Stable Lévy Processes via Lamperti-Type Representations

Stable Lévy processes lie at the intersection of Lévy processes and self-similar Markov processes. Processes in the latter class enjoy a Lamperti-type representation as the space-time path transformation of so-called Markov additive processes (MAPs). This completely new mathematical treatment takes advantage of the fact that the underlying MAP for stable Lévy processes can be explicitly described in one dimension and semi-explicitly described in higher dimensions, and uses this approach to catalogue a large number of explicit results describing the path fluctuations of stable Lévy processes in one and higher dimensions.

Written for graduate students and researchers in the field, this book systematically establishes many classical results as well as presenting many recent results appearing in the last decade, including previously unpublished material. Topics explored include first hitting laws for a variety of sets, path conditionings, law-preserving path transformations, the distribution of extremal points, growth envelopes and winding behaviour.

ANDREAS E. KYPRIANOU was educated at the University of Oxford and University of Sheffield and is currently a professor of mathematics at the University of Bath. He has spent over 25 years working on the theory and application of path-discontinuous stochastic processes and has over 130 publications, including a celebrated graduate textbook on Lévy processes. During his time in Bath, he co-founded and directed the Prob-L@B (Probability Laboratory at Bath), was PI for a multi-million-pound EPSRC Centre for Doctoral Training, and is currently the Director of the Bath Institute for Mathematical Innovation.

JUAN CARLOS PARDO is a full professor at the department of Probability and Statistics at Centro de Investigación en Matemáticas (CIMAT). He was educated at the Universidad Nacional Autónoma de México (UNAM) and Université de Paris VI (Sorbonne Université). He has spent over 13 years working on the theory and application of path-discontinuous stochastic processes and has more than 50 publications in these areas. During the academic year 2018–2019, he held the David Parkin visiting professorship at the University of Bath.

INSTITUTE OF MATHEMATICAL STATISTICS MONOGRAPHS

IMS Monographs are concise research monographs of high quality on any branch of statistics or probability of sufficient interest to warrant publication as books. Some concern relatively traditional topics in need of up-to-date assessment. Others are on emerging themes. In all cases the objective is to provide a balanced view of the field.

Other Books in the Series

1. *Large-Scale Inference*, by Bradley Efron
2. *Nonparametric Inference on Manifolds*, by Abhishek Bhattacharya and Rabi Battacharya
3. *The Skew-Normal and Related Families*, by Adelchi Azzalini
4. *Case-Control Studies*, by Ruth H. Keogh and D. R. Cox
5. *Computer Age Statistical Inference*, by Bradley Efron and Trevor Hastie
6. *Computer Age Statistical Inference (Student Edition)*, by Bradley Efron and Trevor Hastie

Stable Lévy Processes via Lamperti-Type Representations

ANDREAS E. KYPRIANOU
University of Bath

JUAN CARLOS PARDO
Centro de Investigación en Matemáticas, A.C.

CAMBRIDGE
UNIVERSITY PRESS

CAMBRIDGE
UNIVERSITY PRESS

University Printing House, Cambridge CB2 8BS, United Kingdom

One Liberty Plaza, 20th Floor, New York, NY 10006, USA

477 Williamstown Road, Port Melbourne, VIC 3207, Australia

314–321, 3rd Floor, Plot 3, Splendor Forum, Jasola District Centre, New Delhi – 110025, India

103 Penang Road, #05–06/07, Visioncrest Commercial, Singapore 238467

Cambridge University Press is part of the University of Cambridge.

It furthers the University's mission by disseminating knowledge in the pursuit of education, learning, and research at the highest international levels of excellence.

www.cambridge.org
Information on this title: www.cambridge.org/9781108480291
DOI: 10.1017/9781108648318

First published 2022

Printed in the United Kingdom by TJ Books Limited, Padstow Cornwall

A catalogue record for this publication is available from the British Library.

Library of Congress Cataloging-in-Publication Data
Names: Kyprianou, Andreas E., author. | Pardo, Juan Carlos, 1976- author.
Title: Stable Lévy processes via Lamperti-type representations / Andreas E. Kyprianou and Juan Carlos Pardo.
Description: New York : Cambridge University Press, 2022. | Series: Institute of mathematical statistics monographs | Includes bibliographical references and index.
Identifiers: LCCN 2021037625 (print) | LCCN 2021037626 (ebook) | ISBN 9781108480291 (hardback) | ISBN 9781108648318 (ebook)
Subjects: LCSH: Lévy processes. | Stochastic processes.
Classification: LCC QA274.73 .K975 2022 (print) | LCC QA274.73 (ebook) | DDC 519.2/82–dc23
LC record available at https://lccn.loc.gov/2021037625
LC ebook record available at https://lccn.loc.gov/2021037626

ISBN 978-1-108-48029-1 Hardback

This book is dedicated to crossing barriers,
not erecting them

Contents

	Notation		*page* xii
	Preface		xix
	Acknowledgements		xx
1	**Stable distributions**		1
	1.1	One-dimensional stable distributions	1
	1.2	Characteristic exponent of a one-dimensional stable law	4
	1.3	Moments	9
	1.4	Normalised one-dimensional stable distributions	11
	1.5	Distributional identities	13
	1.6	Stable distributions in higher dimensions	20
	1.7	Comments	25
2	**Lévy processes**		27
	2.1	Lévy–Itô decomposition	27
	2.2	Killing	30
	2.3	Path variation and asymmetry	31
	2.4	Feller and strong Markov property	34
	2.5	Infinitesimal generator	35
	2.6	Drifting and oscillating	35
	2.7	Moments	36
	2.8	Exponential change of measure	37
	2.9	Donsker-type convergence	39
	2.10	Transience and recurrence	39
	2.11	Duality	41
	2.12	Hitting points	42
	2.13	Regularity of the half-line	44
	2.14	Excursions and the Wiener–Hopf factorisation	44
	2.15	Reflection	48

	2.16	Creeping	49
	2.17	First passage problems	49
	2.18	Lévy processes in higher dimensions	54
	2.19	Comments	56
3	**Stable processes**		**58**
	3.1	One-dimensional stable processes	58
	3.2	Normalised one-dimensional stable processes	60
	3.3	Path variation, asymmetry and moments	62
	3.4	Path properties in one dimension	64
	3.5	Wiener–Hopf factorisation and the first passage problem	66
	3.6	Isotropic d-dimensional stable processes	70
	3.7	Resolvent density	72
	3.8	Comments	76
4	**Hypergeometric Lévy processes**		**78**
	4.1	β-subordinators	78
	4.2	Hypergeometric processes	81
	4.3	The subclass of Lamperti-stable processes	87
	4.4	The first passage problem	89
	4.5	Exponential functionals	93
	4.6	Distributional densities of exponential functionals	103
	4.7	Distributional tails of exponential functionals	111
	4.8	Comments	113
5	**Positive self-similar Markov processes**		**115**
	5.1	The Lamperti transform	115
	5.2	Starting at the origin	117
	5.3	Stable processes killed on entering $(-\infty, 0)$	122
	5.4	Stable processes conditioned to stay positive	128
	5.5	Stable processes conditioned to limit to 0 from above	133
	5.6	Censored stable processes	136
	5.7	The radial part of an isotropic stable process	144
	5.8	Comments	151
6	**Spatial fluctuations in one dimension**		**153**
	6.1	First exit from an interval	153
	6.2	Hitting points in an interval	160
	6.3	First entrance into a bounded interval	161
	6.4	Point of closest and furthest reach	166
	6.5	First hitting of a two-point set	168
	6.6	First hitting of a point	172
	6.7	First exit for the reflected process	176
	6.8	Comments	181

7 Doney–Kuznetsov factorisation and the maximum 183

7.1 Kuznetsov's factorisation 183
7.2 Quasi-periodicity 185
7.3 The Law of the maximum at a finite time 190
7.4 Doney's factorisation 196
7.5 Comments 204

8 Asymptotic behaviour for stable processes 206

8.1 Stable subordinators 206
8.2 Upper envelopes for $\rho \in (0, 1)$ 214
8.3 Lower envelopes for $\rho \in (0, 1)$ 219
8.4 Comments 226

9 Envelopes of positive self-similar Markov processes 227

9.1 Path decompositions for pssMp 227
9.2 Lower envelopes 234
9.3 Upper envelopes 240
9.4 Comments 250

10 Asymptotic behaviour for path transformations 252

10.1 More on hypergeometric Lévy processes 252
10.2 Distributions of pssMp path functionals 260
10.3 Stable processes conditioned to stay positive 265
10.4 Stable processes conditioned to limit to 0 from above 272
10.5 Censored stable processes 275
10.6 Isotropic stable processes 279
10.7 Comments 285

11 Markov additive and self-similar Markov processes 286

11.1 MAPs and the Lamperti–Kiu transform 286
11.2 Distributional and path properties of MAPs 288
11.3 Excursion theory for MAPs 292
11.4 Matrix Wiener–Hopf factorisation 295
11.5 Self-similar Markov processes in $\mathbb{R}^d$ 300
11.6 Starting at the origin 303
11.7 Comments 304

12 Stable processes as self-similar Markov processes 306

12.1 Stable processes and their h-transforms as ssMp 306
12.2 Stable processes conditioned to avoid or hit 0 314
12.3 One-dimensional Riesz–Bogdan–Żak transform 316
12.4 First entrance into a bounded interval revisited 318
12.5 First hitting of a point revisited 322

12.6 Riesz–Bogdan–Żak transformation in dimension $d \geq 2$ 324
12.7 Radial asymptotics for $d \geq 2$ 328
12.8 Comments 329

13 Radial reflection and the deep factorisation 330
13.1 Radially reflected stable processes when $\alpha \in (0, 1)$ 330
13.2 Deep inverse factorisation of the stable process 332
13.3 Ladder MAP matrix potentials 335
13.4 Stationary limit of the radially reflected process 344
13.5 Deep factorisation of the stable process 348
13.6 Comments 349

14 Spatial fluctuations and the unit sphere 351
14.1 Sphere inversions 351
14.2 Sphere inversions with reflection 354
14.3 First hitting of a sphere 355
14.4 First entrance and exit of a ball 365
14.5 Walk-on-spheres and first exit of general domains 374
14.6 Comments 381

15 Applications of radial excursion theory 383
15.1 Radial excursions 383
15.2 The Point of closest reach to the origin 389
15.3 Deep factorisation in d-dimensions 403
15.4 Radial reflection 406
15.5 Comments 411

16 Windings and up-crossings of stable processes 412
16.1 Polar decomposition of planar stable processes 412
16.2 Windings at infinity 414
16.3 Windings at the origin 417
16.4 Upcrossings of one-dimensional stable processes 421
16.5 Comments 428

***Appendix* 429**
A.1 Useful results from complex analysis 429
A.2 Mellin and Laplace–Fourier inversion 430
A.3 Gamma and beta functions 431
A.4 Double gamma function 432
A.5 Double sine function 434
A.6 Hypergeometric functions 435
A.7 Additive and subadditive functions 436
A.8 Random difference equations 437

A.9 A generalisation of the Borel–Cantelli Lemma 438
A.10 Skorokhod space 439
A.11 Feller processes 440
A.12 Hunt–Nagasawa duality 441
A.13 Poisson point processes 443
References 446
Index 458

Notation

Below is some of the more commonly used notation that appears throughout the text, which has been thematically grouped for convenience. Reference page numbers are presented in the right-hand column.

Stable distributions

α	stability index	3, 61
ρ	positivity index	12, 61
$\mathcal{A}$	parameter set (α, ρ) for stable distributions	12
$p(x, \alpha, \rho)$	pdf of stable distribution with parameters (α, ρ)	14
$M(z)$	Mellin transform of $p(x, \alpha, \rho)$	17

Lévy processes

(Y, P)	general Lévy process	27
$(\hat{Y}, \mathrm{P})$, $(Y, \hat{\mathrm{P}})$	dual of the Lévy process (Y, P)	41
$\Pi(\mathrm{d}x)$	Lévy measure	5, 28
$\pi(x)$	Lévy density	84
$N(\mathrm{d}t, \mathrm{d}x)$	Poisson point process of jumps	29
$\mathcal{L}$	infinitesimal generator	35
$\mathcal{F}_t$	natural filtration	34
P_t	semigroup	41
Ψ	characteristic exponent	28,
ψ	Laplace exponent	94
$\overline{Y}_t$, $\underline{Y}_t$	running supremum and running infimum	41
Ψ_q^+, Ψ_q^-	Wiener–Hopf factors	183
H, $\hat{H}$	ascending and descending ladder height process	46

ς, $\hat{\varsigma}$	lifetime of the ascending and descending ladder height processes	50, 81
κ, $\hat{\kappa}$	ascending and descending ladder height Laplace exponents	46
$U^{(q)}[f]$	q-resolvent	42
$u^{(q)}$	density of q-resolvent	43
U	subordinator resolvent	50
τ^B	first passage time of a Lévy process into B	42
τ_x^-, τ_x^+	first passage times below and above x	34, 50
ζ	lifetime of killed process	30
$\mathcal{E}_t(\beta)$	exponential martingale	37
$\mathfrak{U}$, W	variables characterising asymptotic overshoot of a Lévy process at first passage over a threshold tending to infinity	53, 233
$\mathcal{H}_1, \mathcal{H}_2, \mathcal{H}_3, \mathcal{H}_4$	parametric classes of hypergeometric Lévy processes	84. 88, 93
ξ^*, Ψ^*	Lévy process underlying stable process killed on entering $(-\infty, 0)$ and its characteristic exponent	123
$\xi^\uparrow$, $\Psi^\uparrow$	Lévy process underlying stable process conditioned to stay positive and its characteristic exponent	131, 132
$\xi^\downarrow$, $\Psi^\downarrow$	Lévy process underlying stable process conditioned to limit to 0 from above and its characteristic exponent	135, 135
$\vec{\xi}_t$, $\vec{\Psi}$	Lévy process underlying censored stable process and its characteristic exponent	137, 143
ξ, Ψ	Lévy process underlying the radial part of a stable process and its characteristic exponent	145, 147
η	a constant defined from the hypergeometric Lévy process parameters, equal to $1 - \beta + \gamma + \hat{\beta} + \hat{\gamma}$	84
$\hat{\theta}$	Cramér number	94
$I(\delta, Y)$	integrated exponential functional of Y	93
χ	shorthand for $1/\delta$	94
$p(x)$	pdf of $I(\delta, Y)$	104, 105
$\mathtt{M}(s)$	Mellin transform of $I(\delta, Y)$	94

Stable Processes

$(X, \mathbb{P})$	stable process	58
$\mathcal{A}$	parameter set (α, ρ) for stable processes	61

$\mathscr{A}^+$	parametric set (α,ρ) for stable processes with positive jumps	214
$\mathrm{p}_t(x)$	density of the stable process issued from 0	60, 361
$\lvert X\rvert$	radial distance from the origin	144
X_t^*	running maximum of absolute value	216
$\mathbb{P}^\uparrow$	law of the stable process conditioned to stay positive	129
$\mathbb{P}^\downarrow$	law of the stable process conditioned to limit to 0 from above	133
$\mathbb{P}^\circ$	stable process conditioned to approach 0 continuously (for $\alpha < d$) or conditioned to avoid the origin (for $d < \alpha$)	308, 314
U^A, u^A	resolvent up to exiting the interval A and its density	157, 165
$U^A_{\{z\}}$, $u^A_{\{z\}}$	resolvent up to exiting the interval $A\backslash\{z\}$ and its density	161, 320
R_t	process reflected in its infimum	176
γ_a	first passage time over threshold γ_a of reflected process	177
$\bar{\mathrm{R}}_t$	running supremum of reflected process	179
$\mathtt{J}_t$	future infimum of stable process	240
$\mathrm{p}_{\overline{X}}(x)$	pdf of the maximum at time 1	194
$\mathcal{M}(x)$	Mellin transform of the maximum at time 1	192
$\eta(t)$	time change in Riesz–Bogdan–Żak transformation	316, 324
$\mathsf{G}(\infty)$	time of closest radial reach to the origin	386
$\overline{m}$	time of furthest radial reach from the origin before hitting the origin	340
$\tau_a^{\odot}$	first hitting of the sphere of radius a	355
$\mathtt{D}_a$	last passage time of radial distance below a	419
$\tau_a^{\oplus}$	first entry into the sphere of radius a	365
$\tau_a^{\ominus}$	first exit from the sphere of radius a	365
θ_t, $\theta_{[a,b]}$	winding numbers of planar stable processes	413
U_t, $\mathrm{U}_{[a,b]}$	upcrossings of one-dimensional stable processes	422

Markov additive processes

E	state space of modulator	287, 300
(ξ, J)	MAP with discrete Markov modulator	287
(ξ, Θ)	MAP with general Markov modulator (usually a $\mathbb{S}^{d-1}$-valued modulator)	300
$\mathcal{G}_t$	MAP filtration	300

$\mathbf{P}_{x,i}$	law of MAP with discrete modulator	287
$\mathbf{P}_{x,\theta}$	law of MAP with continuous modulator	300
$\boldsymbol{Q}$	intensity matrix of the discrete modulator	289
$\boldsymbol{G}$	matrix of Laplace transforms of inter-modulator jumps	289
π	stationary distribution of $\boldsymbol{Q}$	289
$\boldsymbol{\Delta}_\pi$	diagonal matrix populated with $\boldsymbol{\pi}$	290
$\boldsymbol{\Psi}$, $\hat{\boldsymbol{\Psi}}$	matrix exponent of MAP and MAP dual	289, 290
$\overline{m}_t$	time at which the MAP ordinate last visits its past maximum before time t	295
(H^+, J^+)	ascending ladder MAP	293
$\boldsymbol{\kappa}(\gamma, \lambda)$	matrix of exponent of space-time ascending ladder MAP	294
$\Phi_i(\gamma, \lambda)$	space-time Laplace exponents of pure subordinator states of ascending ladder MAP	294
$\chi(z)$, $\boldsymbol{v}(z)$	eigenvalue and right eigenvector of matrix exponent of $\boldsymbol{\Psi}(z)$	291
$\boldsymbol{\kappa}(\lambda)$	matrix exponent of the ascending ladder MAP	299
$U_{i,j}(x)$	ascending and descending ladder MAP resolvent for discrete modulator	335
$\mathbf{R}_z[f](\theta)$	MAP resolvent	403
$\boldsymbol{\rho}_z[f](\theta)$, $\hat{\boldsymbol{\rho}}_z[f](\theta)$	ascending and descending ladder MAP resolvent for continuous modulator	403

Self-similar Markov processes

I_t and I_∞	integrated exponential Lévy process underlying a self-similar Markov process	116
$\varphi(t)$	right inverse of integrated exponential Lévy process	116
(Z, P)	self-similar Markov process	116, 227
ζ	lifetime of process	115
$(\Xi, \mathbf{P})$	Lévy process underlying a positive self-similar Markov process	116, 227
$\mathtt{D}_x$	last passage time below x	120
$\overleftarrow{Z}_t$	time reversed process from last passage time	228
$\overleftarrow{S}_y$	first passage time below y of reversed pssMp	228
S_y	first passage time of pssMp above y	249
Γ_1	left limit of positive self-similar Markov process at last passage at level x_1	229

$\mathtt{J}_t$	future infimum of pssMp after time t	240
Υ	scaled left limit of positive self-similar Markov process at last passage at any level	240
$\tilde{p}$	density of distribution of $\Upsilon^\delta I(\delta, \Xi)$	255
F, F_q	right tail distribution of integrated exponential and partially integrated exponential dual Lévy process	234
$\overline{F}_\Upsilon$	left tail distribution of $\Upsilon^\alpha \hat{I}_\infty$	241
G	left tail distribution of S_1 under P_0	248

Excursions

$L,\ \ell$	local time	44, 178, 383
ϵ_t	canonical excursion at local time t	45, 178
$(\epsilon, J^\epsilon),\ (\epsilon, \Theta^\epsilon)$	canonical radial excursion at local time t	294, 384
ζ	excursion lifetime	45
$\overline{\mathcal{U}}(\mathbb{R}),\ \underline{\mathcal{U}}(\mathbb{R})$	space of excursion paths of Lévy processes from maximum and minimum	45
$\overline{\mathcal{U}}(\mathbb{R} \times E)$	space of MAP excursions from ordinate maximum	294
$\underline{\mathcal{U}}(\mathbb{R} \times \mathbb{S}^{d-1})$	space of MAP excursions from ordinate minimum	385
$\overline{n}$ resp. $\underline{n}$	excursion measure of a Lévy process from its maximum resp. minimum	45, 177
$\mathbf{n}_i$	MAP excursion measure when left end point begins with modulator in state i	294
$\mathbb{N}_\theta$	radial excursion measure when left end point begins with modulator in state θ	384

Other notation

$\mathbf{e}_q$	independent and exponentially distributed random variable	30
$\Gamma(z)$	gamma function	431
${}_2F_1(a, b; c; z)$	hypergeometric function	435
$G(z; \tau)$	double gamma function	95, 432
$S_2(z; \tau)$	double sine function	185

$F(s)$	special function derived from double gamma functions	96
C_0, C_∞	families of positive increasing functions with growth at 0 and ∞, respectively	241, 248
$\mathbb{S}^{d-1}(b, r)$	sphere in $\mathbb{R}^d$ of radius r, centred at b	352
$\sigma_a(\mathrm{d}z)$	surface measure on $\mathbb{S}^{d-1}(0, a)$ normalised to have unit mass	356
$K(x)$	spatial inversion through unit sphere (Kelvin transform)	324
x^*, $x^\diamond$	non-centred sphere inversion and non-centred sphere inversion with reflection	352, 354

Preface

There have been a number of developments in the theory of α-stable Lévy processes in recent years. This is largely thanks to a better understanding of their connection to self-similar Markov processes in conjunction with a revised view on the complex analysis that can subsequently be brought into play. We mention in this respect the paper of Caballero and Chaumont [43] as well as the work of Kuznetsov [115, 116], both of which present seminal perspectives in terms of the underlying Wiener–Hopf theory that has stimulated a large base of literature. Among this literature, the PhD theses of Alex Watson in 2013 and Weerapat Satitkanitkul in 2018 stand out.

The basic idea of this book is to give an introductory account of these developments and, accordingly, expose the new techniques that have appeared in the literature since the mid-2000s. The majority of the mathematical computations that are developed in the following chapters pertain either to recent material or to a new approach for classical results. At the end of each chapter, a section is devoted to referencing all material presented in the main body of the chapter. An appendix is also included, and referred to throughout the text, to record some of the more specialist facts from complex analysis, special functions and the theory of Markov processes that are used in the text.

We hope that this text will serve as a standard reference for those interested in the modern theory of α-stable Lévy processes as well as suitable material for a graduate course. Indeed, some of the material in this text has been used in conjunction with lectures given by AEK at the University of Zurich, the National Technical University of Athens, University of Jyväskylä, The Chinese Academy of Sciences and at Prob-L@B in Bath, as well as by JCP at UNAM in Mexico City, CIMAT in Guanajuato and Kyoto University.

Andreas E. Kyprianou
Bath, UK

Juan Carlos Pardo
Guanajuato, Mexico

Acknowledgements

We were inspired to write this text by our mutual friend and collaborator Alexey Kuznetsov, many of whose contributions to the theory of stable processes can be found in this book. During the writing of this book JCP and AEK were in receipt of a Royal Society Advanced Newton International Fellowship, JCP was in receipt of supporting CONACYT grant 250590 and AEK in receipt of supporting EPSRC grants EP/M001784/1 and EP/L002442/1, as well as a Royal Society Wolfson Merit Award. Both authors are grateful for this support, some of which assisted with visits between the UK and Mexico. JCP would especially like to thank the University of Bath for hosting him as David Parkin Visiting Professor for the 2018/19 academic year.

In the final stages of writing, we sent a draft of the manuscript to several people who agreed to act as proofreaders. Predictably, we obtained a shameful amount of corrections. We are immeasurably grateful to the following people in equal measure: Larbi Alili, Sam Baguley, Jean Bertoin, Gabriel Berzunza, Natalia Cardona, Hector Chang, Loic Chaumont, Benjamin Dadoun, Niklas Dexheimer, Ron Doney, Dorottya Fekete, Diana Gillooly, Camilo González, Emma Horton, Sara Klein, Takis Konstantopoulos, Alexey Kuznetsov, Sandra Palau, Helmut Pitters, Tsogzolmaa Saizmaa, Weerapat (Pite) Satitkanitkul, Quan Shi, Lukas Trottner, Stavros Vakeroudis, Matija Vidmar, Alex Watson, Philip Weißmann. The last months of writing took place during the 2020-2021 pandemic lockdown period and we learned the robustness of virtual communication as needs dictated.

Finally, and most importantly, we would like to thank our families for their understanding and patience during the writing of this text.

1

Stable distributions

The starting point of this monograph is the notion of distributional stability and infinite divisibility. Stable distributions are the celebrated class which exhibit both of the aforesaid properties and, accordingly, offer a number of remarkably explicit formulae and identities. We therefore begin our journey by addressing the robust mathematical theory that supports the characterisation of stable distributions in preparation for later chapters.

1.1 One-dimensional stable distributions

We begin our discussion by first restricting ourselves to the one-dimensional setting. The following definition, for which we use $\overset{(d)}{=}$ to mean equality in distribution, is key to the notion of distributional stability.

Definition 1.1 A non-degenerate random variable X has a *stable* distribution if, for any $a > 0$ and $b > 0$, there exists $c > 0$, such that

$$aX_1 + bX_2 \overset{(d)}{=} cX, \tag{1.1}$$

where X_1 and X_2 are independent and $X_1 \overset{(d)}{=} X_2 \overset{(d)}{=} X$. We exclude from this definition the possibility that $X \equiv 0$.

The experienced reader will immediately spot that Definition 1.1 pertains to what is more broadly known in the literature as a *strictly stable* random variable. The notion of a *stable* random variable is reserved for a slightly broader concept. Since we will never have occasion in this book to distinguish the difference, we will depart from the traditional convention and refer only to stable random variables, using this definition.

Observe that (1.1) implies

$$X \stackrel{(\mathrm{d})}{=} \frac{X_1 + X_2}{d_2},$$

for some constant $d_2 > 0$. By induction, it is easy to see that, for any $n \geq 0$, there exists a constant $d_n > 0$ and n independent random variables X_i, $1 \leq i \leq n$, with the same distribution as X, such that

$$X \stackrel{(\mathrm{d})}{=} \frac{X_1 + X_2 + \cdots + X_n}{d_n}. \tag{1.2}$$

Said another way, any stable random variable X is infinitely divisible.

For convenience, let us recall the so-called Lévy–Khintchine representation, which provides a complete characterisation of infinitely divisible distributions. We first introduce some notation. Let μ be the probability distribution of a real-valued random variable and define its characteristic function by

$$\hat{\mu}(z) = \int_{\mathbb{R}} \mathrm{e}^{\mathrm{i}zx} \mu(\mathrm{d}x), \qquad z \in \mathbb{R}.$$

If μ is an infinitely divisible distribution, then it is known that its characteristic function never vanishes. As a consequence, there exists a continuous function $\Psi\colon \mathbb{R} \mapsto \mathbb{C}$, called the *characteristic exponent* of μ, such that $\Psi(0) = 0$, and

$$\exp\{-\Psi(z)\} := \hat{\mu}(z), \quad \text{for } z \in \mathbb{R}. \tag{1.3}$$

Theorem 1.2 (Lévy–Khintchine representation) *A function $\Psi\colon \mathbb{R} \mapsto \mathbb{C}$ is the characteristic exponent of an infinitely divisible random variable if and only if there exists a triple (a, σ, Π), where $a \in \mathbb{R}$, $\sigma \geq 0$ and Π is a measure concentrated on $\mathbb{R} \setminus \{0\}$ satisfying $\int_{\mathbb{R}}(1 \wedge x^2)\Pi(\mathrm{d}x) < \infty$, such that*

$$\Psi(z) = \mathrm{i}az + \frac{1}{2}\sigma^2 z^2 + \int_{\mathbb{R}} \left(1 - \mathrm{e}^{\mathrm{i}zx} + \mathrm{i}zx\mathbf{1}_{(|x|<1)}\right) \Pi(\mathrm{d}x), \tag{1.4}$$

for every $z \in \mathbb{R}$. Moreover, the triple (a, σ^2, Π) is unique within the given arrangement on the right-hand side of (1.4).

The measure Π is called *the Lévy measure* of the distribution μ and σ its *Gaussian coefficient*. Whilst the triple (a, σ, Π) defining $\Psi(z)$ is unique as described, in various situations one may prefer to use a different *regularising* function $h(x)$, in which case (1.4) is written as

$$\Psi(z) = \mathrm{i}\tilde{a}z + \frac{1}{2}\sigma^2 z^2 + \int_{\mathbb{R}} \left(1 - \mathrm{e}^{\mathrm{i}zx} + \mathrm{i}zh(x)\right) \Pi(\mathrm{d}x), \qquad z \in \mathbb{R}, \tag{1.5}$$

where

$$\tilde{a} = a - \int_{\mathbb{R}} (h(x) - x\mathbf{1}_{(|x|<1)})\,\Pi(\mathrm{d}x),$$

which is finite.

In this chapter, we shall interchange between the two equivalent representations given by (1.4) and (1.5). For example, when the measure Π satisfies the stronger condition

$$\int_{\mathbb{R}} (1 \wedge |x|)\,\Pi(\mathrm{d}x) < \infty,$$

we may choose $h(x) \equiv 0$. If the distribution μ has finite mean, we may choose $h(x) \equiv x$. In some cases, it will be convenient to choose $h(x) = \sin(x)$ or $h(x) = x/(1+x^2)$. Everywhere in this book, when we say that the distribution μ has characteristic triple (a, σ, Π) without specifying the regularising function h, we assume that the characteristic exponent is given via (1.4), otherwise we will say that the distribution μ has characteristic triple (a, σ, Π) with the regularising function h, in which case Ψ will be given by (1.5).

The following main result provides the explicit characteristic exponent of stable distributions. As part of its proof, which will be provided in the next section, we will also be obliged to understand the structure of the underlying triple (a, σ, Π) in the associated Lévy–Khintchine formula.

Theorem 1.3 *A stable random variable X has a characteristic exponent satisfying*

$$\Psi(z) = c|z|^{\alpha}\Big(1 - \mathrm{i}\beta \tan(\pi\alpha/2)\,\mathrm{sgn}(z)\Big), \qquad z \in \mathbb{R}, \tag{1.6}$$

where

$$\alpha \in (0,1) \cup (1,2],\ c > 0 \quad \textit{and} \quad \beta \in [-1,1]$$

or

$$\alpha = 1, \beta = 0 \quad \textit{and we understand} \quad \beta \tan(\pi\alpha/2) := 0.$$

The latter case is known as the symmetric Cauchy distribution.

Remark 1.4 Note that the symmetric Cauchy distribution with drift $\delta \in \mathbb{R}$, that is,

$$\Psi(z) = c|z| + \delta z, \qquad z \in \mathbb{R},$$

also belongs to the class of one-dimensional stable distributions. Nonetheless, we will henceforth only deal with the case that $\delta = 0$ when $\alpha = 1$.

Remark 1.5 We also note that the case $\alpha = 2$ corresponds to the case where X has a Gaussian distribution. As we shall see in Chapter 2, associated to each of the distributions discussed in this chapter is a Lévy process. As one might expect, the case $\alpha = 2$ leads to Brownian motion. For other values of α, we will find an association with Lévy processes that do not have continuous paths, the so-called α-stable processes (also referred to as just stable processes). It is the case of processes with path discontinuities that forms the primary concern of this book. For this reason, the overwhelming majority of this text will be restricted to the setting that $\alpha \in (0, 2)$.

1.2 Characteristic exponent of a one-dimensional stable law

We dedicate this section entirely to the proof of Theorem 1.3. As part of this process, we need to establish two key intermediary results.

Lemma 1.6 *The sequence $(d_k)_{k\geq 1}$ defined by (1.2) is strictly increasing and satisfies $d_k = k^{1/\alpha}$ for some $\alpha > 0$, $k \geq 1$.*

Proof Recall that $\hat{\mu}$ denotes the characteristic function of a stable distribution X and, thanks to the infinite divisibility of X, $\hat{\mu}(z) \neq 0$ for $z \in \mathbb{R}$. From the definition of the sequence $(d_k)_{k\geq 1}$, the scaling property in (1.2) can be reworded to say

$$\Psi(d_k z) = k\Psi(z), \qquad z \in \mathbb{R},\ k \geq 1. \tag{1.7}$$

In turn, this implies $|\hat{\mu}(d_{k+1}z)| = |\hat{\mu}(z)||\hat{\mu}(d_k z)| \leq |\hat{\mu}(d_k z)|$ and hence

$$\left|\hat{\mu}\left(\frac{d_{k+1}}{d_k}z\right)\right| \leq |\hat{\mu}(z)|, \qquad k \geq 1.$$

We are now forced to conclude that $d_{k+1} \geq d_k$, for $k \geq 1$. To see why, note that

$$\left|\hat{\mu}\left(\left(\frac{d_{k+1}}{d_k}\right)^n z\right)\right| \leq |\hat{\mu}(z)|, \qquad \text{for any} \quad n \geq 1,$$

with $(d_{k+1}/d_k)^n \to 0$ as $n \to \infty$, which would imply that $1 \leq |\hat{\mu}(z)|$, leading to a contradiction.

Next, we observe that for all $m, n \geq 1$ and $z \in \mathbb{R}$,

$$|\Psi(d_{mn}z)| = mn|\Psi(z)| = n\Big|m\Psi(z)\Big| = \Big|n\Psi(d_m z)\Big| = |\Psi(d_n d_m z)|,$$

implying that $d_{mn} = d_n d_m$. In particular, for any positive integer j, $d_{m^j} = d_m^j$. If $1 < n < m$, there is a positive integer p such that $m^j \leq n^p < m^{j+1}$. Using these inequalities and the established monotonicity of $(d_k)_{k\geq 1}$, we have

$$\frac{j}{j+1}\frac{\log d_m}{\log m} \le \frac{\log d_n}{\log n} \le \frac{j+1}{j}\frac{\log d_m}{\log m}.$$

Hence, taking $j \to \infty$, we get

$$\frac{\log d_m}{\log m} = \frac{\log d_n}{\log n} =: \frac{1}{\alpha},$$

for some strictly positive constant α. Therefore $\log d_n = \log n^{1/\alpha}$ or equivalently $d_n = n^{1/\alpha}$, for $n \ge 1$ and $\alpha > 0$. □

Our second intermediary result characterises the form of the underlying Lévy measure of any stable distribution.

Proposition 1.7 *If X is a stable random variable, then necessarily $\alpha \in (0, 2]$. In the case that $\alpha = 2$, X is Gaussian distributed. Otherwise when $\alpha \in (0, 2)$, then there exist $c_1, c_2 \ge 0$ such that $c_1 + c_2 > 0$ and the underlying Lévy measure Π satisfies*

$$\Pi(\mathrm{d}x) = |x|^{-1-\alpha}\Big(c_1\mathbf{1}_{(x>0)} + c_2\mathbf{1}_{(x<0)}\Big)\,\mathrm{d}x, \qquad x \in \mathbb{R}. \tag{1.8}$$

Proof Recall that identity (1.7) and Lemma 1.6 imply $k\Psi(z) = \Psi(k^{1/\alpha}z)$, for $z \in \mathbb{R}$ and $k \ge 1$. More precisely, we observe

$$\begin{aligned} &\mathrm{i}kaz + \frac{1}{2}k\sigma^2 z^2 + \int_{\mathbb{R}} \Big(1 - \mathrm{e}^{\mathrm{i}zx} + \mathrm{i}zx\mathbf{1}_{(|x|<1)}\Big)\, k\,\Pi(\mathrm{d}x) \\ &\quad = \mathrm{i}ak^{1/\alpha}z + \frac{1}{2}\sigma^2 z^2 k^{2/\alpha} + \int_{\mathbb{R}} \Big(1 - \mathrm{e}^{\mathrm{i}zk^{1/\alpha}x} + \mathrm{i}zk^{1/\alpha}x\mathbf{1}_{(|x|<1)}\Big)\, \Pi(\mathrm{d}x), \end{aligned} \tag{1.9}$$

for any $k \ge 1$ and $z \in \mathbb{R}$. Hence if $\sigma > 0$, we are forced to take $\alpha = 2$. Moreover, still in the setting $\alpha = 2$, if we then let k tend to ∞, the latter identity implies $a = 0$ and $\Pi \equiv 0$. In conclusion, the case that $\alpha = 2$ corresponds to a Gaussian random variable.

Next, we assume $\sigma = 0$. Again from identity (1.9), by changing variables in the integral on the right-hand side, we deduce

$$k\Pi(\mathrm{d}x) = \Pi(k^{-1/\alpha}\mathrm{d}x), \qquad x \neq 0.$$

Therefore, for the functions $\overline{\Pi}^{(+)}(x) := \Pi([x, \infty))$, $x > 0$, and $\overline{\Pi}^{(-)}(x) := \Pi((-\infty, x))$, $x < 0$, we have

$$\overline{\Pi}^{(+)}(x) = \frac{1}{k}\overline{\Pi}^{(+)}\left(k^{-1/\alpha}x\right) \qquad \text{and} \qquad \overline{\Pi}^{(-)}(x) = \frac{1}{k}\overline{\Pi}^{(-)}\left(k^{-1/\alpha}x\right).$$

From the first of these two, we have, for all $k, n \ge 1$,

$$\frac{1}{n}\overline{\Pi}^{(+)}\left(\frac{k^{1/\alpha}}{n^{1/\alpha}}\right) = \overline{\Pi}^{(+)}\left(k^{1/\alpha}\right) = \frac{1}{k}\overline{\Pi}^{(+)}(1).$$

Since $\{(k/n)^{1/\alpha}; k, n \in \mathbb{N}\}$ is dense in $[0, \infty)$ and the function $\overline{\Pi}^{(+)}$ is non-increasing, we deduce $\overline{\Pi}^{(+)}(x) = x^{-\alpha}\overline{\Pi}^{(+)}(1)$, for $x > 0$. Similarly, we may deduce $\overline{\Pi}^{(-)}(x) = |x|^{-\alpha}\overline{\Pi}^{(-)}(1)$, for $x < 0$.

Now taking $c_1 := \alpha\overline{\Pi}^{(+)}(1)$ and $c_2 := \alpha\overline{\Pi}^{(-)}(1)$, we obtain

$$\Pi(\mathrm{d}x) = |x|^{-1-\alpha}\left(c_1\mathbf{1}_{(x>0)} + c_2\mathbf{1}_{(x<0)}\right)\mathrm{d}x, \qquad x \in \mathbb{R},$$

as required. As Π is a Lévy measure, in particular, it must satisfy the integral condition

$$\int_{\mathbb{R}} (1 \wedge |x|^2)\,\Pi(\mathrm{d}x) < \infty.$$

We thus deduce that $\alpha \in (0, 2)$. □

Finally, we are ready to compute the characteristic exponent Ψ as stated in Theorem 1.3.

Proof of Theorem 1.3 Since the case $\alpha = 2$ has already been characterised as Gaussian in the proof of Proposition 1.7, we set $\sigma = 0$ and focus on the case $\alpha \in (0, 2)$.

We first observe that, when $\alpha \in (0, 1)$ the function $x \mapsto |x|^{-(\alpha+1)}$ is integrable near 0 and hence we may take the regularising function in (1.5) to satisfy $h(x) = 0$. From identity (1.7), we deduce that $\tilde{a} = 0$ in (1.5), or in other words,

$$a = -\int_{(|x|<1)} x\,\Pi(\mathrm{d}x).$$

Using the well-known integral identity for the gamma function, see for instance (A.7) in the Appendix, we have

$$\int_0^\infty \mathrm{e}^{\mathrm{i}zx} x^{s-1}\,\mathrm{d}x = z^{-s}\Gamma(s)\mathrm{e}^{\pi\mathrm{i}s/2}, \quad z > 0,\ 0 < s < 1, \tag{1.10}$$

and, appealing to integration by parts, we find that

$$\int_0^\infty \left(\mathrm{e}^{\mathrm{i}zx} - 1\right) x^{-1-\alpha}\,\mathrm{d}x = z^{\alpha}\mathrm{e}^{-\pi\mathrm{i}\alpha/2}\Gamma(-\alpha), \qquad z > 0. \tag{1.11}$$

Making the change of variable $x \mapsto -x$ and taking the complex conjugate of both sides we find

$$\int_{-\infty}^0 \left(\mathrm{e}^{\mathrm{i}zx} - 1\right)|x|^{-1-\alpha}\,\mathrm{d}x = \int_0^\infty \left(\mathrm{e}^{-\mathrm{i}zx} - 1\right) x^{-1-\alpha}\,\mathrm{d}x = z^{\alpha}\mathrm{e}^{\pi\mathrm{i}\alpha/2}\Gamma(-\alpha), \tag{1.12}$$

for $z > 0$. Note also that when z takes negative values, we can similarly make use of the computations leading to (1.12). Then, we apply the following simple identity

$$c_1 e^{-\pi i\alpha/2} + c_2 e^{\pi i\alpha/2} = (c_1 + c_2)\cos(\pi\alpha/2)\left(1 - i\frac{c_1 - c_2}{c_1 + c_2}\tan(\pi\alpha/2)\right),$$

and observe that

$$c = -(c_1 + c_2)\Gamma(-\alpha)\cos(\pi\alpha/2) > 0,$$

since $-\Gamma(-\alpha)$ is positive for $\alpha \in (0, 1)$. This completes the proof of the case $\alpha \in (0, 1)$.

When $\alpha \in (1, 2)$, the function $x \mapsto |x|^{-(\alpha+1)}$ integrates x^2 in a neighbourhood of 0 and hence we may take the regularising function in (1.5) as $h(x) = x$. Again identity (1.7) implies $\tilde{a} = 0$ in (1.5), and therefore

$$a = \int_{(|x|\geq 1)} x\,\Pi(\mathrm{d}x).$$

Similarly, we use (1.10) and apply integration by parts twice to find

$$\int_0^\infty \left(e^{izx} - 1 - izx\right) x^{-1-\alpha}\,\mathrm{d}x = z^\alpha e^{-\pi i\alpha/2}\Gamma(-\alpha), \tag{1.13}$$

for $z > 0$, and the rest of the proof proceeds in the same way as in the case $\alpha \in (0, 1)$.

Finally, the case $\alpha = 1$ must be treated differently. In this case, we observe

$$\begin{aligned}\int_0^\infty \left(1 - e^{izx} + izx\mathbf{1}_{(|x|<1)}\right)\frac{\mathrm{d}x}{x^2} &= \int_0^\infty (1 - \cos(zx))\,\frac{\mathrm{d}x}{x^2}\\ &\quad - i\int_0^\infty \left(\sin(zx) - zx\mathbf{1}_{(|x|<1)}\right)\frac{\mathrm{d}x}{x^2}.\end{aligned} \tag{1.14}$$

A change of variables followed by integration by parts gives us

$$\int_0^\infty (1 - \cos(zx))\,\frac{\mathrm{d}x}{x^2} = |z|\int_0^\infty \frac{\sin(x)}{x}\,\mathrm{d}x = |z|\int_0^\infty\int_0^\infty \sin(x)e^{-xu}\,\mathrm{d}u\,\mathrm{d}x.$$

Since

$$\int_0^\infty e^{-xu}\sin(x)\,\mathrm{d}x = \frac{1}{u^2+1}, \tag{1.15}$$

we get

$$\int_0^\infty (1 - \cos(zx))\,\frac{\mathrm{d}x}{x^2} = \frac{|z|\pi}{2}. \tag{1.16}$$

Next, for simplicity, we assume that $z > 0$. Observe that

$$\begin{aligned}
\int_0^\infty (\sin(zx) - zx\mathbf{1}_{(|x|<1)}) \frac{\mathrm{d}x}{x^2}
&= \int_0^{1/z} (\sin(zx) - zx) \frac{\mathrm{d}x}{x^2} \\
&\quad + \int_{1/z}^\infty \sin(zx)\frac{\mathrm{d}x}{x^2} - z\log z \\
&= z\left(\int_0^1 (\sin(x) - x) \frac{\mathrm{d}x}{x^2} + \int_1^\infty \sin(x) \frac{\mathrm{d}x}{x^2}\right) - z\log z. \qquad (1.17)
\end{aligned}$$

Hence by defining

$$K := \int_0^1 (\sin(x) - x) \frac{\mathrm{d}x}{x^2} + \int_1^\infty \sin(x) \frac{\mathrm{d}x}{x^2},$$

and putting all the pieces in (1.16) and (1.17) back into (1.14), we deduce

$$\int_0^\infty \left(1 - \mathrm{e}^{\mathrm{i}zx} + \mathrm{i}zx\mathbf{1}_{(|x|<1)}\right) \frac{\mathrm{d}x}{x^2} = \frac{|z|\pi}{2} - \mathrm{i}Kz + \mathrm{i}z\log|z|, \quad z \in \mathbb{R} \setminus \{0\}.$$

Therefore, from Proposition 1.7 and the above reasoning, the characteristic exponent Ψ satisfies

$$\Psi(z) = iaz + (c_1 - c_2)\mathrm{i}Kz + (c_1 + c_2)|z|\frac{\pi}{2} + (c_1 - c_2)\mathrm{i}z\log|z|, \quad z \in \mathbb{R} \setminus \{0\}.$$

As we must have $\Psi(k^{1/\alpha}z) = k\Psi(z)$, $z \in \mathbb{R}$, $k \in \mathbb{N}$, albeit now $\alpha = 1$, from Lemma 1.6, we deduce that $c_1 = c_2$ and then

$$\Psi(z) = \mathrm{i}az + (c_1 + c_2)|z|\frac{\pi}{2}, \quad z \in \mathbb{R}.$$

Taking note of Remark 1.4, by taking $a = 0$, we get the desired result. □

Reviewing the proof here, we also get some information about the constants c_1 and c_2, appearing in Proposition 1.7, in relation to the parameters c and β in (1.6).

Corollary 1.8 *When $\alpha \in (0, 2)$, the constants c_1, c_2 appearing in the Lévy measure* (1.8) *satisfy*

$$c = -(c_1 + c_2)\Gamma(-\alpha)\cos(\pi\alpha/2) \quad \text{and} \quad \beta = \frac{c_1 - c_2}{c_1 + c_2}, \qquad (1.18)$$

when $\alpha \in (0, 1) \cup (1, 2)$. Moreover, $c_1 = c_2$ with $c = c_1\pi$, when $\alpha = 1$.

We also get from the proof of Theorem 1.3 the values of a in the Lévy–Khintchine triple 1.4. As such, the following corollary completes the statement of Proposition 1.7.

Corollary 1.9 *When $\alpha \in (0,1)$, the constant a in the Lévy–Khintchine triple is equal to $-\int_{(|x|<1)} x\Pi(\mathrm{d}x)$, when $\alpha \in (1,2)$, we have $a = \int_{(|x|\geq 1)} x\Pi(\mathrm{d}x)$ and when $\alpha = 1$, we have $a = 0$.*

1.3 Moments

An important feature of stable distributions when $\alpha \in (0,2)$, which is one of their signature properties that differs from the setting that $\alpha = 2$, is that they do not possess second moments (and hence no other greater moments). The precise cut-off where positive moments exist is the concern of the next main result.

Theorem 1.10 *Suppose that X is a stable distribution with index $\alpha \in (0,2)$. Then $\mathbb{E}[|X|^\beta] < \infty$, for $0 \leq \beta < \alpha$, and for $\beta \geq \alpha$, we have $\mathbb{E}[|X|^\beta] = \infty$.*

Proof We start by noting that, irrespective of the symmetry in the distribution of X, thanks to the shape of Π given in Theorem 1.7, we have

$$\int_{(|x|\geq 1)} |x|^\beta \, \Pi(\mathrm{d}x) < \infty,$$

for $\beta \in [0,\alpha)$ and infinite for $\beta \in [\alpha, \infty)$.

Next, note that the Lévy–Khintchine exponent (1.4), written here as Ψ, has $\sigma = 0$ and can be decomposed in the form $\Psi = \Psi^{(1)} + \Psi^{(2)}$, where

$$\Psi^{(1)}(z) = \mathrm{i}az + \int_{(|x|\geq 1)} \left(1 - \mathrm{e}^{\mathrm{i}zx}\right) \Pi(\mathrm{d}x), \qquad z \in \mathbb{R},$$

and

$$\Psi^{(2)}(z) = \int_{(|x|<1)} \left(1 - \mathrm{e}^{\mathrm{i}zx} + \mathrm{i}zx\right) \Pi(\mathrm{d}x), \qquad z \in \mathbb{R},$$

with

$$a = \begin{cases} -\int_{(|x|<1)} x\Pi(\mathrm{d}x) & \text{if } \alpha \in (0,1), \\ 0 & \text{if } \alpha = 1, \\ \int_{(|x|\geq 1)} x\Pi(\mathrm{d}x) & \text{if } \alpha \in (1,2). \end{cases}$$

For the first of these two, we note that it corresponds to the characteristic exponent of a compound Poisson random variable, say

$$X^{(1)} = -a + \sum_{i=1}^{\mathtt{N}} \Xi_i,$$

where $\mathtt{N}$ is an independent Poisson distributed random variable with rate $\Pi(|x| \geq 1)$ and $(\Xi_i, i \geq 1)$ are i.i.d. with distribution $\Pi(|x| \geq 1)^{-1}\Pi(\mathrm{d}x)\mathbf{1}_{(|x|\geq 1)}$.

(We use the usual convention that $\sum_{i=1}^{0} := 0$.) We want to consider the moments of $X^{(1)}$. It is already clear from the tail of Π that Ξ_1 has a finite β-moment if $\beta \in [0, \alpha)$ and infinite β-moment if $\beta \geq \alpha$. In particular, Ξ_1 has a first moment (and hence all smaller positive moments) if and only if $\alpha \in (1, 2)$.

When Ξ_1 has a first moment, that is, $\alpha \in (1, 2)$, we observe that $X^{(1)}$ can be rewritten as

$$X^{(1)} = \sum_{i=1}^{\mathtt{N}} \widetilde{\Xi}_i,$$

where each of the $\widetilde{\Xi}_i$ has zero mean. In that case, we may appeal to an inequality for martingale differences, which states that, for $\beta \in [1, \alpha)$ and $n \geq 1$,

$$\mathbb{E}\left[\left|\sum_{i=1}^{n} \widetilde{\Xi}_i\right|^{\beta}\right] \leq 2^{\beta} \sum_{i=1}^{n} \mathbb{E}[|\widetilde{\Xi}_i|^{\beta}]. \tag{1.19}$$

As the right-hand side is equal to $2^{\beta} n \mathbb{E}[|\widetilde{\Xi}_1|^{\beta}]$, it follows by an independent randomisation of n by the Poisson distribution of $\mathtt{N}$ that $\mathbb{E}[|X^{(1)}|^{\beta}] < \infty$.

When $X^{(1)}$ has no first moment, that is, $\alpha \in (0, 1]$, we can use the inequality

$$\left(\sum_{i=1}^{n} u_i\right)^{q} \leq \sum_{i=1}^{n} u_i^{q}, \qquad u_1, \ldots, u_n \geq 0, \tag{1.20}$$

for $q \in (0, 1]$, to deduce that

$$\mathbb{E}\left[\left|\sum_{i=1}^{n} \Xi_i\right|^{\beta}\right] \leq \mathbb{E}\left[\left(\sum_{i=1}^{n} |\Xi_i|\right)^{\beta}\right] \leq \sum_{i=1}^{n} \mathbb{E}\left[|\Xi_i|^{\beta}\right] = n\mathbb{E}\left[|\Xi_1|^{\beta}\right] < \infty,$$

for $\beta \in [0, \alpha)$. Hence, again following an independent randomisation of n by the distribution of $\mathtt{N}$, $\mathbb{E}[|X^{(1)}|^{\beta}] < \infty$, for $\beta \in [0, \alpha)$.

Next, we want to show that $\mathbb{E}[|X^{(2)}|^{\beta}] < \infty$, for $\beta \in [0, \alpha)$ and $\alpha \in (0, 2)$, where $X^{(2)}$ is the random variable whose characteristic exponent is given by $\Psi^{(2)}$. To this end, we write

$$\Psi^{(2)}(z) = -\int_{(|x|<1)} \sum_{k\geq 0} \frac{(\mathrm{i}zx)^{k+2}}{(k+2)!} \Pi(\mathrm{d}x). \tag{1.21}$$

The sum and the integral may be exchanged using Fubini's Theorem and the estimate

$$\sum_{k\geq 0} \int_{(|x|<1)} \frac{|zx|^{k+2}}{(k+2)!} \Pi(\mathrm{d}x) \leq \sum_{k\geq 0} \frac{|z|^{k+2}}{(k+2)!} \int_{(|x|<1)} x^2 \Pi(\mathrm{d}x) < \infty.$$

Hence, the right-hand side of (1.21) can be written as a power series for all $z \in \mathbb{C}$ and is thus entire. In turn this guarantees that $\hat{\mu}^{(2)}(z) := \exp\{-\Psi^{(2)}(z)\}$

is also an entire function. Note that $\hat{\mu}^{(2)}(z)$ is nothing more than the Fourier transform of the measure $\mu^{(2)}(\mathrm{d}x) = \mathbb{P}(X^{(2)} \in \mathrm{d}x)$, for $x \in \mathbb{R}$. Since $\hat{\mu}^{(2)}(z)$ is an entire function, it follows that all the moments of $\mu^{(2)}$, and hence of $X^{(2)}$, exist.

To complete the proof for the case $\beta \in [0, \alpha)$, we can appeal again to (1.19), when $\alpha \in (1, 2)$ and (1.20) when $\alpha \in (0, 1]$ to ensure that $X = X^{(1)} + X^{(2)}$ has the required moment structure.

For the case $\beta \geq \alpha$, suppose that X has β-moments. Without loss of generality we may assume that $\beta \in (0, 2)$, as we will shortly rule out any moments for $\beta \geq \alpha$. Recalling that $X^{(2)}$ always has finite moments, using the inequalities (1.19), when $\beta \geq 1$, and (1.20), when $\beta \in (0, 1]$, together with the simple relation $X^{(1)} = X - X^{(2)}$, we have that the β-moment of $X^{(1)}$ exists. As $X^{(1)} \geq \Xi_1$ on the event $\{\mathtt{N} \geq 1\}$, it follows that Ξ_1 has β-moments. We have already concluded that this can happen when $\beta \in [0, \alpha)$ and hence the required condition follows. □

1.4 Normalised one-dimensional stable distributions

In the sequel, we denote by $\mathcal{S}(\alpha, \beta, c)$ a stable distribution, meaning that its characteristic exponent satisfies (1.6). It appears that there are three parameters naturally associated with stable distributions. However, we want to work with a normalised version of such distributions, reducing the number of parameters from three down to two.

Definition 1.11 Let $\tilde{X}$ be distributed according to $\mathcal{S}(\alpha, \beta, c)$. Define

$$b := c\sqrt{1 + \beta^2 \tan(\pi\alpha/2)^2}, \quad \rho := \frac{1}{2} + \frac{1}{\pi\alpha} \tan^{-1}\left(\beta \tan(\pi\alpha/2)\right), \tag{1.22}$$

where $\tan^{-1}(\cdot)$ denotes the inverse function of $\tan(\cdot)$ restricted to its principal branch $(-\pi/2, \pi/2)$. Then we say that the random variable $X := b^{-\frac{1}{\alpha}}\tilde{X}$ is distributed as a normalised stable distribution with parameters (α, ρ) or simply $X \sim \mathcal{S}_{\mathtt{norm}}(\alpha, \rho)$.

Observe from (1.22) that β, and hence b, can be written in terms of ρ as follows

$$\beta = \cot\left(\frac{\pi\alpha}{2}\right) \tan\left(\pi\alpha\left(\rho - \frac{1}{2}\right)\right), \quad b = \frac{c}{\cos\left(\pi\alpha\left(\rho - \frac{1}{2}\right)\right)}. \tag{1.23}$$

When $\alpha \in (0, 1)$, by varying $\beta \in [-1, 1]$, the parameter ρ ranges over $[0, 1]$, where the boundary points $\rho = 0$ and $\rho = 1$ correspond to the cases $\beta = -1$ and $\beta = 1$, respectively.

The case $\alpha \in (1,2)$ is slightly different. In order to deduce the range of ρ, we first recall the following trigonometric identity

$$\cot\left(\frac{\pi\alpha}{2}\right) = \tan\left(\frac{\pi}{2} - \frac{\pi\alpha}{2}\right),$$

which implies that, by varying $\beta \in [-1,1]$, the range of ρ is $[1 - 1/\alpha, 1/\alpha]$. Note that the boundary points $\rho = 1 - 1/\alpha$ and $\rho = 1/\alpha$ correspond to the cases $\beta = 1$ and $\beta = -1$, respectively.

When $\alpha = 1$, we know that X is symmetric and hence $\rho = 1/2$. Therefore, we introduce the set of admissible parameters

$$\begin{aligned}\mathcal{A} := \{&\alpha \in (0,1),\ \rho \in [0,1]\} \\ &\cup \{\alpha = 1,\ \rho = 1/2\} \\ &\qquad \cup \left\{\alpha \in (1,2),\ \rho \in [1-\alpha^{-1}, \alpha^{-1}]\right\}. \end{aligned} \tag{1.24}$$

Proposition 1.12 *Let $(\alpha,\rho) \in \mathcal{A}$ and assume that X is distributed as $\mathcal{S}_{\text{norm}}(\alpha,\rho)$. Then its characteristic exponent is given by*

$$\Psi(z) = |z|^\alpha \left(e^{\pi i\alpha(\frac{1}{2}-\rho)}\mathbf{1}_{(z>0)} + e^{-\pi i\alpha(\frac{1}{2}-\rho)}\mathbf{1}_{(z<0)}\right). \tag{1.25}$$

The Lévy measure of X satisfies (1.8) *with*

$$c_1 = \Gamma(1+\alpha)\frac{\sin(\pi\alpha\rho)}{\pi}, \quad c_2 = \Gamma(1+\alpha)\frac{\sin(\pi\alpha\hat{\rho})}{\pi}, \tag{1.26}$$

where $\hat{\rho} = 1 - \rho$.

Proof We first prove identity (1.25). In order to do so, we take $\tilde{X}$ with the same distribution as $\mathcal{S}(\alpha,\beta,c)$ and define $X = b^{-\frac{1}{\alpha}}\tilde{X}$, where b was defined in (1.22). We also let $\tilde{\Psi}$ and Ψ denote their respective characteristic exponents. It is then clear that

$$\Psi(z) = \tilde{\Psi}(b^{-\frac{1}{\alpha}}z) = b^{-1}\tilde{\Psi}(z), \qquad z \in \mathbb{R}.$$

Using (1.23), we note that

$$\begin{aligned} &c\big(1 - i\beta\tan(\pi\alpha/2)\,\mathrm{sgn}(z)\big) \\ &= c\left(1 - i\tan\left(\pi\alpha\left(\rho - \frac{1}{2}\right)\right)\mathrm{sgn}(z)\right) \\ &= \frac{c}{\cos\left(\pi\alpha\left(\rho - \frac{1}{2}\right)\right)}\left(\cos\left(\pi\alpha\left(\rho - \frac{1}{2}\right)\right) - i\sin\left(\pi\alpha\left(\rho - \frac{1}{2}\right)\right)\mathrm{sgn}(z)\right) \\ &= \frac{c}{\cos\left(\pi\alpha\left(\rho - \frac{1}{2}\right)\right)}\left(e^{\pi i\alpha(\frac{1}{2}-\rho)}\mathbf{1}_{(z>0)} + e^{-\pi i\alpha(\frac{1}{2}-\rho)}\mathbf{1}_{(z<0)}\right). \end{aligned}$$

Using (1.6) and (1.23) again, we deduce that $\Psi(z)$ is given in the form of (1.25), up to a multiplicative constant.

For the given expressions of c_1 and c_2 in (1.26), using standard trigonometric identities and the reflection formula for the gamma function (see identity (A.12) in the Appendix), we obtain that β and c, defined in (1.18), satisfy

$$\beta = \frac{c_1 - c_2}{c_1 + c_2} = \cot\left(\frac{\pi\alpha}{2}\right)\tan\left(\pi\alpha\left(\rho - \frac{1}{2}\right)\right),$$

and

$$c = -(c_1 + c_2)\Gamma(-\alpha)\cos(\pi\alpha/2) = \cos\left(\pi\alpha\left(\rho - \frac{1}{2}\right)\right), \tag{1.27}$$

as required. With the choices of c_1 and c_2 in (1.26), it is obvious that the first equation in (1.22) gives us $b = 1$, while the second equation in (1.22) becomes an identity, that is, both the left- and right-hand sides are equal to ρ. In conclusion, the choices in (1.26) necessarily hold if $\mathcal{S}_{\mathrm{norm}}(\alpha,\rho)$. □

1.5 Distributional identities

We now consider the probability distribution of stable random variables. This includes understanding where the distribution is supported for the different parameter regimes of α and ρ.

Let $p(x,\alpha,\rho)$ denote the density of $\mathcal{S}_{\mathrm{norm}}(\alpha,\rho)$, where x belongs to its support. Note that, because stable random variables are infinitely divisible but do not belong to the class of compound Poisson distributions, their support is either in the positive half-line, the negative half-line or in the whole real line. Moreover, it is easy to verify that the density $p(x,\alpha,\rho)$ exists and that it is infinitely differentiable. Indeed, observe, for example, from (1.27) that for all values of admissible parameters (α,ρ), we have

$$\left|\alpha\left(\frac{1}{2} - \rho\right)\right| < \frac{1}{2}.$$

Since for $z \in \mathbb{R}$, we necessarily have that

$$\mathrm{Re}(\Psi(z)) = \cos\left(\pi\alpha\left(\frac{1}{2} - \rho\right)\right)|z|^{\alpha},$$

one can deduce that the function $\exp\{-\Psi(z)\}$ is integrable and decays to zero faster than $|z|^{-n}$ for any $n \geq 2$. Therefore, the inverse Fourier transform, which gives $p(\cdot,\alpha,\rho)$, is well defined as follows

$$p(x,\alpha,\rho) = \frac{1}{2\pi}\int_{\mathbb{R}} \mathrm{e}^{-\Psi(z)-\mathrm{i}zx}\,\mathrm{d}z = \frac{1}{\pi}\mathrm{Re}\left[\int_0^\infty \mathrm{e}^{-\Psi(z)-\mathrm{i}zx}\,\mathrm{d}z\right], \tag{1.28}$$

for x in the support of the distribution of X. Moreover, with the given decay of $\exp\{-\Psi(z)\}$, one can similarly write the derivatives of p as inverse Fourier transforms. The next theorem provides the Mellin transform of the positive part of a stable random variable. This identity will be very useful in the sequel.

Theorem 1.13 *Assume that* $X \sim \mathcal{S}_{\mathrm{norm}}(\alpha,\rho)$. *Then for all* $s \in \mathbb{C}$ *in the strip* $-1 < \mathrm{Re}(s) < \alpha$, *we have*

$$\mathbb{E}\left[X^s \mathbf{1}_{(X>0)}\right] = \frac{\sin(\pi\rho s)}{\sin(\pi s)}\frac{\Gamma(1-s/\alpha)}{\Gamma(1-s)}. \tag{1.29}$$

Proof Assume that $-1 < \mathrm{Re}(s) < 0$ and $\alpha \neq 1$. Using (1.28), we obtain

$$\begin{aligned}
\mathbb{E}\left[X^s \mathbf{1}_{(X>0)}\right] &= \int_0^\infty x^s p(x)\,\mathrm{d}x \\
&= \frac{1}{\pi}\mathrm{Re}\left(\int_0^\infty x^s \int_0^\infty \mathrm{e}^{-\Psi(z)-\mathrm{i}zx}\,\mathrm{d}z\,\mathrm{d}x\right) \\
&= \frac{\Gamma(s+1)}{\pi}\mathrm{Re}\left(\mathrm{e}^{-\pi\mathrm{i}(s+1)/2}\int_0^\infty \mathrm{e}^{-\Psi(z)} z^{-s-1}\,\mathrm{d}z\right) \\
&= \frac{\Gamma(s+1)}{\pi\alpha}\Gamma\left(-\frac{s}{\alpha}\right)\mathrm{Re}\left(\mathrm{e}^{-\frac{\pi\mathrm{i}}{2}(s+1)+\pi\mathrm{i}(\frac{1}{2}-\rho)s}\right) \\
&= -\frac{\Gamma(s+1)}{\pi\alpha}\Gamma\left(-\frac{s}{\alpha}\right)\sin(\pi\rho s).
\end{aligned}$$

The last expression is equivalent to the right-hand side of (1.29), after applying the recursion formulae (Eq. A.8 in the Appendix) and the reflection formula for the gamma function (see identity (A.12) in the Appendix). We have proved (1.29) for $\mathrm{Re}(s) \in (-1,0)$ and now we need to use an analytic continuation argument to ensure that it holds for $\mathrm{Re}(s) \in [0,\alpha)$. To this end, we first observe from Theorem 1.10 that

$$\mathbb{E}[|X|^s] < \infty \qquad \text{when} \quad 0 \le s < \alpha.$$

This implies that $\mathbb{E}\left[X^s \mathbf{1}_{(X>0)}\right]$ is analytic in the strip $-1 < \mathrm{Re}(s) < \alpha$. It is not difficult to see that the right-hand side of (1.29) is also an analytic function in the aforesaid domain. The identity thus holds by a standard analytic continuation argument. The case $\alpha = 1$ also follows from continuity properties of both sides of (1.29) in the parameters α and ρ. □

Remark 1.14 As there is explosion on the right-hand side of (1.29) at the critical values $s = \alpha$ and $s = -1$, the above result, in fact, gives us necessary and sufficient conditions for the existence of finite absolute moments. Indeed, $\mathbb{E}[|X|^s] < \infty$ if and only if $-1 < s < \alpha$. This extends the result of Theorem 1.10.

The following corollary to Theorem 1.13 determines whether the support is the negative or positive half-line or the whole real line.

Corollary 1.15 *Assume that X has the same distribution as* $\mathcal{S}_{\mathtt{norm}}(\alpha,\rho)$. *Then*

$$\mathbb{P}(X > 0) = \rho.$$

In particular, if $\alpha \in (0,1)$ *and* $\rho = 1$ *(resp.* $\rho = 0$*), the support of X is the positive half-line (resp. negative half-line) and, in any other case, the support of any stable law is* $\mathbb{R}$.

Proof The first conclusion follows by taking limits, as s goes to 0, in (1.29). Combining it with the comments before Proposition 1.12, the remaining statement in the corollary follows. □

The following result, known as *Zolotarev's duality*, relates the density of a stable distribution with parameters (α,ρ) to the density of a stable distribution with parameters $(1/\alpha, \alpha\rho)$ whenever they are admissible; cf. (1.24).

Theorem 1.16 (Zolotarev's duality) *Assume that both pairs* (α,ρ) *and* $(1/\alpha, \alpha\rho)$ *are admissible. Then for X and* $\tilde{X}$ *which are distributed as* $\mathcal{S}_{\mathtt{norm}}(\alpha,\rho)$ *and* $\mathcal{S}_{\mathtt{norm}}(1/\alpha, \alpha\rho)$, *respectively, we have*

$$\mathbb{P}(X^{-\alpha} \in B, X > 0) = \frac{1}{\alpha}\mathbb{P}(\tilde{X} \in B, \tilde{X} > 0), \tag{1.30}$$

for all Borel sets B.

Proof Let us denote the function on the right-hand side of (1.29) by $m(s;\alpha,\rho)$. By applying the reflection formula (A.12) and the recursion formula (A.8) for the gamma function, it is easy to see that, for all s in the strip $-1 < \mathrm{Re}(s) < \alpha$, we have

$$m(-\alpha s; \alpha,\rho) \equiv \alpha^{-1} m(s; \alpha^{-1}, \alpha\rho),$$

which implies the statement of the Theorem. □

The result of Theorem 1.16 can also be expressed in terms of the density functions as follows

$$z^{-\frac{1}{\alpha}} p\left(z^{-\frac{1}{\alpha}}, \alpha, \rho\right) = z p\left(z, \tfrac{1}{\alpha}, \alpha\rho\right), \tag{1.31}$$

for z in the support of X.

Next, observe that, if the support of the distribution $\mathcal{S}_{\text{norm}}(1/\alpha, \alpha\rho)$ is the real line, then

$$p(-x, \alpha, \rho) = p(x, \alpha, 1 - \rho),$$

thus it is enough to study this function for $x > 0$ in this case. Below we give expressions for the density $p(x, \alpha, \rho)$. We start by treating the case of $\alpha = 1$, the Cauchy distribution, separately.

Theorem 1.17 *When $\alpha = 1$ and $\rho = 1/2$, we have*

$$p(x, 1, 1/2) = \frac{1}{\pi(1 + x^2)}, \qquad x \in \mathbb{R}.$$

Proof Recalling that the distribution is symmetric, we can appeal to Theorem 1.13 and check that the Mellin transform of $p(x, 1, 1/2)$ on the positive half-line is equal to $\sin(\pi s/2)/\sin(\pi s)$, for $-1 < s < 1$. To this end, we note that, for $-1 < s < 1$,

$$\begin{aligned}
\frac{\sin(\pi s/2)}{\sin(\pi s)} &= 2\frac{\Gamma(s)\Gamma(-s)}{\Gamma(s/2)\Gamma(-s/2)} \\
&= \frac{1}{2\pi}\Gamma\left(\frac{1}{2} + \frac{s}{2}\right)\Gamma\left(\frac{1}{2} - \frac{s}{2}\right) \\
&= \frac{1}{2}\int_0^\infty \frac{y^{\frac{s}{2}-\frac{1}{2}}}{\pi(1 + y)}\,\mathrm{d}y \\
&= \int_0^\infty \frac{x^s}{\pi(1 + x^2)}\,\mathrm{d}x,
\end{aligned}$$

where the first equality uses the recursion formula (A.8) and the reflection formula for gamma functions (A.12), the second follows from the duplication formula for gamma functions (A.14), the third uses the definition of the beta function in (A.18) (all of the last four identities found in the Appendix) and the final equality is the result of a change of variables. □

In all other cases we have a convergent power series representation, as described in the next theorem.

Theorem 1.18 *If $\alpha \in (0, 1)$, then*

$$p(x, \alpha, \rho) = \frac{1}{\pi}\sum_{n\geq 1}(-1)^{n-1}\frac{\Gamma(1 + \alpha n)}{n!}\sin(n\pi\alpha\rho)x^{-n\alpha-1}, \qquad x > 0, \tag{1.32}$$

and if $\alpha \in (1, 2)$ then

$$p(x, \alpha, \rho) = \frac{1}{\pi}\sum_{n\geq 1}(-1)^{n-1}\frac{\Gamma(1 + n/\alpha)}{n!}\sin(n\pi\rho)x^{n-1}, \qquad x > 0. \tag{1.33}$$

Moreover, when $\alpha \in (0,1)$ (resp. $\alpha \in (1,2)$) and $|\beta| \neq 1$ (i.e. $0 < \alpha\rho, \alpha\hat{\rho} < 1$) formula (1.33) (resp. (1.32)) provides complete asymptotic expansion as x goes to 0^+ (resp. as x goes to ∞).

Remark 1.19 Before passing to the proof, it is worth emphasising to the unfamiliar reader that the statement above for the asymptotic expansions in the two regimes $\alpha \in (0,1)$ and $\alpha \in (1,2)$ do indeed rely in the series expansion for the opposite regime.

Proof of Theorem 1.18 According to Theorem 1.13, the Mellin transform of $p(x, \alpha, \rho)$ on $(0, \infty)$, satisfies

$$M(z) := \int_0^\infty p(x, \alpha, \rho) x^{z-1}\,\mathrm{d}x = \frac{\sin(\pi\rho(z-1))}{\sin(\pi(z-1))} \frac{\Gamma(1-(z-1)/\alpha)}{\Gamma(2-z)}. \tag{1.34}$$

Observe that this function has simple poles at points $z = 1 + n\alpha$, $n \geq 1$, and $z = -m$, $m \geq 0$. Then by applying Proposition A.1 and identity (A.11) (both in the Appendix), we find that

$$\begin{aligned}\operatorname{Res}(M, 1+n\alpha) &= \left[\frac{\sin(\pi\rho(z-1))}{\sin(\pi(z-1))\Gamma(2-z)}\right]_{z=1+n\alpha} \\ &\qquad \times \operatorname{Res}(\Gamma(1-(z-1)/\alpha), z = 1+n\alpha) \\ &= \frac{\sin(n\pi\alpha\rho)}{\sin(n\pi\alpha)\Gamma(1-n\alpha)} \times (-1)^n \frac{\alpha}{(n-1)!}.\end{aligned}$$

Finally using the reflection formula for the gamma function (A.12) and simplifying the result, we arrive at

$$\operatorname{Res}(M, 1+n\alpha) = -\frac{1}{\pi}(-1)^{n-1}\frac{\Gamma(1+\alpha n)}{n!}\sin(n\pi\alpha\rho).$$

On the other hand, from (A.16), we deduce

$$|\sin(x+\mathrm{i}y)| \sim \frac{\exp\{|y|\}}{2}, \qquad \text{as} \quad y \to \infty,$$

and

$$\left|\frac{\Gamma((x+\mathrm{i}y)/\alpha)}{\Gamma(x+\mathrm{i}y)}\right| \sim \exp\left\{-\frac{\pi}{2}\left(\frac{1}{\alpha}-1\right)|y|\right\}|y|^{x(\frac{1}{\alpha}-1)}\alpha^{-\frac{x}{\alpha}+\frac{1}{2}}, \qquad \text{as} \quad y \to \infty,$$

uniformly in any finite interval $-\infty < a \leq x \leq b < \infty$. This gives us the estimate

$$|M(x+\mathrm{i}y)| \leq C|y|^{x(\frac{1}{\alpha}-1)}\exp\left\{-\frac{\pi}{2}\left(\frac{1}{\alpha}+1-2\rho\right)|y|\right\}, \tag{1.35}$$

as for all y sufficiently large, where $C > 0$ is an unimportant constant. Note that, when $\alpha \in (0,1)$, the exponential term in (1.35) is decreasing on account of the fact that $1/\alpha > 1 > \rho$. Moreover, when $\alpha \in (1,2)$, the exponential term is again decreasing on account of the fact that $\alpha\rho \leq 1$, in which case

$$\frac{1}{\alpha} + 1 - 2\rho \geq 1 - \frac{1}{\alpha} > 0.$$

As such, $M(z)$ is absolutely integrable on the vertical line $c + \mathrm{i}\mathbb{R}$, where c is a constant in $(0, 1 + \alpha)$, therefore we may use the Mellin transform inversion formula

$$p(x,\alpha,\rho) = \frac{1}{2\pi\mathrm{i}} \int_{c+\mathrm{i}\mathbb{R}} M(z)x^{-z}\,\mathrm{d}z.$$

Let us define $b_k = 1 + \alpha(2k+1)/2$ and set ℓ to be an integer. We also consider the contour $L = L_1 \cup L_2 \cup L_3 \cup L_4$, defined as

$$\begin{aligned} L_1 &:= \{\mathrm{Re}(z) = c,\ -\ell \leq \mathrm{Im}(z) \leq \ell\}, \\ L_2 &:= \{\mathrm{Im}(z) = \ell,\ c \leq \mathrm{Re}(z) \leq b_k\}, \\ L_3 &:= \{\mathrm{Re}(z) = b_k,\ -\ell \leq \mathrm{Im}(z) \leq \ell\}, \\ L_4 &:= \{\mathrm{Im}(z) = -\ell,\ c \leq \mathrm{Re}(z) \leq b_k\}. \end{aligned}$$

It is clear that L is the rectangle bounded by vertical lines $\mathrm{Re}(z) = c$, $\mathrm{Re}(z) = b_k$ and by horizontal lines $\mathrm{Im}(z) = \pm\ell$. We assume that L is oriented counter-clockwise; see Figure 1.1.

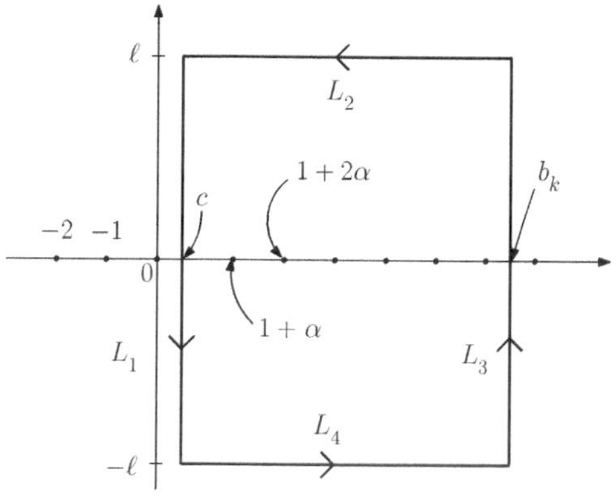

Figure 1.1 The contour $L = L_1 \cup L_2 \cup L_3 \cup L_4$

The function $M(z)$ is analytic in the interior of L, except for simple poles at $s_j = 1 + \alpha j$, for $1 \leq j \leq k$ and is continuous on L. Using the residue theorem we find

$$\frac{1}{2\pi i}\int_L M(z)x^{-z}\,dz = \sum_{j=1}^{k}\text{Res}(M, s_j)\times x^{-s_j}.$$

Next, we estimate the integrals over the horizontal side L_2 as follows

$$\left|\int_{L_2} M(z)x^{-z}\,dz\right| < (b_k - c)\times x^{-b_k}\max_{z\in L_2}|M(z)|.$$

When ℓ increases, we have $\max_{z\in L_2}|M(z)|$ goes to 0. Therefore,

$$\int_{L_2} M(z)x^{-z}\,dz \to 0 \qquad \text{as} \quad \ell\to\infty.$$

Similarly, we deduce that the integral on the contour L_4 goes to 0 as ℓ goes to ∞. Thus putting all the pieces together, we have

$$-\frac{1}{2\pi i}\int_{c+i\mathbb{R}} M(z)x^{-z}\,dz + \frac{1}{2\pi i}\int_{b_k+i\mathbb{R}} M(z)x^{-z}\,dz = \sum_{j=1}^{k}\text{Res}(M, s_j)\times x^{-s_j}.$$

In other words, we have deduced

$$p(x,\alpha,\rho) = -\sum_{n=1}^{k}\text{Res}(M, 1+n\alpha)x^{-(1+\alpha n)} + \frac{1}{2\pi i}\int_{b_k+i\mathbb{R}} M(z)x^{-z}\,dz. \tag{1.36}$$

Now suppose that $\alpha\in(0,1)$. Our aim is to prove that as k goes to ∞, the integral of the right-hand side of (1.36) converges to 0 for $x > 0$. Intuitively this is clear, since the Mellin transform (1.34) can be rewritten with the help of the reflection formula (A.12) as

$$-\frac{\sin(\pi\rho(z-1))}{\sin(\pi(z-1)/\alpha)}\frac{\Gamma(z-1)}{\Gamma((z-1)/\alpha)}.$$

Noting that, for $z = b_k + iu$, $u\in\mathbb{R}$, the ratio of sine functions above, say $H(z)$, is periodic and uniformly bounded in u and b_k as $k\to\infty$. The integral in the right-hand side of (1.36) can thus be estimated as follows

$$\left|\int_{b_k+i\mathbb{R}} M(z)x^{-z}\,dz\right| < x^{-b_k}\int_{\mathbb{R}}\left|\frac{\Gamma(b_k-1+iu)}{\Gamma((b_k-1+iu)/\alpha)}\right| H(b_k-1+iu)\,du. \tag{1.37}$$

To see why the integral on the right-hand side of (1.37) is finite, we can appeal to (A.15) and deduce

$$\frac{\Gamma(z)}{\Gamma(z/\alpha)} = \frac{1}{\sqrt{\alpha}}\exp\left\{-z\left(\frac{1-\alpha}{\alpha}\ln(z) - A\right) + O(z^{-1})\right\}, \quad |z|\to\infty, \text{Re}(z)\geq 0,$$

where $A = (1+\ln(\alpha)+\alpha)/\alpha$, which is negative as we have assumed $\alpha\in(0,1)$. This implies that the right-hand side of (1.37) is finite. Observe, moreover, that $\Gamma(z)/\Gamma(z/\alpha)$ is continuous in the half-plane $\text{Re}(z)\geq 0$, thus

$$x^{-b_k}\left|\frac{\Gamma(b_k-1+\mathrm{i}u)}{\Gamma((b_k-1+\mathrm{i}u)/\alpha)}\right|$$

converges to 0 as k goes to ∞ (uniformly in $u \in \mathbb{R}$), implying that the integral in the right-hand side of (1.37) vanishes as k goes to ∞ and the series representation in (1.32) for the case $\alpha \in (0, 1)$ follows.

We can also pick out elements of the previous arguments to help us prove the asymptotic expansion for $\alpha \in (1, 2)$, as x goes to ∞. More precisely, we take $x \in (1, \infty)$ and observe that the integral in the right-hand side in (1.36) can be estimated as follows

$$\left|\int_{b_k+\mathrm{i}\mathbb{R}} M(z)x^{-z}\,\mathrm{d}z\right| < x^{-b_k}\int_{\mathbb{R}}\left|M(b_k+\mathrm{i}r)\right|\mathrm{d}r = O(x^{-b_k}) \quad \text{as} \quad x \to \infty, \tag{1.38}$$

where the integral on the right-hand side of (1.38) is finite thanks to the estimate (1.35).

The series representation in (1.33) follows from (1.31) (Zolotarev's duality) and (1.32). The asymptotic expansion (1.33) for $\alpha \in (0, 1)$, as x goes to 0 follows from similar arguments for the case $\alpha \in (1, 2)$ as $x \to \infty$. Indeed, one has an identity in the spirit of (1.36), which is constructed from a rectangular contour integral which contains, for example, the first k negative poles. Then an argument similar to (1.38) provides the desired asymptotic. □

1.6 Stable distributions in higher dimensions

Similarly to the one-dimensional case, one can define infinitely divisible distributions in $\mathbb{R}^d$. Just as in one dimension, the characteristic exponent of an $\mathbb{R}^d$ infinitely divisible distribution also has a Lévy–Khintchine representation. Replacing scalar products by Euclidean inner products and understanding $|\cdot|$ as the associated norm, we have, for $z \in \mathbb{R}^d$,

$$\Psi(z) = \mathrm{i}\mathrm{a}\cdot z + \frac{1}{2}z\cdot Qz + \int_{\mathbb{R}^d}\left(1 - \mathrm{e}^{\mathrm{i}z\cdot x} + \mathrm{i}z\cdot x\mathbf{1}_{(|x|<1)}\right)\Pi(\mathrm{d}x), \tag{1.39}$$

where $\mathrm{a} \in \mathbb{R}^d$, Q is a symmetric nonnegative-definite $d \times d$ matrix and Π is a measure on $\mathbb{R}^d$ satisfying

$$\Pi(\{0\}) = 0 \qquad \text{and} \qquad \int_{\mathbb{R}^d}(1 \wedge |x|^2)\,\Pi(\mathrm{d}x) < \infty.$$

We say that X is a d-dimensional *stable* distribution if it takes values on $\mathbb{R}^d$ and satisfies (1.1), where addition is understood in the vectorial sense. For the same reason as in the one-dimensional setting, stable distributions on $\mathbb{R}^d$ are also infinitely divisible distributions. Theorem 1.20 identifies them in terms of a polar decomposition of their Lévy measure.

Theorem 1.20 *An infinitely divisible $\mathbb{R}^d$-valued distribution is a stable distribution if and only if the Lévy–Khintchine representation of its characteristic exponent satisfies either*

(i) $Q \neq 0$, $\Pi \equiv 0$ *and* $\mathrm{a} = 0$, *where* $\alpha = 2$, *or*
(ii) $Q = 0$ *and there is a finite measure* Λ *on* $\mathbb{S}^{d-1} = \{x \in \mathbb{R}^d : |x| = 1\}$ *such that*

$$\Pi(B) = \int_{\mathbb{S}^{d-1}} \Lambda(\mathrm{d}\phi) \int_{(0,\infty)} \mathbf{1}_B(r\phi) \frac{\mathrm{d}r}{r^{\alpha+1}}, \qquad \textit{for} \quad B \in \mathcal{B}(\mathbb{R}^d \setminus \{0\}), \tag{1.40}$$

where $\mathcal{B}(\mathbb{R}^d \setminus \{0\})$ *denotes the Borel* σ*-algebra of* $\mathbb{R}^d \setminus \{0\}$ *and* $\alpha \in (0,2)$. *If* $\alpha \in (0,1) \cup (1,2)$, *then* $\mathrm{a} = 0$.

Proof We first observe that Lemma 1.6 also holds in this case, implying that for $\alpha > 0$ and $k \geq 1$,

$$k\Psi(z) = \Psi(k^{1/\alpha}z), \qquad \text{for} \quad z \in \mathbb{R}^d. \tag{1.41}$$

The above identity and similar arguments to those used in the proof of Proposition 1.7 imply that if $Q \neq 0$, then necessarily $\alpha = 2$, $\Pi \equiv 0$ and $\mathrm{a} = 0$.

If we assume $Q = 0$, then identity (1.41) implies that

$$k\Pi(B) = \Pi(k^{-1/\alpha}B) \qquad \text{for} \quad B \in \mathcal{B}(\mathbb{R}^d \setminus \{0\}).$$

In particular for $\Gamma_\ell = \{x \in \mathbb{R}^d : |x| \geq \ell, x/|x| \in D\}$, where $\ell > 0$ and $D \in \mathcal{B}(\mathbb{S}^{d-1})$, we deduce

$$k\Pi(\Gamma_\ell) = \Pi(k^{-1/\alpha}\Gamma_\ell) = \Pi(\Gamma_{\ell k^{-1/\alpha}}),$$

and by taking $\ell = (k/n)^{1/\alpha}$, we obtain

$$\Pi\left(\Gamma_{(k/n)^{1/\alpha}}\right) = \frac{1}{k}\Pi(\Gamma_{n^{-1/\alpha}}) = \frac{n}{k}\Pi(\Gamma_1).$$

Since the mapping $y \mapsto \Pi(\Gamma_y)$ is monotone decreasing, an argument appealing to denseness of the rational numbers again (see the proof of Proposition 1.7) allows us to obtain

$$\Pi(\Gamma_x) = x^{-\alpha}\Pi(\Gamma_1), \qquad \text{for} \quad x > 0. \tag{1.42}$$

Next, we introduce a finite measure on $\mathbb{S}^{d-1}$ as follows

$$\Lambda(D) = \alpha\Pi(\Gamma_1), \qquad \text{for} \quad D \in \mathcal{B}(\mathbb{S}^{d-1}),$$

and define $\Gamma_{\ell_1,\ell_2} = \{x \in \mathbb{R}^d : \ell_1 \leq |x| < \ell_2,\ x/|x| \in D\}$. Therefore, by identity (1.42), we have

$$\begin{aligned}\Pi(\Gamma_{\ell_1,\ell_2}) &= \Pi(\Gamma_{\ell_1}) - \Pi(\Gamma_{\ell_2})\\ &= \frac{\ell_1^{-\alpha} - \ell_2^{-\alpha}}{\alpha}\Lambda(D)\\ &= \int_D \int_{\ell_1}^{\ell_2} \frac{\mathrm{d}r}{r^{1+\alpha}}\Lambda(\mathrm{d}\phi)\\ &= \int_{\mathbb{S}^{d-1}} \Lambda(\mathrm{d}\phi) \int_{(0,\infty)} \mathbf{1}_{\Gamma_{\ell_1,\ell_2}}(r\phi)\frac{\mathrm{d}r}{r^{\alpha+1}}.\end{aligned}$$

Since the sets Γ_{ℓ_1,ℓ_2} fulfill the conditions of Dynkin's Lemma (or π–λ Theorem) then (1.40) holds for any Borel set of $\mathbb{R}^d \setminus \{0\}$.

Finally, we observe

$$\int_{(|x|<1)} |x|^2\, \Pi(\mathrm{d}x) = \Lambda(\mathbb{S}^{d-1}) \int_0^1 r^{1-\alpha}\, \mathrm{d}r.$$

Since Π is a Lévy measure, we necessarily have in addition to $\alpha > 0$ that $\alpha < 2$. □

Remark 1.21 As in the one-dimensional case, we will proceed ignoring the Gaussian setting in part (i) of Theorem 1.20, allowing us to focus on the regime $\alpha \in (0, 2)$.

Appealing to equation (1.40) and undertaking computations similar in spirit to those done in the proof of Theorem 1.3, one readily finds that the characteristic exponent of a stable distribution on $\mathbb{R}^d$ can be written as follows, for $z \in \mathbb{R}^d$,

$$\Psi(z) = -\Gamma(-\alpha)\cos\left(\frac{\pi\alpha}{2}\right)\int_{\mathbb{S}^{d-1}} |z\cdot\phi|^\alpha \left(1 - \mathrm{i}\tan\left(\frac{\pi\alpha}{2}\right)\mathrm{sgn}(z\cdot\phi)\right)\Lambda(\mathrm{d}\phi),$$

for $\alpha \neq 1$, and

$$\Psi(z) = \mathrm{i}z\cdot a + \frac{\pi}{2}\int_{\mathbb{S}^{d-1}} \big(|z\cdot\phi| + \mathrm{i}z\cdot\phi\log(z\cdot\phi)\big)\Lambda(\mathrm{d}\phi),$$

for $\alpha = 1$, where $a \in \mathbb{R}^d$ and Λ satisfies

$$\int_{\mathbb{S}^{d-1}} \phi\, \Lambda(\mathrm{d}\phi) = 0.$$

The *isotropic* case will be of particular interest in what follows. Recall that a measure μ on $\mathbb{R}^d$ is *symmetric* if $\mu(B) = \mu(-B)$ for $B \in \mathcal{B}(\mathbb{R}^d)$ and *isotropic* if $\mu(B) = \mu(U^{-1}B)$ for every orthogonal matrix U. Observe that when $d = 1$, the notion of isotropy is equivalent to symmetry.

Definition 1.22 We say that X is a *symmetric stable* random variable if it is stable distributed and its law is symmetric.

In particular, if X is symmetric we have that its characteristic function is equal to that of $-X$, that is, $\Psi(z) = \Psi(-z)$ for $z \in \mathbb{R}^d$. By considering the real and imaginary parts of Ψ, the latter clearly implies that

$$\Psi(z) = \int_{\mathbb{S}^{d-1}} |z \cdot \phi|^\alpha \, \Lambda_0(\mathrm{d}\phi) \qquad \text{for} \quad \alpha \in (0, 2), \tag{1.43}$$

where

$$\Lambda_0(\mathrm{d}\phi) = \begin{cases} -\Gamma(-\alpha)\cos\left(\frac{\pi\alpha}{2}\right)\Lambda(\mathrm{d}\phi) & \text{if } \alpha \neq 1, \\ \frac{\pi}{2}\Lambda(\mathrm{d}\phi) & \text{if } \alpha = 1. \end{cases}$$

Using the double angle trigonometric identity and Euler's reflection formula (A.12), we have for $\alpha \neq 1$,

$$\begin{aligned} -\Gamma(-\alpha)\cos\left(\frac{\pi\alpha}{2}\right) &= -\Gamma(-\alpha)\frac{\sin(\pi\alpha)}{2\sin\left(\frac{\pi\alpha}{2}\right)} \\ &= -\frac{\Gamma(-\alpha)}{2}\frac{\Gamma(\alpha/2)\Gamma(1-\alpha/2)}{\Gamma(\alpha)\Gamma(1-\alpha)} \\ &= \frac{\Gamma(\alpha/2)\Gamma(1-\alpha/2)}{2\Gamma(\alpha+1)}. \end{aligned}$$

It is important to note that the right-hand side in the third equality is equal to $\pi/2$ when $\alpha = 1$ thanks to the special values $\Gamma(1/2) = \sqrt{\pi}$ and $\Gamma(2) = 1$. Therefore, in the sequel we define

$$\Lambda_0(\mathrm{d}\phi) = \frac{\Gamma(\alpha/2)\Gamma(1-\alpha/2)}{2\Gamma(\alpha+1)}\Lambda(\mathrm{d}\phi),$$

for $\alpha \in (0, 2)$.

Definition 1.23 We say that X is an *isotropic stable* random variable if it is stable distributed and, for all orthogonal transforms $B\colon \mathbb{R}^d \mapsto \mathbb{R}^d$, BX is equal in law to X.

Observe that if X belongs to this class then X is also symmetric and, consequently, its characteristic function satisfies (1.43) and it can be written as follows

$$\Psi(z) = |z|^\alpha \int_{\mathbb{S}^{d-1}} ||z|^{-1} z \cdot \phi|^\alpha \, \Lambda_0(\mathrm{d}\phi).$$

Since the law of X is isotropic then $\Psi(z) = \Psi(Uz)$, for any orthogonal matrix, U, and therefore, without loss of generality, we may take Λ equal to Lebesgue (surface) measure on $\mathbb{S}^{d-1}$ and, thanks to symmetry,

$$c = \int_{\mathbb{S}^{d-1}} \| |z|^{-1} z \cdot \phi|^{\alpha} \Lambda_0(\mathrm{d}\phi),$$

should be a positive constant, that is, for $\alpha \in (0, 2)$, we have

$$\Psi(z) = c|z|^{\alpha} \qquad \text{for} \quad z \in \mathbb{R}^d. \tag{1.44}$$

Similarly as in the one dimensional case, the explicit computation of the constant c is important for our purposes. In that case, we may take

$$c = \int_{\mathbb{S}^{d-1}} |\mathbf{1} \cdot \phi|^{\alpha} \, \Lambda_0(\mathrm{d}\phi),$$

where $\mathbf{1} = (1, 0, \cdots, 0) \in \mathbb{R}^d$ is the 'North Pole' on $\mathbb{S}^{d-1}$. Hence, using skew product coordinates (also called generalised polar coordinates) in $\mathbb{R}^d$, we deduce

$$\begin{aligned}
\int_{\mathbb{S}^{d-1}} |\mathbf{1} \cdot \phi|^{\alpha} \, \Lambda_0(\mathrm{d}\phi) &= \frac{\Gamma(\alpha/2)\Gamma(1-\alpha/2)}{2\Gamma(\alpha+1)} \frac{2\pi^{(d-1)/2}}{\Gamma((d-1)/2)} \int_0^{\pi} \sin^{d-2}(\theta)|\cos(\theta)|^{\alpha} \mathrm{d}\theta \\
&= 2\frac{\Gamma(\alpha/2)\Gamma(1-\alpha/2)}{\Gamma(\alpha+1)} \frac{\pi^{(d-1)/2}}{\Gamma((d-1)/2)} \int_0^{\pi/2} \sin^{d-2}(\theta) \cos^{\alpha}(\theta) \mathrm{d}\theta \\
&= \pi^{(d-1)/2} \frac{\Gamma(\alpha/2)\Gamma(1-\alpha/2)}{\Gamma(\alpha+1)} \frac{\Gamma((\alpha+1)/2)}{\Gamma((d+\alpha)/2)} \\
&= 2^{1-\alpha}\pi^{d/2} \frac{\Gamma(1-\alpha/2)}{\Gamma(\alpha+1)} \frac{\Gamma(\alpha)}{\Gamma((d+\alpha)/2)} \\
&= 2^{1-\alpha}\pi^{d/2} \frac{\Gamma(1-\alpha/2)}{\alpha\Gamma((d+\alpha)/2)} \\
&= 2^{-\alpha}\pi^{d/2} \frac{|\Gamma(-\alpha/2)|}{\Gamma((d+\alpha)/2)},
\end{aligned}$$

where, from the Appendix, we have used (A.19) in the third equality and the duplication formula (A.14) in the fourth equality. We thus have

$$c = 2^{-\alpha}\pi^{d/2} \frac{|\Gamma(-\alpha/2)|}{\Gamma((d+\alpha)/2)}. \tag{1.45}$$

We may now say the isotropic d-dimensional stable process is *normalised* if

$$\Psi(z) = |z|^{\alpha}, \qquad z \in \mathbb{R}^d.$$

This corresponds to setting

$$\Pi(B) = \frac{1}{c} \int_{\mathbb{S}^{d-1}} \Lambda(\mathrm{d}\phi) \int_0^{\infty} \mathbf{1}_B(r\phi) \frac{\mathrm{d}r}{r^{\alpha+1}}, \qquad \text{for} \quad B \in \mathcal{B}(\mathbb{R}^d \setminus \{0\}),$$

where $\Lambda(\mathrm{d}\phi)$ is the surface measure on $\mathbb{S}^{d-1}$.

It is also worth noting for later that the Lévy measure can be written as absolutely continuous with respect to d-dimensional Lebesgue measure as well as with respect to the skew product measure $\sigma_1(\mathrm{d}\phi)r^{d-1}\mathrm{d}r$, where $\sigma_1(\mathrm{d}\phi)$ is the surface measure on $\mathbb{S}^{d-1}$ normalised to have unit mass.

Theorem 1.24 *For a normalised isotropic stable distribution, that is, having characteristic exponent* $\Psi(z) = |z|^\alpha$, $z \in \mathbb{R}$, *we have*

$$\Pi(B) = 2^\alpha \pi^{-d/2} \frac{\Gamma((d+\alpha)/2)}{|\Gamma(-\alpha/2)|} \int_B \frac{\mathrm{d}z}{|z|^{\alpha+d}} \tag{1.46}$$

and in (generalised) polar coordinates,

$$\Pi(B) = 2^{\alpha-1}\pi^{-d} \frac{\Gamma((d+\alpha)/2)\Gamma(d/2)}{|\Gamma(-\alpha/2)|} \int_{\mathbb{S}^{d-1}} \sigma_1(\mathrm{d}\phi) \int_0^\infty \mathbf{1}_B(r\phi)\frac{1}{r^{\alpha+1}}\,\mathrm{d}r, \tag{1.47}$$

where B is a Borel set in $\mathbb{R}^d$.

Remark 1.25 Also taking inspiration from the one-dimensional case, in particular the computations in the proof of Theorem 1.3, and the statement in Corollary 1.9, we note that, the value of a in (1.39) can be identified explicitly within the choice of normalisation in Theorem 1.24. Specifically, when $\alpha \in (0,1)$ we have $\mathrm{a} = -\int_{\mathbb{R}^d} x\mathbf{1}_{(|x|<1)}\Pi(\mathrm{d}x)$; when $\alpha \in (1,2)$ we have $\mathrm{a} = \int_{\mathbb{R}^d} x\mathbf{1}_{(|x|\geq 1)}\Pi(\mathrm{d}x)$; and when $\alpha = 1$, we have $\mathrm{a} = 0$.

1.7 Comments

The class of stable distributions appeared for the first time in the celebrated monograph of Paul Lévy [143] *Calcul de probabilités*. They were introduced by Lévy as limits for normalised sums of independent identically distributed random variables that do not satisfied a finite second moment condition. The concept of stable distributions was fully developed in 1937 with the appearance of the monographs of Lévy [144] and Khintchine [106]. Both authors characterised them as a class of infinitely divisible distributions. Since their debut, stable distributions have appeared in a vast array of probabilistic models motivated by physics, biology and economics, to name but some of the many areas of influence.

There are several monographs where stable distributions are treated in detail, see for instance Gnedenko and Kolmogorov [80], Linde [146], Sato [190] and Uchaikin and Zolotarev [208]. There are also other monographs that deal specifically with stable distributions or stochastic processes which are associated to them, such as Samorodnitsky and Taqqu [189] and Zolotarev [220].

Zolotarev [220] (see also Uchaikin and Zolotarev [208]) gave a complete treatment to real-valued stable distributions and in particular the series representation and asymptotic expansions of their densities were described in a concise way for the first time. The multidimensional case is treated in Chapter 2 of Samorodnitsky and Taqqu [189] and in Uchaikin and Zolotarev [208]. In Chapter 1 of Samorodnitsky and Taqqu [189], the one-dimensional case is also treated.

The explicit form of the characteristic exponent of stable distributions (Theorem 1.3) was first treated by Lévy [142] and by Khintchine and Lévy [107]. The formulation of Theorem 1.3 follows Gnedenko and Kolmogorov [80] with the corrections of Hall [86]. We stress that the article of Hall corrected a number of different derivations of Theorem 1.3 that have appeared in the literature up to the beginning of the 1980's. Lemma 1.6 is based on similar arguments used in Pitman and Pitman [169]. The proof of Proposition 1.7 and Theorem 1.20 follows from Kuelbs [114], where stable distributions are defined on Hilbert spaces. Theorem 1.10 is based on a more general result for moments of infinitely divisible random variables found in Section 25 of Sato [190]. The moment inequality (1.19) is taken from Lemma 1 of Biggins [30].

The existence of a distributional density thanks to infinite divisibility (that predicates Section 1.5) can be found in Section 24 of Sato [190]. All the results that appear in Section 1.5 can be found in the monographs of Uchaikin and Zolotarev [208] and Zolotarev [220] and we adopt a similar approach, for example in Theorem 1.13. On a final note, we mention that it was shown in Hoffmann-Jørgensen [108] that the densities derived in Theorem 1.18 can be expressed in terms of incomplete hypergeometric functions.

2

Lévy processes

As a precursor to our introduction to stable Lévy processes in the next chapter, as well as a point of reference for our future treatment of self-similar Markov processes, we shall spend some time in this chapter reviewing standard path properties of one- and higher-dimensional Lévy processes.

Lévy processes can be seen as the natural continuous-time analogue of random walks in the sense that they have stationary and independent increments. They are formally defined as follows.

Definition 2.1 (Lévy process) A stochastic process $Y = (Y_t, t \geq 0)$, valued in $\mathbb{R}$ and defined on a probability space $(\Omega, \mathcal{F}, \mathrm{P})$, is said to be a (one-dimensional) Lévy process issued from the origin if it possesses the following properties:

(i) the paths of Y are P-almost surely right-continuous with left limits;
(ii) $\mathrm{P}(Y_0 = 0) = 1$;
(iii) for $0 \leq s \leq t$, $Y_t - Y_s$ is equal in distribution to Y_{t-s};
(iv) for $0 \leq s \leq t$, $Y_t - Y_s$ is independent of $(Y_u, u \leq s)$.

With the exception of Section 2.18, we will concentrate on the one-dimensional case for the rest of this chapter.

2.1 Lévy–Itô decomposition

Thanks to stationary and independent increments and right-continuity, any Lévy process has the property that, for all $t \geq 0$,

$$\mathrm{E}\left[\mathrm{e}^{\mathrm{i}zY_t}\right] = \mathrm{e}^{-t\Psi(z)}, \qquad z \in \mathbb{R}, \tag{2.1}$$

where $\Psi(z) := \Psi_1(z)$ is the *characteristic exponent* of Y_1 in the sense of (1.4).

It turns out that the converse is also true; that is to say, given any infinitely divisible distribution with characteristic exponent Ψ, there exists a Lévy process Y which satisfies (2.1). The conclusion of the previous paragraph and its converse, together give us the following Lévy–Khintchine formula for processes with stationary and independent increments.

Theorem 2.2 (Lévy–Khintchine formula for Lévy processes) *Suppose that $a \in \mathbb{R}$, $\sigma^2 \geq 0$ and Π is a measure concentrated on $\mathbb{R}\backslash\{0\}$ satisfying $\int_{\mathbb{R}}(1 \wedge x^2)\Pi(\mathrm{d}x) < \infty$. From this triple, define for each $z \in \mathbb{R}$*

$$\Psi(z) = \mathrm{i}az + \frac{1}{2}\sigma^2 z^2 + \int_{\mathbb{R}} (1 - \mathrm{e}^{\mathrm{i}zx} + \mathrm{i}zx\mathbf{1}_{(|x|<1)})\,\Pi(\mathrm{d}x). \tag{2.2}$$

If Ψ is the characteristic exponent of a Lévy process in the sense of (2.1), *then it necessarily satisfies (2.2). Conversely, given* (2.2), *there exists a probability space,* $(\Omega, \mathcal{F}, \mathrm{P})$, *on which a Lévy process is defined having characteristic exponent Ψ in the sense of* (2.1).

Remark 2.3 It is important to note that a given Lévy processes is identified by its characteristic exponent only up to a linear dilation in time. If $Y := (Y_t, t \geq 0)$ is a Lévy process with characteristic exponent Ψ then, if $c > 0$ is a constant, the process $Y^c := (Y_{ct}, t \geq 0)$ is the Lévy process with exponent $c\Psi$. The processes Y and Y^c have trajectories which differ only by a linear scaling in time, but are fundamentally the same as they almost surely have the same range.

Two important examples of Lévy processes which help us understand the structure of general Lévy processes are *linear Brownian motion* and *compound Poisson processes with drift*. For the first of these two cases, suppose $\sigma^2 \geq 0$ and $a \in \mathbb{R}$. If we write $B = (B_t, t \geq 0)$ for a standard one-dimensional Brownian motion, then the linear Brownian motion, $\sigma B_t - at$, $t \geq 0$, is a process issued from the origin with stationary and independent increments, continuous paths and Gaussian distributed at each fixed time. Taking account of the Fourier transform of a Gaussian distribution, it is easily verified that the associated characteristic exponent is given by

$$\mathrm{i}az + \frac{1}{2}\sigma^2 z^2, \qquad z \in \mathbb{R}.$$

For the case of a compound Poisson process with drift, suppose that $N(\mathrm{d}t, \mathrm{d}x)$ is a Poisson random measure on $[0, \infty) \times \mathbb{R}$ with intensity $\lambda \mathrm{d}t \times F(\mathrm{d}x)$, where F is a probability distribution concentrated on $\mathbb{R}\backslash\{0\}$. Recall that this means N is a random counting measure on $[0, \infty) \times \mathbb{R}$ such that, for all pairwise disjoint Borel sets $B_1, \ldots, B_n$, $n \in \mathbb{N}$, in $[0, \infty) \times \mathbb{R}$, the counts $N(B_1), \ldots, N(B_n)$ are independent. Moreover they are Poisson distributed with parameter $\int_{B_i} \lambda \mathrm{d}t F(\mathrm{d}x)$, $i = 1, \ldots, n$, respectively. See Appendix A.13 for further details. A compound Poisson process with arrival rate λ, jump distribution F and drift $c \in \mathbb{R}$ can be written in terms of the counting process N via

$$\int_0^t \int_{\mathbb{R}} x N(\mathrm{d}s, \mathrm{d}x) - ct, \qquad t \geq 0.$$

The associated characteristic exponent is then easily computed with the help of Campbell's formula for Poisson random measures and is given by

$$\lambda \int_{\mathbb{R}} (1 - \mathrm{e}^{\mathrm{i}zx}) F(\mathrm{d}x) + \mathrm{i}cz \qquad z \in \mathbb{R}. \tag{2.3}$$

Taking account of the previous two examples, the Lévy–Khintchine formula gives some insight into the path structure of a general Lévy process. Indeed, after some simple reorganisation, we can write the general Lévy–Khintchine exponent in the form

$$\begin{aligned}\Psi(z) = &\left\{ \mathrm{i}az + \frac{1}{2}\sigma z^2 \right\} \\ &+ \left\{ \Pi(x \in \mathbb{R} : |x| \geq 1) \int_{\{|x|\geq 1\}} (1 - \mathrm{e}^{\mathrm{i}zx}) \frac{\Pi(\mathrm{d}x)}{\Pi(x \in \mathbb{R} : |x| \geq 1)} \right\} \\ &+ \left\{ \int_{\{|x|<1\}} (1 - \mathrm{e}^{\mathrm{i}zx} + \mathrm{i}zx) \Pi(\mathrm{d}x) \right\}, \qquad z \in \mathbb{R}.\end{aligned} \tag{2.4}$$

Note that the integrability condition on Π appearing in Theorem 2.2 implies that $\Pi(A) < \infty$ for all Borel $A \subseteq \mathbb{R}$ such that 0 is in the interior of A^c. In particular, it also implies that $\Pi(x \in \mathbb{R} : |x| \geq 1) \in [0, \infty)$. In the case that $\Pi(x \in \mathbb{R} : |x| \geq 1) = 0$, one should think of the second set of curly brackets in (2.4) as absent. Let us name the contents of the three sets of curly brackets in (2.4) as $\Psi^{(1)}(z)$, $\Psi^{(2)}(z)$ and $\Psi^{(3)}(z)$, respectively. The essence of this decomposition revolves around showing that $\Psi^{(1)}(z)$, $\Psi^{(2)}(z)$ and $\Psi^{(3)}(z)$ correspond to the characteristic exponents of three different types of Lévy processes. In this way, Ψ is the characteristic exponent of the independent sum of these three Lévy processes, which is again a Lévy process. Indeed, it is an easy exercise to verify that the characteristic exponent of the sum of independent Lévy processes is equal to the sum of their individual exponents; moreover that, up to a multiplicative constant, Lévy processes are uniquely identified by their characteristic exponents.

It is already clear that $\Psi^{(1)}$ and $\Psi^{(2)}$ correspond, respectively, to a linear Brownian motion, say, $Y^{(1)}$ (which is entirely deterministic if $\sigma = 0$) and an independent compound Poisson process, say $Y^{(2)}$, with arrival rate $\Pi(x \in \mathbb{R} : |x| \geq 1)$ and jump distribution $F(\mathrm{d}x) = \Pi(\mathrm{d}x)/\Pi(x \in \mathbb{R} : |x| \geq 1)$ concentrated on $\{x : |x| \geq 1\}$ (unless $\Pi(x \in \mathbb{R} : |x| \geq 1) = 0$, in which case $Y^{(2)}$ is the process which is identically zero).

The existence of a Lévy process with characteristic exponent given by (2.4) thus boils down to showing the existence of a Lévy process, $Y^{(3)}$, whose characteristic exponent is given by $\Psi^{(3)}$. This can be seen by considering the Poisson random measure, $N(\mathrm{d}t, \mathrm{d}x)$, on $[0, \infty) \times \{x \in \mathbb{R} : |x| < 1\}$ with intensity measure $\mathrm{d}t \times \Pi(\mathrm{d}x)|_{\{x \in \mathbb{R}:|x|<1\}}$. From our computations in (2.3), we can deduce that, for each $\varepsilon > 0$, the process

$$\int_{[0,t]} \int_{\{\varepsilon<|x|<1\}} x N(\mathrm{d}s, \mathrm{d}x) - \left(\int_{\{\varepsilon<|x|<1\}} x \Pi(\mathrm{d}x) \right) t, \qquad t \geq 0, \tag{2.5}$$

is a compound Poisson process with drift (in fact the drift is the mean of the compound Poisson part), whose characteristic exponent is given by

$$\int_{\{\varepsilon<|x|<1\}} (1 - \mathrm{e}^{\mathrm{i}zx} + \mathrm{i}zx) \Pi(\mathrm{d}x).$$

It is straightforward to show that (2.5) is also a square integrable martingale. The process $Y^{(3)}$ should be thought of as the limit, in an appropriate sense, of the compound Poisson process with drift in (2.5) as $\varepsilon \downarrow 0$. Despite the fact that the drift coefficient $\int_{\{\varepsilon<|x|<1\}} x\Pi(\mathrm{d}x)$ may explode, it turns out that the limit does exist and the resulting process remains in the class of Lévy processes. The mathematical sense of the limit requires some care, using the machinery of L^2 convergence in an appropriate Hilbert space of martingales. This L^2 convergence also motivates the necessary and sufficient requirement that $\int_{|x|<1} x^2 \Pi(\mathrm{d}x) < \infty$.

The identification of a general Lévy process, Y, as the independent sum of processes $Y^{(1)}$, $Y^{(2)}$ and $Y^{(3)}$ is known as the *Lévy–Itô decomposition*.

2.2 Killing

When discussing the theory of Lévy processes, in particular in the application to self-similar Markov processes later on in this text, we will need to consider Lévy processes with exponential killing. Suppose that $\tilde{Y}$ is a given Lévy process, then we can additionally define the Lévy process Y killed via

$$Y_t = \begin{cases} \tilde{Y}_t & \text{if } t < \zeta, \\ \partial & \text{otherwise,} \end{cases}$$

where ∂ is a cemetery state annexed to $\mathbb{R}$. As we will discuss below, a special form of killing corresponds to the case that we take $\zeta = \mathbf{e}_q$, an independent and exponentially distributed random variable with rate $q > 0$. Note that the definition still makes sense when $q = 0$, providing we agree with the definition $\mathbf{e}_q := \infty$, so that we understand this case as corresponding to no killing at all.

The lack of memory property of the exponential distribution means that the new process $Y = (Y_t, t < \zeta)$ still has stationary and independent increments in $\mathbb{R}$ on survival. Indeed, one can easy verify that Definition 2.1 is still valid, providing we understand (iii) and (iv) to occur on the event $\{s < \mathbf{e}_q\}$. As such, it is not unnatural to think of the definition of a Lévy process in this slightly broader sense. Accordingly, the characteristic exponent for a general Lévy process needs to be identified in a slightly more general sense.

Suppose we write $\tilde{\Psi}$ for the characteristic exponent of $\tilde{Y}$. Then

$$\mathrm{E}[\mathrm{e}^{\mathrm{i}zY_t}\mathbf{1}_{(t<\zeta)}] = \mathrm{E}[\mathrm{e}^{\mathrm{i}z\tilde{Y}_t}\mathbf{1}_{(t<\mathbf{e}_q)}] = \mathrm{e}^{-(q+\tilde{\Psi}(z))t}, \qquad t \geq 0.$$

Then, $\Psi = q + \tilde{\Psi}$.

For the most part, in this book, what we mean by a 'Lévy process' will conform to Definition 2.1, however, there are occasions when we will need to understand the definition in the killed sense. The context should always be clear in order to distinguish the two scenarios. That said, for the remainder of this chapter, unless otherwise stated, we continue our discussion of Lévy processes without killing.

2.3 Path variation and asymmetry

It is clear from the Lévy–Itô decomposition that the presence of the linear Brownian motion $Y^{(1)}$ would imply that paths of the Lévy process have unbounded variation. On the other hand, should it be the case that $\sigma = 0$, then the Lévy process may or may not have unbounded variation. The term $Y^{(2)}$, being a compound Poisson process, has only bounded variation. Hence, in the case $\sigma = 0$, understanding whether the Lévy process has unbounded variation is an issue determined by the limiting process $Y^{(3)}$.

Reconsidering the definition of $Y^{(3)}$, it is natural to ask: Under what circumstances does

$$\lim_{\varepsilon \downarrow 0} \int_{[0,t]} \int_{\{\varepsilon<|x|<1\}} x\, N(\mathrm{d}s, \mathrm{d}x)$$

exist almost surely without the need for compensation by its mean as in (2.5)? The answer to this question boils down to an analysis of compound Poisson sums through Campbell's formula and it turns out that

$$\int_{[0,t]} \int_{\{|x|<1\}} |x|\, N(\mathrm{d}s, \mathrm{d}x) < \infty \iff \int_{\{|x|<1\}} |x|\, \Pi(\mathrm{d}x) < \infty.$$

In that case, we may identify $Y^{(3)}$ directly via

$$Y_t^{(3)} = \int_{[0,t]} \int_{\{|x|<1\}} x\, N(\mathrm{d}s, \mathrm{d}x) - t \int_{\{|x|<1\}} x\, \Pi(\mathrm{d}x), \qquad t \geq 0.$$

This also tells us that $Y^{(3)}$ will be of bounded variation if and only if $\int_{\{|x|<1\}} |x|\Pi(\mathrm{d}x) < \infty$. This is a stronger integrability condition than $\int_{\{|x|<1\}} x^2\Pi(\mathrm{d}x) < \infty$. In conclusion, we get the following lemma.

Lemma 2.4 *A Lévy process with Lévy–Khintchine exponent corresponding to the triple* (a, σ, Π) *has paths of bounded variation if and only if*

$$\sigma = 0 \quad and \quad \int_{\mathbb{R}} (1 \wedge |x|)\, \Pi(\mathrm{d}x) < \infty. \tag{2.6}$$

Note that the finiteness of the integral in (2.6) also allows for the Lévy–Khintchine exponent of any such bounded variation process to be rewritten in the form

$$\Psi(z) = -\mathrm{i}\mathsf{b}z + \int_{\mathbb{R}} (1 - \mathrm{e}^{\mathrm{i}zx})\, \Pi(\mathrm{d}x), \tag{2.7}$$

where $\mathsf{b} \in \mathbb{R}$ relates to a and Π via

$$\mathsf{b} = -\left(a + \int_{\{|x|<1\}} x\, \Pi(\mathrm{d}x)\right).$$

In this case, we may write the Lévy process in the form

$$Y_t = \mathsf{b}t + \int_{[0,t]} \int_{\mathbb{R}} x\, N(\mathrm{d}s, \mathrm{d}x), \qquad t \geq 0. \tag{2.8}$$

The constant b is referred to as the *drift coefficient*. (The reader should note that the word 'drift' here makes sense because the process has paths of bounded variation and the linear term can be uniquely distinguished from the 'pure jump' component. For other Lévy processes which, for example, have paths of unbounded variation, it is less clear what 'drift' should mean.)

When Ψ is that of a bounded variation Lévy process, to detect the presence of a drift term, it is a straightforward exercise to show that

$$\lim_{|z|\to\infty} \frac{\Psi(z)}{z} = -\mathrm{i}\mathsf{b}. \tag{2.9}$$

The following lemma is now obvious.

Lemma 2.5 *A Lévy process is a compound Poisson process with linear drift if and only if* $\sigma = 0$ *and* $\Pi(\mathbb{R}) < \infty$.

Let us now consider the case of total asymmetry in the Lévy measure and suppose that $\Pi(-\infty, 0) = 0$. From the proof of the Lévy–Itô decomposition, we see that the corresponding Lévy process has no negative jumps. If further we have that $\int_{(0,\infty)}(1 \wedge x)\Pi(\mathrm{d}x) < \infty$, $\sigma = 0$ and, in the representation (2.7) of the characteristic exponent, $\mathrm{b} \geq 0$, then from the representation (2.8) it becomes clear that the Lévy process has non-decreasing paths. In that case, the Lévy process is referred to as a *subordinator*. Conversely, if a Lévy process is a subordinator, then necessarily it has bounded variation. Hence $\int_{(0,\infty)}(1 \wedge x)\Pi(\mathrm{d}x) < \infty$, $\sigma = 0$ and then it is easy to see that in the representation (2.7) of the characteristic exponent, we necessarily have $\mathrm{b} \geq 0$. Summarising, we have the following.

Lemma 2.6 *A Lévy process is a subordinator if and only if* $\Pi(-\infty, 0) = 0$, $\int_{(0,\infty)}(1 \wedge x)\,\Pi(\mathrm{d}x) < \infty$, $\sigma = 0$ *and* $\mathrm{b} = -\left(a + \int_{(0,1)} x\,\Pi(\mathrm{d}x)\right) \geq 0$.

For the sake of clarity, we note that, when Y is a subordinator, further to (2.7), its Lévy–Khintchine formula may be written as

$$\Psi(z) = -\mathrm{i}\mathrm{b}z + \int_{(0,\infty)} (1 - \mathrm{e}^{\mathrm{i}zx})\,\Pi(\mathrm{d}x), \qquad z \in \mathbb{R}.$$

Reconsidering the role of the characteristic exponent Ψ and its analytical representation through infinite divisibility and the Lévy–Itô decomposition, respectively, in the case of a subordinator, it is not difficult to see that one may also work with the Laplace exponent

$$\kappa(\lambda) := -\frac{1}{t} \log \mathrm{E}[\mathrm{e}^{-\lambda Y_t}], \qquad \lambda \geq 0. \tag{2.10}$$

Indeed, one easily verifies that

$$\kappa(\lambda) = \Psi(\mathrm{i}\lambda) = \mathrm{b}\lambda + \int_{(0,\infty)} (1 - \mathrm{e}^{-\lambda x})\,\Pi(\mathrm{d}x), \qquad \lambda \geq 0. \tag{2.11}$$

In general, we say that a Lévy process is *spectrally positive* if $\Pi(-\infty, 0) = 0$ and it does not have monotone paths. A Lévy process, Y, will then be referred to as a *spectrally negative* if $-Y$ is spectrally positive. Together, these two classes of processes are called *spectrally one-sided*. Spectrally one-sided Lévy processes may be of bounded or unbounded variation and, in the latter case, may or may not possess a Gaussian component. Note in particular that when $\sigma = 0$, it is still possible to have paths of unbounded variation. If a spectrally positive Lévy process has bounded variation, then it must take the form

$$Y_t = -\mathrm{b}t + S_t, \qquad t \geq 0, \tag{2.12}$$

where $(S_t, t \geq 0)$ is a pure jump subordinator and, necessarily, $\mathrm{b} > 0$. Note that if $\mathrm{b} \leq 0$, then Y would conform to the definition of a subordinator. The decomposition (2.12) implies that if $\mathrm{E}[Y_1] \leq 0$, then $\mathrm{E}[S_1] < \infty$, as opposed to the case that $\mathrm{E}[Y_1] > 0$, where it is possible that $\mathrm{E}[S_1] = \infty$.

A special feature of spectrally positive processes is that, if

$$\tau_x^- := \inf\{t > 0 : Y_t < x\}, \tag{2.13}$$

where $x < 0$, then $\mathrm{P}(\tau_x^- < \infty) > 0$. Roughly speaking, this probability is strictly positive because, in the setting of bounded variation, the structure (2.12) ensures that there is the possibility of downward movement, whereas, in the setting of unbounded variation, it is the fluctuations of the martingale component of Y that ensures the possibility of downward movement. As there are no downwards jumps,

$$\mathrm{P}(Y_{\tau_x^-} = x | \tau_x^- < \infty) = 1, \tag{2.14}$$

with a similar property for first passage upwards being true for spectrally negative processes.

2.4 Feller and strong Markov property

Lévy processes fit nicely into the class of Feller processes, which, in turn, ensures that they possess the strong Markov property. See the discussion in Section A.11 for a reminder of what these things mean.

It is easy to see that the Markov property is satisfied for all Lévy processes. Indeed, Lévy processes satisfy the stronger condition that the law of $Y_{t+s} - Y_t$ is independent of $\mathcal{F}_t := \sigma(Y_s, s \leq t)$, for all $s, t \geq 0$. As such, we work with the family of probability measures P_x, $x \in \mathbb{R}$, where $\mathrm{P}_x(\cdot) = \mathrm{P}(\cdot \,|\, Y_0 = x)$. In the special case of Lévy processes, because of their stationary independent increments, for each $x \in \mathbb{R}$, we can also define P_x to be the law of $x + Y$, under P.

Next, we introduce the semigroup $(P_t, t \geq 0)$ for Y. For each $t \geq 0$, P_t operates on bounded measurable functions f such that $P_t[f](x) = \mathrm{E}_x[f(Y_t)]$, $x \in \mathbb{R}$. Let $C_0(\mathbb{R})$ be the space of bounded measurable functions which decay to 0 as $|x| \to \infty$. By appealing to the simple fact that, for $f \in C_0(\mathbb{R})$, $P_t[f](x) = \mathrm{E}[f(x + Y_t)]$, together with dominated convergence, it is not difficult to show that the semigroup associated to a Lévy process respects the Feller property (see Definition A.14 in the Appendix). As such, Lévy processes are Feller processes and hence strong Markov processes. In fact, Lévy processes satisfy the stronger condition that they preserve the property of stationary and independent increments over stopping times.

Theorem 2.7 *Suppose that τ is a stopping time. On $\{\tau < \infty\}$, define the process $\widetilde{Y} = \{\widetilde{Y}_t, t \geq 0\}$, where*

$$\widetilde{Y}_t = Y_{T+t} - Y_T, \ t \geq 0.$$

Then, on the event $\{T < \infty\}$, the process $\widetilde{Y}$ is independent of $\mathcal{F}_T$, has the same law as Y and hence in particular is a Lévy process.

In this text, we are largely (but not exclusively) concerned with two types of stopping times for use in conjunction with the strong Markov property. The first type are those taking the form

$$\tau^D = \inf\{t > 0 : Y_t \in D\},$$

where D is either an open or closed set in $\mathbb{R}$. The second will be stopping times of the form

$$\tau_u = \inf\left\{t > 0 : \int_0^t f(Y_s)\,\mathrm{d}s > u\right\}, \tag{2.15}$$

where $f : \mathbb{R} \to (0, \infty)$ is measurable and $u \geq 0$. In both cases, we always work with the standard notion that $\inf \emptyset = \infty$. For example, in (2.15), it may happen that $\int_0^\infty f(Y_s)\,\mathrm{d}s < \infty$, in which case, $\tau_u = \infty$, for all $u > \int_0^\infty f(Y_s)\,\mathrm{d}s$.

2.5 Infinitesimal generator

On account of it being a Feller process, a Lévy process is also in possession of an infinitesimal generator. That is to say, for f belonging to an appropriate class of functions, there exists

$$\mathcal{L}f(x) := \lim_{t\downarrow 0} \frac{P_t[f](x) - f(x)}{t}, \qquad x \in \mathbb{R}. \tag{2.16}$$

The form of the generator can be closely matched to the characteristic exponent Ψ of the Lévy process. In particular, if

$$\Psi(z) = \mathrm{i}az + \frac{1}{2}\sigma^2 z^2 + \int_{\mathbb{R}} (1 - \mathrm{e}^{\mathrm{i}zx} + \mathrm{i}zx\mathbf{1}_{(|x|<1)})\Pi(\mathrm{d}x), \qquad z \in \mathbb{R},$$

then, for all twice continuously differentiable and compactly supported $f : \mathbb{R} \mapsto \mathbb{R}$,

$$\mathcal{L}f(x) = -af'(x) + \frac{1}{2}\sigma^2 f''(x) + \int_{\mathbb{R}} [f(x+y) - f(x) - yf'(x)\mathbf{1}_{(|y|<1)}]\,\Pi(\mathrm{d}y),$$

for $x \in \mathbb{R}$.

2.6 Drifting and oscillating

Thanks to the Lévy–Khintchine formula for Lévy processes, it is straightforward to see that, when the mean of Y is well defined so that $\Psi'(0) := \lim_{z\to 0} \Psi'(z)$ exists, by straightforward differentiation of (2.1), it follows that

$$\mathrm{E}[Y_t] = \mathrm{i}\Psi'(0)t = \mathrm{E}[Y_1]t, \qquad t \geq 0. \tag{2.17}$$

When it exists, we call $\mathrm{E}[Y_1]$ *the mean of the Lévy process* as it characterises $\mathrm{E}[Y_t]$ for all $t \geq 0$. When the mean of the Lévy process is well defined and finite, together with the stationary and independent increments, (2.17) provides compelling evidence that one should expect to see a strong law of large numbers for Y. In turn, this would provide some information about the large-time behaviour of Y. The following theorem gives a relatively complete picture in this respect. In particular it shows that there is a strict trichotomy. Either Y *drifts to* ∞ ($\lim_{s\to\infty} Y_s = \infty$), Y *drifts to* $-\infty$ ($\lim_{s\to\infty} Y_s = -\infty$) or Y *oscillates* ($\limsup_{s\to\infty} Y_s = -\liminf_{s\to\infty} Y_s = \infty$).

Theorem 2.8 *Suppose that Y is a one-dimensional Lévy process with characteristic measure Π.*

(i) If $\mathrm{E}[Y_1]$ is defined and valued in $[-\infty, 0)$, or if $\mathrm{E}[Y_1]$ is undefined and

$$\int_{(1,\infty)} \frac{x\,\Pi(\mathrm{d}x)}{\int_0^x \Pi(-\infty, -y)\,\mathrm{d}y} < \infty, \tag{2.18}$$

then $\lim_{t\to\infty} Y_t/t = \gamma_-$, where $\gamma_- = \mathrm{E}[Y_1]$ in the first case and $\gamma_- = -\infty$ in the second case. In particular, in both cases,

$$\lim_{t\to\infty} Y_t = -\infty.$$

(ii) If $\mathrm{E}[Y_1]$ is defined and valued in $(0, \infty]$, or if $\mathrm{E}[Y_1]$ is undefined and

$$\int_{(-\infty,-1)} \frac{|x|\,\Pi(\mathrm{d}x)}{\int_0^{|x|} \Pi(y, \infty)\,\mathrm{d}y} < \infty, \tag{2.19}$$

then $\lim_{t\uparrow\infty} Y_t/t = \gamma_+$, where $\gamma_+ = \mathrm{E}[Y_1]$ in the first case and $\gamma_+ = \infty$ in the second case. In particular, in both cases,

$$\lim_{t\to\infty} Y_t = \infty.$$

(iii) If $\mathrm{E}[Y_1]$ is defined and equal to zero, or if $\mathrm{E}[Y_1]$ is undefined and both of the integral tests in parts (i) and (ii) fail, then $\lim_{t\to\infty} Y_t/t = 0$ in the first case and $\limsup_{t\to\infty} Y_t/t = -\liminf_{t\to\infty} Y_t/t = \infty$ in the second case. Moreover, in both cases,

$$\limsup_{t\to\infty} Y_t = -\liminf_{t\to\infty} Y_t = \infty.$$

2.7 Moments

We will occasionally be interested to know whether moments of a Lévy process exist. Specifically, if g is a measurable function on $\mathbb{R}$, when can we say that $\mathrm{E}[g(Y_t)] < \infty$? More generally, when can we say that $\mathrm{E}[g(\sup_{s\leq t} |Y_t|)] < \infty$?

An appropriate class of measurable functions to work with are called *submultiplicative*. Specifically, a non-negative measurable function g on $\mathbb{R}$ is submultiplicative if

$$g(x+y) \le g(x)g(y), \qquad x, y \in \mathbb{R}.$$

The class of submultiplicative functions is relatively rich containing some key examples such as $g(x) = |x| \vee 1$, and $g(x) = \exp\{x\}$. More generally, suppose that g is a submultiplicative function, then so is $g(cx+b)^\theta$ where $c, b \in \mathbb{R}$ and $\theta > 0$.

Theorem 2.9 *Suppose that g is a locally bounded, submultiplicative function then* $\mathrm{E}[g(Y_t)] < \infty$*, for some* $t > 0$*, if and only if* $E[g(Y_t)] < \infty$*, for all* $t > 0$*, if and only if*

$$\int_{\{|x|>1\}} g(x)\,\Pi(\mathrm{d}x) < \infty.$$

Moreover, $\mathrm{E}[g(\sup_{s\le t}|Y_s|)] < \infty$*, for some* $t > 0$*, if and only if* $\mathrm{E}[g(\sup_{s\le t}|Y_s|)] < \infty$*, for all* $t > 0$*, if and only if*

$$\int_{\{|x|>1\}} g(|x|)\,\Pi(\mathrm{d}x) < \infty.$$

2.8 Exponential change of measure

Theorem 2.9 gives us a criterion under which we can perform an exponential change of measure, even in the setting of a killed Lévy process. Define the Laplace exponent

$$\psi(\beta) = \frac{1}{t}\log \mathrm{E}[\mathrm{e}^{\beta Y_t}\mathbf{1}_{(t<\zeta)}] = -\Psi(-\mathrm{i}\beta), \tag{2.20}$$

whenever it exists. Theorem 2.9 tells us that the Laplace exponent is finite if and only if

$$\int_{\{|x|\ge 1\}} \mathrm{e}^{\beta x}\,\Pi(\mathrm{d}x) < \infty. \tag{2.21}$$

Define

$$\mathcal{E}_t(\beta) = \mathrm{e}^{\beta Y_t - \psi(\beta)t}, \qquad 0 \le t < \zeta, \tag{2.22}$$

under the assumption (2.21). Appealing to stationary and independent increments and (2.20), we have that, for all $x \in \mathbb{R}$, on $\{t < \zeta\}$,

$$\begin{aligned}\mathrm{E}_x[\mathcal{E}_{t+s}(\beta)\mathbf{1}_{(t+s<\zeta)}|\mathcal{F}_t] &= \mathcal{E}_t(\beta)\mathrm{E}_x\left[\mathrm{e}^{\beta(Y_{t+s}-Y_t)-\psi(\beta)s}\mathbf{1}_{(t+s<\zeta)}\middle|\mathcal{F}_t\right]\\ &= \mathcal{E}_t(\beta)\mathrm{E}_x\left[\mathrm{e}^{\beta Y_s-\psi(\beta)s}\mathbf{1}_{(s<\zeta)}\right]\\ &= \mathcal{E}_t(\beta).\end{aligned}$$

Hence, under (2.21), $\mathcal{E}(\beta) = \{\mathcal{E}_t(\beta)\colon t \geq 0\}$ is a martingale and it may be used to perform a change of measure via

$$\left.\frac{\mathrm{dP}^\beta_x}{\mathrm{dP}_x}\right|_{\mathcal{F}_t} = \mathrm{e}^{-\beta x}\mathcal{E}_t(\beta)\mathbf{1}_{(t<\zeta)}, \qquad t \geq 0, x \in \mathbb{R}. \tag{2.23}$$

The change of measure above is known as an *exponential change of measure*. As the next theorem shows, it has the important property that the (killed) process Y under P^β is still a Lévy process. More importantly, if Y has killing, then, under the change of measure, it will never experience killing.

Theorem 2.10 *Suppose that Y is a (killed) Lévy process with characteristic triple (a, σ, Π) and killing rate $q \geq 0$. Moreover, suppose that $\beta \in \mathbb{R}$ is such that*

$$\int_{\{|x|\geq 1\}} \mathrm{e}^{\beta x}\,\Pi(\mathrm{d}x) < \infty.$$

Under the exponential change of measure P^β, the process Y is still a Lévy process with characteristic triple (a^*, σ^*, Π^*) and killing rate $q^* = 0$, where

$$a^* = a - \beta\sigma^2 + \int_{\{|x|<1\}} (1 - \mathrm{e}^{\beta x})x\,\Pi(\mathrm{d}x), \ \sigma^* = \sigma \ \textit{and} \ \Pi^*(\mathrm{d}x) = \mathrm{e}^{\beta x}\,\Pi(\mathrm{d}x).$$

A compact way of characterising the effect on Y of the exponential change of measure (2.23) is to consider the characteristic exponent of (Y, P^β), denoted here by Ψ_β. In light of the first part of the above theorem, it is straightforward to see that

$$\Psi_\beta(z) := \Psi(z - \mathrm{i}\beta) - \Psi(-\mathrm{i}\beta), \ z \in \mathbb{R}. \tag{2.24}$$

Equivalently, suppose its Laplace exponent ψ is well defined in the interval (a, b). Then under the change of measure described in (2.23), for which necessarily $\beta \in (a, b)$, it maps to

$$\psi_\beta(\theta) = \psi(\theta + \beta) - \psi(\beta), \tag{2.25}$$

for $\theta + \beta \in (a, b)$. The transformations in (2.23) and (2.25) are commonly referred to as the *Esscher transform*. Recall from Section 2.2 that the presence of killing is detected by a constant term in the characteristic exponent. It is clear from the difference of the exponents in (2.24) and (2.25) why the effect of killing disappears.

2.9 Donsker-type convergence

Donsker's Theorem tells us that one may perform a scaling on the paths of random walks in such a way that there is weak convergence on the Skorokhod space $\mathcal{D}([0,\infty),\mathbb{R})$ to Brownian motion (see Section A.10 of the Appendix for details on the Skorokhod space). Similar results may also be obtained for Lévy processes. For example, Lévy processes with second moments can be similarly scaled to limit to Brownian motion on the Skorokhod space.

The following result tells us that controlling convergence on the Skorokhod space for a Lévy process is tantamount to controlling the behaviour of the Lévy–Khintchine exponent.

Theorem 2.11 *Suppose that Ψ_n is a sequence of characteristic exponents of Lévy processes, say $Y^{(n)}$. Suppose that $\Psi_n(z) \to \Psi(z)$, for all $z \in \mathbb{R}$, where Ψ is the characteristic exponent of a Lévy process, say Y, then $(Y^{(n)}_t, t \geq 0)$ converges weakly on the Skorokhod space, as $n \to \infty$, to $(Y_t, t \geq 0)$.*

Returning to the example of scaling Lévy processes in spirit of Donsker's Theorem, suppose that Y is a Lévy process with zero mean and finite second moments. From Section 2.7, to ensure second moments, it is necessary and sufficient to ask that $\int_{\{|x|>1\}} x^2\Pi(\mathrm{d}x) < \infty$, where Π is the Lévy measure of Y. Now note that, for each $r > 0$, $(r^{-1/2}Y_{rt}, t \geq 0)$ is a Lévy process with characteristic exponent $r\Psi(r^{-1/2}z)$, Ψ is the characteristic exponent of Y. Because Y has zero first moment, it is a straightforward exercise to check that we may write its characteristic exponent in the form

$$\Psi(z) = \frac{1}{2}\sigma^2 z^2 + \int_{\mathbb{R}} (1 - \mathrm{e}^{\mathrm{i}zx} + \mathrm{i}zx)\,\Pi(\mathrm{d}x), \qquad z \in \mathbb{R}.$$

Hence, by dominated convergence,

$$\begin{aligned}\lim_{r\to\infty} r\Psi(r^{-1/2}z) &= \frac{1}{2}\sigma^2 z^2 + \lim_{r\to\infty}\int_{\mathbb{R}} r(1 - \mathrm{e}^{\mathrm{i}zr^{-1/2}x} + \mathrm{i}zr^{-1/2}x)\,\Pi(\mathrm{d}x)\\ &= \frac{z^2}{2}\left(\sigma^2 + \int_{\mathbb{R}} x^2\,\Pi(\mathrm{d}x)\right).\end{aligned}$$

Theorem 2.11 now tells us that $(r^{-1/2}Y_{rt}, t \geq 0)$ converges weakly on the Skorokhod space, as $r \to \infty$, to a Brownian motion with variance $\sigma^2 + \int_{\mathbb{R}} x^2\,\Pi(\mathrm{d}x)$.

2.10 Transience and recurrence

As a class of $\mathbb{R}$-valued Markov processes, it is also natural to look at the transience and recurrence properties of Lévy processes. A Lévy process, Y, is said to be transient if, for all $a > 0$,

$$\mathrm{P}\left(\int_0^\infty \mathbf{1}_{(|Y_t|<a)}\,\mathrm{d}t < \infty\right) = 1,$$

and recurrent if, for all $a > 0$,

$$\mathrm{P}\left(\int_0^\infty \mathbf{1}_{(|Y_t|<a)}\,\mathrm{d}t = \infty\right) = 1.$$

In the previous definitions, the requirements for transience and recurrence may appear quite strong as, in principle, the relevant probabilities could be valued in $(0, 1)$. However, the events in the definition belong to the tail sigma-algebra $\bigcap_{t\geq 0}\sigma(Y_s : s \geq t)$. Hence, according to the Hewitt–Savage zero-one law, they cannot have probabilities valued in $(0, 1)$. Nonetheless, we could argue that $\mathrm{P}(\int_0^\infty \mathbf{1}_{(|Y_t|<a)}\mathrm{d}t = \infty) = 0$ for small a, but $\mathrm{P}(\int_0^\infty \mathbf{1}_{(|Y_t|<a)}\mathrm{d}t = \infty) = 1$ for large a. It turns out that Lévy processes always adhere to one of the two cases given in the definition of transience and recurrence above, as is confirmed by the following classic analytic dichotomy.

Theorem 2.12 *Suppose that Y is a Lévy process with characteristic exponent Ψ. Then it is transient if and only if, for some sufficiently small $\varepsilon > 0$,*

$$\int_{\{|z|<\varepsilon\}} \mathrm{Re}\left(\frac{1}{\Psi(z)}\right) \mathrm{d}z < \infty, \tag{2.26}$$

and otherwise it is recurrent.

Straightforward probabilistic reasoning also leads to the following interpretation of the dichotomy.

Theorem 2.13 *Let Y be any Lévy process.*

(i) We have transience if and only if

$$\lim_{t\to\infty} |Y_t| = \infty,$$

almost surely.

(ii) If Y is not a compound Poisson process, then we have recurrence if and only if, for all $x \in \mathbb{R}$,

$$\liminf_{t\to\infty} |Y_t - x| = 0, \tag{2.27}$$

almost surely.

The notion of recurrence thus allows, for each $x \in \mathbb{R}$, the possibility of visiting any arbitrarily small interval around x, say $(x - \varepsilon, x + \varepsilon)$, almost surely infinitely often. The reason for the exclusion of compound Poisson processes in part (ii) of Theorem 2.13 can be seen when one considers the following example. Take Y to be a one-dimensional compound Poisson process, where the jump distribution is supported on a lattice, say $\delta\mathbb{Z}$ for some $\delta > 0$. In that case, it is clear that the set of points visited will be a subset of $\delta\mathbb{Z}$ and (2.27) is no longer valid for, for example, $x \notin \delta\mathbb{Z}$. This example also explains why compound Poisson processes are excluded in other forthcoming results.

2.11 Duality

On account of the fact that a given Lévy process, Y, has stationary independent increments with right-continuous paths having left-limits, it is easy to justify that, when it is time-reversed, the resulting process still has stationary independent increments with paths that are left-continuous having right-limits. The following *Duality Lemma* states that if one adjusts continuity at the jumps of the time-reversed process so that the time reversed path is right-continuous with left limits, then it is easy to identify the resulting object as equal in law to $-Y$.

Lemma 2.14 (Duality Lemma) *For each fixed $t > 0$, define the reversed process*

$$(Y_{(t-s)-} - Y_t, 0 \le s \le t)$$

and the dual process,

$$(-Y_s, 0 \le s \le t).$$

Then the two processes have the same law under P.

It is a straightforward consequence of the Duality Lemma that, when considering deterministic time horizons, the running supremum $\overline{Y}_t := \sup_{s<t} Y_s$ and running infimum $\underline{Y}_t, = \inf_{s\le t} Y_s$ are closely related to one another.

Lemma 2.15 *For each fixed $t > 0$, the pairs $(\overline{Y}_t, \overline{Y}_t - Y_t)$ and $(Y_t - \underline{Y}_t, -\underline{Y}_t)$ have the same distribution under* P.

Suppose we write $\hat{\mathrm{P}}_x$, $x \in \mathbb{R}$, for the probabilities of $\hat{Y} := -Y$. As we shall shortly see, duality can be expressed analytically in terms of the semigroups $(P_t, t \ge 0)$ of Y and $(\hat{P}_t, t \ge 0)$ of $\hat{Y}$, as well as their associated q-resolvents, $U^{(q)}$ and $\hat{U}^{(q)}$, for $q \ge 0$. The latter are defined such that, for bounded measurable f,

$$U^{(q)}[f](x) = \int_0^\infty \mathrm{e}^{-qt} P_t[f](x)\,\mathrm{d}t = \mathrm{E}_x\left[\int_0^\infty \mathrm{e}^{-qt} f(Y_t)\,\mathrm{d}t\right],$$

with the obvious definition holding for $\hat{U}^{(q)}$. When $q = 0$, it is possible that the integral in the first equality above is infinite because Y is recurrent.

The next lemma shows the aforementioned alternative representation of duality in Lemma 2.15.

Lemma 2.16 *Suppose that f and g are non-negative, bounded and measurable functions. Then*

$$\int_{\mathbb{R}} g(x)P_t[f](x)\,\mathrm{d}x = \int_{\mathbb{R}} f(x)\hat{P}_t[g](x)\,\mathrm{d}x$$

and for $q > 0$ *(as well as* $q = 0$ *with f and g compactly supported if Y is transient),*

$$\int_{\mathbb{R}} g(x)U_t^{(q)}[f](x)\,\mathrm{d}x = \int_{\mathbb{R}} f(x)\hat{U}_t^{(q)}[g](x)\,\mathrm{d}x.$$

Lemmas 2.15 and 2.16 scratch the surface of duality theory for semigroups of Lévy processes and their relation to time reversal. We return to this topic for general Markov processes in the Appendix. For now, we mention that the statement in Lemma 2.16 can be extended to the setting where there is killing of the Lévy process on entering a domain.

Suppose that B is an open or closed set and recall that

$$\tau^B := \inf\{t > 0 : Y_t \in B\}.$$

The associated semigroup of Y killed on entering B, say $(P_t^B, t \geq 0)$, is defined by $P_t^B[f](x) = \mathrm{E}_x[f(Y_t);\ t < \tau^B]$, for bounded and measurable f. The associated q-resolvents are written $U_B^{(q)}$, $q \geq 0$. We also use the obvious meaning for $(\hat{P}_t^B, t \geq 0)$ and $\hat{U}_B^{(q)}$, $q \geq 0$. The following result gives what is known as *Hunt's switching identity.*

Lemma 2.17 *Suppose that f and g are non-negative measurable functions and B is an open or closed domain. Then*

$$\int_{\mathbb{R}} g(x)P_t^B[f](x)\,\mathrm{d}x = \int_{\mathbb{R}} f(x)\hat{P}_t^B[g](x)\,\mathrm{d}x$$

and for $q > 0$,

$$\int_{\mathbb{R}} g(x)U_B^{(q)}[f](x)\,\mathrm{d}x = \int_{\mathbb{R}} f(x)\hat{U}_B^{(q)}[g](x)\,\mathrm{d}x.$$

2.12 Hitting points

In Section 2.10, transience and recurrence can be seen as an issue pertaining to the occupation of open intervals. The question of occupying (or, better said, hitting) a single point can also be addressed.

We say that a Lévy process Y *can hit a point* $x \in \mathbb{R}$ if

$$P(Y_t = x \text{ for at least one } t > 0) > 0.$$

Let

$$C = \{x \in \mathbb{R} : P(Y_t = x \text{ for at least one } t > 0) > 0\}$$

be the set of points that a Lévy process can hit. We say that a Lévy process can *hit points* if $C \neq \emptyset$. We have the following classification.

Theorem 2.18 *Suppose that Y is not a compound Poisson process. Then Y can hit points if and only if*

$$\int_{\mathbb{R}} \mathrm{Re}\left(\frac{1}{1+\Psi(z)}\right) \mathrm{d}z < \infty. \tag{2.28}$$

Denote by σ the Gaussian coefficient of Y. Then we have, moreover, that:

(i) If $\sigma > 0$, then Y can hit points and $C = \mathbb{R}$.
(ii) If $\sigma = 0$ but Y is of unbounded variation and Y can hit points then $C = \mathbb{R}$.
(iii) If Y is of bounded variation, then Y can hit points if and only if $\mathrm{b} \neq 0$ where b is the drift in the representation (2.7) of its Lévy–Khintchine exponent Ψ. In that case $C = \mathbb{R}$ unless Y or $-Y$ is a subordinator and then $C = (0, \infty)$ or $C = (-\infty, 0)$ respectively.

The case of a compound Poisson process is excluded above for the same reasons that it was excluded in Theorem 2.13.

It turns out that we can be more specific about hitting probabilities. To this end, let us introduce the q-resolvent measure

$$U^{(q)}(\mathrm{d}x) = \int_0^\infty \mathrm{e}^{-qt} P(Y_t \in \mathrm{d}x)\,\mathrm{d}t, \qquad x \in \mathbb{R}.$$

This notation is consistent with our earlier notion of q-resolvent via the relation $U^{(q)}[f](x) = \int_{\mathbb{R}} f(x+y)U^{(q)}(\mathrm{d}y)$. We can write hitting probabilities in terms of the density, say $\hat{u}^{(q)}$, of the q-resolvent measure of the dual, $\hat{U}^{(q)}$. In that case, we need to work with a 'regular' version of $\hat{u}^{(q)}$, referred to as q-excessive for $-Y$, which ensures that $r\hat{U}^{(r+q)}[f] \leq f$, for all $r > 0$ and measurable $f \geq 0$, and $\lim_{r\to\infty} \hat{U}^{(r+q)}[f] = f$, in the pointwise sense.

Lemma 2.19 *Suppose that Y is not a compound Poisson process and (2.28) holds. Then the resolvent measure of $-Y$ has a q-excessive bounded density, say $\hat{u}^{(q)}$, and, moreover,*

$$E_x\left[\mathrm{e}^{-q\tau^{\{0\}}}\right] = \frac{\hat{u}^{(q)}(x)}{\hat{u}^{(q)}(0)}, \qquad x \in \mathbb{R},$$

where

$$\tau^{\{0\}} = \inf\{t > 0 : Y_t = 0\}.$$

If, moreover, Y is transient, then the result is also valid for $q = 0$. In particular, writing $u = u^{(0)}$, we have

$$\mathrm{P}_x(\tau^{\{0\}} < \infty) = \frac{\hat{u}(x)}{\hat{u}(0)}, \qquad x \in \mathbb{R}.$$

2.13 Regularity of the half-line

For a one-dimensional Lévy process Y (which starts at zero) we say that 0 *is regular for* $(0, \infty)$ (equiv. the upper half-line) if Y enters $(0, \infty)$ immediately. That is to say, if

$$\mathrm{P}(\tau_0^+ = 0) = 1, \quad \text{where} \quad \tau_0^+ = \inf\{t > 0 : Y_t > 0\}.$$

Because of the Blumenthal 0-1 law, the probability $\mathrm{P}(\tau_0^+ = 0)$ is necessarily zero or one. When this probability is zero, we say that 0 is irregular for $(0, \infty)$. We also say that 0 is regular for $(-\infty, 0)$ (equiv. the lower half-line) if $-Y$ is regular for the upper half-line.

Theorem 2.20 *For a Lévy process Y, the point 0 is regular for $(0, \infty)$ if and only if one of the following three situations occurs:*

(i) Y is a process of unbounded variation,
(ii) Y is a process of bounded variation and $\mathrm{b} > 0$ where b is the drift in representation (2.7) of its Lévy–Khintchine exponent Ψ,
(iii) Y is a process of bounded variation, $\mathrm{b} = 0$ (with b as in (ii)) and

$$\int_0^1 \frac{x\,\Pi(\mathrm{d}x)}{\int_0^x \Pi(-\infty, -y)\,\mathrm{d}y} = \infty. \tag{2.29}$$

2.14 Excursions and the Wiener–Hopf factorisation

From the preceding sections, it is apparent that the characteristic exponent, Ψ, and the underlying triple, (a, σ, Π), of any Lévy process encodes a significant amount of information concerning its fine path properties. The Wiener–Hopf factorisation is an analytic factorisation of Ψ that underlines this principle. Through the Wiener–Hopf factorisation, we will see distributional information concerning the local maxima and local minima of the associated Lévy process trajectory can be obtained.

Let us assume henceforth that neither Y nor $-Y$ is a subordinator and not killed. Define the *running maximum* and the *running minimum* processes by

$$\overline{Y}_t = \sup_{s\le t} Y_s \qquad \text{and} \qquad \underline{Y}_t = \inf_{s\le t} Y_s, \qquad t \ge 0,$$

respectively. If we take, for example, the range of the process $\overline{Y} = (\overline{Y}_t, t \ge 0)$, then it corresponds precisely to the range of the first passage points on $(0,\infty)$ given by $\{Y_{\tau_x^+} : x > 0\}$. Therefore, we can expect an understanding of how Lévy processes undergo the process of first passage over a level to be closely related to the distributional properties of $\overline{Y}$ and $\underline{Y}$.

It can be shown that there exists a random measure L on $[0,\infty)$ with the property that its support agrees with the closure of the set $\{t \ge 0 : Y_t = \overline{Y}_t\}$ and with the property that, if T is any $\mathbb{F}$-stopping time such that $Y_T = \underline{Y}_T$ on $\{T < \infty\}$, then $((Y_{T+t} - Y_T, L_{T+t} - L_T), t \ge 0)$ has the same law as $((Y_t, L_t), t \ge 0)$ under $\mathbb{P}$. Moreover, the complement of its support consists of a countable union of open intervals, each one corresponding to an excursion of Y from its running maximum. There is a small technicality to address in the setting that 0 is irregular, for which the process L essentially counts the number of visits to the maximum. To ensure that the forthcoming analysis is valid, it is usual to write L as a sum of auxiliary independent exponential random variables over the number of visits to the maximum instead. In that case, the filtration $\mathbb{F}$ needs to be expanded to include information of the additional randomess.

It is a straightforward exercise to show that the reflected process $\overline{Y} - Y$ is a strong Markov process. The process $L_t := L[0,t]$, $t \ge 0$, is known as the local time at zero for the reflected process, or equivalently the local time of Y at its maximum. It has the special property that $(L_t^{-1}, t \ge 0)$ is a subordinator, which is killed if $L_\infty < \infty$, which, in turn, occurs precisely when $\lim_{t\to\infty} Y_t = -\infty$. For the countable set of times $t > 0$ such that $\Delta L_t^{-1} := L_t^{-1} - L_{t-}^{-1} > 0$, we can identify the excursion of Y from its maximum:

$$\epsilon_t(s) = Y_{L_{t-}^{-1}+s} - Y_{L_{t-}^{-1}}, \qquad 0 \le s \le \Delta L_t^{-1}.$$

As alluded to above, there are a countable number of such excursions due to the fact that there are a countable number of intervals making up the complement of the support of L. The length of each such interval corresponds to a jump of L^{-1}. Together, the excursions $(\epsilon_t, t < L_\infty)$ form a stopped Poisson point process on $[0,\infty) \times \overline{\mathcal{U}}(\mathbb{R})$, where $\overline{\mathcal{U}}(\mathbb{R})$ is the space of paths taking the form $(\epsilon(s) : s \le \zeta)$, where ζ is the path lifetime, which are right-continuous with left limits and which are strictly negative-valued on $(0,\zeta)$. Moreover, when $\zeta < \infty$, $\epsilon(\zeta) \ge 0$. Accordingly, $\zeta = \inf\{t > 0 : \epsilon(t) > 0\}$. The intensity measure of this Poisson point process takes the form $\mathrm{d}t \times \mathrm{d}\overline{n}$, where $\overline{n}$ is a measure on the

Skorokhod space (see Section A.10 in the Appendix), which is concentrated on $\overline{\mathcal{U}}(\mathbb{R})$. It is a consequence of the regenerative nature of the point 0 for $\overline{Y}-Y$ that L_∞ is exponentially distributed (with rate which may be zero, in which case we understand it to be infinite valued with probability one). In the case that $L_\infty < \infty$, the excursion with local time index L_∞ corresponds to the final excursion from the last maximum which never ends as Y drifts to $-\infty$.

We defined the process

$$H_t = Y_{L_t^{-1}}, \qquad t < L_\infty,$$

and otherwise $H_t = \infty$, where ∞ can be considered as a cemetery state. The need for a cemetery state appears if and only if the Lévy process drifts to $-\infty$, that is, $\overline{Y}_\infty := \lim_{t\to\infty} Y_t < \infty$ almost surely, or equivalently $L_\infty < \infty$. The range of the process $H = (H_t, t \geq 0)$ in $[0,\infty)$ agrees with the range of $\overline{Y}$. Moreover, the regenerative property of the state 0 for the reflected process $\overline{Y} - Y$ ensures that H is a subordinator, killed at a constant rate which may be zero (when $L_\infty = \infty$). Clearly the jumps of H correspond to the overshoot at the end of each excursion, that is, the Lévy measure of H is given by $\overline{n}(\epsilon(\zeta) \in \mathrm{d}x)$, $x > 0$. The process H may also possess a drift component, corresponding to there being Lebesgue mass in the range of H. The cause of this is a rather subtle issue and we make no attempt to discuss it here.

By considering the process $-Y$ (which is still a Lévy process), everything we have described above can also be constructed for the process reflected in its infimum, $Y - \underline{Y}$. The range of $\underline{Y}$, or equivalently $(Y_{\tau_x^-}, x < 0) \cup \{0\}$, agrees with the range of $-\hat{H}$, where $\hat{H} = (\hat{H}_t, t \geq 0)$ is a subordinator which is possibly killed at an independent and exponentially distributed time (depending on whether $-\underline{Y}_\infty$ is finite or not) with cemetery state $\{\infty\}$. We call H the *ascending ladder height process* and $\hat{H}$ the *descending ladder height process*.

Suppose that we denote the Laplace exponent of H by κ. To be precise,

$$\kappa(\lambda) = \frac{1}{t}\log \mathrm{E}\left[\mathrm{e}^{-\lambda H_t}\right], \qquad t, \lambda \geq 0.$$

Referring to (2.11), it is easy to deduce that

$$\kappa(\lambda) = q + \mathrm{b}\lambda + \int_{(0,\infty)} (1 - \mathrm{e}^{-\lambda x})\, \Upsilon(\mathrm{d}x), \qquad \lambda \geq 0, \tag{2.30}$$

for some constants $q, \mathrm{b} \geq 0$ and measure Υ concentrated on $(0,\infty)$ satisfying $\int_{(0,\infty)} (1 \wedge x)\Upsilon(\mathrm{d}x) < \infty$. The constant q corresponds to the exponential killing rate when it is strictly positive in value. The Laplace exponent of $\hat{H}$, which we shall henceforth denote by $\hat{\kappa}$, is similarly described.

It is a remarkable fact that its characteristic exponent, Ψ, factorises revealing the two Laplace exponents κ and $\hat{\kappa}$ of the ascending and descending ladder height processes, respectively.

Theorem 2.21 *For $z \in \mathbb{R}$ and constant $c > 0$, we have*

$$\Psi(z) = c\kappa(-\mathrm{i}z)\hat{\kappa}(\mathrm{i}z). \tag{2.31}$$

Equality 2.31 is what is commonly referred to as the *Wiener–Hopf factorisation*. What is interesting is that the factorisation still makes sense when we consider the setting that the underlying Lévy process experiences killing with rate $p > 0$ (the case $p = 0$ being the subject of Theorem 2.21). In such a setting, its characteristic exponent is given by

$$\Psi_p(z) = p + \Psi(z), \qquad z \in \mathbb{R}, \tag{2.32}$$

where Ψ is the characteristic exponent of the underlying Lévy process without killing. The factorisation (2.31) still occurs in this more general setting, only now, κ and $\hat{\kappa}$ are the Laplace exponents of subordinators which depend on the value of p; accordingly we denote them κ_p and $\hat{\kappa}_p$, respectively. In particular, it must be the case that

$$c\kappa_p(0)\hat{\kappa}_p(0) = p, \qquad p > 0. \tag{2.33}$$

The associated subordinators H and $\hat{H}$ are thus both killed when $p > 0$. Their ranges still correspond, respectively, to the range of the running maximum and running minimum of the killed Lévy process.

Keeping with the case of a killed Lévy process, suppose we denote by $\mathbf{e}_p$ the independent and exponentially distributed random variable with rate $p > 0$, which corresponds to the time at which Y is sent to the cemetery state. With a slight abuse of notation, let us write $Y_{\mathbf{e}_p}$ in place of $Y_{\mathbf{e}_p-}$ with a similar meaning for $\underline{Y}_{\mathbf{e}_p}$ and $\overline{Y}_{\mathbf{e}_p}$. As $\mathbf{e}_p$ is independent of Y, it is almost surely not a jump time of Y and hence, for example, $Y_{\mathbf{e}_p} = Y_{\mathbf{e}_p-}$ makes sense. A straightforward computation shows that

$$\mathrm{E}\left[\mathrm{e}^{\mathrm{i}zY_{\mathbf{e}_p}}\right] = \frac{p}{\Psi_p(z)}, \qquad z \in \mathbb{R},$$

where, we recall, Ψ_p satisfies (2.32). The factorisation (2.31), when brought into the above setting, turns out to reveal another remarkable feature of Wiener–Hopf theory for Lévy processes.

Theorem 2.22 *We have that $\overline{Y}_{\mathbf{e}_p}$ and $Y_{\mathbf{e}_p} - \overline{Y}_{\mathbf{e}_p}$ are independent and hence, thanks to duality,*

$$\mathrm{E}\left[\mathrm{e}^{\mathrm{i}zY_{\mathbf{e}_p}}\right] = \mathrm{E}\left[\mathrm{e}^{\mathrm{i}z\overline{Y}_{\mathbf{e}_p}}\right] \times \mathrm{E}\left[\mathrm{e}^{\mathrm{i}z\underline{Y}_{\mathbf{e}_p}}\right], \qquad z \in \mathbb{R}, \tag{2.34}$$

where, additionally,

$$\mathrm{E}\left[\mathrm{e}^{\mathrm{i}z\overline{Y}_{\mathbf{e}_p}}\right] = \frac{\kappa_p(0)}{\kappa_p(-\mathrm{i}z)} \quad \textit{and} \quad \mathrm{E}\left[\mathrm{e}^{\mathrm{i}z\underline{Y}_{\mathbf{e}_p}}\right] = \frac{\hat{\kappa}_p(0)}{\hat{\kappa}_p(\mathrm{i}z)} \tag{2.35}$$

and the terms κ_p *and* $\hat{\kappa}_p$ *are two killed subordinator exponents belonging to the factorisation*

$$\Psi_p(z) = c\kappa_p(-\mathrm{i}z)\hat{\kappa}_p(\mathrm{i}z), \qquad z \in \mathbb{R}. \tag{2.36}$$

Strictly speaking, if $p > 0$, we call (2.36) the *space-time Wiener–Hopf factorisation.* Otherwise, if $p = 0$, we call it the *spatial Wiener–Hopf factorisation.*

The reader should be careful not to confuse κ_p and $\hat{\kappa}_p$ with versions of $p + \kappa$ and $p + \hat{\kappa}$, respectively, where κ and $\hat{\kappa}$ belong to the factorisation in Theorem 2.21. The correct relationship between the two pairs is a little more delicate. Indeed, when κ has the form (2.30), where necessarily $\Upsilon(\mathrm{d}x) = \overline{n}(\epsilon_\zeta \in \mathrm{d}x)$, $x > 0$, (recall that $\overline{n}$ is the excursion measure of the process $\overline{Y} - Y$), then κ_p takes the form

$$\kappa_p(\lambda) = \overline{n}(1 - \mathrm{e}^{-p\zeta}) + \mathrm{b}\lambda + \int_{(0,\infty)} (1 - \mathrm{e}^{-\lambda x})\overline{n}(\mathrm{e}^{-p\zeta}; \epsilon_\zeta \in \mathrm{d}x), \qquad \lambda \geq 0.$$

In, particular,

$$\kappa_p(0) = \overline{n}(1 - \mathrm{e}^{-p\zeta}) = \overline{n}(\zeta = \infty) + \int_{(0,\infty)} (1 - \mathrm{e}^{-px})\overline{n}(\zeta \in \mathrm{d}x).$$

A similar identity will hold for $\hat{\kappa}_p$, albeit in terms of the excursion process of $Y - \underline{Y}$. It is worthy of note that $\kappa_p(\lambda)$ and $\hat{\kappa}_p(\lambda)$ are nonetheless Bernstein functions in λ which are necessarily those of killed subordinators. Moreover, $\kappa_p(0)$ and $\hat{\kappa}_p(0)$ are Bernstein functions in p, which correspond to the Laplace exponents of the inverse local time at the maximum and the same quantity but for the dual, respectively. This makes the factorisation (2.33) all the more remarkable, earning it the name *temporal Wiener–Hopf factorisation.*

Whilst the independence of $\overline{Y}_{\mathbf{e}_p}$ and $Y_{\mathbf{e}_p} - \overline{Y}_{\mathbf{e}_p}$ is a consequence of the decomposition of the path of $(Y_t, t \leq \mathbf{e}_p)$ over its excursions, the identities in (2.35) are a manifestation of the deeper fact that $\overline{Y}_{\mathbf{e}_p}$ and $\overline{Y}_{\mathbf{e}_p} - Y_{\mathbf{e}_p}$ are in fact infinitely divisible. The last fact is again a consequence of the aforementioned excursion decomposition.

2.15 Reflection

An immediate interesting application of the Wiener–Hopf factorisation is that it allows us to deal with the asymptotic behaviour of the Lévy process reflected in its running maximum. Recall that $\overline{Y}_t = \sup_{s\leq t} Y_s$, $t \geq 0$. In Section 2.14 we remarked that $\overline{Y} - Y$ is a strong Markov process.

The trichotomy that all Lévy processes either almost surely drift to $+\infty$, $-\infty$ or oscillate between the two offers a relatively straightforward perspective for the long-term behaviour of the reflected process $\overline{Y} - Y$, when combined with (2.35). Indeed, when $\lim_{t\to\infty} Y_t = \infty$, the global infimum is almost surely bounded away from $-\infty$ and hence $\hat{\kappa}(0) > 0$. Recalling that $-\underline{Y}_t$ is equal in law to $\overline{Y}_t - Y_t$, monotonicity of the former ensures that the limit

$$\lim_{t\to\infty} \mathbb{E}[\mathrm{e}^{-\lambda(\overline{Y}_t - Y_t)}], \qquad \lambda \geq 0,$$

always exists. If we write $p^{-1}\mathbf{e}_1$ in place of $\mathbf{e}_p$ in (2.35) and appeal to an analytic extension to identify the Laplace transform identity

$$\mathrm{E}\left[\mathrm{e}^{-\lambda(\overline{Y}_{p^{-1}\mathbf{e}_1} - Y_{p^{-1}\mathbf{e}_1})}\right] = \frac{\hat{\kappa}_p(0)}{\hat{\kappa}_p(\lambda)}, \qquad \lambda \geq 0,$$

we can take limits as $p \to 0$ and deduce that $\overline{Y} - Y$ convergences in distribution at large times to a non-trivial and non-defective limit.

In contrast, if $\liminf_{t\to\infty} Y_t = -\infty$ (which covers the case of drifting to $-\infty$ or oscillating), we see in a similar way from (2.35) that the limiting distribution of $\overline{Y} - Y$ is defective and concentrated on $+\infty$.

2.16 Creeping

In Section 2.12, for one-dimensional processes, we considered the event that a given point $a \in \mathbb{R}$ lies in the range of a Lévy process, Y. We may consider a refined version of this event in which the point a is visited when a Lévy process, say Y, first enters (a, ∞), that is $\{Y_{\tau_a^+} = a\}$. This event is called *creeping* (upwards over a). Naturally the event of creeping is of no interest when $-Y$ is a subordinator. Let us, therefore, temporarily assume that this is not the case.

Excluding compound Poisson processes, creeping over any $a \in (0, \infty)$ is equivalent to requiring that a belongs to the range of $\overline{Y}$, which is equivalent to requiring that a belongs to the range of H. (In the case that Y is a subordinator, we should understand $H = Y$.) Theorem 2.18 (iii) tells us that this happens if and only if the drift coefficient of the ascending ladder height process, H is strictly positive and, moreover, that creeping occurs for all $a > 0$ with positive probability. In that case, we say simply that Y *creeps upwards.* A similar statement can be made for creeping downwards over $a \in (-\infty, 0)$.

2.17 First passage problems

Understanding the event of creeping is but a part of understanding the bigger picture of the first passage problem for one-dimensional Lévy processes.

Roughly speaking, for a given $a \in \mathbb{R}$, one may think of the first passage problem at a as characterising the law of the so-called *overshoot* $Y_{\tau_a^+} - a$, where

$$\tau_a^+ = \inf\{t > 0 : Y_t > a\}. \tag{2.37}$$

One can be more demanding in this respect and also look at the joint law of the latter together with the so-called *undershoot*, $a - Y_{\tau_a^+ -}$. In fact, the results we shall present will also include the law of the *undershoot of the maximum* prior to first passage, $a - \overline{Y}_{\tau_a^+ -}$. We also allow ourselves the luxury of interpreting the Lévy process Y as having the possibility of being killed in the spirit of Section 2.2, that is, being sent to a cemetery state after an independent and exponentially distributed random time with rate $p \geq 0$.

Just as with the case of creeping, the Wiener–Hopf factorisation plays a central role in the analysis of the more general first passage problem. Part of the information we need to describe the joint law of the overshoots and undershoots described above is contained in the characteristics of the ascending and descending ladder height processes, H and $\hat{H}$.

We define the *renewal measure* associated to H by

$$U(\mathrm{d}x) = \int_0^\infty \mathrm{P}(H_t \in \mathrm{d}x,\, t < \varsigma)\,\mathrm{d}t, \qquad x \geq 0,$$

where we are using ς to denote the lifetime of H, which is necessarily killed if Y is. It is a straightforward computation to deduce that its Laplace transform satisfies

$$\int_{[0,\infty)} \mathrm{e}^{-\lambda x}\, U(\mathrm{d}x) = \frac{1}{\kappa(\lambda)}, \qquad \lambda \geq 0. \tag{2.38}$$

A similar renewal measure, denoted $\hat{U}$, can be defined for the descending ladder height process $\hat{H}$. If the Wiener–Hopf factorisation of a given Lévy process can be identified explicitly and if, further, both U and $\hat{U}$ can be recovered through the relevant inverse Laplace transform, then the following theorem gives an explicit identity for the triple law of the overshoot, undershoot and undershoot of the maximum at first passage over $a > 0$. As usual, the case of compound Poisson processes is excluded to avoid complications in the case that the jump distribution has a lattice support. We also exclude the case that $-Y$ is a subordinator (for obvious reasons) and that Y is a subordinator (which is dealt with later).

Theorem 2.23 *Suppose that Y is a (killed) Lévy process, but not a compound Poisson process, and neither Y nor $-Y$ is a subordinator. Then, for each $a > 0$, we have on $u > 0$, $v \geq y$, $y \in [0, a]$, $s, t \geq 0$,*

$$\mathrm{P}(Y_{\tau_a^+}-a\in \mathrm{d}u,\ a-Y_{\tau_a^+-}\in \mathrm{d}v,\ a-\overline{Y}_{\tau_a^+-}\in \mathrm{d}y\,;\ \tau_a^+<\zeta)\\ = U(a-\mathrm{d}y)\hat{U}(\mathrm{d}v-y)\Pi(\mathrm{d}u+v),$$

where Π is the Lévy measure of Y and ζ is its lifetime. Moreover, if Y creeps upwards, then the renewal measure U has a strictly positive and continuous density with respect to Lebesgue measure, say $u(a)$, such that, for all $a>0$,

$$\mathrm{P}(Y_{\tau_a^+}=a,\ a-Y_{\tau_a^+-}\in \mathrm{d}v,\ a-\overline{Y}_{\tau_a^+-}\in \mathrm{d}y\,;\ \tau_a^+<\zeta) = \mathrm{b}u(a)\delta_0(\mathrm{d}v)\delta_0(\mathrm{d}y),$$

where b *is the drift coefficient of the ascending ladder height process.*

Because of (2.38), the equalities in the theorem above are only valid up to a multiplicative constant on account of the fact that the potentials U and $\hat{U}$ are dependent on the choice of constant that appears in the Wiener–Hopf factorisation (2.31). If this constant is normalised to unity (equivalently, the factorisation can be identified explicitly) then the aforesaid equalities are correct as stated.

Let us also look at the first passage problem in the case that Y is a (killed) subordinator. For this class of processes, it makes no sense to involve the quantity $\overline{Y}$ as $a-\overline{Y}_{\tau_a^+-}=a-Y_{\tau_a^+-}$. We can also understand the potential measure U, defined in (2.38), as now being that of Y, which itself can be taken to be identically equal to H. The relevant result for the first passage problem takes the following form.

Theorem 2.24 *Suppose that Y is a (killed) subordinator. Then for $u>0$ and $y\in[0,a]$,*

$$\mathrm{P}(Y_{\tau_a^+}-a\in \mathrm{d}u,\ a-Y_{\tau_a^+-}\in \mathrm{d}y\,;\ \tau_a^+<\zeta) = U(a-\mathrm{d}y)\Pi(y+\mathrm{d}u), \tag{2.39}$$

where ζ is the lifetime of Y. Moreover, if Y creeps then

$$\mathrm{P}(Y_{\tau_a^+}=a, a-Y_{\tau_a^+-}\in \mathrm{d}v\,;\ \tau_a^+<\zeta) = \mathrm{b}u(a)\delta_0(\mathrm{d}v),$$

where b *is the drift coefficient of Y.*

It is also worth recording a weaker version of this result for future use.

Corollary 2.25 *Suppose that Y is a (killed) subordinator with Laplace exponent $\kappa(\theta):=-\log \mathrm{E}[\exp\{-\theta Y_1\}]$, $\theta\geq 0$. Then, for $q\geq 0$,*

$$\int_0^\infty \mathrm{E}\left[\mathrm{e}^{-qa-\theta(Y_{\tau_a^+}-a)}\mathbf{1}_{(\tau_a^+<\zeta)}\right]\mathrm{d}a = \frac{\kappa(q)-\kappa(\theta)}{(q-\theta)\kappa(q)}.$$

In theory we could marginalise the triple in Theorem 2.23, resp. 2.24, to deduce an expression for $\mathrm{P}(\tau_a^+ < \infty)$ in the case that $\lim_{t\to\infty} Y_t = -\infty$, resp. Y is a killed subordinator. However, there is an alternative shortcut. Note that $\mathrm{P}(\tau_a^+ < \infty) = \mathrm{P}(H_{\mathbf{e}_q-} > a)$, where we should understand $\mathbf{e}_q$ as the independent and exponentially distributed random time at which the subordinator H is killed. Considering (2.38) it is straightforward to deduce the following lemma.

Lemma 2.26 *For $a > 0$,*

$$\mathrm{P}(\tau_a^+ < \zeta) = \kappa(0)U(a, \infty).$$

Clearly renewal measures play an important role in first passage problems, both for subordinators and for non-monotone Lévy processes. To emphasise this point, we conclude with one more theorem which gives us the resolvent of a Lévy process killed on first passage below the origin.

Theorem 2.27 *Suppose that Y is a Lévy process but neither Y nor $-Y$ is a subordinator. For $x > 0$,*

$$\mathrm{E}_x\left[\int_0^{\tau_0^-} f(Y_t)\,\mathrm{d}t\right] = \int_{[0,\infty)} U(\mathrm{d}y) \int_{[0,x]} \hat{U}(\mathrm{d}z)\, f(x+y-z), \qquad x \geq 0,$$

where $\tau_0^- = \inf\{t > 0 : Y_t < 0\}$.

We note that, as many of the above expressions are formulated as convolutions with respect to the renewal measure of the ascending ladder height subordinator, we can appeal to classical renewal theory to obtain results for asymptotic overshoot and undershoot distributions. For example, when Y is a subordinator which is not arithmetic, that is, that the paths of Y do not live in a strict sub-group of $[0, \infty)$ and $m := \mathrm{E}[Y_1] < \infty$, the basic conclusion of renewal theory implies that for bounded measurable sets A, $mU(x + A)/x$ converges to the Lebesgue measure of A as $x \to \infty$. The following corollary to Theorem 2.24 is therefore not too surprising.

Corollary 2.28 *Suppose that Y is a subordinator which is not arithmetic and which has finite mean, say m. Then, for $u > 0$ and $y \geq 0$, in the sense of weak convergence,*

$$\lim_{a\to\infty} \mathrm{P}(Y_{\tau_a^+} - a \in \mathrm{d}u, a - Y_{\tau_a^+ -} \in \mathrm{d}y) = \frac{1}{m}\mathrm{d}y\,\Pi(y + \mathrm{d}u).$$

Corollary 2.28 implies that the subordinator Y converges weakly towards a random variable whose law can be determined explicitly, that is,

$$Y_{\tau_a^+} - a \xrightarrow[a\to\infty]{(w)} \mathfrak{U}W, \qquad (2.40)$$

where '$\xrightarrow{(w)}$' means weak convergence and $\mathfrak{U}$ and W are independent r.v.'s, $\mathfrak{U}$ is uniformly distributed over $[0, 1]$ and the law of W is such that

$$\mathrm{P}(W > u) = \frac{1}{m} \int_{(u,\infty)} s\,\Pi(\mathrm{d}s), \quad \text{for} \quad u \geq 0. \tag{2.41}$$

On a final note, there are a few facts that are worth mentioning for the setting of spectrally negative Lévy processes. The fact that the ascending ladder height process is continuous in the spectrally negative setting means that $\overline{Y}$ is a continuous process. It is also easy to verify that $\overline{Y}$ fulfils the definition of the local time of $\overline{Y} - Y$ at zero, described in Section 2.14. Recalling the property that inverse local time is a (killed) subordinator, and noting that we can also write $\tau_a^+ = \inf\{t > 0 : \overline{Y}_t > a\}$, the result below follows immediately.

Lemma 2.29 *If Y is a spectrally negative Lévy process, then $(\tau_a^+, a \geq 0)$ is a (killed) subordinator.*

It is also straightforward to verify the statement of the above lemma using stationary and independent increments as well as the fact that $Y_{\tau_a^+} = a$ on $\{\tau_a^+ < \infty\}$ thanks to spectral negativity.

The identification of $\overline{Y}$ as a local time also allows for the simplification of a number of calculations in terms of the underlying excursion process of $\overline{Y} - Y$ from 0. In turn, this can be used to develop a number of calculations that reformulate first passage problems in terms of the so called scale function. The scale function $W : [0, \infty) \to [0, \infty)$ is defined via the Laplace transform

$$\int_0^\infty \mathrm{e}^{-\beta x} W(x)\,\mathrm{d}x = \frac{1}{\psi(\beta)}, \qquad \beta \geq \beta_0,$$

where

$$\psi(\beta) = \frac{1}{t} \log \mathrm{E}[\mathrm{e}^{\beta Y_t}], \qquad \beta \geq 0, \tag{2.42}$$

is the Laplace exponent of Y and β_0 is the largest root of the equation $\psi(\beta) = 0$ (there are at most two roots as ψ is convex). The following is a well-used result which shows that the two-sided exit problem has a convenient analytical representation in terms of scale functions.

Lemma 2.30 *For $x \in [0, a]$,*

$$\mathrm{P}_x(\tau_a^+ < \tau_0^-) = \frac{W(x)}{W(a)}, \qquad x \in [0, a]. \tag{2.43}$$

2.18 Lévy processes in higher dimensions

The earlier discussion has focused predominantly on the case of one-dimensional Lévy processes. Whilst some of the concepts, such as regularity for the half-line, creeping and the Wiener–Hopf factorisation, no longer make sense in higher dimensions, a number of the mathematical issues can still be discussed in a meaningful way.

Characteristic exponent and the Lévy–Itô decomposition. Stationary and independent increments still imply that, if we consider $\mathrm{E}[\mathrm{e}^{\mathrm{i}z\cdot Y_t}]$, $t \geq 0$, where, now, we take the parameter $z \in \mathbb{R}^d$ and the product $z \cdot Y_t$ as a Euclidean inner product, then the distribution of Y_t is still infinitely divisible for each $t \geq 0$ and (2.1) holds. From the theory of infinitely divisible random variables in higher dimensions, see (1.39), we recall

$$\Psi(z) = \mathrm{i}a \cdot z + \frac{1}{2} z \cdot Qz + \int_{\mathbb{R}^d} (1 - \mathrm{e}^{\mathrm{i}z\cdot x} + \mathrm{i}z \cdot x\mathbf{1}_{(|x|<1)})\,\Pi(\mathrm{d}x), \qquad z \in \mathbb{R}^d, \quad (2.44)$$

where $a \in \mathbb{R}^d$, Q is a $d \times d$ Gaussian covariance matrix and Π is a uniquely determined measure concentrated on $\mathbb{R}^d \backslash \{0\}$ which satisfies

$$\int_{\mathbb{R}^d} (1 \wedge |x|^2)\,\Pi(\mathrm{d}x) < \infty.$$

Just as in the one-dimensional setting, the structure of the d-dimensional Lévy–Khintchine formula (2.44) pertains to a Lévy–Itô path decomposition which reads almost verbatim (with the obvious adjustments to notation) to the one described in Section 2.1. Note, in particular, that the term $z \cdot Qz/2$ corresponds to the inclusion of an independent d-dimensional Brownian motion with covariance matrix Q.

Bounded versus unbounded variation. Similar arguments to those given in Lemma 2.4 shows that a d-dimensional Lévy process has paths of bounded variation if and only if $\sigma = 0$ and $\int_{\mathbb{R}^d}(1 \wedge |x|)\,\Pi(\mathrm{d}x) < \infty$. Moreover, a simple adaptation of the reasoning leading to Lemma 2.5 shows that a Lévy process is a compound Poisson process with drift if and only if $\sigma = 0$ and $\Pi(\mathbb{R}^d) < \infty$.

Feller property and infinitesimal generator. The notion of the Feller property in Definition A.14 can easily be adapted to cover the case of $\mathbb{R}^d$-processes. As such it is not surprising that Lévy processes remain in the class of Feller processes in higher dimensions. As a consequence, each Lévy process is still equipped with an infinitesimal generator in higher dimensions and the generator can be matched against the characteristic exponent (2.44) similarly as in one dimension. Indeed, for f in the class of continuously differentiable functions with compact support, if Ψ in (2.44) is the exponent of a Lévy process, then its generator is given by

$$\mathcal{L}f(x) = -a \cdot \nabla f(x) + \frac{1}{2} \sum_{i,j \in \{1,\cdots,d\}} Q_{i,j} \frac{\partial^2 f}{\partial x_i \partial x_j}(x) + \int_{\mathbb{R}^d} [f(x+y) - f(x) - y \cdot \nabla f(x) \mathbf{1}_{(|y|<1)}] \Pi(\mathrm{d}y), \quad (2.45)$$

for $x = (x_1, \cdots, x_d) \in \mathbb{R}^d$.

Duality. Duality as described in Section 2.11 is still valid in higher dimensions.

Moments. The notion of moments as discussed in Section 2.7 is equally valid in higher dimensions. The definition of submultiplicative functions and the conclusion of Theorem 2.9 reads verbatim the same in the higher-dimensional setting too.

Transience and recurrence. With the obvious interpretation of $|\cdot|$ for the d-dimensional Euclidean norm, the definition of transience and recurrence remains unchanged in higher dimensions. Moreover, Theorems 2.12 and 2.13 are still valid without adjustment.

Polarity of points. In the case of one dimension, Theorem 2.18 explores points which, with positive probability, are included in the range of a Lévy process (not a compound Poisson process) issued from the origin. It turns out that the set of such points was either $\emptyset$, $(0, \infty)$, $(-\infty, 0)$ or $\mathbb{R}$.

In dimension $d \geq 2$, things are a little more subtle. A point $y \in \mathbb{R}^d$ is polar if for every x,

$$\mathrm{P}_x(Y_t = y \text{ for some } t > 0) = 0.$$

We say that a point $y \in \mathbb{R}^d$ is essentially polar if for Lebesgue almost every x,

$$\mathrm{P}_x(Y_t = y \text{ for some } t > 0) = 0.$$

Theorem 2.31 *For dimension $d \geq 2$, all points are essentially polar.*

Obviously if a point is polar then it is essentially polar. It turns out that, conversely, if the resolvent kernel is absolutely continuous, then any essentially polar points are polar. There are nonetheless examples for which the two classes differ. That said, this is a subtlety that we won't have occasion to work with in this book.

2.19 Comments

The literature on Lévy processes is now vast. There are several books which offer a solid introduction to the core theory of Lévy processes, both in one dimension and higher dimensions. These include Applebaum [6], Bertoin [18], Doney [60], Kyprianou [123] and Sato [190]. There are many more standard texts which contain a chapter or more devoted to introductory material on Lévy processes as an exemplary or favourable stochastic or Markov process; see for example [55, 77, 78, 79, 182, 183] to name but a few.

Sections 2.1 until 2.5 follow a standard approach to be found in many of the aforementioned texts. The complete dichotomy of drifting versus oscillating in Theorem 2.8 is due to Chung and Fuchs [49] and Erickson [67] for random walks, see Bertoin [20] in the Lévy process setting. Section 2.7 offers a short summary of what is otherwise a more thorough handling of moments given in Section 25 of Sato [190]. Theorem 2.11 is a rather elementary application of a classical result for convergence of a sequence of Feller processes to a Feller process on the Skorokhod space. See, for example, the original works of Skorokhod [194, 195] as well as Theorem 2.5 in Ethier and Kurtz [68]. In the case of Lévy processes, the general theory translates directly into the convergence of the characteristic exponents. The exponential change of measure in Section 2.8 is a classical topic rooted in the Esscher transform for random variables and can be found in many texts. In Section 2.9, for the formal definition of the Skorokhod topology, the reader is referred to Chapter VI.1 of Jacod and Shiryaev [95]. Donsker's convergence is a classical result for random walks, which can be found in numerous text books; we mention Billingsley [32] here as but one example.

The notion of transience versus recurrence in Section 2.10 is deeply embedded in the theory of Markov processes. The setting of Lévy processes has been treated by Kingman [110] and Port and Stone [95]. Theorem 2.18 in Section 2.12 can be attributed to Kesten [103] and Bretagnolle [42]. Regularity of the half-line discussed in Theorem 2.20 can be attributed to the work of Rogozin [184], Štatland [202] and Bertoin [20]. The basic notion of duality for Lévy processes is often treated via Lemma 2.14, thanks to a deceptively simple proof which is little different from the analogous result for random walks. The bigger picture when it comes to duality for stochastic processes is much more complex. We have given all but a tiny snapshot in Section A.12. See the further remarks and references therein.

The Wiener–Hopf factorisation has quite a substantial exposure in the history of Lévy processes and random walks. There are many different ways to express the factorisation seen in (2.31) and (2.34), both as an analytical decomposition but also as a probabilistic decomposition. Some of the earliest work in this respect can be found in the setting of random walks from the work of Baxter [13] and Spitzer [197, 198, 200, 201]. Other early contributions can be found in Port [171], Feller [69], Borovkov [41], Pečerskiĭ and Rogozin [168], Gusak and Korolyuk [84] and Fristedt [74]. Bingham [35] gives a comprehensive account of the Wiener–Hopf factorisation for the spectrally one-sided setting. An elegant derivation of the factorisations (2.31) and (2.34) which makes natural use of the underlying excursion theory can be found in the papers of Greenwood and Pitman [83]. See also Chapters IV and VI of Bertoin [18] or Chapter 6 of Kyprianou [123] e.g. where a more detailed handling of local time in the irregular setting is spelled out in detail.

Theorem 2.23 is a simpler version of the so-called quintuple law, proved in Doney and Kyprianou [62]; see also Doney [60], Duquesne [65] and Winkel [214]. Theorem 2.24 is the simpler analogue for subordinators, which is fundamentally based on renewal theory. Together with its consequences in Corollary 2.25 and Lemma 2.26, this result can be traced back to the work of Kesten [104], Bretagnolle [42], Horowitz [88], Bertoin [18] and Andrew [5]. Theorem 2.27 is due to Silverstein [193] and the asymptotic behaviour in Corollary 2.28, although classical for renewal processes, was proved in the context of general subordinators in Bertoin et al. [24]. Lemma 2.30 is one of many identities for spectrally negative Lévy processes that can be written in terms of scale functions; see for example Chapter 8 of Kyprianou [123]. This particular result is originally due to Zolotarev [218] and Takács [204]. Finally the discussion on polarity of points is lifted from Chapter II of Bertoin [18]. Whilst the original literature pertaining to polarity dates back to Orey [157] and Kanda [99, 100, 101]. Bertoin [18] offers a more thorough historical overview.

3

Stable processes

We are now ready to introduce the family of Lévy processes that form the main focus of this book: *Stable processes*. Our first goal in this chapter is to look at the several equivalent definitions that are common in the literature, both in one and higher dimensions, and to explain how one can switch between different parameterisations therein. We will identify, for our own purposes, a normalisation in the definition of the stable processes that will remain in effect throughout the rest of the book. Thereafter we shall revisit the path properties for general Lévy processes that were discussed in Chapter 2, but now within the specific context of stable processes. This will be done first in the one-dimensional setting and then in higher dimensions. In the latter case, we will insist on isotropy just as in the case of stable distributions.

Throughout this and subsequent chapters, the notation $X = (X_t, t \geq 0)$ with probabilities $\mathbb{P}_x$, $x \in \mathbb{R}^d$, for $d \geq 1$, will be reserved for the setting of stable processes. As usual, we generally prefer to write $\mathbb{P}$ in place of $\mathbb{P}_0$.

3.1 One-dimensional stable processes

As we have seen in Chapter 1, stable distributions can be defined in a number of different ways. Naturally, the same is true for stable processes. We give four equivalent definitions here.

Definition 3.1 (Four definitions of a one-dimensional stable process)

(1) The first definition identifies a stable process in terms of its marginals. A Lévy process is called a stable process if its marginal distribution at $t = 1$ is non-Gaussian and stable in the sense of Definition 1.1.

(2) The second definition identifies a stable process as a non-Gaussian Lévy process $X = (X_t, t \geq 0)$ for which there exists an $\alpha > 0$ such that, for all $c > 0$, $(cX_{c^{-\alpha}t}, t \geq 0)$ is equal in law to X.

(3) The third definition of a stable process is via its characteristic triplet. A stable process is a Lévy process whose characteristic triplet takes the form $(0, 0, \Pi)$, where the measure Π is given by

$$\Pi(\mathrm{d}x) = |x|^{-1-\alpha}\left(c_1\mathbf{1}_{(x>0)} + c_2\mathbf{1}_{(x<0)}\right)\mathrm{d}x, \qquad x \in \mathbb{R}, \tag{3.1}$$

such that $c_1 \geq 0$, $c_2 \geq 0$ and $\alpha \in (0, 2)$. Moreover, for the triplet $(0, 0, \Pi)$, we understand there to be a regularisation function $h(x) = 0$ when $\alpha \in (0, 1)$, $h(x) = x$ when $\alpha \in (1, 2)$ and if $\alpha = 1$, then we take $c_1 = c_2$ and $h(x) = x\mathbf{1}_{(|x|<1)}$.

(4) The fourth approach to defining a stable process is through its Lévy–Khintchine exponent. A Lévy process is a stable process if its characteristic exponent takes the form

$$\Psi(z) = c|z|^{\alpha}\left(1 - \mathrm{i}\beta\tan(\pi\alpha/2)\,\mathrm{sgn}(z)\right), \qquad z \in \mathbb{R}, \tag{3.2}$$

where

$$\alpha \in (0, 1) \cup (1, 2],\ c > 0 \quad \text{and} \quad \beta \in [-1, 1],$$

or

$$\alpha = 1 \quad \text{and we understand} \quad \beta\tan(\pi\alpha/2) = 0.$$

Given the exploration of stable distributions in Chapter 1, the reader will note that definitions (1), (3) and (4) are equivalent on account of the fact that the distribution of X_1 entirely determines the law of the Lévy process.

We also note that for the scaling property given in definition (2), it is necessary and sufficient that the scaling holds at time 1, again, on account of the fact that X_1 entirely determines the law of a Lévy process $(X_t, t \geq 0)$. This also means that definitions (1) and (2) are equivalent. In particular, because of this equivalence, definition (2) necessarily implies $\alpha \in (0, 2)$.

For definitions (3) and (4), three parameters are needed to describe the process. In (3), we used the triplet (α, c_1, c_2), in (4), we use the triplet (α, β, c). As noted in Proposition 1.7, the parameters in this pair of triplets are related for $\alpha \in (0, 1) \cup (1, 2)$ by the equalities

$$c = -(c_1 + c_2)\cos(\pi\alpha/2)\Gamma(-\alpha), \quad \beta = \frac{c_1 - c_2}{c_1 + c_2}, \tag{3.3}$$

and for $\alpha = 1$ by

$$c = (c_1 + c_2)\pi/2, \quad \beta = 0. \tag{3.4}$$

For all four definitions, the following fundamental scaling property emerges at the level of the process. We shall use it repeatedly throughout the course of the remainder of this book.

Theorem 3.2 *If $X = (X_t, t \geq 0)$ is a stable process with index $\alpha \in (0,2)$, then for all $c > 0$ and $x \in \mathbb{R}$,*

$$\text{the law of } (cX_{c^{-\alpha}t}, t \geq 0) \text{ under } \mathbb{P}_x \text{ is } \mathbb{P}_{cx}. \tag{3.5}$$

Proof As we have seen from Definition 3.1 (2), the result is true when $x = 0$ simply by definition alone. Stationary and independent increments tells us that $(X, \mathbb{P}_x)$ is equal in law to $(x + X, \mathbb{P})$. Hence, it follows that, for each $c > 0$, $(cX_{c^{-\alpha}t}, t \geq 0)$ under $\mathbb{P}_x$ is equal in law to $(cx + cX_{c^{-\alpha}t}, t \geq 0)$ under $\mathbb{P}$. Moreover the latter is equal in law to $(cx + X_t, t \geq 0)$ under $\mathbb{P}$ and hence X under $\mathbb{P}_{cx}$. □

As with the stable distribution, in the spirit of (1.28), a density for the distribution of X_t exists at all times $t \geq 0$. As such, it is well known that the transition semigroup of X has a density with respect to Lebesgue measure. Taking account of stationary and independent increments, we can write

$$\mathbb{P}_x(X_t \in \mathrm{d}y) = \mathrm{p}_t(y - x)\,\mathrm{d}y, \tag{3.6}$$

for all $x, y \in \mathbb{R}$ and $t \geq 0$, where

$$\mathrm{p}_t(x) := \frac{1}{2\pi} \int_{\mathbb{R}} \mathrm{e}^{-\mathrm{i}zx} \mathrm{e}^{-\Psi(z)t}\,\mathrm{d}z, \qquad x \in \mathbb{R}, t > 0. \tag{3.7}$$

Remark 3.3 We conclude this section by noting that, just as in Chapter 1, we have deliberately excluded the possibility that $\alpha = 2$ from the definitions above. Taking account of the remarks at the end of Section 1.1 regarding stable distributions with index $\alpha = 2$, it is not difficult to see that the corresponding Lévy process is a constant multiple of standard Brownian motion. Since our objective in this book is to deal exclusively with jump processes, we will make no attempt to address this setting.

3.2 Normalised one-dimensional stable processes

In accordance with what was done for one-dimensional stable distributions, we introduce here a normalised version of the stable process. Henceforth, this will be our preferred way of referring to one-dimensional stable processes throughout the remainder of the book. One of the reasons for this is that it has the advantage of reducing the number of parameters from three to two. In essence, this is done by pinning down a value for the constant c in the case that the process is parameterised by the triplet (α, β, c).

For any of the four equivalent definitions given in the previous section, let us consider the scaled version of $(b^{-\frac{1}{\alpha}}X_t, t \geq 0)$ or, equivalently thanks to Theorem 3.2, $(X_{t/b} : t \geq 0)$, where

$$b = c\sqrt{1 + \beta^2 \tan(\pi\alpha/2)^2}. \tag{3.8}$$

This scaling results in a simple form of the characteristic exponent; see Proposition 1.12 and the discussion around the normalised class of stable distributions $\mathcal{S}_{\mathrm{norm}}(\alpha, \rho)$. Specifically, for the aforesaid scaled version of X, we get

$$\Psi(z) = |z|^\alpha \left(e^{\pi i \alpha(\frac{1}{2}-\rho)} \mathbf{1}_{(z>0)} + e^{-\pi i \alpha(\frac{1}{2}-\rho)} \mathbf{1}_{(z<0)} \right), \qquad z \in \mathbb{R}, \tag{3.9}$$

where

$$\rho = \frac{1}{2} + \frac{1}{\pi\alpha} \tan^{-1}(\beta \tan(\pi\alpha/2)) \tag{3.10}$$

is called the positivity parameter and, from what we know of stable distributions, $\rho = \mathbb{P}(X_1 > 0)$ (and hence, thanks to scaling, $\rho = \mathbb{P}(X_t > 0)$ for all $t > 0$). As with the case of normalised stable distributions, with this parameterisation the stable process is defined only by two parameters (α, ρ), which belong to the admissible set (1.24), which we reproduce here for convenience:

$$\begin{aligned} \mathcal{A} := \{\alpha \in (0,1),\ \rho \in [0,1]\} \\ \cup \{\alpha = 1,\ \rho = 1/2\} \\ \cup \{\alpha \in (1,2),\ \rho \in [1-\alpha^{-1}, \alpha^{-1}]\}. \end{aligned} \tag{3.11}$$

The formulas in (3.3), (3.4) and (3.10) show how to obtain an (α, ρ) parameterisation starting from an (α, c_1, c_2) or an (α, β, c) parameterisation. To go in the opposite direction, suppose that X is chosen with normalisation in its characteristic exponent such that (3.9) holds. From Proposition 1.12, the coefficients c_1 and c_2 defining the Lévy measure must be given by

$$c_1 = \Gamma(1+\alpha)\frac{\sin(\pi\alpha\rho)}{\pi}, \quad c_2 = \Gamma(1+\alpha)\frac{\sin(\pi\alpha\hat{\rho})}{\pi}, \tag{3.12}$$

where $\hat{\rho} = 1 - \rho$.

In the one-dimensional setting, we will often find ourselves in future analysis dealing with three different regimes of α; these are the obvious $\alpha \in (0, 1)$, $\alpha = 1$ and $\alpha \in (1, 2)$. Indeed, we already see from (3.11) that the permissible pairs (α, ρ) naturally divide into the aforesaid cases. We shall also repeatedly see this natural partitioning in the next two sections, where we shall review some of the fundamental distributional and path properties introduced in Chapter 2, albeit for the specific setting of stable processes.

3.3 Path variation, asymmetry and moments

Path variation. Referring to (3.1), it is easy to check, using the simple integral, that

$$\int_0^\infty (1 \wedge x)x^{-\alpha-1}\,\mathrm{d}x < \infty,$$

as long as $\alpha \in (0, 1)$, and otherwise the integral diverges. Hence a stable process has paths of bounded variation if and only if $\alpha \in (0, 1)$.

Asymmetry. If $c_2 = 0$, then the process has no negative jumps. Considering (3.3), we see that if $\alpha \in (0, 1)$, then $\beta = 1$, which forces us to conclude in (3.10) that $\rho = 1$. In turn this means that, when $\alpha \in (0, 1)$ and there are no negative jumps, then the associated stable process must be a subordinator i.e. a process with monotone increasing paths. Similarly, if $c_1 = 0$ and $\alpha \in (0, 1)$, then $-X$ is a subordinator. It is easy to check, for example using (2.9), that, in these cases, the associated subordinator has zero drift coefficient.

For a stable subordinator, it is more common to deal with its Laplace exponent than its characteristic exponent. Recall that the Laplace exponent of a subordinator takes the form

$$\kappa(\lambda) := -\frac{1}{t}\log \mathbb{E}\left[\mathrm{e}^{-\lambda X_t}\right], \quad \text{for} \quad t > 0, \lambda \geq 0. \tag{3.13}$$

An expression for κ can be derived from first principles by returning to the computation in (1.11) and performing it again in the context of a positive-valued random variable. One recovers the expression

$$\kappa(\lambda) = \lambda^\alpha, \qquad \lambda \geq 0. \tag{3.14}$$

More generally, when $\alpha \in (0, 1)$ and $\rho \neq \{0, 1\}$, since the associated stable process has paths of bounded variation, it can be written as the difference of two independent subordinators. In this case, on account of the Lévy measure on the positive and negative half-lines, they must be stable subordinators. It is a straightforward exercise to show that

$$X_t = b_1^{1/\alpha}X_t^+ - b_2^{1/\alpha}X_t^-, \qquad t \geq 0, \tag{3.15}$$

where $b_1 = \sin(\pi\alpha\rho)/\sin(\pi\alpha)$ and $b_2 = \sin(\pi\alpha\hat{\rho})/\sin(\pi\alpha)$; and X^+ and X^- are independent (normalised) stable subordinators.

To see why, note first from the scaling property that $X_t \overset{(d)}{=} X^+_{b_1 t} - X^-_{b_2 t}$, $t \geq 0$. In that case, recalling that the characteristic exponent of X^+ and X^- both take the form

$$\tilde{\Psi}(z) = \int_0^\infty (1 - \mathrm{e}^{\mathrm{i}zx})\Gamma(1 + \alpha)\frac{\sin(\pi\alpha)}{\pi}\frac{1}{x^{1+\alpha}}\,\mathrm{d}x,$$

the characteristic exponent of X_1 as defined is equal to

$$\begin{aligned}\Psi(z) &= \tilde{\Psi}(z)b_1 + \tilde{\Psi}(-z)b_2 \\ &= \int_0^\infty (1 - \mathrm{e}^{\mathrm{i}zx})b_1\Gamma(1+\alpha)\frac{\sin(\pi\alpha)}{\pi}\frac{1}{x^{1+\alpha}}\,\mathrm{d}x \\ &\qquad + \int_0^\infty (1 - \mathrm{e}^{-\mathrm{i}zx})b_2\Gamma(1+\alpha)\frac{\sin(\pi\alpha)}{\pi}\frac{1}{x^{1+\alpha}}\,\mathrm{d}x \\ &= \int_{\mathbb{R}} (1 - \mathrm{e}^{\mathrm{i}zx})\left(c_1\frac{1}{x^{1+\alpha}}\mathbf{1}_{(x>0)} + c_2\frac{1}{|x|^{1+\alpha}}\mathbf{1}_{(x<0)}\right)\mathrm{d}x,\end{aligned}$$

where we recall that $c_1 = \Gamma(1+\alpha)\sin(\pi\alpha\rho)/\pi$ and $c_2 = \Gamma(1+\alpha)\sin(\pi\alpha\hat{\rho})/\pi$. It follows from the discussion at the start of this section that the characteristic exponent of X has the required normalised form.

When $\alpha = 1$, the stable process is symmetric, both in its Lévy measure as well as its distribution at all times. One easily verifies from Theorem 1.17 that, for all $t \geq 0$ and $z \in \mathbb{R}$,

$$\mathrm{p}_t(z) = \frac{t}{\pi(t^2 + z^2)}. \tag{3.16}$$

When $\alpha \in (1, 2)$, as the associated stable process has paths of unbounded variation and therefore the case of monotone paths is ruled out. Indeed, from (3.11), we see that there is a restriction on the range of ρ. For example, at its greatest value we have that $\rho = \alpha^{-1}$. From (3.10), noting that

$$\tan^{-1}(\tan(\pi\alpha/2)) = \pi - \frac{\pi\alpha}{2},$$

this corresponds to the case that $\beta = -1$, or equivalently $c_1 = 0$, making the associated stable process spectrally negative. Similarly, at the other extreme we have $\rho = 1 - \alpha^{-1}$, $\beta = 1$ and $c_2 = 0$, making the associated stable process spectrally positive. Spectrally one-sided processes are sometimes described in terms of their Laplace exponents. For example, it is traditional in the literature to define the Laplace exponent of a spectrally positive Lévy process via the relation

$$\exp\{\psi(\lambda)t\} := \mathbb{E}\left[\mathrm{e}^{-\lambda X_t}\right], \qquad t \geq 0,$$

which is well defined for $\lambda \geq 0$. Referring back to the derivation of the characteristic exponent of a general stable process with stability index $\alpha \in (1, 2)$, the computation leading to (1.13), when phrased in terms of a real-valued variable, also delivers us the identity

$$\psi(\lambda) = \lambda^\alpha, \qquad \lambda \geq 0.$$

What lies between the two parameter extremes for ρ, when $\alpha \in (1,2)$, that is,

$$1 - \frac{1}{\alpha} < \rho < \frac{1}{\alpha}$$

must exhibit both positive and negative jumps.

Moments. It is straightforward to check with the help of Theorem 2.9 (see also Theorem 1.10) that stable processes do not possess exponential moments. Indeed, one easily verifies, by considering the integral $\int_1^\infty x^{\theta-\alpha-1}\,\mathrm{d}x$, that, for $\theta > 0$,

$$\mathbb{E}\left[\sup_{s\leq t} |X_s|^\theta\right] < \infty \tag{3.17}$$

if and only if $\theta < \alpha$ and otherwise the above expectation is infinite. In particular this means that stable processes have no moments when $\alpha \in (0,1]$, but a first moment exists when $\alpha \in (1,2)$. Moreover, by differentiating the identity

$$\exp\{-\Psi(z)t\} = \mathbb{E}\left[\mathrm{e}^{\mathrm{i}zX_t}\right], \qquad z \in \mathbb{R}, t \geq 0,$$

at $z = 0$, we see that $\mathbb{E}[X_t] = \mathrm{i}\Psi'(0)t = 0$ for all $t \geq 0$. This means that, unlike the regime $\alpha \in (0,1]$, the associated stable process is also a martingale.

We also learn from (3.17) that stable processes possess no second moments for any $\alpha \in (0,2)$.

3.4 Path properties in one dimension

Drifting and oscillating. As noted above, if $\alpha \in (0,1)$ and either c_2 or c_1 is equal to zero, equivalently $\rho = 1$ or 0, then X is either a subordinator or the negative of a subordinator. Hence, rather trivially, X drifts to $+\infty$ or to $-\infty$, respectively.

When $\alpha \in (0,1]$ and $c_1, c_2 > 0$, because both positive and negative part of the Lévy measure have a density which is proportional to $1/|x|^{1+\alpha}$, it follows that the integrals in (2.18) and (2.19) are either simultaneously finite or simultaneously infinite in value. As Theorem 2.8 forbids them to be simultaneously finite, we are left with the conclusion that X must oscillate and we are forced to conclude that

$$\limsup_{t\to\infty} X_t = -\liminf_{t\to\infty} X_t = \infty. \tag{3.18}$$

Finally, for the case that $\alpha \in (1,2)$, as we have already established in Section 3.3 that $\mathbb{E}[X_1] = 0$, Theorem 2.8 again tells us that (3.18) holds and we have oscillatory behaviour.

index	jumps	path	asymptotic behaviour
$\alpha \in (0,1)$			transient
$\rho = 0$	$-$	monotone decreasing	$\mathbb{P}_x(\tau^{\{0\}} < \infty) = 0, x \in \mathbb{R}$, $\lim_{t\to\infty} X_t = -\infty$
$\rho = 1$	$+$	monotone increasing	$\mathbb{P}_x(\tau^{\{0\}} < \infty) = 0, x \in \mathbb{R}$, $\lim_{t\to\infty} X_t = \infty$
$\rho \in (0,1)$	$+,-$	bounded variation	$\mathbb{P}_x(\tau^{\{0\}} < \infty) = 0, x \in \mathbb{R}$, $\lim_{t\to\infty} \lvert X_t\rvert = \infty$
$\alpha = 1$			recurrent
$\rho = \frac{1}{2}$	$+,-$	unbounded variation	$\mathbb{P}_x(\tau^{\{0\}} < \infty) = 0, x \in \mathbb{R}$, $\limsup_{t\to\infty} \lvert X_t\rvert = \infty$, $\liminf_{t\to\infty} \lvert X_t\rvert = 0$
$\alpha \in (1,2)$			recurrent
$\alpha\rho = 1$	$-$	unbounded variation	$\mathbb{P}_x(\tau^{\{0\}} < \infty) = 1, x \in \mathbb{R}$, $\liminf_{t\to\infty} X_t = -\infty$ $\limsup_{t\to\infty} X_t = \infty$
$\alpha\rho = \alpha - 1$	$+$	unbounded variation	$\mathbb{P}_x(\tau^{\{0\}} < \infty) = 1, x \in \mathbb{R}$, $\liminf_{t\to\infty} X_t = -\infty$ $\limsup_{t\to\infty} X_t = \infty$
$\alpha\rho \in (\alpha - 1, 1)$	$+,-$	unbounded variation	$\mathbb{P}_x(\tau^{\{0\}} < \infty) = 1, x \in \mathbb{R}$, $\liminf_{t\to\infty} X_t = -\infty$ $\limsup_{t\to\infty} X_t = \infty$

Table 3.1 *Path properties of stable processes according to the different parameter regimes of α and ρ*

Transience and recurrence. On account of the fact that, for ε sufficiently small,

$$\int_{-\varepsilon}^{\varepsilon} \mathrm{Re}\left(\frac{1}{\Psi(z)}\right) \mathrm{d}z \approx \int_{-\varepsilon}^{\varepsilon} \frac{1}{|z|^\alpha} \,\mathrm{d}z,$$

where $a \approx b$ means that a can be bounded from above and below by a multiple of b, it follows from Theorem 2.12 that X is transient whenever $\alpha \in (0,1)$ and recurrent when $\alpha \in [1,2)$.

Let us remark that a Lévy process which is recurrent cannot drift to ∞ or $-\infty$, and therefore must oscillate and we see this consistently with stable processes. On the other hand, an oscillating process is not necessarily recurrent. A nice example of this phenomenon is provided by the case of a symmetric stable process of index $0 < \alpha < 1$.

Hitting points. Straightforward computations also show that, for some constant $K > 0$, the ratio of

$$\int_{\mathbb{R}} \mathrm{Re}\left(\frac{1}{1+\Psi(z)}\right) \mathrm{d}z \text{ and } K + \int_{\mathbb{R}\setminus(-1,1)} \frac{1}{|z|^{\alpha}} \mathrm{d}z$$

is bounded by a strictly positive constant if and only if $\alpha \in (1,2)$. Therefore, referring to Theorem 2.18, the process X can hit points almost surely if and only if $\alpha \in (1,2)$.

Regularity of the half line. Taking note of the structure of the Lévy measure for a stable process, it is clear that the integral (2.29) is either finite or infinite for both X and $-X$ simultaneously. In that case, as X is not a compound Poisson process, both integrals must be infinite and 0 is regular for both $(-\infty,0)$ and $(0,\infty)$.

3.5 Wiener–Hopf factorisation and the first passage problem

Wiener–Hopf factorisation. Recall from Theorem 2.21 that the Wiener–Hopf factorisation takes the form

$$\Psi(z) = \kappa(-\mathrm{i}z)\hat{\kappa}(\mathrm{i}z), \qquad z \in \mathbb{R},$$

up to a multiplicative constant, where κ and $\hat{\kappa}$ are the Laplace exponents of the ascending and descending ladder processes. The Wiener–Hopf factorisation only makes sense for Lévy processes that do not have monotone paths. To this end, we assume in this section that $\rho \neq \{0,1\}$ if $\alpha \in (0,1)$.

Stability tells us that, for all $c > 0$, $(c\overline{X}_{c^{-\alpha}t}, t \geq 0)$ has the same law as $\overline{X} := (\overline{X}_t, t \geq 0)$, where $\overline{X}_t = \sup_{s\leq t} X_s$, $t \geq 0$. In particular, for all $c > 0$, the range of $c\overline{X}$ is equal in law to the range of $\overline{X}$. If we write as usual H for the ascending ladder height process, then, for all $c > 0$, the range of cH is equal in law to the range of H. The latter is equivalent to the condition that the Laplace exponent of H is proportional to that of cH, i.e. $\kappa(z) = k_c\kappa(cz)$, for $z \geq 0$, where $k_c > 0$ is a constant that only depends on c. Since $\kappa(1)$ must be a constant, we see that $\kappa(z)/\kappa(1) = \kappa(cz)/\kappa(c)$. Hence, as κ is increasing, one can easily deduce that $\kappa(\lambda) = \kappa(1)\lambda^{\alpha_1}$ for some $\alpha_1 \in [0,1]$. In other words, H is a stable subordinator with parameter α_1. We exclude the case $\alpha_1 = 0$ since it corresponds to the setting where the range of H is the empty set. A similar argument applied to $-X$ shows that the descending ladder height process must also belong to the class of stable subordinators.

We therefore assume that (up to multiplicative constants) $\kappa(z) = z^{\alpha_1}$, $z \geq 0$, and $\hat{\kappa}(z) = z^{\alpha_2}$, $z \geq 0$, for some $\alpha_1, \alpha_2 \in (0,1]$. Appealing to the normalised form of the stable process, (3.9), we must choose the parameters α_1 and α_2 such that, for example, when $z > 0$,

$$z^{\alpha}e^{\pi i\alpha(\frac{1}{2}-\rho)} = z^{\alpha_1}e^{-\frac{1}{2}\pi i\alpha_1} \times z^{\alpha_2}e^{\frac{1}{2}\pi i\alpha_2}.$$

Matching radial and angular parts, this is only possible if α_1 and α_2 satisfy

$$\begin{cases}\alpha_1 + \alpha_2 = \alpha, \\ \alpha_1 - \alpha_2 = -\alpha(1-2\rho),\end{cases} \tag{3.19}$$

which gives us $\alpha_1 = \alpha\rho$ and $\alpha_2 = \alpha\hat{\rho}$. Note from the discussion in Chapter 1, as X does not have monotone paths, it is necessarily the case that $0 < \alpha\rho \leq 1$ and $0 < \alpha\hat{\rho} \leq 1$. In conclusion, for $\theta \geq 0$,

$$\kappa(\theta) = \theta^{\alpha\rho} \text{ and } \hat{\kappa}(\theta) = \theta^{\alpha\hat{\rho}}.$$

When $\alpha\rho = 1$, the ascending ladder height process is a pure linear drift. In that case, the range of the maximum process $\overline{X}$ is $[0,\infty)$. This is consistent with our observations in Section 3.3 that X is spectrally negative. Similarly, when $\alpha\hat{\rho} = 1$, equivalently $\rho = 1 - \alpha^{-1}$, the Wiener–Hopf factorisation concurs with the previous observation that X is spectrally positive.

Creeping. Stable subordinators and the negative thereof cannot creep because they lack a drift component. This is clear in the case that X has monotone paths. Indeed, recalling (2.9), one readily checks that the associated drift coefficient in the decomposition (2.11) is zero. Otherwise, for stable processes with a Wiener–Hopf factorisation, with the exception of the spectrally one-sided setting, as both ascending and descending ladder subordinators are stable subordinators, creeping is not possible. If we take the case of spectrally negative stable processes with a Wiener–Hopf factorisation, that is, with $\alpha \in (1,2)$, as the ascending ladder subordinator is a linear drift but the descending ladder process is a stable subordinator, creeping is possible upwards but not downwards. A similar statement holds for the case of spectrally positive stable processes.

First passage problem. Let us first assume that X is a stable subordinator. Considering the Laplace transform in (2.38) it is straightforward to see, using the expression for the gamma integral (A.7), that the associated renewal measure U satisfies

$$U(\mathrm{d}x) = \frac{1}{\Gamma(\alpha)}x^{\alpha-1}\,\mathrm{d}x, \tag{3.20}$$

for $x \geq 0$. Theorem 2.24 now tells us that

$$\mathbb{P}(X_{\tau_a^+} - a \in \mathrm{d}u, a - X_{\tau_a^+-} \in \mathrm{d}y) = k\cdot(a-y)^{\alpha-1}(y+u)^{-(\alpha+1)}\,\mathrm{d}u\,\mathrm{d}y, \tag{3.21}$$

for $u > 0$, $y \in [0, a]$ and some constant $k > 0$. On account of the fact that its drift coefficient is zero, X cannot creep upwards. Hence the constant k can be obtained by ensuring that the right-hand side of (3.21) integrates to one. Indeed, one readily checks that the following result holds.

Theorem 3.4 *Suppose $\alpha \in (0,1)$ and $\rho = 1$. For all $u, a > 0$ and $y \in [0, a]$,*

$$\mathbb{P}(X_{\tau_a^+} - a \in \mathrm{d}u, a - X_{\tau_a^+ -} \in \mathrm{d}y) = \frac{\alpha \sin(\pi\alpha)}{\pi}(a-y)^{\alpha-1}(y+u)^{-(\alpha+1)}\,\mathrm{d}u\,\mathrm{d}y.$$

It is interesting to note from this further that the distribution of the pair

$$\left(\frac{X_{\tau_a^+} - a}{a}, \frac{a - X_{\tau_a^+ -}}{a}\right) \tag{3.22}$$

is independent of a. This is both clear from Theorem 3.4 as well as easily verified from the scaling property of X. Indeed, for each $c > 0$, suppose we defined $X_t^c := cX_{c^{-\alpha}t}$, $t \geq 0$ and recall that under $\mathbb{P}$, this scaled process is equal in law to X. Then $\tau_x^{+,c} := \inf\{s > 0 : X_s^c > x\}$ is equal in law to $\inf\{s > 0 : X_{c^{-\alpha}s} > x/c\} = c^\alpha \tau_{x/c}^+$. Setting $c = 1/a$, we thus have that, for example,

$$\frac{X_{\tau_a^+} - a}{a} = a^{-1}X_{a^\alpha a^{-\alpha}\tau_a^+} - 1 \overset{(\mathrm{d})}{=} X^{1/a}_{\tau_1^{+,1/a}} - 1 \overset{(\mathrm{d})}{=} X_{\tau_1^+} - 1.$$

The following corollary is also worth recording as it requires a tricky marginalisation of the joint law in Theorem 3.4.

Corollary 3.5 *Suppose $\alpha \in (0,1)$ and $\rho = 1$. For all $u, a > 0$,*

$$\mathbb{P}(X_{\tau_a^+} - a \in \mathrm{d}u) = \frac{\sin(\pi\alpha)}{\pi}\left(\frac{u}{a}\right)^{-\alpha}\left(\frac{1}{a+u}\right)\mathrm{d}u.$$

Proof We have

$$\begin{aligned}\mathbb{P}(X_{\tau_a^+} - a \in \mathrm{d}u) &= \frac{\alpha \sin(\pi\alpha)}{\pi}\int_0^a (a-y)^{\alpha-1}(y+u)^{-(\alpha+1)}\,\mathrm{d}y\,\mathrm{d}u \\ &= \frac{\alpha \sin(\pi\alpha)}{\pi}\int_0^a z^{\alpha-1}(a+u-z)^{-(\alpha+1)}\,\mathrm{d}z\,\mathrm{d}u.\end{aligned}$$

Setting $w = (a+u-z)/z$ so that $z = (a+u)/(1+w)$, we have

$$\begin{aligned}\mathbb{P}(X_{\tau_a^+} - a \in \mathrm{d}u) &= \frac{\alpha \sin(\pi\alpha)}{\pi}\frac{1}{(a+u)}\int_{u/a}^\infty z^{-(\alpha+1)}\,\mathrm{d}z\,\mathrm{d}u \\ &= \frac{\sin(\pi\alpha)}{\pi}\left(\frac{u}{a}\right)^{-\alpha}\frac{1}{(a+u)}\mathrm{d}u,\end{aligned}$$

as required. □

In the case that X does not have monotone paths, and providing that X is not spectrally negative, we can appeal to the Wiener–Hopf factorisation, to help address the first passage problem. From (3.20) we have that $U(\mathrm{d}a) \propto a^{\alpha\rho-1}\mathrm{d}a$ and $\hat{U}(\mathrm{d}a) \propto a^{\alpha\hat{\rho}-1}\mathrm{d}a$, for $a \geq 0$. In that case, Theorem 2.23 tells us that, for $y \in [0,a]$, $v \geq y$ and $u > 0$,

$$\begin{aligned}\mathbb{P}(X_{\tau_a^+} - a \in \mathrm{d}u, a - X_{\tau_a^+-} \in \mathrm{d}v, a - \overline{X}_{\tau_a^+-} \in \mathrm{d}y) \\ = K \cdot \frac{(a-y)^{\alpha\rho-1}(v-y)^{\alpha\hat{\rho}-1}}{(v+u)^{1+\alpha}}\,\mathrm{d}y\,\mathrm{d}v\,\mathrm{d}u,\end{aligned}$$

where K is a strictly positive constant. Again taking account of the assumption that X is not spectrally negative, which excludes the possibility that it can creep upwards, the constant K can be chosen so that the right-hand side above integrates to one. This gives us the following result.

Theorem 3.6 *Suppose that either $\alpha \in (0,1]$ and $\rho \neq \{0,1\}$ or $\alpha \in (1,2)$, $\alpha\rho < 1$ and $\alpha\hat{\rho} \leq 1$. For all $a, u > 0$, $y \in [0,a]$ and $v \leq y$,*

$$\begin{aligned}&\mathbb{P}(X_{\tau_a^+} - a \in \mathrm{d}u, a - X_{\tau_a^+-} \in \mathrm{d}v, a - \overline{X}_{\tau_a^+-} \in \mathrm{d}y) \\ &= \frac{\sin(\pi\alpha\rho)}{\pi}\frac{\Gamma(\alpha+1)}{\Gamma(\alpha\rho)\Gamma(\alpha\hat{\rho})}\frac{(a-y)^{\alpha\rho-1}(v-y)^{\alpha\hat{\rho}-1}}{(v+u)^{1+\alpha}}\,\mathrm{d}y\,\mathrm{d}v\,\mathrm{d}u.\end{aligned}$$

Also, as before, one notes from this triple law, or indeed from the reasoning provided in the discussion following Theorem 3.4, that the distribution of

$$\left(\frac{X_{\tau_a^+} - a}{a}, \frac{a - X_{\tau_a^+-}}{a}, \frac{a - \overline{X}_{\tau_a^+-}}{a}\right)$$

is independent of a. Moreover, below, we also have the easy following corollary to Theorem 3.6, which also follows from Corollary 3.5 and the fact that the ascending ladder height process of X is a stable subordinator with index $\alpha\rho$.

Corollary 3.7 *For $a, u > 0$,*

$$\mathbb{P}(X_{\tau_a^+} - a \in \mathrm{d}u) = \frac{\sin(\pi\alpha\rho)}{\pi}\left(\frac{u}{a}\right)^{-\alpha\rho}\left(\frac{1}{a+u}\right)\mathrm{d}u.$$

Finally we address the special case of the two-sided exit problem for the spectrally negative stable process, that is, the setting that $\alpha \in (1,2)$ and $\alpha\rho = 1$. The scale function in Lemma 2.30 is easy to determine. Indeed, on account of the fact that the Laplace exponent satisfies $\psi(\beta) = \log \mathbb{E}[\mathrm{e}^{\beta Y_1}] = \beta^\alpha$, for $\beta \geq 0$, it is straightforward to check using the definition of a standard gamma integral that

$$W(x) = \frac{x^{\alpha-1}}{\Gamma(\alpha)}, \qquad x \geq 0.$$

This gives us the simple identity from Lemma 2.30.

Lemma 3.8 *For* $x \in [0, a]$,

$$\mathbb{P}_x(\tau_a^+ < \tau_0^-) = \left(\frac{x}{a}\right)^{\alpha-1}.$$

3.6 Isotropic d-dimensional stable processes

Just as with the case of one-dimensional stable processes, it is possible to give a number of equivalent definitions for higher-dimensional stable processes. For example, we could work with the definition of d-dimensional Lévy processes which are self-similar in the spirit of Definition 3.1 (2). The notion of self-similarity here means precisely the same as (3.5) except that the processes concerned are understood in the vectorial sense. One problem with this definition is that it allows for a large variety of directional dependency in the Lévy measure. Mathematically speaking, it is more convenient (and challenging enough from the point of view of the problems we intend to attack) to consider a smaller class, referred to as *isotropic d-dimensional stable processes.* These are self-similar processes with stationary and independent increments such that its law is invariant under any orthogonal transformation. Recall that a measure μ on $\mathbb{R}^d$ is isotropic if for $B \in \mathcal{B}(\mathbb{R}^d)$, $\mu(B) = \mu(\mathrm{U}^{-1}B)$ for every orthogonal d-dimensional matrix U. In one dimension, an isotropic stable process is nothing more than a symmetric process. In other words, a stable process whose positivity parameter satisfies $\rho = 1/2$. In higher dimensions we work with the following equivalent definitions, some of which mirror the equivalent definitions we gave in one dimension.

Definition 3.9 (Four definitions of an isotropic d-dimensional stable process)

(1) A Lévy process, X, is called an isotropic d-dimensional stable process if its marginal distributions at each fixed time are d-dimensional isotropic stable in the sense of Definition 1.23 (see Chapter 1).

(2) The second definition identifies a stable process as a non-Gaussian Lévy process $X = (X_t, t \geq 0)$ such that there exists an $\alpha > 0$ such that, for all $c > 0$, $(cX_{c^{-\alpha}t}, t \geq 0)$ is equal in law to X. Moreover, for all orthogonal transformations U on $\mathbb{R}^d$, the process $(\mathrm{U}X_t, t \geq 0)$ is equal in law to X.

(3) An isotropic d-dimensional stable process is a Lévy process whose characteristic triplet takes the form $(0, 0, \Pi)$, where the measure Π has the following polar decomposition

$$\Pi(B) = 2^{\alpha-1}\pi^{-d}\frac{\Gamma((d+\alpha)/2)\Gamma(d/2)}{|\Gamma(-\alpha/2)|}\int_{\mathbb{S}^{d-1}}\sigma_1(\mathrm{d}\phi)\int_0^\infty \mathbf{1}_B(r\phi)\frac{\mathrm{d}r}{r^{\alpha+1}},$$

for $B \in \mathcal{B}(\mathbb{R}^d)$, where $\sigma_1(\mathrm{d}\phi)$ is the surface measure on $\mathbb{S}^{d-1} := \{x \in \mathbb{R}^{d-1} : |x| = 1\}$, normalised to have unit mass. Stated in Cartesian coordinates, this is equivalent to

$$\Pi(\mathrm{d}z) = 2^{\alpha}\pi^{-d/2}\frac{\Gamma((d+\alpha)/2)}{|\Gamma(-\alpha/2)|}\frac{1}{|z|^{\alpha+d}}\,\mathrm{d}z, \qquad z \in \mathbb{R}^d. \tag{3.23}$$

Here again, the triplet $(0, 0, \Pi)$ is based on the regularisation function $h(x) = 0$ for $\alpha \in (0, 1)$, $h(x) = x$ for $\alpha \in (1, 2)$ and $h(x) = x\mathbf{1}_{(|x|<1)}$ for $\alpha = 1$.

(4) An isotropic d-dimensional Lévy process is a stable process if its characteristic exponent takes the form

$$\Psi(z) = |z|^{\alpha}, \qquad z \in \mathbb{R}^d,$$

for $\alpha \in (0, 2)$.

It is obvious from the proof in Theorem 3.2 that isotropic d-dimensional stable processes are also self-similar in the sense of Theorem 3.2. As with the case of one-dimensional stable processes, we have again deliberately excluded the case that $\alpha = 2$ as, in light of Section 1.1, this corresponds to the case of a Brownian motion. There is however a connection between Brownian motion and isotropic d-dimensional stable processes that we shall make use of later on.

Lemma 3.10 *If $B = (B_t, t \geq 0)$ is a standard d-dimensional Brownian motion and $\Lambda = (\Lambda_t, t \geq 0)$ is an independent stable subordinator with index $\alpha \in (0, 1)$, then $X_t := \sqrt{2}B_{\Lambda_t}$, $t \geq 0$ is an isotropic d-dimensional stable process with index 2α.*

Proof Brownian motion is an isotropic process and hence, since Λ is independent of B, it follows that $X = (X_t, t \geq 0)$ is an isotropic process. As B and Λ are independent and both have stationary and independent increments, we have, for $s, t \geq 0$,

$$\frac{1}{\sqrt{2}}(X_{t+s} - X_s) = B_{\Lambda_{t+s}} - B_{\Lambda_t} \stackrel{(\mathrm{d})}{=} \tilde{B}_{\Lambda_{t+s}-\Lambda_t} \stackrel{(\mathrm{d})}{=} \tilde{B}_{\tilde{\Lambda}_s},$$

where $(\tilde{B}_u, u \geq 0)$ is an independent copy of B and $(\tilde{\Lambda}_t, t \geq 0)$ is an independent copy of Λ. This shows that X has stationary and independent increments. Moreover, as Λ has right-continuous paths with left limits and B has continuous paths, X has right-continuous paths with left limits. We may now say that

X is a Lévy process. To identify X specifically as an isotropic d-dimensional stable process with index 2α, it suffices to consider its characteristic exponent. We have

$$\mathbb{E}\left[e^{iz\cdot X_t}\right] = \mathbb{E}\left[\mathbb{E}\left(e^{iz\sqrt{2}\cdot B_{\Lambda_t}}|\Lambda_t\right)\right] = \mathbb{E}\left[e^{-|z|^2\Lambda_t}\right] = e^{-|z|^{2\alpha}t}, \qquad z \in \mathbb{R}^d,$$

where we have used the fact that the characteristic exponent of a d-dimensional Brownian motion is $|z|^2/2$ and the Laplace exponent of a stable subordinator with index α is given by (3.14). □

Just as for the case of one-dimensional Lévy processes, we can consider some of the finer path properties of isotropic stable processes in higher dimensions.

Transience and recurrence. Following the discussion in Section 2.18, it still makes sense to ask whether such processes are transient or recurrent and whether they can hit points or not. For dimension $d \geq 2$, using polar coordinates, we get for small ε,

$$\begin{aligned}\int_{-\varepsilon}^{\varepsilon} \frac{1}{|z|^\alpha}\, dz &= \frac{2\pi^{(d-1)/2}}{\Gamma((d-1)/2)} \int_0^\varepsilon \frac{1}{r^{\alpha-d+1}} \int_0^\pi (\sin\theta)^{(d-2)}\, d\theta\, dr \\ &= C_d \int_0^\varepsilon \frac{1}{r^{\alpha-d+1}}\, dr,\end{aligned}$$

where C_d is a positive constant that only depends on d, which is proportional to the volume of the d-dimensional sphere. Recalling the integral test (2.26), it follows that there is transience if and only if $\alpha < d$.

Polarity. Theorem 2.31 tells us that all points are essentially polar for dimension $d \geq 2$. In the next section we will show that the resolvent kernel of stable processes are absolutely continuous. As such, from the remarks immediately after Theorem 2.31, we deduce that all points are polar.

3.7 Resolvent density

When $\alpha \in (1, 2)$ and $d = 1$, Lemma 2.19 motivates us to look for the resolvent measure

$$U^{(q)}(dx) = \int_0^\infty e^{-qt}\mathbb{P}(X_t \in dx)\, dt, \qquad x \in \mathbb{R}. \qquad (3.24)$$

In one dimension, by taking Fourier transforms, noting that (2.28) ensures all integrals are well defined, we see that

$$\int_{\mathbb{R}} e^{i\theta x} U^{(q)}(dx) = \frac{1}{q + |\theta|^\alpha \left(e^{\pi i\alpha(\frac{1}{2}-\rho)}\mathbf{1}_{(\theta>0)} + e^{-\pi i\alpha(\frac{1}{2}-\rho)}\mathbf{1}_{(\theta<0)}\right)},$$

for $q \geq 0, \theta \in \mathbb{R}$. No inversion of this transform is known for the case that $q > 0$. It is the case, however, that when $q = 0$, the transform can be inverted giving the resolvent $U := U^{(0)}$ providing X is transient. The case that $\rho = 1$ has already been seen in (3.20). A resolvent cannot exist when X is recurrent simply by the definition of recurrence alone; see Section 2.10. Indeed, this is as much the case for d-dimensional isotropic stable processes as it is one-dimensional processes with $\alpha \in (0, 1)$.

Theorem 3.11 *When X is a transient stable process, that is, $\alpha < d$, its resolvent exists and is absolutely continuous with respect to Lebesgue measure; that is to say $U(dx) = u(x)\,dx$, $x \in \mathbb{R}$. In particular we have the following explicit identities for the resolvent density, u.*

(i) When $d = 1$ and $\alpha \in (0, 1)$,

$$u(x) = \Gamma(1-\alpha)\left(\frac{\sin(\pi\alpha\rho)}{\pi}\mathbf{1}_{(x>0)} + \frac{\sin(\pi\alpha\hat{\rho})}{\pi}\mathbf{1}_{(x<0)}\right)|x|^{\alpha-1}, \tag{3.25}$$

for $x \in \mathbb{R}$. In particular, if $\rho = 1$ (the case of a subordinator), we have

$$u(x) = \frac{1}{\Gamma(\alpha)}x^{\alpha-1}, \qquad x \geq 0.$$

(ii) When $d \geq 2$, (and isotropic),

$$u(x) = 2^{-\alpha}\pi^{-d/2}\frac{\Gamma((d-\alpha)/2)}{\Gamma(\alpha/2)}|x|^{\alpha-d}.$$

Proof (i) Let $x > 0$. From the definition of $U(\cdot)$ on $[0, \infty)$, Fourier inversion for the density of X_t (see for instance (3.7)) together with (3.9) gives us, for bounded measurable $f : [0, \infty) \mapsto [0, \infty)$,

$$\begin{aligned}
\int_0^\infty f(x)U(dx) &= \int_0^\infty \int_0^\infty f(x)\mathbb{P}(X_t \in dx)\,dt \\
&= \frac{1}{2\pi}\mathrm{Re}\left[\int_0^\infty f(x)\int_0^\infty \int_{\mathbb{R}} e^{-\Psi(z)t - izx}\,dz\,dx\,dt\right] \\
&= \mathrm{Re}\left(\frac{e^{-\pi i\alpha(\frac{1}{2}-\rho)}}{2\pi}\int_0^\infty f(x)x^{\alpha-1}\int_0^\infty y^{-\alpha}e^{-iy}\,dy\,dx\right) \\
&\quad - \mathrm{Re}\left(\frac{e^{\pi i\alpha(\frac{1}{2}-\rho)}}{2\pi}\int_0^\infty f(x)x^{\alpha-1}\int_0^\infty y^{-\alpha}e^{iy}\,dy\,dx\right).
\end{aligned}$$

Using identity (1.10) with $s = 1-\alpha$ and $z = 1$, the previous identity is reduced to

$$\int_0^\infty f(x)U(\mathrm{d}x) = \Gamma(1-\alpha)\mathrm{Re}\left(\frac{\mathrm{e}^{-\pi\mathrm{i}(\frac{1}{2}-\alpha\rho)} - \mathrm{e}^{\pi\mathrm{i}(\frac{1}{2}-\alpha\rho)}}{2\pi}\right)\int_0^\infty f(x)x^{\alpha-1}\,\mathrm{d}x$$
$$= \Gamma(1-\alpha)\frac{\sin(\pi\alpha\rho)}{\pi}\int_0^\infty f(x)x^{\alpha-1}\,\mathrm{d}x,$$

as required. Similar argument allow us to deduce the case when $x < 0$ which gives the second term in (3.25).

When $\rho = 1$, we have that the coefficient for $\sin(\pi\alpha\hat{\rho}) = 0$ so that the potential measure is (somewhat obviously) concentrated on the non-negative half-line. The reflection formula for the Gamma function (see (A.12) in the Appendix) now allows us to replace $\Gamma(1-\alpha)\sin(\pi\alpha)/\pi$ by $1/\Gamma(\alpha)$.

(ii) Suppose that $(S_t, t \geq 0)$ is a stable subordinator with index $\alpha/2$. If we write $(B_t, t \geq 0)$ for a standard d-dimensional Brownian motion, then from Lemma 3.10 we have that $X_t := \sqrt{2}B_{S_t}$, $t \geq 0$ is a stable process with index α. Now note that, for bounded and measurable f on $\mathbb{R}^d$, we have

$$\mathbb{E}\left[\int_0^\infty f(X_t)\,\mathrm{d}t\right] = \mathbb{E}\left[\int_0^\infty f(\sqrt{2}B_{S_t})\,\mathrm{d}t\right]$$
$$= \int_0^\infty \mathrm{d}t \int_0^\infty \mathbb{P}(S_t \in \mathrm{d}s)\int_{\mathbb{R}^d} \mathbb{P}(B_s \in \mathrm{d}x) f(\sqrt{2}x)$$
$$= \frac{1}{\Gamma(\alpha/2)\pi^{d/2}2^d}\int_{\mathbb{R}^d}\mathrm{d}y\int_0^\infty \mathrm{d}s\,\mathrm{e}^{-|y|^2/4s}s^{-1+(\alpha-d)/2}f(y)$$
$$= \frac{1}{2^\alpha\Gamma(\alpha/2)\pi^{d/2}}\int_{\mathbb{R}^d}\mathrm{d}y\,|y|^{\alpha-d}\int_0^\infty \mathrm{d}u\,\mathrm{e}^{-u}u^{-1+(d-\alpha/2)}f(y)$$
$$= \frac{\Gamma((d-\alpha)/2)}{2^\alpha\Gamma(\alpha/2)\pi^{d/2}}\int_{\mathbb{R}^d}\mathrm{d}y\,|y|^{\alpha-d}f(y),$$

where in the third equality, we have used the potential for stable subordinators as given in part (i) of this theorem and the standard expression for the d-dimensional Gaussian distribution. This completes the proof. □

In the recurrent cases, that is when $d = 1$ and $\alpha \in [1, 2)$, although a resolvent does not exist, it is still possible to construct an adjusted resolvent. Although we will not use such adjusted resolvents in this book, some of the older literature for stable processes does. Hence we include the following theorem nonetheless.

Theorem 3.12 *For $\alpha \in (1,2)$,*

$$\int_0^\infty (\mathrm{p}_t(x) - \mathrm{p}_t(0))\,\mathrm{d}t = \Gamma(1-\alpha)\left(\frac{\sin(\pi\alpha\rho)}{\pi}\mathbf{1}_{(x>0)} + \frac{\sin(\pi\alpha\hat{\rho})}{\pi}\mathbf{1}_{(x<0)}\right)|x|^{\alpha-1}, \qquad x \in \mathbb{R}.$$

Moreover when $\alpha = 1$,

$$\int_0^\infty (\mathrm{p}_t(x) - \mathrm{p}_t(1))\,\mathrm{d}t = -\frac{1}{\pi}\log|x|, \qquad x \in \mathbb{R}.$$

Proof For the case $\alpha \in (1,2)$, we can appeal to the inverse Fourier transform representation of $\mathrm{p}_t(x)$, $x \in \mathbb{R}$, $t \geq 0$. Assume without loss of generality that $x > 0$. With the help of Fubini's Theorem, using calculations similar to the derivation of (3.25), we deduce

$$\begin{aligned}\int_0^\infty (\mathrm{p}_t(x) - \mathrm{p}_t(0))\,\mathrm{d}t &= -\frac{1}{2\pi}\int_0^\infty \int_{\mathbb{R}} (1 - \mathrm{e}^{-\mathrm{i}zx})\mathrm{e}^{-\Psi(z)t}\,\mathrm{d}z\,\mathrm{d}t \\ &= -\frac{\mathrm{e}^{-\pi\mathrm{i}\alpha(\frac{1}{2}-\rho)}}{2\pi}x^{\alpha-1}\int_0^\infty (1-\mathrm{e}^{-\mathrm{i}y})y^{-\alpha}\,\mathrm{d}y \\ &\quad + \frac{\mathrm{e}^{\pi\mathrm{i}\alpha(\frac{1}{2}-\rho)}}{2\pi}x^{\alpha-1}\int_0^\infty (1-\mathrm{e}^{\mathrm{i}y})y^{-\alpha}\,\mathrm{d}y. \end{aligned} \qquad (3.26)$$

The two integrals above are reminiscent of the characteristic exponent of a stable subordinator with index $\alpha - 1$. Indeed, using identity (1.11) we have

$$\begin{aligned}&\int_0^\infty (\mathrm{p}_t(x) - \mathrm{p}_t(0))\,\mathrm{d}t \\ &= \Gamma(1-\alpha)\frac{x^{\alpha-1}}{2\pi}\left(\mathrm{e}^{-\pi\mathrm{i}\alpha(\frac{1}{2}-\rho)}\mathrm{e}^{\frac{\pi\mathrm{i}}{2}(\alpha-1)} - \mathrm{e}^{\pi\mathrm{i}\alpha(\frac{1}{2}-\rho)}\mathrm{e}^{-\frac{\pi\mathrm{i}}{2}(\alpha-1)}\right) \\ &= \Gamma(1-\alpha)\frac{x^{\alpha-1}}{2\pi}\left(\mathrm{e}^{-\pi\mathrm{i}(\frac{1}{2}-\alpha\rho)} - \mathrm{e}^{\pi\mathrm{i}(\frac{1}{2}-\alpha\rho)}\right) \\ &= \Gamma(1-\alpha)x^{\alpha-1}\frac{\sin(\pi\alpha\rho)}{\pi}.\end{aligned}$$

A similar calculation when $x < 0$ completes the proof in the regime $\alpha \in (1,2)$.

For the case $\alpha = 1$, recall from (3.16) that

$$\mathrm{p}_t(x) = \frac{t}{\pi(x^2+t^2)}, \qquad x \in \mathbb{R}.$$

Hence, using the above and partial fractions, we can evaluate

$$\int_0^\infty (\mathrm{p}_t(x) - \mathrm{p}_t(1))\,\mathrm{d}t$$

$$= -\int_0^\infty \frac{(1-x^2)t}{\pi(1+t^2)(x^2+t^2)}\,\mathrm{d}t$$

$$= -\frac{1}{\pi}\int_0^\infty \left(\frac{1}{2(t-\mathrm{i}x)} + \frac{1}{2(t+\mathrm{i}x)} - \frac{1}{2(t-\mathrm{i})} - \frac{1}{2(t+\mathrm{i})}\right)\mathrm{d}t$$

$$= -\frac{1}{\pi}\log|x|,$$

for $x \in \mathbb{R}$. The proof is now complete. □

3.8 Comments

Around the beginning of the 1930s, Paul Lévy observed that any strictly stable law leads to a random function that can be obtained by an interpolation method, much in the same spirit that Brownian motion is obtained from the Gaussian distribution. This fact motivated Paul Lévy to introduce the general definition of processes with independent increments or additive processes and in particular stable processes. The relevance of strictly stable processes arises from the fact that, along with Brownian motion, they can be obtained as scaling limits of random walks. Moreover, stable processes and Brownian motion are the only Lévy processes satisfying the scaling property.

Similarly to the literature for stable distributions, there are several monographs where stable processes are treated, for instance we mention Samorodnitsky and Taqqu [189], Bertoin [18], Sato [190] and Uchaikin and Zolotarev [208]. The monograph of Janicki and Weron [96] describe approximation and simulation methods for stable processes.

Sections 3.1 and 3.2 follow the same structure as in Chapter 1 where stable distributions are treated. Most of the results presented in Sections 3.3, 3.4 and 3.6, where path properties of stable processes are studied, are simple applications of the general results presented in Chapter 2 to this particular setting. Many of the identities concerning the range of a stable process that appear in its fluctuation theory are analytically tractable since the spatial Wiener–Hopf factorisation presented in Section 3.5 is explicit. This is not the case for the space-time Wiener–Hopf factorisation, which is more more complicated to derive. See for instance Doney [59] as well as Kuznetzov [116], where recent developments can be found on this topic. The space-time Wiener–Hopf factorisation for stable processes will be treated later in Chapter 7. First passage problems for stable processes were first studied by Zolotarev [219], who used analytic methods to prove Corollaries 3.5 and 3.7. The triple law that appears in Theorem 3.6 is an easy corollary of the quintuple law given in Doney and Kyprianou [62]. Finally, the resolvent densities that appear in Theorem 3.11 were obtained by Blumenthal et al. [39], who themselves refer to a method of Kac [98].

4

Hypergeometric Lévy processes

Roughly speaking, hypergeometric Lévy processes are defined by choosing their Wiener–Hopf factors from a special class of (killed) subordinators, called *β-subordinators*. The consequence of having conveniently chosen Wiener–Hopf factors in this way is that many important fluctuation identities become analytically tractable.

It also turns out that different hypergeometric Lévy processes naturally appear through a variety of space-time transformations of the paths of α-stable Lévy processes. In each case, the theory of so-called self-similar Markov processes plays a fundamental role and this connection will play a dominating role in the overwhelming majority of forthcoming results for stable Lévy processes. For this reason, we devote this chapter to introducing the aforementioned class of hypergeometric Lévy processes in detail as well as studying a number of the explicit identities that they offer by way of their path functionals.

4.1 β-subordinators

In Theorem 2.24 we worked with the notion of a killed subordinator. That is, a subordinator that is sent to the cemetery state, $\{\infty\}$, at an independent random time which is exponentially distributed with parameter $q \geq 0$. Taking account of the formula given in (2.11), we recall that the Laplace exponent of a (killed) subordinator is given by

$$\kappa(\lambda) = q + b\lambda + \int_{(0,\infty)} (1 - \mathrm{e}^{-\lambda x})\Upsilon(\mathrm{d}x), \qquad \lambda \geq 0,$$

where $q, b \geq 0$ and $\int_{(0,\infty)}(1 \wedge x)\Upsilon(\mathrm{d}x) < \infty$.

Let us now introduce, by way of a proposition, the family of (killed) β-subordinators with parameters (α, β, γ), which will ultimately be used to build the class of hypergeometric Lévy processes.

Proposition 4.1 *Assume that* $0 \le \alpha \le \beta + \gamma$ *and* $\gamma \in (0,1)$. *Then there exists a subordinator* $Y = (Y_t, 0 \le t \le \zeta)$, *where* ζ *denotes its lifetime and whose Laplace exponent satisfies*

$$\kappa(\lambda) = (\lambda + \alpha)\frac{\Gamma(\lambda + \beta + \gamma)}{\Gamma(\lambda + \beta + 1)}, \qquad \lambda \ge 0. \tag{4.1}$$

In particular, Y has zero drift coefficient and a Lévy measure which has a density with respect to Lebesgue measure given by

$$\upsilon(x) = \frac{1}{\Gamma(1-\gamma)}(1 - \mathrm{e}^{-x})^{-\gamma}\,\mathrm{e}^{-(\beta+\gamma)x}\left[\frac{\gamma}{1-\mathrm{e}^{-x}} + \beta - \alpha\right], \tag{4.2}$$

for $x > 0$. *Moreover, Y has finite lifetime with rate* $\kappa(0) = q = \alpha\Gamma(\beta+\gamma)/\Gamma(\beta+1)$.

Proof Starting with the formula for the standard beta integral (cf. (A.18) in the Appendix), after a straightforward change of variables, we have

$$\frac{\Gamma(\lambda + \beta + \gamma)}{\Gamma(\lambda + \beta + 1)} = \frac{1}{\Gamma(1-\gamma)}\int_0^\infty (1 - \mathrm{e}^{-x})^{-\gamma}\,\mathrm{e}^{-(\beta+\gamma)x}\mathrm{e}^{-\lambda x}\,\mathrm{d}x, \tag{4.3}$$

where $\lambda \ge 0$. Define $w(x) = (1 - \mathrm{e}^{-x})^{-\gamma}\,\mathrm{e}^{-(\beta+\gamma)x}/\Gamma(1-\gamma)$, $x \ge 0$. Performing an integration by parts in (4.3) we get

$$\lambda\frac{\Gamma(\lambda + \beta + \gamma)}{\Gamma(\lambda + \beta + 1)} = -\int_0^\infty w'(x)(1 - \mathrm{e}^{-\lambda x})\,\mathrm{d}x. \tag{4.4}$$

Combining (4.3) and (4.4) we see that

$$\kappa(\lambda) = \int_0^\infty (-\alpha w(x) - w'(x))\left(1 - \mathrm{e}^{-\lambda x}\right)\mathrm{d}x + \alpha\frac{\Gamma(\beta+\gamma)}{\Gamma(\beta+1)},$$

and it is easy to check that the density given in (4.2) is equal to $-\alpha w(x) - w'(x)$ on $(0,\infty)$. □

The Lévy density of the β-subordinator is a completely monotone function. To see this, use the binomial expansion for $(1-\mathrm{e}^{-x})^{-\gamma}$ and $(1-\mathrm{e}^{-x})^{-(1+\gamma)}$ and write (4.2) in the form

$$\upsilon(x) = \frac{1}{\Gamma(1-\gamma)}\sum_{k\ge 0}\frac{(\gamma)_k}{k!}(\beta + k + \gamma - \alpha)\mathrm{e}^{-(\beta+k+\gamma)x}, \qquad x > 0, \tag{4.5}$$

where

$$(\gamma)_k := \gamma(\gamma+1)(\gamma+2)\dots(\gamma+k-1),$$

denotes the Pochhammer symbol.

Recall from the discussion in Section 2.17 that the renewal measure, U, of the subordinator Y is defined by

$$U(\mathrm{d}x) = \int_0^\infty \mathrm{P}(Y_t \in \mathrm{d}x, t < \zeta)\,\mathrm{d}t, \qquad x \geq 0,$$

and satisfies

$$\int_{[0,\infty)} \mathrm{e}^{-\lambda x} U(\mathrm{d}x) = \frac{1}{\kappa(\lambda)}, \qquad \lambda \geq 0.$$

As one might expect, given the straightforward nature of the formula for κ in (4.1), we should expect to obtain a closed form expression for the renewal measure of Y.

Proposition 4.2 *The renewal measure, U, of the β-subordinator satisfies $U(\mathrm{d}x) = u(x)\mathrm{d}x$ on $[0, \infty)$, where*

$$\begin{aligned} u(x) &= \frac{\mathrm{e}^{-\alpha x}}{\Gamma(\gamma)} (1 - \mathrm{e}^{-x})^{\gamma-1}\, \mathrm{e}^{-(1+\beta-\alpha)x} \\ &\quad + \frac{\mathrm{e}^{-\alpha x}}{\Gamma(\gamma)} (\beta + \gamma - \alpha) \int_0^x (1 - \mathrm{e}^{-u})^{\gamma-1}\, \mathrm{e}^{-(1+\beta-\alpha)u}\,\mathrm{d}u, \end{aligned} \tag{4.6}$$

for $x \geq 0$.

Proof We start with the identity

$$\int_{[0,\infty)} \mathrm{e}^{-\lambda x} U(\mathrm{d}x) = \frac{1}{\kappa(\lambda)} = \frac{\Gamma(\lambda + \beta + 1)}{\Gamma(\lambda + \beta + 1 + \gamma)} \times \left(1 + \frac{\beta + \gamma - \alpha}{\lambda + \alpha}\right). \tag{4.7}$$

Again appealing to a straightforward change of variables in the classical beta-integral, we see that $\Gamma(\lambda + \beta + 1)/\Gamma(\lambda + \beta + 1 + \gamma) = \int_{[0,\infty)} \mathrm{e}^{-\lambda x} \mu_1(\mathrm{d}x)$, where

$$\mu_1(\mathrm{d}x) = \frac{1}{\Gamma(\gamma)} (1 - \mathrm{e}^{-x})^{\gamma-1}\, \mathrm{e}^{-(1+\beta)x}\,\mathrm{d}x, \qquad x \geq 0.$$

Moreover, $1 + (\beta + \gamma - \alpha)/(\lambda + \alpha) = \int_{[0,\infty)} \mathrm{e}^{-\lambda x} \mu_2(\mathrm{d}x)$, where

$$\mu_2(\mathrm{d}x) = \delta_0(\mathrm{d}x) + (\beta + \gamma - \alpha)\mathrm{e}^{-\alpha x}\,\mathrm{d}x, \qquad x \geq 0.$$

Since the product of two Laplace transforms corresponds to the Laplace transform of the convolution of the respective underlying measures, we find from (4.7) that $U = \mu_1 * \mu_2$, on $[0, \infty)$, which is equivalent to (4.6). □

The following corollary is an easy consequence of the above propositions and is worth recording for later.

Corollary 4.3 *In the special case that* $\alpha = \beta \geq 0$ *and* $\gamma \in (0,1)$, *the Lévy density of the* β*-subordinator takes the form*

$$\upsilon(x) = \frac{\gamma}{\Gamma(1-\gamma)}(1-\mathrm{e}^{-x})^{-\gamma-1}\,\mathrm{e}^{-(\beta+\gamma)x}, \qquad x \geq 0,$$

and the renewal measure of the β*-subordinator has density given by*

$$u(x) = \frac{1}{\Gamma(\gamma)}\mathrm{e}^{-\beta x}(1-\mathrm{e}^{-x})^{\gamma-1}, \qquad x \geq 0.$$

The β-subordinators described in Corollary 4.3 are also known as *Lamperti-stable subordinators.*

4.2 Hypergeometric processes

As alluded to at the start of this chapter, the basic idea behind the definition of a hypergeometric Lévy process is that they are defined by specifying a particular pair of Wiener–Hopf factors. In order to do so, we want to know when we have the freedom to pick the Laplace exponents of (killed) subordinators and combine them together in the spirit of the formula (2.31). To this end, we take two subordinators H and $\hat{H}$ with lifetimes ς and $\hat{\varsigma}$, respectively, defined by their characteristic exponents, which we assume to take the form

$$\kappa(\lambda) = q + b\lambda + \int_0^\infty \left(1-\mathrm{e}^{-\lambda x}\right)\nu(x)\,\mathrm{d}x, \tag{4.8}$$

$$\hat{\kappa}(\lambda) = \hat{q} + \hat{b}\lambda + \int_0^\infty \left(1-\mathrm{e}^{-\lambda x}\right)\hat{\nu}(x)\,\mathrm{d}x,$$

for $\lambda \geq 0$, where, as usual, we must have that $q, \hat{q}, b, \hat{b} \geq 0$ and $\int_0^\infty (1 \wedge x)\nu(x)\,\mathrm{d}x < \infty$ and $\int_0^\infty (1 \wedge x)\hat{\nu}(x)\,\mathrm{d}x < \infty$. We allow the possibility that one of ν and $\hat{\nu}$ are identically zero and define $V(x) = \int_x^\infty \nu(y)\,\mathrm{d}y$, $x > 0$, and $\hat{V}(x)$, $x > 0$ similarly.

Theorem 4.4 *Assume that*

$$\mathrm{E}\left[H_1^2 \mathbf{1}_{\{1<\varsigma\}}\right] < \infty, \qquad \mathrm{E}\left[\hat{H}_1^2 \mathbf{1}_{\{1<\hat{\varsigma}\}}\right] < \infty,$$

and that $\nu(x)$, $\hat{\nu}(x)$ *are decreasing functions. Define* $\Psi(z) = \kappa(-\mathrm{i}z)\hat{\kappa}(\mathrm{i}z)$, $z \in \mathbb{R}$. *Then* $\Psi(z)$ *is the characteristic exponent of a Lévy process, say* Y, *with lifetime* ζ, *whose parameter is* $\Psi(0) = q\hat{q}$, *and characteristic triplet* $(q\hat{b} - \hat{q}b, 2b\hat{b}, \Pi)$, *where*

$$\Pi(x,\infty) := \int_0^\infty \hat{V}(u)\nu(x+u)\,\mathrm{d}u + \hat{b}\nu(x) + \hat{q}V(x), \tag{4.9}$$

$$\Pi(-\infty,-x) := \int_0^\infty V(u)\hat{\nu}(x+u)\,\mathrm{d}u + b\hat{\nu}(x) + q\hat{V}(x). \tag{4.10}$$

Moreover, we have $\mathbb{E}[Y_1^2 \mathbf{1}_{\{1<\zeta\}}] < \infty$.

We need a preliminary result before proving the above theorem.

Lemma 4.5 *A function $\Psi\colon \mathbb{R} \mapsto \mathbb{C}$ is the characteristic exponent of a Lévy process Y, with lifetime ζ, satisfying $\mathrm{E}[Y_1^2 \mathbf{1}_{\{1<\zeta\}}] < \infty$ if and only if it can be written in the form*

$$\Psi(z) = p + \mathrm{i}az + \frac{1}{2}\sigma^2 z^2 + z^2 \int_{\mathbb{R}} \mathrm{e}^{\mathrm{i}zx}\eta(x)\,\mathrm{d}x, \tag{4.11}$$

where $p > 0$, $a \in \mathbb{R}$, $\sigma \geq 0$, $\eta \in L^1(\mathbb{R})$ is an absolutely continuous function such that its density can be taken as increasing on $(-\infty, 0)$ and decreasing on $(0, \infty)$.

Proof Consider the Lévy–Khintchine formula (1.5) with the regularizing function $h(x) \equiv x$, and, after integrating by parts twice, we can identify it in the form (4.11) with

$$\eta(x) = \begin{cases} \int_x^\infty \Pi(u,\infty)\,\mathrm{d}u, & \text{if } x > 0, \\ \int_{-\infty}^x \Pi(-\infty,u)\,\mathrm{d}u, & \text{if } x < 0. \end{cases} \tag{4.12}$$

This allows us to handle both directions of the proof, with (4.12) acting as the identification of η in one direction of the proof, and as a definition of the Lévy measure Π in the other direction of the proof. □

Proof of Theorem 4.4 The assumption $\mathrm{E}[H_1^2 \mathbf{1}_{\{1<\varsigma\}}] < \infty$ implies (via integration by parts twice) that

$$\int_0^\infty x^2 v(x)\,\mathrm{d}x = \int_0^\infty \left(\int_x^\infty V(y)\,\mathrm{d}y \right) \mathrm{d}x < \infty,$$

and hence $\int_x^\infty V(y)\mathrm{d}y \in L^1(\mathbb{R}_+)$. In particular, we have

$$m := \int_0^\infty x v(x)\,\mathrm{d}x = \int_0^\infty V(y)\mathrm{d}y < \infty.$$

Similarly, we deduce that $\int_x^\infty \hat{V}(y)\mathrm{d}y \in L^1(\mathbb{R}_+)$ and

$$\hat{m} := \int_0^\infty x\hat{v}(x)\,\mathrm{d}x < \infty.$$

Integrating by parts in (4.8), we obtain, for $z \in \mathbb{R}$,

$$\kappa(-\mathrm{i}z) = q - \mathrm{i}bz - \mathrm{i}z \int_0^\infty \mathrm{e}^{\mathrm{i}zx} V(x)\,\mathrm{d}x,$$

and similarly

$$\hat{\kappa}(\mathrm{i}z) = \hat{q} + \mathrm{i}\hat{b}z + \mathrm{i}z \int_0^\infty \mathrm{e}^{-\mathrm{i}zx} \hat{V}(x)\,\mathrm{d}x.$$

Therefore

$$
\begin{aligned}
&\kappa(-\mathrm{i}z)\hat{\kappa}(\mathrm{i}z)\\
&= \left(q - \mathrm{i}bz - \mathrm{i}z\int_0^\infty \mathrm{e}^{\mathrm{i}zx}V(x)\,\mathrm{d}x\right)\left(\hat{q} + \mathrm{i}\hat{b}z + \mathrm{i}z\int_0^\infty \mathrm{e}^{-\mathrm{i}zx}\hat{V}(x)\,\mathrm{d}x\right)\\
&= q\hat{q} + (q\hat{b} - \hat{q}b)\mathrm{i}z + b\hat{b}z^2 + z^2\left(\int_0^\infty \mathrm{e}^{\mathrm{i}zx}V(x)\,\mathrm{d}x\right)\left(\int_0^\infty \mathrm{e}^{-\mathrm{i}zx}\hat{V}(x)\,\mathrm{d}x\right)\\
&\quad + (q - \mathrm{i}bz)\mathrm{i}z\int_0^\infty \mathrm{e}^{-\mathrm{i}zx}\hat{V}(x)\,\mathrm{d}x - (\hat{q} + \mathrm{i}\hat{b}z)\mathrm{i}z\int_0^\infty \mathrm{e}^{\mathrm{i}zx}V(x)\,\mathrm{d}x.
\end{aligned}
$$

Since m and $\hat{m}$ are finite, we rewrite the latter identity as follows

$$
\begin{aligned}
&\kappa(-\mathrm{i}z)\hat{\kappa}(\mathrm{i}z)\\
&= q\hat{q} + \left(q(\hat{b} + \hat{m}) - \hat{q}(b + m)\right)\mathrm{i}z + b\hat{b}z^2 + bz^2\int_0^\infty \mathrm{e}^{-\mathrm{i}zx}\hat{V}(x)\,\mathrm{d}x\\
&\quad + \hat{b}z^2\int_0^\infty \mathrm{e}^{\mathrm{i}zx}V(x)\,\mathrm{d}x + z^2\left(\int_0^\infty \mathrm{e}^{\mathrm{i}zx}V(x)\,\mathrm{d}x\right)\left(\int_0^\infty \mathrm{e}^{-\mathrm{i}zx}\hat{V}(x)\,\mathrm{d}x\right)\\
&\quad - q\mathrm{i}z\int_0^\infty \left(1 - \mathrm{e}^{-\mathrm{i}zx}\right)\hat{V}(x)\,\mathrm{d}x + \hat{q}\mathrm{i}z\int_0^\infty \left(1 - \mathrm{e}^{\mathrm{i}zx}\right)V(x)\,\mathrm{d}x.
\end{aligned}
\tag{4.13}
$$

Next, we integrate by parts and find that

$$
\int_0^\infty \left(1 - \mathrm{e}^{-\mathrm{i}zx}\right)\hat{V}(x)\,\mathrm{d}x = \mathrm{i}z\int_0^\infty \mathrm{e}^{-\mathrm{i}zx}\left[\int_x^\infty \hat{V}(y)\,\mathrm{d}y\right]\mathrm{d}x \tag{4.14}
$$

and similarly

$$
\int_0^\infty \left(1 - \mathrm{e}^{\mathrm{i}zx}\right)V(x)\,\mathrm{d}x = -\mathrm{i}z\int_0^\infty \mathrm{e}^{\mathrm{i}zx}\left[\int_x^\infty V(y)\,\mathrm{d}y\right]\mathrm{d}x. \tag{4.15}
$$

Let us define the function

$$
\eta(x) := \begin{cases} \int_0^\infty \hat{V}(u)V(x+u)\,\mathrm{d}u + \hat{b}V(x) + \hat{q}\int_x^\infty V(y)\,\mathrm{d}y, & \text{if } x > 0,\\ \int_0^\infty V(u)\hat{V}(-x+u)\,\mathrm{d}u + b\hat{V}(-x) + q\int_{-x}^\infty \hat{V}(y)\,\mathrm{d}y, & \text{if } x < 0, \end{cases} \tag{4.16}
$$

which, from our assumptions, belongs to $L^1(\mathbb{R})$. Since the product of Fourier transforms is the Fourier transform of the convolution, we see that formulas (4.13), (4.14), (4.15) and (4.16) imply that

$$
\kappa(-\mathrm{i}z)\hat{\kappa}(\mathrm{i}z) = q\hat{q} + \left(q(\hat{b} + \hat{m}) - \hat{q}(b + m)\right)\mathrm{i}z + b\hat{b}z^2 + z^2\int_{\mathbb{R}} \mathrm{e}^{\mathrm{i}zx}\eta(x)\,\mathrm{d}x.
$$

It is straightforward to see that η is absolutely continuous. To complete the proof, we therefore need to show that the density of η has a version that is increasing on $(-\infty, 0)$ and decreasing $(0, \infty)$. If we take the derivative of both sides in (4.16) and use the fact that the two functions $V'(u) = v(u)$, $\hat{V}'(u) = \hat{v}(u)$ are decreasing, the required monotonicity property follows. □

Now we are ready to introduce the hypergeometric processes which we do through the forthcoming theorem. To this end, we first define the set of admissible parameters which will be used in the definition of hypergeometric Lévy processes: $\mathcal{H} := \mathcal{H}_1 \cup \mathcal{H}_2$, where

$$\mathcal{H}_1 := \{ \beta \le 1,\ \gamma \in (0,1),\ \hat{\beta} \ge 0,\ \hat{\gamma} \in (0,1) \} \tag{4.17}$$

and

$$\mathcal{H}_2 := \left\{ \begin{array}{l} \beta \in [1,2],\ \gamma \in (0,1),\ \hat{\beta} \in [-1,0], \\ \hat{\gamma} \in (0,1) \text{ and } 1 - \beta + \hat{\beta} + (\gamma \wedge \hat{\gamma}) \ge 0 \end{array} \right\}. \tag{4.18}$$

It is important to note that the sets of parameters $\mathcal{H}_1$ and $\mathcal{H}_2$, only coincide in the case $\hat{\beta} = 0$ and $\beta = 1$. As we will see below, each set of parameters codes an individual hypergeometric Lévy process via its Wiener–Hopf factorisation. For each set of parameters, hypergeometric Lévy processes can be killed at an exponential random time, drift to ∞, oscillate or drift to $-\infty$.

Theorem 4.6 (Definition of a hypergeometric Lévy process)

(i) For $(\beta, \gamma, \hat{\beta}, \hat{\gamma}) \in \mathcal{H}$ there exists a Lévy process Y with lifetime ζ, which we henceforth refer to as a hypergeometric Lévy process, having the characteristic function

$$\Psi(z) = \frac{\Gamma(1 - \beta + \gamma - \mathrm{i}z)}{\Gamma(1 - \beta - \mathrm{i}z)} \frac{\Gamma(\hat{\beta} + \hat{\gamma} + \mathrm{i}z)}{\Gamma(\hat{\beta} + \mathrm{i}z)} \qquad z \in \mathbb{R}. \tag{4.19}$$

(ii) The Lévy measure of Y has a density with respect to Lebesgue measure, which is given by

$$\pi(x) = \begin{cases} -\dfrac{\Gamma(\eta)}{\Gamma(\eta - \hat{\gamma})\Gamma(-\gamma)} \mathrm{e}^{-(1-\beta+\gamma)x} {}_2F_1\left(1 + \gamma, \eta; \eta - \hat{\gamma}; \mathrm{e}^{-x}\right), & \text{if } x > 0, \\[2ex] -\dfrac{\Gamma(\eta)}{\Gamma(\eta - \gamma)\Gamma(-\hat{\gamma})} \mathrm{e}^{(\hat{\beta}+\hat{\gamma})x} {}_2F_1\left(1 + \hat{\gamma}, \eta; \eta - \gamma; \mathrm{e}^{x}\right), & \text{if } x < 0, \end{cases} \tag{4.20}$$

where

$$\eta := 1 - \beta + \gamma + \hat{\beta} + \hat{\gamma} \tag{4.21}$$

and, for $z \in \mathbb{R}$ such that $|z| < 1$,

$${}_2F_1(a, b; c; z) := \sum_{k \ge 0} \frac{(a)_k (b)_k}{(c)_k k!} z^k,$$

is the hypergeometric function (see Section A.6 of the Appendix).

(iii) *For* $(\beta,\gamma,\hat{\beta},\hat{\gamma})\in\mathcal{H}_1$ *the ladder height process* H *(resp.* $\hat{H}$*) is a* β*-subordinator with parameters* $(1-\beta,1-\beta,\gamma)$ *(resp.* $(\hat{\beta},\hat{\beta},\hat{\gamma})$*).*

(iv) *For* $(\beta,\gamma,\hat{\beta},\hat{\gamma})\in\mathcal{H}_2$ *the ladder height process* H *(resp.* $\hat{H}$*) is a* β*-subordinator with parameters* $(-\hat{\beta},1-\beta,\gamma)$ *(resp.* $(\beta-1,\hat{\beta},\hat{\gamma})$*).*

(v) *If* Y *is a hypergeometric Lévy process with parameters* $(\beta,\gamma,\hat{\beta},\hat{\gamma})$*, its dual process* $\hat{Y}:=-Y$ *is also a hypergeometric Lévy process with parameters* $(1-\hat{\beta},\hat{\gamma},1-\beta,\gamma)$*.*

Proof The proof of items (i), (iii) and (iv) follows directly from Theorem 4.4, the fact that the Lévy density of the β-subordinators is a decreasing function (see (4.5)) and the identity

$$\Psi(z)=\frac{\Gamma(1-\beta+\gamma-\mathrm{i}z)}{\Gamma(1-\beta-\mathrm{i}z)}\times\frac{\Gamma(\hat{\beta}+\hat{\gamma}+\mathrm{i}z)}{\Gamma(\hat{\beta}+\mathrm{i}z)}$$
$$=(-\hat{\beta}-\mathrm{i}z)\frac{\Gamma(1-\beta+\gamma-\mathrm{i}z)}{\Gamma(2-\beta-\mathrm{i}z)}\times(\beta-1+\mathrm{i}z)\frac{\Gamma(\hat{\beta}+\hat{\gamma}+\mathrm{i}z)}{\Gamma(1+\hat{\beta}+\mathrm{i}z)}.$$

(The first multiplication sign above identifies the Wiener–Hopf factorisation for the regime $\mathcal{H}_1$ and the second for the regime $\mathcal{H}_2$.) Item (v) follows directly from the form of the characteristic exponent of hypergeometric Lévy processes.

Let us now turn to computing the Lévy measure of Y. Consider first the case when $(\beta,\gamma,\hat{\beta},\hat{\gamma})\in\mathcal{H}_1$. Then, according to (4.5) we have

$$v(x)=\frac{1}{\Gamma(1-\gamma)}\sum_{k\geq 0}\frac{(\gamma)_k}{k!}(k+\gamma)\mathrm{e}^{-(k+1-\beta+\gamma)x},\qquad x\geq 0,\tag{4.22}$$

which also gives us an expression for V by integrating each term in the sum. We can now use formula (4.9) to obtain

$$\Pi(x,\infty)=\frac{1}{\Gamma(1-\gamma)}\sum_{k\geq 0}\frac{(\gamma)_k}{k!}(k+\gamma)\mathrm{e}^{-(k+1-\beta+\gamma)x}\int_0^\infty \mathrm{e}^{-(k+1-\beta+\gamma)u}\hat{V}(u)\,\mathrm{d}u$$
$$+\frac{\Gamma(\hat{\beta}+\hat{\gamma})}{\Gamma(\hat{\beta})\Gamma(1-\gamma)}\sum_{k\geq 0}\frac{(\gamma)_k}{k!}\frac{k+\gamma}{k+1-\beta+\gamma}\mathrm{e}^{-(k+1-\beta+\gamma)x},\tag{4.23}$$

for $x>0$. Recalling that $\hat{H}$ is a β-subordinator with parameters $(\hat{\beta},\hat{\beta},\hat{\gamma})$, from formula (4.2) we may compute

$$\int_0^\infty \mathrm{e}^{-\lambda u}\hat{V}(u)\,\mathrm{d}u=\frac{1}{\lambda}\left[\frac{\Gamma(\hat{\beta}+\hat{\gamma}+\lambda)}{\Gamma(\hat{\beta}+\lambda)}-\frac{\Gamma(\hat{\beta}+\hat{\gamma})}{\Gamma(\hat{\beta})}\right],\qquad \lambda\geq 0.$$

Plugging this into (4.23), we get, after a little algebra,

$$\Pi(x,\infty) = \frac{1}{\Gamma(1-\gamma)} \sum_{k\geq 0} \frac{(\gamma)_k}{k!} \mathrm{e}^{-(k+1-\beta+\gamma)x} \frac{k+\gamma}{k+1-\beta+\gamma} \frac{\Gamma(\eta+k)}{\Gamma(\eta+k-\hat{\gamma})}, \tag{4.24}$$

for $x > 0$. Using the fact that

$$\frac{1}{\Gamma(1-\gamma)}(\gamma)_k \times (k+\gamma) = -\frac{1}{\Gamma(-\gamma)}(1+\gamma)_k,$$

and taking derivatives of both sides in (4.24), we get (4.20), for $x > 0$.

The formula for the Lévy density when $x < 0$ follows by considering the dual process, which is also a hypergeometric process with parameters $(1 - \hat{\beta}, \hat{\gamma}, 1-\beta, \gamma)$.

The case when $(\beta, \gamma, \hat{\beta}, \hat{\gamma}) \in \mathcal{H}_2$ follows from similar arguments as those used in the previous case. More precisely, observe that, according to (4.5), the density of the Lévy measure associate to H satisfies

$$\nu(x) = \frac{1}{\Gamma(1-\gamma)} \sum_{k\geq 0} \frac{(\gamma)_k}{k!} (\eta - \hat{\gamma} + k)\mathrm{e}^{-(k+1-\beta+\gamma)x}, \quad x \geq 0, \tag{4.25}$$

which also gives us an expression for V. Again, we use formula (4.9) to obtain

$$\begin{aligned}\Pi(x,\infty) = {} & \frac{1}{\Gamma(1-\gamma)} \sum_{k\geq 0} \frac{(\gamma)_k}{k!} (\eta - \hat{\gamma} + k)\mathrm{e}^{-(k+1-\beta+\gamma)x} \int_0^\infty \mathrm{e}^{-(k+1-\beta+\gamma)u} \hat{V}(u)\, \mathrm{d}u \\ & + (\beta - 1)\frac{\Gamma(\hat{\beta}+\hat{\gamma})}{\Gamma(1+\hat{\beta})\Gamma(1-\gamma)} \sum_{k\geq 0} \frac{(\gamma)_k}{k!} \frac{\eta - \hat{\gamma} + k}{k+1-\beta+\gamma} \mathrm{e}^{-(k+1-\beta+\gamma)x}.\end{aligned}$$

Recalling that $\hat{H}$ is a β-subordinator with parameters $(\beta - 1, \hat{\beta}, \hat{\gamma})$, from formula (4.2) we may compute

$$\int_0^\infty \mathrm{e}^{-\lambda u} \hat{V}(u)\, \mathrm{d}u = \frac{1}{\lambda}\left[(\lambda + \beta - 1)\frac{\Gamma(\hat{\beta}+\hat{\gamma}+\lambda)}{\Gamma(1+\hat{\beta}+\lambda)} - (\beta - 1)\frac{\Gamma(\hat{\beta}+\hat{\gamma})}{\Gamma(1+\hat{\beta})}\right], \qquad \lambda \geq 0.$$

Plutting all the pieces together, we get

$$\Pi(x,\infty) = \frac{1}{\Gamma(1-\gamma)} \sum_{k\geq 0} \frac{(\gamma)_k}{k!} \mathrm{e}^{-(k+1-\beta+\gamma)x} \frac{k+\gamma}{k+1-\beta+\gamma} \frac{\Gamma(\eta+k)}{\Gamma(\eta+k-\hat{\gamma})},$$

which is the same expression as (4.24), thus (4.20) also holds for $x > 0$, in the regime $\mathcal{H}_2$. Similarly, the formula for the density for $x < 0$ follows by using similar arguments as above. This completes the proof. □

Before moving on to analyse some interesting path functionals of hypergeometric Lévy processes, let us identify their long-term behaviour for the different parts of the parameter regimes $\mathcal{H}_1$ and $\mathcal{H}_2$.

More precisely, $\kappa(0) = 0$ if and only if the range of Y is a.s. unbounded above, and $\hat{\kappa}(0) = 0$ if and only if the range of Y is a.s. unbounded below. Accordingly, the corollary below follows directly by examining the Wiener–Hopf factorisation in Theorem 4.6.

Corollary 4.7 *Suppose that Y is a hypergeometric Lévy process. If $\beta < 1$ and $\hat{\beta} > 0$ or $\beta \in (1,2)$ and $\hat{\beta} \in (-1,0)$, the process Y has finite lifetime ζ with rate*

$$\frac{\Gamma(1-\beta+\gamma)}{\Gamma(1-\beta)}\frac{\Gamma(\hat{\beta}+\hat{\gamma})}{\Gamma(\hat{\beta})}.$$

Otherwise the process has an infinite lifetime and

(i) Y oscillates if $\hat{\beta} = 0$ and $\beta = 1$,
(ii) Y drifts to ∞, if either $(\beta,\gamma,\hat{\beta},\hat{\gamma}) \in \mathcal{H}_1$ with $\beta = 1$ and $\hat{\beta} > 0$ or $(\beta,\gamma,\hat{\beta},\hat{\gamma}) \in \mathcal{H}_2$ with $\hat{\beta} = 0$ and $\beta \in (1,2)$,
(iii) Y drifts to $-\infty$, if either $(\beta,\gamma,\hat{\beta},\hat{\gamma}) \in \mathcal{H}_1$ with $\hat{\beta} = 0$ and $\beta < 1$ or $(\beta,\gamma,\hat{\beta},\hat{\gamma}) \in \mathcal{H}_2$ with $\beta = 1$ and $\hat{\beta} < 0$.

Another way of deriving (i) – (iii) is by differentiating (4.19) and taking $z = 0$. Since in any case $\mathrm{i}\Psi'(0) = \mathrm{E}[Y_1]$ is always finite, we can appeal to Theorem 2.8 to conclude. For instance when $(\beta,\gamma,\hat{\beta},\hat{\gamma}) \in \mathcal{H}_1$ with $\hat{\beta} = 0$ and $\beta < 1$, then

$$\mathrm{E}[Y_1] = -\frac{\Gamma(1-\beta+\gamma)\Gamma(\hat{\gamma})}{\Gamma(1-\beta)} < 0.$$

Similarly, when $(\beta,\gamma,\hat{\beta},\hat{\gamma}) \in \mathcal{H}_2$ with $\beta = 1$ and $\hat{\beta} < 0$, we have

$$\mathrm{E}[Y_1] = \hat{\beta}\Gamma(\gamma)\frac{\Gamma(\hat{\beta}+\hat{\gamma})}{\Gamma(1+\hat{\beta})} > 0.$$

4.3 The subclass of Lamperti-stable processes

We continue our analysis of hypergeometric Lévy processes by outlining a subclass of hypergeometric Lévy process which will be of particular interest to us in the forthcoming chapters.

Definition 4.8 A *Lamperti-stable* Lévy process is a hypergeometric process for which the parameters belong to $\mathcal{H}_1$ and for which $\beta = \hat{\beta}$. Said another way, a Lamperti-stable process has characteristic exponent given by

$$\Psi(z) = \frac{\Gamma(1-\beta+\gamma-\mathrm{i}z)}{\Gamma(1-\beta-\mathrm{i}z)}\frac{\Gamma(\beta+\hat{\gamma}+\mathrm{i}z)}{\Gamma(\beta+\mathrm{i}z)} \qquad z \in \mathbb{R},$$

where β, γ and $\hat{\gamma}$ belong to

$$\mathcal{H}_3 := \{\, \beta \in [0,1],\ \gamma \in (0,1),\ \hat{\gamma} \in (0,1)\,\}. \tag{4.26}$$

Consider the formula for the Lévy density of a hypergeometric Lévy process given in (4.20) in the case that $\beta = \hat{\beta}$. Observe that in this case $\eta = 1 + \gamma + \hat{\gamma}$. From identity (A.34) (or from the series representation of the hypergeometric function), for $x > 0$ we have

$$_2F_1\left(1+\gamma, \eta; \eta - \hat{\gamma}; \mathrm{e}^{-x}\right) = \frac{1}{(1-\mathrm{e}^{-x})^{1+\gamma+\hat{\gamma}}}.$$

Similarly, for $x < 0$ we have

$$_2F_1\left(1+\hat{\gamma}, \eta; \eta - \gamma; \mathrm{e}^{x}\right) = \frac{1}{(1-\mathrm{e}^{x})^{1+\gamma+\hat{\gamma}}}.$$

Noting, in addition, that

$$-\frac{\Gamma(1+\gamma+\hat{\gamma})}{\Gamma(1+\gamma)\Gamma(-\gamma)} = \frac{\Gamma(1+\gamma+\hat{\gamma})}{\Gamma(\gamma)\Gamma(1-\gamma)},$$

with a similar result holding when the roles of γ and $\hat{\gamma}$ are interchanged, we draw the conclusion that the Lévy measure of a Lamperti-stable process takes a more welcome compact form.

Lemma 4.9 *The Lévy density of a Lamperti-stable process is given by*

$$\pi(x) = \frac{\Gamma(1+\gamma+\hat{\gamma})}{\Gamma(\gamma)\Gamma(1-\gamma)} \frac{\mathrm{e}^{(\beta+\hat{\gamma})x}}{(\mathrm{e}^{x}-1)^{1+\gamma+\hat{\gamma}}} \mathbf{1}_{(x>0)} + \frac{\Gamma(1+\gamma+\hat{\gamma})}{\Gamma(\hat{\gamma})\Gamma(1-\hat{\gamma})} \frac{\mathrm{e}^{-(1-\beta+\gamma)x}}{(\mathrm{e}^{-x}-1)^{1+\gamma+\hat{\gamma}}} \mathbf{1}_{(x<0)},$$

for $x \in \mathbb{R}$.

The reader will note that the hypergeometric class as well as the subclass of Lamperti-stables do not include the spectrally one-sided cases. The omitted cases are not so difficult to introduce. Indeed by taking $\gamma = 1$ or $\hat{\gamma} = 1$ in the definition of the hypergeometric class, we may obtain respectively the spectrally negative or positive cases. For our purposes, we only consider the subclass of spectrally one-sided Lamperti-stables. Thus, a spectrally negative Lamperti-stable has characteristic exponent given by

$$\Psi(z) = (1-\beta-\mathrm{i}z)\frac{\Gamma(\beta+\hat{\gamma}+\mathrm{i}z)}{\Gamma(\beta+\mathrm{i}z)}, \qquad z \in \mathbb{R},$$

where $\beta \in [0,1]$ and $\hat{\gamma} \in (0,1)$. Moreover, its Lévy density is such that

$$\pi(x) = \frac{\Gamma(2+\hat{\gamma})}{\Gamma(\hat{\gamma})\Gamma(1-\hat{\gamma})} \frac{\mathrm{e}^{-(2-\beta)x}}{(\mathrm{e}^{-x}-1)^{2+\hat{\gamma}}} \mathbf{1}_{(x<0)},$$

for $x \in \mathbb{R}$. The spectrally positive case can be introduced in the same way by taking $\hat{\gamma} = 1$, $\beta \in [0,1]$ and $\gamma \in (0,1)$, or by considering the negative of the spectrally negative case.

4.4 The first passage problem

In this section, we derive the explicit form of the first passage problem for the case of hypergeometric Lévy processes using Theorem 2.23. This will be of particular use at a number of points later on in this book.

Theorem 4.10 *Suppose that Y is a hypergeometric Lévy process. If $(\beta, \gamma, \hat{\beta}, \hat{\gamma}) \in \mathcal{H}_1$, then for $u \geq 0$,*

$$\begin{aligned}
&\mathrm{P}(Y_{\tau_a^+} - a \in \mathrm{d}u, \tau_a^+ < \zeta) \\
&= \frac{\sin \pi\gamma}{\pi} \mathrm{e}^{-(1-\beta+\gamma)(u+a)} \left(\frac{1-\mathrm{e}^{-a}}{\mathrm{e}^{-a} - \mathrm{e}^{-(u+a)}} \right)^{\gamma} (1 - \mathrm{e}^{-(u+a)})^{-1} \,\mathrm{d}u. \qquad (4.27)
\end{aligned}$$

If $(\beta, \gamma, \hat{\beta}, \hat{\gamma}) \in \mathcal{H}_2$, then for $u \geq 0$,

$$\begin{aligned}
&\mathrm{P}(Y_{\tau_a^+} - a \in \mathrm{d}u, \tau_a^+ < \zeta) \\
&= \frac{\sin \pi\gamma}{\pi} \mathrm{e}^{-(2-\beta+\gamma)(u+a)} \left(\frac{1-\mathrm{e}^{-a}}{\mathrm{e}^{-a} - \mathrm{e}^{-(u+a)}} \right)^{\gamma} (1 - \mathrm{e}^{-(u+a)})^{-1} \,\mathrm{d}u \\
&+ \frac{\sin \pi\gamma}{\pi} (1 - \beta + \hat{\beta} + \gamma) \mathrm{e}^{\hat{\beta}a} \mathrm{e}^{-u(1-\beta+\gamma)} (1 - \mathrm{e}^{-u})^{-\gamma} \\
&\times \left(\int_0^a \mathrm{e}^{-y(2-\beta+\hat{\beta})} (1 - \mathrm{e}^{-y})^{\gamma-1} \,\mathrm{d}y \right) \mathrm{d}u.
\end{aligned}$$

Proof For the first statement, we start by recalling that the ascending ladder height process, H, has Laplace exponent

$$\kappa(\lambda) = \frac{\Gamma(1-\beta+\gamma+\lambda)}{\Gamma(1-\beta+\lambda)}, \qquad \lambda \geq 0.$$

Note that the distribution we want to compute is also the overshoot distribution of the (killed) subordinator H over level a. Appealing to Theorem 2.24, together with (4.2) and Corollary 4.3, we have, for $u \geq 0$,

$$\begin{aligned}
&\mathrm{P}(Y_{\tau_a^+} - a \in \mathrm{d}u, \tau_a^+ < \zeta) \\
&= \frac{\gamma \mathrm{e}^{\gamma a} \mathrm{e}^{-(1-\beta+\gamma)(u+a)}}{\Gamma(\gamma)\Gamma(1-\gamma)} \left\{ \int_0^a (1 - \mathrm{e}^{-(a-x)})^{\gamma-1} (1 - \mathrm{e}^{-(u+x)})^{-(\gamma+1)} \mathrm{e}^{-\gamma x} \,\mathrm{d}x \right\} \mathrm{d}u \\
&= \frac{\gamma \mathrm{e}^{-(1-\beta+\gamma)(u+a)}}{\Gamma(\gamma)\Gamma(1-\gamma)} \left\{ \int_0^a (1 - \mathrm{e}^{-w})^{\gamma-1} (1 - \mathrm{e}^{-(u+a-w)})^{-(\gamma+1)} \mathrm{e}^{\gamma w} \,\mathrm{d}w \right\} \mathrm{d}u \\
&= \frac{\gamma \mathrm{e}^{-(1-\beta+\gamma)(u+a)}}{\Gamma(\gamma)\Gamma(1-\gamma)} (1 - \mathrm{e}^{-(u+a)})^{-1} \left\{ \int_0^{\frac{1-\mathrm{e}^{-a}}{1-\mathrm{e}^{-(u+a)}}} s^{\gamma-1} (1-s)^{-(\gamma+1)} \,\mathrm{d}s \right\} \mathrm{d}u,
\end{aligned}$$

where in the third integral we have used the change of variables $s = (1 - \mathrm{e}^{-w})/(1 - \mathrm{e}^{-(u+w)})$. From (A.35) and (A.34) in the Appendix, we can develop the final integral as follows

$$\begin{aligned}
\int_0^{\frac{1-e^{-a}}{1-e^{-(u+a)}}} & s^{\gamma-1}(1-s)^{-(\gamma+1)}\,\mathrm{d}s \\
&= \frac{1}{\gamma}\left(\frac{1-\mathrm{e}^{-a}}{1-\mathrm{e}^{-(u+a)}}\right)^{\gamma} {}_2F_1\left(\gamma, 1+\gamma; 1+\gamma; \frac{1-\mathrm{e}^{-a}}{1-\mathrm{e}^{-(u+a)}}\right) \\
&= \frac{1}{\gamma}\left(\frac{1-\mathrm{e}^{-a}}{1-\mathrm{e}^{-(u+a)}}\right)^{\gamma}\left(1-\left(\frac{1-\mathrm{e}^{-a}}{1-\mathrm{e}^{-(u+a)}}\right)\right)^{-\gamma} \\
&= \frac{1}{\gamma}\left(\frac{1-\mathrm{e}^{-a}}{\mathrm{e}^{-a}-\mathrm{e}^{-(u+a)}}\right)^{\gamma}.
\end{aligned}$$

In conclusion, we have, with the help of the reflection formula for the gamma function, that, for $u \geq 0$,

$$\begin{aligned}
&\mathrm{P}(Y_{\tau_a^+} - a \in \mathrm{d}u, \tau_a^+ < \zeta) \\
&= \frac{\sin \pi\gamma}{\pi} \mathrm{e}^{-(1-\beta+\gamma)(u+a)} \left(\frac{1-\mathrm{e}^{-a}}{\mathrm{e}^{-a}-\mathrm{e}^{-(u+a)}}\right)^{\gamma} (1-\mathrm{e}^{-(u+a)})^{-1}\,\mathrm{d}u.
\end{aligned}$$

Next we turn our attention to the case of the parameter regime $\mathcal{H}_2$. We start by recalling that in this case the ascending ladder height process, H, has Laplace exponent

$$\kappa(\lambda) = (\lambda - \hat{\beta})\frac{\Gamma(1-\beta+\gamma+\lambda)}{\Gamma(2-\beta+\lambda)}, \qquad \lambda \geq 0.$$

In order to compute the overshoot distribution, we will appeal to Corollary 2.25. Specifically, for $q, \theta > 0$,

$$\begin{aligned}
\int_0^\infty \mathrm{e}^{-qa}\mathrm{E}\left[\mathrm{e}^{-\theta(Y_{\tau_a^+}-a)}\mathbf{1}_{\{\tau_a^+<\zeta\}}\right]\mathrm{d}a &= \frac{\kappa(q)-\kappa(\theta)}{(q-\theta)\kappa(q)} \\
&= \frac{1}{q-\theta} - \frac{\kappa(\theta)}{q-\theta}\frac{\Gamma(2-\beta+q)}{\Gamma(2-\beta+\gamma+q)} \\
&\quad - \frac{1}{q-\theta}\kappa(\theta)\frac{1-\beta+\hat{\beta}+\gamma}{q-\hat{\beta}}\frac{\Gamma(2-\beta+q)}{\Gamma(2-\beta+\gamma+q)}.
\end{aligned} \tag{4.28}$$

The first ratio of the right-hand side of the above identity can be easily inverted, because

$$\int_0^\infty \mathrm{e}^{-qx}\mathrm{e}^{\theta x}\,\mathrm{d}x = \frac{1}{q-\theta}.$$

Elementary calculations give us that

$$\frac{1}{\Gamma(\gamma)}\int_0^\infty \mathrm{e}^{-qx}\mathrm{e}^{\theta x}B_{1-\mathrm{e}^{-x}}(\gamma, 2-\beta+\theta)\,\mathrm{d}x = \frac{1}{q-\theta}\frac{\Gamma(2-\beta+q)}{\Gamma(2-\beta+\gamma+q)},$$

where $B_y(\gamma, 2-\beta+\theta)$, $y \in [0,1]$, $\theta \geq 0$, represents the incomplete Beta function, that is,

$$B_y(\gamma, 2-\beta+\theta) = \int_0^y u^{\gamma-1}(1-u)^{1-\beta+\theta}\, du, \qquad y \in [0,1].$$

Similarly, we obtain

$$\begin{aligned}
\frac{1}{\Gamma(\gamma)} \int_0^{\infty} & \mathrm{e}^{-qx}\left(\mathrm{e}^{\theta x} B_{1-\mathrm{e}^{-x}}(\gamma, 2-\beta+\theta) - \mathrm{e}^{\hat{\beta} x} B_{1-\mathrm{e}^{-x}}(\gamma, 2-\beta+\hat{\beta})\right) \mathrm{d}x \\
&= \left(\frac{1}{q-\theta}\frac{\Gamma(2-\beta+q)}{\Gamma(2-\beta+\gamma+q)} - \frac{1}{q-\hat{\beta}}\frac{\Gamma(2-\beta+q)}{\Gamma(2-\beta+\gamma+q)}\right) \\
&= \frac{\theta-\hat{\beta}}{(q-\theta)(q-\hat{\beta})}\frac{\Gamma(2-\beta+q)}{\Gamma(2-\beta+\gamma+q)}.
\end{aligned}$$

Performing the inversion in q in (4.28), we get

$$\begin{aligned}
\mathrm{E}\left[\mathrm{e}^{-\theta(Y_{\tau_a^+}-a)}\mathbf{1}_{\{\tau_a^+<\zeta\}}\right] &= \mathrm{e}^{\theta a} - \frac{\kappa(\theta)}{\Gamma(\gamma)}\mathrm{e}^{\theta a} B_{1-\mathrm{e}^{-a}}(\gamma, 2-\beta+\theta)\left(1+\frac{1-\beta+\hat{\beta}+\gamma}{\theta-\hat{\beta}}\right) \\
&\quad + \frac{\kappa(\theta)}{\Gamma(\gamma)}\frac{1-\beta+\hat{\beta}+\gamma}{\theta-\hat{\beta}}\mathrm{e}^{\hat{\beta} a} B_{1-\mathrm{e}^{-a}}(\gamma, 2-\beta+\hat{\beta}) \\
&= \mathrm{e}^{\theta a} - \frac{\Gamma(2-\beta+\gamma+\theta)}{\Gamma(\gamma)\Gamma(2-\beta+\theta)}\mathrm{e}^{\theta a} B_{1-\mathrm{e}^{-a}}(\gamma, 2-\beta+\theta) \\
&\quad + \frac{\Gamma(1-\beta+\gamma+\theta)(1-\beta+\hat{\beta}+\gamma)}{\Gamma(\gamma)\Gamma(2-\beta+\theta)}\mathrm{e}^{\hat{\beta} a} B_{1-\mathrm{e}^{-a}}(\gamma, 2-\beta+\hat{\beta}) \\
&= \mathrm{e}^{\theta a} I_{\mathrm{e}^{-a}}(\gamma, 2-\beta+\theta) + p(a)\frac{\Gamma(1-\beta+\gamma+\theta)}{\Gamma(2-\beta+\theta)}, \qquad (4.29)
\end{aligned}$$

where

$$I_y(a,b) = \frac{\Gamma(a+b)}{\Gamma(a)\Gamma(b)} \int_0^y u^{a-1}(1-u)^{b-1}\, \mathrm{d}u$$

and

$$p(y) = \frac{1-\beta+\hat{\beta}+\gamma}{\Gamma(\gamma)}\mathrm{e}^{\hat{\beta} y} B_{1-\mathrm{e}^{-y}}(\gamma, 2-\beta+\hat{\beta}).$$

To tackle the right-hand side of (4.29), the following Laplace transform will be useful,

$$\begin{aligned}
\int_0^{\infty} \mathrm{e}^{-u(2-\beta+\theta)}(1-\mathrm{e}^{-(a+u)})^{\gamma-1}\, \mathrm{d}u &= \int_0^{\infty} t^{1-\beta+\theta}(1-\mathrm{e}^{-a}t)^{\gamma-1}\, \mathrm{d}t \\
&= \mathrm{e}^{a(2-\beta+\theta)} B_{\mathrm{e}^{-a}}(\gamma, 2-\beta+\theta), \qquad (4.30)
\end{aligned}$$

where $\mathrm{e}^{-u} = \mathrm{e}^{-a}t$ was used in the second equatily. Let $f(u) = \mathrm{e}^{-u(2-\beta)}(1-\mathrm{e}^{-(a+u)})^{\gamma-1}$. A straightforward integration by parts leads to

$$\int_0^\infty e^{-\theta u}(\gamma-1)e^{-u(2-\beta)}\left(1-e^{-(a+u)}\right)^{\gamma-2} du + (1-e^{-a})^{\gamma-1}\int_0^\infty e^{-\theta u}\delta_0(du)$$
$$= (1-\beta+\gamma+\theta)e^{a(2-\beta+\theta)}B_{e^{-a}}(\gamma, 2-\beta+\theta). \quad (4.31)$$

On the other hand, we also have

$$\frac{\Gamma(1-\beta+\gamma+\theta)}{\Gamma(2-\beta+\theta)} = \frac{1}{\Gamma(1-\gamma)}\int_0^\infty e^{-u\theta}e^{-u(1-\beta+\gamma)}(1-e^{-u})^{-\gamma}\,du. \quad (4.32)$$

For the first term on the right-hand side of (4.29), we will identify a convolution that is the result of the product of (4.31) and (4.32). Performing the changes of variables $t = e^{-v}$, $s = t - e^{-u}$ and $r = s/(e^a - e^{-u})$, for the first, second and third equalities below, we have

$$\frac{\gamma-1}{\Gamma(1-\gamma)}\int_0^u e^{-v(2-\beta)}(1-e^{-(a+v)})^{\gamma-2}e^{-(u-v)(1-\beta+\gamma)}(1-e^{-(u-v)})^{-\gamma}\,dv$$
$$= \frac{\gamma-1}{\Gamma(1-\gamma)}e^{-u(1-\beta+\gamma)}\int_{e^{-u}}^1 (1-e^{-a}t)^{\gamma-2}(t-e^{-u})^{-\gamma}\,dt$$
$$= \frac{\gamma-1}{\Gamma(1-\gamma)}e^{-u(1-\beta+\gamma)}(1-e^{-(a+u)})^{\gamma-2}$$
$$\times\int_0^{1-e^{-u}}\left(1-\frac{e^{-a}s}{1-e^{-(a+u)}}\right)^{\gamma-2}s^{-\gamma}\,ds$$
$$= \frac{\gamma-1}{\Gamma(1-\gamma)}e^{-u(1-\beta+\gamma)}(1-e^{-(a+u)})^{\gamma-2}(e^a-e^{-u})^{1-\gamma}$$
$$\times\int_0^{\frac{1-e^{-u}}{e^a-e^{-u}}} r^{-\gamma}(1-r)^{\gamma-2}\,dr$$
$$= -\frac{1}{\Gamma(1-\gamma)}e^a e^{-u(1-\beta)}(e^a e^u-1)^{-1}(e^u-1)^{1-\gamma}(1-e^{-a})^{\gamma-1}. \quad (4.33)$$

Bearing in mind that $f * \delta_0 = f$ for any function f, we can use (4.33) to decode the product of Laplace transform in (4.31) and (4.32). We get

$$e^{\theta a}I_{e^{-a}}(\gamma, 2-\beta+0)$$
$$= \frac{e^{-a(2-\beta)}}{\Gamma(\gamma)}\frac{\Gamma(1-\beta+\gamma+\theta)}{\Gamma(2-\beta+\theta)}(1-\beta+\gamma+\theta)e^{a(2-\beta+\theta)}B_{e^{-a}}(2-\beta+\theta,\gamma)$$
$$= \frac{e^{-a(2-\beta)}}{\Gamma(\gamma)}\frac{e^a(1-e^{-a})^{\gamma-1}}{\Gamma(1-\gamma)}\int_0^\infty e^{-u\theta}e^{-u(1-\beta)}(e^a e^u-1)^{-1}(e^u-1)^{1-\gamma}\,du$$
$$+ \frac{e^{-a(2-\beta)}}{\Gamma(\gamma)}\frac{(1-e^{-a})^{\gamma-1}}{\Gamma(1-\gamma)}\int_0^\infty e^{-u\theta}e^{-u(1-\beta+\gamma)}(1-e^{-u})^{-\gamma}du.$$

Appealing to the reflection formula for the Gamma function (see (A.12) in the Appendix), together with (4.32) we can complete the inversion of (4.29) and get

$$\begin{aligned}
&\mathrm{P}(Y_{\tau_a^+} - a \in \mathrm{d}u, \tau_a^+ < \zeta) \\
&= \frac{\sin \pi\gamma}{\pi} \mathrm{e}^{-(a+u)(1-\beta)} (\mathrm{e}^u - 1)^{1-\gamma} (1 - \mathrm{e}^{-a})^{\gamma-1} (\mathrm{e}^a \mathrm{e}^u - 1)^{-1} \\
&+ \frac{\sin \pi\gamma}{\pi} \mathrm{e}^{-u(1-\beta+\gamma)} (1 - \mathrm{e}^{-u})^{-\gamma} \Big\{ \mathrm{e}^{-a(2-\beta)} (1 - \mathrm{e}^{-a})^{\gamma-1} \\
&+ (1 - \beta + \hat\beta + \gamma) \mathrm{e}^{\hat\beta a} \int_0^a \mathrm{e}^{-y(2-\beta+\hat\beta)} (1 - \mathrm{e}^{-y})^{\gamma-1}\, \mathrm{d}y \Big\}.
\end{aligned}$$

Straightforward changes of variables in the integrals and the use of the reflection formula for gamma functions (cf. (A.12)) give us the desired result. □

We observe that in the spectrally positive case, that is, when $\hat\gamma = 1$, the first passage problem is also given by the identity (4.27). The latter is justified by the fact that the ascending ladder height of the spectrally one sided case is exactly the same as the one treated in the above result when the parameters lie in $\mathcal{H}_1$. We recall that in this case the parameters β and γ are such that $\beta \in [0, 1]$ and $\gamma \in (0, 1)$.

4.5 Exponential functionals

The exponential of a Lévy process, Y, is defined as the random variable

$$I(\delta, Y) := \int_0^\zeta \mathrm{e}^{\delta Y_t} \mathrm{d}t, \tag{4.34}$$

for $\delta > 0$ and where we recall that ζ denotes the lifetime of Y. We conclude this chapter by studying exponential functionals in the setting that Y belongs to the class of hypergeometric Lévy processes. Moreover, for the remainder of this section. we assume that parameters $(\beta, \gamma, \hat\beta, \hat\gamma)$ belong to

$$\mathcal{H}_4 = (\mathcal{H}_1 \setminus \{\beta = 1\}) \cup (\mathcal{H}_2 \cap \{\eta - \hat\gamma > 0\} \setminus \{\hat\beta = 0\}), \tag{4.35}$$

or, according to Corollary 4.7, that a corresponding hypergeometric Lévy process Y either is killed or drift to $-\infty$.

Under our assumptions, the exponential functional is always finite with probability one. Indeed, when Y is killed, the exponential functional is clearly bounded a.s., and when Y drifts to $-\infty$, the claim follows from the Strong Law of Large Numbers, that is,

$$\lim_{t\to\infty}\frac{Y_t}{t}=\mathrm{E}[Y_1]<0,$$

almost surely, which, in turn, comes from Theorem 2.8. In order to study the law of $I(\delta, Y)$, we introduce its Mellin transform

$$\mathtt{M}(s):=\mathtt{M}(s;\delta,\beta,\gamma,\hat{\beta},\hat{\gamma})=\mathrm{E}\Big[I(\delta,Y)^{s-1}\Big]. \tag{4.36}$$

For convenience it will be better for us to work with the Laplace exponent of Y, defined as

$$\psi(z)=-\Psi(-\mathrm{i}z)=-\frac{\Gamma(1-\beta+\gamma-z)}{\Gamma(1-\beta-z)}\frac{\Gamma(\hat{\beta}+\hat{\gamma}+z)}{\Gamma(\hat{\beta}+z)}, \tag{4.37}$$

for $\mathrm{Re}(z)\in(-\hat{\beta}-\hat{\gamma},1-\beta+\gamma)$. If we agree that the cemetery state is $-\infty$ and that $\exp(-\infty)=0$, so that $\exp(zY_t)$ is well defined for all $t\geq 0$, specifically for $t\geq\zeta$, then

$$\psi(z)=\log\mathrm{E}\Big[\mathrm{e}^{zY_1}\Big],\qquad t\geq 0,\quad \mathrm{Re}(z)\in(-\hat{\beta}-\hat{\gamma},1-\beta+\gamma).$$

To ease the presentation, we also define

$$\chi=1/\delta. \tag{4.38}$$

Proposition 4.11 *Let Y be a hypergeometric Lévy process with parameters $(\beta,\gamma,\hat{\beta},\hat{\gamma})\in\mathcal{H}_4$. Then $\mathtt{M}(s)$ is well defined for* $\mathrm{Re}(s)\in(0,1+\hat{\theta}\chi)$*, where*

$$\hat{\theta}:=\begin{cases}1-\beta & \textit{if } (\beta,\gamma,\hat{\beta},\hat{\gamma})\in\mathcal{H}_1\setminus\{\beta=1\},\\ -\hat{\beta} & \textit{if } (\beta,\gamma,\hat{\beta},\hat{\gamma})\in\mathcal{H}_2\cap\{\eta-\hat{\gamma}>0\}\setminus\{\hat{\beta}=0\}.\end{cases} \tag{4.39}$$

Moreover,

$$\mathtt{M}(s+1)=-\frac{s}{\psi(\delta s)}\mathtt{M}(s),\qquad s\in(0,\hat{\theta}\chi). \tag{4.40}$$

Proof We recall that we have set the cemetery state to be $-\infty$ and accordingly are working with the definition $\exp(-\infty)=0$. Recall that

$$\mathcal{E}_t(z)=\mathrm{e}^{zY_t-\psi(z)t},\qquad t\geq 0,$$

is a positive martingale for any z that $\psi(z)$ is well-defined and real valued. Note that $\psi(0)\leq 0$ and ψ has roots at $1-\beta$ and $-\hat{\beta}$, hence $\psi(\delta s)<0$ for $s\in(0,\hat{\theta}\chi)$. From Doob's L^1-inequality and the Esscher transform (2.23), we observe for $s>0$,

$$\begin{aligned}
\mathrm{E}\left[\left(\int_0^t \mathrm{e}^{\delta Y_u}\mathrm{d}u\right)^s\right] &\le t^s \mathrm{E}\left[\sup_{u\le t} \mathrm{e}^{\delta s Y_u}\right] \\
&\le t^s \mathrm{E}\left[\sup_{u\le t} \mathcal{E}_u(\delta s)\right] \\
&\le \frac{\mathrm{e}}{\mathrm{e}-1} t^s (1 + \mathrm{E}^{\delta s}[\delta s Y_t] - t\psi(\delta s)) \\
&= \frac{\mathrm{e}}{\mathrm{e}-1} t^s (1 + \delta s t\psi'(\delta s) - t\psi(\delta s)),
\end{aligned}$$

which is finite for $s \in (0, \hat{\theta}_\chi)$.

Next, it is easy to see that for all $s > 0$ and $t \ge 0$,

$$\left(\int_0^\infty \mathrm{e}^{\delta Y_u}\mathrm{d}u\right)^s - \left(\int_t^\infty \mathrm{e}^{\delta Y_u}\mathrm{d}u\right)^s = s\int_0^t \mathrm{e}^{\delta s Y_u}\left(\int_0^\infty \mathrm{e}^{\delta(Y_{u+v}-Y_u)}\mathrm{d}v\right)^{s-1}\mathrm{d}u.$$

We take expectations on both sides of the above identity and from the independence of increments, we obtain

$$\mathrm{E}\left[\left(\int_0^\infty \mathrm{e}^{\delta Y_u}\mathrm{d}u\right)^s - \left(\int_t^\infty \mathrm{e}^{\delta Y_u}\mathrm{d}u\right)^s\right] = s\mathtt{M}(s)\int_0^t \mathrm{e}^{u\psi(\delta s)}\mathrm{d}u. \tag{4.41}$$

Using

$$\left||x|^s - |y|^s\right| \le |x-y|^s, \qquad \text{for} \quad x, y \in \mathbb{R},$$

when $0 < s < \hat{\theta}_\chi$, the estimate from the previous paragraph ensures that the left-hand side of (4.41) is bounded by

$$\mathrm{E}\left[\left(\int_0^t \mathrm{e}^{\delta Y_u}\mathrm{d}u\right)^s\right] < \infty.$$

It now follows that $\mathtt{M}(s)$ is also finite when $0 < s < \hat{\theta}_\chi$. In turn, noting the integral on the right-hand side of (4.41) converges as $t \to \infty$, this is sufficient to ensure convergence on the left-hand side of (4.41), giving the identity (4.40).

Since the right-hand side of (4.40) is finite, this functional equation allows us to conclude that $\mathtt{M}(s)$ is well defined for $s \in (0, 1 + \hat{\theta}_\chi)$. It then follows from the general properties of Mellin transforms that $\mathtt{M}(s)$ is finite and analytic for all $s \in \mathbb{C}$ such that $\mathrm{Re}(s) \in (0, 1 + \hat{\theta}_\chi)$. □

We want to identify the Mellin transform $\mathtt{M}$ explicitly. To this end, we must first introduce some special functions that will be of use. The double gamma function is defined by an infinite product in Weierstrass's form, i.e. for $z \in \mathbb{C}$ and $|\arg(\tau)| < \pi$,

$$G(z;\tau) = \frac{z}{\tau}\mathrm{e}^{(az+bz^2/2)/\tau}\prod_{\substack{m,n\ge 0\\ m+n>0}}\left(1 + \frac{z}{m\tau+n}\right)\mathrm{e}^{-(2(m\tau+n)z-z^2)/2(m\tau+n)^2}. \tag{4.42}$$

Note that by definition $G(z;\tau)$ is an entire function in z and if $\tau \notin \mathbb{Q}$ it has simple zeros on the lattice $-m\tau - n$, for $m \geq 0$, $n \geq 0$.

The following function additionally plays an important role in determining $\mathtt{M}(s)$ explicitly and is given in terms of product of double gamma functions.

Definition 4.12 For $s \in \mathbb{C}$, we define

$$\begin{aligned} \mathrm{F}(s) &:= \mathrm{F}(s;\delta,\beta,\gamma,\hat{\beta},\hat{\gamma}) \\ &= C\frac{G(\hat{\beta}\chi + s;\chi)}{G((\hat{\beta}+\hat{\gamma})\chi + s;\chi)} \frac{G((1-\beta+\gamma)\chi + 1 - s;\chi)}{G((1-\beta)\chi + 1 - s;\chi)}, \quad s \in \mathbb{C}, \end{aligned} \tag{4.43}$$

where the constant C is such that $\mathrm{F}(1) = 1$. For simplicity, and as long as it is clear, we use the notation $\mathrm{F}(s)$ instead of $\mathrm{F}(s;\delta,\beta,\gamma,\hat{\beta},\hat{\gamma})$. Whenever the use of the parameters is necessary, we then use the longer form.

Our main result in this section is the following Theorem, which provides an explicit expression for the Mellin transform of the exponential functional $I(\delta, Y)$ in terms of the double gamma function.

Theorem 4.13 *Assume that* $\delta > 0$ *and* $(\beta,\gamma,\hat{\beta},\hat{\gamma}) \in \mathcal{H}_4$.

(i) If $(\beta,\gamma,\hat{\beta},\hat{\gamma}) \in \mathcal{H}_1 \setminus \{\beta = 1\}$, *then*

$$\mathtt{M}(s) \equiv \Gamma(s)\mathrm{F}(s), \qquad \textit{for all} \quad s \in \mathbb{C}.$$

(ii) If $(\beta,\gamma,\hat{\beta},\hat{\gamma}) \in \mathcal{H}_2 \cap \{\eta - \hat{\gamma} > 0\} \setminus \{\hat{\beta} = 0\}$, *then*

$$\mathtt{M}(s) \equiv c\frac{\Gamma(\chi(\hat{\beta}+\hat{\gamma}) + s)}{\Gamma(\chi(\beta - 1) + s)} \frac{\Gamma(-\chi\hat{\beta} + 1 - s)}{\Gamma((1-\beta+\gamma)\chi + 1 - s)}\check{\mathtt{M}}(s), \qquad \textit{for all} \quad s \in \mathbb{C},$$

where $\check{\mathtt{M}}$ *is the Mellin transform of* $I(\delta, \check{Y})$ *such that* $\check{Y}$ *is a hypergeometric process, whose parameters* $(\beta - 1, \gamma, \hat{\beta} + 1, \hat{\gamma})$ *necessarily satisfy the conditions in (i), and the constant* c *is such that* $\mathtt{M}(1) = 1$.

We only prove case (i), the proof of case (ii) follows using exactly the same arguments. For the sake of brevity, we leave the proof of (ii) to the reader. Before proceeding to the proof of Theorem 4.13 (i), we need two preparatory auxiliary results concerning the analytical properties of F. The first pertains to the asymptotic behaviour of F in $\mathbb{C}$ and the second to a recursive equation that it satisfies.

Lemma 4.14 *On the complex plane,*

(i) for $0 < \epsilon < \arg(s) < \pi - \epsilon$, *we have as* $|s| \to \infty$,

$$\log(\mathrm{F}(s)) = -(\gamma + \hat{\gamma})s \log s + s((1 + \log\chi)(\gamma + \hat{\gamma}) + \pi i\gamma) + O(\log s), \tag{4.44}$$

(ii) *in a vertical strip* $-\infty < a < \mathrm{Re}(s) < b < \infty$, *we have, as* $|s| \to \infty$,

$$\left|\mathrm{F}(s)\right| = \exp\left(\frac{\pi}{2}(\gamma - \hat{\gamma})|\mathrm{Im}(s)| + O(\log|\mathrm{Im}(s)|)\right). \tag{4.45}$$

Proof Both asymptotic expansions (4.44) and (4.45) follow from the definition of F and the asymptotic expansion of

$$\log \frac{G(z+a;\tau)}{G(z;\tau)},$$

as $z \to \infty$, in the domain $\arg(z) \le \pi - \epsilon < \pi$, which is provided in (A.24). □

Lemma 4.15 *Write* $\tilde{Y}$ *for the hypergeometric Lévy process with parameters* $(\chi\beta, \chi\gamma, \chi\hat{\beta}, \chi\hat{\gamma})$, *provided that this parameter set belongs to* $\mathcal{H}_1 \backslash \{\beta = 1\}$. *In particular, its Laplace exponent is given by*

$$\tilde{\psi}(z) = -\frac{\Gamma(1-\chi(\beta-\gamma)-z)\Gamma(\chi(\hat{\beta}+\hat{\gamma})+z)}{\Gamma(1-\chi\beta-z)\Gamma(\chi\hat{\beta}+z)}, \tag{4.46}$$

for $\mathrm{Re}(z) \in (-\chi(\hat{\beta}+\hat{\gamma}), 1-\chi(\beta-\gamma))$ *On the complex plane, we have that*

(i) $\mathrm{F}(s)$ *is a real meromorphic function which has zeros*

$$-\hat{\beta}\chi - m\chi - n \text{ and } 1 + (1-\beta+\gamma)\chi + m\chi + n, \tag{4.47}$$

for $m, n \ge 0$ *and poles*

$$z^-_{m,n} := -(\hat{\beta}+\hat{\gamma})\chi - m\chi - n \text{ and } z^+_{m,n} := 1 + (1-\beta)\chi + m\chi + n, \tag{4.48}$$

for $m, n \ge 0$. *All zeros/poles are simple if* $\delta \notin \mathbb{Q}$.

(ii) *Moreover,* $\mathrm{F}(s)$ *satisfies the following functional identities*

$$\mathrm{F}(s+1) = -\frac{1}{\psi(\delta s)}\mathrm{F}(s), \tag{4.49}$$

$$\mathrm{F}(s+\chi) = -\frac{\delta^{-\chi(\hat{\gamma}+\gamma)}}{\tilde{\psi}(s)}\mathrm{F}(s), \tag{4.50}$$

$$\mathrm{F}(s;\delta,\beta,\gamma,\hat{\beta},\hat{\gamma}) = \delta^{(1-s)(\gamma+\hat{\gamma})}\mathrm{F}(\delta s;\chi,\chi\beta,\chi\gamma,\chi\hat{\beta},\chi\hat{\gamma}), \tag{4.51}$$

where ψ *was given in* (4.37).

Proof Part (i) follows from the definition of F and fact that the double gamma function $G(z;\tau)$ has simple zeros on the lattice $m\tau + n$, for $m, n \le 0$ and $\tau \notin \mathbb{Q}$.

The functional identity (4.49) is a straightforward consequence of the quasi-periodic property of period 1 of the double gamma function, see (A.21).

In order to deduce (4.50), we first obtain the functional identity (4.51). We first use the transformation of the double gamma function (A.23) and observe that for all s and x,

$$\frac{G(s+x;\chi)}{G(s;\chi)} = \delta^{\delta s x} C(\delta, x) \frac{G(\delta(s+x);\delta)}{G(\delta s;\delta)}, \tag{4.52}$$

where

$$C(\delta, x) = (2\pi)^{\frac{x}{2}(1-\delta)} \delta^{\frac{x^2\delta}{2} - \frac{x}{2}(1+\delta)}.$$

The identity (4.52) implies that

$$\frac{G(\hat{\beta}\chi + s;\chi)}{G((\hat{\beta}+\hat{\gamma})\chi + s;\chi)} \frac{G((1-\beta+\gamma)\chi + 1 - s;\chi)}{G((1-\beta)\chi + 1 - s;\chi)}$$
$$= \tilde{C}(\delta, \gamma, \hat{\gamma}, \beta, \hat{\beta}) \delta^{-s(\gamma+\hat{\gamma})} \frac{G(\hat{\beta} + \delta s;\delta)}{G(\hat{\beta} + \hat{\gamma} + \delta s;\delta)} \frac{G(1-\beta+\gamma+\delta-\delta s;\delta)}{G(1-\beta+\delta-\delta s;\delta)}, \tag{4.53}$$

where $\tilde{C}(\delta, \gamma, \hat{\gamma}, \beta, \hat{\beta})$ is a constant that does not depend on s. Then, equation (4.53) allows us to deduce

$$C = \frac{G((\hat{\beta}+\hat{\gamma})\chi + 1;\chi)}{G(\hat{\beta}\chi + 1;\chi)} \frac{G((1-\beta)\chi;\chi)}{G((1-\beta+\gamma)\chi;\chi)}$$
$$= \frac{\delta^{\gamma+\hat{\gamma}}}{\tilde{C}(\delta, \gamma, \hat{\gamma}, \beta, \hat{\beta})} \frac{G(\hat{\beta}+\hat{\gamma}+\delta;\delta)}{G(\hat{\beta}+\delta;\delta)} \frac{G(1-\beta;\delta)}{G(1-\beta+\gamma;\delta)},$$

where C is the constant that appears in (4.43). Using the definition of (4.43) and putting all pieces together, we deduce identity (4.51).

The functional identity (4.50) follows from (4.51) and (4.49) but applied to $\mathrm{F}(\delta s;\chi,\chi\beta,\chi\gamma,\chi\hat{\beta},\chi\hat{\gamma})$. Indeed, we observe form (4.51) that

$$\mathrm{F}(s+\chi) = \delta^{(1-s-\chi)(\gamma+\hat{\gamma})} \mathrm{F}(\delta s + 1;\chi,\chi\beta,\chi\gamma,\chi\hat{\beta},\chi\hat{\gamma}).$$

On the other hand, from identity (4.49), we deduce

$$\mathrm{F}(\delta s + 1;\chi,\chi\beta,\chi\gamma,\chi\hat{\beta},\chi\hat{\gamma}) = -\frac{1}{\tilde{\psi}(s)} \mathrm{F}(\delta s;\chi,\chi\beta,\chi\gamma,\chi\hat{\beta},\chi\hat{\gamma}).$$

Using the last two identities together, we deduce (4.50) as expected. □

Proof of Theorem 4.13 (i) Let us introduce $f(s) = \Gamma(s)\mathrm{F}(s)$, where we recall that $\mathrm{F}(s)$ is defined by (4.43). From Lemma 4.15 part (i), we know that $f(s)$ is analytic and zero-free in the strip $\mathrm{Re}(s) \in (0, 1 + (1-\beta)\chi)$. Moreover, from its construction, we have $f(1) = 1$. From identity (4.49), we find that

$$f(s+1) = -\frac{s}{\psi(\delta s)} f(s) \qquad \text{for} \quad s \in (0, (1-\beta)\chi).$$

On the other hand, recall from Proposition 4.11 that $\mathtt{M}(s)$, as the Mellin transform of $I(\delta, Y)$, is well defined for $\mathrm{Re}(s) \in (0, 1 + (1-\beta)\chi)$ and satisfies

$$\mathtt{M}(s+1) = -\frac{s}{\psi(\delta s)}\mathtt{M}(s), \qquad s \in (0, (1-\beta)\chi).$$

Hence, we conclude that the function $H(s) = \mathtt{M}(s)/f(s)$ satisfies

$$H(s+1) = H(s), \text{ for all } s \in (0, (1-\beta)\chi).$$

The rest of the proof is now dedicated to proving that $H(s) \equiv 1$.

Using the assumption that $f(s)$ is analytic and zero-free, we conclude that $H(s)$ is an analytic function in the strip $\mathrm{Re}(s) \in (0, 1 + (1-\beta)\chi)$. Since $H(s)$ is also periodic with period equal to one, it can be extended to an analytic and periodic function in the entire complex plane.

Since $H(s)$ is analytic and periodic in the entire complex plane, it can be represented as a Fourier series

$$H(s) = \sum_{n \in \mathbb{Z}} c_n e^{2\pi i n s},$$

where the series converges in the entire complex plane. This means that the two functions

$$H_1(z) = \sum_{n \geq 1} c_n z^n \quad \text{and} \quad H_2(z) = \sum_{n \geq 1} c_{-n} z^n$$

are analytic in the entire complex plane, and that for all $s \in \mathbb{C}$,

$$H(s) = c_0 + H_1(\exp(2\pi i s)) + H_2(\exp(-2\pi i s)). \tag{4.54}$$

Next, Lemma 4.14 (iii) and the asymptotic for the gamma function given in (A.17) in the Appendix imply that, as $s \to \infty$, in the vertical strip $\mathrm{Re}(s) \in (0, 1 + (1-\beta)\chi)$, we have

$$|f(s)|^{-1} = \exp\left(\frac{\pi}{2}(1 - \gamma + \hat{\gamma})\,|\mathrm{Im}(s)| + o(\mathrm{Im}(s))\right) = o\left(\exp(\pi|\mathrm{Im}(s)|)\right), \tag{4.55}$$

where in the last step we have also used the fact that both γ and $\hat{\gamma}$ belong to the interval $(0, 1)$.

The inequality $|\mathtt{M}(s)| < \mathtt{M}(\mathrm{Re}(s)) = \mathrm{E}[I(\delta, Y)^{\mathrm{Re}(s)-1}]$, (4.55) and the periodicity of $H(s)$, allow us to conclude that uniformly in $\mathrm{Re}(s)$, a sufficiently strong estimate of decay for H can be given by s

$$H(s) = o(\exp(2\pi|\mathrm{Im}(s)|)), \quad \text{as } \mathrm{Im}(s) \to \infty. \tag{4.56}$$

In particular, when $\mathrm{Im}(s) \to \infty$ we have $H_1(\exp(2\pi i s)) \to H_1(0) = 0$, therefore the estimates (4.54) and (4.56) imply that, writing $z = \exp(-2\pi i s)$ so that $\log|z| = 2\pi\mathrm{Im}(s)$, we have

$$H_2(z) = c_0 + H_1(1/z) - H(s) = o(|z|),$$

as $z \to \infty$ in the entire complex plane. Appealing to Cauchy's estimates (see Proposition A.5 in Appendix), for any $R > 0$, we have that, for any z such that $|z| < R$,

$$H_2(z) = \sum_{n\geq 1} c_{-n} z^n = \sum_{n\geq 1} \frac{H^{(n)}(0)}{n!} z^n \leq M_R \sum_{n\geq 1} (z/R)^n, \tag{4.57}$$

where $M_R = \max\{|H_2(z)| : |z| = R\}$. As $H_2(z) = o(|z|)$, it follows that $M_R = o(R)$ and hence, for each fixed $z \in \mathbb{C}$, by choosing R sufficiently large, we can make the right-hand side of (4.57) arbitrarily small. In conclusion, we deduce that $H_2(z) \equiv 0$. Appealing to similar arguments in the setting that $\mathrm{Im}(s) \to -\infty$, we can verify that $H_1(z) \equiv 0$.

In summary, $H(s)$ must be constant, and the value of this constant is equal to one, since $H(1) = \mathtt{M}(1)/f(1) = 1$. We thus conclude that $\mathtt{M}(s) \equiv f(s)$, as required. □

We close this section with the following corollary which computes the Mellin transform $I(\delta, Y)$ for the particular case that $\gamma = \hat{\gamma} = \delta \in (0, 1)$. Its proof follows directly from Theorem 4.13, the quasi-periodic property of period 1 of the double gamma function (see (A.21)) and the reflection formula of the gamma function (see A.12). We leave the details of the proof to the reader.

Corollary 4.16 *Assume that* $\delta \in (0, 1)$ *and* $(\beta, \delta, \hat{\beta}, \delta) \in \mathcal{H}_4$.

(i) If $(\beta, \delta, \hat{\beta}, \delta) \in \mathcal{H}_1 \setminus \{\beta = 1\}$, *then*

$$\mathtt{M}(s) \equiv \frac{\Gamma(\hat{\beta} + \delta)}{\Gamma(1 - \beta)} \frac{\Gamma(s)\Gamma(1 - \beta + \delta(1 - s))}{\Gamma(\hat{\beta} + \delta s)}, \qquad s \in \mathbb{C}.$$

(ii) If $(\beta, \gamma, \hat{\beta}, \hat{\gamma}) \in \mathcal{H}_2 \cap \{\eta - \hat{\gamma} > 0\} \setminus \{\hat{\beta} = 0\}$, *then*

$$\mathtt{M}(s) \equiv c \frac{\sin(\pi(s + \chi(\beta - 1)))}{\sin(\pi(\chi\hat{\beta} + s))} \frac{\Gamma(s)\Gamma(1 - \beta + \delta(1 - s))}{\Gamma(\hat{\beta} + \delta s)}, \qquad s \in \mathbb{C},$$

where the constant c *is such that* $\mathtt{M}(1) = 1$.

Finally, let us consider the spectrally one-sided cases. We start with the spectrally negative case, that is, when the parameters are such that $\gamma = 1, \beta \in [0, 1)$ and $\hat{\gamma} \in (0, 1)$. For this particular case, we have

$$\psi(z) = (z - 1 + \beta)\frac{\Gamma(\beta + \hat{\gamma} + z)}{\Gamma(\beta + z)} = \frac{\Gamma(\beta + \hat{\gamma} + z)}{\Gamma(\beta - 1 + z)}, \tag{4.58}$$

which is well defined for $\mathrm{Re}(z) \in (-(\beta + \hat{\gamma}), \infty)$. Note that the same arguments used in the proof of Proposition 4.11 allow us to deduce that $\mathtt{M}(s)$ is well defined for $\mathrm{Re}(s) \in (0, 1 + (1-\beta)\chi)$ since $\psi(\delta s) \leq 0$, for $s \in (0, 1 + (1-\beta)\chi)$. Moreover, the identity in (4.40) still holds in this case.

We also observe that the arguments used in the proof of Theorem 4.13 are also valid for this particular case and allow us to compute explicitly the Mellin transform $\mathtt{M}$. More precisely, in this setting, the function F is given by

$$\mathrm{F}(s) = c\chi^{s-1}\Gamma\Big((1-\beta)\chi + 1 - s\Big)\frac{G(\beta\chi + s;\chi)}{G((\beta+\hat{\gamma})\chi + s;\chi)}, \qquad s \in \mathbb{C},$$

where c is such that $\mathrm{F}(1) = 1$. We also remark that the function F is meromorphic with zeros at $-\beta\chi - m\chi - n$ and poles

$$z^-_{m,n} := -(\beta + \hat{\gamma})\chi - m\chi - n \quad \text{and} \quad z^+_n := 1 + (1-\beta)\chi + n,$$

for $m, n \geq 0$. As previously, all zeros/poles are simple if $\delta \notin \mathbb{Q}$. It is important to note that Lemma 4.14 and the identities in part (ii) of Lemma 4.15 still hold in this case (replace γ by 1) thanks to the quasi-periodic properties of the double gamma function (see (A.21) in the Appendix). In other words, we may deduce the following result.

Theorem 4.17 *Let $\delta > 0$ and Y be a spectrally negative Lamperti-stable process with parameters $\beta \in [0, 1)$ and $\hat{\gamma} \in (0, 1)$, then*

$$\mathtt{M}(s) = c\chi^{s-1}\Gamma(s)\Gamma\Big((1-\beta)\chi + 1 - s\Big)\frac{G(\beta\chi + s;\chi)}{G((\beta+\hat{\gamma})\chi + s;\chi)}, \qquad s \in \mathbb{C},$$

where c is such that $\mathtt{M}(1) = 1$.

Finally, we consider the spectrally positive case, that is, when $\hat{\gamma} = 1$, $\beta \in [0, 1)$ and $\gamma \in (0, 1)$. As we will see below, this case is slightly different. Here, we have

$$\psi(z) = -(\beta + z)\frac{\Gamma(1-\beta+\gamma-z)}{\Gamma(1-\beta-z)} = \frac{\Gamma(1-\beta+\gamma-z)}{\Gamma(-\beta-z)}, \tag{4.59}$$

which is well defined for $\mathrm{Re}(z) \in (-\infty, 1-\beta+\gamma)$. When the process drifts to $-\infty$, that is $\beta = 0$, we observe that $\mathtt{M}(s)$ can be extended to $\mathrm{Re}(s) \leq 0$.

Proposition 4.18 *Let $\delta > 0$ and Y be a spectrally positive Lamperti-stable process with parameters $\beta \in [0, 1)$ and $\gamma \in (0, 1)$. If*

i) $\beta > 0$, then $\mathtt{M}(s)$ is well defined for $\mathrm{Re}(s) \in (0, 1 + (1-\beta)\chi)$ *and satisfies*

$$\mathtt{M}(s+1) = -\frac{s}{\psi(\delta s)}\mathtt{M}(s), \qquad s \in (0, (1-\beta)\chi).$$

ii) $\beta = 0$, then $\mathtt{M}(s)$ is well defined for $\mathrm{Re}(s) \in (-\infty, 1 + \chi)$ *and satisfies*

$$\mathtt{M}(s+1) = -\frac{s}{\psi(\delta s)}\mathtt{M}(s), \qquad s \in (-\infty, \chi). \tag{4.60}$$

Proof We observe that part (i) follows from the same arguments as in Proposition 4.11, since $\psi(\delta s) \le 0$, for $\mathrm{Re}(s) \in (0, 1 + (1-\beta)\chi)$. The same holds true for part (ii), when $\mathrm{Re}(s) \in (0, 1+\chi)$. Thus, it is enough to prove the result for part (ii), when $\mathrm{Re}(s) \in (-\infty, 0]$.

Observe that, in this case, the process drifts to $-\infty$ and recall that

$$\mathcal{E}_t(z) = \mathrm{e}^{zY_t - \psi(z)t}, \qquad t \ge 0,$$

is a positive martingale for $z \in (-\infty, 0)$. Proceeding similarly as in the proof of Proposition 4.11, we observe that, for $t \in (0, 1]$ and $s > 0$, Doob's L^1-inequality and the Esscher transform (2.23) imply

$$\begin{aligned} \mathrm{E}\left[\left(\int_0^t \mathrm{e}^{\delta Y_u} \mathrm{d}u\right)^{-s}\right] &\le t^{-s} \mathrm{e}^{\psi(-\delta s)\vee 0} \mathrm{E}\left[\sup_{u \le t} \mathcal{E}_u(-\delta s)\right] \\ &\le \frac{\mathrm{e}^{\psi(-\delta s)\vee 0}}{\mathrm{e}-1} t^{-s}(1 - \delta s \psi'(-\delta s) - \psi(-\delta s)), \end{aligned}$$

which is clearly finite. The finiteness for $t > 1$ follows from the fact that the exponential functional $t \mapsto \int_0^t \mathrm{e}^{\delta Y_u} \mathrm{d}u$ is non-decreasing. In consequence, $\mathtt{M}$ is also well defined on $(-\infty, 1)$.

Next, we deduce identity (4.60). Integration by parts gives us

$$\left(\int_t^\infty \mathrm{e}^{\delta Y_u} \mathrm{d}u\right)^{-s} - \left(\int_0^\infty \mathrm{e}^{\delta Y_u} \mathrm{d}u\right)^{-s} = s \int_0^t \mathrm{e}^{-\delta s Y_v} \left(\int_0^\infty \mathrm{e}^{\delta(Y_{u+v} - Y_v)} \mathrm{d}u\right)^{-s-1} \mathrm{d}v,$$

for all $s > 0$ and $t \ge 0$. Hence we take expectations in both sides of the above identity and, since

$$\int_t^\infty \mathrm{e}^{\delta Y_u} \mathrm{d}u = \mathrm{e}^{\delta Y_t} \int_0^\infty \mathrm{e}^{\delta(Y_{u+t} - Y_t)} \mathrm{d}u,$$

stationarity and independent increments imply

$$\mathtt{M}(-s+1)\left(\mathrm{e}^{t\psi(-\delta s)} - 1\right) = -(-s)\mathtt{M}(-s) \int_0^t \mathrm{e}^{u\psi(-\delta s)} \mathrm{d}u,$$

from which identity (4.60) is deduced, for $s \in (-\infty, 0)$. To see that the aforesaid identity holds for $s = 0$, we observe that

$$\lim_{s \to 0} -\frac{\psi(\delta s)}{s} = \Gamma(1+\gamma),$$

and, since $\mathtt{M}$ is finite in $(-\infty, 0)$, we may apply the Dominated Convergence Theorem to dededuce that (4.60) holds for $s \in (-\infty, 0]$.

Finally, we put all pieces together and use general properties of Mellin transforms to get that $\mathtt{M}(s)$ is finite and analytic for all $s \in \mathbb{C}$ such that $\mathrm{Re}(s) \in (-\infty, 1+\chi)$. This completes the proof. □

Similarly to the spectrally negative case, the arguments used in the proof of Theorem 4.13 still holds for the spectrally positive case and allow us to compute explicitly $\mathtt{M}$. In this case, the function F is given by

$$\mathrm{F}(s) = c\frac{\chi^s}{\Gamma(\beta\chi + s)}\frac{G((1-\beta+\gamma)\chi + 1 - s;\chi)}{G((1-\beta+)\chi + 1 - s;\chi)}, \qquad s \in \mathbb{C},$$

where c is such that $\mathrm{F}(1) = 1$. We observe that the function F is meromorphic with zeros at $-\beta\chi - n$ and $1 - (1-\beta+\gamma)\chi + m\chi + n$; and poles

$$z^{+}_{m,n} := 1 + (1-\beta)\chi + m\chi + n,$$

for $m, n \geq 0$. Again, all zeros/poles are simple if $\delta \notin \mathbb{Q}$. Moreover, Lemma 4.14 and the identities in part (ii) of Lemma 4.15 still hold in this case (replace $\hat{\gamma}$ by 1), again thanks to the quasi-periodic properties of the double gamma function. In other words, we may deduce the following result.

Theorem 4.19 *Let $\delta > 0$ and Y be a spectrally positive Lamperti-stable process with parameters $\beta \in [0, 1)$ and $\gamma \in (0, 1)$, then*

$$\mathtt{M}(s) = c\frac{\chi^s\Gamma(s)}{\Gamma(\beta\chi + s)}\frac{G((1-\beta+\gamma)\chi + 1 - s;\chi)}{G((1-\beta)\chi + 1 - s;\chi)}, \qquad s \in \mathbb{C},$$

where the constant c is such that $\mathtt{M}(1) = 1$.

4.6 Distributional densities of exponential functionals

Next, we are interested in inverting the Mellin transform $\mathtt{M}$ in order to deduce an expression for the probability density function of $I(\delta, Y)$, henceforth denoted by

$$p(x) = \frac{\mathrm{d}}{\mathrm{d}x}\mathrm{P}\big(I(\delta, Y) \leq x\big), \quad x \geq 0.$$

We are interested in a convergent series representation as well as a complete asymptotic expansion of the density $p(x)$ as $x \to 0^+$ or $x \to \infty$.

For simplicity of exposition, we only deduce the form of the density $p(x)$ for the case when the parameters $(\beta, \gamma, \hat{\beta}, \hat{\gamma}) \in \mathcal{H}_1 \setminus \{\beta = 1\}$. The other case can be derived using the same arguments.

Recall that ψ, $\tilde{\psi}$ and F denote the functions which were defined in (4.37), (4.46) and (4.43); the sequences $z^{-}_{m,n}$ and $z^{+}_{m,n}$ represent the poles of F and were defined in (4.48). We also recall that $\eta = 1 - \beta + \gamma + \hat{\beta} + \hat{\gamma}$, see (4.21), and that $\chi = 1/\delta$, see (4.38).

Definition 4.20 Assume that $(\beta, \gamma, \hat{\beta}, \hat{\gamma}) \in \mathcal{H}_1 \setminus \{\beta = 1\}$. We define the coefficients a_n, $n \geq 0$, as

$$a_n = -\frac{1}{n!}\prod_{j=0}^{n}\psi(\delta j), \quad n \geq 0. \tag{4.61}$$

Note that if $\hat{\beta} = 0$, we have $\psi(0) = 0$ which implies that $a_n = 0$, for all $n \geq 0$.

The coefficients $b_{m,n}$, $m, n \geq 0$, are defined recursively

$$\begin{cases} b_{0,0} = \chi \dfrac{\Gamma(\eta)\Gamma(-(\hat{\beta}+\hat{\gamma})\chi)}{\Gamma(\eta-\gamma)\Gamma(-\hat{\gamma})} \mathrm{F}(1-(\hat{\beta}+\hat{\gamma})\chi), \\ b_{m,n} = -\dfrac{\psi(\delta z^-_{m,n})}{z^-_{m,n}} b_{m,n-1}, \quad m \geq 0, \ n \geq 1, \\ b_{m,n} = -\delta^{\chi(\gamma+\hat{\gamma})}\tilde{\psi}(z^-_{m,n})\dfrac{\Gamma(z^-_{m,n})}{\Gamma(z^-_{m-1,n})} b_{m-1,n}, \quad m \geq 1, \ n \geq 0. \end{cases} \tag{4.62}$$

Similarly, $c_{m,n}$, $m, n \geq 0$, are defined recursively

$$\begin{cases} c_{0,0} = \chi \dfrac{\Gamma(1+(1-\beta)\chi)\Gamma(1-\beta+\hat{\beta})}{\Gamma(\eta-\gamma)\Gamma(\gamma)} \mathrm{F}((1-\beta)\chi), \\ c_{m,n} = -\dfrac{z^+_{m,n-1}}{\psi(\delta z^+_{m,n-1})} c_{m,n-1}, \quad m \geq 0, \ n \geq 1, \\ c_{m,n} = -\dfrac{\delta^{-\chi(\gamma+\hat{\gamma})}}{\tilde{\psi}(z^+_{m-1,n})}\dfrac{\Gamma(z^+_{m,n})}{\Gamma(z^+_{m-1,n})} c_{m-1,n}, \quad m \geq 1, \ n \geq 0. \end{cases} \tag{4.63}$$

The next result looks at the residues of $\mathtt{M}$. The reader will note that the result excludes the cases that $\delta \in \mathbb{Q}$. This comes about from the fact that, for $\delta \in \mathbb{Q}$, the Mellin transform $\mathtt{M}(s)$, and specifically the double gamma function G, has poles of multiplicity greater than one, which makes the picture much more complicated.

Proposition 4.21 *Assume that $(\beta, \gamma, \hat{\beta}, \hat{\gamma}) \in \mathcal{H}_1 \setminus \{\beta = 1\}$ and $\delta \notin \mathbb{Q}$. For all $m, n \geq 0$, we have*

$$\begin{aligned} \mathrm{Res}(\mathtt{M}(s) : s = -n) &= a_n, \quad \text{if } \hat{\beta} > 0, \\ \mathrm{Res}(\mathtt{M}(s) : s = z^-_{m,n}) &= b_{m,n}, \\ \mathrm{Res}(\mathtt{M}(s) : s = z^+_{m,n}) &= -c_{m,n}. \end{aligned}$$

Proof We start by proving that the residue of $\mathtt{M}(s)$ at $s = z^-_{m,n}$ is equal to $b_{m,n}$. First, use Theorem 4.13 and rearrange the terms in the functional identity (4.49), noting the expression for ψ in (4.37) and the recursion formula for gamma functions (A.8), to find that

$$\mathtt{M}(s) = \frac{\chi \mathrm{F}(s+1)\Gamma(s)}{s+(\hat{\beta}+\hat{\gamma})\chi} \frac{\Gamma(1-\beta+\gamma-\delta s)\Gamma(1+\hat{\beta}+\hat{\gamma}+\delta s)}{\Gamma(1-\beta-\delta s)\Gamma(\hat{\beta}+\delta s)}.$$

The above identity and the definition (4.62) imply that as $s \to -(\hat{\beta} + \hat{\gamma})\chi$

$$\mathtt{M}(s) = \frac{b_{0,0}}{s + (\hat{\beta} + \hat{\gamma})\chi} + O(1),$$

which means that the residue of $\mathtt{M}(s)$ at $z_{0,0}^- = -(\hat{\beta} + \hat{\gamma})\chi$ is equal to $b_{0,0}$.

Next, we show that the residues satisfy the second recursive identity in (4.62). To this end, rewrite (4.49) as

$$\mathtt{M}(s) = -\frac{\psi(\delta s)}{s}\mathtt{M}(s+1). \tag{4.64}$$

We know that $\mathtt{M}(s)$ has a simple pole at $s = z_{m,n}^-$ and $\mathtt{M}(s+1)$ has a simple pole at $z_{m,n}^- + 1 = z_{m,n-1}^-$. One can also check that the function $\psi(\delta s)$ is analytic at $s = z_{m,n}^-$ for $n \geq 1$. Therefore we have, as $s \to z_{m,n}^-$,

$$\mathtt{M}(s) = \mathrm{Res}(\mathtt{M}(z) : z = z_{m,n}^-)\frac{1}{s - z_{m,n}^-} + O(1),$$

$$\mathtt{M}(s+1) = \mathrm{Res}(\mathtt{M}(z) : z = z_{m,n-1}^-)\frac{1}{s - z_{m,n}^-} + O(1),$$

$$-\frac{\psi(\delta s)}{s} = -\frac{\psi(\delta z_{m,n}^-)}{z_{m,n}^-} + O(s - z_{m,n}^-),$$

which, together with (4.64) imply that

$$\mathrm{Res}(\mathtt{M}(s) : s = z_{m,n}^-) = -\frac{\psi(\delta z_{m,n}^-)}{z_{m,n}^-} \times \mathrm{Res}(\mathtt{M}(s) : s = z_{m,n-1}^-).$$

The proof of all remaining cases is very similar and we leave the details to the reader. □

Proposition 4.21 immediately gives us a complete asymptotic expansion of $p(x)$ as $x \to 0^+$ and $x \to \infty$, which we present in the next Theorem.

Theorem 4.22 *Assume that* $(\beta, \gamma, \hat{\beta}, \hat{\gamma}) \in \mathcal{H}_1 \setminus \{\beta = 1\}$ *and* $\delta \notin \mathbb{Q}$. *Then*

$$p(x) \sim \sum_{n\geq 0} a_n x^n + \sum_{m\geq 0}\sum_{n\geq 0} b_{m,n} x^{(m+\hat{\beta}+\hat{\gamma})\chi+n}, \quad x \to 0^+, \tag{4.65}$$

$$p(x) \sim \sum_{m\geq 0}\sum_{n\geq 0} c_{m,n} x^{-(m+1-\beta)\chi-n-1}, \quad x \to \infty. \tag{4.66}$$

Proof The basis of the proof is to identify $p(x)$ as the inverse Mellin transform

$$p(x) = \frac{1}{2\pi \mathrm{i}} \int_{1+\mathrm{i}\mathbb{R}} \mathtt{M}(s) x^{-s} \mathrm{d}s, \quad x > 0. \tag{4.67}$$

In a similar spirit to the proof of (4.55), we can use (A.16) in the Appendix together with (4.45) and the equality $\mathtt{M} = \Gamma\mathtt{F}$ from Theorem 4.13 to deduce that $|\mathtt{M}(x + \mathrm{i}u)|$ decreases exponentially as $u \to \infty$ (uniformly in x in any finite interval). As a consequence, not only does its Fourier inverse exist, but so does the Fourier inverse of its derivatives as well; therefore all exist as continuous functions. As such, $p(x)$ is a smooth function for $x > 0$.

Assume that $c < 0$ satisfying that $c \neq z^-_{m,n}$ and $c \neq -n$ for all m, n, and set ℓ to be an integer. We also consider the contour $L = L_1 \cup L_2 \cup L_3 \cup L_4$ which is defined as

$$\begin{aligned} L_1 &:= \{\mathrm{Re}(z) = c,\ -\ell \le \mathrm{Im}(z) \le \ell\}, \\ L_2 &:= \{\mathrm{Im}(z) = \ell,\ c \le \mathrm{Re}(z) \le 1\}, \\ L_3 &:= \{\mathrm{Re}(z) = 1,\ -\ell \le \mathrm{Im}(z) \le \ell\}, \\ L_4 &:= \{\mathrm{Im}(z) = -\ell,\ c \le \mathrm{Re}(z) \le 1\}. \end{aligned}$$

It is clear that L is the rectangle bounded by vertical lines $\mathrm{Re}(z) = c$, $\mathrm{Re}(z) = 1$ and by horizontal lines $\mathrm{Im}(z) = \pm\ell$. We assume that L is oriented counter-clockwise; see Figure 4.1.

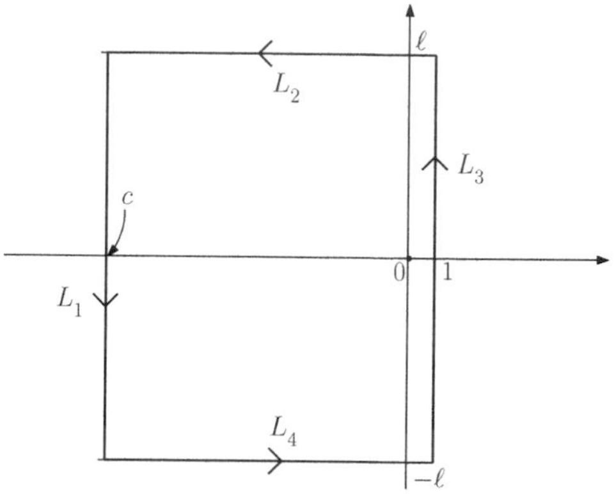

Figure 4.1 The contour $L = L_1 \cup L_2 \cup L_3 \cup L_4$

The function $\mathtt{M}(s)$ is analytic in the interior of L, except for simple poles $z^-_{m,n}$ and $-n$ which lie in $(c, 1)$, and is continuous on L. Using the residue theorem we find

$$\begin{aligned} \frac{1}{2\pi \mathrm{i}} \int_L \mathtt{M}(s) x^{-s}\, \mathrm{d}s = &\sum_{0<|z^-_{m,n}|<|c|} \mathrm{Res}(\mathtt{M}(s) : s = z^-_{m,n}) \times x^{-z^-_{m,n}} \\ &+ \sum_{0\le n<|c|} \mathrm{Res}(\mathtt{M}(s) : s = -n) \times x^n. \end{aligned}$$

Next, we estimate the integrals over the horizontal side L_2 as follows

$$\left|\int_{L_2} \mathtt{M}(s)x^{-s}\,\mathrm{d}s\right| < (\epsilon - c)\times x^{-1}\max_{s\in L_2}|\mathtt{M}(s)|.$$

When ℓ increases, we have $\sup_{s\in L_2}|\mathtt{M}(s)|$ goes to 0. Therefore

$$\int_{L_2} \mathtt{M}(s)x^{-s}\,\mathrm{d}s \to 0, \qquad \text{as} \quad \ell\to\infty.$$

Similarly, we deduce that the integral on the contour L_4 goes to 0 as ℓ goes to ∞. Thus putting all the pieces together, we have

$$-\frac{1}{2\pi\mathrm{i}}\int_{c+\mathrm{i}\mathbb{R}} \mathtt{M}(s)x^{-s}\,\mathrm{d}s + \frac{1}{2\pi\mathrm{i}}\int_{1+\mathrm{i}\mathbb{R}} \mathtt{M}(s)x^{-s}\,\mathrm{d}z$$
$$= \sum_{0<|z^-_{m,n}|<|c|} \mathrm{Res}(\mathtt{M}(s) : s = z^-_{m,n})\times x^{-z^-_{m,n}} + \sum_{0\le n<|c|} \mathrm{Res}(\mathtt{M}(s) : s = -n)\times x^{n}.$$

In other words, we have deduced

$$\begin{aligned} p(x) = &\sum_{0<|z^-_{m,n}|<|c|} \mathrm{Res}(\mathtt{M}(s) : s = z^-_{m,n})\times x^{-z^-_{m,n}} \\ &+ \sum_{0\le n<|c|} \mathrm{Res}(\mathtt{M}(s) : s = -n)\times x^{n} + \frac{1}{2\pi\mathrm{i}}\int_{c+\mathrm{i}\mathbb{R}} \mathtt{M}(s)x^{-s}\mathrm{d}s. \end{aligned} \tag{4.68}$$

Next, we perform a change of variables $s = c + \mathrm{i}u$ and obtain the following estimate

$$\left|\int_{c+\mathrm{i}\mathbb{R}} \mathtt{M}(s)x^{-s}\mathrm{d}s\right| = x^{-c}\left|\int_{\mathbb{R}} \mathtt{M}(s)x^{-\mathrm{i}u}\mathrm{d}u\right| < x^{-c}\int_{\mathbb{R}} |\mathtt{M}(c+\mathrm{i}u)|\mathrm{d}u = O(x^{-c}).$$

Note that $|\mathtt{M}(c+\mathrm{i}u)|$ is integrable thanks to its previously observed exponential decay in u (see remarks below (4.67)). The asymptotic (4.65) now follows.

The proof of (4.66) is identical, except that we have to perform the contour of integration in the opposite direction. In particular, we can build a rectangular anti-clockwise contour similar to the one in Figure 7.1 albeit that the location of left-hand side agrees with that of L_3 in the figure and the other three sides lie in the positive half of the complex plain, capturing the other poles on the positive real line. □

It turns out that, for almost all parameters δ, except for rational numbers and for those real numbers which can be approximated by rational numbers in a certain way, the asymptotic series (4.65) and (4.66) converge to $p(x)$ for all $x > 0$. Unfortunately the proof of such result is rather technical and goes beyond the scope of this manuscript. For that reason, we just state the result without a proof. Nonetheless, in order to state our result the following set of real numbers is needed.

Definition 4.23 Let $\mathcal{L}$ be the set of real irrational numbers x, for which there exists a constant $b > 1$ such that the inequality

$$\left| x - \frac{p}{q} \right| < \frac{1}{b^q} \tag{4.69}$$

is satisfied for infinitely many integers p and q.

The set $\mathcal{L}$ is a proper subset of so-called Liouville numbers and possesses very interesting properties. For instance, as the set of Liouville numbers has zero Hausdorff dimension and hence zero Lebesgue measure, the same is true of $\mathcal{L}$. Moreover, the structure of $\mathcal{L}$ can be described in terms of a continued fraction representation of real numbers and it is closed under addition and multiplication by rational numbers, implying that it is dense in $\mathbb{R}$. As with Proposition 4.21, and Theorem 4.22, and for the same reasons given there, the following exact identity for the density p cannot accommodate for $\delta \in \mathbb{Q}$. Additonally, it cannot accommodate for $\delta \in \mathcal{L}$ for technical reasons that are also beyond the scope of our exposition.

Theorem 4.24 *Assume that* $(\beta, \gamma, \hat{\beta}, \hat{\gamma}) \in \mathcal{H}_1 \setminus \{\beta = 1\}$ *and* $\delta \notin \mathcal{L} \cup \mathbb{Q}$. *Then for all* $x > 0$,

$$p(x) = \begin{cases} \sum_{n\geq 0} a_n x^n + \sum_{m\geq 0} \sum_{n\geq 0} b_{m,n} x^{(m+\hat{\beta}+\hat{\gamma})\chi+n} & \text{if } \gamma + \hat{\gamma} < 1, \\ \sum_{m\geq 0} \sum_{n\geq 0} c_{m,n} x^{-(m+1-\beta)\chi-n-1} & \text{if } \gamma + \hat{\gamma} > 1. \end{cases}$$

For the case $\gamma = \hat{\gamma} = \delta$, there is a simpler representation for $p(x)$ which allows us to remove the assumption $\delta \notin \mathcal{L} \cup \mathbb{Q}$. This representation follows from the fact that the Mellin transform of $p(x)$ can be written exclusively in terms of the gamma function; see Corollary 4.16, noting the use of (A.12).

Theorem 4.25 *Assume that* $(\beta, \delta, \hat{\beta}, \delta) \in \mathcal{H}_1 \setminus \{\beta = 1\}$. *If* $2\delta < 1$, *then*

$$p(x) = \frac{\Gamma(\hat{\beta} + \delta)}{\Gamma(1 - \beta)} \sum_{n\geq 0} \frac{\Gamma(1 - \beta + \delta(1 + n))}{\Gamma(\hat{\beta} - \delta n)} \frac{(-1)^n}{n!} x^n, \tag{4.70}$$

for $x > 0$, *and if* $2\delta > 1$, *then*

$$p(x) = \frac{\chi\Gamma(\hat{\beta} + \delta)}{\Gamma(1 - \beta)} \sum_{n\geq 0} \frac{\Gamma((1 - \beta + n)\chi + 1)}{\Gamma(1 - \beta + \hat{\beta} + \delta + n)} \frac{(-1)^n}{n!} x^{-(1-\beta+n)\chi-1}, \tag{4.71}$$

for $x > 0$. *Moreover, formula* (4.70) *(resp. (4.71)) provides complete asymptotic expansion as* x *goes to* 0^+ *(resp. as* x *goes to* ∞*).*

Proof We give only a brief sketch of the proof. From Corollary 4.16 (i), we see that the Mellin transform $\mathtt{M}(z)$ of $p(x)$ has simple poles at

$$z_n^- = -n \quad \text{and} \quad z_n^+ = 1 + \chi(n + 1 - \beta), \qquad n \geq 0.$$

The residues at these points provide the coefficients in (4.70) and (4.71). Indeed by applying Proposition A.1 and identity (A.11) (both in the Appendix), we find that

$$\operatorname{Res}(\mathtt{M}(s), s = -n) = \frac{\Gamma(\hat{\beta} + \delta)\Gamma(1 - \beta + \delta(1 + n))}{\Gamma(1 - \beta)\Gamma(\hat{\beta} - \delta n)} \frac{(-1)^n}{n!}.$$

Similarly, for the poles at $z_n^+ = 1 + \chi(n + 1 - \beta)$, we get

$$\operatorname{Res}(\mathtt{M}(s) : s = 1 + \chi(n + 1 - \beta)) = \frac{\Gamma(\hat{\beta} + \delta)\Gamma(1 + \chi(n + 1 - \beta))}{\Gamma(1 - \beta)\Gamma(1 - \beta + \hat{\beta} + \delta + n))}(-1)^{n-1}\frac{\chi}{n!}.$$

The rest of the proof follows by using similar ideas to, for example, the proof of Theorem 1.18 (or Theorem 4.22). We first need to guarantee that the Mellin transform inversion can be performed, that is, we need to verify that $\mathtt{M}(z)$ is absolutely integrable on a given vertical line where $\mathtt{M}$ is well-defined. The latter follows from (A.16), which in particular implies the following upper bound

$$\left|\mathtt{M}(x + \mathrm{i}y)\right| \leq C e^{-\frac{\pi}{2}|y|}|y|^{1-\beta-\hat{\beta}+\delta+(1-2\delta)x-\frac{1}{2}}, \qquad \text{as} \quad y \to \infty, \tag{4.72}$$

uniformly in any finite interval $-\infty < a \leq x \leq b < \infty$, where $C > 0$ is an unimportant constant. Thus, we can use an appropriate contour integral which encloses an increasing number of poles as it expands. The poles are chosen in such a way to ensure that one side of the path integral converges to the inverse Mellin transform of $\mathtt{M}(z)$ and the remaining parts of the integral path tend to zero as the contour grows larger. To ensure the latter, one must appeal to the specific form of $\mathtt{M}(z)$ in terms of gamma functions. More precisely, in the spirit of (4.72), albeit with subtle differences in the estimates, the asymptotic relations of the gamma function in (A.15) and the ratio of gamma functions in (A.17) provide an exponential decay for $\mathtt{M}(z)$ when $\operatorname{Re}(z) \leq 0$ with $2\delta < 1$ and when $\operatorname{Re}(z) \geq 0$ with $2\delta > 1$, respectively. Hence, the Residue Theorem gives the desired density as a sum of residues of the captured poles. We leave the details to the interested reader. □

We conclude this section by computing the density of the exponential functional of a hypergeometric Lévy process with a special choice of parameters, which will be of particular interested later on in the book.

Theorem 4.26 *Assume that* $(1,\gamma,\hat{\beta},\hat{\gamma})\in\mathcal{H}_2\cap\{\eta-\hat{\gamma}>0\}\setminus\{\hat{\beta}=0\}$. *If* $2\delta<1$, *then*

$$p(x)=\frac{\delta}{\pi}\sum_{n\geq 1}(-1)^{n-1}\frac{\sin(\pi\chi\hat{\beta})\Gamma(\hat{\beta}+\delta)\Gamma(\delta(n+1)+\hat{\beta})}{\Gamma(1+n+\chi\hat{\beta})\Gamma(-\delta n)}x^{n+\chi\hat{\beta}}, \tag{4.73}$$

for $x>0$, *and if* $2\delta>1$, *then*

$$p(x)=\sum_{n\geq 0}a_n x^{-1+\chi\hat{\beta}-n}+\sum_{m\geq 1}b_m x^{-1-m\chi}, \tag{4.74}$$

for $x>0$, *where*

$$a_n=\frac{\delta}{\pi}\frac{\sin(\pi(\chi\hat{\beta}+1))\Gamma(\hat{\beta}+\delta)\Gamma(\hat{\beta}-\delta n)}{\Gamma(\chi\hat{\beta}-n)\Gamma(\delta(n+1))}(-1)^n$$

and

$$b_m=\frac{\sin(\pi\chi\hat{\beta})\Gamma(\hat{\beta}+\delta)}{\sin(\pi\chi(\hat{\beta}+m))\Gamma(-\chi n)\Gamma(\hat{\beta}+\delta+m)}\frac{(-1)^m}{m!}.$$

Moreover, formula (4.73) *(resp.* (4.74)*) provides complete asymptotic expansion as* x *goes to* 0^+ *(resp. as* x *goes to* ∞*).*

Proof As in Theorem 4.25, we give only a brief sketch of the proof. From Corollary 4.16 (ii), when $\beta=1$, we have

$$\mathtt{M}(s)=\delta\frac{\sin(\pi(\chi\hat{\beta}+1))}{\sin(\pi(\chi\hat{\beta}+s))}\frac{\Gamma(\hat{\beta}+\delta)\Gamma(\delta(1-s))}{\Gamma(1-s)\Gamma(\hat{\beta}+\delta s)}, \qquad s\in\mathbb{C},$$

where we have used (A.10) to determine the normalisation $\mathtt{M}(1)=1$. We see that the Mellin transform $\mathtt{M}(z)$ of $p(x)$ thus has simple poles at

$$z_n^-=-n-\chi\hat{\beta}, \qquad \text{for } n\geq 1,$$

$$z_n^{+,1}=n+1-\chi\hat{\beta}, \qquad \text{for } n\geq 0,$$

and

$$z_n^{+,2}=1+\chi n, \qquad \text{for } n\geq 1.$$

Again, the residues at these points provide the coefficients in (4.73) and (4.74). Indeed, we apply Proposition A.1 and identity (A.11) (both in the Appendix) to find

$$\mathrm{Res}(\mathtt{M}(s):s=-n-\chi\hat{\beta})=\frac{\delta}{\pi}\frac{\sin(\pi(\chi\hat{\beta}+1))\Gamma(\hat{\beta}+\delta)\Gamma(\delta(1+n)+\hat{\beta})}{\Gamma(1+n+\chi\hat{\beta})\Gamma(-\delta n)}(-1)^n.$$

For the poles at $z_n^{+,1}=n+1-\chi\hat{\beta}$, we get

$$\mathrm{Res}(\mathtt{M}(s):s=n+1-\chi\hat{\beta})=\frac{\delta}{\pi}\frac{\sin(\pi(\chi\hat{\beta}+1))\Gamma(\hat{\beta}+\delta)\Gamma(\hat{\beta}-\delta n)}{\Gamma(\chi\hat{\beta}-n)\Gamma(\delta(n+1))}(-1)^{n+1}.$$

Finally, for the poles at $z_n^{+,2} = 1 + \chi n$, we have

$$\mathrm{Res}(\mathtt{M}(s) : s = 1 + \chi n) = \frac{1}{n!} \frac{\sin(\pi(\chi\hat{\beta} + 1))\Gamma(\hat{\beta} + \delta)}{\sin(\pi(\chi(\hat{\beta} + n) + 1)\Gamma(-n\chi)\Gamma(\hat{\beta} + \delta + n)} (-1)^{n+1}.$$

The rest of the proof follows by developing the appropriate contour integrals as in previous theorems and observing, using the asymptotic relations (A.15) and (A.16), that

$$\left|\mathtt{M}(x + \mathrm{i}y)\right| \leq Ce^{-\frac{\pi}{2}|y|}|y|^{\delta-\hat{\beta}+(1-2\delta)x-\frac{1}{2}}, \qquad \text{as} \quad y \to \infty,$$

uniformly in any finite interval $-\infty < a \leq x \leq b < \infty$, where $C > 0$ is an unimportant constant. Moreover, that $\mathtt{M}(z)$ has exponential decay when $\mathrm{Re}(z) \leq 0$ with $2\delta < 1$, and when $\mathrm{Re}(z) \geq 0$ with $2\delta > 1$. We leave the details to the interested reader. □

4.7 Distributional tails of exponential functionals

As we will see for some of the applications later in this text, there are occasions where it suffices to work with the asymptotic upper tail distribution of $I(\delta, Y)$ in place of its probability density. Although the former can be deduced from the latter, it contains marginally less information about the distribution of $I(\delta, Y)$ and so one may be able to derive it with fewer assumptions. A good example of this pertains to one of the assumptions of Theorem 4.24, specifically the requirement there that $\delta \notin \mathcal{L} \cup \mathbb{Q}$. We present a different approach here for studying the tail distribution of $I(\delta, Y)$, which will rule out any restriction on $\delta > 0$.

Proposition 4.27 *Let Y be an hypergeometric Lévy process with parameters $(\beta, \gamma, \hat{\beta}, \hat{\gamma}) \in \mathcal{H}_4$ or a spectrally one-sided Lamperti-stable with parameters $\beta \in [0, 1), \gamma \in (0, 1)$ or $\beta \in [0, 1), \hat{\gamma} \in (0, 1)$. Then*

$$\lim_{t\to\infty} t^{\hat{\theta}\chi}\mathrm{P}\Big(I(\delta, Y) > t\Big) = \frac{\mathtt{M}(\hat{\theta}\chi)}{\psi'(\hat{\theta})}, \tag{4.75}$$

where $\chi = 1/\delta$ and $\hat{\theta}$ is defined in (4.39) when $(\beta, \gamma, \hat{\beta}, \hat{\gamma}) \in \mathcal{H}_4$ and, otherwise, $\hat{\theta}$ is taken as $(1 - \beta)$ in the spectrally one-sided cases.

Proof Let us start by recalling that, in the setting that Y is killed. Its cemetery state is taken to be $-\infty$ and we work with the definition $\exp(-\infty) = 0$. With this in mind, we observe that the exponential functional $I(\delta, Y)$ satisfies

$$\int_0^\infty e^{\delta Y_u}\mathrm{d}u = \int_0^1 e^{\delta Y_u}\mathrm{d}u + e^{\delta Y_1}\int_0^\infty e^{\delta(Y_{u+1}-Y_1)}\mathrm{d}u.$$

Since Y possesses stationary and independent increments, the following identity in law follows

$$I(\delta, Y) \stackrel{d}{=} Q + VI(\delta, Y'),$$

where

$$Q = \int_0^1 e^{\delta Y_u} du, \quad V = e^{\delta Y_1}, \qquad I(\delta, Y') = \int_0^\infty e^{\delta Y'_u} du,$$

and Y' is independent of $(Y_s, 0 \le s \le 1)$, with the same law as Y. Therefore $I(\delta, Y')$ is independent of the pair (Q, V) and has the same law as $I(\delta, Y)$. That is to say, $I(\delta, Y)$ is a random recursive equation in the sense of (A.39) in the Appendix. Hence, if Q and V fulfils the hypothesis of Theorem A.11 (see the Appendix), the result will follow.

Since Y is not arithmetic, the same property is inherited to the random variable V. From the analytical expression for ψ given in (4.37), we not that the quantity $\hat{\theta}_X$ satisfies

$$\mathrm{E}\left[V^{\hat{\theta}_X}\right] = \mathrm{E}\left[e^{\hat{\theta} Y_1}\right] = 1.$$

Next, we introduce

$$\vartheta := \begin{cases} \gamma & \text{if } (\beta, \gamma, \hat{\beta}, \hat{\gamma}) \in \mathcal{H}_1 \setminus \{\beta = 1\}, \\ \eta - \hat{\gamma} & \text{if } (\beta, \gamma, \hat{\beta}, \hat{\gamma}) \in \mathcal{H}_2 \cap \{\eta - \hat{\gamma} > 0\} \setminus \{\hat{\beta} = 0\}. \end{cases}$$

For the spectrally positive case we take $\vartheta = \gamma$ and for the spectrally negative case, ϑ can be taken as ∞. Thus, we observe, from (4.37), (4.58) and (4.59) that for $\epsilon \in (0, \vartheta)$, we have

$$\mathrm{E}\left[V^{\hat{\theta}_X} \ln^+(V)\right] = \mathrm{E}\left[e^{\hat{\theta} Y_1} Y_1^+\right] \le \mathrm{E}\left[e^{(\hat{\theta}+\epsilon) Y_1}\right] < \infty,$$

where $y^+ = \max\{0, y\}$. Finally, from Doob's L^1-inequality and the Esscher transform (2.23), we verify that

$$\mathrm{E}\left[Q^{\hat{\theta}_X}\right] = \mathrm{E}\left[\left(\int_0^1 e^{\delta Y_u} du\right)^{\hat{\theta}_X}\right] \le \mathrm{E}\left[\sup_{u \le 1} \mathcal{E}_u(\hat{\theta})\right] \le \frac{e}{e-1}\left(1 + \hat{\theta}\mathrm{E}\left[e^{\hat{\theta} Y_1} Y_1^+\right]\right) < \infty.$$

Therefore, Theorem A.11 guarantees that

$$\lim_{t \to \infty} t^{\hat{\theta}_X} \mathrm{P}\big(I(\delta, Y) > t\big) = C_+,$$

where

$$C_+ = \frac{\mathrm{E}\left[(Q + VI(\delta, Y'))^{\hat{\theta}_X}\right] - \mathrm{E}\left[(VI(\delta, Y'))^{\hat{\theta}_X}\right]}{\hat{\theta}_X \psi'(\hat{\theta})}.$$

Now, we compute explicitly the constant C_+. Recall the identities in law from the beginning of the proof. Then using integration by parts and the stationary and independent increment property of Y, we get

$$\begin{aligned}
C_+ &= \frac{\mathrm{E}\left[\left(\int_0^\infty \mathrm{e}^{\delta Y_s}\mathrm{d}s\right)^{\hat{\theta}\chi} - \left(\int_1^\infty \mathrm{e}^{\delta Y_s}\mathrm{d}s\right)^{\hat{\theta}\chi}\right]}{\hat{\theta}\chi\psi'(\hat{\theta})} \\
&= \frac{1}{\psi'(\hat{\theta})}\mathrm{E}\left[\int_0^1 \mathrm{e}^{\delta Y_u}\left(\int_u^\infty \mathrm{e}^{\delta Y_s}\mathrm{d}s\right)^{\hat{\theta}\chi-1}\mathrm{d}u\right] \\
&= \frac{1}{\psi'(\hat{\theta})}\mathrm{E}\left[\int_0^1 \mathrm{e}^{\hat{\theta} Y_u}\left(\int_0^\infty \mathrm{e}^{\delta(Y_{s+u}-Y_u)}\mathrm{d}s\right)^{\hat{\theta}\chi-1}\mathrm{d}u\right] \\
&= \frac{1}{\psi'(\hat{\theta})}\left(\int_0^1 \mathrm{e}^{\psi(\hat{\theta})u}\mathrm{d}u\right)\mathrm{E}\left[I(\delta, Y)^{\hat{\theta}\chi-1}\right] \\
&= \frac{\mathtt{M}(\hat{\theta}\chi)}{\psi'(\hat{\theta})},
\end{aligned}$$

as required. □

4.8 Comments

The first example of a β-subordinator appeared in Lamperti [139]. Since then, hypergeometric Lévy processes can be found in a variety of different contexts, too numerous to list here. The reader may wish to consult [26, 115, 131] to name but a few. Hypergeometric Lévy processes were first introduced in Kuznetsov et al. [119] and later developed more thoroughly in Kuznetsov and Pardo [121], for the set of admisible parameters $\mathcal{H}_1$, and then extended by Kyprianou et al. [128] to the set $\mathcal{H}_2$. The basic idea of Hypergeometric Lévy processes is to express them via a Wiener–Hopf factorisation consisting of two β-subordinators, appealing to Vigon's theory of philanthropy to justify that the product of factors makes a Lévy–Khintchine exponent. The latter gives conditions under which two subordinators may be associated as ascending and descending ladder heights in the context of a Wiener–Hopf factorisation. In this respect, Theorem 4.4 is a special version of a more general result given in Vigon [211], where as Theorem 4.6 was proved in Kuznetsov and Pardo [121] and Kyprianou et al. [128].

Hypergeometric Lévy processes have appeared consistently in various literature at the intersection of self-similar Markov processes and stable processes. See for example [44, 129] and Chapter 13 of Kyprianou [123]. We note that hypergeometric processes and extensions thereof belong to a larger family of Lévy processes called *meromorphic* processes, which are described in Kuznetsov et al. [118]; see also [89, 130]. Meromorphic processes have the feature that their Wiener–Hopf factors can be written as infinite products of rational functions.

Exponential functionals of Lévy processes appear in various aspects of probability theory, see for instance the survey of Bertoin and Yor [28], and has been a very active topic of research in the last couple of decades. Indeed, determining distributional features of such functionals is still a very prolific research topic; see for example [2, 14, 15, 16, 70, 97, 145, 148, 158, 162, 163, 165, 166, 179, 180, 188, 203] to name but a few key articles. Proposition 4.11 holds for a bigger class of Lévy process and its general form can be found in [148, 180]. The hypergeometric class seems to be one of the very few examples of Lévy processes with two-sided jumps (outside of the class of stable processes and Gaussian processes with compound Poisson, exponentially distributed jumps) for which the law of its associated exponential functionals can be completely characterised. Proposition 4.11 and Theorem 4.13 are taken from Kuznetsov and Pardo [121] (part (i)) and Kyprianou et al. [128] (part(ii)); the results concerning the distribution of the integrated exponential functional, for example, Theorem 4.24 can largely be derived from Kuznetsov and Pardo [121], with, for example, versions of Proposotion 4.27 appearing earlier in Maulik and Zwart [148] and Rivero [180].

5

Positive self-similar Markov processes

In this chapter we introduce one of the key mathematical tools that we shall use to analyse stable processes: *positive self-similar Markov processes*. We shall often denote this class by the shorthand *pssMp*. In the next section we give the definition of these processes and their pathwise characterisation as space-time-changed Lévy processes through the *Lamperti transform*. Thereafter, we spend the rest of the chapter exploring a number of examples of pssMp which can be constructed through path transformations of stable processes. Each of these examples of pssMp turn out to be intimately connected, through the Lamperti transform, to a different Lévy process belonging to the hypergeometric class.

5.1 The Lamperti transform

Let us begin with a definition of the fundamental class of processes that will dominate our analysis. The reader may first find it useful to refer to Section A.11 in the Appendix for an introduction to Feller processes.

Definition 5.1 A $(0,\infty)$-valued regular Feller process, say $Z = (Z_t, t \geq 0)$, is called a *positive self-similar Markov process* if there exists a constant $\alpha > 0$ such that, for any $x > 0$ and $c > 0$,

$$\text{the law of } (cZ_{c^{-\alpha}t}, t \geq 0) \text{ under } P_x \text{ is } P_{cx}, \tag{5.1}$$

where P_x is the law of Z when issued from x. In that case, we refer to α as the *index of self-similarity*.

There is a natural bijection between the class of exponentially killed Lévy processes and positive self-similar Markov processes, up to a naturally defined lifetime,

$$\zeta = \inf\{t > 0 : Z_t = 0\},$$

that is, the first moment it visits the origin. Roughly speaking, this bijection shows that the property of self-similarity is interchangeable with the property of having stationary and independent increments through an appropriate space-time transformation. Below, we state this bijection as a theorem.

Let us first introduce some more notation. Throughout this section, we shall use $\Xi := (\Xi_t, t \geq 0)$ to denote a one-dimensional Lévy process (not necessarily issued from the origin) that is killed and sent to the cemetery state $-\infty$ at an independent and exponentially distributed random time, $\mathbf{e} = \inf\{t > 0 : \Xi_t = -\infty\}$, with rate in $[0, \infty)$. As usual, we understand $\mathbf{e}$ as an exponential distribution in the *broader sense*, so that if its rate is 0, then $\mathbf{e} = \infty$ with probability one, that is, there is no killing.

We will be interested in applying a time change to the process Ξ by using its integrated exponential process, $I := (I_t, t \geq 0)$, where

$$I_t = \int_0^t e^{\alpha \Xi_s} \, ds, \qquad t \geq 0. \tag{5.2}$$

As the process I is increasing, we may define its almost sure limit, $I_\infty := \lim_{t\uparrow\infty} I_t$. We are also interested in the inverse process

$$\varphi(t) = \inf\{s > 0 : I_s > t\}, \qquad t \geq 0. \tag{5.3}$$

As usual, we work with the convention $\inf \emptyset = \infty$.

The following decomposition describes the celebrated Lamperti transformation, we omit the proof here as it is long and a distraction from our main objectives.

Theorem 5.2 (The Lamperti transform) *Fix $\alpha > 0$.*

(i) If (Z, P_x), $x > 0$, is a positive self-similar Markov process with index of self-similarity α, then up to its first visit of the origin, it can be represented as follows:

$$Z_t \mathbf{1}_{(t<\zeta)} = \exp\{\Xi_{\varphi(t)}\}, \qquad t \geq 0, \tag{5.4}$$

such that $\Xi_0 = \log x$ and either

(1) $P_x(\zeta = \infty) = 1$ for all $x > 0$, in which case, Ξ is a Lévy process satisfying $\limsup_{t\uparrow\infty} \Xi_t = \infty$,

(2) $P_x(\zeta < \infty$ and $Z_{\zeta-} = 0) = 1$ for all $x > 0$, in which case Ξ is a Lévy process satisfying $\lim_{t\uparrow\infty} \Xi_t = -\infty$, or

(3) $P_x(\zeta < \infty$ and $Z_{\zeta-} > 0) = 1$ for all $x > 0$, in which case Ξ is a Lévy process killed at an independent and exponentially distributed random time.

In all cases, we may identify $\zeta = I_\infty$.

(ii) *Conversely, for each $x > 0$, suppose that Ξ is a given (killed) Lévy process, issued from* $\log x$. *Define*

$$Z_t = \exp\{\Xi_{\varphi(t)}\}\mathbf{1}_{(t<I_\infty)}, \qquad t \geq 0.$$

Then Z defines a positive self-similar Markov process up to its absorption time $\zeta = I_\infty$, which satisfies $Z_0 = x$ and which has index α.

In the forthcoming sections, we shall identify a number of different pssMp, all of which are derived from path transformations of the stable process, for which we shall use some notational conventions. We shall generally use a modified version of the letter Ξ to denote individual Lévy processes associated to different pssMp. However, without confusion, φ and ζ will always denote the time change associated to the underlying Lévy process and the lifetime of the particular pssMp at hand, respectively.

Before passing to the promised specific examples of pssMp, we will first address one outstanding issue, namely whether it is possible to develop the notion of a pssMp issued from the origin. That is to say, whether we can find a candidate for P_0 that is consistent with the laws $P = (P_x, x > 0)$ described by Theorem 5.2, for any given pssMp. The Lamperti transform cannot help us in its given format. It would require the Lévy process Ξ to be issued from $-\infty$ and the associated time change to be appropriately well behaved in order for the representation (5.4) to correspond to a pssMp process issued from the origin. In the next section, we state without proof the main results in the literature, which deal with entry from the origin.

5.2 Starting at the origin

Theorem 5.2 (i) indicates that positive self-similar Markov processes naturally divide into two classes. First, *conservative processes*, for which $\zeta = \infty$ almost surely, and, secondly, *non-conservative processes*, for which $\zeta < \infty$ almost surely. Suppose that Z is a conservative positive self-similar Markov process. As alluded to above, we want to find a way to give a meaning to P_0 (and even better would be to give further meaning to this candidate as '$P_0 := \lim_{x\downarrow 0} P_x$', which would offer a sense of uniqueness). One way to do this is to look at the behaviour of the transition semigroup of (Z, P) as its initial value tends to zero. That is to say, to consider whether the weak limit

$$P_0(Z_t \in \mathrm{d}y) := \lim_{x\downarrow 0} P_x(Z_t \in \mathrm{d}y), \qquad t, y > 0, \tag{5.5}$$

exists. In that case, for any sequence of times $0 < t_1 \leq t_2 \leq \cdots \leq t_n < \infty$ and $y_1, \cdots, y_n \in (0, \infty)$, $n \in \mathbb{N}$, the Markov property gives us

$$
\begin{aligned}
&P_0(Z_{t_1} \in \mathrm{d}y_1, \cdots, Z_{t_n} \in \mathrm{d}y_n) \\
&\quad := \lim_{x \downarrow 0} P_x(Z_{t_1} \in \mathrm{d}y_1, \cdots, Z_{t_n} \in \mathrm{d}y_n) \\
&\quad = \lim_{x \downarrow 0} P_x(Z_{t_1} \in \mathrm{d}y_1) P_{y_1}(Z_{t_2 - t_1} \in \mathrm{d}y_2, \cdots, Z_{t_n - t_2} \in \mathrm{d}y_n) \\
&\quad = P_0(Z_{t_1} \in \mathrm{d}y_1) P_{y_1}(Z_{t_2 - t_1} \in \mathrm{d}y_2, \cdots, Z_{t_n - t_2} \in \mathrm{d}y_n).
\end{aligned}
$$

The limit (5.5), when it exists, thus implies the existence of P_0 as limit of P_x as $x \downarrow 0$, in the sense of convergence of finite-dimensional distributions. The following result gives a stronger sense of convergence, which implies the latter, as well as identifying the limiting law of (Z, P_0) at fixed times.

Theorem 5.3 *Assume that Z is a conservative positive self-similar Markov process. Moreover, suppose that the Lévy process $(\Xi, \mathbf{P})$, associated with Z through the Lamperti transform, is not a compound Poisson process and has an ascending ladder height process H which satisfies $\mathbf{E}[H_1] < \infty$. Then $P_0 := \lim_{x \downarrow 0} P_x$ exists, in the sense of convergence of on the Skorokhod space (see Section A.10 in the Appendix). Moreover, under P_0, the process Z leaves the origin continuously. Conversely, if $\mathbf{E}[H_1] = \infty$, then this limit does not exist. Under the additional assumption that $\mathbf{E}[\Xi_1] \in (0, \infty)$, for any positive measurable function f and $t > 0$,*

$$
E_0[f(Z_t)] = \frac{1}{\alpha \hat{\mathbf{E}}[|\Xi_1|]} \hat{\mathbf{E}}\left[\frac{1}{I_\infty} f\left((t/I_\infty)^{1/\alpha}\right)\right], \tag{5.6}
$$

where $I_\infty = \int_0^\infty \exp\{\alpha \Xi_s\} \mathrm{d}s$ and $(\Xi, \hat{\mathbf{P}})$ is equal in law to $(-\Xi, \mathbf{P})$.

Remark 5.4 The reader will note that we could equally have phrased (5.6) in the form

$$
E_0[f(Z_t)] = \frac{1}{\alpha \mathbf{E}[\Xi_1]} \mathbf{E}\left[\frac{1}{\hat{I}_\infty} f\left(\left(t/\hat{I}_\infty\right)^{1/\alpha}\right)\right],
$$

where $\hat{I}_\infty = \int_0^\infty \exp\{-\alpha \Xi_s\} \mathrm{d}s$. However, we have chosen the format as stated in Theorem 5.3 so that, when the reader is confronted with entrance laws for more general self-similar Markov processes later on in this text (see Theorem 11.16), the consistency will be clear.

The form of the entrance law (5.6) suggests that there must be a connection between (Z, P_0) and the law of the positive self-similar Markov process associated to $(\Xi, \hat{\mathbf{P}})$, that is, the dual of Ξ. As we will see below, this connection manifests through time reversal of the paths of (Z, P_0).

Let us assume that the Lévy process Ξ satisfies

$$
0 < \mathbf{E}[\Xi_1] < \infty, \tag{5.7}
$$

and consider the pssMp $(Z, \hat{P}_x)$ with self-similar index $\alpha > 0$, which is associated to $\hat{\Xi} = -\Xi$. In other words, the Lamperti transform of $(Z, \hat{P}_x)$ is equal in law to

$$x \exp\left\{\hat{\Xi}_{\hat{\varphi}(tx^{-\beta})}\right\} \mathbf{1}_{(t<x^{\beta}\hat{I}_\infty)}, \qquad t \geq 0, \tag{5.8}$$

where $\hat{I}_t = \int_0^t \exp(\alpha\hat{\Xi}_u)\mathrm{d}u$, $t \geq 0$, and $\hat{\varphi}$ denotes the right-continuous inverse process of $\hat{I}$. Note, as Ξ drifts to ∞ (cf. Section 2.6), it follows that $\hat{\Xi}$ drifts to $-\infty$. This means that the process $(Z, \hat{P})$ is continuously absorbed at the origin, where $\hat{P} = (\hat{P}_x, x > 0)$.

We want to show that (Z, P) and $(Z, \hat{P})$ are in duality with respect to a given measure, in other words, the latter has the law of the former when time reversed in an appropriate way. This will lead us to Proposition 5.5 below. In order to do this, we will use Hunt–Nagasawa duality theory (see Section A.12 in the Appendix), for which we must introduce the resolvent operators associated to (Z, P) and $(Z, \hat{P})$, respectively. For every $q \geq 0$, and measurable $f, g\colon (0, \infty) \to [0, \infty)$, we introduce the resolvent operators of (Z, P_x) and $(Z, \hat{P}_x)$,

$$V^q f(x) := E_x\left[\int_0^\infty \mathrm{e}^{-qt} f(Z_t)\,\mathrm{d}t\right] \quad \text{and} \quad \hat{V}^q g(x) := \hat{E}_x\left[\int_0^\zeta \mathrm{e}^{-qt} g(Z_t)\,\mathrm{d}t\right],$$

for $x > 0$, where ζ denotes the absorption time of $(Z, \hat{P})$.

Proposition 5.5 *Suppose* (5.7) *holds. The resolvent operators of* (Z, P) *and* $(Z, \hat{P})$, *are in duality with respect to the measure* $\mu(\mathrm{d}x) = x^{\alpha-1}\mathrm{d}x$. *That is to say, for every* $q \geq 0$, *and measurable functions* $f, g\colon (0, \infty) \to [0, \infty)$, *we have*

$$\int_0^\infty g(x) V^q f(x) x^{\alpha-1}\,\mathrm{d}x = \int_0^\infty f(x)\hat{V}^q g(x) x^{\alpha-1}\,\mathrm{d}x\,. \tag{5.9}$$

Proof Fix $q \geq 0$, and consider two measurable functions $f, g\colon (0, \infty) \to [0, \infty)$. From the Lamperti transform, we have

$$\begin{aligned}
\int_0^\infty g(x) V^q f(x) x^{\alpha-1}\,\mathrm{d}x &= \int_0^\infty g(x)\mathbf{E}\left[\int_0^\infty \mathrm{e}^{-qt} f\big(x\exp\{\Xi_{\varphi(tx^{-\alpha})}\}\big)\,\mathrm{d}t\right] x^{\alpha-1}\,\mathrm{d}x \\
&= \mathbf{E}\left[\int_0^\infty \mathrm{d}x\, x^{\alpha-1} g(x) \int_0^\infty \mathrm{d}s\, \mathrm{e}^{-qx^\alpha I_s} f\left(x\mathrm{e}^{\Xi_s}\right) x^\alpha \mathrm{e}^{\alpha\Xi_s}\right] \\
&= \mathbf{E}\left[\int_0^\infty \mathrm{d}s \int_0^\infty \mathrm{d}y\, \mathrm{e}^{-qy^\alpha \mathrm{e}^{-\alpha\Xi_s} I_s} g\left(y\mathrm{e}^{-\Xi_s}\right) f(y) y^{2\alpha-1} \mathrm{e}^{-\alpha\Xi_s}\right] \\
&= \int_0^\infty \mathrm{d}y\, y^{\alpha-1} f(y) \int_0^\infty \mathrm{d}s\, \mathbf{E}\left[\mathrm{e}^{-qy^\alpha \mathrm{e}^{-\alpha\Xi_s} I_s} g\left(y\mathrm{e}^{-\Xi_s}\right) y^\alpha \mathrm{e}^{-\alpha\Xi_s}\right] \\
&= \int_0^\infty \mathrm{d}y\, y^{\alpha-1} f(y) \int_0^\infty \mathrm{d}s\, \mathbf{E}\left[\mathrm{e}^{-qy^\alpha I'_s} g\left(y\mathrm{e}^{-\Xi'_s}\right) y^\alpha \mathrm{e}^{-\Xi'_s}\right],
\end{aligned}$$

where we have used the change of variable $y = x\exp(\Xi_s)$ in the third equality and, in the final equality, $\Xi'_u = \Xi_s - \Xi_{(s-u)^-}$, for $0 < u < s$, and $I'_s = \int_0^s e^{-\alpha\Xi'_u} du$. By the duality Lemma for Lévy processes (see Lemma 2.14), the processes $(\Xi_u, 0 < u < t)$ and $(\Xi'_u, 0 < u < t)$ have the same law. In other words, we have

$$\mathbf{E}\left[e^{-qy^\alpha I'_s} g\left(ye^{-\Xi'_s}\right) y^\alpha e^{-\alpha\Xi'_s}\right] = \hat{\mathbf{E}}\left[e^{-qy^\alpha I_s} g\left(ye^{\Xi_s}\right) y^\alpha e^{\alpha\Xi_s}\right],$$

where we recall that $\hat{\mathbf{P}}$ denotes the law of $\hat{\Xi}$. By using a similar change of variables as above, we deduce

$$\int_0^\infty g(x) V^q f(x) x^{\alpha-1}\, dx = \int_0^\infty f(x) \hat{V}^q g(x) x^{\alpha-1}\, dx,$$

as required. □

The weak duality between the laws of (Z, P) and $(Z, \hat{P})$, in the sense of (5.9), together with Hunt–Nagasawa's theory of time reversal (see Section A.12 in the Appendix), suggest that we can express the law of the time reversal of (Z, P_0) at some specific random times in terms of the law of $(Z, \hat{P})$. Our aim is to study the law of the time reversal of (Z, P_0) from last passage times

$$\mathtt{D}_x = \sup\{t \geq 0 : Z_t \leq x\}, \qquad \text{for} \quad x > 0$$

(with the definition $\sup\emptyset := 0$) and deduce that, under the event $\{Z_{\mathtt{D}_x-} = z\}$, it can be expressed in terms of the law of $(Z, \hat{P}_z)$. As a consequence we obtain a representation of the law of $\mathtt{D}_x$ under P_0, in terms of the law of the exponential functional $\hat{I}_\infty$.

To simplify the notation, we denote by $\mathcal{S}_x$ the support set of the law of $Z_{\mathtt{D}_x-}$. We also make the convention that $Z_{0-} = Z_0$.

Proposition 5.6 *Fix $x > 0$. Suppose* (5.7) *holds. The process $(Z, \hat{P}_z)$ is equal in law to that of the process $(Z_{(\mathtt{D}_x-t)-}, 0 \leq t \leq \mathtt{D}_x)$, under $P_0(\cdot \,|\, Z_{\mathtt{D}_x-} = z)$, for $z \in \mathcal{S}_x$.*

Proof The result follows from Theorem 5.3, Proposition 5.5 and Hunt–Nagasawa's theory of time reversal for Markov processes (see Section A.12 in the Appendix), specifically Theorem A.15.

Let $p_t(dx)$ be the entrance law of (Z, P_0) at time $t > 0$. It follows from the scaling property of Z that, for any $t > 0$, $p_t(dx) = p_1(t^{-\alpha} dx)$. The latter implies

$$\int_0^\infty p_t(dx)\, dt = \left(\int_0^\infty \frac{1}{y^\alpha} p_1(dy)\right) \alpha x^{\alpha-1}\, dx, \qquad x > 0.$$

From Theorem 5.3, we deduce

$$\int_0^\infty \frac{1}{y^\alpha} p_1(dy) = \frac{1}{\alpha\, \mathbf{E}[\Xi_1]}.$$

In other words, the resolvent measure $\int_{[0,\infty)} \delta_{\{0\}}(\mathrm{d}a) E_a \left[\int_0^\infty \mathbf{1}_{(Z_t \in \mathrm{d}x)} \mathrm{d}t\right]$, $x \geq 0$, is proportional to the measure $\mu(\mathrm{d}x)$, that is,

$$\frac{1}{\mathbf{E}[\xi_1]} \int_0^\infty f(x) x^{\alpha-1}\, \mathrm{d}x = E_0 \left[\int_0^\infty f(Z_t)\, \mathrm{d}t\right]. \tag{5.10}$$

Condition **(A)** of Nagasawa's Theorem A.15, specifically (A.43) and (A.42), are satisfied thanks to (5.9) and (5.10). Condition **(B)** of the same theorem can be easily satisfied thanks to the Feller property of $(Z, \hat{P})$. We may thus apply Theorem A.15 with the last exit time $\mathtt{D}_x$, noting in particular that, thanks to the assumed transience of (Z, P_0) in (5.7), $P_0(0 < \mathtt{D}_x < \infty) = 1$, and the proposition is proved. □

Another way to state Proposition 5.6 is as follows. For any $z \in \mathcal{S}_x$ with $x \geq z$, the time reversed process $(Z_{(z^\alpha I_\infty - t)-}, 0 \leq t \leq z^\alpha I_\infty)$ under $\hat{P}_z$ has the same law as $(Z_t, 0 \leq t < \mathtt{D}_x)$, under $P_0(\cdot \,|\, Z_{\mathtt{D}_x -} = z)$.

Now suppose that Z is a non-conservative positive self-similar Markov process. If there is a way to describe how Z can be issued from the origin then, in principle, one should be able to reissue it from the origin at all subsequent hitting times of this point in such a way that the resulting process remains strong Markov, thereby generating what is known as a recurrent extension. To be more precise, we say that a regular Feller process on $[0,\infty)$, $Z' := (Z'_t : t \geq 0)$, with probabilities $(P'_x, x \geq 0)$, is a *recurrent extension* of Z if, for each $x > 0$, the origin is not an exit boundary (i.e. a killing state) P'_x-almost surely and $(Z'_t, t < \zeta')$ under P'_x has the same law as (Z, P_x), where

$$\zeta' = \inf\{t > 0 : Z'_t = 0\}.$$

Showing that a recurrent extension exists is a very technical task and revolves around the theory of excursions. Roughly speaking, instead of constructing an entrance law for the family $(P_x, x > 0)$, it turns out that the correct mathematical procedure is to use $(P_x, x > 0)$ to construct an entrance law for an excursion measure that will describe the sojourns of Z' away from zero. Then with the help of what is known as *Itô synthesis*, one may piece together excursions end to end in an appropriate way to generate the desired recurrent extension.

In theory, one may approach the problem of constructing an excursion entrance law, and hence the problem of constructing a recurrent extension, in two different ways. Either excursions start by leaving the origin with a jump, or they leave the origin continuously. We focus on the case of recurrent extensions which leave the origin continuously, on account of the fact that the construction is unique. Otherwise, in the case of processes which leave the origin with a jump, there is no unique construction.

Theorem 5.7 *Assume that Z is a non-conservative positive self-similar Markov process. Suppose that $(\Xi, \mathbf{P})$ is the (killed) Lévy process associated with Z through the Lamperti transform. Then there exists a unique recurrent extension of Z which leaves 0 continuously if and only if there exists a Cramér number $\beta \in (0, \alpha)$ such that*

$$\mathbf{E}[e^{\beta \Xi_1}] = 1. \tag{5.11}$$

Here, as usual, α is the index of self-similarity.

5.3 Stable processes killed on entering $(-\infty, 0)$

Excluding the case of subordinators, a stable process is not a positive-valued process (albeit strong Markov and self-similar). However, by killing such processes as they enter $(-\infty, 0)$ and sending them to the cemetery state $\{0\}$, we can preserve the strong Markov, right-continuous and quasi-left-continuous properties, whilst introducing the property of positivity. Appealing to our previous notation for positive self-similar Markov processes, let us define, for $x > 0$,

$$Z_t = X_t \mathbf{1}_{(\underline{X}_t \geq 0)}, \qquad t \geq 0, \tag{5.12}$$

where X is a one-dimensional stable process. Our claim is that this process is also self-similar. To this end we need some additional facts about the self-similarity property of stable processes (which will also be of use in other examples).

Lemma 5.8 *Given any stable process $X = (X_t, t \geq 0)$, the pair $((X_t, \underline{X}_t), t \geq 0)$ is a strong Markov process. Moreover, if we denote its probabilities by $\mathbb{P}_{(x,s)}$, $-\infty < x \leq s < \infty$, then, for all $c > 0$ and $-\infty < x \leq s < \infty$,*

$$\textit{the law of } (c(X_{c^{-\alpha}t}, \underline{X}_{c^{-\alpha}t}), t \geq 0) \textit{ under } \mathbb{P}_{(x,s)} \textit{ is } \mathbb{P}_{(cx,cs)}.$$

Proof Note the statement of the theorem is trivial if X has monotone paths. We may thus assume that this is not the case for the remainder of the proof. Start by noticing that the process $((X_t, \underline{X}_t), t \geq 0)$ under $\mathbb{P}_{(x,s)}$ is equal in law to the process

$$(x + X_t, s \wedge (x + \underline{X}_t)), \qquad t \geq 0,$$

under $\mathbb{P}_{(0,0)}$. Now suppose that τ is any stopping time with respect to the filtration $(\mathcal{F}_t, t \geq 0)$ of X, which, by default, we take as naturally enlarged (see

Remark A.13 in the Appendix) so that it is right-continuous. Using the stationary and independent increments of X, it follows that, for any bounded measurable function f, we can write

$$\begin{aligned}
&\mathbb{E}_{(x,s)}[f(X_{T+t}, \underline{X}_{T+t})|\mathcal{F}_T] \\
&= \mathbb{E}_{(x,s)}\left[f\left(X_T + \widetilde{X}_t, \underline{X}_T \wedge (X_T + \underline{\widetilde{X}}_t)\right) \middle| \mathcal{F}_T \right] \\
&= \mathbb{E}_{(X_T, \underline{X}_T)}\left[f(\widetilde{X}_t, \underline{\widetilde{X}}_t) \right],
\end{aligned}$$

where, for $t \geq 0$, $\widetilde{X}_t = X_{T+t} - X_T$, $\underline{\widetilde{X}}_t = \inf_{s \leq t} \widetilde{X}_s$ and the process $(\widetilde{X}, \underline{\widetilde{X}})$ is also independent of $\mathcal{F}_T$. The strong Markov property now follows immediately.

To check self-similarity, we have that, for all $t \geq 0$ and $c > 0$,

$$c\underline{X}_{c^{-\alpha}t} = c \inf_{s \leq c^{-\alpha}t} X_s = \inf_{u \leq t} cX_{c^{-\alpha}u},$$

and hence, thanks to the self-similarity of X, it follows that $(c(X_{c^{-\alpha}t}, \underline{X}_{c^{-\alpha}t}), t \geq 0)$ under $\mathbb{P}_{(0,0)}$ is equal in law to $\mathbb{P}_{(0,0)}$. Next note that $(c(X_{c^{-\alpha}t}, \underline{X}_{c^{-\alpha}t}), t \geq 0)$ under $\mathbb{P}_{(x,s)}$ is equal in law to

$$\left(\left(cx + cX_{c^{-\alpha}t}, cs \wedge \left(cx + c\underline{X}_{c^{-\alpha}t}\right)\right), t \geq 0\right) \quad \text{under} \quad \mathbb{P}_{(0,0)},$$

which, in turn, is equal in law to $\mathbb{P}_{(cx,cs)}$. □

Returning now to the claim that the process (5.12) is self-similar, we note that, for $x, c > 0$, under $\mathbb{P}_x = \mathbb{P}_{(x,x)}$,

$$cZ_{c^{-\alpha}t} = cX_{c^{-\alpha}t}\mathbf{1}_{(\underline{X}_{c^{-\alpha}t} \geq 0)}, \qquad t \geq 0,$$

and, thanks to Lemma 5.8, this is equal in law to $(Z, \mathbb{P}_{cx})$.

The process (5.12) is a particular example of a positive self-similar Markov process which can fall into all three categories of Theorem 5.2. If X has negative jumps, then it is in class (3). If X has only positive jumps, then it falls into category (1) for the case of monotone paths (i.e. $\alpha \in (0,1)$) and otherwise category (2) (i.e. $\alpha \in (1,2)$). Let us now proceed to derive in explicit detail its Lamperti transform for the setting that Z is in category (1).

First, we turn our attention to computing the killing rate of the underlying Lévy process, which we henceforth refer to as $\xi^* = (\xi^*_t, 0 \leq t < \zeta^*)$, where ζ^* is its lifetime. At each moment in time $t > 0$ such that $\{t < \zeta\}$, given the Poisson point process of jumps with intensity given by (3.1) that govern the movement of X, the process Z is killed and sent to the origin at a rate equal to the rate of arrival of a negative jump of size Z_{t-} or greater. The rate at which Z is killed at time t on $\{t < \zeta\}$ is thus equal to

$$\Pi(-\infty, -Z_{t-})\,\mathrm{d}t = \left(\int_{-\infty}^{-Z_{t-}} \frac{c_2}{|x|^{1+\alpha}}\,\mathrm{d}x\right)\mathrm{d}t = \frac{c_2}{\alpha} Z_{t-}^{-\alpha}\,\mathrm{d}t.$$

Now suppose that q^* is the rate at which the underlying process ξ^* in the Lamperti transform is killed. On the probability space which supports the Lamperti representation, we also have that the rate at which Z is killed and sent to the origin is proportional to $q^*\,\mathrm{d}\varphi(t)$. Noting, however,

$$\int_0^{\varphi(t)} \mathrm{e}^{\alpha\xi_u^*}\,\mathrm{d}u = t,$$

it follows that, on $\{t < \zeta\}$,

$$q^*\,\mathrm{d}\varphi(t) = q^*\mathrm{e}^{-\alpha\xi^*_{\varphi(t)}}\,\mathrm{d}t = q^* Z_t^{-\alpha}\,\mathrm{d}t.$$

Comparing these two rates, we come to rest at

$$q^* = \frac{c_2}{\alpha} = \Gamma(\alpha)\frac{\sin(\pi\alpha\hat{\rho})}{\pi}, \tag{5.13}$$

where we have used (3.12). It is worth noting that, when X has only positive jumps, the above calculation rightfully tells us that the process ξ^* does not experience killing.

Next, we turn our attention to computing the Lévy measure and the characteristic exponent, respectively denoted by ν^* and Ψ^*, for the process ξ^*. We do this using a simple computation based on a fluctuation identity. Before proceeding thus, let us momentarily restrict ourselves to the setting that X experiences two-sided jumps. From the definition of Z as a stable process killed on first entering $(-\infty, 0)$, we know that, on the one hand, $Z_{\zeta-} = X_{\tau_0^- -}$, where

$$\tau_0^- = \inf\{t > 0 : X_t < 0\}.$$

On the other hand, thanks to the Lamperti representation, we also know that $Z_{\zeta-}$ has the Mellin transform which satisfies

$$\mathbb{E}_1\left[(Z_{\zeta-})^{\mathrm{i}\theta}\right] = \frac{q^*}{(\Psi^*(\theta) - q^*) + q^*}, \qquad \theta \in \mathbb{R},$$

which is the Fourier transform of the law of ξ^* at the moment before it is killed and sent to its cemetery state. Said another way, the right-hand side above is the Fourier transform of the Lévy process ξ^* stripped of its exponential killing rate and sampled at an independent and exponentially distributed random time with parameter q^*. We thus have that

$$\frac{q^*}{\Psi^*(\theta)} = \mathbb{E}_1\left[(X_{\tau_0^- -})^{\mathrm{i}\theta}\right]. \tag{5.14}$$

Setting

$$C = \frac{\sin(\alpha\hat{\rho}\pi)}{\pi}\frac{\Gamma(\alpha+1)}{\Gamma(\alpha\rho)\Gamma(\alpha\hat{\rho})}, \tag{5.15}$$

we note, with the help of Theorem 3.6, that, for all $v \geq 0$,

$$\begin{aligned}
\mathbb{P}_1(X_{\tau_0^- -} \in \mathrm{d}v) &= \hat{\mathbb{P}}(1 - X_{\tau_1^+ -} \in \mathrm{d}v) \\
&= C\left(\int_0^\infty \int_0^\infty \mathbf{1}_{(y \leq 1 \wedge v)} \frac{(1-y)^{\alpha\hat{\rho}-1}(v-y)^{\alpha\rho-1}}{(v+u)^{1+\alpha}} \mathrm{d}u\, \mathrm{d}y\right) \mathrm{d}v \\
&= \frac{C}{\alpha}\left(\int_0^1 \mathbf{1}_{(y \leq v)} v^{-\alpha}(1-y)^{\alpha\hat{\rho}-1}(v-y)^{\alpha\rho-1}\mathrm{d}y\right) \mathrm{d}v,
\end{aligned}$$

where $\hat{\mathbb{P}}$ is the law of $-X$ and we recall that $\tau_1^+ = \inf\{t > 0 : X_t > 1\}$. It is now a straightforward argument to show that, appealing to (5.14), for all $\theta \in \mathbb{R}$,

$$\begin{aligned}
\frac{q^*}{\Psi^*(\theta)} &= \frac{C}{\alpha}\int_0^1 (1-y)^{\alpha\hat{\rho}-1} \int_0^\infty \mathbf{1}_{(y \leq v)} v^{\mathrm{i}\theta - \alpha\hat{\rho}-1}\left(1 - \frac{y}{v}\right)^{\alpha\rho - 1} \mathrm{d}v\, \mathrm{d}y \\
&= \frac{C}{\alpha}\int_0^1 (1-y)^{\alpha\hat{\rho}-1} y^{\mathrm{i}\theta-\alpha\hat{\rho}}\, \mathrm{d}y \frac{\Gamma(\alpha\hat{\rho} - \mathrm{i}\theta)\Gamma(\alpha\rho)}{\Gamma(\alpha - \mathrm{i}\theta)} \\
&= \frac{C}{\alpha}\frac{\Gamma(1 - \alpha\hat{\rho} + \mathrm{i}\theta)\Gamma(\alpha\hat{\rho})}{\Gamma(1+\mathrm{i}\theta)} \cdot \frac{\Gamma(\alpha\hat{\rho} - \mathrm{i}\theta)\Gamma(\alpha\rho)}{\Gamma(\alpha - \mathrm{i}\theta)},
\end{aligned}$$

where in the first equality Fubini's Theorem has been used, in the second equality a straightforward substitution $w = y/v$ has been used for the inner integral on the preceding line together with the classical Beta integral and, finally, in the third equality, the Beta integral has been used for a second time. Inserting the respective values for the constants q^* and C, we come to rest at the following result.

Theorem 5.9 *Suppose that X has two-sided jumps. For the pssMp constructed by killing a stable process on first entry to $(-\infty, 0)$, the underlying Lévy process, ξ^*, that appears through the Lamperti transform has characteristic exponent given by*

$$\Psi^*(z) = \frac{\Gamma(\alpha - \mathrm{i}z)}{\Gamma(\alpha\hat{\rho} - \mathrm{i}z)}\frac{\Gamma(1+\mathrm{i}z)}{\Gamma(1 - \alpha\hat{\rho} + \mathrm{i}z)}, \qquad z \in \mathbb{R}. \tag{5.16}$$

In particular, the process ξ^* belongs to the class of Lamperti-stable processes with parameters $(\beta, \gamma, \hat{\gamma}) = (1 - \alpha\hat{\rho}, \alpha\rho, \alpha\hat{\rho})$.

Since $\Psi^*(0) = \Gamma(\alpha)/(\Gamma(\alpha\hat{\rho})\Gamma(1-\alpha\hat{\rho})) > 0$, we conclude that ξ^* is a killed Lévy process. Note that, in this case, we have a non-conservative pssMp. From (5.16), one easily verifies that $\Psi(-\mathrm{i}\alpha\hat{\rho}) = 0$, (which means that (5.11) holds with $\alpha\hat{\rho} < \alpha$) and hence, by Theorem 5.7 a recurrent extension at the origin is possible. A little thought reveals that the recurrent extension is nothing more than $X - \underline{X}$.

Let us return to the setting of one-sided jumps, which we excluded in Theorem 5.9. We will shortly see that, in fact, the exponent (5.16) correctly describes the underlying Lévy process ξ^* even in the one-sided jump setting, providing we take account of cancellations of the gamma functions when we insert the special values of ρ corresponding to those cases.

Let us first consider the case that X has only negative jumps. This means either $\alpha \in (0, 1)$ and $\rho = 0$, so that $-X$ is a subordinator which may be started from $x > 0$, or $\alpha \in (1, 2)$ and $\alpha\rho = 1$. The calculations preceding Theorem 5.9, are still meaningful. Note, in the case that $-X$ is a subordinator, then we should appeal to Theorem 3.4 in place of Theorem 3.6 in order to develop the right-hand side of (5.14). We leave the calculations to the reader. Summarising we have the following result.

Theorem 5.10 *Suppose that X has no positive jumps.*

(i) If $\alpha \in (0, 1)$ and $\rho = 0$, then ξ^* is the negative of a β-subordinator, whose Laplace exponent, in the sense of (2.10), *satisfies*

$$\kappa^*(\lambda) = \frac{\Gamma(1 + \lambda)}{\Gamma(1 - \alpha + \lambda)}, \qquad \lambda \geq 0.$$

(i) *If $\alpha \in (1, 2)$ and $\alpha\rho = 1$, then ξ^* is a Lamperti-stable spectrally negative Lévy process with parameters $(\beta, \gamma, \hat{\gamma}) = (2 - \alpha, 1, \alpha - 1)$ and Laplace exponent, in the sense of* (2.42), *given by*

$$\psi^*(\lambda) = (\lambda - \alpha + 1)\frac{\Gamma(1 + \lambda)}{\Gamma(2 - \alpha + \lambda)}, \qquad \lambda \geq 0.$$

As alluded to above, (5.16) is still valid, providing one takes account of the fact that, for the negative subordinator setting, the first ratio of gamma functions is unity and in the other setting the first ratio of gamma functions is linear thanks to the recursion formula. Moreover, as with the setting of two sided jumps, we note in both cases, ξ^* experiences killing, which we can see by setting the variable $\lambda = 0$ in the two Laplace exponents.

When X has no negative jumps, then either $\alpha \in (0, 1)$ and $\rho = 1$, in which case X is a subordinator, or $\alpha \in (1, 2)$ and $\alpha\hat{\rho} = 1$. In the first of these two settings, it is clear that ξ^* is a pure jump subordinator. We can appeal to the overshoot distribution of X past a threshold a, which, up to an exponential change of spatial scale, must coincide with the equivalent overshoot distribution of ξ^* over $\log a$. More precisely, for $a > 1$,

$$\frac{X_{\tau_a^+} - a}{a} = \exp\left(\xi^*_{\tau^{*,+}_{\log a}} - \log a\right) - 1, \tag{5.17}$$

where $\tau^{*,+}_{\log a} = \inf\{t > 0 : \xi^*_t > \log a\}$. Thanks to scaling, the law of the left-hand side of (5.17) under $\mathbb{P}_x$, as $a \to \infty$, is the same as its law as $x \to 0$. Moreover, from the discussion following Theorem 3.4, the left-hand side of (5.17) is invariant under $\mathbb{P}$. Hence by setting $a = 1$ in Corollary 3.5, and recalling the asymptotic overshoot distribution given by Corollary 2.28 we see that, for bounded measurable f on $[0, \infty)$,

$$\mathbb{E}\left[f\left(X_{\tau_1^+} - 1\right)\right] = \int_0^\infty f(u) \frac{\sin(\pi\alpha)}{\pi} \frac{u^{-\alpha}}{1+u} du$$
$$= \int_0^\infty f(e^y - 1) \int_0^\infty \frac{1}{m^*} \pi^*(y+z)\, dz\, dy, \tag{5.18}$$

where π^* is the (preemptively) assumed density of the jump measure associated to ξ^* and m^* is its mean (which is necessarily finite given that a non-trivial limit on the right-hand side of (5.17) exists). By changing variables, setting $u = e^y - 1$ in the second integral of (5.18), its easy to deduce, up to a multiplicative constant, that

$$\pi^*(x) = \alpha \frac{\sin \pi\alpha}{\pi} \frac{e^x}{(e^x - 1)^{\alpha+1}}. \tag{5.19}$$

It is thus clear that ξ^* is a β-subordinator which is characterised by the density given in (5.19).

Let us now turn to the case that X is a spectrally positive stable process with $\alpha \in (1, 2)$ and $\alpha\hat{\rho} = 1$. In this case, we may appeal to the two-sided exit formula in Lemma 3.8 to deduce that, for $x \in (0, 1)$,

$$\mathbb{P}_x(\tau_0^- < \tau_1^+) = (1 - x)^{\alpha-1}.$$

On the other hand, if we write $\mathbf{P}^*_x$, $x \in \mathbb{R}$, for the law of ξ^*, which is obviously a spectrally positive Lévy process, then from the two-sided exit problem in Lemma 2.30, we also have that

$$\mathbb{P}_x(\tau_0^- < \tau_1^+) = \mathbf{P}^*_{\log x}(\tau_0^{*,+} = \infty) = \frac{W^*(-\log x)}{W^*(\infty)},$$

where $\tau_0^{*,+} = \inf\{t > 0 : \xi^*_t > 0\}$ and W^* is the scale function of $-\xi^*$ (the negative sign makes it spectrally negative). In other words, up to a multiplicative constant, we can identify

$$W^*(y) = (1 - e^{-y})^{\alpha-1}, \qquad y \geq 0. \tag{5.20}$$

Now recall that the Laplace transform of W^* is equal to the reciprocal of the Laplace exponent of $-\xi^*$, say ψ^*, in the sense of (2.42), and hence the latter is easy to deduce from (5.20). Summarising the case of no negative jumps, we have the following result.

Theorem 5.11 *Suppose that X has no negative jumps.*

(i) *If* $\alpha \in (0,1)$ *and* $\rho = 1$, *then* ξ^* *is a* β*-subordinator, whose Laplace exponent, in the sense of* (2.10), *satisfies*

$$\kappa^*(\lambda) = \frac{\Gamma(\lambda+\alpha)}{\Gamma(\lambda)}, \qquad \lambda \geq 0.$$

(i) *If* $\alpha \in (1,2)$ *and* $\alpha\hat{\rho} = 1$, *then* $-\xi^*$ *is a Lamperti-stable spectrally negative Lévy process with* $(\beta, \gamma, \hat{\gamma}) = (0, \alpha - 1, 1)$. *and Laplace exponent, in the sense of* (2.42), *given by*

$$\psi^*(\lambda) = \lambda\frac{\Gamma(\lambda+\alpha)}{\Gamma(\lambda+1)}, \qquad \lambda \geq 0.$$

Again, we see that, as predicted above, the identities we obtained for the Laplace exponents in the conclusion of Theorem 5.11 are consistent with the expression (5.16) when we plug in the relevant values of ρ, providing we allow for cancellations of gamma functions in the latter. As one would expect, in both cases, there is no killing. This is obvious in the setting that X is a subordinator. For the case that $\alpha \in (1,2)$ and has no negative jumps, the process X reaches the origin continuously and hence ξ^* is a spectrally positive Lévy process that drifts to $-\infty$.

5.4 Stable processes conditioned to stay positive

Suppose that X is a stable process which does not have monotone paths (this is assumed throughout this section). Consider the process killed on first entry into $(-\infty, 0)$, that is to say (5.12). Writing ζ for the lifetime of the stochastic processes we consider, we have from the Lamperti transform that

$$\zeta = I^*_\infty := \int_0^{\zeta^*} \mathrm{e}^{\alpha\xi^*_t}\,\mathrm{d}t,$$

where ξ^* has characteristic exponent given by (5.16). Noting that ζ can otherwise be understood as τ_0^-, if we are careful to take account of the point of issue of X, then, with the help of Proposition 4.27, we can establish the following tail asymptotic.

Lemma 5.12 *For* $t, x > 0$,

$$\mathbb{P}_x(\tau_0^- > t) \sim \frac{\alpha}{\Gamma(\rho)\Gamma(1+\alpha\hat{\rho})} t^{-\hat{\rho}} x^{-\alpha\hat{\rho}}, \qquad t \to \infty. \tag{5.21}$$

Proof From Theorem 5.9 we note that $(\beta, \gamma, \hat{\beta}, \hat{\gamma}) = (1-\alpha\hat{\rho}, \alpha\rho, 1-\alpha\hat{\rho}, \alpha\hat{\rho})$ and hence ξ^* belongs to the class $\mathcal{H}_4$ defined in (4.35) (more precisely, it belongs to $\mathcal{H}_1\backslash\{\beta = 1\}$). When X experiences one-sided jumps, the process ξ^* is a Lamperti-stable with parameters $(\beta, \gamma, \hat{\gamma}) = (2-\alpha, 1, \alpha-1)$ in the spectrally negative case (see Theorem 5.10), or with parameters $(\beta, \gamma, \hat{\gamma}) = (0, \alpha-1, 1)$, in the spectrally positive case (see Theorem 5.11). As alluded to above, we can appeal to Proposition 4.27, with the observation that $\hat{\theta}_\chi = \hat{\rho}$ and $\hat{\theta} = \alpha\hat{\rho}$, where $\hat{\rho} = 1 - 1/\alpha$ in the spectrally negative case and $\hat{\rho} = 1/\alpha$ in the spectrally positive case. In other words, we write

$$\mathbb{P}_x(\tau_0^- > t) = \mathbb{P}_x(\zeta > t) = \mathbf{P}^*\left(I_\infty^* > tx^{-\alpha}\right) \sim \frac{\mathtt{M}^*(\hat{\rho})}{(\psi^*)'(\alpha\hat{\rho})} t^{-\hat{\rho}} x^{-\alpha\hat{\rho}}, \qquad t \to \infty, \tag{5.22}$$

where $\mathbf{P}^*$ is the law of ξ^* when issued from the origin, $\mathtt{M}^*$ denotes the Mellin transform of I_∞^* and $\psi^*(z) := -\Psi^*(-\mathrm{i}z)$ is the Laplace transform of ξ^*. Moreover, in this case $\mathtt{M}^*(\hat{\rho})$ and $(\psi^*)'(\alpha\hat{\rho})$ can be computed explicitly. The key ingredients are: The identities in Theorems 4.13 (i) and 5.9, for the two-sided jump case, Theorems 4.17 and 5.10 (ii), for the spectrally negative case, and Theorems 4.19 and 5.11 (ii), for the spectrally positive case; some straightforward manipulation using quasi-periodic properties of double gamma functions (see Appendix A.4); and standard properties of gamma functions (see Appendix A.3). We find that

$$\mathtt{M}^*(\hat{\rho}) = \frac{\alpha\Gamma(\alpha\rho)}{\Gamma(\rho)} \qquad \text{and} \qquad (\psi^*)'(\alpha\hat{\rho}) = \Gamma(\alpha\rho)\Gamma(1+\alpha\hat{\rho}),$$

thus concluding the proof. □

With this asymptotic, we can develop the notion of the stable process conditioned to stay positive. For $t \geq 0$, $x > 0$ and $A \in \mathcal{F}_t$, *if we are permitted to exchange limits and expectations* (something we will deal with later), then with the help of the Markov property, we have,

$$\mathbb{P}_x^\uparrow(A, t < \zeta) := \lim_{s\to\infty} \mathbb{P}_x(A, t < \zeta | t + s < \zeta) \tag{5.23}$$

$$= \lim_{s\to\infty} \mathbb{E}_x\left[\mathbf{1}_{(t<\tau_0^-)} \frac{\mathbb{P}_y(s < \tau_0^-)|_{y=X_t}}{\mathbb{P}_x(t + s < \tau_0^-)}\right] \tag{5.24}$$

$$= \mathbb{E}_x\left[\mathbf{1}_{(A,\, t<\tau_0^-)} \frac{X_t^{\alpha\hat{\rho}}}{x^{\alpha\hat{\rho}}}\right]. \tag{5.25}$$

As such, we can define the new probabilities $\mathbb{P}^\uparrow = (\mathbb{P}_x^\uparrow, x > 0)$ via the change of measure

$$\left.\frac{\mathrm{d}\mathbb{P}_x^\uparrow}{\mathrm{d}\mathbb{P}_x}\right|_{\mathcal{F}_t} := \frac{X_t^{\alpha\hat{\rho}}}{x^{\alpha\hat{\rho}}} \mathbf{1}_{(t<\zeta)}, \qquad t \geq 0. \tag{5.26}$$

Note, at this stage, it is unclear whether $(X, \mathbb{P}^\uparrow)$ is conservative or not.

Remark 5.13 An alternative way to define the stable process X conditioned to stay positive is to consider the limiting procedure

$$\mathbb{P}_x^{\uparrow}(A,\, t < \zeta) = \lim_{a\to\infty} \mathbb{P}_x(A,\, t < \zeta \wedge \tau_a^+ | \, \tau_a^+ < \zeta), \tag{5.27}$$

for $t \geq 0$ and $A \in \mathcal{F}_t$. Knowing an asymptotic for $\mathbb{P}_x(\tau_a^+ < \tau_0^-)$ as $a \to \infty$ is needed to compute the limit. This can be done, however, we first need to develop an identity for $\mathbb{P}_x(\tau_a^+ < \tau_0^-)$. Independently of the calculations in this section, this is done in the forthcoming Section 6.1. We leave it as an exercise for the reader to return to (5.27) later and verify that it results in the exact same change of measure as in (5.26). Note, this approach does not suffer the need for a relatively delicate result of the form (5.21), which is typically harder than deriving the two-sided exit probability in the stable setting.

To know whether $(X, \mathbb{P}^{\uparrow})$ is conservative is to know whether the right-hand side of (5.26) is a martingale or not. To this end, we can take advantage of the Lamperti transform in Section 5.3.

Note from (5.16) that $\Psi^*(-\mathrm{i}\alpha\hat{\rho}) = 0$, which implies that $(\exp(\alpha\hat{\rho}\xi_t^*), t \geq 0)$ is a martingale, where it is understood that ξ^* has cemetery state $-\infty$ and $\exp(-\infty) := 0$. In particular, if we write $\mathbf{e}$ for the lifetime of ξ^*, then as the time change in the Lamperti transform, $\varphi(t)$, is a stopping time (which may be infinite with positive probability), it follows that

$$\mathrm{e}^{\alpha\hat{\rho}\xi^*_{\varphi(t)}} = \mathbf{1}_{(\varphi(t)<\mathbf{e})}\mathrm{e}^{\alpha\hat{\rho}\xi^*_{\varphi(t)}} = \mathrm{e}^{\alpha\hat{\rho}\xi^*_{\varphi(t)}}\mathbf{1}_{(t<I^*_\infty)} = \mathbf{1}_{(t<\tau_0^-)}X_t^{\alpha\hat{\rho}}, \qquad t \geq 0.$$

It thus follows that, for $x > 0, t \geq 0$,

$$\mathbb{E}_x[\mathbf{1}_{(t<\tau_0^-)}X_t^{\alpha\hat{\rho}}] = \mathbf{E}^*_{\log x}[\mathrm{e}^{\alpha\hat{\rho}\xi^*_{\varphi(t)}}] = \mathrm{e}^{\alpha\hat{\rho}(\log x)} = x^{\alpha\hat{\rho}},$$

where $(\mathbf{P}^*_x, x \in \mathbb{R})$ are the probabilities of the killed Lévy process ξ^*. The martingale property now follows from the Markov Property as

$$\mathbb{E}_x[\mathbf{1}_{(t+s<\tau_0^-)}X_{t+s}^{\alpha\hat{\rho}}|\mathcal{F}_t] = \mathbf{1}_{(t<\tau_0^-)}\mathbb{E}_y[\mathbf{1}_{(s<\tau_0^-)}X_s^{\alpha\hat{\rho}}]_{y=X_t} = \mathbf{1}_{(t<\tau_0^-)}X_t^{\alpha\hat{\rho}}. \tag{5.28}$$

Now that we know that $x^{\alpha\hat{\rho}}$ is an invariant function in the above sense, we can return to (5.23) and justify the limit. It is easy to deduce that the right-hand side in (5.25) is a lower bound when the liminf is taken in (5.24) and then moved inside the expectation, thanks to Fatou's Lemma. On the other hand, since the last statement is true for all $A \in \mathcal{F}_t$, we can use the martingale property in (5.28) and deduce

$$
\begin{aligned}
&\limsup_{s\to\infty} \mathbb{P}_x(A,\, t<\zeta \,|\, t+s<\zeta)\\
&\qquad = 1 - \liminf_{s\to\infty} \mathbb{P}_x(A^c,\, t<\zeta \,|\, t+s<\zeta)\\
&\qquad \le 1 - \mathbb{E}_x\left[\mathbf{1}_{(A^c,\, t<\tau_0^-)}\frac{X_t^{\alpha\hat\rho}}{x^{\alpha\hat\rho}}\right]\\
&\qquad = \mathbb{E}_x\left[\mathbf{1}_{(t<\tau_0^-)}\frac{X_t^{\alpha\hat\rho}}{x^{\alpha\hat\rho}}\right] - \mathbb{E}_x\left[\mathbf{1}_{(A^c,\, t<\tau_0^-)}\frac{X_t^{\alpha\hat\rho}}{x^{\alpha\hat\rho}}\right]\\
&\qquad = \mathbb{E}_x\left[\mathbf{1}_{(A,\, t<\tau_0^-)}\frac{X_t^{\alpha\hat\rho}}{x^{\alpha\hat\rho}}\right].
\end{aligned}
\tag{5.29}
$$

The equality in (5.23) is thus justified.

Among those trajectories that remain positive, the change of measure (5.26) rewards trajectories that move to large positive values and penalises trajectories that visit close to the origin. Accordingly, we see heuristically that the resulting process should be repelled from the origin. We can make this more precise by examining our next claim that $(\mathbb{P}_x^\uparrow, x>0)$ describes a family of probability measures that belong to a pssMp. Said another way, we claim that the stable process conditioned to stay positive is a pssMp.

We observe, using Lemma 5.8, that for all positive, bounded and measurable functions f of $(X_s, s\le t)$,

$$
\begin{aligned}
&\mathbb{E}_x^\uparrow[f(cX_{c^{-\alpha}s}, s\le t)]\\
&\quad = \mathbb{E}\left[f(c(x+X_{c^{-\alpha}s}), s\le t)\frac{(x+X_{c^{-\alpha}t})^{\alpha\hat\rho}}{x^{\alpha\hat\rho}}\mathbf{1}_{(x+\underline{X}_{c^{-\alpha}t}\ge 0)}\right]\\
&\quad = \mathbb{E}\left[f(c(x+X_{c^{-\alpha}s}), s\le t)\frac{(cx+cX_{c^{-\alpha}t})^{\alpha\hat\rho}}{(cx)^{\alpha\hat\rho}}\mathbf{1}_{(cx+c\underline{X}_{c^{-\alpha}t}\ge 0)}\right]\\
&\quad = \mathbb{E}\left[f(cx+X_s, s\le t)\frac{(cx+X_t)^{\alpha\hat\rho}}{(cx)^{\alpha\hat\rho}}\mathbf{1}_{(cx+\underline{X}_t\ge 0)}\right]\\
&\quad = \mathbb{E}_{cx}^\uparrow[f(X_s, s\le t)],
\end{aligned}
\tag{5.30}
$$

for $c, t>0$. It is automatic that $(X, \mathbb{P}^\uparrow)$ is a Markov process from the change of measure. The remaining limiting properties of the semigroup associated to $\mathbb{P}^\uparrow$ that identify $(X, \mathbb{P}^\uparrow)$ as a Feller process (see Definition A.14 in the Appendix) are easy to verify once using the change of measure and dominated convergence. It follows that the stable process conditioned to stay positive is yet another example of a pssMp.

Now suppose we denote by $\xi^\uparrow = (\xi_t^\uparrow, t\ge 0)$ the Lévy process associated through the Lamperti transform to the conditioned stable process and write

$(\mathbf{P}_x^{\uparrow}, x \in \mathbb{R})$ for its probabilities (reserving, as usual, the special notation $\mathbf{P}^{\uparrow}$ in place of $\mathbf{P}_0^{\uparrow}$). We are again interested in computing the characteristic exponent of $\xi^{\uparrow}$, which we henceforth write $\Psi^{\uparrow}$.

Recalling the previously observed fact that, for each $t \geq 0$ and $x > 0$, the quantity $\varphi(t)$ is a stopping time, we can use the change of measure (5.26) at this stopping time and write, for all bounded measurable function g,

$$\mathbb{E}_1^{\uparrow}[g(X_t^{\uparrow})] = \mathbf{E}^{\uparrow}[g(\mathrm{e}^{\xi^{\uparrow}_{\varphi(t)}})] = \mathbf{E}^{*}\left[g(\mathrm{e}^{\xi^{*}_{\varphi(t)}})\mathrm{e}^{\alpha\hat{\rho}\xi^{*}_{\varphi(t)}}\mathbf{1}_{(t<I^{*}_{\infty})}\right].$$

Recalling that $\Psi^*(-\mathrm{i}\alpha\hat{\rho}) = 0$, we can deduce that $\xi^{\uparrow}$ has the law of the process ξ^* under an Esscher transform and in particular that $\Psi^{\uparrow}(z) = \Psi^*(z - \mathrm{i}\alpha\hat{\rho})$, for $z \in \mathbb{R}$; see Section 2.8 and formula (2.24) in particular. Note that we are applying the Esscher transform to a Lévy process which is killed at an independent and exponentially distributed random time. Nonetheless, Theorem 2.10 still accommodates for this context.

The Lévy process $\xi^{\uparrow}$ experiences no killing as $\Psi^{\uparrow}(0) = \Psi^*(-\mathrm{i}\alpha\hat{\rho})$ and, since

$$\Psi^{\uparrow\prime}(0) = \Psi^{*\prime}(-\mathrm{i}\alpha\hat{\rho}) = -\mathrm{i}\Gamma(\alpha\rho)\Gamma(1 + \alpha\hat{\rho}),$$

it follows that $\xi^{\uparrow}$ drifts to infinity at rate $\mathrm{i}\Psi^{\uparrow\prime}(0) = \Gamma(\alpha\rho)\Gamma(1 + \alpha\hat{\rho}) > 0$. Returning to (5.26), and recalling our earlier remarks that it rewards paths that explore large values and penalises paths that remain close to the origin, we now see that the Lamperti representation and the fact that $\xi^{\uparrow}$ drifts to ∞ actualises this heuristic. Summarising, we have the following result.

Theorem 5.14 *Suppose that X is a stable process which does not have monotone paths. The stable process conditioned to stay positive, defined by* (5.23) *or equivalently* (5.26), *is a conservative positive self-similar Markov process whose underlying Lévy process is given by*

$$\Psi^{\uparrow}(z) = \frac{\Gamma(\alpha\rho - \mathrm{i}z)}{\Gamma(-\mathrm{i}z)}\frac{\Gamma(1 + \alpha\hat{\rho} + \mathrm{i}z)}{\Gamma(1 + \mathrm{i}z)}, \qquad z \in \mathbb{R}. \tag{5.31}$$

In particular, in the spectrally one-sided cases, that is $\alpha \in (1,2), \alpha\rho = 1$ and $\alpha \in (1,2), \alpha\hat{\rho} = 1$, we interpret one of the two ratios of gamma functions as linear. Moreover, the process $\xi^{\uparrow}$ belongs to the class of Lamperti-stable processes with parameters $(1, \alpha\rho, \alpha\hat{\rho})$.

As the stable process conditioned to stay positive is a conservative process we can verify whether we can include the probabilities $\mathbb{P}_0^{\uparrow}$ in its definition or not by appealing to Theorem 5.3. For convenience, we restrict ourselves to the setting of two-sided jumps, the story for one-sided jumps is easily dealt with in a similar fashion. Thanks to the Wiener–Hopf factorisation (5.31), the ascending ladder height process of $\xi^{\uparrow}$ is a β-subordinator with parameters $(0, 0, \alpha\rho)$;

cf. Proposition (4.1). Suppose its Lévy density is denoted by $\upsilon^{\uparrow}$, then Corollary 4.3, gives us

$$\upsilon^{\uparrow}(x) = \frac{\alpha\rho}{\Gamma(1-\alpha\rho)}(1-\mathrm{e}^{-x})^{-\alpha\rho-1}\,\mathrm{e}^{-\alpha\rho x}, \qquad x > 0.$$

Hence, from Theorem 2.9, verifying directly that $\int_1^\infty \upsilon^{\uparrow}(x)\mathrm{d}x < \infty$, we see that the mean of the ascending ladder height is finite. Thus the conditions of Theorem 5.3 are met and we may include $\mathbb{P}_0^{\uparrow}$ in the definition of the process conditioned to stay non-negative. Moreover, we have all the ingredients to explicitly characterise $\mathbb{P}_0^{\uparrow}$. More precisely, we can compute explicitly the Mellin transform of X_1 and therefore determine its density.

Recall $\mathbf{P}^{\uparrow}$ is the law of $\xi^{\uparrow}$. Corollary 4.7 (ii), tells us that $\mathbf{E}^{\uparrow}[\xi_1^{\uparrow}] \in (0,\infty)$. Hence from Theorem 5.3, the Mellin transform of the entrance law of $(X,\mathbb{P}_0^{\uparrow})$ is such that

$$\mathbb{E}_0^{\uparrow}\left[X_t^{s-1}\right] = \frac{t^{\frac{s-1}{\alpha}}}{\alpha\mathbf{E}^{\uparrow}\left[\xi_1^{\uparrow}\right]}\mathbf{E}^{\uparrow}\left[\left(\hat{I}_\infty^{\uparrow}\right)^{\frac{1-s}{\alpha}-1}\right],$$

where $\hat{I}_\infty^{\uparrow} = \int_0^\infty \exp\{-\alpha\xi_s^{\uparrow}\}\mathrm{d}s$. From the explicit form of the characteristic exponent of the process $\xi^{\uparrow}$, see (5.31), we deduce that

$$\mathbf{E}^{\uparrow}\left[\xi_1^{\uparrow}\right] = \Gamma(\alpha\rho)\Gamma(1+\alpha\hat{\rho}).$$

Moreover, from Theorem 4.13 (i), the Mellin transform of the exponential functional $\hat{I}_\infty^{\uparrow}$ can be computed explicitly in terms of the double gamma function since $-\xi^{\uparrow}$ is in the class $\mathcal{H}_4$. In other words, the entrance law of $(X,\mathbb{P}_0^{\uparrow})$ satisfies

$$\mathbb{E}_0^{\uparrow}\left[X_t^{s-1}\right] = C_{\alpha,\rho}t^{\frac{s-1}{\alpha}}\Gamma\left(\frac{1-s}{\alpha}\right)\frac{G\left(\frac{1-s}{\alpha};\frac{1}{\alpha}\right)}{G\left(\rho+\frac{1-s}{\alpha};\frac{1}{\alpha}\right)}\frac{G\left(\frac{\alpha\hat{\rho}+\alpha+s}{\alpha};\frac{1}{\alpha}\right)}{G\left(\frac{\alpha+s}{\alpha};\frac{1}{\alpha}\right)},$$

where the function G is the double gamma function introduced in (4.42) and

$$C_{\alpha,\rho} = \frac{1}{\alpha\Gamma(\alpha\rho)\Gamma(1+\alpha\hat{\rho})}\frac{G\left(\rho+1;\frac{1}{\alpha}\right)}{G\left(1;\frac{1}{\alpha}\right)}\frac{G\left(\frac{1}{\alpha};\frac{1}{\alpha}\right)}{G\left(\frac{1+\alpha\hat{\rho}}{\alpha};\frac{1}{\alpha}\right)}.$$

5.5 Stable processes conditioned to limit to 0 from above

Let us again assume that X is a stable process which does not have monotone paths. There is another type of conditioning for Lévy processes, which also boils down to a change of measure, in the spirit of (5.26), which can be used to identify a family of positive self-similar Markov processes in the special setting that we work with α-stable processes. The conditioning of interest is that of the stable process conditioned to limit continuously to the origin before entering $(-\infty, 0)$. With this, we can immediately further exclude from interest the setting that $\alpha \in (1,2)$ and $\alpha\hat{\rho} = 1$ (spectrally positive processes with non-monotone paths) since they automatically exhibit this behaviour without the need for conditioning.

As in the previous two sections, we work with ζ as the generic notation for the lifetime of the processes we consider so, for example, $\zeta = \tau_0^-$ for $(X, \mathbb{P}_x)$, $x > 0$. Define the family of probabilities $(\mathbb{P}_x^{\downarrow}, x > 0)$ such that, for each $A \in \mathcal{F}_t$, $x, t, \eta > 0$,

$$\mathbb{P}_x^{\downarrow}(A,\, t < \zeta) = \lim_{\eta \downarrow 0} \mathbb{P}_x^{\downarrow}(A,\, t < \tau_\eta^-) := \lim_{\eta \downarrow 0} \lim_{\varepsilon \downarrow 0} \mathbb{P}_x(A,\, t < \tau_\eta^- | \underline{X}_{\tau_0^- -} \leq \varepsilon), \quad (5.32)$$

where $\underline{X}_t = \inf_{s \leq t} X_s$.

The limit can be computed once we recall the identity in Theorem 3.4 applied to the descending ladder subordinator of X, say H, (which is a stable subordinator with index $\alpha\hat{\rho}$) at first passage time $T_x^+ = \inf\{t > 0 : H_t > x\}$. From this identity we can derive, for any $0 < \varepsilon < x$,

$$\begin{aligned}\mathbb{P}_x(\underline{X}_{\tau_0^- -} \leq \varepsilon) &= \mathbb{P}_0(x - H_{T_x^+ -} \leq \varepsilon)\\ &= \frac{\sin(\pi\alpha\hat{\rho})}{\pi} \int_0^{\varepsilon} (x-u)^{\alpha\hat{\rho}-1} u^{-\alpha\hat{\rho}} \mathrm{d}u.\end{aligned}$$

From the above equality, L'Hôpital's rule thus gives us that

$$\lim_{\varepsilon \to 0} \varepsilon^{\alpha\hat{\rho}-1} \mathbb{P}_x(\underline{X}_{\tau_0^- -} \leq \varepsilon) = \frac{\sin(\pi\alpha\hat{\rho})}{\pi(1-\alpha\hat{\rho})} x^{\alpha\hat{\rho}-1}.$$

With this in hand, *again assuming that limits can be exchanged with expectations*, we have with the help of L'Hôpital's rule that

$$\begin{aligned}\mathbb{P}_x^{\downarrow}(A,\, t < \tau_\eta^-) &= \lim_{\varepsilon \downarrow 0} \mathbb{E}_x\left[\mathbf{1}_{(A,\, t<\tau_\eta^-)} \frac{\mathbb{P}_y(\underline{X}_{\tau_0^- -} \leq \varepsilon)|_{y=X_t}}{\mathbb{P}_x(\underline{X}_{\tau_0^- -} \leq \varepsilon)}\right] \quad (5.33)\\ &= \mathbb{E}_x\left[\mathbf{1}_{(A,\, t<\tau_\eta^-)} \frac{X_t^{\alpha\hat{\rho}-1}}{x^{\alpha\hat{\rho}-1}}\right].\end{aligned}$$

The conditioning (5.32) thus corresponds to the change of measure

$$\left.\frac{\mathrm{d}\mathbb{P}_x^{\downarrow}}{\mathrm{d}\mathbb{P}_x}\right|_{\mathcal{F}_t} = \frac{X_t^{\alpha\hat{\rho}-1}}{x^{\alpha\hat{\rho}-1}} \mathbf{1}_{(t<\zeta)}, \qquad t \geq 0. \quad (5.34)$$

Define $\mathbb{P}^{\downarrow} = (\mathbb{P}_x^{\downarrow}, x > 0)$. As in the previous section, we can verify, through careful analysis of (5.34), that the resulting process $(X, \mathbb{P}^{\downarrow})$ is both well defined (in particular that the limit (5.33) is justified) and that it is a pssMp. The key detail is that the right-hand side of (5.34) is a martingale. Once again, this a consequence of an exponential change of measure for the underlying Lévy process ξ^*. It is straightforward to observe that $\Psi^*(-\mathrm{i}(\alpha\hat{\rho}-1)) = 0$ from (5.16) and hence $\exp((\alpha\hat{\rho}-1)\xi^*_t)\mathbf{1}_{(t<\zeta)}$, $t \geq 0$, is a martingale. In a similar way to the calculations in the previous section, this translates to the right-hand side of (5.34) being a martingale as desired.

With our martingale in hand, we leave to the reader the details of the justification of the limit in (5.33), as well as the fact that $(X, \mathbb{P}^\downarrow)$ is a Feller process. They are essentially the same as in the previous section.

The process $(X, \mathbb{P}^\downarrow)$ is referred to as the stable process conditioned to limit to 0 from above. Unlike the construction of the stable process conditioned to stay positive, on account of the fact that $\alpha\hat{\rho} < 1$, amongst those paths that remain positive, the change of measure in (5.34) rewards paths that pass close to the origin and penalise those paths that explore large values.

Suppose now that we defined $\xi^\downarrow = (\xi^\downarrow_t, t \geq 0)$ as the Lévy process associated through the Lamperti transform to the stable process conditioned to limit to 0 from above. Similarly to the case of $\xi^\uparrow$, it turns out relatively simple to compute its characteristic exponent, which we shall henceforth denote by $\Psi^\downarrow$. Indeed, as before, for all bounded measurable g, we have from (5.34) and the Lamperti transform that

$$\mathbb{E}^\downarrow_1\left[g(X^\downarrow_t)\right] = \mathbf{E}^\downarrow\left[g(\xi^\downarrow_{\varphi(t)})\right] = \mathbf{E}^*\left[g(\xi^*_{\varphi(t)})\mathrm{e}^{(\alpha\hat{\rho}-1)\xi^*_{\varphi(t)}}\mathbf{1}_{(t<I^*_\infty)}\right],$$

where $\mathbf{P}^\downarrow$ is the law of $\xi^\downarrow$ issued from the origin, $\varphi(t)$ is the time change in the Lamperti transform. Note this also shows that the right-hand side of (5.34) is a martingale similarly to the setting for $\mathbb{P}^\uparrow$. In essence, $\xi^\downarrow$ has the same law as ξ^* under an Esscher transform. More precisely, one easily sees that $\Psi^\downarrow(z) = \Psi^*(z-\mathrm{i}(\alpha\hat{\rho}-1))$. The process $\xi^\downarrow$ belongs to the class of Lamperti-stable processes with parameters $(\beta, \gamma, \hat{\gamma}) = (0, \alpha\rho, \alpha\hat{\rho})$. Moreover, this process has no killing, that is, $\Psi^\downarrow(0) = 0$. Since

$$\Psi^{\downarrow\prime}(0) = \Psi^{*\prime}(-\mathrm{i}(\alpha\hat{\rho}-1)) = \mathrm{i}\Gamma(1+\alpha\rho)\Gamma(\alpha\hat{\rho}),$$

we have that $\xi^\downarrow_1$ has mean rate of increment $i\Psi^{\downarrow\prime}(0) = -\Gamma(1+\alpha\rho)\Gamma(\alpha\hat{\rho})$, which in turn ensures that $\xi^\downarrow$ drifts to $-\infty$. Indeed, by comparing their characteristic exponents, one may note that $\xi^\downarrow$ has the law of $-\xi^\uparrow$ with the roles of ρ and $\hat{\rho}$ interchanged. The long term drift of $\xi^\downarrow$ to $-\infty$ also emphasises the heuristic given earlier concerning the interpretation of the change of measure (5.34) in terms of how it rewards paths.

Theorem 5.15 *Suppose that X is a stable process without monotone paths, which has the possibility of negative jumps. The stable process conditioned to limit to* 0 *from above, defined by* (5.32) *or equivalently* (5.34), *is* *a non-conservative pssMp such that the underlying Lévy process, $\xi^{\downarrow}$, has characteristic exponent given by*

$$\Psi^{\downarrow}(z) = \frac{\Gamma(1 + \alpha\rho - \mathrm{i}z)}{\Gamma(1 - \mathrm{i}z)} \frac{\Gamma(\mathrm{i}z + \alpha\hat{\rho})}{\Gamma(\mathrm{i}z)}, \qquad z \in \mathbb{R}.$$

In other words, $\xi^{\downarrow}$ is a Lamperti-stable Lévy process with parameters $(\beta, \gamma, \hat{\gamma}) = (0, \alpha\rho, \alpha\hat{\rho})$.

Clearly the stable process conditioned to conditioned to limit to 0 from above is non-conservative. The only strictly positive Cramér number associated to $\xi^{\downarrow}$ is given by 1, that is, $\Psi(-\mathrm{i}) = 0$. It thus follows from Theorem 5.7 that the condition (5.11) is satisfied with $\beta = 1$. Hence a recurrent extension is possible if and only if $\alpha \in (1, 2)$.

5.6 Censored stable processes

Suppose that X is a stable process which has two-sided jumps. Define the occupation time of $(0, \infty)$ for X,

$$A_t = \int_0^t \mathbf{1}_{(X_s > 0)}\,\mathrm{d}s, \qquad t \geq 0,$$

and let

$$\gamma(t) = \inf\{s \geq 0 : A_s > t\}, \qquad t \geq 0, \tag{5.35}$$

be its right-continuous inverse. Define a process $(\check{Z}_t, t \geq 0)$ by setting $\check{Z}_t = X_{\gamma(t)}$, $t \geq 0$. This is the process formed by erasing the negative components of the space-time trajectory of X and shunting together the remaining positive sections of path.

We now turn zero into a cemetery state. Define the stopping time

$$T_0 = \inf\{t > 0 : \check{Z}_t = 0\}, \tag{5.36}$$

and the process

$$Z_t = \check{Z}_t \mathbf{1}_{(t < T_0)}, \qquad t \geq 0,$$

which is killed and absorbed at its cemetery state zero. We call the process Z the *censored stable process*. Our claim is that, up to its killing time T_0, this process is a positive self-similar Markov process. It is perhaps worth noting that we have *a priori* excluded the case that X has one-sided jumps as the resulting path description above produces the corresponding stable process killed on entering the lower half-line.

We now consider the scaling property. For each $c > 0$, define the rescaled process $(\check{Z}_t^c, t \geq 0)$ by $\check{Z}_t^c = c\check{Z}_{c^{-\alpha}t}$, and, correspondingly, let γ^c be defined such that

$$\int_0^{\gamma^c(t)} \mathbf{1}_{(X_s^c>0)}\,\mathrm{d}s = t, \tag{5.37}$$

where $X_t^c = cX_{c^{-\alpha}t}$, $t \geq 0$. By changing variable with $u = c^{-\alpha}s$ in (5.37) and noting that $A_{\gamma(c^{-\alpha}t)} = c^{-\alpha}t$, a short calculation shows that

$$c^\alpha \gamma(c^{-\alpha}t) = \gamma^c(t).$$

For each $x, c > 0$, we have under $\mathbb{P}_x$,

$$c\check{Z}_{c^{-\alpha}t} = cX_{\gamma(c^{-\alpha}t)} = cX_{c^{-\alpha}\gamma^c(t)} = X^c_{\gamma^c(t)}, \qquad t \geq 0.$$

The right-hand side above is equal in law to the process $(\check{Z}, \mathbb{P}_{cx})$, which establishes self-similarity of $\check{Z}$. Note, moreover, that, for all $c > 0$, if T_0^c is the time to absorption in $\{0\}$ of $\check{Z}$, then

$$T_0^c = \inf\{t > 0 : \check{Z}_{c^{-\alpha}t} = 0\} = c^\alpha \inf\{s > 0 : \check{Z}_s = 0\} = c^\alpha T_0. \tag{5.38}$$

It follows that, for all $x, c > 0$, under $\mathbb{P}_x$, $cZ_{c^{-\alpha}t} = c\check{Z}_{c^{-\alpha}t}\mathbf{1}_{(c^{-\alpha}t<T_0)}$, $t \geq 0$, which is equal in law to Z under $\mathbb{P}_{cx}$.

It remains to show that Z has the Feller property. This is easily verified through the Feller property of X and left as an exercise to the reader.

We now consider the pssMp Z more closely for different values of $\alpha \in (0,2)$. Denote by $\vec{\xi} = (\vec{\xi}_t, t \geq 0)$ the Lévy process associated to the censored stable process through the Lamperti transform. From the exposition in Chapter 3 we know that, for $\alpha \in (0,1]$, the stable process X cannot hit points. This implies that $T_0 = \infty$ almost surely, and so, in this case, $Z = \check{Z}$ and $\vec{\xi}$ experiences no killing. Moreover, when $\alpha \in (0,1)$, the process X is transient, meaning $\lim_{t\to\infty} |X_t| = \infty$, which implies that Z has almost surely finite occupancy of any bounded interval, and hence $\lim_{t\to\infty} \vec{\xi}_t = \infty$. When $\alpha = 1$, the process X is recurrent, meaning $\liminf_{t\to\infty} |X_t| = 0$ and $\limsup_{t\to\infty} |X_t| = \infty$, and so, $\limsup_{t\to\infty} \vec{\xi}_t = -\liminf_{t\to\infty} \vec{\xi}_t = \infty$. Meanwhile, for $\alpha \in (1,2)$, X can hit every point. Hence, we have, in particular, that $T_0 < \infty$. However, on account of the fact that creeping is not possible, the process X must make infinitely many jumps across zero during any arbitrarily small period of time immediately prior to hitting zero. Therefore, for $\alpha \in (1,2)$, Z approaches zero continuously.

Unlike the previous examples of pssMp, understanding the Lamperti transform of the censored stable process is a much less straightforward procedure. Nonetheless we shall proceed through a number of steps to achieve this goal. We start with the following result.

Theorem 5.16 *Suppose that X has two-sided jumps. The Lévy process $\vec{\xi}$ can be identified as follows.*

(i) *It is equal in law to the sum of two independent Lévy processes ξ^{L} and ξ^{C_1}.*

(ii) *The Lévy process ξ^{L} has characteristic exponent*

$$\Psi^*(z) - q^*, \qquad z \in \mathbb{R},$$

where we recall that Ψ^* is the characteristic exponent of the process ξ^* defined in Section 5.3 and $q^* = \Psi^*(0)$ is the killing rate of ξ^*, see (5.13). Said another way, ξ^{L} is formed by removing the effect of independent killing from ξ^*.

(iii) *The process ξ^{C_1} is a compound Poisson process with jump rate q^* and jump distribution, F^{C_1} on $\mathbb{R}$, given by its characteristic function,*

$$\int_{\mathbb{R}} \mathrm{e}^{\mathrm{i}\theta x} F^{C_1}(\mathrm{d}x) = \frac{\sin(\pi\alpha\rho)}{\pi\Gamma(\alpha)}\Gamma(1-\alpha\rho+\mathrm{i}\theta)\Gamma(\alpha\rho-\mathrm{i}\theta)\Gamma(1+\mathrm{i}\theta)\Gamma(\alpha-\mathrm{i}\theta), \tag{5.39}$$

for $\theta \in \mathbb{R}$.

Before beginning the proof, let us make some preparatory remarks. Recall that $\tau_0^- = \inf\{t > 0 : X_t < 0\}$ and let

$$\sigma = \inf\{t > \tau_0^- : X_t > 0\}$$

be the return time to $(0, \infty)$ for X. Note that, due to the continuous nature of the time-change γ,

$$Z_{\tau_0^-} = X_\sigma \quad \text{and} \quad Z_{\tau_0^- -} = X_{\tau_0^- -}.$$

In order to prove Theorem 5.16, we require the following lemma.

Lemma 5.17 *Suppose that X has two-sided jumps. For each $x > 0$, the joint law of $(X_{\tau_0^-}, X_{\tau_0^- -}, X_\sigma)$ under $\mathbb{P}_x$ is equal to the joint law of $(xX_{\tau_0^-}, xX_{\tau_0^- -}, xX_\sigma)$ under $\mathbb{P}_1$.*

Proof Recall that, for each $c > 0$, we defined the rescaled process $(X_t^c, t \geq 0)$ by $X_t^c = cX_{c^{-\alpha}t}$, $t \geq 0$. Let $\tau_0^{c,-} = \inf\{t > 0 : X_t^c < 0\}$ and $\sigma^c = \inf\{t > \tau_0^{c,-} : X_t^c > 0\}$. Then

$$c^\alpha \tau_0^- = \inf\{c^\alpha t > 0 : X_t < 0\} = \inf\{s > 0 : cX_{c^{-\alpha}s} < 0\} = \tau_0^{c,-},$$

and, similarly,

$$c^\alpha \sigma = \inf\{c^\alpha t > c^\alpha \tau_0^- : X_t > 0\} = \inf\{s > \tau_0^{c,-} : cX_{c^{-\alpha}s} < 0\} = \sigma^c.$$

With the classical scaling property of X, this implies that for every $c, x > 0$, under $\mathbb{P}_x$,

$$(cX_{\tau_0^-}, cX_{\tau_0^- -}, cX_\sigma) = (cX_{c^{-\alpha}\tau_0^{c,-}}, cX_{c^{-\alpha}\tau_0^{c,-}-}, cX_{c^{-\alpha}\sigma^c}) \stackrel{d}{=} (X_{\tau_0^-}, X_{\tau_0^- -}, X_\sigma),$$

under $\mathbb{P}_{cx}$. The claim follows by setting $c = 1/x$. □

Proof of Theorem 5.16 (i) and (ii) Applying the strong Markov property at the stopping time τ_0^-, it is sufficient to study the process $(Z_t, t \le \tau_0^-)$. It is clear that the path section $(Z_t, t < \tau_0^-)$ (note the strict inequality) agrees with $(X_t, t < \tau_0^-)$; however, rather than being killed at time τ_0^-, the process Z jumps to a positive state. Recalling that ξ^* is the Lévy process that describes, through the Lamperti transform, the process X killed on entering $(-\infty, 0)$, it follows that the dynamics of $\tilde{\xi}$ agree with those of ξ^* up to, but not including, the moment of killing of the latter. Instead of being killed at rate q^*, the process $\tilde{\xi}$ experiences an additional jump at rate q^*. This yields the decomposition of $\tilde{\xi}$ into the sum of $\xi^{\mathrm{L}} := (\xi_t^{\mathrm{L}}, t \ge 0)$ and $\xi^{C_1} := (\xi_t^{C_1}, t \ge 0)$, where ξ^{C_1} is a process which jumps at the times of a Poisson process with rate q^*, but whose jumps may depend on the position of $\tilde{\xi}$ prior to this jump. What remains to be shown is that the value of the first jump (and hence all subsequent jumps) of ξ^{C_1} is also independent of the previous path of ξ^{L}.

Let T_1 be the time of the first jump of the process ξ^{C_1} and note from above that it is an independent exponentially distributed random variable with parameter q^*. Using only the independence of the jump times of ξ^{L} and ξ^{C_1}, we can compute

$$\begin{aligned}\Delta Z_{\tau_0^-} := Z_{\tau_0^-} - Z_{\tau_0^- -} &= \exp(\xi_{T_1}^{\mathrm{L}} + \xi_{T_1}^{C_1}) - \exp(\xi_{T_1 -}^{\mathrm{L}} + \xi_{T_1 -}^{C_1}) \\ &= \exp(\tilde{\xi}_{T_1 -})[\exp(\Delta\xi_{T_1}^{C_1}) - 1] \\ &= X_{\tau_0^- -}[\exp(\Delta\xi_{T_1}^{C_1}) - 1],\end{aligned}$$

where $\Delta\xi_s^{C_1} = \xi_s^{C_1} - \xi_{s-}^{C_1}$. It follows that

$$\exp(\Delta\xi_{T_1}^{C_1}) = 1 + \frac{\Delta Z_{\tau_0^-}}{X_{\tau_0^- -}} = 1 + \frac{X_\sigma - X_{\tau_0^- -}}{X_{\tau_0^- -}} = \frac{X_\sigma}{X_{\tau_0^- -}}.$$

Hence, it is sufficient to show that $X_\sigma / X_{\tau_0^- -}$ is independent of $(X_t, t < \tau_0^-)$. To this end, observe that one consequence of Lemma 5.17 is that, for a bounded measurable function g and $x > 0$,

$$\mathbb{E}_x\left[g\left(\frac{X_\sigma}{X_{\tau_0^- -}}\right)\right] = \mathbb{E}_1\left[g\left(\frac{X_\sigma}{X_{\tau_0^- -}}\right)\right].$$

Now, fix $n \in \mathbb{N}$, take bounded, measurable functions f and g and $0 \le s_1 < s_2 < \cdots < s_n \le t$. Then, using the Markov property and the above equality,

$$\begin{aligned}
&\mathbb{E}_1\left[f(X_{s_1},\ldots,X_{s_n})g\left(\frac{X_\sigma}{X_{\tau_0^- -}}\right)\mathbf{1}_{(t<\tau_0^-)}\right] \\
&\quad = \mathbb{E}_1\left[f(X_{s_1},\ldots,X_{s_n})\mathbf{1}_{(t<\tau_0^-)}\mathbb{E}_{X_t}\left[g\left(\frac{X_\sigma}{X_{\tau_0^- -}}\right)\right]\right] \\
&\quad = \mathbb{E}_1\left[f(X_{s_1},\ldots,X_{s_n})\mathbf{1}_{(t<\tau_0^-)}\right]\mathbb{E}_1\left[g\left(\frac{X_\sigma}{X_{\tau_0^- -}}\right)\right].
\end{aligned}$$

We have now shown that ξ^{L} and ξ^{C_1} are independent. □

For the proof of Theorem 5.16 (iii), we need to have some understanding of the jump distribution of the compound Poisson process ξ^{C_1}. Let us introduce some additional notation in order to meet this goal. Let $\hat{X}$ be an independent copy of the dual process $-X$ and let

$$\hat{\tau}_0^- = \inf\{t > 0 : \hat{X}_t < 0\}.$$

Note that $\hat{X}$ is also a stable process. Furthermore, we shall denote by $\Delta\xi^{C_1}$ a random variable whose law is the same as the jump distribution of ξ^{C_1}. Before proving part (iii) of Theorem 5.16, we first need an intermediate lemma.

Lemma 5.18 *Suppose that X has two-sided jumps. The distribution of $\exp(\Delta\xi^{C_1})$ is equal to that of*

$$\left(-\frac{X_{\tau_0^-}}{X_{\tau_0^- -}}\right)\left(-\hat{X}_{\hat{\tau}_0^-}\right),$$

where X and $\hat{X}$ are taken to be independent, such that $X_0 = \hat{X}_0 = 1$.

Proof In the proof of Theorem 5.16, we saw that

$$\exp(\Delta\xi^{C_1}_{T_1}) = \frac{X_\sigma}{X_{\tau_0^- -}}. \tag{5.40}$$

Applying the Markov property at time τ_0^-, and then using Lemma 5.17 but applied to the stable process $\hat{X}$, we obtain for bounded, measurable f,

$$\begin{aligned}
\mathbb{E}_1[f(X_\sigma)|\mathcal{F}_{\tau_0^-}] &= \mathbb{E}_{-y}[f(-\hat{X}_{\hat{\tau}_0^-})]\Big|_{y=X_{\tau_0^-}} \\
&= \mathbb{E}_1[f(y\hat{X}_{\hat{\tau}_0^-})]\Big|_{y=X_{\tau_0^-}}.
\end{aligned}$$

Then, by disintegration,

$$\begin{aligned}
\mathbb{E}_1\left[f\left(\frac{X_\sigma}{X_{\tau_0^- -}}\right)\right] &= \mathbb{E}_1\left[\int_{(0,\infty)} f\left(\frac{x}{X_{\tau_0^- -}}\right)\mathbb{P}_1(X_\sigma \in \mathrm{d}x|\mathcal{F}_{\tau_0^-})\right] \\
&= \mathbb{E}_1\left[\int_{(0,\infty)} f\left(\frac{x}{X_{\tau_0^- -}}\right)\mathbb{P}_1[y\hat{X}_{\hat{\tau}_0^-} \in \mathrm{d}x]\Big|_{y=X_{\tau_0^-}}\right] \\
&= \mathbb{E}_1\left[\mathbb{E}_1\left[f\left(\frac{y\hat{X}_{\hat{\tau}_0^- -}}{z}\right)\right]\Big|_{y=X_{\tau_0^-},\, z=X_{\tau_0^- -}}\right].
\end{aligned}$$

The claim now follows. □

Proof of Theorem 5.16 (iii) The jump rate being equal to q^* has already been established. The characteristic function of $\Delta\xi^{C_1}$ can be evaluated by using the explicit distributional details of overshoots and undershoots of stable Lévy processes in the context of Lemma 5.18. For this Theorem 3.6 will be essential.

Recall that, for $a \in \mathbb{R}$,

$$\tau_a^+ = \inf\{t > 0 : X_t > a\},$$

and let $\hat{\tau}_a^+$ be defined similarly for $\hat{X}$. We have with the help of the beta function and Theorem 3.6,

$$\begin{aligned}
\mathbb{E}_1[(-\hat{X}_{\hat{\tau}_0^-})^{\mathrm{i}\theta}] &= \mathbb{E}[(X_{\tau_1^+} - 1)^{\mathrm{i}\theta}] \\
&= \frac{\sin(\pi\alpha\rho)}{\pi}\int_0^\infty t^{\mathrm{i}\theta-\alpha\rho}(1+t)^{-1}\,\mathrm{d}t \\
&= \frac{\sin(\pi\alpha\rho)}{\pi}\Gamma(1-\alpha\rho+\mathrm{i}\theta)\Gamma(\alpha\rho-\mathrm{i}\theta), \qquad \theta \in \mathbb{R}. \qquad (5.41)
\end{aligned}$$

Furthermore,

$$\begin{aligned}
\mathbb{E}_1\left[\left(-\frac{X_{\tau_0^-}}{X_{\tau_0^- -}}\right)^{\mathrm{i}\theta}\right] &= \mathbb{E}\left[\left(\frac{\hat{X}_{\hat{\tau}_1^+} - 1}{1 - \hat{X}_{\hat{\tau}_1^+ -}}\right)^{\mathrm{i}\theta}\right] \\
&= \frac{\sin(\pi\alpha\hat{\rho})}{\pi}\frac{\Gamma(\alpha+1)}{\Gamma(\alpha\rho)\Gamma(\alpha\hat{\rho})} \\
&\quad \times \int_0^1\int_y^\infty\int_0^\infty \frac{u^{\mathrm{i}\theta}(1-y)^{\alpha\hat{\rho}-1}(v-y)^{\alpha\rho-1}}{v^{\mathrm{i}\theta}(v+u)^{1+\alpha}}\,\mathrm{d}u\,\mathrm{d}v\,\mathrm{d}y. \qquad (5.42)
\end{aligned}$$

For the innermost integral above, substituting $w = v/u$ and appealing to the integral representation of the beta function in Appendix (A.18), we have

$$\int_0^\infty \frac{u^{\mathrm{i}\theta}}{(u+v)^{1+\alpha}}\,\mathrm{d}u = v^{\mathrm{i}\theta-\alpha}\int_0^\infty \frac{w^{\alpha-\mathrm{i}\theta-1}}{(1+w)^{1+\alpha}}\,\mathrm{d}w = v^{\mathrm{i}\theta-\alpha}\frac{\Gamma(\mathrm{i}\theta+1)\Gamma(\alpha-\mathrm{i}\theta)}{\Gamma(\alpha+1)}.$$

Substituting $z = v/y - 1$, the next iterated integral in (5.42) becomes

$$\int_y^\infty v^{-\alpha}(v-y)^{\alpha\rho-1}\,\mathrm{d}v = y^{-\alpha\hat{\rho}}\int_0^\infty \frac{z^{\alpha\rho-1}}{(1+z)^\alpha}\,\mathrm{d}z = y^{-\alpha\hat{\rho}}\frac{\Gamma(\alpha\rho)\Gamma(\alpha\hat{\rho})}{\Gamma(\alpha)}.$$

Finally, it remains to calculate the resulting outer integral of (5.42),

$$\int_0^1 y^{-\alpha\hat{\rho}}(1-y)^{\alpha\hat{\rho}-1}\,\mathrm{d}y = \Gamma(1-\alpha\hat{\rho})\Gamma(\alpha\hat{\rho}).$$

Multiplying together these expressions and using the reflection identity for the gamma function, see (A.12) in the Appendix, we obtain

$$\mathbb{E}_1\left[\left(-\frac{X_{\tau_0^-}}{X_{\tau_0^- -}}\right)^{\mathrm{i}\theta}\right] = \frac{\Gamma(\mathrm{i}\theta+1)\Gamma(\alpha-\mathrm{i}\theta)}{\Gamma(\alpha)}. \tag{5.43}$$

The result now follows from Lemma 5.18 by multiplying (5.41) and (5.43) together. □

Finally we are ready to fulfil our objective of computing the characteristic exponent $\vec{\Psi}$ of $\vec{\xi}$. First we compute Ψ^{C_1} and Ψ^{L}, the characteristic exponents of ξ^{C_1} and ξ^{L}. As ξ^{C_1} is a compound Poisson process with jump rate q^* given by (5.13) and jump distribution given by (5.39), we obtain, after re-writing with the help of the reflection formula (cf. (A.12) in the Appendix),

$$q^* = \frac{\Gamma(\alpha)}{\Gamma(\alpha\hat{\rho})\Gamma(1-\alpha\hat{\rho})},$$

that, for $\theta \in \mathbb{R}$,

$$\Psi^{C_1}(\theta) = \frac{\Gamma(\alpha)}{\Gamma(\alpha\hat{\rho})\Gamma(1-\alpha\hat{\rho})}\left(1 - \frac{\Gamma(1-\alpha\rho+\mathrm{i}\theta)\Gamma(\alpha\rho-\mathrm{i}\theta)\Gamma(1+\mathrm{i}\theta)\Gamma(\alpha-\mathrm{i}\theta)}{\Gamma(\alpha\rho)\Gamma(1-\alpha\rho)\Gamma(\alpha)}\right).$$

On the other hand, Theorem 5.9 provides an expression for the characteristic exponent Ψ^* of the Lamperti-stable process ξ^*, and removing the killing constant q^*, Theorem 5.16 (i) gives us

$$\Psi^L(\theta) = \frac{\Gamma(\alpha-\mathrm{i}\theta)}{\Gamma(\alpha\hat{\rho}-\mathrm{i}\theta)}\frac{\Gamma(1+\mathrm{i}\theta)}{\Gamma(1-\alpha\hat{\rho}+\mathrm{i}\theta)} - \frac{\Gamma(\alpha)}{\Gamma(\alpha\hat{\rho})\Gamma(1-\alpha\hat{\rho})}.$$

Applying the reflection formula twice, we compute

$$\begin{aligned}
\vec{\Psi}(\theta) &= \Psi^L(\theta) + \Psi^{C_1}(\theta)\\
&= \Gamma(\alpha-\mathrm{i}\theta)\Gamma(1+\mathrm{i}\theta)\\
&\quad\times\left(\frac{1}{\Gamma(\alpha\hat{\rho}-\mathrm{i}\theta)\Gamma(1-\alpha\hat{\rho}+\mathrm{i}\theta)} - \frac{\Gamma(1-\alpha\rho+\mathrm{i}\theta)\Gamma(\alpha\rho-\mathrm{i}\theta)}{\Gamma(\alpha\rho)\Gamma(1-\alpha\rho)\Gamma(\alpha\hat{\rho})\Gamma(1-\alpha\hat{\rho})}\right)\\
&= \Gamma(\alpha-\mathrm{i}\theta)\Gamma(1+\mathrm{i}\theta)\Gamma(1-\alpha\rho+\mathrm{i}\theta)\Gamma(\alpha\rho-\mathrm{i}\theta)\\
&\quad\times\left(\frac{\sin(\pi(\alpha\hat{\rho}-\mathrm{i}\theta))\sin(\pi(\alpha\rho-\mathrm{i}\theta))}{\pi^2} - \frac{\sin(\pi\alpha\hat{\rho})\sin(\pi\alpha\rho)}{\pi^2}\right).
\end{aligned} \tag{5.44}$$

Manipulations of the classical product and sum identities for trigonometric functions give us

$$\sin(\pi(\alpha\hat{\rho} - \mathrm{i}\theta))\sin(\pi(\alpha\rho - \mathrm{i}\theta)) + \sin(\pi\mathrm{i}\theta)\sin(\pi(\alpha - \mathrm{i}\theta)) = \sin(\pi\alpha\hat{\rho})\sin(\pi\alpha\rho).$$

Hence we can push (5.44) a little further and write

$$\overset{\rightarrow}{\Psi}(\theta) = \Gamma(\alpha - \mathrm{i}\theta)\Gamma(1 + \mathrm{i}\theta)\Gamma(1 - \alpha\rho + \mathrm{i}\theta)\Gamma(\alpha\rho - \mathrm{i}\theta)\frac{\sin(\pi\mathrm{i}\theta)\sin(\pi(\alpha - \mathrm{i}\theta))}{\pi^2}.$$

Again, using the reflection formula for gamma functions twice (see (A.12) in the Appendix), this leads to the following main result.

Theorem 5.19 *Suppose that X has two-sided jumps. For the pssMp constructed by censoring the stable process in $(-\infty, 0)$, the underlying Lévy process $\overset{\rightarrow}{\xi}$ that appears through the Lamperti transform has characteristic exponent given by*

$$\overset{\rightarrow}{\Psi}(z) = \frac{\Gamma(\alpha\rho - \mathrm{i}z)}{\Gamma(-\mathrm{i}z)}\frac{\Gamma(1 - \alpha\rho + \mathrm{i}z)}{\Gamma(1 - \alpha + \mathrm{i}z)}, \qquad z \in \mathbb{R}. \tag{5.45}$$

In other words, the process $\overset{\rightarrow}{\xi}$ belongs to the class of hypergeometric Lévy processes with $(\beta, \gamma, \hat{\beta}, \hat{\gamma}) = (1, \alpha\rho, 1 - \alpha, \alpha\hat{\rho})$.

We may now deduce from (5.45) that $\overset{\rightarrow}{\xi}$ drifts to ∞, oscillates, drifts to $-\infty$, respectively, as $\alpha \in (0, 1)$, $\alpha = 1$ and $\alpha \in (1, 2)$, with two-sided jumps in all cases. In other words, its associated pssMp Z is transient for $\alpha \in (0, 1)$ and recurrent for $\alpha \in [1, 2)$. We also deduce that, in accordance with the behaviour of X, the process Z drift to ∞, oscillates or hits 0 continuously, respectively when $\alpha \in (0, 1)$, $\alpha = 1$ and $\alpha \in (1, 2)$.

In the conservative setting, that is, if $\alpha \in (0, 1]$, from Theorem 4.6 the ascending ladder process is a β-subordinator with Laplace exponent $\lambda \mapsto \Gamma(\alpha\rho + \lambda)/\Gamma(\lambda)$. As in the setting of the stable process conditioned to stay positive, it is easy to verify that the conditions of Theorem 5.3 are thus met and hence the censored stable process is well defined when issued from the origin. Moreover, we can compute explicitly the Mellin transform of Z_1 and therefore determine its density when $\alpha \in (0, 1)$.

Let us assume that $\alpha \in (0, 1)$ and write $\overset{\rightarrow}{\mathbf{P}}$ for the law of $\overset{\rightarrow}{\xi}$. Recall from Corollary 4.7 part (ii), that $\overset{\rightarrow}{\mathbf{E}}[\overset{\rightarrow}{\xi}_1] \in (0, \infty)$. Hence from Theorem 5.3, the Mellin transform of the entrance law of $(Z, \mathbb{P}_0)$ is such that

$$\mathbb{E}_0\left[Z_t^{s-1}\right] = \frac{t^{\frac{s-1}{\alpha}}}{\alpha\,\overset{\rightarrow}{\mathbf{E}}\left[\overset{\rightarrow}{\xi}_1\right]}\,\overset{\rightarrow}{\mathbf{E}}\left[\left(\vec{I}_\infty\right)^{\frac{1-s}{\alpha}-1}\right],$$

where $\vec{I}_\infty = \int_0^\infty \exp\{-\alpha\overset{\rightarrow}{\xi}_s\}\mathrm{d}s$. From the explicity form of the characteristic exponent of the process $\overset{\rightarrow}{\xi}$, see (5.45), we deduce that

$$\overset{\rightarrow}{\mathbf{E}}\left[\overset{\rightarrow}{\xi}_1\right] = \Gamma(\alpha\rho)\Gamma(1-\alpha\rho).$$

Moreover, from Theorem 4.13 (i), the Mellin transform of the exponential functional $\vec{I}{}^{\uparrow}_\infty$ can be computed explicitly in terms of the double gamma function since $-\overset{\rightarrow}{\xi}$ is in the class $\mathcal{H}_4$. In other words, the entrance law of $(Z, \mathbb{P}_0)$ satisfies

$$\mathbb{E}_0\left[Z_t^{s-1}\right] = C_{\alpha,\rho} t^{\frac{s-1}{\alpha}} \Gamma\left(\frac{1-s}{\alpha}\right) \frac{G\left(\frac{1-s}{\alpha};\frac{1}{\alpha}\right)}{G\left(\rho+\frac{1-s}{\alpha};\frac{1}{\alpha}\right)} \frac{G\left(\frac{\alpha\hat{\rho}+s}{\alpha};\frac{1}{\alpha}\right)}{G\left(\frac{s}{\alpha};\frac{1}{\alpha}\right)},$$

where the function G is the double gamma function introduced in (4.42) and

$$C_{\alpha,\rho} = \frac{1}{\alpha\Gamma(\alpha\rho)\Gamma(1-\alpha\rho)} \frac{G\left(\rho+1;\frac{1}{\alpha}\right)}{G\left(1;\frac{1}{\alpha}\right)} \frac{G\left(\frac{1-\alpha}{\alpha};\frac{1}{\alpha}\right)}{G\left(\frac{1-\alpha\rho}{\alpha};\frac{1}{\alpha}\right)}.$$

In the non-conservative setting, that is, when $\alpha \in (1,2)$, we note that the non-negative Cramér number is equal to $0 < \alpha - 1 < 1 < \alpha$. The conditions of Theorem 5.7 thus holds and a recurrent extension is possible. The recurrent extension is, of course, the continued censoring of X.

5.7 The radial part of an isotropic stable process

In this section, we consider an isotropic d-dimensional stable process X with index $\alpha \in (0,2)$. In particular, we are interested in the process defined by its radial part, that is, $(|X_t|, t \geq 0)$, where $|\cdot|$ denotes the Euclidean norm.

Similar to the case of the censored stable process, we turn zero into a cemetery state since the process X may be recurrent and hit zero. Define the stopping time

$$\tau^{\{0\}} = \inf\{t > 0 : |X_t| = 0\} \tag{5.46}$$

and the process

$$Z_t = |X_t|\mathbf{1}_{(t<\tau^{\{0\}})}, \qquad t \geq 0, \tag{5.47}$$

which is killed and absorbed at its cemetery state whenever X hits 0 for the first time.

Recall that X is isotropic. That is to say, for any orthogonal matrix U on $\mathbb{R}^d$, the processes X and UX have the same law under $\mathbb{P}_0$. It follows that the process

$Z = (Z_t, t \geq 0)$ is Markovian. Indeed, suppose we identify $(X_t^{(x)}, t \geq 0)$ as equal in law to $(X, \mathbb{P}_x)$. Distributional rotational invariance implies that

$$(|X_t^{(x)}|, t \geq 0) \overset{(d)}{=} (|X_t^{(|x|1)}|, t \geq 0), \tag{5.48}$$

where $\overset{(d)}{=}$ means equality in law and $1 = (1, 0, \cdots, 0) \in \mathbb{R}^d$ is the 'North Pole' on $\mathbb{S}^{d-1}$. Moreover, for any bounded and measurable function g, appealing to the Markov property of $X^{(x)}$ and (5.48), we have

$$\begin{aligned}\mathbb{E}\left[g\left(|X_{t+s}^{(x)}|\right) \middle| \sigma(|X_u^{(x)}|, u \leq s)\right] &= \mathbb{E}\left[g\left(|\tilde{X}_t^{(X_s^{(x)})}|\right) \middle| \sigma(|X_u^{(x)}|, u \leq s)\right] \\ &= \mathbb{E}\left[g\left(|\tilde{X}_t^{(|X_s^{(x)}|1)}|\right) \middle| \sigma(|X_u^{(x)}|, u \leq s)\right] \\ &= \mathbb{E}\left[g\left(|\tilde{X}_t^{(y1)}|\right)\right]_{y=|X_s^{(x)}|},\end{aligned}$$

where $\tilde{X}_t^{(y)}$, $t \geq 0$ is an independent copy of $(X, \mathbb{P}_y)$, $y \in \mathbb{R}^d$. It follows that $|X|$ and hence Z is a Markov process. This argument can easily be developed to deduce that Z is a Feller process (see Definition A.14 in the Appendix) by using the Feller property of X and dominated convergence.

The process Z also inherits the scaling property from X and the scaling of $\tau^{\{0\}}$, similar in spirit to (5.38). This implies that the radial part of an isotropic stable process killed when it hits zero is a pssMp with index α.

We now consider the process Z more closely for different values of α and d, and denote by $\xi = (\xi_t, t \geq 0)$ its associated Lévy process through the Lamperti transform. From the exposition in Chapter 3 we know that, for $d \geq \alpha$, the stable process X cannot hit points. This implies that $\tau^{\{0\}} = \infty$ almost surely, and so, in this case, Z and ξ experience no killing. Moreover, when $\alpha < d$, the process X is transient implying that Z and ξ drift to ∞. When $d = \alpha = 1$, the process X is recurrent which implies that the Lévy process ξ oscillates. In the remaining case, that is, $d = 1$ and $\alpha \in (1, 2)$, the process X is recurrent and can hit every point, in other words, $\tau^{\{0\}} < \infty$ almost surely. Since X must make infinitely many jumps across zero during any arbitrarily small period of time immediately prior to hitting zero, the process Z approaches zero continuously implying that ξ drifts to $-\infty$.

Towards the goal of computing the characterisitic exponent of ξ, we first show that in the one-dimensional setting there is a similar decomposition of the Lévy process which is associated to the censored stable process. More precisely, we have the following result.

Proposition 5.20 *Assume that $d = 1$. The Lévy process ξ can be decomposed as follows*

(i) *It is equal in law to the sum of two independent Lévy processes* ξ^{L} *and* ξ^{C_2}.

(ii) *The Lévy process* ξ^{L} *has characteristic exponent*

$$\Psi^*(z) - q^*, \qquad z \in \mathbb{R},$$

where we recall that Ψ^* *is the characteristic exponent of the process* ξ^* *defined in Section 5.3 and* $q^* = \Psi^*(0)$ *is the killing rate of* ξ^**, see (5.13). Said another way,* ξ^{L} *is formed by removing the effect of independent killing from* ξ^*.

(iii) *The process* ξ^{C_2} *is a compound Poisson process with jump rate* q^* *and jump distribution,* F^{C_2} *on* $\mathbb{R}$*, given by its characteristic function,*

$$\int_{\mathbb{R}} \mathrm{e}^{\mathrm{i}\theta x} F^{C_2}(\mathrm{d}x) = \frac{\Gamma(1+\mathrm{i}\theta)\Gamma(\alpha-\mathrm{i}\theta)}{\Gamma(\alpha)}. \tag{5.49}$$

Proof In order to prove this result, we use similar arguments as those used in the censored stable process case that we repeat for the sake of completeness. Recall that $\tau_0^- = \inf\{t > 0 : X_t < 0\}$ and observe that it is almost surely finite since the process X is symmetric. Again from symmetry and the strong Markov property at τ_0^-, one can deduce that it is sufficient to study the process $(Z_t, t \le \tau_0^-)$. It is clear that the path section $(Z_t, t < \tau_0^-)$ agrees with $(X_t, t < \tau_0^-)$; however, rather than being killed at time τ_0^-, the process Z jumps to a positive state. Recalling that ξ^* is the Lévy process that describes, through the Lamperti transform, the process X killed on entering $(-\infty, 0)$, it follows that the dynamics of ξ agree with those of ξ^* up to, but not including, the moment of killing of the latter. Instead of being killed at rate q^*, the process ξ experiences an additional jump at rate q^*. This yields the decomposition of ξ into the sum of $\xi^{\mathrm{L}} := (\xi_t^{\mathrm{L}}, t \ge 0)$ and $\xi^{C_2} := (\xi_t^{C_2}, t \ge 0)$ a process which jumps at the times of a Poisson process with rate q^*, but whose jumps may depend on the position of ξ prior to this jump.

Let T_1 be the time of the first jump of the process ξ^{C_2} and note that it is an independent exponentially distributed random variable with parameter q^*. Using only the independence of the jump times of ξ^{L} and ξ^{C_2}, we can compute

$$\begin{aligned}\Delta Z_{\tau_0^-} := Z_{\tau_0^-} - Z_{\tau_0^- -} &= \exp(\xi_{T_1}^{\mathrm{L}} + \xi_{T_1}^{C_2}) - \exp(\xi_{T_1 -}^{\mathrm{L}} + \xi_{T_1 -}^{C_2}) \\ &= \exp(\xi_{T_1 -})[\exp(\Delta\xi_{T_1}^{C_2}) - 1] \\ &= X_{\tau_0^- -}[\exp(\Delta\xi_{T_1}^{C_2}) - 1],\end{aligned}$$

where $\Delta\xi_s^{C_2} = \xi_s^{C_2} - \xi_{s-}^{C_2}$. It follows that

$$\exp(\Delta\xi_{T_1}^{C_2}) = 1 + \frac{\Delta Z_{\tau_0^-}}{X_{\tau_0^- -}} = 1 + \frac{-X_{\tau_0^-} - X_{\tau_0^- -}}{X_{\tau_0^- -}} = -\frac{X_{\tau_0^-}}{X_{\tau_0^- -}},$$

which is independent of $(X_t, t < \tau_0^-)$ according to the proof of Theorem 5.16, thus implying that ξ^{L} and ξ^{C_2} are independent. Moreover, from the proof of Proposition 5.20 and (5.43), the characteristic function of the jump distribution F^{C_2} satisfies (5.49). □

We now proceed with the main result in this section which determines the characteristics of ξ.

Theorem 5.21 *For the pssMp constructed using the radial part of an isotropic d-dimensional stable process, the underlying Lévy process, ξ that appears through the Lamperti transform has characteristic exponent given by*

$$\Psi(z) = 2^{\alpha}\frac{\Gamma(\frac{1}{2}(-\mathrm{i}z+\alpha))}{\Gamma(-\frac{1}{2}\mathrm{i}z)}\frac{\Gamma(\frac{1}{2}(\mathrm{i}z+d))}{\Gamma(\frac{1}{2}(\mathrm{i}z+d-\alpha))}, \qquad z \in \mathbb{R}. \tag{5.50}$$

In other words, the process 2ξ belongs to the class of hypergeometric Lévy processes with $(\beta, \gamma, \hat{\beta}, \hat{\gamma}) = (1, \alpha/2, (d-\alpha)/2, \alpha/2)$.

Proof We need to appeal to two different tactics according to whether $\alpha < d$ or $d \le \alpha$. Note that the first case covers dimensions $d \ge 2$ and dimension $d = 1$ with $\alpha \in (0,1)$, whereas the second case covers the setting that $\alpha \in [1,2)$ and $d = 1$. Essentially these cases distinguish the setting that X is transient and recurrent, respectively. The methods for each of these settings are quite different.

Assume that $d > \alpha$, that is, that the process X is transient. In this case, the process ξ drifts towards ∞ since the process Z is also transient. Write $(\mathbf{P}_x, x \in \mathbb{R})$ for the probabilities of ξ, reserving, as usual, the notation $\mathbf{P}$ to mean $\mathbf{P}_0$. Our strategy is to compute the integral

$$\int_0^\infty \mathbb{E}_1[Z_t^u]\,\mathrm{d}t = \int_0^\infty \mathbf{E}[\mathrm{e}^{u\xi_{\varphi(t)}}]\,\mathrm{d}t = \int_0^\infty \mathbf{E}[\mathrm{e}^{(u+\alpha)\xi_s}]\,\mathrm{d}s, \tag{5.51}$$

where we have used the fact that

$$\int_0^{\varphi(t)} \mathrm{e}^{\alpha\xi_s}\mathrm{d}s = t \quad \text{and hence} \quad \frac{\mathrm{d}\varphi(t)}{\mathrm{d}t}\mathrm{e}^{\alpha\xi_{\varphi(t)}} = 1.$$

Once we have an identity for (5.51), then it must be the case that this equals $1/\psi(\alpha+u)$, where ψ is the Laplace exponent of ξ, given by

$$\mathbf{E}[\mathrm{e}^{z\xi_t}] = \mathrm{e}^{\psi(z)t}, \qquad t \ge 0,$$

for any $z \in \mathbb{R}$ such that the right-hand side of (5.51) is well defined. It will turn out that this will restrict us to $\mathrm{Re}(z) \in (-d, -\alpha)$.

Recall from Lemma 3.10, that the process X can be seen as a subordinated d-dimensional Brownian motion. This implies that the process Z starting from

1, has the same law as $(\sqrt{2}R^{(\nu)}_{\Lambda_t}, t \geq 0)$, where $\nu = (d/2) - 1$, $R^{(\nu)}$ denotes a Bessel process of dimension ν with $R^{(\nu)}_0 = 1/\sqrt{2}$, and $\Lambda = (\Lambda_t, t \geq 0)$ is an independent stable subordinator with index $\alpha/2$. Thus,

$$\int_0^\infty \mathbb{E}_1[Z_t^u]\,\mathrm{d}t = 2^{u/2}\int_0^\infty\int_0^\infty \mathbb{E}\left[\left(R^{(\nu)}_s\right)^u \,\Big|\, R^{(\nu)}_0 = \frac{1}{\sqrt{2}}\right]\mathbb{P}(\Lambda_t \in \mathrm{d}s)\,\mathrm{d}t$$
$$= \frac{2^{u/2}}{\Gamma(\alpha/2)}\int_0^\infty \mathbb{E}\left[\left(R^{(\nu)}_s\right)^u \,\Big|\, R^{(\nu)}_0 = \frac{1}{\sqrt{2}}\right] s^{\frac{\alpha}{2}-1}\,\mathrm{d}s,$$

where in the last identity, in the spirit of (3.20), we have used that the renewal measure for Λ satisfies

$$U(\mathrm{d}s) = \frac{1}{\Gamma(\alpha/2)} s^{\alpha/2-1}\,\mathrm{d}s, \qquad s \geq 0.$$

On the other hand, recall that $R^{(\nu)}_t$ starting from x, has transition probability density given by

$$\mathbb{P}\left(R^{(\nu)}_t \in \mathrm{d}y \mid R^{(\nu)}_0 = x\right) = \frac{1}{t}x^{-\nu}y^{\nu+1}\mathrm{e}^{-(x^2+y^2)/2t} I_\nu\left(\frac{xy}{t}\right)\mathrm{d}y,$$

where I_ν denotes the modified Bessel function of the first kind, which is given by

$$I_\nu(z) = \sum_{k\geq 0}\left(\frac{z}{2}\right)^{2k+\nu}\frac{\Gamma(\nu+k+1)}{k!}, \qquad z > 0.$$

Hence,

$$\mathbb{E}\left[\left(R^{(\nu)}_s\right)^u \,\Big|\, R^{(\nu)}_0 = \frac{1}{\sqrt{2}}\right] = \frac{2^{\nu/2}}{s}\int_0^\infty y^{u+\nu+1}\mathrm{e}^{-(\frac{1}{2}+y^2)/2s} I_\nu\left(\frac{y}{\sqrt{2}s}\right)\mathrm{d}y$$
$$= \mathrm{e}^{-1/4s}(2s)^{u/2}\frac{\Gamma\left(\frac{u}{2}+\nu+1\right)}{\Gamma(\nu+1)}\,{}_1F_1\left(\frac{u}{2}+\nu+1, \nu+1, \frac{1}{4s}\right),$$

where we have used the identity

$$\int_0^\infty y^{\rho-1}\mathrm{e}^{-py^2} I_\nu(cy)\,\mathrm{d}y = 2^{-\nu-1}c^\nu p^{-(\rho+\nu)/2}\frac{\Gamma\left(\frac{\rho+\nu}{2}\right)}{\Gamma(\nu+1)}\,{}_1F_1\left(\frac{\rho+\nu}{2}, \nu+1, \frac{c^2}{4p}\right),$$

for $c > 0$. The formula from above is valid for $u/2 + \nu + 1 > 0$, which is equivalent to the condition $u > -d$. Now, applying the identity

$$\mathrm{e}^{-x}\,{}_1F_1(a, b, x) = {}_1F_1(b-a, b, -x),$$

we observe, for $u > -d$,

$$\mathbb{E}\left[\left(R^{(\nu)}_s\right)^u \,\Big|\, R^{(\nu)}_0 = \frac{1}{\sqrt{2}}\right] = (2s)^{u/2}\frac{\Gamma\left(\frac{u}{2}+\nu+1\right)}{\Gamma(\nu+1)}\,{}_1F_1\left(-\frac{u}{2}, \nu+1, -\frac{1}{4s}\right).$$

Therefore, for $u \in (-d, -\alpha)$, we deduce

$$\begin{aligned}
\int_0^\infty \mathbb{E}_1[Z_t^u]\,\mathrm{d}t &= \frac{2^u\Gamma\left(\frac{u}{2}+\nu+1\right)}{\Gamma(\alpha/2)\Gamma(\nu+1)}\int_0^\infty {}_1F_1\left(-\frac{u}{2},\nu+1,-\frac{1}{4s}\right)s^{\frac{\alpha+u}{2}-1}\,\mathrm{d}s\\
&= \frac{2^{-\alpha}\Gamma\left(\frac{u}{2}+\nu+1\right)}{\Gamma(\alpha/2)\Gamma(\nu+1)}\int_0^\infty {}_1F_1\left(-\frac{u}{2},\nu+1,-x\right)x^{-\frac{\alpha+u}{2}-1}\,\mathrm{d}x\\
&= \frac{2^{-\alpha}\Gamma\left(\frac{u}{2}+\nu+1\right)}{\Gamma(\alpha/2)\Gamma(\nu+1)}\times\frac{\Gamma(-(\alpha+u)/2)\Gamma(\nu+1)\Gamma(\alpha/2)}{\Gamma(-u/2)\Gamma(\nu+1+(\alpha+u)/2)}\\
&= 2^{-\alpha}\frac{\Gamma((d+u)/2)}{\Gamma((d+\alpha+u)/2)}\frac{\Gamma(-(\alpha+u)/2)}{\Gamma(-u/2)},
\end{aligned}$$

where we have used the following identity

$$\int_0^\infty x^{b-1}{}_1F_1(a,c,-x)\,\mathrm{d}x = \frac{\Gamma(b)\Gamma(c)\Gamma(a-b)}{\Gamma(a)\Gamma(c-b)}, \qquad \text{for } 0<b<a. \tag{5.52}$$

We may now restate (5.51) more carefully and, with the addition of Fubini's Theorem, we can conclude that, for $u \in (-d, -\alpha)$,

$$\int_0^\infty \mathbb{E}_1[Z_t^u]\,\mathrm{d}t = \mathbf{E}\left[\int_0^\infty \mathrm{e}^{(u+\alpha)\xi_s}\,\mathrm{d}s\right] = -\frac{1}{\psi(\alpha+u)}. \tag{5.53}$$

Putting all the pieces together, we obtain that the Laplace exponent of ξ satisfies

$$\psi(z) = -2^\alpha\frac{\Gamma((\alpha-z)/2)}{\Gamma(-z/2)}\frac{\Gamma((z+d)/2)}{\Gamma((z+d-\alpha)/2)}, \qquad z\in(-d,\alpha).$$

A simple argument of analytic extension provides the characteristic exponent of ξ as stated in (5.50).

Next, assume that $d=1$ and $\alpha\in[1,2)$, that is, that the process X is recurrent. In order to compute the characteristic exponent of ξ, we use the decomposition stated in Proposition 5.20. Recall that the characteristic exponent ξ^{L} was computed in the previous section and satisfies, for $\theta\in\mathbb{R}$,

$$\Psi^L(\theta) = \frac{\Gamma(\alpha-\mathrm{i}\theta)}{\Gamma(\frac{\alpha}{2}-\mathrm{i}\theta)}\frac{\Gamma(1+\mathrm{i}\theta)}{\Gamma(1-\frac{\alpha}{2}+\mathrm{i}\theta)} - \frac{\Gamma(\alpha)}{\Gamma(\alpha/2)\Gamma(1-\frac{\alpha}{2})}.$$

In order to finish the proof, the computation of the characteristic exponent of ξ^{C_2} is needed. Since ξ^{C_2} is a compound Poisson process with jump rate q^* and jump with characteristic function given by (5.49), we obtain

$$\Psi^{C_2}(\theta) = \frac{\Gamma(\alpha)}{\Gamma(\alpha/2)\Gamma(1-\frac{\alpha}{2})}\left(1-\frac{\Gamma(1+\mathrm{i}\theta)\Gamma(\alpha-\mathrm{i}\theta)}{\Gamma(\alpha)}\right), \qquad \theta\in\mathbb{R}.$$

Hence summing up the characteristic exponents of ξ^{L} and ξ^{C_2}, and applying the reflection formula (A.12), we deduce

$$\begin{aligned}\Psi(\theta) &= \Gamma(1+\mathrm{i}\theta)\Gamma(\alpha-\mathrm{i}\theta)\left(\frac{1}{\Gamma(\frac{\alpha}{2}-\mathrm{i}\theta)\Gamma(1-\frac{\alpha}{2}+\mathrm{i}\theta)}-\frac{1}{\Gamma(\alpha/2)\Gamma(1-\frac{\alpha}{2})}\right)\\ &= \frac{\Gamma(1+\mathrm{i}\theta)\Gamma(\alpha-\mathrm{i}\theta)}{\pi}\left(\sin\left(\pi\left(\frac{\alpha}{2}-\mathrm{i}\theta\right)\right)-\sin\left(\frac{\pi\alpha}{2}\right)\right),\end{aligned} \tag{5.54}$$

for $\theta\in\mathbb{R}$.

Manipulations of the classical product and sum identities for trigonometric functions and the reflection formula for the gamma function (A.12), give us

$$\begin{aligned}\sin\left(\pi\left(\frac{\alpha}{2}-\mathrm{i}\theta\right)\right)-\sin\left(\frac{\pi\alpha}{2}\right) &= 2\sin\left(-\frac{i\theta\pi}{2}\right)\sin\left((1-\alpha+i\theta)\frac{\pi}{2}\right)\\ &= 2\pi^2\frac{1}{\Gamma(-i\theta/2)\Gamma\left(1+\frac{i\theta}{2}\right)}\frac{1}{\Gamma((1-\alpha+i\theta)/2)\Gamma((1+\alpha-i\theta)/2)}.\end{aligned}$$

On the other hand from the duplication formula (A.14), we see

$$\Gamma(1+\mathrm{i}\theta)=\Gamma\left(\frac{1+\mathrm{i}\theta}{2}\right)\Gamma\left(1+\frac{\mathrm{i}\theta}{2}\right)\frac{2^{\mathrm{i}\theta}}{\sqrt{\pi}}$$

and

$$\Gamma(\alpha-\mathrm{i}\theta)=\Gamma\left(\frac{\alpha-\mathrm{i}\theta}{2}\right)\Gamma\left(\frac{1+\alpha-\mathrm{i}\theta}{2}\right)\frac{2^{\alpha-1-\mathrm{i}\theta}}{\sqrt{\pi}}.$$

Finally putting all the pieces together in (5.54), we deduce

$$\Psi(\theta)=2^{\alpha}\frac{\Gamma(\frac{1}{2}(-\mathrm{i}\theta+\alpha))}{\Gamma(-\frac{1}{2}\mathrm{i}\theta)}\frac{\Gamma(\frac{1}{2}(\mathrm{i}\theta+1))}{\Gamma(\frac{1}{2}(\mathrm{i}\theta+1-\alpha))},\qquad \theta\in\mathbb{R},$$

which completes the proof. □

As one might expect, there is no issue with defining $|X|$ when issued from the origin. We need only distinguish between the two cases of the representation of $|X|$ as a conservative pssMp or a recurrent extension of the pssMp (5.47). Given the known behaviour of the stable process in one and higher dimensions (cf. Chapter 3), it is straightforward to declare that recurrent extension occurs when $\alpha\in(1,2)$ and $d=1$, whereas, in all other cases of (α,d), the point 0 is an entrance boundary and (5.47) is otherwise a conservative process on $(0,\infty)$. Inspecting (5.50), the previous remarks correlate precisely with the conclusions of Theorem 5.3 and 5.7. Indeed, we note that, when $\alpha\in(1,2)$ and $d=1$, $\Psi(-\mathrm{i}(\alpha-1))=0$, that is, there is a Cramér number $0<\alpha-1<1<\alpha$, hence Theorem 5.7 correctly predicts a recurrent extension of (5.47). Moreover, in the other cases, when $0<\alpha<d$, we note that the ascending ladder process

is again a β-subordinator and hence has finite mean. Accordingly Theorem 5.3 correctly predicts entrance at 0.

Similarly to the case of the stable process conditioned to stay positive and the censored stable process for $\alpha \in (0,1)$, we have all the ingredients to compute explicitly the Mellin transform of Z_1, and therefore determine its density, for the radial part of an isotropic stable process when $d > \alpha$.

Let us assume that $d > \alpha$ and recall from Corollary 4.7 part (ii), that $\mathbf{E}[\xi_1] \in (0, \infty)$. Hence from Theorem 5.3, the Mellin transform of the entrance law of $(Z, \mathbb{P}_0)$ is such that

$$\mathbb{E}_0\left[Z_t^{s-1}\right] = \frac{t^{\frac{s-1}{\alpha}}}{\alpha \mathbf{E}\left[\xi_1\right]} \mathbf{E}\left[\hat{I}_\infty^{\frac{1-s}{\alpha}-1}\right],$$

where $\hat{I}_\infty = \int_0^\infty \exp\{-\alpha \xi_s\} \mathrm{d}s$. From the explicit form of the characteristic exponent of the process ξ, see (5.50), we deduce that

$$\mathbf{E}\left[\xi_1\right] = 2^{\alpha-1} \frac{\Gamma\left(\frac{\alpha}{2}\right)\Gamma\left(\frac{d}{2}\right)}{\Gamma\left(\frac{d-\alpha}{2}\right)}.$$

Moreover, from part (i) of Corollary 4.16, the Mellin transform of the exponential functional $\hat{I}_\infty$ can be computed explicitly in terms of the gamma function. In other words, the entrance law of $(Z, \mathbb{P}_0)$ satisfies

$$\mathbb{E}_0\left[Z_t^{s-1}\right] = \frac{t^{\frac{s-1}{\alpha}}}{\alpha 2^{\alpha-1}\Gamma\left(\frac{d}{2}\right)} \Gamma\left(\frac{1-s}{\alpha}\right) \frac{\Gamma\left(\frac{d-1+s}{2}\right)}{\Gamma(1-s)}.$$

5.8 Comments

The notion that a pssMp can be expressed as the exponential of a time changed Lévy process was first described in the foundational work of Lamperti [139]. Section 5.1 summarises Lamperti's main findings. The proof of Theorem 5.2 is due to Lamperti [139], however a more complete version of the proof is found in Chapter 13 of Kyprianou [123]. Also in [139], Lamperti computed the characteristics of the underlying Lévy process embedded in the Lamperti transform of a stable subordinator, which turned out to be an example of what we now refer to as a β-subordinator. Interest in positive self-similar Markov processes was rekindled around the turn of the Millennium with various works concerning the problem of the existence of an entrance law at the origin (cf. Bertoin and Caballero [22], Bertoin and Savov [23], Bertoin and Yor [27], Caballero and Chaumont [44], and Chaumont et al. [46]). The conclusion of Theorem 5.3 gives the union of the aforesaid literature. Bertoin and Savov

[23] go further and give a pathwise construction of the process (Z, P_0). The duality property that appears in Proposition 5.5, was studied by Bertoin and Yor [27]. The time-reversal of (Z, P_0) at last passage times of Proposition 5.6 is taken from Chaumont and Pardo [48]. The existence of a recurrent extension from the origin was dealt with by Rivero [180, 181] and Fitzsimmons [71] and Theorem 5.7 is a summary of their work.

In this text, for any stable process, we always work with the natural enlargement of the filtration generated by the process itself; see Warning 1.3.39 of Bichteler [29] for further elaboration on this issue. It was a landmark observation of in Caballero and Chaumont [43], which noted that further concrete examples of pssMp could be studied for which their Lamperti transform could be characterised explicitly. Identification of the underlying Lévy processes that Lamperti transform to the positive self-similar Markov processes described in Sections 5.3, 5.4, 5.5 was undertaken in Caballero and Chaumont [43] using a method that examined their infinitesimal generators. A different approach that uses fluctuation identities associated to the aforesaid pssMp was used in Chapter 13 of Kyprianou [123] to identify the underlying Lévy processes using the Wiener–Hopf factorisation. As alluded to in the latter, we take a more economical approach, taking advantage of the fact that all three underlying Lévy processes are related by an Esscher transform. Lemma 5.12 that was used in constructing the stable process to stay positive in Section 5.4 was previously known up to a constant, see for example Chapter XIII of Bertoin [18]. The representation of the entrance law of $\mathbb{P}_0^{\uparrow}$, discussed following Theorem 5.14, is implicit in existing results of Kuznetsov and Pardo [121].

Section 5.6 is taken from the recent work of Kyprianou et al. [129]. The entrance law discussed after Theorem 5.19 is presented here for the first time. Some of the results in Section 5.7 are to be found in Caballero et al. [45], albeit, there, they used a different method, again appealing to infinitesimal generators. More precisely, Theorem 5.21 can be found in Caballero et al. [45] in the case that $\alpha < d$ and $\alpha = 1 = d$. The approach we take to Theorem 5.7 is new and was suggested to us by Alexey Kuznetsov. The characterisation of the entrance law of the radial part of a d-dimensional stable process given at the end of Section 5.7 is new. On a final note, the identity (5.52) for the hypergeometric function ${}_1F_1$ is one of the many identities that can be found for hypergeometric functions, see for example, formula 10, page 273 of the book of Bateman and Erdélyi [11].

6

Spatial fluctuations in one dimension

Having developed the relationship between several path functionals of stable processes and pssMp, we shall go to work and show how an explicit understanding of each of their Lamperti transforms leads, in a relatively straightforward way, to a suite of fluctuation identities. In essence, we will see that all of the identities we are interested in can be rephrased in terms of an underlying hypergeometric Lévy process. The specific nature of the Wiener–Hopf factorisation for this class, together with the identities in Section 2.17 is what gives us access to explicit results. Throughout this chapter, we keep to our usual notation that $X = (X_t, t \geq 0)$ is a one-dimensional stable process with probabilities $\mathbb{P}_x$, $x \in \mathbb{R}$ (reserving the special notation $\mathbb{P}$ in place of $\mathbb{P}_0$).

6.1 First exit from an interval

Theorem 3.6 deals with the event of first exit of a stable process from the interval $(-\infty, x)$, for some $x > 0$. A natural problem to consider thereafter is the event of first exit of a stable process from an interval, say $[0, a]$, for some $a > 0$. To this end, let us write as usual

$$\tau_a^+ = \inf\{t > 0 \colon X_t > a\} \quad \text{and} \quad \tau_0^- = \inf\{t > 0 \colon X_t < 0\},$$

where X is a stable process. As with many of the results in this chapter, we must be careful on occasion to distinguish whether or not the process X is spectrally negative. Recall that $0 < \alpha\hat{\rho}, \alpha\rho < 1$ if and only if X has jumps in both directions. Moreover, when $\alpha \in (1, 2)$, $\alpha\rho = 1$, corresponds to the case that X is spectrally negative without monotone paths and $\alpha\hat{\rho} = 1$ corresponds to the case that X is spectrally positive without monotone paths. The reader may also assume throughout that the setting of monotone paths, that is, $\alpha \in (0, 1)$ and $\rho \in \{0, 1\}$, are always excluded.

As a warm-up to the main result in this section, let us start by computing the two-sided exit probabilities. The reader will note that the result is consistent with Lemma 3.8, which deals with the case that $\alpha\rho = 1$ (spectrally negativity).

Lemma 6.1 *Suppose that $\alpha \in (0, 2)$ and $0 < \alpha\rho < 1$. For $a > 0$ and $x \in [0, a]$,*

$$\mathbb{P}_x(\tau_a^+ < \tau_0^-) = \frac{\Gamma(\alpha)}{\Gamma(\alpha\rho)\Gamma(\alpha\hat{\rho})} \int_0^{x/a} t^{\alpha\hat{\rho}-1}(1-t)^{\alpha\rho-1}\,\mathrm{d}t.$$

Proof Recall that ξ^* was defined in Section 5.3. Denote by $\mathbf{P}^*$ the law of ξ^* and, for $b > 0$, let

$$\tau_b^{*,+} = \inf\{t > 0 \colon \xi_t^* > b\}.$$

Recalling that the range of the stable process killed on exiting $[0, \infty)$ agrees with the range of the exponential of the process ξ^*, we have, with the help of Lemma 2.26, Theorems 5.9 and 5.11 as well as Proposition 4.3,

$$\begin{aligned}\mathbb{P}_x(\tau_a^+ < \tau_0^-) &= \mathbf{P}^*(\tau_{\log(a/x)}^{*,+} < \infty)\\ &= \frac{\Gamma(\alpha)}{\Gamma(\alpha\rho)\Gamma(\alpha\hat{\rho})} \int_{\log(a/x)}^{\infty} \mathrm{e}^{-\alpha\hat{\rho}y}(1-\mathrm{e}^{-y})^{\alpha\rho-1}\,\mathrm{d}y\\ &= \frac{\Gamma(\alpha)}{\Gamma(\alpha\rho)\Gamma(\alpha\hat{\rho})} \int_0^{x/a} t^{\alpha\hat{\rho}-1}(1-t)^{\alpha\rho-1}\,\mathrm{d}t,\end{aligned}$$

where in the final equality we have applied the change of variable $t = \mathrm{e}^{-y}$. □

Now we turn to a more general identity around the event of two-sided exit. The reader will note that Lemma 6.1 is, in principle, a corollary to Theorem 6.2 below. However, as it will shortly become apparent, the marginalisation of the more general fluctuation identity in Theorem 6.2 to derive Lemma 6.1 is not necessarily the most convenient way of doing things. We also note that we exclude the case of spectral negativity in the next theorem as the right-hand side of the identity would otherwise be zero.

Theorem 6.2 *Suppose that $\alpha \in (0, 2)$ and $0 < \alpha\rho < 1$. For $a, u > 0$, $x \in [0, a]$, $y \in [0, a-x]$ and $v \in [y, a]$,*

$$\begin{aligned}&\mathbb{P}_x(X_{\tau_a^+} - a \in \mathrm{d}u, a - X_{\tau_a^+-} \in \mathrm{d}v, a - \overline{X}_{\tau_a^+-} \in \mathrm{d}y\,;\, \tau_a^+ < \tau_0^-)\\ &= \frac{\sin(\pi\alpha\rho)}{\pi}\frac{\Gamma(\alpha+1)}{\Gamma(\alpha\rho)\Gamma(\alpha\hat{\rho})}\frac{x^{\alpha\hat{\rho}}(a-x-y)^{\alpha\rho-1}(v-y)^{\alpha\hat{\rho}-1}(a-v)^{\alpha\rho}}{(a-y)^{\alpha}(u+v)^{\alpha+1}}\,\mathrm{d}u\,\mathrm{d}v\,\mathrm{d}y.\end{aligned}$$

Proof The overshoot and undershoots at first passage over the level a for X on the event $\{\tau_a^+ < \tau_0^-\}$ are, up to a logarithmic change of spatial variable, equal to the overshoot and undershoots at first passage over the level $\log a$ for ξ^* on the event this first passage occurs before ξ^* is killed. Note that, for

$u \geq 0$, $y \in [0, a - x]$ and $v \in [y, a]$, with the help of Theorem 2.23, up to a multiplicative constant,

$$\mathbb{P}_x\left(\frac{X_{\tau_a^+}}{a} - 1 > u/a,\ 1 - \frac{X_{\tau_a^+ -}}{a} > v/a\,,1 - \frac{\overline{X}_{\tau_a^+ -}}{a} > y/a,\ \tau_a^+ < \tau_0^-\right)$$

$$= \mathbf{P}^*\left(\xi^*_{\tau^{*,+}_{\log(a/x)}} - \log(a/x) > \log\left(\frac{a+u}{a}\right),\right.$$
$$\log(a/x) - \xi^*_{\tau^{*,+}_{\log(a/x)}-} > -\log\left(\frac{a-v}{a}\right),$$
$$\left.\log(a/x) - \overline{\xi^*}_{\tau^{*,+}_{\log(a/x)}-} > -\log\left(\frac{a-y}{a}\right), \tau^{*,+}_{\log(a/x)} < \infty\right)$$
$$= \int_{-\log(\frac{a-y}{a})}^{\log(a/x)} \int_{-\log(\frac{a-v}{a})}^{\infty} \int_{\log(\frac{a+u}{a})}^{\infty} u^*(\log(a/x) - r)$$
$$\times \hat{u}^*(z - r)\pi^*(w + z)\mathbf{1}_{(z \geq r)}\mathrm{d}w\mathrm{d}z\mathrm{d}r,$$

where π^* is the Lévy density of ξ^* and, moreover, u^* and $\hat{u}^*$ are the densities of the renewal measures of the ascending and descending ladder height processes, respectively. Taking derivatives, we get

$$\mathbb{P}_x(X_{\tau_a^+} - a \in \mathrm{d}u,\ a - X_{\tau_a^+ -} \in \mathrm{d}v,\ a - \overline{X}_{\tau_a^+ -} \in \mathrm{d}y,\ \tau_a^+ < \tau_0^-)$$
$$= u^*\left(\log\left(\frac{a-y}{x}\right)\right)\hat{u}^*\left(\log\left(\frac{a-y}{a-v}\right)\right)\pi^*\left(\log\left(\frac{a+u}{a-v}\right)\right)\frac{\mathrm{d}u\,\mathrm{d}v\,\mathrm{d}y}{(a-y)(a-v)(a+u)}.$$

Given that the Wiener–Hopf factorisation of ξ^* has been described in explicit detail in Theorem 5.9, we can now develop the right-hand side above further. To this end, recall that the process ξ^* belongs to the class of Lamperti-stable processes with characteristic exponent (and hence Wiener–Hopf factorisation) given by

$$\Psi^*(z) = \frac{\Gamma(\alpha - \mathrm{i}z)}{\Gamma(\alpha\hat{\rho} - \mathrm{i}z)} \times \frac{\Gamma(1 + \mathrm{i}z)}{\Gamma(1 - \alpha\hat{\rho} + \mathrm{i}z)}, \qquad z \in \mathbb{R}.$$

From Lemma 4.9 we have

$$\pi^*(x) = \frac{\Gamma(1+\alpha)}{\Gamma(\alpha\rho)\Gamma(1-\alpha\rho)} \frac{\mathrm{e}^x}{(\mathrm{e}^x - 1)^{1+\alpha}}, \qquad x > 0,$$

and from Corollary 4.3 the renewal measures of the ascending and descending ladder height processes (which are clearly β-subordinators given the expression for Ψ^*) have densities given by

$$u^*(x) = \frac{1}{\Gamma(\alpha\rho)}\mathrm{e}^{-\alpha\hat{\rho}x}(1 - \mathrm{e}^{-x})^{\alpha\rho - 1},$$

and

$$\hat{u}^*(x) = \frac{1}{\Gamma(\alpha\hat{\rho})}\mathrm{e}^{-(1-\alpha\hat{\rho})x}(1 - \mathrm{e}^{-x})^{\alpha\hat{\rho} - 1},$$

for $x \geq 0$, respectively. In the special case that X (and hence ξ^*) is spectrally positive, the descending ladder height process of ξ^* is a pure drift and hence its potential measure is equal to Lebesgue measure restricted to $[0,\infty)$. In this sense we understand $u^*(x) \equiv 1$ when $\alpha\hat{\rho} = 1$. Putting everything together, straightforward algebra yields the desired result. □

In principle, one can marginalise the identity in Theorem 6.2 to give both the joint law of $(X_{\tau_a^+} - a, a - X_{\tau_a^+-})$ and the law of $X_{\tau_a^+} - a$ on the event $\{\tau_a^+ < \tau_0^-\}$. Whilst this is possible, albeit clumsy, we approach the matter in a different way. We deal with the law of $X_{\tau_a^+} - a$ on $\{\tau_a^+ < \tau_0^-\}$ by noting that it is also the law of the exponential of the overshoot of the ascending ladder height process of ξ^*. To establish the law of the pair $(X_{\tau_a^+} - a, a - X_{\tau_a^+-})$ on $\{\tau_a^+ < \tau_0^-\}$, we will appeal to a method based around the compensation formula which involves first computing the resolvent of X up to exiting $[0,a]$. Once again, spectral negativity is excluded.

Corollary 6.3 *Suppose that $\alpha \in (0,2)$ and $0 < \alpha\rho < 1$. For $x \in [0,a]$ and $u > 0$,*

$$\mathbb{P}_x(X_{\tau_a^+} - a \in du\,;\, \tau_a^+ < \tau_0^-)$$
$$= \frac{\sin \pi\alpha\rho}{\pi}(a-x)^{\alpha\rho}x^{\alpha\hat{\rho}}u^{-\alpha\rho}(u+a)^{-\alpha\hat{\rho}}(u+a-x)^{-1}\,du.$$

Proof Inspired by the proof of Theorem 6.2, we note that

$$\mathbb{P}_x\left(\frac{X_{\tau_a^+}}{a} - 1 > u/a\,;\, \tau_a^+ < \tau_0^-\right)$$
$$= \mathbf{P}^*\left(\xi^*_{\tau^{*,+}_{\log(a/x)}} - \log(a/x) > \log\left(\frac{a+u}{a}\right);\, \tau^{*,+}_{\log(a/x)} < \zeta^*\right),$$

where $u > 0$ and $x \in [0,a]$ and ζ^* is the lifetime of ξ^*. Note, however, that the law of the overshoot of ξ^* above $\log(a/x)$ is also equal to the law of the overshoot of its ascending ladder process over the same level. Accordingly, referring to Theorem 2.24 we have that, if υ^* is the Lévy density of the ascending ladder height process of ξ^*, then

$$\mathbf{P}^*\left(\xi^*_{\tau^{*,+}_{\log(a/x)}} - \log(a/x) > \log\left(\frac{a+u}{a}\right);\, \tau^{*,+}_{\log(a/x)} < \zeta^*\right)$$
$$= \int_0^{\log(a/x)} u^*(y)\,dy \int_{\log(a/x)-y+\log(\frac{a+u}{a})}^{\infty} \upsilon^*(z)\,dy.$$

Taking derivatives give us

$$\mathbb{P}_x\left(X_{\tau_a^+} - a \in du\,;\, \tau_a^+ < \tau_0^-\right)$$
$$= \left(\int_0^{\log(a/x)} u^*(y)\upsilon^*\left(\log\left(\frac{a+u}{x}\right) - y\right)dy\right)\frac{1}{(a+u)}\,du.$$

Now recall the expression for u^* given in the proof of Theorem 6.2 as well as the identity for v^* (taken from Corollary 4.3), noting in particular that $\alpha\hat{\rho} = 1$ and $\alpha\rho = \alpha - 1$ in the spectrally positive setting. We have

$$\begin{aligned}
&\mathbb{P}_x\left(X_{\tau_a^+} - a \in \mathrm{d}u\,;\ \tau_a^+ < \tau_0^-\right)\\
&= \frac{\alpha\rho x^\alpha (a+u)^{-\alpha\hat{\rho}}}{\Gamma(\alpha\rho)\Gamma(1-\alpha\rho)}\left(\int_0^{\log(a/x)} (\mathrm{e}^y - 1)^{\alpha\rho-1}\,(a+u-x\mathrm{e}^y)^{-\alpha\rho-1}\,\mathrm{e}^y\,\mathrm{d}y\right) du\\
&= \frac{\alpha\rho x^\alpha (a+u)^{-\alpha\hat{\rho}}}{\Gamma(\alpha\rho)\Gamma(1-\alpha\rho)}\left(x^{-\alpha\rho}(a+u-x)^{-1}\int_0^{(a-x)/u} \theta^{\alpha\rho-1}\,\mathrm{d}\theta\right) \mathrm{d}u,
\end{aligned}$$

where in the final equality, we have applied the change of variable

$$\mathrm{e}^y - 1 = \frac{(a+u-x)}{x}\frac{\theta}{\theta+1}.$$

It is now a minor amount of algebra to establish the identity given in the statement of the corollary. □

As promised, let us consider the resolvent of the stable process up to exiting the interval $[0, a]$,

$$U^{[0,a]}(x, \mathrm{d}y) = \int_0^\infty \mathbb{P}_x(X_t \in \mathrm{d}y,\ t < \tau_a^+ \wedge \tau_0^-)\,\mathrm{d}t,$$

for $y \in [0, a]$.

Theorem 6.4 *For $0 \le x, y \le a$, the measure $U^{[0,a]}(x, \mathrm{d}y)$ has a density with respect to Lebesgue measure which is almost everywhere equal to*

$$u^{[0,a]}(x,y) := \begin{cases} \dfrac{(y-x)^{\alpha-1}}{\Gamma(\alpha\rho)\Gamma(\alpha\hat{\rho})}\displaystyle\int_0^{\frac{x(a-y)}{a(y-x)}} (s+1)^{\alpha\rho-1}s^{\alpha\hat{\rho}-1}\,\mathrm{d}s, & x < y,\\[2ex] \dfrac{(x-y)^{\alpha-1}}{\Gamma(\alpha\rho)\Gamma(\alpha\hat{\rho})}\displaystyle\int_0^{\frac{y(a-x)}{a(x-y)}} s^{\alpha\rho-1}(s+1)^{\alpha\hat{\rho}-1}\,\mathrm{d}s, & x > y. \end{cases}$$

Proof We exclude from the proof the case of spectral negativity, that is $\alpha\rho = 1$. This is not really a restriction as the spectrally negative case can be established by applying it to the dual in the spectrally positive case.

On account of the fact that X cannot creep upwards, it follows that, for $u > 0$, $v \in [0, a]$ and $y \in [v \vee x, a]$,

$$\begin{aligned}
&\mathbb{P}_x(X_{\tau_a^+} - a \in \mathrm{d}u, X_{\tau_a^+-} \in \mathrm{d}v,\ \overline{X}_{\tau_a^+-} \le y\,;\ \tau_a^+ < \tau_0^-)\\
&= \mathbb{E}_x\left[\sum_{t<\infty} \mathbf{1}_{(X_{t-}\in \mathrm{d}v,\, \overline{X}_{t-}\le y,\, t<\tau_a^+\wedge\tau_0^-)}\mathbf{1}_{(X_{t-}+\Delta X_t - a\in\,\mathrm{d}u)}\right], \qquad (6.1)
\end{aligned}$$

where $\Delta X_t = X_t - X_{t-}$ and the sum is over the Poisson point process $((t, \Delta X_t), t \geq 0$ and $\Delta X_t \neq 0)$ which has intensity $\mathrm{d}t \times \Pi(\mathrm{d}x)$ (cf. Appendix A.13), representing the arrival of jumps in the stable process, and Π is the Lévy measure given by (3.1). It follows from the classical compensation formula for Poisson integrals of this type that

$$\begin{aligned}
&\mathbb{P}_x(X_{\tau_a^+} - a \in \mathrm{d}u, X_{\tau_a^+-} \in \mathrm{d}v, \overline{X}_{\tau_a^+} \leq y, ; \tau_a^+ < \tau_0^-) \\
&= \mathbb{E}_x\left[\int_0^\infty \mathbf{1}_{(X_{t-} \in \mathrm{d}v, \overline{X}_{t-} \leq y, t < \tau_a^+ \wedge \tau_0^-)} \mathrm{d}t\right] \Pi(a - v + \mathrm{d}u) \\
&= \Gamma(1+\alpha)\frac{\sin(\pi\alpha\rho)}{\pi}\mathbb{E}_x\left[\int_0^\infty \mathbf{1}_{(X_t \in \mathrm{d}v, \overline{X}_t \leq y, t < \tau_a^+ \wedge \tau_0^-)} \mathrm{d}t\right] \frac{1}{(a - v + u)^{1+\alpha}} \mathrm{d}u \\
&= \Gamma(1+\alpha)\frac{\sin(\pi\alpha\rho)}{\pi} U^{[0,y]}(x, \mathrm{d}v) \frac{1}{(a - v + u)^{1+\alpha}} \mathrm{d}u. \qquad (6.2)
\end{aligned}$$

From Theorem 6.2, we also have that, for $u > 0$, $v \in [0, a]$ and $y \in [v \vee x, a]$,

$$\begin{aligned}
&\mathbb{P}_x(X_{\tau_a^+} - a \in \mathrm{d}u, X_{\tau_a^+-} \in \mathrm{d}v, \overline{X}_{\tau_a^+-} \leq y, ; \tau_a^+ < \tau_0^-) \\
&= \frac{\sin(\pi\alpha\rho)}{\pi} \frac{\Gamma(\alpha+1)}{\Gamma(\alpha\rho)\Gamma(\alpha\hat{\rho})} \left\{\int_{v \vee x}^y \frac{x^{\alpha\hat{\rho}}(z - x)^{\alpha\rho - 1}(z - v)^{\alpha\hat{\rho} - 1} v^{\alpha\rho}}{z^\alpha (a - v + u)^{1+\alpha}} \mathrm{d}z\right\} \mathrm{d}u\, \mathrm{d}v.
\end{aligned}$$

The reader will note that in the spectrally positive case, we simply interpret the above expression with $\alpha\hat{\rho} = 1$. The consequence of this last observation is that, for $0 \leq v \vee x \leq y$, $U^{[0,y]}(x, \mathrm{d}v)$ is absolutely continuous with respect to Lebesgue measure and its density is given by

$$u^{[0,y]}(x, v) = \frac{x^{\alpha\hat{\rho}} v^{\alpha\rho}}{\Gamma(\alpha\rho)\Gamma(\alpha\hat{\rho})} \left\{\int_{v \vee x}^y \frac{(z - x)^{\alpha\rho - 1}(z - v)^{\alpha\hat{\rho} - 1}}{z^\alpha} \mathrm{d}z\right\}. \qquad (6.3)$$

To evaluate the integral in (6.3), we must consider two cases according to the value of x in relation to v. To this end, we first suppose that $x \leq v$. We have

$$\begin{aligned}
&x^{\alpha\hat{\rho}} v^{\alpha\rho} \int_v^y \frac{(z - x)^{\alpha\rho - 1}(z - v)^{\alpha\hat{\rho} - 1}}{z^\alpha} \mathrm{d}z \\
&= (v - x)^{\alpha - 2} \int_v^y \left[\frac{v(z - x)}{z(v - x)}\right]^{\alpha\rho - 1} \left[\frac{x(z - v)}{z(v - x)}\right]^{\alpha\hat{\rho} - 1} \frac{xv}{z^2} \mathrm{d}z \\
&= (v - x)^{\alpha - 1} \int_0^{\frac{x(y-v)}{y(v-x)}} (s + 1)^{\alpha\rho - 1} s^{\alpha\hat{\rho} - 1} \mathrm{d}s,
\end{aligned}$$

where in the final equality we have changed variables using $s = x(z - v)/z(v - x)$. To deal with the case $x > v$, one proceeds as above except that the lower delimiter on the integral in (6.3) is equal to x, we multiply and divide by $(x-v)^{\alpha-2}$ and one makes the change of variable $s = v(z-x)/z(x-v)$. This completes the proof. □

By taking limits as $a \to \infty$ in the expression for the resolvent $U^{[0,a]}$, one can appeal to monotone convergence to obtain an expression for the resolvent killed on exiting $[0,\infty]$. Note that this resolvent has a density with respect to Lebesgue measure which we denote by $u^{[0,\infty]}$. The corollary below gives an expression for $u^{[0,\infty]}$. Note, that the same result can also be derived from Theorem 2.27 by taking account of the fact that the ascending and descending ladder height processes of a stable processes are both stable subordinators of index $\alpha\rho$ and $\alpha\hat{\rho}$; see the discussion on the Wiener–Hopf factorisation in Section 3.4.

Corollary 6.5 *For $x, z \geq 0$,*

$$u^{[0,\infty]}(x,z) = \begin{cases} \dfrac{(y-x)^{\alpha-1}}{\Gamma(\alpha\rho)\Gamma(\alpha\hat{\rho})} \displaystyle\int_0^{\frac{x}{(y-x)}} (s+1)^{\alpha\rho-1} s^{\alpha\hat{\rho}-1}\,\mathrm{d}s, & x \leq y, \\[2ex] \dfrac{(x-y)^{\alpha-1}}{\Gamma(\alpha\rho)\Gamma(\alpha\hat{\rho})} \displaystyle\int_0^{\frac{y}{(x-y)}} s^{\alpha\rho-1}(s+1)^{\alpha\hat{\rho}-1}\,\mathrm{d}s, & x > y. \end{cases}$$

The computation of the resolvent in Theorem 6.4 allows us to write down the joint law of $(X_{\tau_a^+} - a, a - X_{\tau_a^+ -})$ on $\{\tau_a^+ < \tau_0^-\}$ without having to perform a marginalisation of the identity in Theorem 6.2. Indeed, the aforesaid marginalisation has already implicitly taken place when computing the identity for the resolvent in Theorem 6.4. Once again, spectral negativity is excluded to avoid a trivial result.

Corollary 6.6 *Suppose that $\alpha \in (0,2)$ and $0 < \alpha\rho < 1$. For $a, u > 0$, $x \in [0,a]$ and $v \in [0,a]$,*

$$\begin{aligned} &\mathbb{P}_x(X_{\tau_a^+} - a \in \mathrm{d}u, X_{\tau_a^+ -} \in \mathrm{d}v\,;\, \tau_a^+ < \tau_0^-) \\ &= \frac{\sin(\pi\alpha\rho)}{\pi}\frac{\Gamma(1+\alpha)}{\Gamma(\alpha\rho)\Gamma(\alpha\hat{\rho})} \times \\ &\qquad \begin{cases} (v-x)^{\alpha-1}\left(\displaystyle\int_0^{\frac{x(a-v)}{a(v-x)}} (s+1)^{\alpha\rho-1}s^{\alpha\hat{\rho}-1}\,\mathrm{d}s\right)\mathrm{d}u\,\mathrm{d}v, & x \leq v, \\[2ex] (x-v)^{\alpha-1}\left(\displaystyle\int_0^{\frac{v(a-x)}{a(x-v)}} s^{\alpha\rho-1}(s+1)^{\alpha\hat{\rho}-1}\,\mathrm{d}s\right)\mathrm{d}u\,\mathrm{d}v, & x > v. \end{cases} \end{aligned}$$

Proof Following the reasoning in Theorem 6.4, we can write the desired probability in terms of the Poisson point process of jumps. In that case, the compensation formula gives us that

$$\begin{aligned}
&\mathbb{P}_x(X_{\tau_a^+} - a \in \mathrm{d}u, X_{\tau_a^+-} \in \mathrm{d}v\,;\ \tau_a^+ < \tau_0^-)\\
&= \mathbb{E}_x\left[\sum_{t>0} \mathbf{1}_{(X_{t-}+\Delta X_t - a\,\in\,\mathrm{d}u)} \mathbf{1}_{(X_{t-}\in \mathrm{d}v,\, t<\tau_a^+\wedge\tau_0^-)}\right]\\
&= \mathbb{E}_x\left[\int_0^\infty \mathbf{1}_{(X_{t-}\in \mathrm{d}v,\, t<\tau_a^+\wedge\tau_0^-)}\mathrm{d}t\right]\Pi(a-v+\mathrm{d}u)\\
&= U^{[0,a]}(x,\mathrm{d}v)\Pi(a-v+\mathrm{d}u)\\
&= \Gamma(1+\alpha)\frac{\sin(\pi\alpha\rho)}{\pi}u^{[0,a]}(x,v)\frac{1}{(a-v+u)^{1+\alpha}}\,\mathrm{d}v\,\mathrm{d}u.
\end{aligned}$$

The result now follows. □

6.2 Hitting points in an interval

In the spirit of Lemma 2.19, we can develop an identity concerning the probability of hitting individual points in $(0,a)$, for $a > 0$, before exiting the interval, when $\alpha \in (1,2)$. The restriction on α ensures that points can be hit. To this end, let us introduce the notation

$$\tau^{\{y\}} = \inf\{t > 0 \colon X_t = y\},$$

for $y \in \mathbb{R}$.

Theorem 6.7 *For $\alpha \in (1,2)$ and $x, y \in (0,a)$,*

$$\mathbb{P}_x(\tau^{\{y\}} < \tau_a^+ \wedge \tau_0^-) = (\alpha-1)\begin{cases} \dfrac{a^{\alpha-1}(y-x)^{\alpha-1}}{y^{\alpha-1}(a-y)^{\alpha-1}}\displaystyle\int_0^{\frac{x(a-y)}{a(y-x)}} (s+1)^{\alpha\rho-1}s^{\alpha\hat{\rho}-1}\,\mathrm{d}s, & x \le y,\\[2ex] \dfrac{a^{\alpha-1}(x-y)^{\alpha-1}}{y^{\alpha-1}(a-y)^{\alpha-1}}\displaystyle\int_0^{\frac{y(a-x)}{a(x-y)}} s^{\alpha\rho-1}(s+1)^{\alpha\hat{\rho}-1}\,\mathrm{d}s, & x > y.\end{cases}$$

Proof We appeal to a standard technique and note that, for $x, y \in (0,a)$,

$$u^{[0,a]}(x,y) = \mathbb{P}_x(\tau^{\{y\}} < \tau_a^+ \wedge \tau_0^-)u^{[0,a]}(y,y),$$

where we may use L'Hôpital's rule to compute $u^{[0,a]}(y,y) = \lim_{x\uparrow y} u^{[0,a]}(x,y)$. The details are straightforward and left to the reader. □

Corollary 6.8 *For $\alpha \in (1,2)$ and $0 < \alpha\rho < 1$, $u > 0$ and $0 < y < x < a$,*

$$
\begin{aligned}
&\mathbb{P}_x(X_{\tau_a^+} - a \in \mathrm{d}u, \tau_a^+ < \tau_0^- \wedge \tau^{\{y\}}) \\
&= \Big\{ \frac{\sin \pi\alpha\rho}{\pi} (a-x)^{\alpha\rho} x^{\alpha\hat{\rho}} u^{-\alpha\rho} (u+a)^{-\alpha\hat{\rho}} (u+a-x)^{-1} \\
&\quad - (\alpha - 1)\frac{\sin \pi\alpha\rho}{\pi} (a-y)^{1-\alpha\hat{\rho}} y^{1-\alpha\rho} u^{-\alpha\rho} (u+a)^{-\alpha\hat{\rho}} (u+a-y)^{-1} \,\mathrm{d}u \\
&\qquad\qquad \times a^{\alpha-1} (x-y)^{\alpha-1} \int_0^{\frac{y(a-x)}{a(x-y)}} s^{\alpha\rho-1} (s+1)^{\alpha\hat{\rho}-1} \Big\} \,\mathrm{d}u.
\end{aligned}
$$

Proof The proof follows from the following observation which is the result of counting paths. For $u > 0$,

$$
\begin{aligned}
&\mathbb{P}_x(X_{\tau_a^+} - a \in \mathrm{d}u, \tau_a^+ < \tau_0^- \wedge \tau^{\{y\}}) \\
&\quad = \mathbb{P}_x(X_{\tau_a^+} - a \in \mathrm{d}u, \tau_a^+ < \tau_0^-) \\
&\qquad\qquad - \mathbb{P}_x(\tau^{\{y\}} < \tau_a^+ \wedge \tau_0^-) \mathbb{P}_y(X_{\tau_a^+} - a \in \mathrm{d}u, \tau_a^+ < \tau_0^-).
\end{aligned}
$$

The result now follows using the conclusions in Corollary 6.3 and Theorem 6.7. □

A similar identity to the one in Corollary 6.8 can be written in the case that $0 < x < y < a$, we leave the details to the reader. Another exercise for the reader is to consider the resolvent

$$
U_{\{z\}}^{[0,a]}(x, \mathrm{d}y) = \int_0^\infty \mathbb{P}_x(X_t \in \mathrm{d}y,\ t < \tau^{\{z\}} \wedge \tau_a^+ \wedge \tau_0^-), \quad x, y \in (0,a)\backslash\{z\}, z \in (0,a).
$$

By the strong Markov property, it is not difficult to see that this resolvent has a density with respect to Lebesgue measure, say $u_{\{z\}}^{[0,a]}(x,y)$, $x, y \in (0,a)\backslash\{z\}$, $z \in (0,a)$, where

$$
u_{\{z\}}^{[0,a]}(x,y) = u^{[0,a]}(x,y) - \mathbb{P}_x(\tau^{\{z\}} < \tau_a^+ \wedge \tau_0^-) u^{[0,a]}(z,y).
$$

6.3 First entrance into a bounded interval

In Section 6.1 we looked at the law of the stable process as it first exits an interval. In this section, we shall look at the law of the stable process as it first *enters* a interval. Accordingly, we introduce the first hitting time of the interval $(0,a)$,

$$
\tau^{(0,a)} = \inf\{t > 0 \colon X_t \in (0,a)\}.
$$

The next result provides the law of $X_{\tau^{(0,a)}}$. Because of the issue of creeping when X is spectrally one sided it is necessary to consider the cases of one-sided and two-sided jumps separately when $\alpha \in (1,2)$. Recall that we exclude the setting of monotone paths. Our first result deals exclusively processes with two-sided jumps.

Theorem 6.9 *Let $x > a > 0$. Then, when $\alpha, \rho \in (0,1)$ or $\alpha = 1, \rho = 1/2$, then*

$$\begin{aligned}&\mathbb{P}_x(X_{\tau^{(0,a)}} \in \mathrm{d}y,\ \tau^{(0,a)} < \infty)\\ &\quad = \frac{\sin(\pi\alpha\hat{\rho})}{\pi} x^{\alpha\rho} y^{-\alpha\rho}(x-a)^{\alpha\hat{\rho}}(a-y)^{-\alpha\hat{\rho}}(x-y)^{-1}\mathrm{d}y,\end{aligned} \tag{6.4}$$

for $y \in (0,a)$. When $\alpha \in (1,2)$ and $0 < \alpha\hat{\rho}, \alpha\rho < 1$,

$$\begin{aligned}&\mathbb{P}_x(X_{\tau^{(0,a)}} \in \mathrm{d}y)\\ &\quad = \frac{\sin(\pi\alpha\hat{\rho})}{\pi} y^{-\alpha\rho}(a-y)^{-\alpha\hat{\rho}}\Bigg((x-a)^{\alpha\hat{\rho}}x^{\alpha\rho}(x-y)^{-1}\\ &\qquad -(\alpha-1)\left(\frac{a}{2}\right)^{\alpha-1}\int_1^{\frac{2x}{a}-1}(t-1)^{\alpha\hat{\rho}-1}(t+1)^{\alpha\rho-1}\,\mathrm{d}t\Bigg)\mathrm{d}y,\end{aligned} \tag{6.5}$$

for $y \in (0,a)$.

Proof Just as with the proof of Theorem 6.2, the proof here relies on reformulating the problem at hand in terms of an underlying positive self-similar Markov process. In this case, we will appeal to the censored stable process defined in Section 5.6. The key observation that drives the proof is that, when $X_0 = x > a > 0$, on $\{\tau^{(0,a)} < \infty\}$,

$$X_{\tau^{(0,a)}} \equiv x\exp\{\vec{\xi}_{\vec{\tau}^-_{\log(a/x)}}\},$$

where $\vec{\xi}$ is the Lévy process described in Theorem 5.19 and

$$\vec{\tau}^-_{\log(a/x)} = \inf\{t > 0 : \ \vec{\xi}_t < \log(a/x)\}.$$

Note, moreover, that $\{\tau^{(0,a)} < \infty\}$ and $\{\vec{\tau}^-_{\log(a/x)} < \infty\}$ are corresponding events. If we denote the law of $\vec{\xi}$ when issued from the origin by $\vec{\mathbf{P}}$, then, for $\alpha \in (0,2)$ and $y \in (0,a)$,

$$\begin{aligned}&\mathbb{P}_x(X_{\tau^{(0,a)}} \le y,\ \tau^{(0,a)} < \infty)\\ &\quad = \vec{\mathbf{P}}\left(\log(a/x) - \vec{\xi}_{\vec{\tau}^-_{\log(a/x)}} \ge \log(a/y),\ \vec{\tau}^-_{\log(a/x)} < \infty\right),\end{aligned}$$

and hence

$$\begin{aligned}&\mathbb{P}_x(X_{\tau^{(0,a)}} \in \mathrm{d}y,\ \tau^{(0,a)} < \infty)\\ &\quad = \frac{1}{y}\frac{\mathrm{d}}{\mathrm{d}z}\vec{\mathbf{P}}\left(\log(a/x) - \vec{\xi}_{\vec{\tau}^-_{\log(a/x)}} \le z,\ \vec{\tau}^-_{\log(a/x)} < \infty\right)\mathrm{d}y\Bigg|_{z=\log(a/y)},\end{aligned} \tag{6.6}$$

where we have pre-emptively assumed that the overshoot distribution of $\vec{\xi}$ has a density. Note that the dual of the process $\vec{\xi}$ has characteristic exponent given by

$$z \mapsto \frac{\Gamma(1-\alpha\rho-\mathrm{i}z)}{\Gamma(1-\alpha-\mathrm{i}z)}\frac{\Gamma(\alpha\rho+\mathrm{i}z)}{\Gamma(\mathrm{i}z)}, \qquad z \in \mathbb{R},$$

which is an $(\alpha, \alpha\hat{\rho}, 0, \alpha\rho)$-hypergeometric Lévy process.

We may now appeal to the two parts of Theorem 4.10, accordingly as $\alpha \in (0,1]$ and $\alpha \in (1,2)$, to develop the right-hand side of (6.6) by considering the first passage problem of the dual of $\vec{\xi}$ over the threshold $\log(x/a)$. After a straightforward computation, the identity (6.4) for $\alpha \in (0,1]$ emerges from the first part of Theorem 4.10 (in particular it becomes clear that the overshoot distribution of $\vec{\xi}$ has a density as was assumed in the previous paragraph). The case $\alpha \in (1,2)$ requires the evaluation of an extra term. More precisely, from the second part of Theorem 4.10, we get

$$\begin{aligned}
&\mathbb{P}_x(X_{\tau^{(0,a)}} \in \mathrm{d}y)\\
&= \frac{\sin(\pi\alpha\hat{\rho})}{\pi} y^{-\alpha\rho}(a-y)^{-\alpha\hat{\rho}}\Bigg((x-a)^{\alpha\hat{\rho}}x^{\alpha\rho-1}y(x-y)^{-1}\\
&\qquad - a^{\alpha-1}(\alpha\rho-1)\int_0^{1-\frac{a}{x}} t^{\alpha\hat{\rho}-1}(1-t)^{1-\alpha}\,\mathrm{d}t\Bigg)\mathrm{d}y. \qquad (6.7)
\end{aligned}$$

By the substitution $t = (s-1)/(s+1)$, we deduce

$$\begin{aligned}
&\int_0^{1-\frac{a}{x}} t^{\alpha\hat{\rho}-1}(1-t)^{1-\alpha}\,\mathrm{d}t\\
&\quad = 2^{1-\alpha}\Bigg(\int_1^{\frac{2x}{a}-1}(s-1)^{\alpha\hat{\rho}-1}(s+1)^{\alpha\rho-1}\,\mathrm{d}s\\
&\qquad - \int_1^{\frac{2x}{a}-1}(s-1)^{\alpha\hat{\rho}}(s+1)^{\alpha\rho-2}\,\mathrm{d}s\Bigg). \qquad (6.8)
\end{aligned}$$

Now evaluating the second term on the right-hand side above via integration by parts and substituting back into (6.7) yields the required law. □

Recall that the ascending ladder height of a stable process is a stable subordinator with index $\alpha\rho$. As one might expect, the analogue of the statement in Theorem 6.9 for the spectrally negative case agrees with (6.5) in the limit as $\alpha\rho \to 1$, albeit that this does not constitute a proof. We take a more rigorous approach below.

Proposition 6.10 *Let $\alpha \in (1,2)$, and suppose that X is spectrally negative, that is, $\alpha\rho = 1$. Then, the hitting distribution of $(0,a)$ is given by*

$$\mathbb{P}_x(X_{\tau^{(0,a)}} \in \mathrm{d}y) = \frac{\sin(\pi(\alpha-1))}{\pi}(x-a)^{\alpha-1}(a-y)^{1-\alpha}(x-y)^{-1}\mathrm{d}y + \frac{\sin(\pi(\alpha-1))}{\pi}\int_0^{\frac{x-a}{x}} t^{\alpha-2}(1-t)^{1-\alpha}\,\mathrm{d}t\,\delta_0(\mathrm{d}y), \quad (6.9)$$

for $x > a$, $y \in [0,a]$, where δ_0 is the unit point mass at 0.

Proof Since the process has only negative jumps, we have two possibilities: either the process jumps below the level a and hits the interval $(0,a)$ or the process jumps below the level 0 and then hits 0 continuously. In other words,

$$\mathbb{P}_x(X_{\tau^{(0,a)}} \in \mathrm{d}y) = \mathbb{P}_x(X_{\tau_a^-} \in \mathrm{d}y) + \mathbb{P}_x(X_{\tau_a^-} < 0)\delta_0(\mathrm{d}y).$$

Now, we observe

$$\mathbb{P}_x(X_{\tau_a^-} \in \mathrm{d}y) = \hat{\mathbb{P}}\Big(X_{\tau_{x-a}^+} - (x-a) \in a - \mathrm{d}y\Big) = \frac{\sin(\pi(\alpha-1))}{\pi}(x-a)^{\alpha-1}(a-y)^{1-\alpha}(x-y)^{-1}\mathrm{d}y, \quad (6.10)$$

where $\hat{\mathbb{P}}$ is the law of $-X$ when issued from the origin and the second equality above follows from Corollary 3.7. Writing $\mathbb{P}_x(X_{\tau_a^-} < 0)$ as an integral with respect to the density in (6.10), after a change of variable similar to the one in (6.8), the identity in (6.9) follows. □

We also have the following straightforward corollary that gives the probability that the process never hits the interval $(0,a)$, in the case when $\alpha \in (0,1)$ (but X is not a subordinator). The result can be deduced from Theorem 6.9 by integrating out y in expression (6.4), however, we present a more straightforward proof.

Corollary 6.11 *When $\alpha, \rho \in (0,1)$, for $x > a$,*

$$\mathbb{P}_x(\tau^{(0,a)} = \infty) = \frac{\Gamma(1-\alpha\rho)}{\Gamma(\alpha\hat{\rho})\Gamma(1-\alpha)}\int_0^{\frac{x-a}{x}} t^{\alpha\hat{\rho}-1}(1-t)^{-\alpha}\,\mathrm{d}t.$$

Proof From Theorem 5.19 we know that the descending ladder height process has Laplace exponent given by $\Gamma(1-\alpha\rho+\lambda)/\Gamma(1-\alpha+\lambda)$, $\lambda \geq 0$. By Corollary 4.3 there is an associated potential density given by

$$\frac{1}{\Gamma(\alpha\hat{\rho})}\mathrm{e}^{-(1-\alpha)x}(1-\mathrm{e}^{-x})^{\alpha\hat{\rho}-1}, \qquad x \geq 0.$$

Now appealing to Corollary 2.26, we have that

$$\begin{aligned}\mathbb{P}_x(\tau^{(0,a)} = \infty) &= \vec{\mathbf{P}}_{\log(x/a)}(\vec{\tau}_0^- = \infty)\\ &= 1 - \frac{\Gamma(1-\alpha\rho)}{\Gamma(1-\alpha)}\int_{\log(x/a)}^{\infty}\frac{1}{\Gamma(\alpha\hat{\rho})}\mathrm{e}^{-(1-\alpha)y}(1-\mathrm{e}^{-y})^{\alpha\hat{\rho}-1}\mathrm{d}y\\ &= 1 - \frac{\Gamma(1-\alpha\rho)}{\Gamma(\alpha\hat{\rho})\Gamma(1-\alpha)}\int_{(x-a)/a}^{1} t^{\alpha\hat{\rho}-1}(1-t)^{-\alpha}\mathrm{d}t,\end{aligned}$$

where in the last equality we have performed the change of variable $t = 1-\mathrm{e}^{-y}$. The desired probability now follows as a straightforward consequence of the beta integral. □

To remain consistent with the previous sections in this chapter, our next point of interest is the resolvent

$$U^{(0,a)^c}(x,\mathrm{d}y) = \int_0^{\infty}\mathbb{P}_x(X_t\in\mathrm{d}y,\ t<\tau^{(0,a)})\,\mathrm{d}t, \qquad y\in(0,a)^c,$$

where $a > 0$. The theorem below gives us an identity for the above resolvent in the case of two-sided jumps, that is, for $0 < \alpha\rho, \alpha\hat{\rho} < 1$. We will have to defer its proof however until we have built up more machinery. In particular, we will have to wait until Chapter 12 where we will introduce the Riesz–Bogdan–Żak transform. This space-time transformation will play a crucial role as well as exemplifying a methodology which is robust enough to develop related identities in higher dimensions.

Theorem 6.12 *Suppose that X has two-sided jumps. For $y > x > a$, the measure $U^{(0,a)^c}(x,\mathrm{d}y)$ has a density given by*

$$\begin{aligned}&u^{(0,a)^c}(x,y)\\ &= \frac{2^{1-\alpha}}{\Gamma(\alpha\rho)\Gamma(\alpha\hat{\rho})}\Bigg(|y-x|^{\alpha-1}\int_1^{\left|\frac{x+y-2xy/a}{y-x}\right|}(s+1)^{\alpha\rho-1}(s-1)^{\alpha\hat{\rho}-1}\,\mathrm{d}s\\ &\qquad - (\alpha-1)^+\left(\frac{a}{2}\right)^{\alpha-1}\int_1^{(2x/a)-1}(s+1)^{\alpha\rho-1}(s-1)^{\alpha\hat{\rho}-1}\,\mathrm{d}s\\ &\qquad \times\int_1^{(2y/a)-1}(s+1)^{\alpha\hat{\rho}-1}(s-1)^{\alpha\rho-1}\,\mathrm{d}s\Bigg),\end{aligned}$$

where $(\alpha-1)^+ = \max\{0,\alpha-1\}$. Moreover, if $x > y > a$ then appealing to duality (cf. Lemma 2.16),

$$u^{(0,a)^c}(x,y) = u^{(0,a)^c}(y,x)|_{\rho\leftrightarrow\hat{\rho}},$$

where $\rho\leftrightarrow\hat{\rho}$ is understood to mean that the roles of ρ and $\hat{\rho}$ are interchanged. If $x > a$, $y < 0$, then

$$u^{(0,a)^c}(x,y)$$

$$= \frac{\sin(\alpha\hat{\rho})}{\sin(\alpha\rho)} \frac{2^{1-\alpha}}{\Gamma(\alpha\rho)\Gamma(\alpha\hat{\rho})} \Bigg(|y-x|^{\alpha-1} \int_1^{\left|\frac{x+y-2xy/a}{y-x}\right|} (s+1)^{\alpha\rho-1}(s-1)^{\alpha\hat{\rho}-1}\, \mathrm{d}s$$

$$- (\alpha-1)^+ \int_1^{(2x/a)-1} (s+1)^{\alpha\rho-1}(s-1)^{\alpha\hat{\rho}-1}\, \mathrm{d}s$$

$$\times \int_1^{|(2y/a)-1|} (s+1)^{\alpha\rho-1}(s-1)^{\alpha\hat{\rho}-1}\, \mathrm{d}s \Bigg).$$

Finally, if $x < 0, y < 0$ *or* $x < 0, y > a$, *then* $u^{(0,a)^c}(x,y) = u^{(0,a)^c}(a-x, a-y)|_{\rho\leftrightarrow\hat{\rho}}$.

6.4 Point of closest and furthest reach

We are interested in the 'point of closest reach' to the origin for stable processes with index $\alpha \in (0,1)$. Recall that for this index range, the stable process does not hit points and, moreover, $\lim_{t\to\infty} |X_t| = \infty$. Hence, either on the positive or negative side of the origin, the path of the stable process has a minimal radial distance. Moreover, thanks to path regularity (as discussed in Section 3.4), this distance is achieved at the unique time $\underline{m}$ such that $|X_t| \geq |X_{\underline{m}}|$ for all $t \geq 0$. Note, uniqueness follows thanks to regularity of X for both $(0,\infty)$ and $(-\infty, 0)$.

Theorem 6.13 *Suppose that* $\alpha, \rho \in (0,1)$, *then for* $x > 0$ *and* $|z| \leq x$,

$$\mathbb{P}_x(X_{\underline{m}} \in \mathrm{d}z) = \frac{\Gamma(1-\alpha\rho)}{\Gamma(1-\alpha)\Gamma(\alpha\hat{\rho})} \frac{x+z}{|2z|^\alpha} (x-|z|)^{\alpha\hat{\rho}-1} (x+|z|)^{\alpha\rho-1}\, \mathrm{d}z.$$

Proof Following our standard notation, let $\tau^{(-1,1)} := \inf\{t \geq 0 \colon |X_t| < 1\}$. From Corollary 6.11, after shifting and scaling, we have that, for $x > 1$,

$$\mathbb{P}_x(\tau^{(-1,1)} = \infty) = \frac{\Gamma(1-\alpha\rho)}{\Gamma(\alpha\hat{\rho})\Gamma(1-\alpha)} \int_0^{(x-1)/(x+1)} t^{\alpha\hat{\rho}-1}(1-t)^{-\alpha}\, \mathrm{d}t.$$

Now fix $x > 0$. Let $\underline{m}^+$ be the unique time such that $X_{\underline{m}^+} > 0$ and $X_t \geq X_{\underline{m}^+}$ for all $t \geq 0$ such that $X_t > 0$. Similarly let $\underline{m}^-$ be the unique time such that $X_{\underline{m}^-} < 0$ and $X_t \leq X_{\underline{m}^-}$ for all $t \geq 0$ such that $X_t < 0$. In words, $\underline{m}^+$ and $\underline{m}^-$ are the times when X is at the closest point to the origin on the positive and negative side of the origin, respectively. Consequently, we have that $X_{\underline{m}} > 0$ if and only if $X_{\underline{m}^+} < |X_{\underline{m}^-}|$.

Define

$$G(x) = \frac{\Gamma(1-\alpha\rho)}{\Gamma(\alpha\hat{\rho})\Gamma(1-\alpha)} \int_0^{(x-1)/(x+1)} t^{\alpha\hat{\rho}-1}(1-t)^{-\alpha}\,\mathrm{d}t, \qquad x > 1.$$

In fact $G(x) = \mathbb{P}_x(\tau^{(-1,1)} = \infty)$. We now have that

$$\mathbb{P}_x(|X_{\underline{m}^-}| > u, X_{\underline{m}^+} > v) = \mathbb{P}_x(\tau^{(-u,v)} = \infty) = G\left(\frac{2x+u-v}{u+v}\right), \qquad u, v > 0,$$

where, in the second equality, we have scaled space and used the self-similarity of X.

Next we have that for $z \geq 0$,

$$\begin{aligned}
\frac{\mathrm{d}}{\mathrm{d}z}\mathbb{P}_x(X_{\underline{m}} \leq z) &= -\frac{\partial}{\partial v}\mathbb{P}_x(|X_{\underline{m}^-}| > z; X_{\underline{m}^+} > v)\Big|_{v=z} \\
&= -\frac{\partial}{\partial v} G\left(\frac{2x+z-v}{z+v}\right)\Big|_{v=z} \\
&= \frac{x+z}{2z^2} G'\left(\frac{x}{z}\right).
\end{aligned} \qquad (6.11)$$

The proposition for $z > 0$ now follows from an easy computation. The result for $z < 0$ follows similarly. □

In the case that $\alpha = 1$, the stable process does not hit points and we have that $\limsup_{t\to\infty} |X_t| = \infty$ and $\liminf_{t\to\infty} |X_t| = 0$ and hence it is not possible to produce a result in the spirit of Theorem 6.13. However, when $\alpha \in (1,2)$ and there are two-sided jumps, the stable process will hit all points almost surely, in particular $\tau^{\{0\}} := \inf\{t > 0 : X_t = 0\}$ is $\mathbb{P}_x$-almost surely finite for all $x \in \mathbb{R}$. This allows us to talk about the 'point of furthest' reach until first hitting the origin. To this end, we define $\overline{m}$ to be the unique time such that $|X_t| \leq |X_{\overline{m}}|$ for all $t \leq \tau^{\{0\}}$. Once again, uniqueness follows by path regularity of X for both the upper and lower half lines.

Theorem 6.14 *Suppose that $\alpha \in (1,2)$ and $0 < \alpha\hat{\rho}, \alpha\rho < 1$, then for each $x > 0$ and $|z| > x$,*

$$\mathbb{P}_x(X_{\overline{m}} \in \mathrm{d}z) = \frac{\alpha-1}{2|z|^\alpha}\Bigg(|x+z|(|z|-x)^{\alpha\rho-1}(|z|+x)^{\alpha\hat{\rho}-1} - (\alpha-1)x\int_1^{|z|/x}(t-1)^{\alpha\rho-1}(t+1)^{\alpha\hat{\rho}-1}\,\mathrm{d}t\Bigg)\,\mathrm{d}z.$$

Proof The proof here is very similar to that of the case when $\alpha < 1$, thus we skip some of the details.

From the discussion following Theorem 5.19, we know that the positively censored (and similarly negatively) censored stable process must hit the origin continuously when $\alpha \in (1,2)$. Hence, without censoring, we also have that $\tau^{\{0\}} := \inf\{t \geq 0: X_t = 0\} < \infty$ and $X_{\tau^{\{0\}}-} = 0$ almost surely.

From Theorem 6.7, after shifting and scaling, we have that, for every $x \in (0,1)$ and $y \in (x,1)$,

$$\mathbb{P}_x(\tau^{\{y\}} < \tau_1^+ \wedge \tau_{-1}^-) = (\alpha - 1)\left(\frac{x-y}{1-y^2}\right)^{\alpha-1} \bar{G}\left(\left|\frac{1-xy}{x-y}\right|\right), \tag{6.12}$$

where

$$\bar{G}(z) = \int_1^z (t-1)^{\alpha\rho-1}(t+1)^{\alpha\hat{\rho}-1}\,\mathrm{d}t.$$

In the spirit of the proof of Proposition 6.13, we apply a linear spatial transformation to the probability $\mathbb{P}_x(\tau_v^+ \wedge \tau_{-u}^- < \tau^{\{0\}})$ and, using (6.12), write it in terms of $\bar{G}$. Similarly to the derivation of (6.11) in the proof of Proposition 6.13, for each $x > 0$ and $|z| > x$,

$$\mathbb{P}_x(X_{\overline{m}} \in \mathrm{d}z) = \frac{\alpha - 1}{2x^{2-\alpha}|z|^{\alpha}}\left(|x+z|\bar{G}'\left(\frac{|z|}{x}\right) - (\alpha-1)x\bar{G}\left(\frac{|z|}{x}\right)\right)\mathrm{d}z.$$

The result now follows from straight forward computations. □

6.5 First hitting of a two-point set

Let us define the hitting times

$$\tau^{\{b\}} = \inf\{t > 0: X_t = b\},$$

for $b \in \mathbb{R}$, and consider the two point hitting problem of evaluating $\mathbb{P}_x(\tau^{\{b\}} < \tau^{\{a\}})$ for $a, b, x \in \mathbb{R}$ with $a < b$. Naturally for this problem to be distinct from other problems we have considered, we need to assume, as in the previous section, that $\alpha \in (1,2)$ and $0 < \alpha\hat{\rho}, \alpha\rho < 1$. The requirement that $\alpha \in (1,2)$ ensures that points can be hit and the second requirement, that is, that there are two-sided jumps removes the reduction of the problem to existing exit problems in the spectrally one-sided case. The two point hitting problem is a classical problem for Brownian motion. However, for the case of a stable process, on account of the fact that it may wander either side of the points a and b before hitting one of them, the situation is significantly different.

It turns out that censoring the stable process is a useful way to analyse this problem. Indeed if we write $\vec{\xi}$ for the Lévy process which drives the Lamperti transformation of the censored stable process (cf. Section 5.6) and denote its probabilities by $\vec{\mathbf{P}}_x$, $x \in \mathbb{R}$, then by choosing $a = 0 < b$ and $x > 0$,

$$\mathbb{P}_x(\tau^{\{b\}} < \tau^{\{0\}}) = \vec{\mathbf{P}}_{\log x}(\vec{\tau}^{\{\log b\}} < \infty),$$

where

$$\overset{\leftarrow}{\tau}^{\{\log b\}} = \inf\{t > 0 : \overset{\leftarrow}{\xi}_t = \log b\}.$$

According to Lemma 2.19, the previous probability can be written

$$\mathbb{P}_x(\tau^{\{b\}} < \tau^{\{0\}}) = \frac{\overset{\leftarrow}{u}(-\log(x/b))}{\overset{\leftarrow}{u}(0)}, \tag{6.13}$$

where

$$\begin{aligned}\int_{\mathbb{R}} e^{zx}\,\overset{\leftarrow}{u}(x)\mathrm{d}x &= \int_0^\infty \overset{\leftarrow}{\mathbf{E}}[e^{z\overset{\leftarrow}{\xi}_t}]\,\mathrm{d}t \\ &= \frac{1}{\overset{\leftarrow}{\Psi}(-\mathrm{i}z)} \\ &= \frac{\Gamma(-z)}{\Gamma(\alpha\rho - z)}\frac{\Gamma(1-\alpha+z)}{\Gamma(1-\alpha\rho+z)},\end{aligned} \tag{6.14}$$

for $\mathrm{Re}(z) \in (0, \alpha - 1)$. More generally, $\overset{\leftarrow}{\psi}(z) := -\overset{\leftarrow}{\Psi}(-\mathrm{i}z)$ is well defined as a Laplace exponent for $\mathrm{Re}(z) \in (\alpha\rho - 1, \alpha\rho)$, having roots at 0 and $\alpha - 1$. As $\overset{\leftarrow}{\psi}(z)$ is convex for real z, recalling from the discussion following Theorem 5.19 that $\overset{\leftarrow}{\mathbf{E}}[\overset{\leftarrow}{\xi}_1] < 0$, we can deduce that $\mathrm{Re}(\Psi(-\mathrm{i}z)) > 0$ for $\mathrm{Re}(z) \in (0, \alpha - 1)$.

It turns out that the potential density $\overset{\leftarrow}{u}$ can be explicitly identified by inverting (6.14).

Theorem 6.15 *Suppose that $\alpha \in (1,2)$ and $0 < \alpha\hat{\rho}, \alpha\rho < 1$. For $x > 0$, we have*

$$\overset{\leftarrow}{u}(x) = -\frac{1}{\pi}\Gamma(1-\alpha)\left(\frac{\sin(\pi\alpha\rho)}{\pi}\left[1 - (1 - e^{-x})^{\alpha-1}\right] + \frac{\sin(\pi\alpha\hat{\rho})}{\pi}e^{-(\alpha-1)x}\right),$$

and for $x < 0$,

$$\overset{\leftarrow}{u}(x) = -\frac{1}{\pi}\Gamma(1-\alpha)\left(\frac{\sin(\pi\alpha\rho)}{\pi} + \frac{\sin(\pi\alpha\hat{\rho})}{\pi}\left[1 - (1 - e^{x})^{\alpha-1}\right]e^{-(\alpha-1)x}\right).$$

Moreover,

$$\overset{\leftarrow}{u}(0) = -\frac{1}{\pi}\Gamma(1-\alpha)\left(\frac{\sin(\pi\alpha\rho)}{\pi} + \frac{\sin(\pi\alpha\hat{\rho})}{\pi}\right).$$

Proof Appealing to (A.17), we have that

$$\frac{1}{\overset{\leftarrow}{\Psi}(-\mathrm{i}z)} = z^{-\alpha}(1 + o(1)), \quad \mathrm{Im}(z) \to \infty, \tag{6.15}$$

which is valid uniformly in any sector $|\mathrm{Arg}(z)| < \pi - \epsilon$. This and the fact that there are no poles along the vertical line $c + \mathrm{i}\mathbb{R}$, for $c \in (0, \alpha - 1)$, allows us to invert (6.14) via the integral

$$\overleftrightarrow{u}(x) = \frac{1}{2\pi \mathrm{i}} \int_{c+\mathrm{i}\mathbb{R}} \frac{1}{\overleftrightarrow{\Psi}(-\mathrm{i}z)} \mathrm{e}^{-zx}\, \mathrm{d}z. \tag{6.16}$$

We can give a concrete value to the above integral by appealing to a standard contour integration argument in connection with Cauchy's residue theory.

The function $1/\overleftrightarrow{\Psi}(-\mathrm{i}z)$ has simple poles at the points

$$\{0, 1, 2, \ldots\} \cup \{\alpha - 1, \alpha - 2, \alpha - 3, \ldots\}.$$

Suppose that γ_R is the contour described in Figure 6.1. That is, $\gamma_R = \{c + \mathrm{i}x \colon |x| \le R\} \cup \{c + R\mathrm{e}^{\mathrm{i}\theta} \colon \theta \in (-\pi/2, \pi/2)\}$, where we recall $c \in (0, \alpha - 1)$.

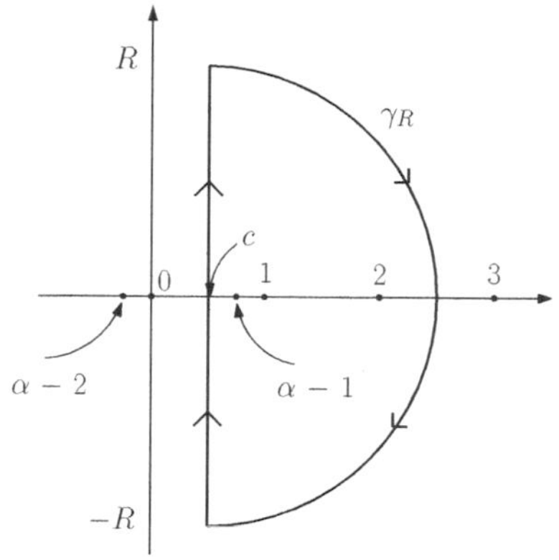

Figure 6.1 The contour γ_R

Residue calculus gives us

$$\begin{aligned}
&\frac{1}{2\pi \mathrm{i}} \int_{c+\mathrm{i}x \colon |x| \le R} \frac{1}{\overleftrightarrow{\Psi}(-\mathrm{i}z)} \mathrm{e}^{-zx}\, \mathrm{d}z \\
&\quad = -\frac{1}{2\pi \mathrm{i}} \int_{c+R\mathrm{e}^{\mathrm{i}\theta} \colon \theta \in (-\pi/2, \pi/2)} \frac{1}{\overleftrightarrow{\Psi}(-\mathrm{i}z)} \mathrm{e}^{-zx}\, \mathrm{d}z \\
&\qquad - \mathrm{Res}(1/\overleftrightarrow{\Psi}(-\mathrm{i}z) \colon z = \alpha - 1)\mathrm{e}^{-(\alpha-1)x} \\
&\qquad - \sum_{1 \le k \le \lfloor R \rfloor} \mathrm{Res}(1/\overleftrightarrow{\Psi}(-\mathrm{i}z) \colon z = k)\mathrm{e}^{-kx}.
\end{aligned} \tag{6.17}$$

Now fix $x \ge 0$. The uniform estimate (6.15), the positivity of x and the arc $\{c + R\mathrm{e}^{\mathrm{i}\theta} \colon \theta \in (-\pi/2, \pi/2)\}$ having length πR allows us to estimate

$$\left|\int_{c+Re^{i\theta}:\,\theta\in(-\pi/2,\pi/2)} \frac{1}{\overleftarrow{\Psi}(-iz)} e^{-zx}\,dz\right| \le CR^{-(\alpha-1)},$$

for some constant $C > 0$, and hence

$$\lim_{R\to\infty} \int_{c+Re^{i\theta}:\,\theta\in(-\pi/2,\pi/2)} \frac{1}{\overleftarrow{\Psi}(-iz)} e^{-zx}\,dz = 0.$$

Together with (6.16), we can use this convergence and take limits as $R \to \infty$ in (6.17) to conclude that

$$\begin{aligned}\overleftarrow{u}(x) = &- \mathrm{Res}(1/\overleftarrow{\Psi}(-iz)\colon z = \alpha - 1)e^{-(\alpha-1)x} \\ &- \sum_{k=1}^{\infty} \mathrm{Res}(1/\overleftarrow{\Psi}(-iz)\colon z = k)e^{-kx}.\end{aligned}$$

To compute the residues, we make straightforward use of the fact that $\mathrm{Res}(\Gamma(z)\colon z = -n) = (-1)^n/n!$, for $n \ge 0$; see (A.11). Hence, with the help of the binomial series identity, we finally obtain

$$\begin{aligned}\overleftarrow{u}(x) &= -\frac{1}{\pi}\sin(\pi\alpha\hat{\rho})\Gamma(1-\alpha)e^{-(\alpha-1)x} + \frac{1}{\pi}\sin(\pi\alpha\rho)\sum_{k=1}^{\infty}\frac{\Gamma(1-\alpha+k)}{k!}e^{-kx} \\ &= -\frac{1}{\pi}\sin(\pi\alpha\hat{\rho})\Gamma(1-\alpha)e^{-(\alpha-1)x} \\ &\qquad + \frac{1}{\pi}\sin(\pi\alpha\rho)\Gamma(1-\alpha)\left[(1-e^{-x})^{\alpha-1} - 1\right],\end{aligned}$$

which is valid for $x > 0$.

The proof in the case $x < 0$ is identical, except that we need to shift the arc of the contour γ_R to extend into the negative part of the complex plane, catching the poles $\{0, \alpha-2, \alpha-3, \cdots\}$. The details are left to the reader. □

Theorem 6.16 *Suppose that $\alpha \in (1,2)$ and $0 < \alpha\hat{\rho}, \alpha\rho < 1$. For any two distinct points x, b in $\mathbb{R}$ that are also different from the origin,*

$$\mathbb{P}_x(\tau^{\{b\}} < \tau^{\{0\}}) = \frac{|b|^{\alpha-1}s(b) - |b-x|^{\alpha-1}s(b-x) + |x|^{\alpha-1}s(-x)}{|b|^{\alpha-1}(\sin(\pi\alpha\rho) + \sin(\pi\alpha\hat{\rho}))}, \tag{6.18}$$

where $s(y) = \mathbf{1}_{(y>0)}\sin(\pi\alpha\rho) + \mathbf{1}_{(y<0)}\sin(\pi\alpha\hat{\rho})$.

Proof There are only two cases to consider: $x < b$ and $x > b$. In the first case, note that $-\log(x/b) > 0$. We therefore use the first of the two expressions for $\overleftarrow{u}(x)$ in Theorem 6.15 for the identity (6.13). We have

$$
\begin{aligned}
&\mathbb{P}_x(\tau^{\{b\}} < \tau^{\{0\}}) \\
&= \frac{\sin(\pi\alpha\rho)\left[1-(1-x/b)^{\alpha-1}\right] + \sin(\pi\alpha\hat{\rho})(x/b)^{\alpha-1}}{(\sin(\pi\alpha\rho)+\sin(\pi\alpha\hat{\rho}))} \\
&= \frac{b^{\alpha-1}\sin(\pi\alpha\rho) - (b-x)^{\alpha-1}\sin(\pi\alpha\rho) + x^{\alpha-1}\sin(\pi\alpha\hat{\rho})}{b^{\alpha-1}(\sin(\pi\alpha\rho)+\sin(\pi\alpha\hat{\rho}))},
\end{aligned}
$$

as required. When $x > b$, we have $-\log(x/b) < 0$ and we use the second of the two expressions in Theorem 6.15 for the identity (6.13). In that case, we have

$$
\begin{aligned}
&\mathbb{P}_x(\tau^{\{b\}} < \tau^{\{0\}}) \\
&= \frac{\sin(\pi\alpha\hat{\rho})\left[1-(1-b/x)^{\alpha-1}\right](x/b)^{\alpha-1} + \sin(\pi\alpha\rho)}{(\sin(\pi\alpha\rho)+\sin(\pi\alpha\hat{\rho}))} \\
&= \frac{b^{\alpha-1}\sin(\pi\alpha\rho) - (x-b)^{\alpha-1}\sin(\pi\alpha\hat{\rho}) + x^{\alpha-1}\sin(\pi\alpha\hat{\rho})}{b^{\alpha-1}(\sin(\pi\alpha\rho)+\sin(\pi\alpha\hat{\rho}))}.
\end{aligned}
$$

Now note that for $x, b > 0$, we have

$$
\mathbb{P}_x(\tau^{\{0\}} < \tau^{\{b\}}) = 1 - \mathbb{P}_x(\tau^{\{b\}} < \tau^{\{0\}})
$$

and hence, for $b < 0 < x$, by spatial homogeneity,

$$
\begin{aligned}
\mathbb{P}_x(\tau^{\{b\}} < \tau^{\{0\}}) &= \mathbb{P}_{x-b}(\tau^{\{0\}} < \tau^{\{|b|\}}) \\
&= \frac{|b|^{\alpha-1}\sin(\pi\alpha\hat{\rho}) + x^{\alpha-1}\sin(\pi\alpha\hat{\rho}) - (x-b)^{\alpha-1}\sin(\pi\alpha\hat{\rho})}{|b|^{\alpha-1}(\sin(\pi\alpha\rho)+\sin(\pi\alpha\hat{\rho}))} \\
&= \frac{|b|^{\alpha-1}s(b) - |b-x|^{\alpha-1}s(b-x) + x^{\alpha-1}s(-x)}{|b|^{\alpha-1}(\sin(\pi\alpha\rho)+\sin(\pi\alpha\hat{\rho}))}.
\end{aligned}
$$

When $b > 0 > x$ or $x, b < 0$, we have

$$
\begin{aligned}
\mathbb{P}_x(\tau^{\{b\}} < \tau^{\{0\}}) &= \mathbb{P}_{-x}(\tau^{\{-b\}} < \tau^{\{0\}})\Big|_{\rho\leftrightarrow\hat{\rho}} \\
&= \frac{|-b|^{\alpha-1}s(-b) - |b-x|^{\alpha-1}s(-b+x) + (-x)^{\alpha-1}s(x)}{|b|^{\alpha-1}(\sin(\pi\alpha\rho)+\sin(\pi\alpha\hat{\rho}))}\Bigg|_{\rho\leftrightarrow\hat{\rho}} \\
&= \frac{|b|^{\alpha-1}s(b) - |b-x|^{\alpha-1}s(b-x) + |x|^{\alpha-1}s(-x)}{|b|^{\alpha-1}(\sin(\pi\alpha\rho)+\sin(\pi\alpha\hat{\rho}))},
\end{aligned}
$$

where $\rho \leftrightarrow \hat{\rho}$ means that the roles of ρ and $\hat{\rho}$ are interchanged. □

6.6 First hitting of a point

Recall that, for $x \in \mathbb{R}$, $\tau^{\{x\}} = \inf\{t > 0: X_t = x\}$. We are interested in characterising the distribution of $\tau^{\{x\}}$. As mentioned earlier, we know that $\mathbb{P}(\tau^{\{x\}} < \infty) = 1$ (see e.g. Table 3.4) and, for example, in the case that X has two-sided jumps, we can confirm this from the conclusion of Theorem 6.16. We have,

$$\lim_{x\to\infty} \mathbb{P}(\tau^{\{b\}} < \tau^{\{-x\}}) = \lim_{x\to\infty} \frac{(b+x)^{\alpha-1}\sin(\pi\alpha\rho) - |b|^{\alpha-1}s(b) + x^{\alpha-1}\sin(\pi\alpha\hat{\rho})}{(b+x)^{\alpha-1}(\sin(\pi\alpha\rho)+\sin(\pi\alpha\hat{\rho}))} = 1,$$

where $s(b) = \mathbf{1}_{(b>0)}\sin(\pi\alpha\rho) + \mathbf{1}_{(b<0)}\sin(\pi\alpha\hat{\rho})$.

For $\alpha \in (1,2)$, Lemma 2.19 and duality tell us that, for $q > 0$,

$$\mathbb{E}\left[\mathrm{e}^{-q\tau^{\{x\}}}\right] = \frac{u^{(q)}(x)}{u^{(q)}(0)}, \qquad x \in \mathbb{R},$$

where $u^{(q)}$ is the density of the occupation measure

$$U^{(q)}(\mathrm{d}x) = \int_0^\infty \mathrm{e}^{-qt}\mathbb{P}(X_t \in \mathrm{d}x)\,\mathrm{d}t = \left(\int_0^\infty \mathrm{e}^{-qt}\mathrm{p}_t(x)\,\mathrm{d}t\right)\mathrm{d}x, \qquad x \in \mathbb{R},$$

and $\mathrm{p}_t(x)$ is the probability density function of X_t under $\mathbb{P}$, for $t > 0$. Observe from self-similarity that $\mathrm{p}_t(x) = t^{-1/\alpha}\mathrm{p}_1(t^{-1/\alpha}x)$, where $\mathrm{p}_1(x) = p(x,\alpha,\rho)$, which was defined in Section 1.4. It follows that

$$\mathbb{E}\left[\mathrm{e}^{-q\tau^{\{x\}}}\right] = \frac{1}{u^{(q)}(0)}\int_0^\infty \mathrm{e}^{-qt}t^{-1/\alpha}\mathrm{p}_1(t^{-1/\alpha}x)\,\mathrm{d}t.$$

Note that by setting $x = 0$, since the left-hand side above is equal to unity, this tells us that

$$u^{(q)}(0) = \mathrm{p}_1(0)\int_0^\infty \mathrm{e}^{-qt}t^{-1/\alpha}\,\mathrm{d}t = \frac{\sin(\pi\rho)}{\pi}\Gamma(1+1/\alpha)\Gamma(1-1/\alpha)q^{-1+1/\alpha},$$

where the second equality follows by appealing to the integral representation of the gamma function and (1.33).

In a similar spirit to (5.38), we have by scaling that

$$\tau^{\{x\}} \overset{(\mathrm{d})}{=} x^\alpha\tau^{\{1\}}, \tag{6.19}$$

thus we can proceed with our computations for the special case that $x = 1$. To this end, let $\Lambda = X_1^{-\alpha}\mathbf{1}_{(X_1>0)}$. With a change of variable $u = t^{-1/\alpha}$, we can now write

$$\begin{aligned}\mathbb{E}\left[\mathrm{e}^{-q\tau^{\{1\}}}\right] &= \frac{\pi}{\sin(\pi\rho)}\frac{\alpha q^{(\alpha-1)/\alpha}}{\Gamma(1+1/\alpha)\Gamma(1-1/\alpha)}\int_0^\infty \mathrm{e}^{-q/u^\alpha}\mathrm{p}_1(u)u^{-\alpha}\,\mathrm{d}u\\ &= \frac{\pi}{\sin(\pi\rho)}\frac{\alpha q^{(\alpha-1)/\alpha}}{\Gamma(1+1/\alpha)\Gamma(1-1/\alpha)}\mathbb{E}[\mathrm{e}^{-q\Lambda}\Lambda].\end{aligned}$$

On the other hand, we also have, for $s \in (0,1)$, that

$$
\begin{aligned}
\mathbb{E}[(\tau^{\{1\}})^{-s}] &= \frac{1}{\Gamma(s)}\int_0^\infty \mathbb{E}[\mathrm{e}^{-q\tau^{\{1\}}}]q^{s-1}\,\mathrm{d}q \\
&= \frac{\pi}{\sin(\pi\rho)}\frac{\alpha}{\Gamma(1+1/\alpha)\Gamma(1-1/\alpha)}\mathbb{E}\left[\Lambda\int_0^\infty \mathrm{e}^{-q\Lambda}q^{s-1/\alpha}\,\mathrm{d}q\right] \\
&= \frac{\pi}{\sin(\pi\rho)}\frac{\alpha}{\Gamma(1+1/\alpha)\Gamma(1-1/\alpha)}\frac{\Gamma(1-1/\alpha+s)}{\Gamma(s)}\mathbb{E}[\Lambda^{\frac{1}{\alpha}-s}] \\
&= \frac{\pi}{\sin(\pi\rho)}\frac{\alpha}{\Gamma(1+1/\alpha)\Gamma(1-1/\alpha)}\frac{\Gamma(1-1/\alpha+s)}{\Gamma(s)}\mathbb{E}[X_1^{\alpha s-1}\mathbf{1}_{(X_1>0)}].
\end{aligned}
$$

From Theorem 1.13, we note that

$$
\mathbb{E}\left[X_1^{\alpha s-1}\mathbf{1}_{(X_1>0)}\right] = \frac{\sin(\pi\rho(\alpha s-1))}{\sin(\pi(\alpha s-1))}\frac{\Gamma(1-s+1/\alpha)}{\Gamma(2-\alpha s)}.
$$

All together, we have

$$
\mathbb{E}[(\tau^{\{1\}})^{-s}] = \frac{\pi\alpha}{\sin(\pi\rho)}\frac{\Gamma(1-1/\alpha+s)\Gamma(1-s+1/\alpha)}{\Gamma(1+1/\alpha)\Gamma(1-1/\alpha)\Gamma(2-\alpha s)\Gamma(s)}\frac{\sin(\pi\rho(\alpha s-1))}{\sin(\pi(\alpha s-1))}.
$$

After some algebra, involving repeated use of the reflection formula for the gamma function, see (A.12), the above expression can be simplified to

$$
\mathbb{E}[(\tau^{\{1\}})^{-s}] = \frac{\sin(\pi/\alpha)}{\sin(\pi\rho)}\frac{\sin(\pi\rho(\alpha s-1))}{\sin(\pi(s-1/\alpha))}\frac{\Gamma(1+\alpha s)}{\Gamma(1+s)},
$$

and we get the following theorem.

Theorem 6.17 *Suppose that $\alpha \in (1,2)$. For $s \in (0,1)$,*

$$
\mathbb{E}[(\tau^{\{1\}})^{s-1}] = \frac{\sin(\pi/\alpha)}{\sin(\pi\rho)}\frac{\sin(\pi\rho(\alpha-1-\alpha s))}{\sin(\pi(1-1/\alpha-s))}\frac{\Gamma(\alpha+1-\alpha s)}{\Gamma(2-s)}. \tag{6.20}
$$

In principal one may use the ideas found in the proofs of, for example, Theorems 1.18 and 4.22 to perform an inverse Mellin transform of (6.20). It turns out that, for (6.20), there are some complications that one needs to be wary of in doing so, which we address in the comments section at the end of this chapter. We skip the details in preference of continuing with our exposition of fluctuation identities.

We complete this section by returning to Theorem 6.16 from which we can draw a simple conclusion about the resolvent of the stable process with killing at the origin,

$$
U^{\{0\}}(x,\mathrm{d}b) = \int_0^\infty \mathbb{P}_x(X_t \in \mathrm{d}b,\, t < \tau^{\{0\}})\,\mathrm{d}t = \int_0^\infty \mathbb{P}_x^{\{0\}}(X_t \in \mathrm{d}b)\,\mathrm{d}t.
$$

Let us momentarily restrict our discussion to the setting that $\alpha \in (1,2)$ and $0 < \alpha\hat{\rho}, \alpha\rho < 1$. Without loss of generality, assume that $x, b > 0$. Using the strong Markov property, we have

$$U^{\{0\}}(x, \mathrm{d}b) = \mathbb{P}_x^{\{0\}}(\tau^{\{b\}} < \infty)U^{\{0\}}(b, \mathrm{d}b) = \mathbb{P}_x(\tau^{\{b\}} < \tau^{\{0\}})U^{\{0\}}(b, \mathrm{d}b). \quad (6.21)$$

If we pre-emptively assume that $U^{\{0\}}(x, \mathrm{d}b)$ has a density, written $u^{\{0\}}(x, b)$, then we can use (6.18) to deduce the result

$$u^{\{0\}}(x, b) = c_b \left(|b|^{\alpha-1} s(b) - |b - x|^{\alpha-1} s(b - x) + |x|^{\alpha-1} s(-x)\right)$$

for some constant $c_b \in (0, \infty)$ (which could in principle depend on b), where $s(x) = \sin(\pi\alpha\rho)\mathbf{1}_{(x\geq 0)} + \sin(\pi\alpha\hat{\rho})\mathbf{1}_{(x<0)}$. Pinning down the constant c_b is difficult. Formally speaking,

$$\int_{\mathbb{R}} u^{\{0\}}(x, b)\mathrm{d}b = \int_0^\infty \mathbb{P}_x(t < \tau^{\{0\}})\, \mathrm{d}t = \mathbb{E}_x[\tau^{\{0\}}]. \quad (6.22)$$

However, the right-hand side turns out to be infinite. To see why, note that

$$\mathbb{E}_x[\tau^{\{0\}}] = \mathbb{E}_x[\vec{\tau}^{\{0\}}] + \hat{\mathbb{E}}_{-x}[\vec{\tau}^{\{0\}}],$$

where $\vec{\tau}^{\{0\}}$ is the lifetime of the censored stable process and, as usual, $(\hat{\mathbb{P}}_x, x \in \mathbb{R})$ are the probabilities of $-X$. Recall from Theorem 5.19 that the (positively) censored stable has underlying Lévy process $\vec{\xi}$ which belongs to the class of hypergeometric Lévy processes with characteristic exponent

$$\vec{\Psi}(z) = \frac{\Gamma(\alpha\rho - \mathrm{i}z)}{\Gamma(-\mathrm{i}z)} \frac{\Gamma(1 - \alpha\rho + \mathrm{i}z)}{\Gamma(1 - \alpha + \mathrm{i}z)}, \qquad z \in \mathbb{R}.$$

Accordingly, for $x = 1$, thanks to Fubini's Theorem, we have

$$\mathbb{E}_1[\vec{\tau}^{\{0\}}] = \vec{\mathbf{E}}\left[\int_0^\infty \mathrm{e}^{\alpha\vec{\xi}_t}\mathrm{d}t\right] = \int_0^\infty \vec{\mathbf{E}}\left[\mathrm{e}^{\alpha\vec{\xi}_t}\right]\mathrm{d}t, \quad (6.23)$$

where $\vec{\mathbf{P}}$ is the law of $\vec{\xi}$ issued from the origin. The expression above for $\vec{\Psi}$, indicates that the Laplace exponent of $\vec{\xi}$ (which can be seen as an analytic extension of $\vec{\Psi}$) only exists on the strip $\{z \in \mathbb{C}\colon \mathrm{Re}(z) \in (\alpha\rho, \alpha\rho - 1)\}$. Accordingly, we see in (6.23) that the right-hand side blows up.

Nonetheless, it is possible to find a precise formula for the above resolvent. As was the case for the proof of Theorem 6.12, we will have to wait to develop some additional technology in Chapter 12 before we can prove the following result. (Its proof can be found in Section 12.5.)

Theorem 6.18 *Suppose that $\alpha \in (1,2)$ and $0 < \alpha\hat{\rho}, \alpha\rho < 1$. The resolvent with killing at the origin is absolutely continuous such that, for x, y in $\mathbb{R}$ and distinct from the origin,*

$$u^{\{0\}}(x,y) = -\frac{1}{\pi^2}\Gamma(1-\alpha)\left(|y|^{\alpha-1}s(y) - |y-x|^{\alpha-1}s(y-x) + |x|^{\alpha-1}s(-x)\right),$$

where $s(x) = \sin(\pi\alpha\rho)\mathbf{1}_{(x\geq 0)} + \sin(\pi\alpha\hat{\rho})\mathbf{1}_{(x<0)}$.

6.7 First exit for the reflected process

In this section, we will consider the reflected stable process

$$\mathrm{R}_t = X_t - (0 \wedge \underline{X}_t), \qquad t \geq 0,$$

where, under $\mathbb{P}_z$, X is a stable process issued from $z \geq 0$ and $\underline{X}_t = \inf_{s\leq t} X_s$. Note that $\mathbb{P}_z(\mathrm{R}_0 = z) = 1$. To avoid trivialities, we exclude the setting that X has monotone paths throughout this section. The process R is valued in $[0,\infty)$ and, roughly speaking, its paths evolve as follows. When R, is issued from $z > 0$, it follows the same trajectory as X until time $\tau_0^- = \inf\{t > 0 : X_t < 0\}$. At this moment in time, the process R takes the value zero and thereafter, its excursions away from zero correspond precisely to the excursions of X away from its previous infimum.

As we shall see shortly, R possesses the strong Markov property and hence there are natural first passage problems which, inevitably, are very closely related to those of X. Noting that $\mathrm{R} = (\mathrm{R}_t, t \geq 0)$ is valued in $[0,\infty)$, our main concern in this section is to look at the problem of first exit from an interval of the form $[0,a]$, where $a > 0$.

We start by returning to our earlier claim that R is strong Markovian. As before $(\mathcal{F}_t : t \geq 0)$ is the filtration generated by X which is naturally enlarged.

Lemma 6.19 *The process* R *is a strong Markov process in* $(\mathcal{F}_t : t \geq 0)$, *under the family of measures* $(\mathbb{P}_z, z \geq 0)$.

Proof Suppose that τ is any stopping time with respect to $(\mathcal{F}_t : t \geq 0)$. Let $\widetilde{X}_s = X_{\tau+s} - X_\tau$, $s \geq 0$, and recall that, on $\{\tau < \infty\}$, $(\widetilde{X}_s, s \geq 0)$ is independent of $\mathcal{F}_\tau$ with the same law as $(X,\mathbb{P})$. Note that, for $s \geq 0$, on $\{\tau < \infty\}$,

$$\begin{aligned} X_{\tau+s} - (0 \wedge \underline{X}_{\tau+s}) &= X_\tau + \widetilde{X}_s - \left(0 \wedge \underline{X}_\tau \wedge \inf_{u\in[\tau,\tau+s]} X_u\right) \\ &= \widetilde{X}_s - \left[(X_\tau - (0 \wedge \underline{X}_\tau)) \wedge \left(X_\tau - \inf_{u\in[\tau,\tau+s]} X_u\right)\right] \\ &= \widetilde{X}_s - \left[\mathrm{R}_\tau \wedge \inf_{u\in[0,s]} \widetilde{X}_u\right]. \end{aligned}$$

From the right-hand side above, it is clear that the law of $\mathrm{R}_{\tau+s}$ depends only on $(\mathrm{R}_u, u \leq \tau)$ through the value of R_τ. It follows that $(\mathrm{R}_t, t \geq 0)$ is a strong Markov process. □

We are now ready to state the main result of this section, for which we define the first exit time

$$\gamma_a = \inf\{t > 0 \colon \mathrm{R}_t > a\},$$

where $a > 0$. In order for the result to be non-trivial, we further exclude the setting that X is spectrally negative.

Theorem 6.20 *Suppose $\alpha \in (0,2)$ and $0 < \alpha\rho < 1$. For $u > 0$ and $v \in [0,a]$,*

$$\mathbb{P}(\mathrm{R}_{\gamma_a} - a \in \mathrm{d}u,\ a - \mathrm{R}_{\gamma_a -} \in \mathrm{d}v) = \frac{\alpha \sin(\pi\alpha\rho)}{\pi} \frac{v^{\alpha\hat{\rho}}(a-v)^{\alpha\rho-1}}{(u+v)^{\alpha+1}} \mathrm{d}u\, \mathrm{d}v. \tag{6.24}$$

It is worth remarking that, as X cannot creep upwards over a, then it is also not possible for R to creep upwards over a; indeed it is easy to verify that the distribution in (6.24) is proper.

The proof of this result depends heavily on the understanding of the process $(R, \mathbb{P})$ in terms of its excursions from the origin. The latter is nothing more than the process of excursions of X from its running infimum $\underline{X}$. This was discussed for general Lévy processes in Section 2.14 (albeit in the context of excursions from the running maximum, which maps to the current setting by considering $-X$).

For convenience, let us transfer some of the notation across. We will write $(L_t, t \geq 0)$ for the local time of X at its minimum. From Table 3.4, we know that, since X always oscillates, we have that $L_\infty = \infty$. Denote by $\underline{\mathcal{U}}(\mathbb{R})$ the space of paths taking the form $\epsilon = (\epsilon(s), s \leq \zeta)$ which are right-continuous with left limits and which are strictly positive-valued on $(0, \zeta)$. When $\zeta < \infty$, we have the terminal value $\epsilon(\zeta) \leq 0$. Accordingly $\zeta = \inf\{t > 0 \colon \epsilon(t) \leq 0\}$. Then for all $t > 0$ such that $\Delta L_t^{-1} := L_t^{-1} - L_{t-}^{-1} > 0$, we can identify the excursion of R from $\{0\}$

$$\epsilon_t(s) = \mathrm{R}_{L_{t-}^{-1}+s} - \mathrm{R}_{L_{t-}^{-1}}, \qquad 0 \leq s \leq \Delta L_t^{-1}.$$

The key feature of excursion theory in the current setting that we will use is that

$$\left((t, \epsilon_t), t \geq 0 \text{ and } \Delta L_t^{-1} > 0\right)$$

is a Poisson point process with values in $[0,\infty) \times \underline{\mathcal{U}}(\mathbb{R})$ and intensity measure $\mathrm{d}t \times \mathrm{d}\underline{n}$, where $\underline{n}$ is a measure on the Skorokhod space (see Section A.10 in the Appendix), which is concentrated on $\underline{\mathcal{U}}(\mathbb{R})$.

Proof of Theorem 6.20 We start by noting that the statement of the theorem dictates that $\mathrm{R}_0 = 0$. As such, we do not need to deal with the occasion that the process exceeds a from its point of issue before reflecting at the origin. Using the thinning theorem for Poisson point processes, the first excursion that exceeds a threshold a has law given by

$$\mathbb{Q}(\cdot) := \frac{\underline{n}(\cdot \cap \{\gamma_a < \zeta\})}{\underline{n}(\gamma_a < \zeta)}, \tag{6.25}$$

where, for the generic excursion $\epsilon \in \mathcal{U}(\mathbb{R})$, we make a slight abuse of notation and use $\gamma_a = \inf\{t > 0\colon \epsilon(t) > a\}$. Next, suppose that $f\colon (0,\infty) \times [0,a] \to [0,\infty)$ is measurable and uniformly bounded and consider the martingale

$$M_\theta := \mathbb{Q}\Big(f(\epsilon(\gamma_a) - a,\ a - \epsilon(\gamma_a-))\,\mathbf{1}_{(\gamma_a<\zeta)}\,|\,\mathcal{G}_\theta\Big),$$

for $\theta \in (0,a]$, where $\mathcal{G}_\theta$ is the sigma-algebra generated by $(\epsilon_u, u \le \gamma_\theta)$ and $\overline{\epsilon}(t) = \sup_{s\le t}\epsilon(s)$, for $0 \le t \le \zeta$. The martingale property follows as a consequence of the tower property for the excursion measure and the fact that $\underline{n}(\gamma_a < \zeta) < \infty$. (If the latter were false, then the otherwise infinite rate of excursions exceeding height a would imply that $\mathbb{P}(\gamma_a = 0) = 1$, which contradicts the fact that the paths of Y are right-continuous.)

We can otherwise write

$$M_\theta = \frac{\underline{n}\Big(f(\epsilon(\gamma_a) - a,\ a - \epsilon(\gamma_a-))\,\mathbf{1}_{(\gamma_a<\zeta)}|\,\mathcal{G}_\theta\Big)}{\underline{n}(\gamma_a < \zeta|\,\mathcal{G}_\theta)}, \tag{6.26}$$

for $\theta \in (0,a]$, where the equality is the result of Bayes formula. Moreover, by the Markov property of excursions, on the event $\{\epsilon(\gamma_\theta) < a\}$, we can write

$$\begin{aligned}\underline{n}\Big(f(\epsilon(\gamma_a) - a,\ a - \epsilon(\gamma_a-))\,\mathbf{1}_{(\gamma_a<\zeta)}|\,\mathcal{G}_\theta\Big)\\ = \mathbb{E}_{\epsilon(\gamma_\theta)}\Big[f(X_{\tau_a^+} - a,\ a - X_{\tau_a^+-})\,\mathbf{1}_{(\tau_a^+<\tau_0^-)}\Big]\end{aligned} \tag{6.27}$$

and

$$\underline{n}(\gamma_a < \zeta\,|\,\mathcal{G}_\theta) = \mathbb{P}_{\epsilon(\gamma_\theta)}(\tau_a^+ < \tau_0^-). \tag{6.28}$$

Next, observing that $\epsilon(\gamma_\theta) \to 0$ as $\theta \downarrow 0$ almost surely thanks to regularity of the upper half line of X (see Section 3.4), we have, on the one hand, from (6.26), that

$$\begin{aligned}\lim_{\theta\downarrow 0} M_\theta &= \mathbb{Q}\Big[f(\epsilon(\gamma_a) - a,\ a - \epsilon(\gamma_a-))\,\mathbf{1}_{(\gamma_a<\zeta)}\Big]\\ &= \mathbb{E}\Big[f(\mathrm{R}_{\gamma_a} - a,\ a - \mathrm{R}_{\gamma_a-})\Big].\end{aligned}$$

The second equality above follows on account of the fact that the joint law of the overshoot and undershoot of the first excursion exceeding threshold a agrees with the joint law of the analogous quantities for R. Taking note of

(6.27) and (6.28), we have, with the help of Lemma 6.1, Theorem 6.2 and an application of L'Hôpital's rule, that

$$\mathbb{E}\left[f(\mathrm{R}_{\gamma_a} - a,\, a - \mathrm{R}_{\gamma_a -})\right]$$

$$= \lim_{x\downarrow 0} \frac{\int_0^\infty \int_0^a \int_0^{(a-x)\wedge v} f(u,v) \frac{\alpha \sin(\pi\alpha\rho)}{\pi} \frac{x^{\alpha\hat\rho}(a-v)^{\alpha\rho}}{(u+v)^{\alpha+1}} \frac{(a-x-y)^{\alpha\rho-1}(v-y)^{\alpha\hat\rho-1}}{(a-y)^\alpha}}{\int_0^{x/a} t^{\alpha\hat\rho-1}(1-t)^{\alpha\rho-1}\,\mathrm{d}t} \mathrm{d}y\,\mathrm{d}v\,\mathrm{d}u$$

$$= \alpha\hat\rho a^{\alpha\hat\rho} \frac{\alpha\sin(\pi\alpha\rho)}{\pi}$$

$$\times \int_0^\infty \int_0^a \int_0^v f(u,v) \frac{(a-v)^{\alpha\rho}}{(u+v)^{\alpha+1}} (a-y)^{-(1+\alpha\hat\rho)} (v-y)^{\alpha\hat\rho-1}\,\mathrm{d}y\,\mathrm{d}v\,\mathrm{d}u.$$

Setting $w = (v-y)/(a-y)$ and marginalising, the statement of the theorem follows. □

Define $\overline{\mathrm{R}}_t = \sup_{s\le t} \mathrm{R}_s$, for $t \ge 0$. If we note that, for $y \in [0,a]$, $v \le y$ and $u \ge 0$,

$$\{\overline{\mathrm{R}}_{\gamma_a -} \le y,\, a - \mathrm{R}_{\gamma_a -} \in \mathrm{d}v,\, \mathrm{R}_{\gamma_a} - a \in \mathrm{d}u\}$$
$$= \{a - \mathrm{R}_{\gamma_y -} \in \mathrm{d}v,\, \mathrm{R}_{\gamma_y} - a \in \mathrm{d}u\},$$

we easily derive the following corollary of Theorem 6.20.

Corollary 6.21 *Suppose* $\alpha \in (0,2)$ *and* $0 < \alpha\rho < 1$. *For* $u > 0$, $v \in [0,a]$ *and* $y \in [0,v]$,

$$\mathbb{P}(\overline{\mathrm{R}}_{\gamma_a -} \le y,\, \mathrm{R}_{\gamma_a} - a \in \mathrm{d}u,\, a - \mathrm{R}_{\gamma_a -} \in \mathrm{d}v)$$
$$= \frac{\alpha\sin(\pi\alpha\rho)}{\pi} \frac{(a-v)^{\alpha\rho-1}(v-a+y)^{\alpha\hat\rho}}{(u+v)^{\alpha+1}}\,\mathrm{d}y\,\mathrm{d}v\,\mathrm{d}u.$$

As noted in its proof, Theorem 6.20 does not consider the case that $\mathrm{R}_0 = z \in [0,a]$. This case may be dealt with in a relatively straightforward way, although it delivers an identity which is no object of beauty. Note that, for $z \in [0,a]$, $u > 0$ and $v \in [0,a]$,

$$\mathbb{P}_z(\mathrm{R}_{\gamma_a} - a \in \mathrm{d}u,\, a - \mathrm{R}_{\gamma_a -} \in \mathrm{d}v)$$
$$= \mathbb{P}_z(X_{\tau_a^+} - a \in \mathrm{d}u,\, a - X_{\tau_a^+ -} \in \mathrm{d}v;\, \tau_a^+ < \tau_0^-)$$
$$+ \mathbb{P}_z(\tau_0^- < \tau_a^+)\mathbb{P}(\mathrm{R}_{\gamma_a} - a \in \mathrm{d}u,\, a - \mathrm{R}_{\gamma_a -} \in \mathrm{d}v), \qquad (6.29)$$

where we recall that

$$\tau_a^+ = \inf\{t > 0 \colon X_t > a\} \qquad \text{and} \qquad \tau_0^- = \inf\{t > 0 \colon X_t < 0\}.$$

Again referring to Lemma 6.1 and also to Corollary 6.6, as well as the expression in (6.24), the relevant expressions for each of the probabilities in (6.29) may be plugged in.

We can also obtain an identity for the resolvent of the reflected process up to exiting the interval $[0, a]$,

$$V^{[0,a]}(x, \mathrm{d}y) = \int_0^\infty \mathbb{P}_x(\mathrm{R}_t \in \mathrm{d}y,\ t < \gamma_a)\mathrm{d}t, \qquad y \in [0, a].$$

Corollary 6.22 *Suppose $\alpha \in (0, 2)$ and $0 < \alpha\rho < 1$. Fix $a > 0$. For each $x \in [0, a]$, the resolvent $V^{[0,a]}(0, \mathrm{d}y)$ has a density, denoted by $v^{[0,a]}(0, y)$, $y \in [0, a]$, which satisfies*

$$v^{[0,a]}(0, y) = \frac{1}{\Gamma(\alpha)}(a - y)^{\alpha\hat{\rho}} y^{\alpha\rho - 1}\,\mathrm{d}y. \tag{6.30}$$

Proof From (6.24), we have that, for $u > 0$ and $y \in [0, a]$,

$$\mathbb{P}(\mathrm{R}_{\gamma_a} - a \in \mathrm{d}u,\ \mathrm{R}_{\gamma_a-} \in \mathrm{d}y) = \frac{\alpha \sin(\pi\alpha\rho)}{\pi} \frac{(a - y)^{\alpha\hat{\rho}} y^{\alpha\rho - 1}}{(u + a - y)^{\alpha + 1}}\,\mathrm{d}u\,\mathrm{d}v. \tag{6.31}$$

On the other hand, appealing to the use of the compensation formula, in the spirit of the computations in (6.1) and (6.2), we have that, for positive, bounded and measurable functions $f : [0, \infty) \times [0, a] \to [0, \infty)$,

$$\begin{aligned}
&\mathbb{E}\left[f(\mathrm{R}_{\gamma_a} - a,\ \mathrm{R}_{\gamma_a-})\right] \\
&= \mathbb{E}\left[\sum_{t>0} \mathbf{1}_{(t<\gamma_a)} f(\mathrm{R}_{t-} + \Delta X_t - a,\ \mathrm{R}_{t-}) \mathbf{1}_{(\mathrm{R}_{t-} + \Delta X_t > a)}\right] \\
&= \mathbb{E}\left[\int_0^\infty \int_{\mathbb{R}} \mathbf{1}_{(t<\gamma_a)} f(\mathrm{R}_{t-} + x - a,\ \mathrm{R}_{t-}) \mathbf{1}_{(\mathrm{R}_{t-} + x > a)} \Pi(\mathrm{d}x)\,\mathrm{d}t\right] \\
&= \Gamma(1 + \alpha)\frac{\sin(\pi\alpha\rho)}{\pi} \int_0^\infty \int_{[0,a]} V^{[0,a]}(0, \mathrm{d}y) f(y + x - a,\ y) \mathbf{1}_{(y + x - a > 0)} \frac{1}{x^{1+\alpha}}\,\mathrm{d}x \\
&= \Gamma(1 + \alpha)\frac{\sin(\pi\alpha\rho)}{\pi} \int_0^\infty \int_{[0,a]} V^{[0,a]}(0, \mathrm{d}y) f(u,\ y) \frac{1}{(u + a - y)^{1+\alpha}}\,\mathrm{d}u,
\end{aligned} \tag{6.32}$$

where the sum in the first equality is taken over the Poisson point process of jumps with intensity $\mathrm{d}t \times \Pi(\mathrm{d}x)$ (cf. Appendix A.13), where Π is the Lévy measure given by (3.1). Now comparing the two joint laws in (6.31) and (6.32), the result follows. □

Just as with the discussion surrounding (6.29) we can push the conclusion of Corollary 6.22 just a little further to get an explicit identity for $v^{[0,a]}(x, \mathrm{d}y)$, albeit being a little ugly (and therefore we refrain from writing it out in detail).

Note that, for $x, y \in (0, a)$, the measure $V^{[0,a]}(x, \mathrm{d}y)$ is absolutely continuous with density

$$v^{[0,a]}(x, \mathrm{d}y) = u^{[0,a]}(x, \mathrm{d}y) + \mathbb{P}_x(\tau_0^- < \tau_a^+)v^{[0,a]}(0, \mathrm{d}y),$$

where $U^{[0,a]}$ was given in Theorem 6.4 and $\mathbb{P}_x(\tau_0^- < \tau_a^+)$ can be recovered from Lemma 6.1.

6.8 Comments

Lemma 6.1 and Corollary 6.3 first appeared in Rogozin [185]. Using the approach given here, Theorem 6.2 was established in Kyprianou et al. [127]. Theorem 6.4 was first proved in the symmetric case in Blumenthal et al. [39] and in the more general setting by Kyprianou and Watson [137] as well as Profeta and Simon [175]. Corollaries 6.5 and 6.6 can similarly be deduced or found directly in the aforementioned literature. Theorem 6.7 and Corollary 6.8 can be found in both Kyprianou [124] and Profeta and Simon [175]. Theorem 6.9 and Corollary 6.11 were first proved in the symmetric case in Blumenthal et al. [39], then later in Kyprianou et al. [129] and Profeta and Simon [175]. Proposition 6.10 can again be found in Kyprianou et al. [129], but first appeared in Port [171]. An incomplete version of Theorem 6.12 can be found in Kyprianou et al. [129], however a full statement was proved in Profeta and Simon [175]. An alternative proof of Theorem 6.12 was put forward in Kyprianou [125] and this is the one that will be give later in the forthcoming Section 12.4.

Theorems 6.13 and 6.14 are taken from Kyprianou et al. [132]. Theorem 6.16 was first proved in Getoor [76], however the proof given here is new, based on self-similarity rather than a potential analytic approach. Theorem 6.17 was first proved in Kuznetsov et al. [120], together with an asymptotic expansion of the density of the law of $\tau^{\{1\}}$. The shorter proof of Theorem 6.17 that we offer here comes from Letemplier and Simon [141]. As alluded to earlier, care is needed when performing the inverse Mellin transform using standard techniques. One of the problems highlighted in Kuznetsov et al. [120] is the multiplicity of poles when α is rational. As such the results obtained there place restrictions on α. In the spectrally positive case, a full expansion for the aforesaid density was given in Peskir [167] using a methodology different to Mellin inverse transform, which holds for all $\alpha \in (1, 2)$. Theorem 6.18 is a relatively new result, found in the review article Kyprianou [125], as are many of the proofs given in this chapter. The proof of Theorem 6.18 could in principle be derived from Theorem 6.12 by taking limits as $a \to 0$ and invoking monotonicity, however this seems analytically rather difficult to execute on account of the limit resulting in the difference of two infinities. Instead, we opted for a more elegant proof based on the Riesz–Bogdan–Żak transformation, given in Section 12.5.

The method for Theorem 6.20 using excursion theory in the stable setting was developed in Kyprianou [122] alongside similar ideas appearing in the related setting of spectrally one-sided Lévy processes in Avram et al. [9]. Baurdoux [12] exploited other excursion theoretic techniques to analyse resolvents for general reflected Lévy processes. His results include Corollary 6.22 for the case of symmetric stable processes. Appealing to Baurdoux's approach Kyprianou and Watson [137] proved Corollary 6.21.

7

Doney–Kuznetsov factorisation and the maximum

In this chapter, we are interested in describing the law of the running supremum of a stable process at an independent and exponentially distributed random time, as well as at a fixed time. As alluded to in Theorem 2.22, this boils down to a better understanding of the space-time Wiener–Hopf factorisation.

Stable processes provide a rare example of a Lévy process with discontinuous paths for which a full space-time Wiener–Hopf can be developed in explicit detail. Overwhelmingly, the majority of what is known in this arena is due to the work of R. A. Doney and A. Kuznetsov. We present each of their perspectives here. Accordingly, we collectively discuss the Doney–Kuznetsov factorisation. Additional remarks regarding the precise layout of the literature are found in the comments at the end of this chapter as usual.

Thanks to self-similarity, it will turn out that the Mellin transform of one of the Wiener–Hopf factors is associated with the Mellin transform of the law of the running supremum at a fixed time. We are able to provide a series representation of the density of the latter by making a connection with the earlier theory we have developed in Section 4.5 on the law of integrated exponential hypergeometric Lévy processes.

7.1 Kuznetsov's factorisation

Recall that X denotes a stable process with parameters (α,ρ) that are admissible, that is, $(\alpha,\rho)\in\mathcal{A}$ where the set $\mathcal{A}$ is defined as in (3.11). We also recall that $\overline{X}_t := \sup_{s\le t} X_s$ and $\underline{X}_t := \inf_{s\le t} X_s$, denote its running supremum and infimum, respectively, at time $t\ge 0$. Let $\mathbf{e}_q$ be an exponentially distributed random time with rate $q\ge 0$, which is independent of X, and we denote by Ψ_q^+ and Ψ_q^- the Laplace transforms of $\overline{X}_{\mathbf{e}_q}$ and $\underline{X}_{\mathbf{e}_q}$, respectively. That is to say,

$$\Psi_q^+(z) = \mathbb{E}\left[\mathrm{e}^{-z\overline{X}_{\mathbf{e}_q}}\right] \quad \text{and} \quad \Psi_q^-(z) = \mathbb{E}\left[\mathrm{e}^{z\underline{X}_{\mathbf{e}_q}}\right] \quad \text{for} \quad z\ge 0. \tag{7.1}$$

From Theorem 2.22, the functions Ψ_q^+ and Ψ_q^- are known as the (space-time) Wiener–Hopf factors. Their characterisation is one of the aims of this chapter. We call such characterisation the *Doney–Kuznetsov factorisation*.

In the spectrally one-sided cases, but excluding the case of monotone paths, characterising the Wiener–Hopf factors (7.1) is straightforward. Indeed, let us assume that X is spectrally negative with index $\alpha \in (1,2)$ and hence positivity parameter $\rho = \alpha^{-1}$. Observe from Lemma 2.29 that

$$\tau_x^+ = \inf\left\{t > 0 : \overline{X}_t > x\right\}, \qquad x \geq 0,$$

is a subordinator and, moreover, from the discussion of creeping in Section 3.5, it is necessarily stable. Since the stable process is self-similar with index α, the subordinator $(\tau_x^+, x \geq 0)$ is also self-similar with index α^{-1}. In other words, we have

$$\mathbb{E}\left[\mathrm{e}^{-\lambda\tau_x^+}\right] = \mathrm{e}^{-x\lambda^{\frac{1}{\alpha}}}, \qquad \lambda > 0.$$

On the other hand, we see

$$\mathbb{P}\left(\overline{X}_{\mathbf{e}_q} > x\right) = \mathbb{P}\left(\tau_x^+ < \mathbf{e}_q\right) = \mathbb{E}\left[\mathrm{e}^{-q\tau_x^+}\right] = \mathrm{e}^{-xq^{\frac{1}{\alpha}}},$$

implying that $\overline{X}_{\mathbf{e}_q}$ has the same distribution as an exponential random variable with parameter $q^{\frac{1}{\alpha}}$. Recall that the Laplace exponent satisfies $\psi(z) = \log\mathbb{E}[\mathrm{e}^{zX_1}] = z^\alpha$. Using this expression for ψ together with the Wiener–Hopf factorisation, we deduce

$$\Psi_q^+(z) = \frac{q^{\frac{1}{\alpha}}}{q^{\frac{1}{\alpha}} + z} \quad \text{and} \quad \Psi_q^-(z) = \frac{q\left(q^{\frac{1}{\alpha}} - z\right)}{q^{\frac{1}{\alpha}}\left(q - z^\alpha\right)} \quad \text{for} \quad z \geq 0.$$

Thanks to duality, we also obtain the following expressions in the spectrally positive case

$$\Psi_q^+(z) = \frac{q\left(q^{\frac{1}{\alpha}} + z\right)}{q^{\frac{1}{\alpha}}\left(q + z^\alpha\right)} \quad \text{and} \quad \Psi_q^-(z) = \frac{q^{\frac{1}{\alpha}}}{q^{\frac{1}{\alpha}} - z} \quad \text{for} \quad z \geq 0.$$

For the remainder of this section we will thus keep our attention on the two-sided jump setting, that is, $0 < \alpha\hat{\rho}, \alpha\rho < 1$. From duality, the Laplace transform Ψ_q^- has exactly the same form as Ψ_q^+ albeit the role of ρ is played by $\hat{\rho}$. The scaling property implies

$$\Psi_q^+(z) = \Psi_1^+(zq^{-1/\alpha}), \qquad z \geq 0, \tag{7.2}$$

with the same property holding for Ψ_q^-. In other words, in order to describe the Wiener–Hopf factors it is enough to study Ψ_1^+.

Before we can state Kuznetsov's identification of these two factors, we need to introduce a family of special functions known as *double sine functions*. For

$\tau > 0$, the double sine function $S_2(z;\tau)$ is defined in terms of the double gamma function $G(z,\tau)$, which was introduced in (4.42), as follows

$$S_2(z;\tau) = (2\pi)^{(1+\tau)/2-z} \frac{G(z;\tau)}{G(1+\tau-z;\tau)}, \qquad z \in \mathbb{C}. \tag{7.3}$$

The double sine function also has a Weierstrass product representation, see (A.26), which can be simplified when $\tau = 1$ (see (A.27)). Moreover, the function $z \mapsto S_2(z;\tau)$ is meromorphic, which has zeros at points $\{-m\tau - n\colon m, n \geq 0\}$ and poles at points $\{m\tau + n\colon m, n \geq 1\}$. All zeros and poles are simple if and only if τ is irrational.

The double sine function also satisfies

$$S_2(z+1;\tau) = \frac{S_2(z;\tau)}{2\sin(\pi z/\tau)}, \quad S_2(z+\tau) = \frac{S_2(z;\tau)}{2\sin(\pi z)}, \tag{7.4}$$

and has the normalising condition $S_2((1+\tau)/2;\tau) = 1$. From the asymptotic behaviour of the double sine function (A.29), we deduce that for every $b, c \in \mathbb{R}$,

$$\begin{aligned} &\left| S_2(b + \mathrm{i}\tau \log(\mathrm{e}^{\mathrm{i}c} y)/(2\pi);\tau) S_2(b - \mathrm{i}\tau \log(\mathrm{e}^{\mathrm{i}c} y)/(2\pi);\tau) \right| \\ &\qquad = \begin{cases} y^{1/2+\tau/2-b}(1+o(1)), & \text{as } y \to +\infty, \\ y^{-1/2-\tau/2+b}(1+o(1)), & \text{as } y \to 0^+. \end{cases} \end{aligned} \tag{7.5}$$

Moreover, the above asymptotic result holds uniformly in b and c on compact subsets of $\mathbb{R}$. From identity (7.3), we also observe that the function $S_2(z;\tau)$ satisfies the following reflection formula

$$S_2(z;\tau) S_2(1+\tau-z;\tau) = 1. \tag{7.6}$$

For further details and properties of the double sine function we refer to the Appendix A.5. We are otherwise now ready to state Kuznetsov's factorisation.

Theorem 7.1 (Kuznetsov's factorisation) *Assume that* $(\alpha,\rho) \in \mathcal{A}$. *For* $\mathrm{Re}(z) \geq 0$, *we have*

$$\begin{aligned} \Psi_1^+(z) = {}& z^{-\alpha\rho/2} S_2((1+\alpha(1+\rho))/2 + \mathrm{i}\alpha \log(z)/(2\pi);\alpha) \\ & \times S_2((1+\alpha(1+\rho))/2 - \mathrm{i}\alpha \log(z)/(2\pi);\alpha), \end{aligned} \tag{7.7}$$

and $\Psi_1^-(z)$ *can be obtained from the above identity by replacing* ρ *by* $\hat{\rho}$.

7.2 Quasi-periodicity

A key observation in proving Theorem 7.1 is the identification of a quasi-periodic property associated to the factors Ψ_1^+ and Ψ_1^-. In this section, we develop a proposition (below), which characterises the aforesaid quasi-periodicity. We can then use what we know about double sine functions to guess and rigorously verify the factorisation in Theorem 7.1 using this notion of quasi-periodicity.

To this end, let us introduce the following notation, $\Lambda_{a,b} := \{z \in \mathbb{C} : a < \mathrm{Im}(z) < b\}$, the horizontal open strip, $\overline{\Lambda_{a,b}} := \{z \in \mathbb{C} : a \le \mathrm{Im}(z) \le b\}$, the closed horizontal strip, and $\gamma_a := \{z \in \mathbb{C} : \mathrm{Im}(z) = a\}$, the horizontal line.

On account of the fact that both $\Psi_1^+(z)$ and $\Psi^-(z)$ are Laplace transforms of positive random variables, it is immediately obvious that they are analytic in the half-plane $\mathrm{Re}(z) > 0$ and continuous in the closed half-plane $\mathrm{Re}(z) \ge 0$. On the other hand, from the Wiener–Hopf factorisation (2.34) and the explicit form of the characteristic exponent of the stable process (see (3.9)), the characteristic exponent of X, we have, for example, that

$$\Psi_1^+(-\mathrm{i}z)\Psi_1^-(\mathrm{i}z) = \frac{1}{1 + \mathrm{e}^{\pi\mathrm{i}\alpha(1/2-\rho)}z^\alpha}, \qquad \text{for} \quad z > 0. \tag{7.8}$$

In the next proposition, identity (7.8), together with Schwartz's reflection principle, will allow us to play the analytic properties of Ψ_1^+ off against those of Ψ_1^- (and vice versa). This results in analytic continuation and a functional equation satisfied by each of the Wiener–Hopf factors. The aforesaid functional equation demonstrates the aforementioned quasi-periodicity.

Proposition 7.2 *For $|\mathrm{Im}(w)| < \pi/2$, we define $f(w) = \Psi_1^+(\mathrm{e}^w)$ and $\hat{f}(w) = \Psi_1^-(\mathrm{e}^w)$. The functions $f(w)$ and $\hat{f}(w)$ can be analytically continued to meromorphic functions and satisfy the quasi-periodic functional equations*

$$f(w + 2\pi\mathrm{i}) = \mathrm{e}^{-\pi\mathrm{i}\alpha\rho}\frac{\cos(\mathrm{i}\alpha w/2 - \pi\alpha\hat{\rho}/2)}{\cos(\mathrm{i}\alpha w/2 - \pi\alpha(1+\rho)/2)}f(w), \tag{7.9}$$

$$\hat{f}(w + 2\pi\mathrm{i}) = \mathrm{e}^{-\pi\mathrm{i}\alpha\hat{\rho}}\frac{\cos(\mathrm{i}\alpha w/2 - \pi\alpha\rho/2)}{\cos(\mathrm{i}\alpha w/2 - \pi\alpha(1+\hat{\rho})/2)}\hat{f}(w), \tag{7.10}$$

for $w \in \mathbb{C}$.

Proof We begin by applying a change of variable of the form $z = \mathrm{e}^w$, in the Wiener–Hopf factorisation (7.8), to deduce

$$f(w - \pi\mathrm{i}/2)\hat{f}(w + \pi\mathrm{i}/2) = \frac{1}{1 + \mathrm{e}^{\pi\mathrm{i}\alpha(1/2-\rho)+\alpha w}}, \qquad w \in \mathbb{R}. \tag{7.11}$$

From its representation as an expectation, cf. (7.1), the function $f(w - \pi\mathrm{i}/2)$ is analytic in the open strip $\Lambda_{0,\pi}$ and is continuous in $\overline{\Lambda_{0,\pi}}$. Similarly, the function $\hat{f}(w + \pi\mathrm{i}/2)$ is analytic in the open strip $\Lambda_{-\pi,0}$ and is continuous in $\overline{\Lambda_{-\pi,0}}$. Rewriting equation (7.11) in the form

$$f(w - \pi\mathrm{i}/2) = \frac{1}{1 + \mathrm{e}^{\pi\mathrm{i}\alpha(1/2-\rho)+\alpha w}} \times \frac{1}{\hat{f}(w + \pi\mathrm{i}/2)}, \quad w \in \mathbb{R}, \tag{7.12}$$

and applying Morera's Theorem (see Theorem A.4 in the Appendix) we can analytically (or meromorphically, depending on where the poles of the factor $(1 + \mathrm{e}^{\pi\mathrm{i}\alpha(1/2-\rho)+\alpha w})^{-1}$ lie) continue $f(w - \pi\mathrm{i}/2)$ into the wider strip $\Lambda_{-\pi,0}$. Since $f(w - \pi\mathrm{i}/2)$ takes real values on the line $\gamma_{\pi/2}$, we can apply Corollary A.3 in the Appendix, and we see that $f(w - \pi\mathrm{i}/2)$ is a meromorphic function in the strip $\Lambda_{-\pi,2\pi}$.

Next, rewriting equation (7.11) as follows

$$\hat{f}(w + \pi\mathrm{i}/2) = \frac{1}{1 + \mathrm{e}^{\pi\mathrm{i}\alpha(1/2-\rho)+\alpha w}} \times \frac{1}{f(w - \pi\mathrm{i}/2)}, \tag{7.13}$$

we obtain an analytic (or meromorphic) continuation of $\hat{f}(w+\pi\mathrm{i}/2)$ in the strip $\Lambda_{-\pi,2\pi}$. Since $\hat{f}(w + \pi\mathrm{i}/2)$ takes real values on the line $\gamma_{-\pi/2}$, we apply again Corollary A.3 and we see that $\hat{f}(w + \pi\mathrm{i}/2)$ is a meromorphic function in the wider strip $\Lambda_{-3\pi,2\pi}$.

We repeat this procedure and observe that formula (7.12) and Corollary A.3 guarantee analytic continuation of $f(w-\pi\mathrm{i}/2)$ in the strip $\Lambda_{-3\pi,4\pi}$, then formula (7.13) and the Corollary A.3 allow us to continue $\hat{f}(w+\pi\mathrm{i}/2)$ into strip $\Lambda_{-5\pi,4\pi}$, and so on. In other words, repeating the steps from above repeatedly, we have deduced that both functions $f(w)$ and $\hat{f}(w)$ can be analytically continued to meromorphic functions in $\mathbb{C}$.

In order to finish our proof, it remains to show that these functions satisfy the functional equations (7.9) and (7.10). From formula (7.13), we have

$$\hat{f}(w)^{-1} = (1 + \mathrm{e}^{\pi\mathrm{i}\alpha(w-\rho)})f(w - \pi\mathrm{i}).$$

Since $\hat{f}(w)$ is real for $w \in \mathbb{R}$, we deduce that for $w \in \mathbb{R}$,

$$(1 + \mathrm{e}^{\pi\mathrm{i}\alpha(w-\rho)})f(w - \pi\mathrm{i}) = \hat{f}(w)^{-1} = \overline{\hat{f}(w)^{-1}} = (1 + \mathrm{e}^{-\pi\mathrm{i}\alpha(w-\rho)})f(w + \pi\mathrm{i}),$$

where we have used the fact that, if a and b are two complex numbers, then $\overline{ab} = \overline{a} \times \overline{b}$ and that, as f is a meromorphic function taking real values on the real axis, $\overline{f(z)} = f(\overline{z})$. By analytic continuation, we observe

$$(1 + \mathrm{e}^{\pi\mathrm{i}\alpha(w-\rho)})f(w - \pi\mathrm{i}) = (1 + \mathrm{e}^{-\pi\mathrm{i}\alpha(w-\rho)})f(w + \pi\mathrm{i}), \qquad \text{for} \quad w \in \mathbb{C}.$$

The above identity is equivalent to the functional equation (7.9). The functional equation (7.10) is established similarly. The proof is now complete. □

Let us now see how the quasi-periodic relations in Proposition 7.2 can be used to prove Theorem 7.1. Using the second functional equation in (7.4), we can guess the following solution to equation (7.9),

$$f(w) = e^{-\alpha\rho w/2} S_2((1+\alpha(1+\rho))/2 + i\alpha w/(2\pi);\alpha) \times S_2((1+\alpha(1+\rho))/2 - i\alpha w/(2\pi);\alpha). \tag{7.14}$$

Indeed, let us verify that the above function is a solution to (7.9),

$$\begin{aligned} f(w+2\pi i) &= e^{-\alpha\rho w/2 - \pi i\alpha\rho} S_2((1+\alpha(1+\rho))/2 + i\alpha w/(2\pi) - \alpha;\alpha) \\ &\quad \times S_2((1+\alpha(1+\rho))/2 - i\alpha w/(2\pi) + \alpha;\alpha) \\ &= e^{-\alpha\rho w/2 - \pi i\alpha\rho} S_2((1+\alpha(1+\rho))/2 + i\alpha w/(2\pi);\alpha) \\ &\quad \times 2\sin(\pi((1+\alpha(1+\rho))/2 + i\alpha w/(2\pi) - \alpha)) \\ &\quad \times \frac{S_2((1+\alpha(1+\rho))/2 - i\alpha w/(2\pi);\alpha)}{2\sin(\pi((1+\alpha(1+\rho))/2 - i\alpha w/(2\pi)))} \\ &= e^{-\pi i\alpha\rho} \frac{\cos(i\alpha w/2 - \pi\alpha\hat{\rho}/2)}{\cos(i\alpha w/2 - \pi\alpha(1+\rho)/2)} f(w). \end{aligned}$$

This suggests candidates for the Wiener–Hopf factors Ψ_1^+ and Ψ_1^-. To be more precise, we have $\Psi_1^+(z) = f(\log(z))$, where $f(w)$ is given in (7.14), and $\Psi_1^-(z)$ can be obtained from Ψ_1^+ by exchanging the roles of ρ and $\hat{\rho}$. Solutions to equation (7.9) are certainly not unique since we can multiply any solution by an arbitrary periodic function $F(w)$ satisfying $F(w+2\pi i) = F(w)$, and the result would still be a solution. Thus, we need to verify that our guess in (7.14) is correct via means other than (7.9).

Proof of Theorem 7.1 Let us define

$$H(z) = z^{-\alpha\rho/2} S_2((1+\alpha(1+\rho))/2 + i\alpha\log(z)/(2\pi);\alpha) \times S_2((1+\alpha(1+\rho))/2 - i\alpha\log(z)/(2\pi);\alpha),$$

that is, the function in the right-hand side of (7.7) and, by $\widehat{H}(z)$, the same function, but with $\hat{\rho}$ in place of ρ. Our first goal is to verify that the functions H and $\widehat{H}$ satisfy the Wiener–Hopf factorisation (7.8). Assume that $z > 0$ and to ease the presentation, write $w = \alpha\log(z)/(2\pi i)$. Then, recalling $\rho + \hat{\rho} = 1$,

$$\begin{aligned} H(-iz)\widehat{H}(iz) &= z^{-\alpha/2} e^{\pi i\alpha(\rho-1/2)/2} S_2(1/2 + \alpha/4 + \alpha\rho/2 + w;\alpha) \\ &\quad \times S_2(1/2 + 3\alpha/4 + \alpha\rho/2 - w;\alpha) S_2(1/2 + 5\alpha/4 - \alpha\rho/2 + w;\alpha) \\ &\quad \times S_2(1/2 + 3\alpha/4 - \alpha\rho/2 - w;\alpha) \\ &= z^{-\alpha/2} e^{\pi i\alpha(\rho-1/2)/2} S_2(1/2 + 3\alpha/4 + \alpha\rho/2 - w;\alpha) \\ &\quad \times S_2(1/2 + 5\alpha/4 - \alpha\rho/2 + w;\alpha), \end{aligned}$$

where we have used the reflection formula (7.6) in the form

$$S_2(1/2+\alpha/4+\alpha\rho/2+w;\alpha)S_2(1/2+3\alpha/4-\alpha\rho/2-w;\alpha)=1.$$

Next, according to the second functional equation in (7.4), we have

$$S_2(1/2+5\alpha/4-\alpha\rho/2+w)=\frac{S_2(1/2+\alpha/4-\alpha\rho/2+w)}{2\sin(\pi(1/2+\alpha/4-\alpha\rho/2+w))}.$$

Using the above identity and the reflection formula (7.6) again, but in the form

$$S_2(1/2+\alpha/4-\alpha\rho/2+w;\alpha)S_2(1/2+3\alpha/4+\alpha\rho/2-w;\alpha)=1,$$

we obtain

$$\begin{aligned} H(-\mathrm{i}z)\widehat{H}(\mathrm{i}z) &= \frac{z^{-\alpha/2}e^{\pi\mathrm{i}\alpha(\rho-1/2)/2}}{2\sin(\pi(1/2+\alpha/4-\alpha\rho/2+w))} \\ &= \frac{1}{1+z^{\alpha}e^{\pi\mathrm{i}\alpha(1/2-\rho)}}=\frac{1}{1+\Psi(z)}, \end{aligned}$$

where we have used the identity

$$2\sin(\pi(1/2+\alpha/4-\alpha\rho/2+w))=z^{\alpha/2}e^{\pi\mathrm{i}(1/2-\rho)/2}+z^{-\alpha/2}e^{\pi\mathrm{i}(\rho-1/2)/2}.$$

In other words, the functions H and $\widehat{H}$ satisfy the Wiener–Hopf factorisation (7.8) for $z>0$. The proof for $z<0$ follows by taking the complex conjugate in the computations above.

To prove that our candidate solutions H and $\widehat{H}$ are in fact the Wiener–Hopf factors, we need to apply a uniqueness argument. First, let us establish some properties of the functions H and $\widehat{H}$.

The function $S_2(z;\alpha)$ is analytic and zero-free in the strip $0<\mathrm{Re}(z)<1+\alpha$. Observe that

$$\mathrm{Re}((1+\alpha(1+\rho))/2\pm\mathrm{i}\alpha\log(z)/(2\pi))=\frac{1}{2}(1+\alpha(1+\rho))\mp\frac{\alpha}{2\pi}\arg(z)$$

and, hence noting that $(1+\alpha\rho)/2\le 1$, the functions H and $\widehat{H}$ are analytic and zero-free in the half-plane $\mathrm{Re}(z)\ge 0$ (i.e. $\arg(z)\in[-\pi/2,\pi/2]$). Moreover, we have

$$z^{-1}\log(H(z))\to 0 \qquad \text{and} \qquad z^{-1}\log(\widehat{H}(z))\to 0,$$

as z goes to ∞ (uniformly in the half-plane $\mathrm{Re}(z)\ge 0$). The latter asymptotics can be derived from the asymptotic result (7.5).

On the other hand, recall from (2.35) that the Wiener–Hopf factor $\Psi_1^+(z)$ can be written as

$$\Psi_1^+(z)=\frac{\kappa_1(0)}{\kappa_1(z)}, \tag{7.15}$$

where $\kappa_q(z)$, for $q, z \geq 0$, is the bivariate Laplace exponent of the ascending ladder process (see (2.30)). From this representation and from the well-known fact that the Laplace exponent of a subordinator is analytic in the half-plane $\mathrm{Re}(z) > 0$ and is zero-free in the half-plane $\mathrm{Re}(z) \geq 0$, we deduce that Ψ_1^+ also inherits such properties in the half-plane $\mathrm{Re}(z) \geq 0$. Since Ψ_1^- satisfies a similar representation, then we deduce that both functions, Ψ_1^+ and Ψ^-, are analytic and zero-free in the half-plane $\mathrm{Re}(z) \geq 0$. Moreover, from (7.15), recalling that, as a Bernstein function (and, in particular, a concave function), κ_p will grow no faster than linearly, as z goes to ∞ on the right half-plane of $\mathbb{C}$,

$$z^{-1}\log(\Psi_1^+(z)) \to 0 \qquad \text{and} \qquad z^{-1}\log(\Psi_1^-(z)) \to 0$$

(uniformly in the half-plane $\mathrm{Re}(z) \geq 0$).

Next, we define the function $F(z)$ as follows:

$$F(z) := \begin{cases} H(z)/\Psi_1^+(z), & \text{if } \mathrm{Re}(z) \geq 0, \\ \Psi_1^-(-z)/\widehat{H}(-z), & \text{if } \mathrm{Re}(z) \leq 0. \end{cases} \tag{7.16}$$

Note that the function $F(z)$ is well defined for $\mathrm{Re}(z) = 0$ since both pairs Ψ_1^+, Ψ_1^- and H, $\widehat{H}$ satisfy (7.8), thus $H(z)/\Psi_1^+(z) = \Psi_1^+(-z)/\widehat{H}(-z)$ for $\mathrm{Re}(z) = 0$.

From the properties satisfied by H and $\widehat{H}$, we see that the function $F(z)$ is analytic and zero-free in the two half-planes $\mathrm{Re}(z) > 0$ and $\mathrm{Re}(z) < 0$ and from our discussion in the previous paragraph, it follows that $F(z)$ is continuous in the entire complex plane. Therefore, by Morera's theorem (see Theorem A.4), the function $F(z)$ must be analytic in the entire complex plane.

In other words, we have an analytic and zero-free function $F(z)$ and therefore its logarithm $\log(F(z))$ is also an entire function. The properties satisfied by Ψ_1^+ and Ψ_1^- imply that $z^{-1}\log(F(z)) \to 0$ as $z \to \infty$ uniformly in the entire complex plane. Liouville's theorem (cf. Theorem A.6) now implies that $\log(F(z))$ must be constant. The value of this constant is easily seen to be zero, since $F(0) = 1$ (this follows from $H(0) = \widehat{H}(0) = \Psi_1^+(0) = \Psi_1^-(0) = 1$). Thus, $F(z) = 1$ for all $z \in \mathbb{C}$, which implies $\Psi_1^+(z) = H(z)$ and $\Psi_1^-(z) = \widehat{H}(z)$ for all z in the half-plane $\mathrm{Re}(z) \geq 0$. This completes the proof. □

7.3 The Law of the maximum at a finite time

The Wiener–Hopf factor Ψ_q^+ provides information about the distribution of the supremum up to an independent exponentially distributed random time with parameter $q > 0$, written $\mathbf{e}_q$. In principle, a Laplace–Fourier inversion applied to $q \mapsto \Psi_q^+$ will give the law of $\overline{X}_t$, for $t \geq 0$.

On the other hand, recalling from (7.2) that $\Psi_q^+(z) = \Psi_1^+(zq^{-1/\alpha})$, for $q > 0$, we see that there is the possibility of converting an inversion with respect to the variable q with an inverse with respect to the variable z. This observation can otherwise be seen as equivalent to the noting that, thanks to the scaling property and the independence of $\mathbf{e}_q$ and X, the law of $\overline{X}_{\mathbf{e}_q}$ is equal to the law of $\mathbf{e}_q^{1/\alpha}\overline{X}_1$, when $X_0 = 0$. Indeed, the following computation, which considers the Mellin transform of Ψ_q^+, connects these two random variables,

$$\begin{aligned}
\int_0^\infty z^{s-1}\Psi_q^+(z)\mathrm{d}z &= q^{s/\alpha}\int_0^\infty u^{s-1}\Psi_1^+(u)\mathrm{d}u \\
&= q^{s/\alpha}\mathbb{E}\left[\int_0^\infty u^{s-1}\mathrm{e}^{-u\overline{X}_{\mathbf{e}_1}}\,\mathrm{d}u\right] \\
&= q^{s/\alpha}\Gamma(s)\mathbb{E}\left[\mathbf{e}_1^{-s/\alpha}\left(\overline{X}_1\right)^{-s}\right] \\
&= q^{s/\alpha}\Gamma(s)\Gamma\left(1-\frac{s}{\alpha}\right)\mathcal{M}(1-s),
\end{aligned}$$

where in the fourth equality we have used the integral representation of the gamma function (see (A.7) in the Appendix) and $\mathcal{M}$ denotes the Mellin transform of $\overline{X}_1$, that is,

$$\mathcal{M}(s) := \mathbb{E}\left[\left(\overline{X}_1\right)^{s-1}\right].$$

Our approach in determining $\mathcal{M}$ relies on the Lamperti transform instead of computing directly the Mellin transform of Ψ_q^+, which can be done thanks to the explicit expression given in Theorem 7.1. Indeed, the Lamperti transform provides a natural relationship between the law of $\overline{X}_1$ and the law of an exponential functional of the Lamperti-stable process ξ^*, described in Section 5.3, that we explain as follows.

Recall that $\tau_0^- = \inf\{t > 0 : X_t < 0\}$ and also that $\hat{\mathbb{P}}_1$ denotes the law of the dual process $-X$ issued from 1. Then spatial homogeneity, duality and the scaling property give us

$$\hat{\mathbb{P}}_1(\tau_0^- < t) = \hat{\mathbb{P}}(\tau_{-1}^- < t) = \hat{\mathbb{P}}\left(\underline{X}_t < -1\right) = \mathbb{P}\left(\overline{X}_t > 1\right) = \mathbb{P}\left(\overline{X}_1 > t^{-\frac{1}{\alpha}}\right). \quad (7.17)$$

In other words, the law of $(\overline{X}_1)^{-\alpha}$, under $\mathbb{P}$, is the same as the law of τ_0^-, under $\hat{\mathbb{P}}_1$. The latter can be identified as the life time of the stable process killed on entering $(-\infty, 0)$, under $\hat{\mathbb{P}}_1$.

On the other hand, recall from Section 5.3 that the stable process killed on entering $(-\infty, 0)$ is defined as follows

$$Z_t = X_t\mathbf{1}_{(\underline{X}_t \geq 0)}, \qquad t \geq 0,$$

and that Z is a positive self-similar Markov process. Its underlying Lévy process, $(\xi^*, \mathbf{P})$, that appears through the Lamperti transform is a Lamperti-stable process with parameters $(1 - \alpha\hat{\rho}, \alpha\rho, \alpha\hat{\rho})$. In other words, its characteristic exponent is given by

$$\Psi^*(z) = \frac{\Gamma(\alpha - \mathrm{i}z)}{\Gamma(\alpha\hat{\rho} - \mathrm{i}z)} \frac{\Gamma(1 + \mathrm{i}z)}{\Gamma(1 - \alpha\hat{\rho} + \mathrm{i}z)}, \qquad z \in \mathbb{R}.$$

Under $\hat{\mathbb{P}}_1$, the stable process killed on entering $(-\infty, 0)$ and implicitly its associated Lamperti-stable process, only experiences a change in its parameters by replacing ρ by $\hat{\rho}$. In other words, under the law $\hat{\mathbb{P}}_1$, the underlying Lévy process is a Lamperti-stable process with parameters $(1 - \alpha\rho, \alpha\hat{\rho}, \alpha\rho)$ that we denote by $(\hat{\xi}^*, \hat{\mathbf{P}})$. It is with remembering from Section 5.3 that everything we have said regarding $\hat{\xi}^*$ applies to both the two-sided and one-sided jump cases (albeit that we are excluding the setting of monotone paths). In particular, in the spectrally negative case, that is, $\rho = \alpha^{-1}$, the Lamperti-stable process $\hat{\xi}^*$ is spectrally positive with parameters $(0, \alpha - 1, 1)$ and in the spectrally positive case, the Lamperti-stable process $\hat{\xi}^*$ is spectrally negative with parameters $(2 - \alpha, 1, \alpha - 1)$. From the Lamperti transform, under $\hat{\mathbb{P}}_1$, we can also identify the life time of Z with the exponential functional associated to $\alpha\hat{\xi}^*$, which implies that the law of τ_0^- is the same as the law of the exponential functional of $\alpha\hat{\xi}^*$. In other words, the law of τ_0^-, under $\hat{\mathbb{P}}_1$, is the same as $I(\alpha, \hat{\xi}^*)$, under $\hat{\mathbf{P}}$, where

$$I(\alpha, \hat{\xi}^*) = \int_0^{\zeta^*} \mathrm{e}^{\alpha\hat{\xi}^*_s} \mathrm{d}s,$$

and ζ^* is the lifetime of ξ^*. Hence, as a consequence of (7.17), we deduce that the Mellin transform of $\overline{X}_1$ satisfies

$$\mathcal{M}(s) := \mathbb{E}\left[\left(\overline{X}_1\right)^{s-1}\right] = \hat{\mathbf{E}}\left[I(\alpha, \hat{\xi}^*)^{-(s-1)/\alpha}\right]. \qquad (7.18)$$

Our main result of this section provides a complete characterisation of the Mellin transform $\mathcal{M}$ in terms of the so-called double gamma $G(z; \tau)$. The latter follows from the above identity and Theorems 4.13, 4.17 and 4.19.

Theorem 7.3 *For $s \in \mathbb{C}$ and if (α, ρ) is such that*

(i) $0 < \alpha\rho, \alpha\hat{\rho} < 1$, *we have*

$$\mathcal{M}(s) = \alpha^{s-1} \frac{G(\alpha\rho; \alpha)}{G(\alpha\hat{\rho} + 1; \alpha)} \frac{G(\alpha\hat{\rho} + 2 - s; \alpha)}{G(\alpha\rho - 1 + s; \alpha)} \frac{G(\alpha - 1 + s; \alpha)}{G(\alpha + 1 - s; \alpha)},$$

(ii) $\alpha\rho = 1$, *we have*

$$\mathcal{M}(s) = \frac{\Gamma(s - 1)}{\Gamma\left(\frac{s-1}{\alpha}\right)},$$

(iii) $\alpha\hat{\rho} = 1$, *we have*

$$\mathcal{M}(s) = \frac{\sin\left(\frac{\pi}{\alpha}\right)}{\sin\left(\pi\left(\frac{2-s}{\alpha}\right)\right)} \frac{\Gamma\left(1 - \frac{s-1}{\alpha}\right)}{\Gamma(2-s)}.$$

Proof We first prove part (i). Recall that $\hat{\xi}^*$ is a Lamperti-stable process with parameters $(1 - \alpha\rho, \alpha\hat{\rho}, \alpha\rho)$ and hence belongs to the hypergeometric family of Lévy processes whose parameters lie in $\mathcal{H}_1 \setminus \{\beta = 1\}$ (cf. (4.17)). That is to say, we can use Theorem 4.13 and deduce

$$\begin{aligned}\mathcal{M}(s) &= \mathtt{M}\left(1 - \frac{s-1}{\alpha}; \alpha, 1 - \alpha\rho, \alpha\hat{\rho}, 1 - \alpha\rho, \alpha\rho\right) \\ &= \Gamma\left(1 - \frac{s-1}{\alpha}\right) \mathrm{F}\left(1 - \frac{s-1}{\alpha}; \alpha, 1 - \alpha\rho, \alpha\hat{\rho}, 1 - \alpha\rho, \alpha\rho\right).\end{aligned}$$

From the definition of F, we get

$$\mathcal{M}(s) = \Gamma\left(\frac{\alpha - s + 1}{\alpha}\right) \frac{G\left(\frac{\alpha\rho}{\alpha}; \frac{1}{\alpha}\right) G\left(\frac{\alpha+1}{\alpha}; \frac{1}{\alpha}\right)}{G\left(1; \frac{1}{\alpha}\right) G\left(\frac{1+\alpha\hat{\rho}}{\alpha}; \frac{1}{\alpha}\right)} \frac{G\left(\frac{2-\alpha\hat{\rho}-s}{\alpha}; \frac{1}{\alpha}\right) G\left(\frac{\alpha+s+1}{\alpha}; \frac{1}{\alpha}\right)}{G\left(\frac{2+\alpha-s}{\alpha}; \frac{1}{\alpha}\right) G\left(\frac{\alpha\rho+s-1}{\alpha}; \frac{1}{\alpha}\right)}.$$

Using the transformation (A.23), we deduce

$$\frac{G\left(\frac{\alpha\rho}{\alpha}; \frac{1}{\alpha}\right) G\left(\frac{\alpha+1}{\alpha}; \frac{1}{\alpha}\right)}{G\left(1; \frac{1}{\alpha}\right) G\left(\frac{1+\alpha\hat{\rho}}{\alpha}; \frac{1}{\alpha}\right)} = c(\alpha) \frac{G(\alpha\rho; \alpha) G(\alpha + 1; \alpha)}{G(\alpha; \alpha) G(1 + \alpha\hat{\rho}; \alpha)},$$

where

$$c(\alpha) = (2\pi)^{\frac{\alpha(\hat{\rho}-\rho)}{2}(1-\frac{1}{\alpha})} \alpha^{\frac{(\hat{\rho}-\rho)}{2} + \frac{\alpha}{2}((\hat{\rho}-\rho)-(\hat{\rho}^2-\rho^2))+\rho}.$$

Similarly, we have

$$\frac{G\left(\frac{2-\alpha\hat{\rho}-s}{\alpha}; \frac{1}{\alpha}\right) G\left(\frac{\alpha+s+1}{\alpha}; \frac{1}{\alpha}\right)}{G\left(\frac{2+\alpha-s}{\alpha}; \frac{1}{\alpha}\right) G\left(\frac{\alpha\rho+s-1}{\alpha}; \frac{1}{\alpha}\right)} = \frac{G(2 - \alpha\hat{\rho} - s; \alpha) G(\alpha + s + 1; \alpha)}{G(2 + \alpha - s; \alpha) G(\alpha\rho + s - 1; \alpha)} \alpha^{s-1} c(\alpha)^{-1}.$$

Finally, we use the quasi-periodic identity (A.21) of G and get

$$G(2 + \alpha - s; \alpha) = \Gamma\left(\frac{1 + \alpha - s}{\alpha}\right) G(1 + \alpha - s; \alpha) \text{ and } G(1 + \alpha; \alpha) = G(\alpha; \alpha).$$

Putting all pieces together, we conclude that

$$\mathcal{M}(s) = \alpha^{s-1} \frac{G(\alpha\rho; \alpha)}{G(1 + \alpha\hat{\rho}; \alpha)} \frac{G(2 - \alpha\hat{\rho} - s; \alpha) G(\alpha + s + 1; \alpha)}{G(1 + \alpha - s; \alpha) G(\alpha\rho + s - 1; \alpha)},$$

as expected.

For the remaining two cases, we proceed similarly as above but using the explicit expressions for the Mellin transform of $I(\alpha, \hat{\xi}^*)$ provided in Theorem 4.19, for part (ii), and in Theorem 4.17, for part (iii). □

Finally, we use identity (7.18), knowledge of the density of the exponential functional $I(\alpha, \hat{\xi}^*)$ and its tail behaviour near 0 and ∞ to deduce similar distributional properties of $\overline{X}_1$. Denote the distributional densities of $I(\alpha, \hat{\xi}^*)$ and $\overline{X}_1$ by

$$p_\alpha(x) = \frac{\mathrm{d}}{\mathrm{d}x}\hat{\mathbf{P}}\big(I(\alpha, \hat{\xi}^*) \le x\big) \quad \text{and} \quad \mathrm{p}_{\overline{X}}(x) = \frac{\mathrm{d}}{\mathrm{d}x}\mathbb{P}\big(\overline{X}_1 \le x\big), \quad \text{for} \quad x \ge 0.$$

From identity (7.18), these two densities are related by

$$\mathrm{p}_{\overline{X}}(x) = \alpha x^{-\alpha-1} p_\alpha(x^\alpha), \qquad x > 0.$$

First, we consider the case of two-sided jumps, that is, when $0 < \alpha\rho, \alpha\hat{\rho} < 1$. The spectrally negative and positive cases will be treated separately in the next section. As alluded to previously, in the case of two-sided jumps, the Lévy process $\hat{\xi}^*$ is a hypergeometric Lévy process in the class $\mathcal{H}_1 \setminus \{\beta = 1\}$. Accordingly, we can apply directly Theorems 4.22 and 4.24.

Let us first introduce the sequences $a_{m,n}$, $m, n \ge 0$, and $b_{m,n}$, $m \ge 0, n \ge 1$, as follows

$$\begin{aligned} a_{m,n} &= \frac{(-1)^{m+n}}{\Gamma\left(1 - \rho - n - \frac{m}{\alpha}\right)\Gamma(\alpha\rho + m + \alpha n)} \\ &\quad \times \prod_{j=1}^{m} \frac{\sin\left(\frac{\pi}{\alpha}(\alpha\rho + j - 1)\right)}{\sin\left(\frac{\pi j}{\alpha}\right)} \prod_{j=1}^{n} \frac{\sin(\pi\alpha(\rho + j - 1)}{\sin(\pi\alpha j)}, \end{aligned} \tag{7.19}$$

and

$$b_{m,n} = \frac{\Gamma\left(1 - \rho - n - \frac{m}{\alpha}\right)\Gamma(\alpha\rho + m + \alpha n)}{\Gamma\left(1 + n + \frac{m}{\alpha}\right)\Gamma(-m - \alpha n)} a_{m,n}. \tag{7.20}$$

We also recall from Definition 4.23 that $\mathcal{L}$ denotes the set of real irrational numbers x, for which there exists a constant $b > 1$ such that the inequality

$$\left| x - \frac{p}{q} \right| < \frac{1}{b^q}$$

is satisfied for infinitely many integers p and q. For further details about this set, we refer to the comments after Definition 4.23.

Theorem 7.4 *Assume that* $0 < \alpha\rho, \alpha\hat{\rho} < 1$. *For* $\alpha \notin \mathbb{Q}$, *we have*

$$\mathrm{p}_{\overline{X}}(x) \sim x^{\alpha\rho - 1} \sum_{m \ge 0} \sum_{n \ge 0} a_{m,n} x^{m + \alpha n}, \qquad x \to 0^+, \tag{7.21}$$

$$\mathrm{p}_{\overline{X}}(x) \sim x^{-1-\alpha} \sum_{m \ge 0} \sum_{n \ge 0} b_{m,n+1} x^{-m - \alpha n}, \qquad x \to \infty. \tag{7.22}$$

For $\alpha \notin \mathcal{L} \cup \mathbb{Q}$., we have for all $x > 0$,

$$
\mathrm{p}_{\overline{X}}(x) = \begin{cases} x^{1-\alpha} \sum_{m\geq 0} \sum_{n\geq 0} b_{m,n+1} x^{-m-\alpha n} & \text{if } \alpha \in (0,1), \\ x^{\alpha\rho-1} \sum_{m\geq 0} \sum_{n\geq 0} a_{m,n} x^{m+\alpha n} & \text{if } \alpha \in (1,2). \end{cases} \tag{7.23}
$$

There are two ways of getting explicitly the coefficients that appear in (7.21), (7.22) and (7.23). One is performing the computations directly from (4.20) or by observing that the Mellin transform $\mathcal{M}(s)$ has simple poles at $\{m + \alpha n \colon m, n \geq 1\}$ and $\{1 - \alpha\rho - m - \alpha n \colon m, n \geq 0\}$ with residues

$$
\mathrm{Res}(\mathcal{M}(s) \colon s = 1 - \alpha\rho - m - \alpha n) = -b_{m-1,n},
$$

and

$$
\mathrm{Res}(\mathcal{M}(s) \colon s = m + \alpha n) = a_{m,n}.
$$

The latter identities can be deduced by iterating the following quasi-periodic properties of $\mathcal{M}$,

$$
\mathcal{M}(s+1) = \frac{\alpha}{\pi} \sin\left(\pi\left(\rho - \frac{1-s}{\alpha}\right)\right) \Gamma\left(1 - \frac{s}{\alpha}\right) \Gamma\left(1 - \frac{1-s}{\alpha}\right) \mathcal{M}(s),
$$

and

$$
\mathcal{M}(s+\alpha) = \frac{\alpha}{\pi} \sin\left(\pi\left(\alpha\rho - 1 + s\right)\right) \Gamma\left(1 - s\right) \Gamma\left(\alpha - 1 + s\right) \mathcal{M}(s),
$$

which follows directly from the quasi-periodic properties of the double gamma function found in (A.21) of the Appendix.

We observe that the behaviour of the density $\mathrm{p}_{\overline{X}}$ from Theorem 7.4 only provides the asymptotic behaviour of the tail distribution of $\overline{X}_1$ at 0 and ∞ for $\alpha \notin \mathbb{Q}$ and $0 < \alpha\rho, \alpha\hat{\rho} < 1$. As we will see in the next chapter, tail asymptotic behaviour is desirable for all values of α. To avoid the aforementioned restriction in α, we use Theorem 1.18 to deduce the upper tail behaviour of the distribution of $\overline{X}_1$ for any $\alpha \in (0, 2)$.

Proposition 7.5 *Suppose that X possesses positive jumps. Then*

$$
\mathbb{P}\left(\overline{X}_1 > x\right) \sim \mathbb{P}(X_1 > x) \sim \frac{\Gamma(\alpha)}{\pi} \sin(\pi\alpha\rho) x^{-\alpha} \qquad \textit{as} \quad x \to \infty.
$$

Proof From the asymptotic expansion in Theorem 1.18, we get the following estimate

$$
\mathbb{P}(X_1 > x) \sim \frac{\Gamma(\alpha)}{\pi} \sin(\pi\alpha\rho) x^{-\alpha} \qquad \text{as} \quad x \to \infty. \tag{7.24}
$$

Hence, it is clear that

$$
\liminf_{x\to\infty} x^{\alpha} \mathbb{P}\left(\overline{X}_1 > x\right) \geq \frac{\Gamma(\alpha)}{\pi} \sin(\pi\alpha\rho).
$$

For the upper bound, we fix $\epsilon > 0$ and observe from the Markov and scaling properties that

$$\begin{aligned}\mathbb{P}\big(X_1 > (1-\epsilon)x\big) &\geq \mathbb{P}\big(\overline{X}_1 > x, X_1 > (1-\epsilon)x\big)\\ &\geq \int_{[0,1]} \mathbb{P}\big(\tau_x^+ \in du\big)\mathbb{P}(X_{1-u} > -\epsilon x)\\ &= \int_{[0,1]} \mathbb{P}\big(\tau_x^+ \in du\big)\mathbb{P}\left(X_1 > \frac{-\epsilon x}{(1-u)^{1/\alpha}}\right)\\ &\geq \mathbb{P}\big(\overline{X}_1 > x\big)\mathbb{P}(X_1 > -\epsilon x),\end{aligned}$$

where we recall the usual notation $\tau_x^+ = \inf\{t > 0\colon X_t > x\}$.

Next, we use again the estimate in (7.24), together with the fact that

$$\lim_{x\to\infty} \mathbb{P}(X_1 > -\epsilon x) = 1,$$

to deduce

$$\limsup_{x\to\infty} x^\alpha \mathbb{P}\big(\overline{X}_1 > x\big) \leq \frac{\Gamma(\alpha)}{\pi}\sin(\pi\alpha\rho)(1-\epsilon)^{-\alpha}.$$

The desired result follows by choosing ϵ arbitrarily close to 0. □

Scaling tells us that τ_x^+ is equal in distribution to $x^\alpha\tau_1^+$ and hence, in a similar spirit to (7.17), we have that

$$\mathbb{P}(\overline{X}_1 < x) = \mathbb{P}(\tau_x^+ > 1) = \mathbb{P}(\tau_1^+ > x^{-\alpha}) = \hat{\mathbb{P}}_1(\tau_0^- > x^{-\alpha}).$$

The asymptotic behaviour in Lemma 5.12 (which is ultimately rooted in Proposition 4.27) now gives us the lower tail behaviour of the distribution of $\overline{X}_1$, albeit that we must interchange the roles of ρ and $\hat{\rho}$.

Proposition 7.6 *Suppose that $|X|$ is not a subordinator. Then*

$$\mathbb{P}\big(\overline{X}_1 < x\big) \sim \frac{\alpha}{\Gamma(\hat{\rho})\Gamma(1+\alpha\rho)} x^{\alpha\rho}, \qquad \text{as} \quad x \to 0.$$

7.4 Doney's factorisation

It turns out that the Mellin transform $\mathcal{M}$ given in Theorem 7.3 can be simplified to a more explicit form for special parameter choices.

Definition 7.7 (Doney classes) For $k, l \in \mathbb{Z}$, we say that a stable process X with admissible parameters, $(\alpha,\rho) \in \mathcal{A}$ (cf. (3.11)), belongs to the *Doney class* $C_{k,l}$ if the following identity holds

$$\rho + k = \frac{l}{\alpha}. \tag{7.25}$$

Note that the spectrally one-sided cases are included in the Doney classes. Indeed setting $k = 0$, $l = 1$ in (7.25) gives us the spectrally negative case and $k = -1$, $l = -1$ gives us the spectrally positive case. Moreover, if (α, ρ) satisfy (7.25), then we observe that $\alpha \in \mathbb{Q}$ if and only if $\rho \in \mathbb{Q}$. If $\alpha \notin \mathbb{Q}$ there exists a unique pair of integers k, l such that identity (7.25) holds. If $\alpha = m/n$ for some coprime integers m, n and $X \in C_{k,l}$ then identity (7.25) holds for any pair $(\hat{k}, \hat{l}) = (k + jn, l + jm)$, for integer $j \in \mathbb{Z}$. In this case, we assume that $0 \leq k < n$ and $1 \leq l < m$.

Our next result provides the explicit form of the Mellin transform $\mathcal{M}$ for the Doney classes.

Theorem 7.8 *If $X \in C_{k,l}$ and $l > 0$, then, for $s \in \mathbb{C}$,*

$$\mathcal{M}(s) = \frac{\Gamma(s)}{\Gamma(1 - (1 - s)/\alpha)} \prod_{i=1}^{l-1} \frac{\sin(\pi(s - 1 + i)/\alpha)}{\sin(\pi i/\alpha)} \prod_{j=1}^{k} \frac{\sin(\pi\alpha j)}{\sin(\pi(1 - s + j\alpha))}.$$

If $X \in C_{k,l}$ and $l < 0$, then

$$\mathcal{M}(s) = \frac{\Gamma(1 - (1 - s)/\alpha)}{\Gamma(2 - s)} \prod_{i=1}^{|k|-1} \frac{\sin(\pi(s - 1 + i\alpha))}{\sin(\pi\alpha i)} \prod_{j=1}^{|l|} \frac{\sin(\pi j/\alpha)}{\sin(\pi(1 - s + j)/\alpha)}.$$

Proof Since both identities use similar arguments, we only provided the arguments to show the first identity. Assume that $l > 0$ and recall that under our assumptions $\alpha\rho = l - \alpha k$. From Theorem 7.3, we have on $\mathbb{C}$

$$\mathcal{M}(s) = \alpha^{s-1} \frac{G(l - \alpha k; \alpha)}{G(\alpha(k + 1) - l + 1; \alpha)} \frac{G(\alpha(k + 1) - l + 2 - s; \alpha)}{G(l - \alpha k - 1 + s; \alpha)} \frac{G(\alpha - 1 + s; \alpha)}{G(\alpha + 1 - s; \alpha)}.$$

Using the quasi-periodic properties for the double gamma function (A.21), we deduce the following identities

$$\frac{G(\alpha(k + 1) - l + 2 - s; \alpha)}{G(\alpha + 1 - s; \alpha)} = (2\pi)^{\frac{\alpha-1}{2}k} \prod_{i=1}^{l-1} \frac{1}{\Gamma\left(\frac{\alpha+1-s-i}{\alpha}\right)} \times \prod_{j=1}^{k} \alpha^{-(\alpha j-s-l+2)+\frac{1}{2}} \Gamma(\alpha j - s - l + 2),$$

and

$$\frac{G(\alpha - 1 + s; \alpha)}{G(s + l - 1 - \alpha k; \alpha)} = (2\pi)^{\frac{\alpha-1}{2}(k+1)} \alpha^{-(s-1)+\frac{1}{2}} \frac{\Gamma(s - 1)}{\Gamma(1 - (1 - s)/\alpha)} \prod_{i=1}^{l-1} \frac{1}{\Gamma\left(\frac{s-1+i}{\alpha}\right)} \times \prod_{j=1}^{k} \alpha^{-(s+l-1-\alpha j)+\frac{1}{2}} \Gamma(s + l - \alpha j - 1).$$

Multiplying both identities and using the reflection formula (A.12), we observe

$$\frac{G(\alpha(k+1)-l+2-s;\alpha)}{G(\alpha+1-s;\alpha)}\frac{G(\alpha-1+s;\alpha)}{G(s+l-1-\alpha k;\alpha)} = (2\pi)^{(\alpha-1)(k+1/2)}\alpha^{-(s-1)+\frac{3}{2}}$$
$$\times\frac{\Gamma(s)}{\Gamma\left(1-\frac{1-s}{\alpha}\right)}\prod_{i=1}^{l-1}\frac{\sin(\pi(s-1+i)/\alpha)}{\pi}\prod_{j=1}^{k}\frac{\pi}{\sin(\pi(1-s-(l-1)+j\alpha))}.$$

In particular for $s=1$, we obtain

$$\frac{G(l-\alpha k;\alpha)}{G(\alpha(k+1)-l+1;\alpha)} = (2\pi)^{-(\alpha-1)(k+1/2)}\alpha^{-\frac{3}{2}}$$
$$\times\prod_{i=1}^{l-1}\frac{\pi}{\sin(\pi i/\alpha)}\prod_{j=1}^{k}\frac{\sin(\pi(-(l-1)+j\alpha))}{\pi}.$$

Finally, using the reflection identity of the sine function,

$$\sin(\theta-\pi) = -\sin(\theta),$$

in the last two identities, we deduce the form of the Mellin transform $\mathcal{M}$ as required. □

Similar to Theorem 7.4, the explicit form of the Mellin transport $\mathcal{M}$ for the Doney classes allows us to describe explicitly the density of $\overline{X}_1$. In order to do so, we introduce the sequence of coefficients as follows. If $l>0$, then for $n\in\{0,1,2,\dots,k\}$ and $m\in\mathbb{Z}$, we define

$$c^+_{m,n} = \frac{(-1)^{m(k+1)+nl+1}}{\Gamma(1+n+m/\alpha)\Gamma(-m-n\alpha)}$$
$$\times\prod_{j=1}^{l-1}\frac{\sin(\pi(j+m)/\alpha)}{\sin(\pi j/\alpha)}\prod_{i=1}^{k-n}\frac{\sin(\pi\alpha(i+n))}{\sin(\pi i\alpha)},$$

while if $l<0$, then for $m\in\{0,1,\dots,|l|\}$ and $n\in\mathbb{Z}$, we define

$$c^-_{m,n} = \frac{(-1)^{mk+n(l+1)+1}}{\Gamma(1+n+m/\alpha)\Gamma(-m-n\alpha)}$$
$$\times\prod_{i=1}^{|k|-1}\frac{\sin(\pi\alpha(i+n))}{\sin(\pi i\alpha)}\prod_{j=1}^{|l|-m}\frac{\sin(\pi(j+m)/\alpha)}{\sin(\pi j/\alpha)}.$$

Theorem 7.9 *Assume that $X\in C_{k,l}$. Then if $\alpha\in(0,1)$ and $l>0$, we have a convergent series representation*

$$p_{\overline{X}}(x) = -\sum_{n=1}^{k}\sum_{m\geq 0}c^+_{m,n}x^{-m-\alpha n-1}, \qquad x\in\mathbb{R}^+, \tag{7.26}$$

and an asymptotic expansion

$$\mathrm{p}_{\overline{X}}(x) \sim \sum_{n=0}^{k} \sum_{m \le -l} c^{+}_{m,n} x^{-m-\alpha n-1}, \qquad x \to 0^{+}. \tag{7.27}$$

Similarly, if $\alpha \in (0,1)$ and $l < 0$, we have a convergent series representation

$$\mathrm{p}_{\overline{X}}(x) = -\sum_{m=0}^{|l|} \sum_{n \ge 1} c^{-}_{m,n} x^{-m-\alpha n-1}, \qquad x \in \mathbb{R}^{+}, \tag{7.28}$$

and an asymptotic expansion

$$\mathrm{p}_{\overline{X}}(x) \sim \sum_{m=1}^{|l|} \sum_{n \le k} c^{-}_{m,n} x^{-m-\alpha n-1}, \qquad x \to 0^{+}. \tag{7.29}$$

If $\alpha \in (1,2)$ and $l > 0$, we have a convergent series representation

$$\mathrm{p}_{\overline{X}}(x) = \sum_{n=0}^{k} \sum_{m \le -l} c^{+}_{m,n} x^{-m-\alpha n-1}, \qquad x \in \mathbb{R}^{+}, \tag{7.30}$$

and an asymptotic expansion

$$\mathrm{p}_{\overline{X}}(x) \sim -\sum_{n=1}^{k} \sum_{m \ge 0} c^{+}_{m,n} x^{-m-\alpha n-1}, \qquad x \to \infty. \tag{7.31}$$

Similarly, if $\alpha \in (1,2)$ and $l < 0$, we have a convergent series representation

$$\mathrm{p}_{\overline{X}}(x) = \sum_{m=1}^{|l|} \sum_{n \le k} c^{-}_{m,n} x^{-m-\alpha n-1}, \qquad x \in \mathbb{R}^{+}, \tag{7.32}$$

and an asymptotic expansion

$$\mathrm{p}_{\overline{X}}(x) \sim -\sum_{m=0}^{|l|} \sum_{n \ge 1} c^{-}_{m,n} x^{-m-\alpha n-1}, \qquad x \to \infty. \tag{7.33}$$

Proof We follow the same arguments used in the proof of Theorem 4.22 and observe that, in order to deduce the asymptotic expansions (7.27) and (7.29), only the knowledge of the residues at the poles of $\mathcal{M}(s)$ are needed. The rest of the proof follows exactly the same arguments as in the proof of the aforementioned result.

From the explicit form of $\mathcal{M}(s)$ given in Theorem 7.8, we deduce for the case $l > 0$ that $\mathcal{M}(s)$ has simple poles at $s_{m,n} = m + n\alpha$ for $m \le 1 - l$ and $n \in \{0, 1, 2, \ldots, k\}$ or $m \ge 1$ and $n \in \{1, 2, \ldots, k\}$. The associated residues are such that,

$$\mathrm{Res}(\mathcal{M}(s)\colon s = m + \alpha n) = c^{+}_{m-1,n}.$$

Similarly, for $l < 0$, the Mellin transform $\mathcal{M}(s)$ has simple poles at $s_{m,n} = m + n\alpha$, for $m \in \{1, 2, 3, \ldots, |l| + 1\}$ and $n \geq 1$, or $m \in \{2, 3, \ldots, |l| + 1\}$ and $n \leq k$. The associated residues are such that,

$$\mathrm{Res}(\mathcal{M}(s)\colon\ s = m + \alpha n) = c^{-}_{m-1,n}.$$

The above characterisation of the residues of $\mathcal{M}(s)$ completes the proof of (7.27) and (7.29).

Next, we establish the convergence of (7.26) and (7.28). With this aim in mind, we assume that $\alpha \in (1, 2)$ and $l > 0$. We choose $c \in (0, 1)$ such that $c \neq m + \alpha n$, for $n \in \{0, 1, 2, \ldots, k\}$ and $m \in \mathbb{Z}$. Our starting point is the expression of $p(x)$ as the inverse Mellin transform

$$\mathrm{p}_{\overline{X}}(x) = \frac{1}{2\pi \mathrm{i}} \int_{1+\mathrm{i}\mathbb{R}} \mathcal{M}(s) x^{-s} \mathrm{d}s, \quad x > 0. \tag{7.34}$$

In a similar spirit to the proof of Theorem 4.22, recalling that we are using the identity (7.18) for $\mathcal{M}$, where the underlying Lévy process belongs to the hypergeometric family of Lévy processes with parameters $(1 - \alpha\rho, \alpha\hat{\rho}, 1 - \alpha\rho, \alpha\rho)$, we can use (A.16) in the Appendix together with (4.45) and the explicit form of $\mathcal{M}$ from Theorem 7.3 to deduce that $|\mathcal{M}(x + \mathrm{i}u)|$ decreases exponentially as $u \to \infty$ (uniformly in x in any finite interval). To be more precise, for $x \in \mathbb{R}$, we have as $u \to \infty$,

$$\log\left(|\mathcal{M}(x + iu)|\right) = -\frac{\pi|u|}{2\alpha}\left(\alpha(1 - \rho) + 1 - \alpha\rho\right) + o(u). \tag{7.35}$$

As a consequence, not only does its Fourier inverse exist but so does the Fourier inverse of its derivatives, and therefore all of them are continuous. As such, $\mathrm{p}_{\overline{X}}(x)$ is a smooth function for $x > 0$.

We take N to be a large positive number and assume that ℓ is an integer. Define the contour $L = L_1 \cup L_2 \cup L_3 \cup L_4$, where

$$\begin{aligned}
L_1 &:= \{\mathrm{Re}(z) = c - N,\ -\ell \leq \mathrm{Im}(z) \leq \ell\},\\
L_2 &:= \{\mathrm{Im}(z) = \ell,\ c - N \leq \mathrm{Re}(z) \leq 1\},\\
L_3 &:= \{\mathrm{Re}(z) = 1,\ -\ell \leq \mathrm{Im}(z) \leq \ell\},\\
L_4 &:= \{\mathrm{Im}(z) = -\ell,\ c - N \leq \mathrm{Re}(z) \leq 1\}.
\end{aligned}$$

It is clear that L is the rectangle bounded by vertical lines $\mathrm{Re}(z) = c - N$, $\mathrm{Re}(z) = 1$ and by horizontal lines $\mathrm{Im}(z) = \pm\ell$. We assume that L is oriented counter-clockwise; see Figure 7.1.

The function $\mathcal{M}(s)$ is analytic in the interior of L, except for simple poles $s_{m,n}$, which lie in $(c - N, 1)$, moreover, it is continuous on L. Using the residue

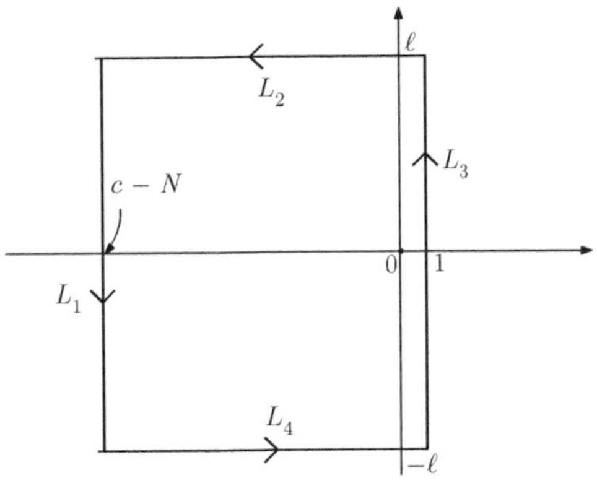

Figure 7.1 The contour $L = L_1 \cup L_2 \cup L_3 \cup L_4$

theorem (cf. Section A.1 in the Appendix), we find

$$\frac{1}{2\pi i}\int_L \mathcal{M}(s)x^{-s}\,ds = \sum_{c-N<s_{m,n}<1} \operatorname{Res}(\mathcal{M}(s)\colon s = s_{m,n}) \times x^{-s_{m,n}},$$

where the summation is over all m, n, such that $c - N < s_{m,n} < 1$. Next, we estimate the integrals over the horizontal side L_2 as follows

$$\left|\int_{L_2} \mathcal{M}(s)x^{-s}\,ds\right| < (1 - c + N) \times x^{-1} \max_{s\in L_2} |\mathcal{M}(s)|.$$

When ℓ increases, we see from (7.35) that $\max_{s\in L_2} |\mathcal{M}(s)|$ tends to 0. Therefore

$$\int_{L_2} \mathcal{M}(s)x^{-s}\,ds \to 0 \qquad \text{as} \quad \ell \to \infty.$$

Similarly, we deduce that the integral on the contour L_4 goes to 0, as ℓ goes to ∞. Thus, putting all the pieces together, we have

$$-\frac{1}{2\pi i}\int_{c-N+i\mathbb{R}} \mathcal{M}(s)x^{-s}\,ds + \frac{1}{2\pi i}\int_{1+i\mathbb{R}} \mathcal{M}(s)x^{-s}\,dz = \sum_{c-N<s_{m,n}<1} \operatorname{Res}(\mathcal{M}(s)\colon s = s_{m,n}) \times x^{-s_{m,n}}.$$

In other words, we have deduced

$$p_{\overline{X}}(x) = \sum_{c-N<s_{m,n}<1} \operatorname{Res}(\mathcal{M}(s)\colon s = s_{m,n})x^{-s_{m,n}} + \frac{1}{2\pi i}\int_{c-N+i\mathbb{R}} \mathcal{M}(s)x^{-s}ds. \tag{7.36}$$

Now, we use the explicit form of $\mathcal{M}$ from Theorem 7.8 and the reflection formula for the gamma function (see (A.12) in the Appendix) to find that, for some constant $C \in \mathbb{R}$,

$$\mathcal{M}(s) = C\frac{\Gamma\left(\frac{1-s}{\alpha}\right)}{\Gamma(1-s)}\frac{\prod_{j=0}^{l-1}\sin(\pi(s-1+j)/\alpha)}{\prod_{j=0}^{k}\sin(\pi(1-s+\alpha j))}$$

on $\mathbb{C}$. Finally, we prove that as N increases, the integral on the right-hand side of (7.36) converges to zero for any $x > 0$. Indeed, we observe

$$\left|\int_{c-N+\mathrm{i}\mathbb{R}} \mathcal{M}(s)x^{-s}\mathrm{d}s\right| < Cx^{N-c}\int_{\mathbb{R}}\left|\frac{\Gamma(\frac{N-c+\mathrm{i}t}{\alpha})}{\Gamma(N-c+\mathrm{i}t)}\right| g(t)\mathrm{d}t,$$

where

$$g(t) = \mathrm{e}^{\frac{\pi}{\alpha}l|t|}\prod_{j=0}^{k}|\mathrm{cosech}(\pi(t+i(\alpha j - c)))|\,.$$

Using the asymptotic (A.16) for the gamma function, we deduce that for any $x > 0$, the function

$$x^{N-1}\frac{\Gamma(\frac{N-c+\mathrm{i}t}{\alpha})}{\Gamma(N-c+\mathrm{i}t)}$$

converges to zero as N increases (uniformly in $t \in \mathbb{R}$), thus the integral in the right-hand side of (7.36) vanishes as N increases and implicitly the convergent series for $\alpha \in (1,2)$ is such that

$$\mathrm{p}_{\overline{X}}(x) = \sum_{n=0}^{k}\sum_{m\le -l} c^{+}_{m,n}x^{-m-\alpha n-1}, \qquad x > 0.$$

The convergence of the series for the case $l < 0$ can be established in the same way, except that now we have to change the rectangular contour so that one side lies along the line $\mathrm{Re}(z) = 1$, but the other three sides are arranged so that the contour encloses an increasing number of poles on the positive real line. The case $\alpha \in (0,1)$ can be deduced exactly in the same way as above. The details are left to the reader. □

It is worthy of note that there is another way to characterise the coefficients $c^{\pm}_{m,n}$, by making use of the sequences $\{a_{m,n}\}_{n,m\ge 0}$ and $\{b_{m,n}\}_{m\ge 0, n\ge 1}$, defined in (7.19) and (7.20), respectively. Indeed, by performing straightforward computations (similar to those used in the proof of Theorem 7.8), we can deduce that $a_{m,n} = c^{\pm}_{-l-m,k-n}$ and $b_{m,n} = -c^{\pm}_{m,n}$ accordingly as $\pm l > 0$.

We also note by way of a corollary that the spectrally negative case ($k = 0$, $l = 1$) and the spectrally positive case ($k = -1$, $l = -1$) offer more convenient expressions than already apparent from Theorem 7.9.

Corollary 7.10 *When X is spectrally negative, that is, $\alpha \in (1,2)$ and $\rho = \alpha^{-1}$,*

$$\mathrm{p}_{\overline{X}}(x) = \sum_{m\geq 1} \frac{(-1)^{m-1}}{\Gamma\left(1-\frac{m}{\alpha}\right)(m-1)!} x^{m-1}, \qquad x > 0. \tag{7.37}$$

When X is spectrally positive, that is, $\alpha \in (1,2)$ and $\rho = 1-\alpha^{-1}$, we have

$$\mathrm{p}_{\overline{X}}(x) = \sum_{n\geq 1} \frac{1}{\Gamma\left(1-n+\frac{1}{\alpha}\right)\Gamma(\alpha n - 1)} x^{\alpha n-2}, \qquad x > 0. \tag{7.38}$$

Finally, we state the Wiener–Hopf factor for stable processes in the Doney classes. In order to do so, we first introduce the q-Pochhammer symbol as follows

$$(a\colon q)_n := \prod_{k=0}^{n-1}(1-aq^k), \qquad \text{for} \quad n \in \mathbb{N},$$

and $(a\colon q)_0 = 1$. If $|q| < 1$, we define

$$(a\colon q)_\infty := \prod_{k\geq 0}(1-aq^k).$$

Theorem 7.11 (Doney's factorisation) *Assume that $X \in \mathcal{C}_{k,l}$. Then, for $|\arg(z)| < \pi$,*

$$\Psi_1^+(z) = \begin{cases} \dfrac{\left(z^\alpha(-1)^{1-l}\mathrm{e}^{(1-k)\pi\mathrm{i}\alpha} \ :\ \mathrm{e}^{2\pi\mathrm{i}\alpha}\right)_k}{\left(z(-1)^{1-k}\mathrm{e}^{-(1-l)\pi\mathrm{i}/\alpha} \ :\ \mathrm{e}^{-2\pi\mathrm{i}/\alpha}\right)_l} & \text{if} \quad l > 0, \\[3em] \dfrac{\left(z(-1)^{1+k}\mathrm{e}^{-(1+l)\pi\mathrm{i}/\alpha} \ :\ \mathrm{e}^{-2\pi\mathrm{i}/\alpha}\right)_{|l|}}{\left(z^\alpha(-1)^{1+l}\mathrm{e}^{(1+k)\pi\mathrm{i}\alpha} \ :\ \mathrm{e}^{2\pi\mathrm{i}\alpha}\right)_{|k|}} & \text{if} \quad l < 0. \end{cases}$$

We will not provide the complete proof here but will offer instead a sketch of the main arguments. The complete and formal arguments go beyond the scope of our exposition.

First, we observe that the form of the Wiener–Hopf factor Ψ_1^+ in (7.7) can be written in terms of the double gamma function G using the identity (7.3). For the second step, we also observe that the resulting identity also holds for $\alpha \in \{w \in \mathbb{C}\colon \mathrm{Re}(w) > 0\}$. After this, we develop the identity further using the analogue of the reflection formula for the double gamma function (A.22), which, in our particular case, reads as

$$-2\pi i\alpha G\left(\frac{1}{2}+z;\alpha\right)G\left(\frac{1}{2}-z;-\alpha\right)=\frac{\left(-\,\mathrm{e}^{2\pi iz}\,:\,\mathrm{e}^{2\pi i\alpha}\right)_{\infty}}{\left(\mathrm{e}^{2\pi i\alpha}\,:\,\mathrm{e}^{2\pi i\alpha}\right)_{\infty}},$$

for $\mathrm{Im}(\alpha)>0$, together with the transformation that appears in (A.23), which in particular is written as

$$G(z;\alpha)=(2\pi)^{\frac{z}{2}\left(1-\frac{1}{\alpha}\right)}\alpha^{\frac{z-z^2}{2\alpha}+\frac{z}{2}-1}G\left(\frac{z}{\alpha};\frac{1}{\alpha}\right).$$

To complete our arguments, the following identity is needed

$$\frac{(a:\,q)_\infty}{(aq^n:\,q)_\infty}=(a;q)_n,$$

which can be easily deduced from the definition of the q-Pochhammer symbol. For the last step, in the resulting identity that we have developed from (7.7) with the help of the above identities, we let $\mathrm{Im}(\alpha)$ go to 0 and use analytic continuation, to give the desired expression for the Wiener–Hopf factor when $\alpha\in(0,2)$.

7.5 Comments

Darling [56] and Heyde [87] are the first authors interested in the Wiener–Hopf factorisation for stable processes. They observed that evaluating the Wiener–Hopf factors analytically is equivalent to the evaluation of a certain definite integral, usually referred to as Darling's integral. This integral was explicitly computed in the case of symmetric Cauchy process by Darling [56] and, in the case of spectrally negative Lévy processes, by Bingham [34]. In particular, Darling found a simple expression and Bingham obtained an absolutely convergent series representation for the density of the supremum; see identity (7.37).

In his seminal paper, Doney [59] obtained a closed-form expression for the Wiener–Hopf factors for a dense set of parameters that we introduce here as the Doney classes, and which include the one-sided cases. The method employed to evaluate Darling's integral by Doney was similar to that used by Bingham. Theorem 7.11 is the main result in Doney [59] and is included for completeness in spite of the cases there being covered by Theorem 7.1. The sketch proof that we highlighted is not Doney's original proof, but rather a method suggested by Kuznetsov [116].

After the result of Doney and following an absence of any progress for around 30 years, many other significant results related to the supremum at a fixed time for stable processes and its density appeared in the literature. For instance, an absolutely convergent series representation was obtained for the spectrally positive case by Bernyk et al. [17], see identity (7.38). Doney [61] found the first asymptotic term of the density at infinity in the spectrally positive case and, in the same setting, Patie [164] provided a complete asymptotic expansion. In the setting of two-sided jumps, the first term of the asymptotic expansion of the density at 0 or at infinity was obtained by Doney and Savov [63]. Graczyk and Jakubowski [81] have also discovered a series representation for the logarithm of the Wiener–Hopf factor.

Kuznetsov [116] observed from Doney's main result in [59] that certain properties were reminiscent of aspects of the theory of elliptic functions. Motivated by this observation, Kuznetsov used the theory of elliptic functions to deduce the Wiener–Hopf factors as described in Theorem 7.1. Theorem 7.3, which describes the Mellin transform of the supremum at a fixed time, was also established in [116]. In this paper, he also establishes the asymptotic expansion of its density, which is the first part of our Theorem 7.4. Moreover, he constructs the asymptotic expansion and the convergent series representation of the density of the supremum at a fixed time for the Doney classes, which was given in Theorem 7.9. The tail distribution in Theorem 7.5 is based on a similar result in Chapter VIII of Bertoin [18].

In [116], Kuznetsov also points out that a convergent series for the density of the distribution of $\overline{X}_1$ is not so easy to deduce in full generality and a conjecture was established. This conjecture, which corresponds to the second part of Theorem 7.4, was proved shortly after by Hubalek and Kuznetsov [90]. However, this is not the end of the story. Further developments include Kuznetsov [117], who showed that there exists an uncountable dense set of irrational α's, for which the series representation that appears in the second part of Theorem 7.4 does not converge absolutely, for almost all ρ. Moreover, Kuznetsov [117] gives an infinite series representation for the density in the setting that α is rational. Finally, Hackmann and Kuznetsov [85] show that, for every irrational α, there is a way to rearrange the terms of the double series in Theorem 7.4 so that the series converges to the density of the supremum.

8

Asymptotic behaviour for stable processes

Now that we have a suitable knowledge of the distribution of a stable process and its extrema, we are in a position to develop integral tests that determine lower and upper envelopes of their sample paths as $t \to 0$. Similar arguments can be applied for $t \to \infty$, so we only state and prove our results for small times. The results for large times are similar, albeit that integral tests at 0 must be replaced by the same integral tests at ∞.

In keeping with our standard notation, $(X_t, t \geq 0)$ with probabilities $\mathbb{P}_x$, $x \in \mathbb{R}$, will always denote a stable process. Recall that $\mathcal{A}$ denotes the admissible set of parameters defined in (3.11). In the following sections, we will work our way around different regimes of the fundamental parameters $(\alpha, \rho) \in \mathcal{A}$, establishing path envelope properties for each. For some parameter combinations, the results on envelopes will be strong enough to state a law of iterated logarithm.

8.1 Stable subordinators

We begin by considering the simplest scenario, that is to say, the case when the sample paths of the stable process X are monotone increasing, that is, the case of a subordinator. Recall that, for this case, the parameters (α, ρ) are such that $\alpha \in (0, 1)$ and $\rho = 1$.

Our first result describes the upper envelope of stable subordinators.

Theorem 8.1 *Suppose* $\alpha \in (0, 1)$ *and* $\rho = 1$. *Let* $f\colon (0, \infty) \to (0, \infty)$ *be an increasing function such that* $\lim_{t\to 0} t^{-1/\alpha} f(t) = \infty$. *Then*

$$\limsup_{t\to 0} \frac{X_t}{f(t)} = 0 \quad \text{or} \quad \infty, \quad \mathbb{P}\text{-a.s.},$$

accordingly as the integral

$$\int_{0+} \frac{dt}{f(t)^\alpha} \qquad \textit{converges or diverges.} \tag{8.1}$$

Proof We first deal with the setting that the integral in (8.1) is convergent. Let $r \in (0, 1)$ and define, for $n \geq 1$, the events

$$A_n = \left\{X_{r^n} > r^{-2/\alpha} c f(r^{n+1})\right\},$$

where c is any positive constant smaller than 1.

Next, we observe that the following inequality holds

$$\begin{aligned}\sum_{n\geq 1} \mathbb{P}(A_n) &= \sum_{n\geq 1} \mathbb{P}\Big(X_{r^{n+2}} > cf(r^{n+1})\Big)\\ &\leq \int_1^\infty \mathbb{P}\Big(X_{r^t} > cf(r^t)\Big)\,dt,\end{aligned}$$

where in the equality we have used the scaling property and, in the inequality, we used the monotonicity of X and f; in particular, for $t \in [k, k+1]$, $k \geq 1$, we have $\{X_{r^{k+1}} > cf(r^k)\} \subseteq \{X_{r^t} > cf(r^t)\}$. Now appealing to the change of variable $u = r^t$ followed by the scaling property, we get

$$\sum_{n\geq 1} \mathbb{P}(A_n) \leq -\frac{1}{\log r} \int_0^r \mathbb{P}\left(X_1 > c\frac{f(u)}{u^{1/\alpha}}\right) \frac{du}{u}. \tag{8.2}$$

From (1.32), it is not difficult to deduce the asymptotic

$$\mathbb{P}(X_1 > x) \sim k(\alpha) x^{-\alpha}, \qquad \text{as} \quad x \to \infty, \tag{8.3}$$

where $\kappa(\alpha)$ is a constant that depends only on α. Putting the pieces together, we have

$$\sum_{n\geq 1} \mathbb{P}(A_n) < \infty \qquad \text{whenever} \qquad \int_{0+} \frac{dt}{f(t)^\alpha} < \infty.$$

Then, from Borel–Cantelli's Lemma, it follows that

$$X_{r^n} \leq c r^{-2/\alpha} f(r^{n+1}) \quad \text{eventually as } n \to \infty, \qquad \mathbb{P}\text{-a.s.}$$

In a similar spirit to remarks above, monotonicity of f and X tells us that, for $t \in [r^{n+1}, r^n]$, $n \geq 1$, we have $\{X_{r^n} \leq c r^{-2/\alpha} f(r^{n+1})\} \subseteq \{X_t \leq c r^{-2/\alpha} f(t)\}$. In other words, with probability one,

$$\limsup_{t\to 0} \frac{X_t}{f(t)} \leq r^{-2/\alpha} c, \qquad \text{for all} \quad r \in (0, 1).$$

The result now follows since c can be taken arbitrarily small.

For the setting that the integral in (8.1) diverges, we use again Borel–Cantelli's Lemma for the independent sequence of events

$$B_n = \left\{X_{r^{n+1}} - X_{r^{n+2}} > r^{2/\alpha} c f(r^n)\right\},$$

where c is any positive constant bigger than 1 and $r \in (0, 1)$. Following similar steps to those that led to (8.2), we have

$$\begin{aligned}\sum_{n\geq 1} \mathbb{P}(B_n) &= \sum_{n\geq 1} \mathbb{P}\left(X_{r^{n+1}(1-r)} > r^{2/\alpha} c f(r^n)\right)\\ &= \sum_{n\geq 1} \mathbb{P}\left(X_{r^{n-1}} > \left(\frac{1}{1-r}\right)^{1/\alpha} c f(r^n)\right)\\ &\geq \int_1^\infty \mathbb{P}\left(X_{r^t} > \left(\frac{1}{1-r}\right)^{1/\alpha} c f(r^t)\right) \mathrm{d}t\\ &= -\frac{1}{\log r} \int_0^r \mathbb{P}\left(X_1 > \left(\frac{1}{1-r}\right)^{1/\alpha} c \frac{f(u)}{u^{1/\alpha}}\right) \frac{\mathrm{d}u}{u},\end{aligned}$$

where in the first equality, we used stationary and independent increments, for the second and last equalities, we used the scaling property and for the inequality, we used monotonicity of f and X.

Using the asymptotic (8.3) and taking r close to 0, we get

$$\sum_{n\geq 1} \mathbb{P}(B_n) = \infty \qquad \text{whenever} \qquad \int_{0+} \frac{\mathrm{d}t}{f(t)^\alpha} = \infty.$$

From Borel–Cantelli's Lemma for independent events, it follows that

$$X_{r^{n+1}} - X_{r^{n+2}} > r^{2/\alpha} c f(r^n) \quad \text{i.o., as } n \to \infty, \qquad \mathbb{P}\text{-a.s.}$$

Again, noting in a similar spirit to earlier remarks, $\{X_{r^{n+1}} > X_{r^{n+2}} + r^{2/\alpha} c f(r^n)\} \subseteq \{X_t > r^{2/\alpha} c f(t)\}$, for $t \in [r^{n+1}, r^n]$, $n \geq 1$, thanks to the monotonicity of f and X. In other words, with probability one, we have

$$\limsup_{t\to 0} \frac{X_t}{f(t)} \geq r^{2/\alpha} c, \qquad \text{for all} \quad r < 1.$$

Then the result follows since c can be taken arbitrarily large. The proof is now complete. □

Next, we study the lower envelope. In this case, one is able to get a precise result in the form of a law of the iterated logarithm.

Theorem 8.2 *For $\alpha \in (0, 1)$ and $\rho = 1$, we have*

$$\liminf_{t\to 0} \frac{X_t (\log|\log t|)^{(1-\alpha)/\alpha}}{t^{1/\alpha}} = \alpha(1-\alpha)^{(1-\alpha)/\alpha}, \quad \mathbb{P}\text{-a.s.}$$

We prove Theorem 8.2 in two steps. First, we obtain an integral test in terms of the law of X_1 that allows us to identify lower envelopes of stable subordinators. Thereafter, we use a sharp estimate for the lower tail of the distribution of X_1 to develop the aforementioned integral test into an explicit form, which gives the desired result.

Lemma 8.3 *Suppose* $\alpha \in (0,1)$ *and* $\rho = 1$. *Let* $f\colon (0,\infty) \to (0,\infty)$ *be an increasing function such that* $\lim_{t\to 0} t^{-1/\alpha} f(t) = 0$.

(i) If

$$\int_{0+} \mathbb{P}\left(X_1 < \frac{f(u)}{u^{1/\alpha}}\right)\frac{du}{u} < \infty,$$

then

$$\liminf_{t\to 0} \frac{X_t}{f(t)} \geq 1, \quad \mathbb{P}\text{-}a.s.$$

(ii) If

$$\int_{0+} \mathbb{P}\left(X_1 < \frac{f(u)}{u^{1/\alpha}}\right)\frac{du}{u} = \infty,$$

then

$$\liminf_{t\to 0} \frac{X_t}{f(t)} \leq 1, \qquad \mathbb{P}\text{-}a.s.$$

Proof We first deduce part (i). Similarly to the proof of Theorem 8.1, we introduce, for $r \in (0,1)$ and $n \geq 2$, the events

$$A_n = \left\{X_{r^{n+1}} < r^{2/\alpha} f(r^n)\right\}$$

and use scaling and monotonicity to deduce that

$$\sum_{n\geq 2} \mathbb{P}(A_n) \leq -\frac{1}{\log r}\int_0^r \mathbb{P}\left(X_1 < \frac{f(u)}{u^{1/\alpha}}\right)\frac{du}{u}.$$

That is to say, if the previous integral is finite, then from Borel–Cantelli's Lemma, it follows, with probability one, that

$$X_{r^{n+1}} \geq r^{2/\alpha} f(r^n) \quad \text{eventually as } n \to \infty, \qquad \mathbb{P}\text{-a.s.}$$

Again, a monotonicity argument allows us to deduce that with probability one, we have

$$\liminf_{t\to 0} \frac{X_t}{f(t)} \geq r^{2/\alpha}, \qquad \text{for all} \quad r < 1.$$

As r may be taken arbitrarily close to 1, the result follows.

Now, we prove part (ii). Let us take $r \in (0,1)$ and define, for $n \geq 1$, the events

$$C_n = \left\{X_t < r^{-2/\alpha} f(t), \text{ for some } t \in (0, r^n)\right\}.$$

Note that the family of events $(C_n)_{n\geq 1}$ is decreasing. Moreover,

$$C := \bigcap_{n\geq 1} C_n = \left\{X_t < r^{-2/\alpha} f(t) \text{ i.o., as } t \to 0\right\}.$$

As such, if we prove that

$$\lim_{n\to\infty} \mathbb{P}(C_n) = 1, \tag{8.4}$$

it follows that

$$\mathbb{P}\left(X_t < r^{-2/\alpha} f(t) \text{ i.o., as } t \to 0\right) = 1, \tag{8.5}$$

which, in turn, implies the claim. In order to show (8.4), we introduce the sets,

$$B_n = \left\{X_{r^{n-1}} < r^{-2/\alpha} f(r^n)\right\}, \quad \text{for } n \geq 1.$$

Following arguments already used several times, it is clear from scaling and monotonicity that

$$\sum_{n\geq 1} \mathbb{P}(B_n) \geq -\frac{1}{\log r} \int_0^r \mathbb{P}\left(X_1 < \frac{f(u)}{u^{1/\alpha}}\right) \frac{du}{u},$$

so that the hypothesis of the claim in (ii) implies that $\sum_{n\geq 1} \mathbb{P}(B_n) = \infty$.

Since $\lim_{n\to\infty} \mathbb{P}(C_n)$ exists by monotonicity, to prove (8.4), it suffices for us to find increasing sequences $(m_k)_{k\geq 0}$ and $(n_k)_{k\geq 0}$ tending to infinity such that $0 \leq m_k \leq n_k - 1$,

$$\mathbb{P}(C_{m_k}) \geq 1 - G(m_k, n_k),$$

and $\lim_{k\to\infty} G(m_k, n_k) = 0$, where

$$G(m,n) = \mathbb{P}\left(X_{r^{j-1}} > r^{-2/\alpha} f(r^j), \text{ for all } m \leq j \leq n-1\right).$$

To this end, let us introduce

$$H(n,m) = \mathbb{P}\left(X_{r^{j-1}} - X_{r^{n-1}} > r^{-2/\alpha} f(r^j), \text{ for all } m \leq j \leq n-1\right),$$

and

$$\rho_{m,n}(x) = \mathbb{P}\left(X_{r^{j-1}} - X_{r^{n-2}} > r^{-2/\alpha} f(r^j) - x, \text{ for all } m \leq j \leq n-2\right).$$

Observe that the map $x \mapsto \rho_{m,n}(x)$ is increasing. From stationary and independent increments and the scaling property, $H(m,n)$ and $G(m,n)$ can be expressed as follows

$$H(m,n) = \int_{b_k}^{\infty} \rho_{m,n}(a_k x)\,\mathbb{P}(X_{1-r} \in \mathrm{d}x),$$

and

$$G(m,n) = \int_{b_k}^{\infty} \rho_{m,n}(a_k x)\mathbb{P}(X_1 \in \mathrm{d}x),$$

where $a_k = r^{(k-2)/\alpha}$ and $b_k = r^{-k/\alpha} f(r^{k-1})$. In particular, it follows that, for m, n sufficiently large,

$$H(n,m) \geq \rho_{n,m}(a_n N)\,\mathbb{P}(X_{1-r} \geq N), \qquad \text{for} \quad N \geq C, \tag{8.6}$$

where $C = \sup_{x \leq r^{-1}} x^{-1/\alpha} f(x)$, which can be assumed to be finite without loss of generality.

Our objective is now to show that there exist increasing sequences $(m_k)_{k\geq 0}$ and $(n_k)_{k\geq 0}$ tending to infinity such that $0 \leq m_k \leq n_k - 1$ and

$$H(m_k, n_k) \underset{k\to\infty}{\longrightarrow} 0. \tag{8.7}$$

Indeed, with (8.7) in hand, noting (8.6), it implies that $\rho_{n_k,m_k}(a_{n_k} N)$ also converges to 0, for every $N \geq C$. On the other hand, we have

$$G(n_k, m_k) \leq \rho_{n_k,m_k}(a_{n_k} N)\mathbb{P}(X_1 \leq N) + \mathbb{P}(X_1 > N),$$

and hence letting k and N tend to infinity, we get that $\lim_{k\to\infty} G(n_k, m_k) = 0$.

We thus complete the proof by proving (8.7). To this end, suppose the contrapositive, that is, that there exists $\delta > 0$ such that $H(m,n) \geq \delta$, for all sufficiently large integers m and n. From the independence of increments, we deduce

$$\begin{aligned}
1 &\geq \mathbb{P}\left(\bigcup_{n=m+1}^{\infty} B_n\right) \\
&\geq \sum_{n=m+1}^{\infty} \mathbb{P}\left(\left(\bigcup_{j=m}^{n-1} B_j\right)^c \bigcap B_j\right) \\
&\geq \sum_{n=m+1}^{\infty} \mathbb{P}(B_n) H(m,n) \\
&\geq \delta \sum_{n=m+1}^{\infty} \mathbb{P}(B_n),
\end{aligned}$$

but since the last term diverges, we deduce that the limit in (8.7) holds. □

For the next Lemma, we introduce

$$g(t) := (\log|\log t|)^{\frac{\alpha-1}{\alpha}}, \qquad 0 \leq t \leq \mathrm{e}^{-1}.$$

Lemma 8.4 *Suppose $\alpha \in (0,1)$ and $\rho = 1$. For every $c > 0$, we have*

$$-\log \mathbb{P}(X_1 \le cg(t)) \sim (1-\alpha)\left(\frac{\alpha}{c}\right)^{\frac{\alpha}{1-\alpha}} \log|\log t|, \quad \text{as} \quad t \to 0. \tag{8.8}$$

Proof We first deduce the lower bound. From Chebyshev's inequality, we have

$$\mathbb{P}(X_1 \le cg(t)) \le \exp\left\{\lambda cg(t) - \lambda^\alpha\right\}, \qquad \text{for any} \quad \lambda > 0.$$

A straightforward optimisiation of the exponent on the right-hand side above gives us

$$-\log \mathbb{P}(X_1 \le cg(t)) \ge (1-\alpha)\left(\frac{\alpha}{c}\right)^{\alpha/(1-\alpha)} \log|\log t|. \tag{8.9}$$

To complete the proof, we need an upper bound in the spirit of (8.9) as $t \to 0$. Recall from (2.22) that $\mathcal{E}_t(\lambda) := \exp(-\lambda X_t + \lambda^\alpha t)$, $t \ge 0$, is a martingale that induces the following change of measure

$$\mathbb{P}^\lambda(\Lambda) = \mathbb{E}[\mathcal{E}_1(\lambda)\mathbf{1}_\Lambda], \qquad \Lambda \in \mathcal{F}_1, \tag{8.10}$$

see (2.23). Under $\mathbb{P}^\lambda$, the process X is still a subordinator with Laplace transform given by

$$\Phi_\lambda(\theta) = (\lambda + \theta)^\alpha - \lambda^\alpha, \qquad \theta \ge 0.$$

In particular,

$$\mathbb{E}^\lambda[X_1] = \alpha\lambda^{\alpha-1}, \qquad \text{and} \qquad \mathbb{E}^\lambda\left[\left(X_1 - \alpha\lambda^{\alpha-1}\right)^2\right] = \alpha(1-\alpha)\lambda^{\alpha-2}. \tag{8.11}$$

Using (8.10) transform with $\lambda = \lambda(t)$, where

$$\lambda(t) = \left(\frac{c(1-\epsilon)}{\alpha}\right)^{\frac{1}{\alpha-1}} (\log|\log t|)^{1/\alpha}, \quad \text{for} \quad 0 < t \le \mathrm{e}^{-1},$$

we observe

$$\begin{aligned}
\mathbb{P}(X_1 \le cg(t)) &= \mathbb{E}^{\lambda(t)}\left[\mathcal{E}_1(\lambda(t))^{-1}\mathbf{1}_{\{X_1 \le cg(t)\}}\right] \\
&\ge \exp\left\{\lambda(t)c(1-2\epsilon)g(t) - \lambda^\alpha(t)\right\}\mathbb{P}^{\lambda(t)}\left(c(1-2\epsilon)g(t) \le X_1 \le cg(t)\right) \\
&= \exp\left\{-C_{\alpha,\epsilon}\log|\log t|\right\}\mathbb{P}^{\lambda(t)}\left(|X_1 - c(1-\epsilon)g(t)| \le \epsilon cg(t)\right),
\end{aligned} \tag{8.12}$$

where $\epsilon > 0$ is arbitrarily small and

$$C_{\alpha,\epsilon} = \left(\frac{\alpha}{c(1-\epsilon)}\right)^{\frac{\alpha}{1-\alpha}}\left(1 - \alpha\frac{1-2\epsilon}{1-\epsilon}\right).$$

Noting that $\mathbb{E}^{\lambda(t)}[X_1] = c(1-\epsilon)g(t)$, we have from Chebyshev's inequality and (8.11), that the following inequality holds

$$\mathbb{P}^{\lambda(t)}\Big(|X_1 - c(1-\epsilon)g(t)| > \epsilon c g(t)\Big) \le \widetilde{C}_{\alpha,\epsilon}\frac{1}{\log|\log t|}, \tag{8.13}$$

where $\widetilde{C}_{\alpha,\epsilon}$ is a positive constant that only depends on α and ϵ. In other words, as t goes to 0, the probability in (8.13) goes to 0, implying that

$$\lim_{t\to 0}\mathbb{P}^{\lambda(t)}\Big(|X_1 - c(1-\epsilon)g(t)| \le \epsilon c g(t)\Big) = 1,$$

and, hence, from (8.12), we deduce

$$\limsup_{t\to 0}\frac{-\log\mathbb{P}(X_1 \le cg(t))}{\log|\log t|} \le \left(\frac{\alpha}{c(1-\epsilon)}\right)^{\frac{\alpha}{1-\alpha}}\left(1 - \alpha\frac{1-2\epsilon}{1-\epsilon}\right).$$

Since ϵ was taken arbitrarily small, the required asymptotic upper bound follows. This completes the proof. □

Proof of Theorem 8.2 In order to deduce the result, we take

$$f(t) = ct^{1/\alpha}\left(\log|\log t|\right)^{(\alpha-1)/\alpha}, \qquad 0 < t < \mathrm{e}^{-1},$$

with $0 < c < \alpha(1-\alpha)^{(1-\alpha)/\alpha}$. From the estimate in Lemma 8.4, we note that

$$\int_{0+}\mathbb{P}\left(X_1 < \frac{f(u)}{u^{1/\alpha}}\right)\frac{\mathrm{d}u}{u} \le \int_{0+}|\log u|^{-C_\alpha}\frac{\mathrm{d}u}{u} < \infty,$$

with

$$C_\alpha = (1-\alpha)\left(\frac{\alpha}{c}\right)^{\frac{\alpha}{1-\alpha}} > 1. \tag{8.14}$$

Using part (i) in Lemma 8.3, we deduce

$$\liminf_{t\to 0}\frac{X_t}{f(t)} \ge 1, \quad \mathbb{P}\text{-a.s.} \tag{8.15}$$

The proof is completed by establishing an asymptotic upper bound for the limsup that complements (8.15). We take $f(t)$ as before but with $c > \alpha(1-\alpha)^{(1-\alpha)/\alpha}$, which from (8.14) means $C_\alpha < 1$. We get that

$$\int_{0+}\mathbb{P}\left(X_1 < \frac{f(u)}{u^{1/\alpha}}\right)\frac{\mathrm{d}u}{u} \ge \int_{0+}|\log u|^{-C_\alpha}\frac{\mathrm{d}u}{u} = \infty.$$

From part (ii) in Lemma 8.3, we deduce

$$\liminf_{t\to 0}\frac{X_t}{f(t)} \le 1, \quad \mathbb{P}\text{-a.s.},$$

which completes the proof. □

8.2 Upper envelopes for $\rho \in (0,1)$

Next, we proceed to study the upper envelope when the sample paths of stable processes are not necessarily monotone. We first assume that the stable process X possesses positive jumps and is not a subordinator. In other words, $(\alpha,\rho) \in \mathcal{A}^+$ where

$$\mathcal{A}^+ := \Big\{\alpha \in (0,1),\ \rho \in (0,1)\} \cup \{\alpha = 1,\ \rho = 1/2\} \\ \cup \{\alpha \in (1,2),\ \rho \in [1-\alpha^{-1}, \alpha^{-1})\Big\}. \tag{8.16}$$

We also recall that for any $t \geq 0$, $\overline{X}_t$ denotes the running supremum of X, i.e.

$$\overline{X}_t = \sup_{s \leq t} X_s.$$

Theorem 8.5 *Suppose* $(\alpha,\rho) \in \mathcal{A}^+$. *Let* $f\colon (0,\infty) \to (0,\infty)$ *be an increasing function such that* $\lim_{t\to 0} t^{-1/\alpha} f(t) = \infty$. *Then*

$$\limsup_{t\to 0} \frac{X_t}{f(t)} = \limsup_{t\to 0} \frac{\overline{X}_t}{f(t)} = 0 \quad or \quad \infty, \quad \mathbb{P}\text{-a.s.}, \tag{8.17}$$

accordingly as the integral

$$\int_{0+} \frac{\mathrm{d}t}{f(t)^\alpha} \qquad \textit{converges or diverges.}$$

Proof Let us start by noting from the obvious inequality $\overline{X}_t \geq X_t$, $t \geq 0$, that

$$\limsup_{t\to 0} \frac{X_t}{f(t)} \leq \limsup_{t\to 0} \frac{\overline{X}_t}{f(t)}, \tag{8.18}$$

almost surely. On the other hand, there is a sequence of times at which $X_t = \overline{X}_t$, specifically, when t belongs to the range of the inverse local time at the maximum. Moreover, since f is increasing, this means we have a collection of times, that is, the left end points of excursion of X from $\overline{X}$, say $\mathcal{T} \subseteq [0,\infty)$, which is such that $[0,\infty)\backslash\mathcal{T}$ is the countable union of open intervals on which $\overline{X}$ is constant. In addition, for each $t \in \mathcal{T}$, $X_t = \overline{X}_t$ and $f(t) \leq f(s)$, for all $s \geq t$ satisfying $\overline{X}_s = \overline{X}_t$. It follows that

$$\limsup_{t\to 0} \frac{X_t}{f(t)} \geq \limsup_{t\in\mathcal{T}} \frac{X_t}{f(t)} = \limsup_{t\in\mathcal{T}} \frac{\overline{X}_t}{f(t)} \geq \limsup_{s\to\infty} \frac{\overline{X}_s}{f(s)},$$

the converse inequality to (8.18) holds, and the first equality of (8.17) is automatic.

We now proceed similarly as in the proof of Theorem 8.1 to deduce the convergent part of the integral test. Indeed, let $r \in (0,1)$ and define, for $n \geq 1$, the events

$$A_n = \left\{\overline{X}_{r^n} > r^{-2/\alpha} c f(r^{n+1})\right\},$$

where c is any positive constant smaller than 1. Then, using the scaling property and the monotonicity of $\overline{X}$ and f as before, we deduce

$$\sum_{n\geq 1} \mathbb{P}(A_n) \leq -\frac{1}{\log r} \int_0^r \mathbb{P}\left(\overline{X}_1 > c\frac{f(u)}{u^{1/\alpha}}\right) \frac{\mathrm{d}u}{u}.$$

On the other hand, from Proposition 7.5, we have that there exists a constant $\kappa_1 > 0$, such that

$$\mathbb{P}\left(\overline{X}_1 > x\right) \sim k_1 x^{-\alpha} \qquad \text{as} \quad x \to \infty. \tag{8.19}$$

In other words, for r close to 0, we have

$$\sum_n \mathbb{P}(A_n) < \infty \qquad \text{whenever} \qquad \int_{0+} \frac{\mathrm{d}t}{f(t)^\alpha} < \infty.$$

Then, from Borel–Cantelli's Lemma and the monotonicity of $\overline{X}$ and f, we obtain, with probability one, that

$$\limsup_{t\to 0} \frac{\overline{X}_t}{f(t)} \leq r^{-2/\alpha} c, \qquad \text{for all} \quad r < 1.$$

The result then follows since c can be taken arbitrarily small.

For the divergent part of the integral test, let $r \in (0,1)$ and introduce, for $n \geq 1$, the following sets

$$B_n = \left\{|X_{r^n}| < N r^{n/\alpha}\right\} \quad \text{and} \quad C_n = \left\{\overline{X}^{(n)}_{r^{n-1}(1-r)} > c f(r^{n-1}) + N r^{n/\alpha}\right\},$$

where $\overline{X}^{(n)}_t = \sup_{s\leq t} X_{r^n+s} - X_{r^n}$ and N, c are two positive arbitrary constants. Let $D_n = B_n \cap C_n$, from stationary and independent increments and the scaling property, we see

$$\mathbb{P}(D_n) = \mathbb{P}(B_n)\mathbb{P}(C_n) \tag{8.20}$$

$$= \mathbb{P}(|X_1| < N)\mathbb{P}\left(\overline{X}_1 > c\frac{f(r^{n-1})}{r^{\frac{n}{\alpha}}(\frac{1}{r}-1)^{1/\alpha}} + \frac{N}{(\frac{1}{r}-1)^{1/\alpha}}\right). \tag{8.21}$$

It follows that, for $t \in [n-1, n]$,

$$\mathbb{P}(D_n) \geq \kappa_N \mathbb{P}\left(\overline{X}_1 > c\frac{f(r^t)}{r^{\frac{t}{\alpha}}(\frac{1}{r}-1)^{1/\alpha}} + \frac{N}{(\frac{1}{r}-1)^{1/\alpha}}\right), \tag{8.22}$$

where $\kappa_N = \mathbb{P}(|X_1| < N) > 0$. Recalling that $\lim_{u\to 0} f(u)/u^{1/\alpha} = \infty$ and the polynomial decay in the tail distribution of $\overline{X}_1$, cf. (8.19), we may asymptotically, for n sufficiently large, replace the lower bound in (8.22) by

$$\mathbb{P}(D_n) \geq \kappa_N \mathbb{P}\left(\overline{X}_1 > c\frac{f(r^l)}{r^{\frac{l}{\alpha}}(\frac{1}{r}-1)^{1/\alpha}}\right). \tag{8.23}$$

Putting these pieces together, we note that

$$\sum_{n\geq 1} \mathbb{P}(D_n) = \infty \quad \text{if} \quad -\frac{\kappa_N}{\log r}\int_0^{r^{-1}} \mathbb{P}\left(\overline{X}_1 > \frac{c}{(\frac{1}{r}-1)^{1/\alpha}}\frac{f(u)}{u^{1/\alpha}}\right)\frac{\mathrm{d}u}{u} = \infty,$$

which is the case thanks to the upper tail distribution of $\overline{X}_1$ in (8.19) and the assumed divergent integral test.

On the other hand, we observe that for $m < n$, using stationary and independent increments together with (8.20),

$$\mathbb{P}(D_m \cap D_n) \leq \mathbb{P}(D_m \cap C_n) = \mathbb{P}(D_m)\frac{\mathbb{P}(D_n)}{\mathbb{P}(B_n)} \leq \frac{1}{\kappa_N}\mathbb{P}(D_m)\mathbb{P}(D_n). \tag{8.24}$$

Clearly the case $n > m$ holds by symmetry. The inequality (8.24) fulfils the conditions of Lemma A.12 in Appendix A.9, which is of a Borel–Cantelli type, and we deduce

$$\mathbb{P}\left(\limsup_{k\geq 1} D_k\right) \geq \kappa_N.$$

Next, it is clear that on the event D_n, the inequality $\overline{X}_{r^{n-1}} > cf(r^{n-1})$ holds. Moreover, on $\limsup_{n\geq 1} D_n$,

$$\limsup_{t\to 0}\frac{\overline{X}_t}{f(t)} \geq \limsup_{t\to 0}\frac{X_t}{f(t)} \geq \frac{c}{(\frac{1}{r}-1)^{1/\alpha}},$$

with probability equal to at least $\mathbb{P}(|X_1| < N)$. Since N is arbitrary, we deduce that the latter holds with probability one. The result then follows since c can be taken arbitrarily large. □

Now let us introduce the notation

$$X_t^* = \sup_{s\leq t}|X_s|, \qquad \text{for} \quad t \geq 0. \tag{8.25}$$

Theorem 8.6 *Assume that* $(\alpha,\rho) \in \mathcal{A}$ *and let* $f\colon (0,\infty) \to (0,\infty)$ *be an increasing function such that* $\lim_{t\to 0} t^{-1/\alpha}f(t) = \infty$. *Then*

$$\limsup_{t\to 0}\frac{|X_t|}{f(t)} = \limsup_{t\to 0}\frac{X_t^*}{f(t)} = 0 \qquad \text{or} \quad \infty, \qquad \mathbb{P}\text{-a.s.},$$

accordingly as the integral

$$\int_{0+}\frac{\mathrm{d}t}{f(t)^\alpha} \qquad \text{converges or diverges.}$$

Proof Let us assume that the process has positive jumps, then if the integral diverges we use Theorem 8.5 and obtain

$$\infty = \limsup_{t\to 0} \frac{X_t}{f(t)} \le \limsup_{t\to 0} \frac{|X_t|}{f(t)} = \limsup_{t\to 0} \frac{X_t^*}{f(t)}, \qquad \mathbb{P}\text{-a.s.}$$

If the process has no positive jumps, we consider its dual process, that is, $\hat{X} = -X$. Since the dual process $\hat{X}$ possesses positive jumps, the previous argument guarantees that

$$\limsup_{t\to 0} \frac{|X_t|}{f(t)} = \limsup_{t\to 0} \frac{X_t^*}{f(t)} = \infty, \qquad \mathbb{P}\text{-a.s.}$$

Next, we assume that the integral test converges. If $|X|$ is a subordinator, then the result follows from Theorem 8.1. If the process X has two-sided jumps, then Theorem 8.5 guarantees

$$\limsup_{t\to 0} \frac{\hat{X}_t}{f(t)} = \limsup_{t\to 0} \frac{X_t}{f(t)} = 0, \qquad \mathbb{P}\text{-a.s,}$$

which implies our assertion.

If the process is spectrally negative, we recall that stationary and independent increments imply that the first passage time process $(\tau_x^+, x \ge 0)$ is a subordinator, where $\tau_x^+ = \inf\{s \ge 0 \colon X_s > x\}$; see Lemma 2.29. In addition, it is not difficult to verify that $(\tau_x^+, x \ge 0)$ respects the scaling property with self-similar index α^{-1}. From the Markov inequality, one easily deduces

$$\mathbb{P}\left(\overline{X}_1 > x\right) = \mathbb{P}(\tau_x^+ > 1) \le \mathrm{e}^{-1}\mathbb{E}\left[\mathrm{e}^{-\tau_x^+}\right] \le \mathrm{e}^{-1}\mathrm{e}^{-x}.$$

The latter inequality, together with the asymptotic (8.19), implies

$$\mathbb{P}(X_1^* > x) = \mathbb{P}\left(\overline{X}_1 > x\right) + \mathbb{P}\left(-\underline{X}_1 > x\right) - \mathbb{P}(\overline{X}_1 > x, -\underline{X}_1 > x) \sim \kappa_1 x^{-\alpha},$$

as $x \to \infty$, where $\underline{X}_1 = \inf_{s\le 1} X_s$, which is equal in law to the law of the maximum until time 1 of the dual process. Using the same arguments as in the proof of Theorem 8.5 for the case of the convergent integral test, we deduce our claim. The proof is now complete. □

In the special case of a spectrally negative stable process, that is, $\alpha \in (1,2)$ and $\rho = \alpha^{-1}$, we again recover a law of the iterated logarithm.

Theorem 8.7 *For $\alpha \in (1,2)$ and $\rho = \alpha^{-1}$, the following law of the iterated logarithm holds*

$$\limsup_{t\to 0} \frac{X_t}{t^{1/\alpha}(\log|\log t|)^{1-1/\alpha}} = \alpha^{1/\alpha}\left(\frac{\alpha}{\alpha-1}\right)^{\frac{\alpha-1}{\alpha}},$$

almost surely.

Proof Let $r \in (0,1)$ and, for $n \geq 0$, we define the events

$$A_n = \left\{\overline{X}_{r^n} > c_1(R,\alpha) f(r^{n+1})\right\},$$

where

$$f(t) = t^{1/\alpha}(\log|\log t|)^{1-1/\alpha}, \qquad t > 0,$$

and, for $R > 1$,

$$c_1(R,\alpha) := R^{\frac{\alpha-1}{\alpha}} r^{-1/\alpha} \alpha^{1/\alpha} \left(\frac{\alpha}{\alpha-1}\right)^{\frac{\alpha-1}{\alpha}}.$$

Note in particular that $r \in (0,1)$ and $\alpha \in (1,2)$ implies that $c_1(R,\alpha) > 1$. In the absence of positive jumps, we observe

$$A_n = \left\{\tau^+_{c_1(R,\alpha) f(r^{n+1})} < r^n\right\},$$

where $(\tau^+_x, x \geq 0)$ is a stable subordinator with scaling parameter α^{-1} (see earlier remarks and Lemma 2.29). Thus we use the scaling property and Lemma 8.4 to deduce that, for n sufficiently large, after some rather tedious algebra, we have the estimates

$$\sum_{k \geq n} \mathbb{P}(A_k) = \sum_{k \geq n} \mathbb{P}\left(\tau^+_1 < r^{-1} c_1(R,\alpha)^{-\alpha} (\log|\log r^{k+1}|)^{1-\alpha}\right) \leq \kappa_1(r,\alpha) \sum_{k \geq n} k^{-C_{1,\alpha}},$$

for n sufficiently large, where $\kappa_1(r,\alpha)$ is an unimportant constant and

$$C_{1,\alpha} := \left(1 - \frac{1}{\alpha}\right)\left(\frac{r c_1(R,\alpha)^\alpha}{\alpha}\right)^{\frac{1}{\alpha-1}} = R > 1.$$

We thus have that $\sum_{n=1}^\infty \mathbb{P}(A_n) < \infty$ and hence, since we may take r and R as close to 1 as we like, it follows from Borel–Cantelli's Lemma that

$$\limsup_{t \to 0} \frac{X_t}{t^{1/\alpha}(\log|\log t|)^{1-1/\alpha}} \leq \alpha^{1/\alpha}\left(\frac{\alpha}{\alpha-1}\right)^{\frac{\alpha}{\alpha-1}}, \qquad \mathbb{P}\text{-a.s.}$$

To show that

$$\limsup_{t \to 0} \frac{X_t}{t^{1/\alpha}(\log|\log t|)^{1-1/\alpha}} \geq \alpha^{1/\alpha}\left(\frac{\alpha}{\alpha-1}\right)^{\frac{\alpha}{\alpha-1}}, \qquad \mathbb{P}\text{-a.s.,} \qquad (8.26)$$

we define

$$B_n = \left\{\tau^+_{c_2(r,\alpha) f(r^n)} < r^n\right\},$$

where, now, we take the different definition

$$c_2(R,\alpha) := R^{\frac{\alpha-1}{\alpha}} \alpha^{1/\alpha}\left(\frac{\alpha}{\alpha-1}\right)^{\frac{\alpha-1}{\alpha}},$$

for some $R \in (0,1)$. Again, we use self-similarity, Lemma 8.4 applied to the subordinator $(\tau_s^+, s \geq 0)$, remembering that it has self-similarity index $1/\alpha$, and similar arguments as above to deduce that, for n sufficiently large,

$$\sum_{k\geq n} \mathbb{P}(B_k) = \sum_{n\geq 0} \mathbb{P}\left(\tau_1^+ < c_2(R,\alpha)^{-\alpha}(\log|\log r^k|)^{1-\alpha}\right) \geq \kappa_2(r,\alpha) \sum_{k\geq n} k^{-C_{2,\alpha}},$$

where, again, $\kappa_2(r,\alpha)$ is an unimportant constant, but this time,

$$C_{2,\alpha} = \left(1 - \frac{1}{\alpha}\right)\left(\frac{c_2(R,\alpha)^\alpha}{\alpha}\right)^{\frac{1}{\alpha-1}} = R < 1.$$

In consequence, $\sum_{n=1}^\infty \mathbb{P}(B_n) = \infty$. Similar reasoning to the proof of Theorem 8.3 (ii) leads us to the conclusion

$$\liminf_{s\to 0} \frac{\tau^+_{c_2(R,\alpha)f(s)}}{s} \leq 1, \qquad \mathbb{P}\text{-a.s.}$$

This implies that the set $\{s\colon \tau^+_{c_2(R,\alpha)f(s)} < s\}$ is an unbounded set $\mathbb{P}$-a.s. Spectral negativity means that the latter is equivalent to $\{t\colon \overline{X}_t > c_2(R,\alpha)f(t)\}$ is also an unbounded set $\mathbb{P}$-a.s. In other words,

$$\limsup_{t\to 0} \frac{\overline{X}_t}{f(t)} \geq R^{\frac{\alpha-1}{\alpha}} \alpha^{1/\alpha} \left(\frac{\alpha}{\alpha-1}\right)^{\frac{\alpha}{\alpha-1}}, \qquad \mathbb{P}\text{-a.s.,} \tag{8.27}$$

and we are free to choose R as close to 1 from below as we like. Similar reasoning to that found in the proof of Lemma 8.5 tells us that we may replace $\overline{X}_t$ by X_t in (8.27), thus giving us (8.26). This completes the proof. □

8.3 Lower envelopes for $\rho \in (0,1)$

Finally, we describe the lower envelope of the sample paths of stable processes. We first present an integral test that describes the lower envelope of the running supremum of stable processes.

Theorem 8.8 *Suppose $(\alpha,\rho) \in \mathcal{A}$ and $\rho \in (0,1)$. Let $f\colon (0,\infty) \to (0,\infty)$ be an increasing function, then*

$$\liminf_{t\to 0} \frac{\overline{X}_t}{t^{1/\alpha} f(t)} = 0 \quad \text{or} \quad \infty, \qquad \mathbb{P}\text{-a.s.,}$$

accordingly as the integral

$$\int_{0+} \frac{f(t)^{\alpha\rho}}{t} \mathrm{d}t \qquad \text{diverges or converges.}$$

Proof The integral test follows the same arguments as in the proof of Theorem 8.5. Indeed, the convergent part uses the events

$$A_n = \left\{\overline{X}_{r^{n+1}} < r^{1/\alpha} c r^{n/\alpha} f(r^n)\right\}, \qquad n \geq 1,$$

with $r \in (0,1)$ and c a positive constant bigger than 1. Then, it is enough to deduce the behaviour of $\mathbb{P}\left(\overline{X}_1 < x\right)$ for x small enough. From Proposition 7.6, we have

$$\mathbb{P}\left(\overline{X}_1 < x\right) \sim \frac{\alpha}{\Gamma(\hat{\rho})\Gamma(1+\alpha\rho)} x^{\alpha\rho}, \qquad \text{as} \quad x \to 0. \tag{8.28}$$

In other words, for r close to 0, we have

$$\sum_{n\geq 1} \mathbb{P}(A_n) < \infty \qquad \text{whenever} \qquad \int_{0+} \frac{f(t)^{\alpha\rho}}{t} \mathrm{d}t < \infty.$$

Then, from Borel–Cantelli's Lemma and the monotonicity of $\overline{X}$ and f, we obtain the result.

For the divergent part of the integral test, let $r \in (0,1)$ and introduce, for $n \geq 1$, the following sets

$$B_n = \left\{X_{r^n} < -\epsilon r^{n/\alpha}\right\} \text{ and } C_n = \left\{ \sup_{t \in (r^{n-1}, r^n]} (X_t - X_{r^n}) < c r^{\frac{n-1}{\alpha}} f(r^{n-1}) + \epsilon r^{n/\alpha} \right\},$$

where ϵ, c are two positive arbitrary constants. Let $D_n = B_n \cap C_n$, from the independence of increments and the scaling property, following similar reasoning to the proof of Theorem 8.5, we see

$$\begin{aligned}
\mathbb{P}(D_n) &= \mathbb{P}(B_n)\mathbb{P}(C_n) \\
&= \mathbb{P}(X_1 < -\epsilon)\mathbb{P}\left(\overline{X}_1 < c\frac{f(r^{n-1})}{(1-r)^{1/\alpha}} + \frac{\epsilon r^{\frac{1}{\alpha}}}{(1-r)^{1/\alpha}}\right) \\
&\geq \mathbb{P}(X_1 < -\epsilon)\mathbb{P}\left(\overline{X}_1 < c\frac{f(r^{n-1})}{(1-r)^{1/\alpha}}\right).
\end{aligned} \tag{8.29}$$

In particular, noting that $\mathbb{P}(X_1 < -\epsilon) \to 1 - \rho$ as $\epsilon \to 0$, we deduce that

$$\sum_{n\geq 1} \mathbb{P}(D_n) \geq -\frac{\kappa_\epsilon}{\log r} \int_0^{r^{-1}} \mathbb{P}\left(\overline{X}_1 < cf(u)\right)\frac{\mathrm{d}u}{u} = \infty,$$

where, recalling that the law of X_1 is supported on $\mathbb{R}$ as $\rho \in (0,1)$,

$$\kappa_\epsilon := \mathbb{P}(X_1 < -\epsilon) > 0.$$

Again, following the reasoning in the proof of Theorem 8.5 for $m \neq n$, we have

$$\mathbb{P}(D_m \cap D_n) \leq \kappa_\epsilon^{-1}\mathbb{P}(D_m)\mathbb{P}(D_n),$$

which, from Lemma A.12 (cf. Appendix A.9), implies

$$\mathbb{P}\left(\limsup_{k\geq 1} D_k\right) \geq \kappa_\epsilon.$$

Next, it is clear that under the event $\limsup_{n\geq 1} D_n$, the following inequality holds

$$\limsup_{t\to 0} \frac{\overline{X}_t}{t^{1/\alpha} f(t)} \leq c, \tag{8.30}$$

with probability at least equal to $\mathbb{P}(X_1 < -\epsilon) > 0$. Since c is arbitrary and the event (8.30) is in the tail sigma algebra of X (and therefore has a 0 or 1 probability), we deduce our result. □

Our last result describes the lower envelope of X^*, defined in (8.25), by a law of the iterated logarithm.

Theorem 8.9 *Suppose* $(\alpha, \rho) \in \mathcal{A}$ *and* $\rho \in (0,1)$. *There exists a constant* $k > 0$ *such that*

$$\liminf_{t\to 0} \frac{X_t^*(\log|\log(t)|)^{1/\alpha}}{t^{1/\alpha}} = k^{1/\alpha}.$$

The proof of this result relies on the following estimate.

Lemma 8.10 *Suppose that* $|X|$ *is not a subordinator, then there exists a constant* $k \in (0,\infty)$ *such that*

$$\log \mathbb{P}(X_t^* < 1) \sim -kt, \qquad \textit{as} \quad t \to \infty. \tag{8.31}$$

Proof Let us consider the function

$$f(t) = \sup_{|x|<1} \mathbb{P}_x(X_t^* < 1).$$

From the Markov property, we deduce

$$\mathbb{P}_x(X_{t+s}^* < 1) = \mathbb{E}_x\left[\mathbf{1}_{\{X_s^*<1\}} \mathbb{P}_{X_s}(X_t^* < 1)\right].$$

The previous identity implies the inequality $f(t+s) \leq f(t)f(s)$ and thus the function $\log f$ is subadditive. From Theorem A.10 (see the Appendix A.7) this tells us that there exists a constant $k \in (0,\infty]$ such that

$$\lim_{t\to\infty} \frac{1}{t} \log f(t) = -k. \tag{8.32}$$

In particular,

$$\limsup_{t\to\infty} \frac{1}{t} \log \mathbb{P}(X_t^* < 1) \leq -k. \tag{8.33}$$

Let us now turn our attention to showing

$$\liminf_{t\to\infty} \frac{1}{t} \log \mathbb{P}(X_t^* < 1) \geq -k. \tag{8.34}$$

Fix $\epsilon > 0$ arbitrarily small and deduce by the scaling property and (8.32) that, for every $t > 0$ large enough, there exists $y(t) \in [\epsilon - 1, 1 - \epsilon]$ such that

$$\mathbb{P}_{y(t)}(X_t^* < 1 - \epsilon) \geq \frac{1}{2} \sup_{|x|<1-\epsilon} \mathbb{P}_x(X_t^* \leq 1 - \epsilon) \geq \frac{1}{2} \exp\{-k(1-\epsilon)^{-\alpha} t\}.$$

The above inequality entails that for $y \in [y(t) - \epsilon, y(t) + \epsilon]$ and t sufficiently large, we have

$$\mathbb{P}_y(X_t^* < 1) \geq \frac{1}{2} \exp\{-k(1-\epsilon)^{-\alpha} t\}.$$

Next, we claim that we can find a deterministic $T > 0$ for which

$$g_T(x) := \inf_{|y|<1-\epsilon} \mathbb{P}_x\big(X_T \in [y - \epsilon, y + \epsilon], X_T^* < 1\big) > 0. \tag{8.35}$$

In order to deduce the previous claim, we proceed by contradiction. Let us assume that, for each $T > 0$, $g_T(x) = 0$, that is,

$$\inf_{|y|<1-\epsilon} \mathbb{P}_x\big(X_T \in (y - \epsilon, y + \epsilon], X_T^* < 1\big) = 0. \tag{8.36}$$

Define $F_T(z) := \mathbb{P}_x(X_T \leq z, X_T^* < 1)$, and note that (8.36) can otherwise be written as

$$\inf_{|y|<1-\epsilon} \big(F_T(y+\epsilon) - F_T(y-\epsilon)\big) = 0,$$

for every $T > 0$. Since the distribution function F_T is càdlàg, we deduce that there exists $\hat{y} \in [\epsilon - 1, 1 - \epsilon]$ such that $F_T(\hat{y}+\epsilon) - F_T(\hat{y}-\epsilon) = 0$. In other words, for all $T > 0$,

$$\mathbb{P}_x\big(X_T \in (\hat{y} - \epsilon, \hat{y} + \epsilon], X_T^* < 1\big) = 0. \tag{8.37}$$

On the other hand, recall from Theorem 6.4 that

$$U^{[-1,1]}(x, [\hat{y} - \epsilon, \hat{y} + \epsilon]) > 0,$$

where, for any Borel set A in $[-1, 1]$,

$$U^{[-1,1]}(x, A) := \int_0^\infty \mathbb{P}_x(X_t^* < 1, X_t \in A)\,\mathrm{d}t.$$

However, this implies that there exists a $\hat{T} > 0$ such that

$$\mathbb{P}_x(X_{\hat{T}}^* < 1, X_{\hat{T}} \in [\hat{y} - \epsilon, \hat{y} + \epsilon]) > 0,$$

which contradicts (8.37).

Putting all the pieces together and, using that (8.35) holds for $T > 0$ together with the Markov property, we deduce

$$\mathbb{P}_x(X^*_{t+T} < 1) \geq g_T(x) \inf_{y \in [y(t)-\epsilon, y(t)+\epsilon]} \mathbb{P}_y(X^*_t < 1) \geq \frac{1}{2} g_T(x) \exp\{-k(1-\epsilon)^{-\alpha} t\}.$$

This gives us the liminf in (8.34) as ϵ may be taken arbitrarily small in our reasoning.

In order to finish the proof, we need to verify that $k < \infty$. Recall that $p_1(x)$, the density of X_1, is continuous and positive. Hence we can take $c, c_1, \kappa > 0$ such that

$$\mathbb{P}_x(X^*_1 < c, |X_1| < c_1) \geq \kappa, \qquad \text{for all} \quad |x| < c_1.$$

From the scaling property, we have that for every integer $n > 0$,

$$\inf_{|x| < c_1 n^{-1/\alpha}} \mathbb{P}_x\left(X^*_{1/n} < cn^{-1/\alpha}, |X_{1/n}| < c_1 n^{-1/\alpha}\right) \geq \kappa.$$

Then applying the Markov property twice, we see

$$\begin{aligned} &\mathbb{P}(X^*_1 < cn^{-1/\alpha}) \\ &= \mathbb{E}\left[\mathbf{1}_{(X^*_{1/n} < cn^{-1/\alpha})} \mathbb{P}_{X_{1/n}}\left(X^*_{(n-1)/n} < cn^{-1/\alpha}\right)\right] \\ &\geq \mathbb{P}\left(X^*_{1/n} < c_1 n^{-1/\alpha},\ |X_{1/n}| < c_1 n^{-1/\alpha}\right) \inf_{|x| < c_1 n^{-1/\alpha}} \mathbb{P}_x\left(X^*_{(n-1)/n} < cn^{-1/\alpha}\right), \end{aligned}$$

which, by a recursive argument, implies

$$\mathbb{P}(X^*_1 < cn^{-1/\alpha}) \geq \left(\inf_{|x| < c_1 n^{-1/\alpha}} \mathbb{P}_x\left(X^*_{1/n} < cn^{-1/\alpha}, |X_{1/n}| < c_1 n^{-1/\alpha}\right)\right)^n \geq \kappa^n,$$

and, after applying scaling to the probability on the left-hand side above, this shows that k must be finite. □

Proof of Theorem 8.9 Let us introduce

$$f(t) = \frac{t}{\log|\log t|}, \qquad \text{for} \quad t > 0.$$

For the lower bound, we take $r \in (0,1)$ and $0 < c < c_1 < k^{1/\alpha}$. If we take n sufficiently large and r close enough to 1, we get

$$\mathbb{P}\left(X^*_{r^n} \leq cf(r^{n-1})^{1/\alpha}\right) \leq \mathbb{P}\left(X^*_{r^n} \leq c_1 f(r^n)^{1/\alpha}\right).$$

From the scaling property and the asymptotic behaviour in (8.31), we obtain for any $k' > k$,

$$\log \mathbb{P}\left(X^*_{r^n} \leq cf(r^{n-1})^{1/\alpha}\right) \leq \log \mathbb{P}\left(X^*_{\frac{r^n}{c_1^\alpha f(r^n)}} \leq 1\right) \leq -\frac{k'}{c_1^\alpha} \log(n|\log r|),$$

as $n \to \infty$. Hence for n large enough, we have

$$\sum_{m \geq n} \mathbb{P}\left(X^*_{r^m} \leq cf(r^{m-1})^{1/\alpha}\right) \leq C \sum_{m \geq n} m^{-k/c_1^\alpha}.$$

Since $k > c_1^\alpha$, one can deduce

$$\sum_{n\geq 1} \mathbb{P}\left(X^*_{r^n} \leq cf(r^{n-1})^{1/\alpha}\right) < \infty,$$

and the lower bound

$$\liminf_{t\to 0} \frac{X^*_t(\log|\log(t)|)^{1/\alpha}}{t^{1/\alpha}} \geq k^{1/\alpha}$$

thus follows from Borel–Cantelli's Lemma and taking c arbitrarily close to $k^{1/\alpha}$.

For the upper bound

$$\liminf_{t\to 0} \frac{X^*_t(\log|\log(t)|)^{1/\alpha}}{t^{1/\alpha}} \leq k^{1/\alpha}, \tag{8.38}$$

we take $r > 1$ and $k^{1/\alpha} < c_2$. Let $t_n = \exp\{-n^r\}$. Since

$$\sup_{s\in[t_{n+1},t_n]} |X_s - X_{t_{n+1}}| \overset{(d)}{=} X^*_{t_n - t_{n+1}} \leq X^*_{t_n}, \tag{8.39}$$

we have that $\{X^*_{t_n} \leq x\}$ is contained in $\{X^*_{t_n-t_{n+1}} \leq x\}$ and hence

$$\mathbb{P}\left(\sup_{s\in[t_{n+1},t_n]} |X_s - X_{t_{n+1}}| \leq x\right) \geq \mathbb{P}\left(X^*_{t_n} \leq x\right).$$

For r close enough to 1, we now have

$$\begin{aligned}\sum_{m\geq n} \mathbb{P}\left(\sup_{s\in[t_{m+1},t_m]} |X_s - X_{t_{m+1}}| \leq c_2 f(t_m)^{1/\alpha}\right) &\geq \sum_{m\geq n} \mathbb{P}\left(X^*_{t_m} \leq c_2 f(t_m)^{1/\alpha}\right)\\ &\geq C \sum_{m\geq n+1} m^{-kc_2^{-\alpha} r}\\ &= \infty,\end{aligned}$$

where the second inequality follows from scaling and the asymptotic (8.31) and the infinite sum follows since we can choose $kc_2^{-\alpha} r < 1$ on account of the fact that $kc_2^{-\alpha} < 1$ and we can choose r as close to 1 as we like from above. Using Borel–Cantelli's Lemma, we deduce

$$\liminf_{n\to\infty} \frac{\sup_{s\in[t_{n+1},t_n]} |X_s - X_{t_{n+1}}|}{f(t_n)^{1/\alpha}} \leq c_2 < k^{1/\alpha}, \qquad \mathbb{P}\text{-a.s.} \tag{8.40}$$

Next, note that, since $t_n > 2t_{n+1}$, we have in contrast to (8.39) that

$$\sup_{s\in[t_{n+1},t_n]} |X_s - X_{t_{n+1}}| \overset{(d)}{=} X^*_{t_n - t_{n+1}} \geq X^*_{t_{n+1}}.$$

Hence since

$$\frac{X^*_{t_n}}{f(t_n)^{1/\alpha}} = \frac{X^*_{t_{n+1}}}{f(t_n)^{1/\alpha}} \vee \Big(\frac{\sup_{s\in[t_{n+1},t_n]}|X_s|}{f(t_n)^{1/\alpha}}\Big)$$
$$\leq \frac{X^*_{t_{n+1}}}{f(t_n)^{1/\alpha}} \vee \Big(\frac{\sup_{s\in[t_{n+1},t_n]}|X_s - X_{t_{n+1}}|}{f(t_n)^{1/\alpha}} + \frac{|X_{t_{n+1}}|}{f(t_n)^{1/\alpha}}\Big),$$

it easily follows from the independent comparison of $X^*_{t_{n+1}}$ and $\sup_{s\in[t_{n+1},t_n]}|X_s - X_{t_{n+1}}|$, if we can show that

$$\lim_{n\to\infty} \frac{|X_{t_{n+1}}|}{f(t_n)^{1/\alpha}} = 0, \qquad \mathbb{P}\text{-a.s.,} \tag{8.41}$$

then

$$\frac{X^*_{t_n}}{f(t_n)^{1/\alpha}} \leq \frac{\sup_{s\in[t_{n+1},t_n]}|X_s - X_{t_{n+1}}|}{f(t_n)^{1/\alpha}},$$

infinitely often. Together with (8.40) this is sufficient to deduce

$$\liminf_{n\to\infty} \frac{X^*_{t_n}}{f(t_n)^{1/\alpha}} \leq k^{1/\alpha}, \qquad \mathbb{P}\text{-a.s.,}$$

in other words (8.38) holds.

To show (8.41), the coarse estimate $X_1 \leq \overline{X}_1$ and $-X_1 \leq -\underline{X}_1$, we use the scaling property and the asymptotic (8.19) to get

$$\begin{aligned}\mathbb{P}\left(|X_{t_{n+1}}| \geq \epsilon f(t_n)^{1/\alpha}\right) &= \mathbb{P}\left(|X_1| \geq \epsilon t_{n+1}^{-1/\alpha} f(t_n)^{1/\alpha}\right)\\ &= \mathbb{P}\left(X_1 \geq \epsilon t_{n+1}^{-1/\alpha} f(t_n)^{1/\alpha}\right) + \mathbb{P}\left(-X_1 \geq \epsilon t_{n+1}^{-1/\alpha} f(t_n)^{1/\alpha}\right)\\ &\leq C\epsilon^{-\alpha} t_{n+1} f(t_n)^{-1}\\ &= C\epsilon^{-\alpha}\exp\{-(n+1)^r + n^r\} r \log n,\end{aligned}$$

for some constant $C \in (0,\infty)$. In other words,

$$\sum_{m\geq n} \mathbb{P}\left(|X_{t_{n+1}}| \geq \epsilon f(t_n)^{1/\alpha}\right) \leq C\epsilon^{-\alpha} \sum_{k\geq n} \exp\{-n^r((1+1/n)^r - 1)\} \log n^r < \infty.$$

From the classical Borel–Cantelli Lemma, we have that

$$\limsup_{n\to\infty} \frac{|X_{t_{n+1}}|}{f(t_n)} \leq \epsilon, \qquad \mathbb{P}\text{-a.s.,}$$

which implies (8.41) as we may take ϵ as small as we like. This completes the proof. □

8.4 Comments

There is a huge variety of results concerning the upper and lower envelopes of stable processes. Most of them have been extended to the setting of more general classes of Lévy processes. One of the earliest works in this arena, if not the first, is Khintchine [105], who considered the upper envelope of stable subordinators. The description of the lower envelope of stable subordinators, manifesting in a law of the iterated logarithm, was derived by Fristedt [73]. We refer to Fristedt [74] for a survey about the asymptotic behaviour of subordinators. The arguments used in Theorems 8.1 and 8.2 follow similar reasoning to those used by Watanabe [213], where the sample path behaviour of increasing self-similar processes with independent increments is studied. The asymptotic behaviour of the lower tail of the law of a stable subordinator described in Lemma 8.4 is taken from Bertoin [18].

The lower and upper envelopes of the supremum of stable processes, Theorems 8.5 and 8.8, were noted by Bertoin [18]; the versions we present here are taken from Fourati [72]. The upper envelope of the radial part of one-dimensional stable processes, Theorem 8.6, was noted by Khintchine [105]. Chapter 12, the d-dimensional analogue will be treated. The law of the iterated logarithm in the spectrally negative case presented in Theorem 8.7 was first noted by Zolotarev [217], but our approach is from Bertoin [21]. Theorem 8.9 is from Taylor [207] and Lemma 8.10 is from Bertoin [19]. An additional reference which contains summary results concerning upper and lower envelopes, albeit now dated, is the PhD thesis of Mijnheer [150].

9

Envelopes of positive self-similar Markov processes

In the spirit of Chapter 8, we are interested in developing integral tests that describe the lower and upper envelopes at $t \to 0$ and as $t \to \infty$, but now for general positive self-similar Markov processes starting from the origin. Given Lamperti's characterisation of pssMps (cf. Theorem 5.2), such integral tests should ideally be written in terms of the Lévy process that underlies the Lamperti transform. It turns out that the law of its exponential functional is the natural quantity that serves a purpose to that end. This emerges from a path decomposition at last passage, which we discuss in Section 9.1. Similarly to the previous chapter, we only give proofs of the integral tests associated with upper and lower envelopes as $t \to 0$. The proofs of the asymptotic behaviour as $t \to \infty$ are essentially the same with minor modifications.

Ultimately, our aim is to develop the aforementioned integral tests into an explicit form for the setting of the pssMps associated with the path functionals of stable processes discussed in Chapter 5. This is done in the next chapter.

9.1 Path decompositions for pssMp

Let us consider a pssMp with self-similar index $\alpha > 0$, as usual written (Z, P), where $P = (P_x, x \geq 0)$ is its family of laws. We denoted by Ξ its associated Lévy process via the Lamperti transform (cf. Theorem 5.2). In other words, for every $t \geq 0$,

$$Z_t = x \exp\left\{\Xi_{\varphi(tx^{-\alpha})}\right\} \mathbf{1}_{(t<x^\alpha I_\infty)}, \tag{9.1}$$

where

$$I_t = \int_0^t e^{\alpha \Xi_s}\, ds \qquad \text{with} \qquad I_\infty := \lim_{t\uparrow\infty} I_t$$

and

$$\varphi(t) = \inf\{s > 0 \colon I_s > t\}.$$

We also consider its dual (cf. Proposition 5.5), $(Z, \hat{P})$, where $\hat{P} = (\hat{P}_x, x \geq 0)$ are its probabilities, which is also a pssMp with self-similar index $\alpha > 0$, and which is associated with the dual of the Lévy process Ξ, that is, $\hat{\Xi} = -\Xi$. In other words, the Lamperti transform of $(Z, \hat{P}_x)$, is equal in law to

$$x \exp\left\{\hat{\Xi}_{\hat{\varphi}(tx^{-\alpha})}\right\} \mathbf{1}_{(t<x^{\alpha}\hat{I}_\infty)}, \qquad t \geq 0, \tag{9.2}$$

where $\hat{I}_{\cdot}$ and $\hat{\varphi}(\cdot)$ are defined as above but with $\hat{\Xi}$ instead of Ξ.

In what follows, we always assume that the Lévy process Ξ, with law $\mathbf{P}$, is not arithmetic (i.e. that its paths do not almost surely live in a strict sub-group of $\mathbb{R}$) and satisfies

$$0 < \mathbf{E}[\Xi_1] < \infty. \tag{9.3}$$

We will also use $\hat{\mathbf{P}}$ to denote the law of $\hat{\Xi}$.

Assumption (9.3) implies that $\hat{\Xi}_t \to -\infty$ almost surely and hence $(Z, \hat{P})$ hits 0 continuously so that $x^{\alpha}\hat{I}_\infty$ corresponds to the first hitting time at 0 of $(Z, \hat{P})$. Moreover, (9.3) tells us that the process (Z, P) is conservative and that the conditions of Theorem 5.3 are fulfilled. In other words, we have that $P_0 := \lim_{x\downarrow 0} P_x$ exists, in the sense of weak convergence on the Skorokhod space (cf. Theorem 5.3).

Let us denote the last passage time of (Z, P_0) by

$$\mathtt{D}_y = \sup\{t \geq 0 : Z_t \leq y\}, \qquad \text{for} \quad y > 0, \tag{9.4}$$

with the convention that $\sup \emptyset = 0$. We also recall from Proposition 5.6 that the law of $(Z, \hat{P}_z)$ is a regular version of the law of the process

$$\overleftarrow{Z}_t := Z_{(\mathtt{D}_x - t)-}, \qquad 0 \leq t \leq \mathtt{D}_x, \tag{9.5}$$

under $P_0(\cdot | Z_{\mathtt{D}_x-} = z)$, for $z \in \mathcal{S}_x$, where $\mathcal{S}_x$ denotes the support of the law of $Z_{\mathtt{D}_x-}$.

Fix a decreasing sequence $(x_n)_{n\geq 1}$, of positive real numbers such that $x_n \downarrow 0$ as n goes to ∞. For purposes that will soon become clear, we need to decompose the paths of (Z, P_0) at the sequence of last passage times $(\mathtt{D}_{x_n})_{n\geq 1}$.

To this end, we introduce the first passage time of $\overleftarrow{Z}$ below y, as follows

$$\overleftarrow{S}_y = \inf\{t > 0 : \overleftarrow{Z}_t \leq y\} \qquad y > 0.$$

Proposition 9.1 *Between the first passage times $\overleftarrow{S}_{x_n}$ and $\overleftarrow{S}_{x_{n+1}}$, the process $\overleftarrow{Z}$ has the pathwise description*

$$\left(\overleftarrow{Z}_{\overleftarrow{S}_{x_n}+t}, \ 0 \leq t \leq \overleftarrow{S}_{x_{n+1}} - \overleftarrow{S}_{x_n}\right) = \left(\Gamma_n \exp\left\{\hat{\Xi}^{(n)}_{\hat{\varphi}^{(n)}(t/\Gamma_n^{\alpha})}\right\}, \ 0 \leq t \leq \Delta_n\right), \quad n \geq 1,$$

where the processes $(\hat{\Xi}^{(n)})_{n\geq 1}$ *are also independent copies of* $\hat{\Xi}$, *which are independent of*

$$\Gamma_1 := Z_{\mathrm{D}_{x_1}-}.$$

For $n \geq 1$, *we have*

$$\hat{\varphi}^{(n)}(t) = \inf\left\{s > 0\colon \hat{I}^{(n)}_s \geq t\right\}, \qquad \hat{I}^{(n)}_s = \int_0^s \mathrm{e}^{\alpha\hat{\Xi}^{(n)}_u}\,\mathrm{d}u$$

and, iteratively,

$$\Gamma_{n+1} := \Gamma_n \exp\left(\hat{\Xi}^{(n)}_{\hat{T}^{(n)}(\log(x_{n+1}/\Gamma_n))}\right) \quad \text{and} \quad \Delta_n := \Gamma_n^\alpha\, \hat{I}^{(n)}_{\hat{T}^{(n)}(\log(x_{n+1}/\Gamma_n))},$$

with

$$\hat{T}^{(n)}(z) = \inf\{t > 0\colon \hat{\Xi}^{(n)}_t \leq z\}.$$

Moreover, for each $n \geq 1$, Γ_n *is independent of* $\hat{\Xi}^{(n)}$ *and*

$$x_n^{-1}\Gamma_n \overset{(\mathrm{d})}{=} x_1^{-1}\Gamma_1, \tag{9.6}$$

where $\overset{(\mathrm{d})}{=}$ *means identity in distribution.*

Before moving to the proof, we should note that the above definitions may be degenerate, depending on the spacing of the sequence $(x_n)_{n\geq 1}$. Indeed, by definition, $\hat{T}^{(n)}(\log(x_{n+1}/\Gamma_n)) = 0$ on the event that $\{\Gamma_n \leq x_{n+1}\}$, in which case we have, for example, $\Gamma_{n+1} = \Gamma_n$.

Proof of Proposition 9.1 From (9.2) and Proposition 5.6, recalling the notation (9.5), the process $\overleftarrow{Z}$, with point of issue $Z_0 = x_1$, may be described as

$$\left(\Gamma_1 \exp\left\{\hat{\Xi}^{(1)}_{\hat{\varphi}^{(1)}(t/\Gamma_1^\alpha)}\right\},\ 0 \leq t \leq \Gamma_1^\alpha \hat{I}^{(1)}_\infty\right),$$

where $\hat{\Xi}^{(1)}$ has the same law as $\hat{\Xi}$, is independent of Γ_1 and

$$\hat{\varphi}^{(1)}(t) = \inf\left\{s > 0\colon \hat{I}^{(1)}_s \geq t\right\}, \qquad \text{with} \qquad \hat{I}^{(1)}_s = \int_0^s \mathrm{e}^{\alpha\hat{\Xi}^{(1)}_u}\,\mathrm{d}u.$$

On the other hand, we observe that $\Gamma_1 \leq x_1$, almost surely, so between the first passage times $\overleftarrow{S}_{x_1} = 0$ and $\overleftarrow{S}_{x_2}$, the process $\overleftarrow{Z}$ is clearly described as in the statement with $\hat{\Xi}^{(1)} = \hat{\Xi}$ and

$$\overleftarrow{S}_{x_2} - \overleftarrow{S}_{x_1} = \Delta_1 = \Gamma_1^\alpha\, \hat{I}^{(1)}_{\hat{T}^{(1)}(\log(x_2/\Gamma_1))}.$$

Now, if we set

$$\hat{\Xi}^{(2)}_t := \hat{\Xi}^{(1)}_{\hat{T}^{(1)}(\log(x_2/\Gamma_1))+t} - \hat{\Xi}^{(1)}_{\hat{T}^{(1)}(\log(x_2/\Gamma_1))}, \qquad t \geq 0,$$

then with the definitions of the statement, we see in the pathwise sense,

$$\left(\overleftarrow{Z}_{\overleftarrow{S}_{x_2}+t},\ t \ge 0\right) = \left(\Gamma_2 \exp\left\{\hat{\Xi}^{(2)}_{\hat{\varphi}^{(2)}(t/\Gamma_2^\alpha)}\right\},\ t \ge 0\right), \tag{9.7}$$

and

$$\overleftarrow{S}_{x_3} - \overleftarrow{S}_{x_2} = \inf\{t > 0:\ \overleftarrow{Z}_{\overleftarrow{S}_{x_2}+t} \le x_3\} = \Delta_2.$$

Thanks to stationary and independent increments, the process $\hat{\Xi}^{(2)}$ is independent of the couple $((\hat{\Xi}^{(1)}_t,\ 0 \le t \le \hat{T}^{(1)}(\log(x_2/\Gamma_1))),\ \Gamma_1)$.

From the scaling property, we have that the processes

$$\left(\frac{x_2}{x_1} Z_{(x_1/x_2)^\alpha t},\ 0 \le t \le \left(\frac{x_2}{x_1}\right)^\alpha \mathrm{D}_{x_1}\right) \quad \text{and} \quad \left(Z_t,\ 0 \le t \le \mathrm{D}_{x_2}\right),$$

have the same law under P_0, which implies that the couples

$$\left(x_1^{-1} Z_{\mathrm{D}_{x_1}-}, x_1^{-\alpha}\mathrm{D}_{x_1}\right) \quad \text{and} \quad \left(x_2^{-1} Z_{\mathrm{D}_{x_2}-}, x_2^{-\alpha}\mathrm{D}_{x_2}\right), \tag{9.8}$$

have the same law, under P_0. On the other hand, we see from the definition of $\overleftarrow{Z}$ in Proposition 5.6 that, in the pathwise sense,

$$\left(\overleftarrow{Z}_{S_{x_2}+t},\ 0 \le t \le \mathrm{D}_{x_1} - S_{x_2}\right) \text{ and } \left(Z_{(\mathrm{D}_{x_2}-t)-},\ 0 \le t \le \mathrm{D}_{x_2}\right)$$

are equal. This gives us (9.6) for $n = 2$ from this identity, (9.7) and the identity in law in (9.8). The remainder of the proof follows by a straightforward inductive argument, which we leave to the reader. □

Corollary 9.2 *With the same notation as in Proposition 9.1, for each $n \ge 1$, the last passage time D_{x_n} can be written as the decomposition*

$$\mathrm{D}_{x_n} = \sum_{k \ge n} \Gamma_k^\alpha \hat{I}^{(k)}_{\hat{T}^{(k)}(\log(x_{k+1}/\Gamma_k))}. \tag{9.9}$$

In particular, for all $z > 0$, we have the almost sure inequality

$$z^\alpha \mathbf{1}_{\{\Gamma_n \ge z\}} \hat{I}^{(n)}_{\hat{T}^{(n)}(\log(x_{n+1}/z))} \le \mathrm{D}_{x_n} \le x_n^\alpha \overline{I}^{(n)}_\infty, \tag{9.10}$$

where $\overline{\Xi}^{(n)}$, $n \ge 1$, live on the same space to and are equal in law to $\hat{\Xi}$ and

$$\overline{I}^{(n)}_\infty = \int_0^\infty \exp\left\{\alpha \overline{\Xi}^{(n)}_t\right\} \mathrm{d}t.$$

In the spirit of an earlier remark, we note that, by definition, since $\Gamma_n \le x_n$ and $\hat{T}^{(n)}(y) = 0$ for $y \ge 0$, the first inequality in (9.10) is relevant only when $x_{n+1} < z < x_n$.

Proof of Corollary 9.2 The identity (9.9) follows from Proposition 9.1 and the fact that

$$\mathrm{D}_{x_n} = \sum_{k \ge n} \left(\overleftarrow{S}_{x_{k+1}} - \overleftarrow{S}_{x_k}\right).$$

From (9.9), we deduce

$$\Gamma_n^\alpha \hat{I}^{(n)}_{\hat{T}^{(n)}(\log(x_{n+1}/\Gamma_n))} \leq \mathtt{D}_{x_n},$$

which clearly implies the first inequality in (9.10).

In order to deduce the second inequality in (9.10), let us start by identifying the processes $\bar{\Xi}^{(n)}$, $n \geq 0$. To this end, define

$$\Sigma_k^{(n)} = \sum_{j=n}^{n+k-1} \hat{T}^{(j)}(\log(x_{j+1}/\Gamma_j))$$

and use these times to define, iteratively,

$$\overline{\Xi}_t^{(n)} = \begin{cases} \hat{\Xi}_t^{(n)} & \text{if } t \in [0, \Sigma_1^{(n)}), \\ \hat{\Xi}^{(n+1)}_{t-\Sigma_1^{(n)}} & \text{if } t \in [\Sigma_1^{(n)}, \Sigma_2^{(n)}), \\ \vdots & \\ \hat{\Xi}^{(n+k)}_{t-\Sigma_k^{(n)}} & \text{if } t \in [\Sigma_k^{(n)}, \Sigma_{k+1}^{(n)}), \\ \vdots & \end{cases} \tag{9.11}$$

which is independent of Γ_n and has the same law as $\hat{\Xi}$.

With the definition (9.11), we note that the process $(\hat{\Xi}_t^{(n)}, 0 \leq t \leq \hat{T}^{(n)}(\log(x_{n+1}/\Gamma_n))$ is the same as the process $\overline{\Xi}^{(n)}$ killed at $\overline{T}^{(n)}(\log(x_{n+1}/\Gamma_n))$, where

$$\overline{T}^{(n)}(x) = \inf\{t \colon \overline{\Xi}_t^{(n)} \leq x\}, \qquad x \leq 0.$$

Moreover, from the definition (9.5), Proposition 5.6 and the strong Markov property, for any $n \geq 1$, in the pathwise sense,

$$\left(\overleftarrow{Z}_{\overleftarrow{S}_{x_n}+t}, \ 0 \leq t \leq \mathtt{D}_{x_1} - \overleftarrow{S}_{x_n}\right) \text{ and } \left(\Gamma_n \exp\left\{\overline{\Xi}^{(n)}_{\overline{\varphi}^{(n)}(t/\Gamma_n^\alpha)}\right\}, \ 0 \leq t \leq \mathtt{D}_{x_1} - \overleftarrow{S}_{x_n}\right)$$

are equal, where

$$\overline{\varphi}^{(n)}(t) = \inf\left\{s > 0 \colon \overline{I}_s^{(n)} > t\right\}, \quad \overline{I}_s^{(n)} = \int_0^s \mathrm{e}^{\alpha \overline{\Xi}_u^{(n)}} \, du \text{ and } \Gamma_n = \overleftarrow{Z}_{\overleftarrow{S}_{x_n}}$$

(cf. Proposition 9.1). It remains to note that

$$\mathtt{D}_{x_1} - \overleftarrow{S}_{x_n} = \mathtt{D}_{x_n} = \Gamma_n^\alpha \overline{I}_\infty^{(n)} \tag{9.12}$$

and that, by definition, $\Gamma_n \leq x_n$, which gives us the second inequality of (9.10), as required. □

Combining (9.6) with (9.12), we also have the following corollary.

Corollary 9.3 *For all $n \geq 1$,*

$$\mathtt{D}_{x_n} \stackrel{(d)}{=} \Gamma_n^\alpha I_\infty^{(n)}.$$

In order to study the envelope of a pssMp as $t \to \infty$, we also need to understand the law of the time $\mathtt{D}_x$ when x is large. A similar decomposition as for the decreasing-to-zero sequence $(x_n)_{n\geq 1}$, can be provided for an increasing sequence $(y_n)_{n\geq 1}$, which tends to ∞, as it is stated below. Once again, the reader is alerted to the degeneracies of some of the statements, depending on the spacings of $(y_n)_{n\geq 1}$.

Corollary 9.4 *Let $(y_n)_{n\geq 1}$ be an increasing sequence of positive real numbers which increases to ∞. There exist sequences of processes $(\check{\Xi}^{(n)})_{n\geq 1}$, $(\tilde{\Xi}^{(n)})_{n\geq 1}$ and variables $(\check{\Gamma}_n)_{n\geq 1}$, on the same probability space such that, for each $n \geq 1$, the processes $\check{\Xi}^{(n)}$ and $\tilde{\Xi}^{(n)}$ have the same law as $\hat{\Xi}$ and the random variables $\check{\Gamma}_n$ have the same law as Γ_1. Moreover, $\check{\Gamma}_n$ and $\check{\Xi}^{(n)}$ are independent, the Lévy processes $(\check{\Xi}^{(n)})_{n\geq 1}$ are mutually independent and we have, for all $z > 0$,*

$$z^\alpha \mathbf{1}_{\{\check{\Gamma}_n \geq z\}} \int_0^{\check{T}^{(n)}(\log(y_{n-1}/z))} e^{\alpha \check{\Xi}_s^{(n)}} \, ds \leq \mathtt{D}_{y_n} \leq y_n^\alpha \tilde{I}_\infty^{(n)}, \tag{9.13}$$

almost surely, where

$$\tilde{I}_\infty^{(n)} := \int_0^\infty e^{\alpha \tilde{\Xi}_u^{(n)}} du \quad and \quad \check{T}_z^{(n)} = \inf\{t > 0 : \check{\Xi}_t^{(n)} \leq z\}.$$

Proof Fix an integer $n \geq 1$ and define the decreasing sequence $x_1, \ldots, x_n$ by $x_n = y_1, x_{n-1} = y_2, \ldots, x_1 = y_n$. Using the sequence $(x_1, \cdots, x_n)$, we can use the definitions of $\hat{\Xi}^{(1)}, \ldots, \hat{\Xi}^{(n)}, \Gamma_1, \ldots, \Gamma_n$ from $x_1, \ldots, x_n$ and $\overline{\Xi}^{(1)}, \ldots, \overline{\Xi}^{(n)}$ from Proposition 9.1 and Corollary 9.2.

Now, define $\check{\Xi}^{(1)} = \hat{\Xi}^{(n)}, \check{\Xi}^{(2)} = \hat{\Xi}^{(n-1)}, \ldots, \check{\Xi}^{(n)} = \hat{\Xi}^{(1)}$ and $\tilde{\Xi}^{(1)} = \overline{\Xi}^{(n)}, \tilde{\Xi}^{(2)} = \overline{\Xi}^{(n-1)}, \ldots, \tilde{\Xi}^{(n)} = \overline{\Xi}^{(1)}$ and $\check{\Gamma}_1 = \Gamma_n, \check{\Gamma}_2 = \Gamma_{n-1}, \ldots, \check{\Gamma}_n = \Gamma_1$. With this new notation, it is implicit from (9.10), that for any $k = 2, \ldots, n$, for all $z > 0$,

$$z^\alpha \mathbf{1}_{\{\check{\Gamma}_k \geq z\}} \int_0^{\check{T}^{(k)}(\log(y_{k-1}/z))} e^{\alpha \check{\Xi}_s^{(k)}} \, ds \leq \mathtt{D}_{y_k} \leq y_k^\alpha \tilde{I}_\infty^{(n)},$$

almost surely, The sequences $(\tilde{\Xi}^{(n)})_{n\geq 1}$, $(\check{\Xi}^{(n)})_{n\geq 1}$ and $(\check{\Gamma}_n)_{n\geq 1}$ are thus well defined with the desired properties. □

We now identify the law of Γ_1 in terms of the stationary ascending overshoot distribution of Ξ. Let us write $H = (H_t, t \geq 0)$ for ascending ladder height process of Ξ; see Section 2.14 for its formal definition. Since the process Ξ

is not arithmetic and has positive finite mean, the Wiener–Hopf factorisation (2.31) implies that its ladder height process H is also not arithmetic (i.e. that its range not live in a strict sub-group of $[0,\infty)$) and that $\mathbf{E}[H_1] < \infty$. Suppose we denote by Π the Lévy measure of the subordinator H and by τ_x its first passage time above the level $x > 0$. The family of overshoots of the ascending ladder height process $(H_{\tau_x} - x, x \geq 0)$ agrees with the family of the overshoots of Ξ, that is, $(\Xi_{T_x^+} - x, x \geq 0)$, where $T_x^+ = \inf\{t > 0 \colon \Xi_t > x\}$.

Corollary 2.28 tells us that, when treated as a stochastic process, the aforementioned family of overshoots converges in distribution towards a random variable whose law can be determined explicitly. More precisely,

$$H_{\tau_x} - x \xrightarrow[x\to\infty]{(w)} \mathfrak{U}W, \tag{9.14}$$

where '$\xrightarrow{(w)}$' means weak convergence, $\mathfrak{U}$ and W are independent random variables, $\mathfrak{U}$ is uniformly distributed over $[0,1]$ and the law of W is such that

$$\mathbf{P}(W > u) = \frac{1}{\mathbf{E}[H_1]}\int_{(u,\infty)} s\,\Pi(\mathrm{d}s), \quad \text{for} \quad u \geq 0. \tag{9.15}$$

The relationship between the above limiting distribution and Γ_1 can now be given in the next result.

Lemma 9.5 *Assume that Ξ is not arithmetic and that $0 < \mathbf{E}[\Xi_1] < \infty$, then the law of Γ_1 is characterised as follows*

$$\log(x_1^{-1}\Gamma_1) \stackrel{(\mathrm{d})}{=} -\mathfrak{U}W,$$

where $\mathfrak{U}$ and W are defined as above. In particular, for all $x < x_1$, we have $\mathbf{P}(\Gamma_1 > x) > 0$.

Proof We showed in Proposition 9.1, that

$$\begin{aligned} x_1^{-1}\Gamma_1 &\stackrel{(\mathrm{d})}{=} x_{n+1}^{-1}\Gamma_{n+1} \\ &= \exp\left\{\hat{\Xi}^{(n)}_{\hat{T}^{(n)}(\log(x_{n+1}/\Gamma_n))} - \log(x_{n+1}/\Gamma_n)\right\} \\ &\stackrel{(\mathrm{d})}{=} \exp\left\{\hat{\Xi}_{\hat{T}(\log(x_{n+1}/x_n)+\log(x_1^{-1}\Gamma_1))} - \log(x_{n+1}/x_n) - \log(x_1^{-1}\Gamma_1)\right\}. \end{aligned}$$

Note that, since $(x_n)_{n\geq 0}$ is a decreasing sequence, $\log(x_{n+1}/x_n) \leq 0$, and, by definition, $\log(x_1^{-1}\Gamma_1) \leq 0$. As such,

$$\hat{\Xi}_{\hat{T}(\log(x_{n+1}/x_n)+\log(x_1^{-1}\Gamma_1))} - \log(x_{n+1}/x_n) - \log(x_1^{-1}\Gamma_1) = \hat{\Xi}_{\hat{T}(z_n)} + z_n = -(\Xi_{T_{z_n}^+} - z_n),$$

where $z_n = -\log(x_{n+1}/x_n) - \log(x_1^{-1}\Gamma_1)$. Then, by taking, for example, $x_n = \mathrm{e}^{-n^2}$ (any choice for which $z_n \to \infty$ will do), we deduce from the equalities (9.16) that $\log(x_1^{-1}\Gamma_1)$ has the same law as the limiting overshoot distribution of Ξ as claimed. □

9.2 Lower envelopes

In order to describe the lower envelope of a pssMp, we need to study the behaviour of the upper tail distributions of $\hat{I}_\infty$ and

$$\hat{I}_{\hat{T}^-_{-q}} = \int_0^{\hat{T}^-_{-q}} e^{\alpha \hat{\Xi}_u}\, du, \qquad q > 0,$$

where

$$\hat{T}^-_x = \inf\left\{t > 0 : \hat{\Xi}_t < x\right\}.$$

To this end, let us introduce, for $t > 0$,

$$F(t) := \mathbf{P}(\hat{I}_\infty > t) \qquad \text{and} \qquad F_q(t) := \mathbf{P}\left(\hat{I}_{\hat{T}^-_{-q}} > t\right), \tag{9.16}$$

for $t \geq 0$.

It will turn out that the two distributions F and F_q are the natural quantities from which to develop integral tests for the lower envelope. Before we engage in the main conclusions, for which the aforesaid integral tests are given, let us first develop some analytical results for F and F_q. The following result will be used to show that, for particular cases, knowing F suffices to describe the lower envelope of (Z, P_0).

Lemma 9.6 *Assume that there exists $\gamma > 1$ such that,*

$$\limsup_{t\to\infty} \frac{F(\gamma t)}{F(t)} < 1.$$

For any $q > 0$ and $\delta > \gamma e^{-\alpha q}$, we have that

$$\liminf_{t\to\infty} \frac{F_q((1-\delta)t)}{F(t)} > 0.$$

Proof It follows from the decomposition of Ξ into the two independent processes $(\hat{\Xi}_s,\ s \leq \hat{T}^-_{-q})$ and $\hat{\Xi}' := (\hat{\Xi}_{s+\hat{T}^-_{-q}} - \hat{\Xi}_{\hat{T}^-_{-q}},\ s \geq 0)$ that

$$\hat{I}_\infty = \hat{I}_{\hat{T}^-_{-q}} + e^{\alpha \hat{\Xi}_{\hat{T}^-_{-q}}} \hat{I}'_\infty \leq \hat{I}_{\hat{T}^-_{-q}} + e^{-\alpha q} \hat{I}'_\infty,$$

where $\hat{I}'_\infty = \int_0^\infty e^{\alpha \hat{\Xi}'_s}\, ds$ is a copy of $\hat{I}_\infty$, which is independent of $\hat{I}_{\hat{T}^-_{-q}}$. Then we can write, for any $q > 0$ and $\delta \in (0, 1)$, the inequalities

$$\begin{aligned}\hat{\mathbf{P}}(I_\infty > t) &\leq \hat{\mathbf{P}}\left(I_{T^-_{-q}} + e^{-\alpha q} I'_\infty > t\right)\\ &\leq \hat{\mathbf{P}}\left(I_{T^-_{-q}} > (1-\delta)t\right) + \hat{\mathbf{P}}\left(e^{-\alpha q} I_\infty > \delta t\right).\end{aligned}$$

Therefore, for $\delta > \gamma e^{-\alpha q}$, we deduce

$$1 - \frac{\hat{\mathbf{P}}(I_\infty > \gamma t)}{\hat{\mathbf{P}}(I_\infty > t)} \le 1 - \frac{\hat{\mathbf{P}}(I_\infty > e^{\alpha q}\delta t)}{\hat{\mathbf{P}}(I_\infty > t)} \le \frac{\hat{\mathbf{P}}\left(I_{T^-_{-q}} > (1-\delta)t\right)}{\hat{\mathbf{P}}(I_\infty > t)},$$

from which the result follows. □

With all these tools in hand, we are ready to state the main result of this section which provides an integral test as $t \to 0$ for the lower envelope of (Z, P_0). As alluded to earlier, this theorem asserts that the asymptotic behaviour of (Z, P_0) depends only on the analytic properties of the distributions F and F_q.

Theorem 9.7 *The lower envelope of (Z, P_0) at 0 is described as follows. Let f be an increasing function.*

(*i*) *If*

$$\int_{0+} F\left(\frac{t}{f(t)^\alpha}\right)\frac{dt}{t} < \infty,$$

then, for all $\varepsilon > 0$,

$$P_0\left(Z_t < (1-\varepsilon)f(t) \text{ i.o. as } t \to 0\right) = 0.$$

(*ii*) *If, for all $q > 0$,*

$$\int_{0+} F_q\left(\frac{t}{f(t)^\alpha}\right)\frac{dt}{t} = \infty,$$

then, for all $\varepsilon > 0$,

$$P_0\left(Z_t < (1+\varepsilon)f(t) \text{ i.o. as } t \to 0\right) = 1.$$

(*iii*) *Suppose that $t \mapsto t^{-1}f(t)^\alpha$ is increasing. If there exists $\gamma > 1$ such that,*

$$\limsup_{t\to\infty} \frac{F(\gamma t)}{F(t)} < 1 \quad \text{and} \quad \int_{0+} F\left(\frac{t}{f(t)^\alpha}\right)\frac{dt}{t} = \infty,$$

then, for all $\varepsilon > 0$,

$$P_0\left(Z_t < (1+\varepsilon)f(t) \text{ i.o. as } t \to 0\right) = 1.$$

Proof We start with some general remarks before proving each of the three parts of the theorem individually.

Let $(x_n)_{n\ge 1}$ be a decreasing sequence such that $x_n \downarrow 0$ as n goes to ∞. We define the events

$$A_n = \left\{\text{there exists } t \in [D_{x_{n+1}}, D_{x_n}] \text{ such that } Z_t < f(t)\right\}.$$

Since $\mathtt{D}_{x_n}$ tends to 0 a.s., as $n \to \infty$, we have

$$\left\{Z_t < f(t) \text{ i.o. as } t \to 0\right\} = \limsup_{n\to\infty} A_n\,. \tag{9.17}$$

We also introduce $(z_n)_{n\geq 1}$ a decreasing sequence satisfying $x_{n+1} < z_n < x_n$. Since f is increasing and $Z_t \geq x_{n+1}$ for $t \in [\mathtt{D}_{x_{n+1}}, \mathtt{D}_{x_n}]$, we deduce the following inclusions

$$\left\{Z_{\mathtt{D}_{x_n}-} > z_n, x_n < f(\mathtt{D}_{x_n}-)\right\} \subset A_n \subset \left\{x_{n+1} \leq f(\mathtt{D}_{x_n})\right\}. \tag{9.18}$$

For the first inclusion, we observe that the event $\{Z_{\mathtt{D}_{x_n}-} > z_n\}$ guarantees that $\mathtt{D}_{x_{n+1}} < \mathtt{D}_{x_n}$. To see why, note that the contrapositive (i.e. $\mathtt{D}_{x_{n+1}} = \mathtt{D}_{x_n}$) would entail that $z_n < Z_{\mathtt{D}_{x_n}-} = Z_{\mathtt{D}_{x_{n+1}}-} < x_{n+1}$, which violates the assumption that $z_n > x_{n+1}$. Hence if we also have that $x_n \leq f(\mathtt{D}_{x_n}-)$ in addition to $Z_{\mathtt{D}_{x_n}-} > z_n$, the right continuity of Z and f allow us to find $\epsilon > 0$ such that for all $s \in (\mathtt{D}_{x_n} - \epsilon, \mathtt{D}_{x_n})$, $Z_s < f(s)$. The second inclusion always holds even for those cases when $\mathtt{D}_{x_{n+1}} = \mathtt{D}_{x_n}$.

(i) Let us choose $x_n = r^{-n}$ for $r > 1$, and recall from relation (9.10) above that $\mathtt{D}_{r^{-n}} \leq r^{-\alpha n}\overline{I}^{(n)}_\infty$. From this inequality, (9.18) and the monotone property of f, we observe that

$$A_n \subset \left\{r^{-(n+1)} \leq f\left(r^{-\alpha n}\overline{I}^{(n)}_\infty\right)\right\}. \tag{9.19}$$

Therefore the classical version of the Borel–Cantelli Lemma, (9.19) and (9.17) imply that, if

$$\sum_{n\geq 1} \hat{\mathbf{P}}\left(r^{-(n+1)} \leq f(r^{-\alpha n} I_\infty)\right) < \infty,$$

then

$$P_0\left(Z_t < f(t), \;\text{ i.o. as } t \to 0\right) = 0. \tag{9.20}$$

Using the change of variable $s = \mathrm{e}^{-\alpha t} I_\infty$, we observe that

$$\int_1^\infty \hat{\mathbf{P}}\left(r^{-t} \leq f(r^{-\alpha t} I_\infty)\right) \mathrm{d}t = \int_0^\infty \frac{\hat{\mathbf{P}}\left(s < f(s)^\alpha I_\infty,\; s < I_\infty r^{-\alpha}\right)}{s\alpha \log(r)} \mathrm{d}s.$$

Since f is increasing, the following inequalities hold

$$\begin{aligned}\sum_{n=1}^\infty \hat{\mathbf{P}}\left(r^{-n} \leq f(r^{-\alpha(n+1)} I_\infty)\right) &\leq \int_0^\infty \hat{\mathbf{P}}\left(\frac{s}{f(s)^\alpha} < I_\infty,\; s < I_\infty r^{-\alpha}\right) \frac{\mathrm{d}s}{s\alpha\log(r)} \\ &\leq \sum_{n=1}^\infty \hat{\mathbf{P}}\left(r^{-(n+1)} \leq f(r^{-\alpha n} I_\infty)\right).\end{aligned} \tag{9.21}$$

(We note that the first inequality is relevant to the current proof of part (i) but the second inequality will be relevant to the proof of part (ii).) With no loss of generality, we can restrict ourselves to the case $f(0) = 0$. It is not difficult to check that for any $r > 1$,

$$\int_{0+} \hat{\mathbf{P}}\left(\frac{s}{f(s)^\alpha} < I_\infty,\ s < I_\infty r^{-\alpha}\right) \frac{\mathrm{d}s}{s} < \infty \ \text{ iff } \int_{0+} F\left(\frac{s}{f(s)^\alpha}\right) \frac{\mathrm{d}s}{s} < \infty. \tag{9.22}$$

Suppose the latter condition holds, then from (9.21), for all $r > 1$,

$$\sum_{n=2}^{\infty} \hat{\mathbf{P}}\Big(r^{-(n+1)} \le r^{-2} f(r^{-\alpha n} I_\infty)\Big) < \infty,$$

and from (9.20), for all $r > 1$,

$$P_0\Big(Z_t < r^{-2} f(t),\ \text{ i.o. as } t \to 0\Big) = 0,$$

which proves the desired result for part (i).

(ii) Again, we choose $x_n = r^{-n}$ for $r > 1$, and $z_n = kr^{-n}$, where $k = 1-\varepsilon+\varepsilon/r$, for $0 < \varepsilon < 1$ (so that $x_{n+1} < z_n < x_n$). We set

$$B_n = \left\{r^{-n} \le f_{r,\varepsilon}\left(k^\alpha r^{-\alpha n} \mathbf{1}_{\{\Gamma_n \ge kr^{-n}\}} \hat{I}^{(n)}_{\hat{T}^{(n)}(\log(x_{n+1}/z_n))}\right)\right\},$$

where, $f_{r,\varepsilon}(t) = rf(t/k^\alpha)$ and observe that, for each n,

$$\hat{I}^{(n)}_{\hat{T}^{(n)}(\log(x_{n+1}/z_n))} \stackrel{(\mathrm{d})}{=} \int_0^{\hat{T}^-(\log(1/rk))} \mathrm{e}^{\alpha \hat{\Xi}_s}\, \mathrm{d}s, \tag{9.23}$$

which is independent of Γ_n, and Γ_n is such that $x_n^{-1}\Gamma_n \stackrel{(\mathrm{d})}{=} x_1^{-1}\Gamma_1$. Moreover the random variables

$$\hat{I}^{(n)}_{\hat{T}^{(n)}(\log(x_{n+1}/z_n))}, \qquad n \ge 1,$$

are mutually independent and identity (9.23) shows that they have the same law as $\hat{I}_{\hat{T}^-(-q)}$ defined before Lemma 9.6, with $q = -\log(1/rk)$. Without loss of generality, we may assume that $f(0) = 0$, so that we rewrite

$$B_n = \left\{r^{-n} \le f_{r,\varepsilon}\left(k^\alpha r^{-\alpha n} \hat{I}^{(n)}_{\hat{T}^{(n)}(\log(x_{n+1}/z_n))}\right),\ \Gamma_n \ge kr^{-n}\right\}$$

and from the above remarks we deduce

$$\mathbf{P}(B_n) = \hat{\mathbf{P}}\Big(r^{-n} \le f_{r,\varepsilon}\left(k^\alpha r^{-\alpha n} I_{T^-_{-q}}\right)\Big)\mathbf{P}\left(\Gamma_1 \ge kr^{-1}\right). \tag{9.24}$$

The arguments which are developed above to show (9.21) and (9.22), are also valid if we replace $\hat{I}_\infty$ by $\hat{I}_{\hat{T}^-_{-q}}$. Hence from the hypothesis in the statement of part (ii), since

$$\int_{0+} \hat{\mathbf{P}}\Big(s < f(s)^\alpha I_{T^-_{-q}}\Big)\frac{\mathrm{d}s}{s} = \infty,$$

and from (9.21) and (9.22) with $\hat{I}_\infty$ replaced by $\hat{I}_{\hat{T}^-_{-q}}$, we have

$$\sum_{n=1}^{\infty} \hat{\mathbf{P}}\left(r^{-(n+1)} \le f_{r,\varepsilon}\left(r^{-\alpha n} I_{T_{-q}^-}\right)\right) = \sum_{n=1}^{\infty} \hat{\mathbf{P}}\left(r^{-n} \le f\left(k^\alpha r^{-\alpha n} I_{T_{-q}^-}\right)\right) = \infty,$$

and from (9.24) we deduce $\sum_{n\ge1} \mathbf{P}(B_n) = \infty$. Therefore, remembering the mutual independence of $\Xi^{(n)}$ and $\Xi^{(m)}$, the definitions of B_n and B_m, we have for any $n \neq m$,

$$\mathbf{P}(B_n \cap B_m) \le \hat{\mathbf{P}}\left(r^{-n} \le f_{r,\varepsilon}\left(k^\alpha r^{-\alpha n} I_{T_{-q}^-}\right)\right)\hat{\mathbf{P}}\left(r^{-m} \le f_{r,\varepsilon}\left(k^\alpha r^{-\alpha m} I_{T_{-q}^-}\right)\right).$$

Using identity (9.24), we get

$$\mathbf{P}(B_n \cap B_m) \le \mathbf{P}(\Gamma_1 \ge kr^{-1})^{-2}\mathbf{P}(B_n)\mathbf{P}(B_m),$$

where $\mathbf{P}(\Gamma_1 \ge kr^{-1}) > 0$, from Lemma 9.5. From the generalisation of Borel–Cantelli's lemma given in Lemma A.12, we obtain

$$\mathbf{P}\left(\limsup_{n\to\infty} B_n\right) \ge \mathbf{P}(\Gamma_1 \ge kr^{-1})^2 > 0. \tag{9.25}$$

Then, we recall from Corollary 9.2 the inequality

$$k^\alpha r^{-\alpha n} \mathbf{1}_{\{\Gamma_n \ge kr^{-n}\}} \hat{I}^{(n)}_{\hat{T}^{(n)}(\log(x_{n+1}/z_n))} \le \mathsf{D}_{r^{-n}},$$

which implies from (9.18) that $B_n \subset A_n$, (where in the definition of A_n we replaced f by $f_{r,\varepsilon}$). So, from (9.25), $P_0(\limsup_n A_n) > 0$, but since (Z, P_0) is a Feller process and $\limsup_n A_n$ is a tail event, we have

$$P_0\left(\limsup_{n\ge1} A_n\right) = 1.$$

We deduce from the scaling property of (Z, P_0) and (9.17) that

$$\begin{aligned} P_0\left(Z_t \le f_{r,\varepsilon}(t) \text{ i.o. as } t \to 0\right) &= P_0(Z_{k^\alpha t} \le rf(t) \text{ i.o. as } t \to 0) \\ &= P_0(Z_t \le k^{-1} rf(t) \text{ i.o. as } t \to 0) \\ &= 1. \end{aligned}$$

Since $k = 1 - \varepsilon + \varepsilon/r$, with $r > 1$ and $0 < \varepsilon < 1$ are arbitrary, we obtain the statement of part (ii).

(iii) The sequences $(x_n)_{n\ge1}$ and $(z_n)_{n\ge1}$ are defined as in the proof of part (ii). Recall that $q = -\log(1/rk)$ and take $\delta > \gamma e^{-\alpha q}$ as in Lemma 9.6. Without loss of generality, we may assume that $t^{-1} f(t)^\alpha \to 0$, as $t \to 0$ Then from our hypothesis and Lemma 9.6, we have

$$\int_{0+} F_q\left(\frac{(1-\delta)t}{f(t)^\alpha}\right) \frac{\mathrm{d}t}{t} = \infty.$$

As already noticed above, this is equivalent to

$$\int_1^\infty \hat{\mathbf{P}}\left((1-\delta)r^{-t} \le f\left(r^{-\alpha t} I_{T^-_{-q}}\right)\right) \mathrm{d}t = \infty .$$

Since $t \mapsto t^{-1} f(t)^\alpha$ increases,

$$\int_1^\infty \hat{\mathbf{P}}\left((1-\delta)r^{-t} \le f\left(r^{-\alpha t} I_{T^-_{-q}}\right)\right) \mathrm{d}t \le \sum_{n=1}^\infty \hat{\mathbf{P}}\left((1-\delta)r^{-n} \le f\left(r^{-\alpha n} I_{T^-_{-q}}\right)\right) = \infty .$$

Set $f_r^{(\delta)}(t) = (1-\delta)^{-1} f(t/k^\alpha)$, then

$$\sum_1^\infty \hat{\mathbf{P}}\left(r^{-n} \le f_r^{(\delta)}\left(k^\alpha r^{-\alpha n} I_{T^-_{-q}}\right)\right) = \infty .$$

Similarly to the proof of part (ii), we define

$$B'_n = \left\{ r^{-n} \le f_r^{(\delta)}\left(k^\alpha r^{-\alpha n} \hat{I}^{(n)}_{\hat{T}^{(n)}(\log(x_{n+1}/z_n))}\right),\ \Gamma_n \ge k r^{-n} \right\}.$$

Then $B'_n \subset A_n$, (where in the definition of A_n we replaced f by $f_r^{(\delta)}$). From the same arguments as above, since $\sum_{n\ge1} \mathbf{P}(B'_n) = \infty$, we have $P_0(\limsup_n A_n) = 1$. Hence from the scaling property of (Z, P_0) and (9.17)

$$\begin{aligned} P_0(Z_t \le f_r^{(\delta)}(t) \text{ i.o. as } t \to 0) &= P_0\left(Z_{k^\alpha t} \le (1-\delta)^{-1} f(t) \text{ i.o. as } t \to 0\right) \\ &= P_0\left(Z_t \le k^{-1}(1-\delta)^{-1} f(t) \text{ i.o. as } t \to 0\right) \\ &= 1 . \end{aligned}$$

Since $k = 1 - \varepsilon + \varepsilon/r$ with $r > 1$, $0 < \varepsilon < 1$ and $\delta > \gamma e^{-q} = \gamma/(r + \varepsilon(1-r))$, by choosing r sufficiently large and ε sufficiently small, δ can be taken sufficiently small so that $k^{-1}(1-\delta)^{-1}$ is arbitrary close to 1. This completes the proof of part (iii). □

We can use the same arguments of Theorem 9.7 to address similar results for the lower envelope as $t \to \infty$ of (Z, P_x), but now for any point of issue $x \ge 0$. We do not give a proof of the result below as the arguments go through almost verbatim as in Theorem 9.7 with some minor technical variation.

Theorem 9.8 *Let $x \ge 0$. The lower envelope of (Z, P_x) at ∞ is described as follows. Let f be an increasing function.*

(i) If

$$\int^\infty F\left(\frac{t}{f(t)^\alpha}\right) \frac{\mathrm{d}t}{t} < \infty ,$$

then for all $\varepsilon > 0$,

$$P_x\Big(Z_t < (1-\varepsilon)f(t)\ \ i.o.\ as\ t \to \infty\Big) = 0\,.$$

(ii) *If for all* $q > 0$,

$$\int^{\infty} F_q\left(\frac{t}{f(t)^{\alpha}}\right)\frac{dt}{t} = \infty\,,$$

then for all $\varepsilon > 0$,

$$P_x\Big(Z_t < (1+\varepsilon)f(t)\ \ i.o.\ as\ t \to \infty\Big) = 1\,.$$

(iii) *Suppose that* $t \mapsto t^{-1}f(t)^{\alpha}$ *is decreasing. If there exists* $\gamma > 1$ *such that*

$$\limsup_{t\to\infty}\frac{F(\gamma t)}{F(t)} < 1 \qquad and \qquad \int^{\infty} F\left(\frac{t}{f(t)^{\alpha}}\right)\frac{dt}{t} = \infty\,,$$

then for all $\varepsilon > 0$, *and for all* $x \geq 0$,

$$P_x\Big(Z_t < (1+\varepsilon)f(t)\ \ i.o.\ as\ t \to \infty\Big) = 1\,.$$

9.3 Upper envelopes

For the study of the upper envelope of (Z, P_0), we proceed in two steps. First we describe the upper envelope of its future infimum, which is defined as follows

$$\mathtt{J}_t = \inf_{s\geq t} Z_s, \qquad \text{for} \quad t \geq 0,$$

and then we compare it with the upper envelope of the process itself.

Observe that the process $\mathtt{J} = (\mathtt{J}_t, t \geq 0)$ is an increasing self-similar process with the same index of self-similarity as Z. It is clear that when the process Z starts at 0, the process $\mathtt{J}$ also starts from 0. When the process Z starts from $x > 0$, the future infimum $\mathtt{J}$ starts from its global infimum, that is to say

$$\mathtt{J}_0 = \underline{Z}_{\infty} = \inf_{t\geq 0} Z_t.$$

We also introduce the random variable Υ which is independent of $\hat{I}_{\infty}$ and has the same distribution as $x_1^{-1}\Gamma_1$. That is to say,

$$\Upsilon \overset{(d)}{=} x^{-1} Z_{\mathrm{D}_x -}, \text{ for all } x > 0, \tag{9.26}$$

where we recall that there is no dependency on x thanks to scaling. We also note that the support of the law of Υ is a subset of the interval [0,1].

Moreover, with this notation Corollary 9.3 has a more convenient statement of the law of the last passage time at $x > 0$.

Corollary 9.9 *The last passage time at $x > 0$ under P_0 satisfies*

$$\mathtt{D}_x \stackrel{(d)}{=} x^\alpha \Upsilon^\alpha \hat{I}_\infty. \tag{9.27}$$

In order to state our results, we set

$$\overline{F}_\Upsilon(t) = P_0\Big(\Upsilon^\alpha \hat{I}_\infty \leq t\Big), \qquad t \geq 0. \tag{9.28}$$

Definition 9.10 Denote by C_0 the set of positive increasing functions $h(t)$ on $(0, \infty)$ satisfying

(i) $h(0) = 0$, and
(ii) there exists $\delta \in (0, 1)$ such that $\sup_{t<\delta} th(t)^{-\alpha} < \infty$.

We will also need the following integral condition in all the main results of this section:

$$\mathbf{E}\left[\log^+\left(\frac{1}{\Upsilon^\alpha \hat{I}_\infty}\right)\right] < \infty. \tag{9.29}$$

This condition may come across as a concern as it may appear to be difficult to verify. Later, in Section 10.1, we demonstrate that, for a class of pssMps driven by hypergeometric Lévy processes, (9.29) can indeed be verified. This is of particular importance in the next chapter, where we develop further the integral tests for the upper and lower envelopes of this chapter for particular examples of pssMp, which correspond to a number of the path functionals of stable processes that we have seen earlier in this book.

Our first result of this section provides integral tests for the upper envelope at 0 of the future infimum of pssMps.

Theorem 9.11 *Let $h \in C_0$ and assume that* (9.29) *holds.*

(i) If

$$\int_{0^+} \overline{F}_\Upsilon\left(\frac{t}{h(t)^\alpha}\right) \frac{\mathrm{d}t}{t} < \infty,$$

then for all $\epsilon > 0$,

$$P_0\Big(\mathtt{J}_t > (1+\epsilon)h(t) \text{ i.o. as } t \to 0\Big) = 0.$$

(ii) If

$$\int_{0^+} \overline{F}_\Upsilon\left(\frac{t}{h(t)^\alpha}\right) \frac{\mathrm{d}t}{t} = \infty,$$

then for all $\epsilon > 0$,

$$P_0\Big(\mathtt{J}_t > (1-\epsilon)h(t) \text{ i.o. as } t \to 0\Big) = 1.$$

Proof As with the proof of Theorem 9.7, we start again with some general remarks before proving each of the two parts of the theorem individually. Let $(x_n)_{n\geq 1}$ be a decreasing sequence which converges to 0. Recalling that $\mathtt{D}_x$ is the last passage time below $x > 0$, see (9.4), we define the events

$$A_n = \Big\{\text{there exists } t \in [\mathtt{D}_{x_{n+1}}, \mathtt{D}_{x_n}) \text{ such that } \mathtt{J}_t > h(t)\Big\}.$$

From the fact that $\mathtt{D}_{x_n}$ tends to 0, P_0-a.s., when n goes to ∞, we see

$$\Big\{\mathtt{J}_t > h(t) \text{ i.o. as } t \to 0\Big\} = \limsup_{n\geq 1} A_n.$$

(i) Since h is a non-decreasing function and $x_n > \mathtt{J}_t$ for $t \in [\mathtt{D}_{x_{n+1}}, \mathtt{D}_{x_n})$, the following inclusion holds

$$A_n \subset \Big\{x_n > h(\mathtt{D}_{x_{n+1}})\Big\}. \tag{9.30}$$

We choose $x_n = r^n$, for $r < 1$, and introduce $h_r(t) = r^{-2}h(t)$. Appealing again to monotonicity, we deduce that

$$\begin{aligned}\sum_{n\geq 0} P_0\Big(r^n > h_r(\mathtt{D}_{r^{n+1}})\Big) &\leq \int_1^\infty P_0\Big(r^s > h(\mathtt{D}_{r^s})\Big)\mathrm{d}s \\ &= -\frac{1}{\log(r)}\int_0^r P_0\Big(t > h(\mathtt{D}_t)\Big)\frac{\mathrm{d}t}{t},\end{aligned} \tag{9.31}$$

where the last identity follows from the change of variable $s = r^t$. Hence, if we replace h by h_r in (9.30), since r may be taken arbitrarily close to 1 from below, we see from (9.31) that the result follows if

$$\int_0^r P_0\Big(t > h(\mathtt{D}_t)\Big)\frac{\mathrm{d}t}{t} < \infty.$$

From Corollary 9.3, Fubini's Theorem and the change of variable $s = t^\alpha \Upsilon^\alpha \hat{I}_\infty$, we have

$$\begin{aligned}\int_0^r P_0\Big(t > h(\mathtt{D}_t)\Big)\frac{\mathrm{d}t}{t} &= \int_0^r \mathbf{P}\Big(t > h\big(t^\alpha \Upsilon^\alpha \hat{I}_\infty\big)\Big)\frac{\mathrm{d}t}{t} \\ &= \frac{1}{\alpha}\mathbf{E}\left[\int_0^{g(r)} \mathbf{1}_{\{sr^{-\alpha} < \Upsilon^\alpha \hat{I}_\infty < sh(s)^{-\alpha}\}}\frac{\mathrm{d}s}{s}\right],\end{aligned} \tag{9.32}$$

where $g(s) := \inf\{t > 0, h(t) > s\}$ denotes the right-continuous inverse function of h. Then, this integral on the right-hand side of (9.32) converges if

$$\int_0^{g(r)} \mathbf{P}\left(\Upsilon^\alpha \hat{I}_\infty < \frac{t}{h(t)^\alpha}\right)\frac{\mathrm{d}t}{t} < \infty.$$

By choosing r suitably close to 1 and using similar reasoning to the proof of part (i) of Theorem 9.7, we deduce part (*i*).

(ii) We use an adaptation of the proof of part (ii) of Lemma 8.3. We suppose that h satisfies

$$\int_{0^+} \overline{F}_\Upsilon\left(\frac{t}{h(t)^\alpha}\right)\frac{\mathrm{d}t}{t} = \infty. \tag{9.33}$$

Again we take $x_n = r^n$, for $r < 1$, and introduce $h_r(t) = rh(t)$. By replacing h by h_r in the definition of A_n, we note

$$B_n = \bigcup_{j=n-1}^{\infty} A_j = \left\{\text{there exist } t \in (0, \mathtt{D}_{r^{n-1}}) \text{ such that } \mathtt{J}_t > h_r(t)\right\}. \tag{9.34}$$

Hence, if we prove that

$$\lim_{n\to\infty} P_0(B_n) > 0, \tag{9.35}$$

we can use the fact that (Z, P_0) is a Feller process, invoke Blumenthal's 0-1 law and deduce that

$$P_0(\mathtt{J}_t > h_r(t) \text{ i.o., as } t \to 0) = 1.$$

The latter identity, clearly implies the statement in part (ii).

Since $\mathtt{J}_{\mathtt{D}_x} \geq x$, for $x \geq 0$, we deduce that for $m \geq n + 1$, the inclusion below holds

$$\begin{aligned} \bigcap_{j=n-1}^{m} A_j^c &= \left\{\mathtt{J}_t \leq h(t) \text{ for all } t \in [\mathtt{D}_{x_{m+1}}, \mathtt{D}_{x_{n-1}})\right\} \\ &\subset \bigcap_{j=n}^{m} \left\{x_j \leq h(\mathtt{D}_{x_j})\right\}. \end{aligned} \tag{9.36}$$

From the inclusion in (9.36), (9.34) and Corollary 9.3, we see

$$P_0(B_n) \geq 1 - \mathbf{P}\left(r^j \leq h_r\left(\Gamma_j^\alpha \overline{I}_\infty^{(j)}\right), \text{ for all } n \leq j \leq m\right), \tag{9.37}$$

where m is chosen arbitrarily such that $m \geq n + 1$.

The remainder of the proof is quite long. Before embarking on it, we give a brief summary first of the two main objectives.

Step 1: Define the events

$$C_n = \left\{r^n > h_r\left(\Gamma_n^\alpha \overline{I}_\infty^{(n)}\right)\right\}.$$

We will prove that

$$\sum_{n\geq 1} \mathbf{P}(C_n) = \infty. \tag{9.38}$$

Step 2: Following the same strategy as in the proof of part (ii) in Lemma 8.3, having established (9.38), we can use it to show the existence of two subsequences $(n_l)_{l\geq 1}$ and $(m_l)_{l\geq 1}$, both tending to infinity, such that $0 \leq n_l \leq m_l - 1$, and

$$\lim_{l\to\infty} \mathbf{P}\left(r^j \leq h_r\left(\Gamma_j^\alpha \overline{I}_\infty^{(j)}\right), \text{ for all } n_l \leq j \leq m_l\right) < 1. \tag{9.39}$$

From (9.39) and (9.37), we may thus conclude that (9.35) holds, which, in turn, gives us the desired result.

Proof of Step 1: Recall from the identity in law in (9.6) that Γ_j has the same law as $r^j\Upsilon$. Since the function h is increasing, it is straightforward to see

$$\begin{aligned}\sum_{n\geq 1} \mathbf{P}(C_n) &\geq \int_0^\infty \mathbf{P}\left(r^t > h\left(r^{\alpha t}\Upsilon^\alpha \hat{I}_\infty\right)\right) \mathrm{d}t \\ &= -\frac{1}{\log(r)} \int_0^1 \mathbf{P}\left(s > h(s^\alpha \Upsilon^\alpha \hat{I}_\infty)\right) \frac{\mathrm{d}s}{s},\end{aligned} \tag{9.40}$$

where the first inequality follows by splitting the integral in the right-hand side along the positive integers and the last identity by the change of variables $s = r^t$. Hence, it is enough for our purposes to prove that this last integral is infinite. Using arguments similar to those in the proof of part (i), we have

$$\int_0^r \mathbf{P}\left(t > h(t^\alpha \Upsilon^\alpha \hat{I}_\infty)\right) \frac{\mathrm{d}t}{t} = \mathbf{E}\left[\int_0^{g(r)} \mathbf{1}_{\{tr^{-\alpha} < \Upsilon^\alpha \hat{I}_\infty < th(t)^{-\alpha}\}} \frac{\mathrm{d}t}{t}\right],$$

where $g(s) := \inf\{t > 0, h(t) > s\}$ denotes the right-continuous inverse function of h. On the other hand, we see

$$\begin{aligned}\int_0^{g(r)} \mathbf{P}\left(\Upsilon^\alpha \hat{I}_\infty < \frac{t}{h(t)^\alpha}\right) \frac{\mathrm{d}t}{t} &= \int_0^{g(r)} \mathbf{P}\left(\frac{t}{r^\alpha} < \Upsilon^\alpha \hat{I}_\infty < \frac{t}{h(t)^\alpha}\right) \frac{\mathrm{d}t}{t} \\ &\quad + \int_0^{g(r)} \mathbf{P}\left(\Upsilon^\alpha \hat{I}_\infty < \frac{t}{r^\alpha}\right) \frac{\mathrm{d}t}{t}.\end{aligned}$$

Next, we observe

$$\int_0^{g(r)} \mathbf{P}\left(\Upsilon^\alpha \hat{I}_\infty < \frac{t}{r^\alpha}\right) \frac{\mathrm{d}t}{t} - \hat{\mathbf{E}}\left[\log^+\left(\frac{g(r)}{r^\alpha \Upsilon^\alpha \hat{I}_\infty}\right)\right],$$

which is finite from the assumption in (9.29). In other words, we have deduced that

$$\mathbf{E}\left[\int_0^{g(r)} \mathbf{1}_{\{tr^{-\alpha} < \Upsilon^\alpha \hat{I}_\infty < th(t)^{-\alpha}\}} \frac{\mathrm{d}t}{t}\right] = \infty,$$

where the latter follows from our assumption. Hence, from (9.40), we have $\sum_{n\geq 1} \mathbf{P}(C_n) = \infty$.

Proof of Step 2: For $n \le m-1$, we introduce

$$D_{(n,m)} = \left\{ r^j \le h(\Gamma_j^\alpha \overline{I}_{(j,m)}) \text{ for all } n \le j \le m \right\},$$

and, for $r < \kappa < 1$ and $n \le m-2$, we also define

$$E_{n,m-1}^{(\kappa)} = \left\{ r^j \le h\left(\Gamma_j^\alpha \overline{I}_{(j,m-1)} + \Gamma_{m-1}^\alpha \overline{I}^{(m-1),\kappa}\right) \text{ for all } n \le j \le m-1 \right\},$$

where

$$\overline{I}_{(j,m-1)} = \int_0^{\overline{T}^{(j)}(\log(r^{m-1}/\Gamma_j))} \mathrm{e}^{\alpha \overline{\Xi}_s^{(j)}} \, \mathrm{d}s, \qquad \overline{I}^{(m-1),\kappa} = \int_0^{\overline{T}^{(m-1)}(\log(r/\kappa))} \mathrm{e}^{\alpha \overline{\Xi}_s^{(m-1)}} \, \mathrm{d}s,$$

and for $n \le j \le m-1$, $\overline{\Xi}^{(j)}$ is the Lévy process defined as in (9.11). From the definition of $\overline{\Xi}^{(j)}$, we note that, for $j < m$,

$$\overline{\Xi}_t^{(m-1)} = \overline{\Xi}^{(j)}_{\overline{T}^{(j)}(\log(r^{m-1}/\Gamma_j))+t} - \overline{\Xi}^{(j)}_{\overline{T}^{(j)}(\log(r^{m-1}/\Gamma_j))}, \qquad t \ge 0$$

and

$$\Gamma_{m-1} = \Gamma_j \mathrm{e}^{\overline{\Xi}^{(j)}_{\overline{T}^{(j)}(\log(r^{m-1}/\Gamma_j))}}.$$

Furthermore, we have the following identity

$$\begin{aligned} \overline{T}^{(j)}(\log(r^m/\Gamma_j)) &= \overline{T}^{(j)}(\log(r^{m-1}/\Gamma_j)) \\ &\quad + \inf\left\{ t \ge 0 : \overline{\Xi}_t^{(m-1)} \le \log(r^m/\Gamma_{m-1}) \right\}. \end{aligned}$$

The above decomposition allows us to determine the following identity

$$\overline{I}_{(j,m)} = \overline{I}_{(j,m-1)} + \mathrm{e}^{\alpha \overline{\Xi}^{(j)}_{\overline{T}^{(j)}(\log(r^{m-1}/\Gamma_j))}} \overline{I}_{(m-1,m)}. \tag{9.41}$$

By Proposition 9.1 and the decomposition in (9.11), it follows that $\overline{I}_{(m-1,m)}$ is independent of

$$\overline{I}_{(j,m-1)} \text{ and } \mathrm{e}^{\alpha \overline{\Xi}^{(j)}_{\overline{T}^{(j)}(\log(r^{m-1}/\Gamma_j))}}. \tag{9.42}$$

Similarly, we can also deduce that

$$\overline{I}_\infty^{(j)} = \overline{I}_{(j,m-1)} + \mathrm{e}^{\alpha \overline{\Xi}^{(j)}_{\overline{T}^{(j)}(\log(r^{m-1}/\Gamma_j))}} \overline{I}_\infty^{(m-1)}. \tag{9.43}$$

Again, by Proposition 9.1 and the decomposition in (9.11), it follows that $\overline{I}_\infty^{(m-1)}$ is also independent of the pair in (9.42) and has the same distribution as $\hat{I}_\infty$, under $\mathbf{P}$. Moreover, since Γ_j is independent of $\overline{\Xi}^{(j)}$ (see Proposition 9.1), we deduce that

$$\Gamma_j^\alpha \overline{I}_\infty^{(j)} = \Gamma_j^\alpha \overline{I}_{(j,m-1)} + \Gamma_{m-1}^\alpha \overline{I}_\infty^{(m-1)}, \tag{9.44}$$

and that $\overline{I}_\infty^{(m-1)}$ is independent of $(\Gamma_j, \overline{I}_{(j,m-1)}, \Gamma_{m-1})$.

On the other hand, it is clear that

$$D_{(n,m)} = D_{(n,m-1)} \bigcap \left\{ r^{m-1} \le h\left(\Gamma^{\alpha}_{m-1} \overline{I}_{(m-1,m)} \right) \right\},$$

for $n \le m - 1$, and we also introduce

$$H(n,m) = \mathbf{P}\left(E^{(\kappa)}_{n,m-1}, r^{m-1} \le h\left(\Gamma^{\alpha}_{m-1} \overline{I}^{(m-1),\kappa} \right), \Gamma_{m-1} > r^{m-1}\kappa \right).$$

On the event $\{\Gamma_{m-1} > r^{m-1}\kappa\}$, we have $\overline{I}^{(m-1),\kappa} \le \overline{I}_{(m-1,m)}$. Thus, since $\kappa > r$ and h is increasing, we may deduce that $\mathbf{P}(D_{(n,m)}) \ge H(n,m)$.

Next, we prove that there exist two increasing sequences, $(n_l)_{l\ge1}$ and $(m_l)_{l\ge1}$, tending to infinity such that $0 \le n_l \le m_l - 1$ and $H(n_l, m_l)$ tends to 0 as l goes to infinity. The previous claim is important to deduce (9.39).

Arguing by contradiction, we suppose the contrapositive is true, that is, that there exist $\delta > 0$ such that $H(n,m) \ge \delta$ for all sufficiently large integers m and n. Hence from identity (9.44) and the independence of $\overline{I}^{(m)}_{\infty}$ and $(\Gamma_j, \overline{I}_{(j,m)}, \Gamma_m)$, we see

$$\begin{aligned}
1 &\ge \mathbf{P}\left(\bigcup_{m=n+1}^{\infty} C_m \right) \\
&\ge \sum_{m=n+1}^{\infty} \mathbf{P}\left(C_m \bigcap \left(\bigcap_{j=n}^{m-1} C_j^c \right) \right) \\
&\ge \sum_{m=n+1}^{\infty} \mathbf{P}\left(C_m \bigcap \left(\bigcap_{j=n}^{m-1} \left\{ r^j \le rh\left(\Gamma^{\alpha}_j \overline{I}_{(j,m)} + \Gamma^{\alpha}_m \overline{I}^{(m)}_{\infty} \right) \right\} \right) \right) \\
&\ge \sum_{m=n+1}^{\infty} \mathbf{P}\left(r^m > rh\left(\Gamma^{\alpha}_m \overline{I}^{(m)}_{\infty} \right) \right) \mathbf{P}(D_{(n,m)}) \\
&\ge \sum_{m=n+1}^{\infty} \mathbf{P}\left(r^m > rh\left(\Gamma^{\alpha}_m \overline{I}^{(m)}_{\infty} \right) \right) H(n,m) \\
&\ge \delta \sum_{m=n+1}^{\infty} \mathbf{P}(C_m),
\end{aligned}$$

but since $\sum_{n\ge1} \mathbf{P}(C_n)$ diverges, we see that our assertion concerning $H(n_l, m_l)$ is true.

Next, let us deduce (9.39). With this purpose in mind, we define

$$\rho_{n_l,m_l}(x) = \mathbf{P}\left(\bigcap_{j=n_l}^{m_l-2} \left\{ r^j \le rh(\Gamma^{\alpha}_j \overline{I}_{(j,m_l-1)} + \Gamma_{m_l-1} x) \right\}, \Gamma_{m_l-1} > \kappa r^{m_l-1} \right)$$

and

$$G(n_l, m_l) = \mathbf{P}\left(\bigcap_{j=n_l}^{m_l-1} \left\{r^j \le rh\left(\Gamma_j^\alpha \overline{I}_\infty^{(j)}\right)\right\}, \Gamma_{m_l-1} > \kappa r^{m_l-1}\right).$$

Since h is increasing, we see that $\rho_{n_l,m_l}(x)$ is increasing in x. Moreover, $H(n_l, m_l)$ and $G(n_l, m_l)$ satisfy

$$H(n_l, m_l) \ge \int_0^\infty \mathbf{1}_{\{h(r^{\alpha(m_l-1)}\kappa^\alpha x) \ge r^{m_l-2}\}} \rho_{n_l,m_l}(x) \mathbf{P}\left(\hat{I}_{\hat{T}^-(\log(r/\kappa))} \in \mathrm{d}x\right) \qquad (9.45)$$

and

$$G(n_l, m_l) \le \int_0^\infty \mathbf{1}_{\{h(r^{\alpha(m_l-1)}x) \ge r^{m_l-2}\}} \rho_{n_l,m_l}(x) \mathbf{P}\left(\hat{I}_\infty \in \mathrm{d}x\right). \qquad (9.46)$$

The inequality (9.45) follows by virtue of the fact that h is increasing and the independence of $\overline{I}^{(m_l-1),\kappa}$ and the ensemble

$$\Gamma_{m_l-1} \text{ and } \left(\overline{I}_{(j,m_l-1)}, \Gamma_j \text{ for } n_l \le j \le m_l - 2\right).$$

To show (9.46), we use (9.44) and the independence of $\overline{I}_\infty^{(m_l-1)}$ and the ensemble

$$\Gamma_{m_l-1} \text{ and } \left(\overline{I}_{(j,m_l-1)}, \Gamma_j \text{ for } n_l \le j \le m_l - 2\right).$$

In particular, we get that, for l sufficiently large,

$$H(n_l, m_l) \ge \rho_{n_l,m_l}(N) \int_N^\infty \mathbf{P}\left(\hat{I}_{\hat{T}^-(\log(r/\kappa))} \in \mathrm{d}x\right), \qquad \text{for} \quad N \ge rC,$$

where $C = \sup_{x \le \delta} xh(x)^{-\alpha}$. Since $H(n_l, m_l)$ converges to 0, as l goes to ∞ and the law of $\hat{I}_{\hat{T}^-(\log(r/\kappa))}$, under $\mathbf{P}$, does not depend on l, it follows that $\rho_{n_l,m_l}(N)$ also converges to 0 as l tends to ∞, for every $N \ge rC$. On the other hand, we have

$$G(n_l, m_l) \le \rho_{n_l,m_l}(N) \int_0^N \mathbf{P}\left(\hat{I}_\infty \in \mathrm{d}x\right) + \int_N^\infty \mathbf{P}\left(\hat{I}_\infty \in \mathrm{d}x\right).$$

By letting l and N tend to infinity, we get that $G(n_l, m_l)$ tends to 0.

Finally, recall from (9.37) that

$$P_0(B_{n_l}) \ge 1 - \mathbf{P}\left(r^j < rh\left(\Gamma_j^\alpha \overline{I}_\infty^{(j)}\right), \text{ for all } n_l \le j \le m_l - 1\right).$$

It is not difficult to see

$$\mathbf{P}\left(r^j < rh\left(\Gamma_j^\alpha \overline{I}_\infty^{(j)}\right), \text{ for all } n_l \le j \le m_l - 1\right) \le \mathbf{P}(\Gamma_{m_l-1} \le \kappa r^{m_l-1}) + G(n_l, m_l),$$

which implies (9.39) on account of the fact that $G(n_l, m_l)$ tends to 0 and

$$\mathbf{P}(\Gamma_{m_l-1} > kr^{m_l-1}) = \mathbf{P}(\Gamma_1 > kr) > 0,$$

cf. (9.6) and Lemma 9.5. The latter allow us to conclude that $\lim P_0(B_n) > 0$ and hence, from the discussion following (9.35) the proof is complete. □

Next we will look at integral tests for upper envelopes at $t \to \infty$. Before doing so, we will introduce another class of functions.

Definition 9.12 Define C_∞, the class of positive increasing functions $h(t)$ on $(0, \infty)$ satisfying

(i) $\lim_{t\to\infty} h(t) = \infty$, and
(ii) there exists $\delta > 1$ such that $\sup_{t>\delta} th(t)^{-\alpha} < \infty$.

Then the upper envelope of $\mathtt{J}$ as $t \to \infty$, under P_x for $x \geq 0$, is given by Theorem 9.13 below; we omit its proof on account of it being similar to that of Theorem 9.11.

Theorem 9.13 *Let $h \in C_\infty$ and assume that* (9.29) *holds.*

(i) If

$$\int^\infty \overline{F}_\Upsilon\left(\frac{t}{h(t)^\alpha}\right)\frac{\mathrm{d}t}{t} < \infty,$$

then for all $\epsilon > 0$ and for all $x \geq 0$,

$$P_x\Big(\mathtt{J}_t > (1+\epsilon)h(t) \text{ i.o. as } t \to \infty\Big) = 0.$$

(ii) If

$$\int^\infty \overline{F}_\Upsilon\left(\frac{t}{h(t)^\alpha}\right)\frac{\mathrm{d}t}{t} = \infty,$$

then for all $\epsilon > 0$ and for all $x \geq 0$,

$$P_x\Big(\mathtt{J}_t > (1-\epsilon)h(t) \text{ i.o. as } t \to \infty\Big) = 1.$$

Our next result provides integral tests for the upper envelope at 0 of the pssMp (Z, P_0) under the same hypotheses as above, that is, that Ξ is not arithmetic, $0 < \mathbf{E}[\Xi] < \infty$ and that (9.29) is satisfied.

Let S_y be the first passage time of the pssMp Z above the level $y > 0$, that is,

$$S_y = \inf\Big\{t \geq 0 \colon Z_t \geq y\Big\}.$$

We also introduce

$$G(t) := P_0(S_1 \leq t). \tag{9.47}$$

Proposition 9.14 *Let $h \in \mathcal{C}_0$ and assume that* (9.29) *holds.*

(i) If

$$\int_{0^+} G\left(\frac{t}{h(t)^\alpha}\right)\frac{dt}{t} < \infty,$$

then for all $\epsilon > 0$,

$$P_0\Big(Z_t > (1+\epsilon)h(t) \text{ i.o. as } t \to 0\Big) = 0.$$

ii) If

$$\int^{\infty} \overline{F}_\Upsilon\left(\frac{t}{h(t)^\alpha}\right)\frac{dt}{t} = \infty,$$

then for all $\epsilon > 0$,

$$P_0\Big(Z_t < (1-\epsilon)h(t) \text{ i.o. as } t \to 0\Big) = 1.$$

Proof We begin with some general comments. Let $(x_n)_{n\geq 1}$ be a decreasing sequence which converges to 0. We define the events

$$A_n = \{\text{there exists } t \in [S_{x_{n+1}}, S_{x_n}) \text{ such that } Z_t > h(t)\}.$$

Appealing to the fact that S_{x_n} tends to 0 as n tends to ∞, we have

$$\Big\{Z_t > h(t) \text{ i.o. as } t \to 0\Big\} = \limsup_{n\geq 1} A_n.$$

Since h is an increasing function and $Z_{S_{x_{n+1}}} \leq x_n$ when $S_{x_n} > S_{x_{n+1}}$, the following inclusion holds

$$A_n \subset \Big\{x_n > h(S_{x_{n+1}})\Big\}. \tag{9.48}$$

We observe that when $S_{x_n} = S_{x_{n+1}}$, the event A_n is an empty set and hence (9.48) is trivial.

(i) We choose $x_n = r^n$, for $r < 1$ and define $h_r(t) = r^{-2}h(t)$. Since h is increasing, in a similar spirit to (9.31), we deduce

$$\sum_{n\geq 0} P_0\Big(r^n > h_r(S_{r^{n+1}})\Big) \leq -\frac{1}{\log r}\int_0^r P_0\Big(t > h(S_t)\Big)\frac{dt}{t}.$$

Replacing h by h_r in (9.48) and recalling that r can be chosen arbitrarily close to 1 from below, we obtain the desired result if

$$\int_0^r P_0\Big(t > h(S_t)\Big)\frac{dt}{t} < \infty.$$

Also, in the spirit of (9.32), we can similarly derive

$$\int_0^r P_0\Big(t > h(S_t)\Big)\frac{dt}{t} = E_0\left[\int_0^{g(r)} \mathbf{1}_{\{tr^{-\alpha} < S_1 < th(t)^{-\alpha}\}}\frac{dt}{t}\right],$$

where $g(s) = \inf\{t > 0, h(t) > s\}$, denotes the right inverse function of h. This integral converges if

$$\int_0^{g(r)} P_0\left(S_1 < \frac{t}{h(t)^\alpha}\right)\frac{dt}{t} < \infty.$$

This proves part (i).

(ii) The statement of this part follows from part (ii) of Theorem 9.11 on account of the fact that $Z_t \geq J_t$, for $t \geq 0$. □

Below we give the corresponding result for the upper envelope of Z at ∞, under P_x for $x \geq 0$ as $t \to \infty$. As with other envelopes as $t \to \infty$, we omit the proofs on account of their similarity to the case that $t \to 0$.

Proposition 9.15 *Let $h \in C_\infty$ and assume that* (9.29) *holds.*

(i) If

$$\int^{\infty} G\left(\frac{t}{h(t)^\alpha}\right)\frac{dt}{t} < \infty,$$

then for all $\epsilon > 0$ and for all $x \geq 0$,

$$P_x\Big(Z_t > (1+\epsilon)h(t) \text{ i.o. as } t \to \infty\Big) = 0.$$

ii) If

$$\int^{\infty} \overline{F}_\Upsilon\left(\frac{t}{h(t)^\alpha}\right)\frac{dt}{t} = \infty,$$

then for all $\epsilon > 0$ and for all $x \geq 0$,

$$P_x\Big(Z_t < (1-\epsilon)h(t) \text{ i.o. as } t \to \infty\Big) = 1.$$

9.4 Comments

There exist several results on the lower and upper envelopes for particular families of pssMp, the oldest of which are due to Dvoretzky and Erdős [66] and Motoo [153] who studied the special case of transient Bessel processes. The asymptotic behaviour of the future infimum of transient Bessel processes was studied by Khoshnevisan et al. [109]. Xiao [215] considers a bigger class of self-similar Markov processes whose transition functions admit some special bounds.

The first author to use Lamperti's representation of pssMp to study their asymptotic behaviour starting from a positive state is Lamperti himself, [139]. Rivero [179] provided the first integral test for the lower envelope of pssMp starting at 0 in the case when the sample paths are increasing. He was the first to observe that the exponential functional of the underlying Lévy process was crucial to describe the lower envelope of pssMp.

The path decomposition at last passage times for pssMp in Section 9.1 is taken from Chaumont and Pardo [48]. It is based on the construction of the entrance law of pssMp starting at the origin given in Caballero and Chaumont [44]. The integral tests that describe the lower envelopes of pssMp at the origin and at infinity (i.e. Theorems 9.7 and 9.8) also originate from Chaumont and Pardo [48]. The upper envelopes of pssMp and their future infimum are taken from Pardo [159, 160]. Theorems 9.11 and 9.13 are modified versions of those presented in Pardo [159]. In Pardo [159], the integral tests of the divergent parts of both results are given only in terms of the probability distribution of the exponential functional of the underlying Lévy processes. This is different to the presentation here which additionally relies on the distribution of the random variable Υ, that is, the undershoot at last passage. We account for this difference as, in the next Chapter, it will allow us to produce more precise integral tests for upper and lower envelopes for some of the path transformations of stable processes introduced in Chapter 5.

10

Asymptotic behaviour for path transformations

In the spirit of the previous two chapters, we develop integral tests that describe the lower and upper envelopes as $t \to 0$ and $t \to \infty$, for the path functionals of stable processes introduced in Chapter 5. Our arguments are based on the integral tests developed in Chapter 9 together with explicit knowledge of the upper and lower tail behaviour of the relevant distribution functions, for example, those of integrated exponential functionals. All of the path transformations of stable processes that we consider boil down to positive self-similar Markov processes driven by hypergeometric Lévy processes. In this respect, we rely on some of the results in Chapter 4 and more elaborate versions thereof, in addition to some new distributional results for positive self-similar Markov processes. With these in hand, we will work our way around different regimes of the parameters $(\alpha, \rho) \in \mathcal{A}$, establishing asymptotic envelopes for conditioned stable processes, censored stable processes and radial stable processes.

10.1 More on hypergeometric Lévy processes

Let us recall some notation from the previous chapter, which deals with general positive self-similar Markov processes. The process (Z, P), denotes a general pssMp, with index of self-similarity α, and $(\Xi, \mathbf{P})$ is the underlying Lévy process in the Lamperti transform (9.1). With $\hat{\Xi} = -\Xi$, the integrated exponential functional of the dual is denoted by

$$I(\delta, \hat{\Xi}) := \int_0^\infty e^{\delta \hat{\Xi}_t} \mathrm{d}t, \tag{10.1}$$

for $\delta > 0$. We will mostly be interested in the case that $\delta = \alpha$ but when we consider the setting of the radius of d-dimensional stable processes at the end of this chapter, we will need to deal with the setting $\delta = \alpha/2$. From (9.26), without loss of generality thanks to scaling, we may take

$$\Upsilon \stackrel{(d)}{=} Z_{\mathtt{D}_1-}, \tag{10.2}$$

which is bounded from above by unity, where we recall

$$\mathtt{D}_1 = \sup\{t > 0 : Z_t \leq 1\} \tag{10.3}$$

is the last passage time of Z below 1. Importantly, the variable Υ is always taken to be independent of $I(\delta, \hat{\Xi})$. Theorem 9.11 requires us to work with an assumption of the form

$$\mathbf{E}\left[\log^+\left(\frac{1}{\Upsilon^\delta I(\delta, \hat{\Xi})}\right)\right] < \infty, \tag{10.4}$$

which is core to a number of results in Chapter 9. As alluded to in the discussion preceding Theorem 9.11, (10.4) may be difficult to verify. In this section, we show that this is not a problem for pssMps that are driven by a suitable subclass of hypergeometric Lévy processes. As we have already seen in Chapter 5, the class of pssMp driven by hypergeometric Lévy processes contain several examples of path functionals of stable processes. Removing the condition (10.4) for at least these processes is thus a first step to developing more concrete statements than those offered in the main results of Chapter 9.

For the following results, we recall that the parametric regimes $\mathcal{H}_1$ and $\mathcal{H}_2$ for hypergeometric Lévy processes are given in (4.17) and (4.18).

Lemma 10.1 *Suppose that $\hat{\Xi}$ is a hypergeometric Lévy process whose parameters $(\beta, \gamma, \hat{\beta}, \hat{\gamma})$ belong to $\mathcal{H}_1$ with $\hat{\beta} = 0$ and $\beta < 1$ or to $\mathcal{H}_2$ with $\beta = 1$ and $-\hat{\beta} < \gamma$, then the condition* (10.4) *is satisfied.*

Proof Recall from Proposition 4.11 that, for $s \in (0, \hat{\theta}_\chi)$, where

$$\hat{\theta} := \begin{cases} 1-\beta & \text{if } (\beta, \gamma, \hat{\beta}, \hat{\gamma}) \in \mathcal{H}_1 \cap \{\hat{\beta} = 0\} \setminus \{\beta = 1\}, \\ -\hat{\beta} & \text{if } (\beta, \gamma, \hat{\beta}, \hat{\gamma}) \in \mathcal{H}_2 \cap \{\beta = 1\} \setminus \{\hat{\beta} = 0\}, \end{cases}$$

we have

$$\mathbf{E}\left[I(\delta, \hat{\Xi})^s\right] = -\frac{s}{\psi(\delta s)}\mathbf{E}\left[I(\delta, \hat{\Xi})^{s-1}\right]. \tag{10.5}$$

Here,

$$\psi(z) = \log \mathbf{E}\left[e^{z\hat{\Xi}_1}\right] = -\frac{\Gamma(1-\beta+\gamma-z)}{\Gamma(1-\beta-z)}\frac{\Gamma(\hat{\beta}+\hat{\gamma}+z)}{\Gamma(\hat{\beta}+z)},$$

for $\mathrm{Re}(z) \in (-\hat{\beta}-\hat{\gamma}, 1-\beta+\gamma)$. According to the classification in Corollary 4.7 (iii) the assumed parameter restriction for $\hat{\Xi}$ ensures that

$$\lim_{s\to 0}\frac{\psi(s)}{s} = \psi'(0+) = \mathbf{E}[\hat{\Xi}_1],$$

implying, via (10.5), that $\mathbf{E}[I(\delta,\hat{\Xi})^{-1}] < \infty$ and therefore, since $\log(1+x) \leq x$, for $x > 0$,

$$\mathbf{E}\left[\log\left(1+\frac{1}{I(\delta,\hat{\Xi})}\right)\right] < \infty. \tag{10.6}$$

On the other hand, from Lemma 9.5, the random variable $-\log \Upsilon$ is equal in distribution to $\mathfrak{U}W$, where $\mathfrak{U}$ is a uniform random variable on $[0,1]$, independent of W, which is given as in (9.15). In the case of the hypergeometric Lévy processes of interest here, the law of W can be computed explicitly.

To this end, recall from Theorem 4.6 that the ascending ladder height of Ξ, denoted here by H, is a β-subordinator with parameters

$$\begin{cases} (0,0,\hat{\gamma}) & \text{if } (\beta,\gamma,\hat{\beta},\hat{\gamma}) \in \mathcal{H}_1 \cap \{\hat{\beta}=0\} \setminus \{\beta = 1\}, \\ (0,\hat{\beta},\hat{\gamma}) & \text{if } (\beta,\gamma,\hat{\beta},\hat{\gamma}) \in \mathcal{H}_2 \cap \{\beta=1\} \setminus \{\hat{\beta} = 0\}. \end{cases} \tag{10.7}$$

From Proposition 4.1, the associated Lévy measure of H has a density that satisfies

$$\pi(x) = \frac{1}{\Gamma(1-\hat{\gamma})}(1-\mathrm{e}^{-x})^{-\hat{\gamma}}\,\mathrm{e}^{-(\rho+\hat{\gamma})x}\left[\frac{\hat{\gamma}}{1-\mathrm{e}^{-x}}+\rho\right], \qquad x > 0,$$

where ρ equals 0 or $\hat{\beta}$, respectively, according to the two parameter choices in (10.7). In both cases, it is not so difficult to deduce that

$$-\mathbf{E}[\log \Upsilon] = \frac{1}{2\mathbf{E}[H_1]}\int_0^\infty x^2\pi(x)\mathrm{d}x < \infty.$$

Recalling that $\Upsilon \leq 1$, it is easy to see that (10.4) is satisfied if and only (10.6) and $\mathbf{E}[\log(1/\Upsilon^\delta)] = -\delta\mathbf{E}[\log \Upsilon] < \infty$. Hence the proof is complete. □

For many of the results in the previous chapter, knowing the behaviour of the density of $\Upsilon^\delta I(\delta,\hat{\Xi})$ near 0 is important. We spend the rest of this section providing an explicit computation giving the Mellin transform of $\Upsilon^\delta I(\delta,\hat{\Xi})$, again, for a special class of hypergeometric Lévy processes. Moreover, we deduce the asymptotic expansion of its density.

Proposition 10.2 *Suppose that $\hat{\Xi}$ is a hypergeometric Lévy process with parameters $(\beta,\gamma,0,\hat{\gamma})$ belonging to $\mathcal{H}_1 \setminus \{\beta = 1\}$ and let $\chi = 1/\delta$. Then the Mellin transform of $\Upsilon^\delta I(\delta,\hat{\Xi})$ satisfies, for $s \in \mathbb{C}$,*

$$\tilde{\mathtt{M}}(s) := \mathbf{E}\left[(\Upsilon^\delta I(\delta,\hat{\Xi}))^{s-1}\right] = c\frac{\Gamma(s-1)G(s-1;\chi)G((1-\beta+\gamma)\chi+1-s;\chi)}{G(s-1+\hat{\gamma}\chi;\chi)G((1-\beta)\chi+1-s;\chi)},$$

where

$$c := \chi\frac{G((1-\beta)\chi;\chi)G(\hat{\gamma}\chi;\chi)}{G((1-\beta+\gamma)\chi;\chi)}.$$

Proof From the explicit form of the Mellin transform $\mathtt{M}$ of $I(\delta, \hat{\Xi})$ in Theorem 4.13 (i) and the independence of $I(\delta, \hat{\Xi})$ and the pair $(\mathfrak{U}, W)$, it is enough to compute the Laplace transform of $\mathfrak{U}W$, that is,

$$
\begin{aligned}
\mathbf{E}\left[\mathrm{e}^{-\delta s \mathfrak{U} W}\right] &= \frac{1}{\delta s}\mathbf{E}\left[\frac{1-\mathrm{e}^{-\delta s W}}{W}\right] \\
&= \frac{1}{\delta s \mathbf{E}[H_1]}\int_0^\infty \left(1-\mathrm{e}^{-\delta s w}\right)\nu(w)\mathrm{d}w \\
&= \frac{1}{\mathbf{E}[H_1]}\frac{\Gamma(\delta s+\hat{\gamma})}{\Gamma(\delta s+1)}, \qquad (10.8)
\end{aligned}
$$

where the last identity follows from Proposition 4.1. Moreover from the explicit form of the Laplace exponent of H, see (4.1), we get

$$\mathbf{E}[H_1] = \Gamma(\hat{\gamma}).$$

Hence putting the expression for $\mathtt{M}$ given in Theorem 4.13 (i) and the previous computation together, and appealing to the quasi-periodic properties of the double gamma function G, see (A.21), we get the desired result. □

Next, we invert the Mellin transform $\tilde{\mathtt{M}}$ of Proposition 10.2 in order to deduce an asymptotic expression for the probability density function of $\Upsilon^\delta I(\delta, \hat{\Xi})$, henceforth denoted by

$$\tilde{p}(x) = \frac{\mathrm{d}}{\mathrm{d}x}\mathbf{P}\left(\Upsilon^\delta I(\delta, \hat{\Xi}) \le x\right), \qquad \text{for} \quad x \ge 0.$$

We are interested in the existence of an asymptotic expansion of the density $\tilde{p}(x)$ as $x \to 0^+$ or $x \to \infty$. Our approach follows a similar computational philosophy to that of Section 4.6 for integrated exponential Lévy processes.

We first observe that the zeros of $\tilde{\mathtt{M}}$ are positioned at

$$-m\chi - n + 1 \quad (m \ge 1) \quad \text{and} \quad 1 + (1-\beta+\gamma)\chi + m\chi + n,$$

and its poles at

$$z^-_{m,n} := 1 - \hat{\gamma}\chi - m\chi - n \quad \text{and} \quad z^+_{m,n} := 1 + (1-\beta)\chi + m\chi + n, \qquad (10.9)$$

for $m, n \ge 0$. We recall from standard properties of the double gamma function in the Appendix A.4 that all zeros/poles are simple if $\delta \notin \mathbb{Q}$.

Next, we introduce the functions

$$\phi(s) := \delta\frac{\Gamma(\delta(s-1)+\hat{\gamma})\Gamma(1-\beta+\gamma-\delta s)}{\Gamma(\delta(s-1)+1)\Gamma(1-\beta-\delta s)}, \qquad (10.10)$$

$$\tilde{\phi}(s) := \frac{\Gamma(s-1+\hat{\gamma}\chi)\Gamma(1-\beta\chi+\gamma\chi-s)}{\Gamma(s-1+\chi)\Gamma(1-\beta\chi-s)}, \qquad (10.11)$$

and recall that $\eta = 1-\beta+\gamma+\hat{\gamma}$.

Definition 10.3 Assume that $\hat{\Xi}$ is a hypergeometric Lévy process with parameters $(\beta, \gamma, 0, \hat{\gamma}) \in \mathcal{H}_1 \setminus \{\beta = 1\}$. We define the coefficients recursively

$$\begin{cases} b_{0,0} = \dfrac{\Gamma(\eta - \delta)}{\Gamma(\eta - \gamma - \delta)\Gamma(1 - \hat{\gamma})} \tilde{\mathtt{M}}(2 - \hat{\gamma}\chi), \\ b_{m,n} = \phi(z^-_{m,n}) b_{m,n-1}, \quad m \geq 0,\ n \geq 1, \\ b_{m,n} = \delta^{\chi(\gamma+\hat{\gamma})} \tilde{\phi}(z^-_{m,n}) b_{m-1,n}, \quad m \geq 1,\ n \geq 0, \end{cases} \tag{10.12}$$

where $\chi = 1/\delta$. Similarly, $c_{m,n}$, $m, n \geq 0$, are defined recursively

$$\begin{cases} c_{0,0} = \chi^2 \dfrac{\Gamma(2 - \beta - \delta)}{\Gamma(\eta - \gamma - \delta)\Gamma(\gamma)} \tilde{\mathtt{M}}((1 - \beta)\chi), \\ c_{m,n} = \dfrac{1}{\phi(z^+_{m,n-1})} c_{m,n-1}, \quad m \geq 0,\ n \geq 1, \\ c_{m,n} = \dfrac{\chi^{\chi(\gamma+\hat{\gamma})}}{\tilde{\phi}(z^+_{m-1,n})} c_{m-1,n}, \quad m \geq 1,\ n \geq 0. \end{cases} \tag{10.13}$$

The next result computes the residues of $\tilde{\mathtt{M}}$. As with other calculations in Section 4.6, we keep the assumption that $\delta \notin \mathbb{Q}$ to ensure that G has simple poles. The situation for $\delta \in \mathbb{Q}$ is more complicated.

Proposition 10.4 *Assume that $\hat{\Xi}$ is a hypergeometric Lévy process with parameters $(\beta, \gamma, 0, \hat{\gamma}) \in \mathcal{H}_1 \setminus \{\beta = 1\}$ and $\delta \notin \mathbb{Q}$. For all $m, n \geq 0$, we have*

$$\operatorname{Res}(\tilde{\mathtt{M}}(s) \colon s = z^-_{m,n}) = b_{m,n},$$
$$\operatorname{Res}(\tilde{\mathtt{M}}(s) \colon s = z^+_{m,n}) = -c_{m,n}.$$

Proof We start by observing that the quasi-periodicity of the double gamma function at 1 and at χ (see (A.21) in the Appendix) implies that

$$\tilde{\mathtt{M}}(s+1) = \frac{1}{\phi(s)} \tilde{\mathtt{M}}(s) \qquad \text{and} \qquad \tilde{\mathtt{M}}(s+\chi) = \frac{\chi^{(\gamma+\hat{\gamma})\chi}}{\tilde{\phi}(s)} \tilde{\mathtt{M}}(s). \tag{10.14}$$

Next, we prove that the residue of $\tilde{\mathtt{M}}(s)$ at $s = z^-_{m,n}$ is equal to $b_{m,n}$. We use the explicit form of $\tilde{\mathtt{M}}$ in Proposition 10.2 and rearrange the terms in the first functional identity in (10.14), making use of the expression for ϕ in (10.10) and the recursion formula for gamma functions (A.8), to find that

$$\tilde{\mathtt{M}}(s) = \frac{\tilde{\mathtt{M}}(s+1)}{s - 1 + \hat{\gamma}\chi} \frac{\Gamma(1 + \hat{\gamma} + \delta(s-1))\Gamma(1 - \beta + \gamma - \delta s)}{\Gamma(\delta(s-1) + 1)\Gamma(1 - \beta - \delta s)}.$$

The above identity and the definition (10.12) imply that, as $s \to 1 - \hat{\gamma}\chi$,

$$\tilde{\mathtt{M}}(s) = \frac{b_{0,0}}{s - 1 + \hat{\gamma}\chi} + O(1),$$

which means that the residue of $\tilde{\mathtt{M}}(s)$ at $z^-_{0,0} = 1 - \hat{\gamma}\chi$ is equal to $b_{0,0}$.

Next, we show that the residues satisfy the second recursive identity in (10.12). To this end, rewrite the first identity in (10.14) as

$$\tilde{\mathtt{M}}(s) = \phi(s)\tilde{\mathtt{M}}(s+1). \tag{10.15}$$

We know that $\tilde{\mathtt{M}}(s)$ has a simple pole at $s = z^-_{m,n}$ and $\tilde{\mathtt{M}}(s+1)$ has a simple pole at $z^-_{m,n} + 1 = z^-_{m,n-1}$. One can also check that the function $\phi(s)$ is analytic at $s = z^-_{m,n}$ for $n \geq 1$. Therefore we have, as $s \to z^-_{m,n}$,

$$\tilde{\mathtt{M}}(s) = \mathrm{Res}(\tilde{\mathtt{M}}(s)\colon s = z^-_{m,n})\frac{1}{s - z^-_{m,n}} + O(1),$$

$$\tilde{\mathtt{M}}(s+1) = \mathrm{Res}(\tilde{\mathtt{M}}(s)\colon s = z^-_{m,n-1})\frac{1}{s - z^-_{m,n}} + O(1),$$

$$\phi(s) = \phi(z^-_{m,n}) + O(s - z^-_{m,n}),$$

which, together with (10.15), imply that

$$\mathrm{Res}(\tilde{\mathtt{M}}(s)\colon s = z^-_{m,n}) = \phi(z^-_{m,n})\mathrm{Res}(\tilde{\mathtt{M}}(s)\colon s = z^-_{m,n-1}).$$

The proof of all remaining cases is very similar, and we leave the details to the reader. □

As with the Mellin transform of the exponential functional of hypergeometric Lévy processes in section 4.6, Proposition 10.4 immediately provides a complete asymptotic expansion of $\tilde{p}(x)$, as $x \to 0^+$ and $x \to \infty$, which we present in the next theorem.

Theorem 10.5 *Assume that $\hat{\Xi}$ is a hypergeometric Lévy process with parameters $(\beta, \gamma, 0, \hat{\gamma}) \in \mathcal{H}_1 \setminus \{\beta = 1\}$ and $\delta \notin \mathbb{Q}$. Then*

$$\tilde{p}(x) \sim \sum_{m\geq 0}\sum_{n\geq 0} b_{m,n} x^{(m+\hat{\gamma})\chi+n-1}, \qquad x \to 0^+, \tag{10.16}$$

$$\tilde{p}(x) \sim \sum_{m\geq 0}\sum_{n\geq 0} c_{m,n} x^{-(m+1-\beta)\chi-n-1}, \qquad x \to \infty. \tag{10.17}$$

The proof of the above result follows exactly the same style of reasoning as in the proof of Theorem 4.22. We thus omit it and leave the details to the reader. Instead, we continue with the presentation of some special cases where the density $\tilde{p}$ can be written as a convergent series for which the restriction $\delta \notin \mathbb{Q}$ is no longer required.

The Mellin transform $\tilde{\mathtt{M}}$ is simplified when γ and $\hat{\gamma}$ take the specific value of $\delta \in (0, 1)$, in the sense that it can be written purely in terms of gamma functions.

Corollary 10.6 *Let $\gamma = \hat{\gamma} = \delta$ and assume that $\hat{\Xi}$ is a hypergeometric Lévy process with parameters $(\beta, \delta, 0, \delta) \in \mathcal{H}_1 \setminus \{\beta = 1\}$. Then the Mellin transform of $\Upsilon^\delta I(\delta, \hat{\Xi})$ satisfies*

$$\tilde{\mathtt{M}}(s) = \frac{1}{\Gamma(1-\beta)} \frac{\Gamma(s)\Gamma(1-\beta+\delta(1-s))}{\Gamma(\delta(s-1)+1)}.$$

Hence, if $2\delta < 1$, then

$$\tilde{p}(x) = \frac{1}{\Gamma(1-\beta)} \sum_{n\geq 0} \frac{\Gamma(1-\beta+\delta(1+n))}{\Gamma(1-\delta(n+1))} \frac{(-1)^n}{n!} x^n, \tag{10.18}$$

for $x > 0$ and, if $2\delta > 1$, then

$$\tilde{p}(x) = \frac{\chi}{\Gamma(1-\beta)} \sum_{n\geq 0} \frac{\Gamma((1-\beta+n)\chi+1))}{\Gamma(2-\beta+n)} \frac{(-1)^n}{n!} x^{-(1-\beta+n)\chi-1}, \tag{10.19}$$

for $x > 0$. Moreover, formula (10.18) *(resp.* (10.19)*) provides a complete asymptotic expansion as $x \to 0^+$ (resp. as $x \to \infty$).*

Proof We give only a brief sketch of the proof. The explicit form of $\tilde{\mathtt{M}}$ follows from Proposition 10.2 and the quasi-periodicity at 1 of the double gamma function (see (A.21) in the Appendix).

On the other hand, it is clear that the Mellin transform $\tilde{\mathtt{M}}(z)$ of $\tilde{p}(x)$ has simple poles at

$$z_n^- = -n \qquad \text{and} \qquad z_n^+ = 1 + \chi(n+1-\beta), \quad \text{for } n \geq 0.$$

The residues at these points provide the coefficients in (10.18) and (10.19). Indeed, by applying Proposition A.1 and identity (A.11) (both in the Appendix), we find that

$$\mathrm{Res}(\tilde{\mathtt{M}}(s)\colon s = -n) = \frac{\Gamma(1-\beta+\delta(1+n))}{\Gamma(1-\beta)\Gamma(1-\delta(n+1))} \frac{(-1)^n}{n!}.$$

Similarly, for the poles at $z_n^+ = 1 + \chi(n+1-\beta)$, we get

$$\mathrm{Res}(\tilde{\mathtt{M}}(s)\colon s = 1+\chi(n+1-\beta)) = \frac{\Gamma(1+\chi(n+1-\beta))}{\Gamma(1-\beta)\Gamma(2-\beta+n)} (-1)^{n-1} \frac{\chi}{n!}.$$

The rest of the proof follows by using similar ideas to, for example, the proof of Theorem 4.22. In particular, this pertains to the use of an appropriate contour integral which encloses an increasing number of poles as it expands. The poles are chosen in such a way to ensure that one side of the contour integral converges to the inverse Mellin transform of $\tilde{\mathtt{M}}(z)$, and the remaining parts of the contour integral tend to zero as the contour grows larger. The Residue Theorem gives the desired density as a sum of residues of the captured poles. We should note that the cases $2\delta < 1$ and $2\delta > 1$ coincide with the need for two different contour integrals that capture a different set of poles in a similar spirit to the two cases in Theorem 4.24. We leave the details to the interested reader. □

We conclude this section with another important case where $\tilde{\mathtt{M}}$ has an explicit form and which is not covered by the previous results. Let us assume that the parameters $(1, \delta, \hat{\beta}, \delta)$ lie in $\mathcal{H}_2 \setminus \{\hat{\beta} = 0\}$. The result below is needed later on in this chapter.

Proposition 10.7 *Let $\gamma = \hat{\gamma} = \delta \in (0, 1)$ and assume that $\hat{\Xi}$ is a hypergeometric Lévy process with parameters $(1, \delta, \hat{\beta}, \delta) \in \mathcal{H}_2 \setminus \{\hat{\beta} = 0\}$. Then the Mellin transform of $\Upsilon^\delta I(\delta, \hat{\Xi})$ satisfies*

$$\tilde{\mathtt{M}}(s) = \frac{\delta \sin(-\pi\chi\hat{\beta})\Gamma(1 + \hat{\beta})\Gamma(\delta(1 - s))}{\Gamma(\delta(s - 1) + 1 + \hat{\beta}) \sin(\pi(\chi\hat{\beta} + s))\Gamma(1 - s)}. \tag{10.20}$$

Hence, if $2\delta < 1$, we have

$$\tilde{p}(x) = \frac{\delta}{\pi} \sum_{n \geq 0} (-1)^n \frac{\sin(-\pi\chi\hat{\beta})\Gamma(1 + \hat{\beta})\Gamma(\delta(n + 1) + \hat{\beta})}{\Gamma(1 + n + \chi\hat{\beta})\Gamma(1 - \delta(n + 1))} x^{n+\chi\hat{\beta}}, \tag{10.21}$$

for $x > 0$, and, if $2\delta > 1$, we have

$$\tilde{p}(x) = \sum_{n \geq 0} a_n x^{-1+\chi\hat{\beta}-n} + \sum_{m \geq 1} b_m x^{-1-m\chi}, \qquad x > 0, \tag{10.22}$$

where

$$a_n = \frac{\delta}{\pi} \frac{\sin(-\pi\chi\hat{\beta})\Gamma(1 + \hat{\beta})\Gamma(\hat{\beta} - \delta n)}{\Gamma(\chi\hat{\beta} - n)\Gamma(\delta n + 1)} (-1)^n$$

and

$$b_m = \frac{\sin(-\pi\chi\hat{\beta})\Gamma(1 + \hat{\beta})}{\sin(\pi(\chi(\hat{\beta} + m) + 1))\Gamma(-\chi n)\Gamma(\hat{\beta} + 1 + m)} \frac{(-1)^m}{m!}.$$

Moreover, when formula (10.21) *(resp. (10.22)) provides complete asymptotic expansion as x goes to 0^+ (resp. as x goes to ∞).*

Proof For the explicit form of $\tilde{\mathtt{M}}$, we proceed similarly as in the proof of Proposition 10.2. Using the style of reasoning in (10.8) and the nature of the ascending ladder height process, cf. Proposition 4.1, we can easily deduce that

$$\mathbf{E}\left[e^{-\delta s \mathfrak{U} W}\right] = \frac{\Gamma(1 + \hat{\beta})\,\Gamma(\delta s + \hat{\beta} + \hat{\gamma})}{\Gamma(\hat{\beta} + \hat{\gamma})\,\Gamma(\delta s + \hat{\beta} + 1)}.$$

The independence of $\hat{\Xi}$ and the pair $(\mathfrak{U}, W)$, the Mellin transform of $I(\delta, \hat{\Xi})$ in Corollary 4.16 and the quasi-periodic properties of the double gamma function G, see (A.21) now gives us the identity (10.20).

For the convergent series representation, we proceed as in the proof of Corollary 10.6. We first observe from (10.20) that the Mellin transform $\tilde{\mathtt{M}}(z)$ of $\tilde{p}(x)$ has simple poles at

$$z_n^- = -n - \chi\hat{\beta}, \qquad z_n^{+,1} = n + 1 - \chi\hat{\beta}, \qquad \text{for} \quad n \geq 0,$$

and

$$z_n^{+,2} = 1 + \chi n, \qquad \text{for} \quad n \geq 1.$$

The residues at these points provide the coefficients in (10.21) and (10.22). To find them, we again apply Proposition A.1 and identity (A.11) (both in the Appendix) to get

$$\mathrm{Res}(\mathtt{M}(s)\colon s = -n - \chi\hat{\beta}) = \frac{\delta}{\pi}\frac{\sin(-\pi\chi\hat{\beta})\Gamma(1+\hat{\beta})\Gamma(\delta(1+n)+\hat{\beta})}{\Gamma(1+n+\chi\hat{\beta})\Gamma(1-\delta(n+1))}(-1)^n,$$

$$\mathrm{Res}(\mathtt{M}(s)\colon s = n + 1 - \chi\hat{\beta}) = \frac{\delta}{\pi}\frac{\sin(-\pi\chi\hat{\beta})\Gamma(1+\hat{\beta})\Gamma(\hat{\beta}-\delta n)}{\Gamma(\chi\hat{\beta}-n)\Gamma(\delta n+1)}(-1)^{n+1},$$

and

$$\mathrm{Res}(\mathtt{M}(s)\colon s = 1 + \chi n) = \frac{1}{n!}\frac{\sin(-\pi\chi\hat{\beta})\Gamma(1+\hat{\beta})}{\sin(\pi(\chi(\hat{\beta}+n)+1)\Gamma(-n\chi)\Gamma(\hat{\beta}+1+n)}(-1)^{n+1}.$$

The rest of the proof follows by developing the appropriate contour integrals as in Chapter 4. We leave the details to the interested reader. □

10.2 Distributions of pssMp path functionals

Let us put ourselves into the general setting of Chapter 9. That is to say we will consider a positive self-similar Markov process $Z = (Z_t, t \geq 0)$, with index of self-similarity $\alpha > 0$ and probabilities P_x, $x > 0$. The Lévy processes associated to it via the Lamperti transformation is denoted again Ξ, with law $\mathbf{P}$, which is assumed non-arithmetic, and its mean at time 1 satisfies

$$0 < \mathbf{E}[\Xi_1] < \infty.$$

This ensures that $\hat{\Xi}$, the dual of Ξ drifts to $-\infty$ and hence its integrated exponential function $I(\delta, \hat{\Xi})$, defined in (10.1) is almost surely finite. We are specifically interested in the setting that $\delta = \alpha$ in this section and accordingly define $\hat{I}_\infty = I(\alpha, \hat{\Xi})$, that is,

$$\hat{I}_\infty = \int_0^\infty \mathrm{e}^{\alpha\hat{\Xi}_s}\mathrm{d}s.$$

The setting that $\hat{I}_\infty < \infty$ almost surely is sufficient for us to assume that the process Z may be extended to include entrance from the origin, with law P_0, such that $\lim_{z\to 0} P_z = P_0$ in the weak sense on the Skorokhod space.

Under additional assumptions, we provide three results in this section concerning the distributional tail behaviour of certain path functionals of Z that appear in the integral tests of Chapter 9.

Recall that $\hat{T}_x^- = \inf\left\{t > 0 : \hat{\Xi}_t < x\right\}$ and, for $q > 0$,

$$\hat{I}_{\hat{T}_{-q}^-} = \int_0^{\hat{T}_{-q}^-} e^{\alpha\hat{\Xi}_u}\, du.$$

For our first result, we are interested in tail properties of the distributions

$$F(t) := \mathbf{P}(\hat{I}_\infty > t) \qquad \text{and} \qquad F_q(t) := \mathbf{P}\left(\hat{I}_{\hat{T}_{-q}^-} > t\right),$$

for $t \geq 0$. The following result shows that, for any $q > 0$, functions F_q and F are asymptotically equivalent, that is, $F_q(t) \asymp F(t)$, as $t \to \infty$, as soon as F has polynomial decay.

Lemma 10.8 *Assume that*

$$F(t) \sim Ct^{-\gamma}, \qquad \textit{as} \quad t \to \infty, \tag{10.23}$$

where C and γ are strictly positive constants. If (10.23) *holds then for all $q > 0$,*

$$(1 - e^{-\gamma q})F(t) \leq F_q(t) \leq F(t), \tag{10.24}$$

for all t large enough.

Proof Recall from the proof of Lemma 9.6, that $(\hat{\Xi}_s,\ s \leq \hat{T}(-q))$ and $\hat{\Xi}' = (\hat{\Xi}_{s+\hat{T}(-q)} - \hat{\Xi}_{\hat{T}(-q)},\ s \geq 0)$ are independent and also that the following inequality holds

$$\hat{I}_\infty = \hat{I}_{\hat{T}^-(-q)} + e^{\alpha\hat{\Xi}_{\hat{T}(-q)}}\hat{I}'_\infty \leq \hat{I}_{\hat{T}^-(-q)} + e^{-\alpha q}\hat{I}'_\infty, \tag{10.25}$$

where $\hat{I}'_\infty$ is a copy of $\hat{I}_\infty$ that is independent of $\hat{I}_{\hat{T}^-(-q)}$. From (10.25), it is clear that the second inequality of the lemma holds. To deduce the first inequality in (10.24), we write, for all $\varepsilon > 0$,

$$\begin{aligned}
F((1+\varepsilon)t) &= \mathbf{P}\left(\hat{I}_\infty > (1+\varepsilon)t\right) \\
&\leq \mathbf{P}\left(\hat{I}_{\hat{T}^-(-q)} + e^{-\alpha q}\hat{I}'_\infty \geq (1+\varepsilon)t\right) \\
&\leq \mathbf{P}\left(\hat{I}_{\hat{T}^-(-q)} > t\right) + \mathbf{P}\left(e^{-\alpha q}\hat{I}_\infty > t\right) + \mathbf{P}\left(\hat{I}_{\hat{T}^-(-q)} > \varepsilon t\right)\mathbf{P}\left(e^{-\alpha q}\hat{I}_\infty > \varepsilon t\right) \\
&\leq \mathbf{P}\left(\hat{I}_{\hat{T}^-(-q)} > t\right) + \mathbf{P}\left(e^{-\alpha q}\hat{I}_\infty > t\right) + \mathbf{P}\left(\hat{I}_\infty > \varepsilon t\right)\mathbf{P}\left(e^{-\alpha q}\hat{I}_\infty > \varepsilon t\right),
\end{aligned}$$

so that

$$\liminf_{t\to\infty} \frac{\mathbf{P}\left(\hat{I}_{\hat{T}^-(-q)} > t\right)}{\mathbf{P}\left(\hat{I}_\infty > t\right)} \geq (1+\varepsilon)^{-\gamma} - e^{-\alpha q \gamma}.$$

The result now follows since ε can be chosen arbitrary small. □

In the spirit of Lemma 10.8, the next result shows asymptotic equivalence of the two distributions

$$\overline{F}_\Upsilon(t) = P_0\left(\Upsilon^\alpha \hat{I}_\infty \leq t\right) \quad \text{and} \quad G(t) = P_0(S_1 \leq t), \qquad t \geq 0,$$

when the former of the two is assumed to have polynomial decay. Here, we recall the definitions of Υ and $\mathtt{D}_1$ given in (10.2) and (10.3) as well as

$$S_y = \inf\{t \geq 0 \colon Z_t > y\}. \tag{10.26}$$

Proposition 10.9 *Assume that*

$$\overline{F}_\Upsilon(t) \sim Ct^\gamma, \qquad as \quad t \to 0, \tag{10.27}$$

where C and γ are strictly positive constants. Under condition (10.27), we have that

$$C_1 t^\gamma \leq G(t) \leq C_2 t^\gamma, \qquad as \quad t \to 0,$$

where c and C_1 are two positive constants such that $0 < C_1 \leq C \leq C_2$.

Proof In the notation of Chapter 9, recall that $Z = (Z_t, t \geq 0)$ is a pssMp with self-similar index α and its future infimum process is denoted by $(\mathtt{J}_t, t \geq 0)$, where $\mathtt{J}_t = \inf_{s\geq t} Z_s$. Let us introduce the running supremum of Z, written $M = (M_t, t \geq 0)$, where

$$M_t = \sup_{0\leq s\leq t} Z_s, \qquad t \geq 0.$$

Since $\mathtt{J}_t \leq M_t$ for all $t \geq 0$, we clearly have that $S_1 \leq \mathtt{D}_1$. Moreover, since $\mathtt{D}_1$, under P_0, has the same law as $\Upsilon^\alpha \hat{I}_\infty$, under $\mathbf{P}$, cf. Corollary 9.9, we deduce that $\overline{F}_\Upsilon(t) \leq G(t)$, for all $t \geq 0$.

For the upper bound, we fix $\epsilon > 0$. Then, by the Markov property and the fact that $\mathtt{J}$ is an increasing process, we have

$$\begin{aligned} P_0\left(\mathtt{J}_1 > \frac{1-\epsilon}{t}\right) &\geq P_0\left(\mathtt{J}_1 > \frac{1-\epsilon}{t}, M_1 \geq \frac{1}{t}\right) \\ &= E_0\left[P_{Z_{S_{1/t}}}\left(\mathtt{J}_{1-S_{1/t}} > \frac{1-\epsilon}{t}\right)\mathbf{1}_{\{S_{1/t}\leq 1\}}\right] \\ &\geq E_0\left[P_{Z_{S_{1/t}}}\left(\mathtt{J}_0 > \frac{1-\epsilon}{t}\right)\mathbf{1}_{\{S_{1/t}\leq 1\}}\right]. \end{aligned} \tag{10.28}$$

Since $Z_{S_{1/t}} \geq 1/t$, P_0-almost surely, using the Lamperti transform (see Theorem 5.2), we deduce that

$$E_0\left[P_{Z_{S_{1/t}}}\left(\mathtt{J}_0 > \frac{1-\epsilon}{t}\right)\mathbf{1}_{\{S_{1/t}\leq 1\}}\right] \geq P_0(S_{1/t} < 1)\mathbf{P}\left(\inf_{s\geq 0}\Xi_s > \log(1-\epsilon)\right). \tag{10.29}$$

On the other hand, since Ξ drifts towards ∞, we have from the Wiener–Hopf factorisation (see Theorem 2.21) that the descending ladder height $\hat{H}$ is equal in law to a subordinator which is killed at an independent time $\mathbf{e}_q$, which is exponentially distributed with some parameter $q > 0$. Following similar ideas to those in the discussion above Lemma 2.26, we deduce

$$\mathbf{P}\left(\inf_{s\geq 0}\Xi_s > \log(1-\epsilon)\right) = \mathbf{P}\left(T^-_{\log(1-\epsilon)} = \infty\right) = \mathbf{P}\left(\hat{H}_{\mathbf{e}_q-} \leq -\log(1-\epsilon)\right),$$

where $T^-_x = \inf\{t > 0 \colon \Xi_t < x\}$. Since the process Ξ is not arithmetic, the descending ladder height $\hat{H}$ has support on $[0,\infty)$, implying that, for all $\epsilon > 0$,

$$K_\epsilon := \mathbf{P}\left(\inf_{s\geq 0}\Xi_s > \log(1-\epsilon)\right) > 0.$$

Hence, using the scaling property of Z which ensures that $S_{1/t}$ is equal in law to $t^{-\alpha}S_1$, we have from (10.28) and (10.29),

$$K_\epsilon^{-1}P_0\left(\mathtt{J}_1 > \frac{1-\epsilon}{t}\right) \geq P_0\left(S_1 < t^\alpha\right). \tag{10.30}$$

Note that

$$P_0\left(\mathtt{J}_1 > \frac{1-\epsilon}{t}\right) = P_0\left(\mathtt{D}_{(1-\epsilon)/t} < 1\right) = P_0\left(\mathtt{D}_1 < \left(\frac{t}{1-\epsilon}\right)^\alpha\right),$$

where we have used a similar scaling property for $\mathtt{D}_{(1-\epsilon)/t}$. Feeding this back in (10.30) and recalling that $\mathtt{D}_1$ is equal in law to $\Upsilon^\alpha \hat{I}_\infty$, for which we have assumed (10.27), we now get

$$CK_\epsilon^{-1}\left(\frac{t}{1-\epsilon}\right)^{\alpha\gamma} \geq K_\epsilon^{-1}P_0\left(\mathtt{D}_1 < \left(\frac{t}{1-\epsilon}\right)^\alpha\right) \geq P_0\left(S_1 < t^\alpha\right), \quad \text{as} \quad t \to 0.$$

This completes the proof. □

Our third and final result for this section concerns the setting that the pssMp $(Z_t, t \geq 0)$ has no positive jumps, or equivalently that Ξ has no positive jumps, albeit that we exclude the case of monotone paths. In this setting, we necessarily have that $\Upsilon = 1$, P_0-almost surely. In other words,

$$\overline{F}_\Upsilon(t) = \overline{F}(t) := \mathbf{P}(\hat{I}_\infty \leq t), \qquad \text{for} \quad t \geq 0.$$

Proposition 10.10 *Excluding monotone paths, assume that Z has no positive jumps and that*

$$-\log \overline{F}(1/t) \sim Ct^{\gamma}, \qquad \text{as } t \to \infty, \tag{10.31}$$

where C and γ are strictly positive constants. Then

$$-\log G(1/t) \sim Ct^{\gamma}, \qquad \text{as} \quad t \to \infty. \tag{10.32}$$

Proof First, we prove an upper bound for (10.32). As noted earlier, the future infimum process $(\mathtt{J}_t, t \geq 0)$ and the process of last passage times $(\mathtt{D}_x, x > 0)$ are conveniently related via $\{\mathtt{J}_1 > t\} = \{\mathtt{D}_t < 1\}$. Corollary 9.9 states that $\mathtt{D}_t$ is equal in law to $t^{\alpha}\Upsilon^{\alpha}\hat{I}_{\infty}$. Since $\Upsilon = 1$, we have that

$$P_0\big(\mathtt{J}_1 > t\big) = \mathbf{P}(\hat{I}_{\infty} < 1/t^{\alpha}).$$

Also recalling from the proof of Proposition 10.9 that $\mathtt{J}_1 \leq M_1 := \sup_{0 \leq s \leq 1} Z_s$, we have

$$-\log \mathbf{P}\big(\hat{I}_{\infty} < 1/t^{\alpha}\big) = -\log P_0\big(\mathtt{J}_1 > t\big) \geq -\log P_0\big(M_1 > t\big),$$

which implies

$$1 \geq \limsup_{t \to \infty} \frac{-\log P_0\big(M_1 > t^{1/\alpha}\big)}{Ct^{\gamma}}.$$

Since $P_0(M_1 > t^{1/\alpha}) = P_0(S_{t^{1/\alpha}} < 1)$, the scaling property of Z, which tells us that $S_{t^{1/\alpha}}$ is equal in law to tS_1, implies that

$$1 \geq \limsup_{t \to \infty} \frac{-\log P_0(S_1 < 1/t)}{Ct^{\gamma}}.$$

Now, fix $\epsilon > 0$. Decomposing the path of Z at time 1, we have

$$P_0(\mathtt{J}_1 > (1-\epsilon)t^{1/\alpha}) \geq P_0(S_{t^{1/\alpha}} < 1)\mathbf{P}\left(\inf_{s \geq 0} \Xi_s > \log(1-\epsilon)\right). \tag{10.33}$$

Since

$$K_{\epsilon} = \mathbf{P}\Big(\inf_{s \geq 0} \Xi_s > \log(1-\epsilon)\Big) > 0,$$

again using the scaling property, we deduce from (10.33) that

$$-\log P_0(\mathtt{J}_1 > (1-\epsilon)t^{1/\alpha}) \leq -\log P_0\big(S_1 < 1/t\big) - \log K_{\epsilon}.$$

Recalling that $\{\mathtt{J}_1 > (1-\epsilon)t^{1/\alpha}\} = \{\mathtt{D}_{(1-\epsilon)t^{1/\alpha}} < 1\}$ and that $\mathtt{D}_t$ is equal in law to $t^{\alpha}\Upsilon^{\alpha}\hat{I}_{\infty}$ (with $\Upsilon = 1$ in the current setting), we have from the assumption on $\overline{F}$, the lower bound

$$(1-\epsilon)^{\alpha\gamma} \leq \liminf_{t \to \infty} \frac{-\log P_0\big(S_1 < 1/t\big)}{Ct^{\gamma}}.$$

Since ϵ can be chosen arbitrarily small, (10.32) is proved. □

Now, we are ready to apply the results of Chapter 9 and identify explicit integral tests that describe the lower and upper envelopes, as $t \to 0$ and $t \to \infty$, for the path functionals of stable processes discussed in Chapter 5. We start with the case of stable processes conditioned to stay positive.

10.3 Stable processes conditioned to stay positive

Recall from Section 5.4 that the stable process conditioned to stay positive X has probabilities $\mathbb{P}^{\uparrow} = (\mathbb{P}^{\uparrow}_x, x > 0)$, where $\mathbb{P}^{\uparrow}_x$ is the law starting from x. The process $(X, \mathbb{P}^{\uparrow})$ is defined by (5.23) or equivalently by the change of measure (5.26).

According to Theorem 5.14, the process $(X, \mathbb{P}^{\uparrow})$ is a conservative positive self-similar Markov process, with self-similar index α, whose underlying Lévy process $\xi^{\uparrow}$, with law $\mathbf{P}^{\uparrow}$, belongs to the class of hypergeometric processes with parameters $(\beta, \gamma, \hat{\beta}, \hat{\gamma}) = (1, \alpha\rho, 1, \alpha\hat{\rho})$. Moreover its dual, $-\xi^{\uparrow}$, is a hypergeometric Lévy process with parameters $(\beta, \gamma, \hat{\beta}, \hat{\gamma}) = (0, \alpha\hat{\rho}, 0, \alpha\rho)$, which places it in the class $\mathcal{H}_1 \backslash \{\beta = 1\}$. We also have that $-\xi^{\uparrow}$ fulfils the conditions of Theorem 5.3 which guarantees that $\mathbb{P}^{\uparrow}_0 := \lim_{x \downarrow 0} \mathbb{P}^{\uparrow}_x$ exists, in the sense of weak convergence on the Skorokhod space (see Section A.10 in the Appendix). In other words, we can apply the results of the previous section to describe the lower and upper envelope of $(X, \mathbb{P}^{\uparrow}_0)$ at 0 and at ∞.

We start by describing the lower envelope. In order to do so, we introduce the integrated exponential functional of $-\xi^{\uparrow}$, namely

$$\hat{I}^{\uparrow}_{\infty} = \int_0^{\infty} \mathrm{e}^{-\alpha\xi^{\uparrow}_s} \mathrm{d}s.$$

From Proposition 4.27, we deduce that the upper tail behaviour of the distribution of $\hat{I}^{\uparrow}_{\infty}$ has polynomial decay. Recall that $\mathcal{A}$ is the set of (α, ρ) parameter combinations for stable processes; cf. (3.11).

Lemma 10.11 *For $(\alpha, \rho) \in \mathcal{A}$ and $\rho \in (0, 1)$, we have*

$$F^{\uparrow}(t) := \mathbf{P}^{\uparrow}\left(\hat{I}^{\uparrow}_{\infty} > t\right) \sim \frac{\Gamma\big(1/\alpha\big)}{\Gamma(\rho + 1)\Gamma(\hat{\rho})} t^{-\frac{1}{\alpha}}, \qquad \text{as} \quad t \to \infty. \tag{10.34}$$

Proof Recall from the discussion preceding the statement of this lemma that $-\xi^{\uparrow}$ is a hypergeometric Lévy process whose parameters belong to $\mathcal{H}_1 \backslash \{\beta = 1\}$. As alluded to above, we can appeal to Proposition 4.27, with the observation that $\hat{\theta} = 1$ and $\chi = 1/\alpha$, to write

$$\mathbf{P}^{\uparrow}\left(\hat{I}^{\uparrow}_{\infty} > t\right) \sim \frac{\hat{\mathtt{M}}^{\uparrow}\left(1/\alpha\right)}{\hat{\psi}^{\uparrow\prime}(1)} t^{-\frac{1}{\alpha}}, \qquad t \to \infty, \tag{10.35}$$

where $\hat{\mathtt{M}}^{\uparrow}$ denotes the Mellin transform of $\hat{I}^{\uparrow}_{\infty}$ and $\hat{\psi}^{\uparrow}(z)$ is the Laplace exponent of $-\xi^{\uparrow}$. Moreover, in this case, $\hat{\mathtt{M}}^{\uparrow}(1/\alpha)$ and $\hat{\psi}^{\uparrow\prime}(1)$ can be computed explicitly. Indeed, starting from the identities in Theorem 4.13 (i) and 4.37, by appealing to the quasi-periodic properties of double gamma functions (see (A.21) in the Appendix) as well as standard properties of gamma functions (see Appendix Section A.3), we find

$$\hat{\mathtt{M}}^{\uparrow}\left(1/\alpha\right) = \Gamma\left(1/\alpha\right) \frac{\alpha\Gamma(\alpha\hat{\rho})\Gamma(\alpha\rho)}{\Gamma(\rho)\Gamma(\hat{\rho})} \qquad \text{and} \qquad \hat{\psi}^{\uparrow\prime}(1) = \Gamma(\alpha\hat{\rho})\Gamma(1+\alpha\rho),$$

thus concluding the proof. □

For $q \geq 0$, let us now define $\hat{T}^{\uparrow,-}(-q) := \inf\{t \geq 0 : \ -\xi^{\uparrow}_t \leq -q\}$,

$$\hat{I}^{\uparrow}_{\hat{T}^{\uparrow,-}(-q)} := \int_0^{\hat{T}^{\uparrow,-}(-q)} e^{-\alpha\xi^{\uparrow}_s} ds \quad \text{and} \quad F^{\uparrow}_q(t) := \mathbf{P}^{\uparrow}\left(\hat{I}^{\uparrow}_{\hat{T}^{\uparrow,-}(-q)} > t\right).$$

For the current setting, the above lemma allows us to develop more precise integral tests for lower and upper path envelopes than those described in Theorems 9.7 and 9.8. In particular, the appearance of the factors $1-\varepsilon$ and $1+\varepsilon$ can be absorbed into the envelope with no change to the integral test thanks to the polynomial asymptotic of $F^{\uparrow}$. The proof of the next result follows from a simple application of Theorems 9.7 and 9.8, Lemma 10.30 and the equivalence between $F^{\uparrow}$ and $F^{\uparrow}_q$, for $q > 0$, given in Lemma 10.8.

Theorem 10.12 *The lower envelope at* 0 *and at* ∞ *of the stable process conditioned to stay positive is characterised as follows.*

(i) Let f be an increasing function such that either

$$\lim_{t\to 0} \frac{f(t)^{\alpha}}{t} = 0 \qquad \text{or} \qquad \liminf_{t\to 0} \frac{f(t)^{\alpha}}{t} > 0,$$

then,

$$\mathbb{P}^{\uparrow}_0\left(X_t < f(t) \text{ i.o. as } t \to 0\right) = 0 \text{ or } 1,$$

accordingly as

$$\int_{0+} \frac{f(t)}{t^{\frac{1}{\alpha}+1}} dt \quad \text{is finite or infinite.}$$

ii) Let g be an increasing function such that either

$$\lim_{t\to\infty} \frac{g(t)^\alpha}{t} = 0 \qquad or \qquad \liminf_{t\to\infty} \frac{g(t)^\alpha}{t} > 0,$$

then, for all $x \geq 0$,

$$\mathbb{P}_x^{\uparrow}\Big(X_t < g(t) \text{ i.o. as } t \to \infty\Big) = 0 \text{ or } 1,$$

accordingly as

$$\int^{\infty} \frac{g(t)}{t^{\frac{1}{\alpha}+1}} \mathrm{d}t \quad \text{is finite or infinite.}$$

Next, we proceed to describe the upper envelope of $(X, \mathbb{P}^\uparrow)$ at 0 and at ∞. As an intermediate step, we describe the upper envelope of the future infimum at 0 and at ∞ and, thereafter, we compare it with the upper envelope of the original process.

We first consider the case when the process has positive jumps since the spectrally negative case possesses a completely different behaviour, as we will see later. For the spectrally positive case, the reader should implicitly assume that we are excluding the case that X is a stable subordinator. The conditioned process in the subordinator case is nothing more than itself again and envelopes of stable subordinators were already considered in Chapter 8.

Let us recall some notation. The future infimum process is denoted by $\mathtt{J} = (\mathtt{J}_t, t \geq 0)$, where

$$\mathtt{J}_t = \inf_{s\geq t} Z_s, \qquad t \geq 0,$$

and recall the last passage time below the level 1, denoted $\mathtt{D}_1$, was defined in (10.3). We also consider the random variable $\Upsilon^\uparrow$ which is independent of $\hat{I}_\infty^\uparrow$ and has the same law, under $\mathbf{P}^\uparrow$, as $X_{\mathtt{D}_1-}$ under $\mathbb{P}_0^\uparrow$.

Under the assumption that $\xi^\uparrow$ has positive jumps, the following result gives the behaviour for the lower tail distribution of $(\Upsilon^\uparrow)^\alpha \hat{I}_\infty^\uparrow$ near 0, with the restriction that $\alpha \notin \mathbb{Q}$. Its proof follows directly from the asymptotic expansion of $\tilde{p}(x)$, for x close to 0, given in Theorem 10.5, when the parameters $(\beta, \gamma, \hat{\beta}, \hat{\gamma})$ take the specific values $(0, \alpha\hat{\rho}, 0, \alpha\rho)$. (The reader will recall from (8.16) that $(\alpha,\rho) \in \mathcal{A}^+$ is the parameter regime for stable processes with positive jumps.)

Lemma 10.13 *Let us assume that $\alpha \notin \mathbb{Q}$. For $(\alpha,\rho) \in \mathcal{A}^+$ and $\rho \in (0,1)$, we have*

$$\overline{F}^\uparrow_{\Upsilon^\uparrow}(t) := \mathbf{P}^\uparrow\left((\Upsilon^\uparrow)^\alpha \hat{I}_\infty^\uparrow < t\right) \sim \frac{\tilde{\mathtt{M}}^\uparrow(1+\hat{\rho})}{\Gamma(1-\alpha\hat{\rho})\Gamma(1-\alpha\rho)} t^\rho, \qquad as \quad t \to 0, \tag{10.36}$$

where $\tilde{\mathtt{M}}^\uparrow$ denotes the Mellin transform of $(\Upsilon^\uparrow)^\alpha \hat{I}_\infty^\uparrow$.

As with Theorem 10.12, the polynomial behaviour of the tail of $\overline{F}^{\uparrow}_{\Upsilon^\uparrow}(t)$ allows us to drop the factors $1-\varepsilon$ and $1+\varepsilon$ in the envelope of Theorems 9.11 and 9.13, whilst keeping the same integral test. Recall that C_0 and C_∞ denote the set of positive increasing functions defined in Definitions 9.10 and 9.12. In a similar manner to Theorem 10.12, the following result follows from simple applications of Theorems 9.11 and 9.13 and the estimate in 10.41.

Theorem 10.14 *Let $(\alpha,\rho) \in \mathcal{A}^+$ and assume that $\alpha \notin \mathbb{Q}$. The upper envelope at 0 and at ∞ of the future infimum of the stable process conditioned to stay positive is as follows.*

(i) Let $f \in C_0$ such that either

$$\lim_{t\to 0} \frac{t}{f(t)^\alpha} = 0 \qquad \textit{or} \qquad \liminf_{t\to 0} \frac{t}{f(t)^\alpha} > 0,$$

then,

$$\mathbb{P}^{\uparrow}_0\left(\mathtt{J}_t > f(t) \text{ i.o. as } t \to 0\right) = 0 \text{ or } 1,$$

accordingly as

$$\int_{0+} \frac{t^{\rho-1}}{f(t)^{\alpha\rho}} \mathrm{d}t \quad \textit{is finite or infinite.}$$

ii) Let $g \in C_\infty$ such that either

$$\lim_{t\to \infty} \frac{t}{g(t)^\alpha} = 0 \qquad \textit{or} \qquad \liminf_{t\to \infty} \frac{t}{g(t)^\alpha} > 0,$$

then, for all $x \geq 0$,

$$\mathbb{P}^{\uparrow}_x\left(\mathtt{J}_t > g(t) \text{ i.o. as } t \to \infty\right) = 0 \text{ or } 1,$$

accordingly as

$$\int^{\infty} \frac{t^{\rho-1}}{g(t)^{\alpha\rho}} \mathrm{d}t \quad \textit{is finite or infinite.}$$

In order to deduce the upper envelope of the stable process conditioned to stay positive, we first need to compare the behaviour near 0 of $\overline{F}^{\uparrow}_{\Upsilon^\uparrow}(t)$ with that of

$$G^\uparrow(t) := \mathbb{P}^{\uparrow}_0\left(S_1 < t\right),$$

where we recall from (10.26) that $S_1 = \inf\{t > 0\colon X_u > 1\}$. Proposition 10.9 and Lemma 10.13 ensures that $G^\uparrow$ has the same polynomial behaviour as $\overline{F}^{\uparrow}_{\Upsilon^\uparrow}$.

As we have seen with Theorems 10.12 and 10.14, the polynomial behaviour of $\overline{F}^{\uparrow}_{\Upsilon^\uparrow}$ and $G^\uparrow$ affords us the development of a cleaner version of the upper envelope integral tests given in Propositions 9.14 and 9.15.

Theorem 10.15 *Let $\alpha \notin \mathbb{Q}$ and assume $(\alpha, \rho) \in \mathcal{A}^+$. Hence the upper envelope of the stable process conditioned to stay positive at 0 and at ∞ is as follows.*

(i) Let $f \in C_0$ such that either

$$\lim_{t\to 0} \frac{t}{f(t)^\alpha} = 0 \qquad or \qquad \liminf_{t\to 0} \frac{t}{f(t)^\alpha} > 0,$$

then,

$$\mathbb{P}_0^\uparrow\Big(X_t > f(t) \text{ i.o. as } t \to 0\Big) = 0 \text{ or } 1,$$

accordingly as

$$\int_{0+} \frac{t^{\rho-1}}{f(t)^{\alpha\rho}} \mathrm{d}t \quad \text{is finite or infinite.}$$

ii) Let $g \in C_\infty$ such that either

$$\lim_{t\to \infty} \frac{t}{g(t)^\alpha} = 0 \qquad or \qquad \liminf_{t\to \infty} \frac{t}{g(t)^\alpha} > 0,$$

then, for all $x \geq 0$,

$$\mathbb{P}_x^\uparrow\Big(X_t > g(t) \text{ i.o. as } t \to \infty\Big) = 0 \text{ or } 1,$$

accordingly as

$$\int^{\infty} \frac{t^{\rho-1}}{g(t)^{\alpha\rho}} \mathrm{d}t \quad \text{is finite or infinite.}$$

It is interesting to note from Theorems 10.14 and 10.15 that, in the parameter regime $\mathcal{A}^+$, the stable process conditioned to stay positive and its future infimum have the same upper functions.

Finally, we consider the case when the stable process conditioned to stay positive has no positive jumps, that is, when $\alpha \in (1,2)$ and $\rho = \alpha^{-1}$. As we will see below, this assumption allows us to obtain a law of the iterated logarithm for the upper envelope of $(X, \mathbb{P}_0^\uparrow)$ as well as for its future infimum and for the process reflected at its future infimum.

First, we observe that $\Upsilon^\uparrow = 1$, $\mathbf{P}^\uparrow$-almost surely because of spectral negativity. Under the assumption that $\xi^\uparrow$ has no positive jumps, we have the following behaviour for the lower tail distribution of $\hat{I}_\infty^\uparrow$ near 0.

Lemma 10.16 *Let $\alpha \in (1,2)$ and $\rho = \alpha^{-1}$, then*

$$-\log \overline{F}^\uparrow(1/t) \sim \frac{\alpha-1}{\alpha}\left(\frac{1}{\alpha}\right)^{1/(\alpha-1)} t^{1/(\alpha-1)}, \qquad \text{as } t \to \infty. \tag{10.37}$$

Proof Recall from Theorem 5.14 that $\xi^\uparrow$ is a spectrally negative Lamperti-stable process with parameters $(\beta,\gamma,\hat{\gamma}) = (1,1,\alpha-1)$. Moreover its dual $-\xi^\uparrow$ is a spectrally positive Lamperti-stable Lévy process with parameters $(\beta,\gamma,\hat{\gamma}) = (0,\alpha-1,1)$. According to Theorem 5.11, the process $-\xi^\uparrow$ has the same law as the Lévy process associated with the spectrally positive stable process killed on entering $(-\infty,0)$. In other words, the spectrally negative stable process conditioned to stay positive is in duality with the spectrally positive stable process killed at entering at $(-\infty,0)$. Indeed, since

$$0 < \mathbf{E}^\uparrow[\xi_1^\uparrow] < \infty,$$

we deduce from Proposition 5.6 that

$$X_t \mathbf{1}_{(\underline{X}_t \geq 0)}, \qquad t \geq 0,$$

under $\hat{\mathbb{P}}_z$, is equal in law to the process $\tilde{X} = (X_{(\mathtt{D}_z - t)-}, 0 \leq t \leq \mathtt{D}_z)$, under $\mathbb{P}_0^\uparrow$, for $z > 0$. That is to say, the laws of $(\mathtt{D}_z, \mathbb{P}_0^\uparrow)$ and $(\tau_0^-, \hat{\mathbb{P}}_z)$ are the same, where $\tau_y^- = \inf\{s \geq 0 \colon X_t < y\}$. Recall from Corollary 9.9 that $\mathtt{D}_1 \overset{(d)}{=} \hat{I}_\infty^\uparrow$ and so, with the above duality in mind, for $t \geq 0$,

$$\overline{F}^\uparrow(t) = \mathbb{P}_0^\uparrow(D_1 \leq t) = \hat{\mathbb{P}}_1\left(\tau_0^- \leq t\right) = \hat{\mathbb{P}}_0\left(\tau_{-1}^- \leq t\right).$$

Since $(\tau_{-x}^-, x \geq 0)$ is a stable subordinator under $\hat{\mathbb{P}}_0$, from Lemma 8.4, we get the desired estimate. □

The following result provides laws of the iterated logarithm for the stable process conditioned to stay positive with no positive jumps as well as for its future infimum process. It takes advantage of the comparison between the behaviour near 0, of $\overline{F}^\uparrow(t)$ with that of $G^\uparrow(t)$ given by Proposition 10.10.

Theorem 10.17 *Assume that $\alpha \in (1,2)$ and $\rho = \alpha^{-1}$. Then the upper envelope of the stable process conditioned to stay positive with no positive jumps, as well as its future infimum, are described by the following law of the iterated logarithms,*

$$\limsup_{t\to 0} \frac{X_t}{t^{1/\alpha}(\log|\log t|)^{1-1/\alpha}} = \alpha(\alpha-1)^{-\frac{\alpha-1}{\alpha}}, \qquad \mathbb{P}_0^\uparrow - a.s.$$

and

$$\limsup_{t\to 0} \frac{\mathtt{J}_t}{t^{1/\alpha}(\log|\log t|)^{1-1/\alpha}} = \alpha(\alpha-1)^{-\frac{\alpha-1}{\alpha}}, \qquad \mathbb{P}_0^\uparrow - a.s.$$

The same law of the iterated logarithm holds, as $t \to \infty$, under $\mathbb{P}_x^\uparrow$, for any $x > 0$.

This result follows from the integral tests of Theorem 9.11 and Proposition 9.14 together with the asymptotic behaviour of $\overline{F}^{\uparrow}$ and $G^{\uparrow}$, which follows from (10.37) and Proposition 10.10, respectively. The arguments to deduce the laws of the iterated logarithm are similar to those used in the proof of Theorem 8.2. It is also interesting to note that Theorem 10.17 indicates that, for no positive jumps, the stable process conditioned to stay positive and its future infimum satisfy the same law of the iterated logarithm despite the fact that they do not necessarily have the same integral tests.

Our last result of this section tells us that the stable process conditioned to stay positive with no positive jumps reflected at its future infimum also satisfies the same law of the iterated logarithm.

Theorem 10.18 *Let us assume that $\alpha \in (1,2)$ and $\rho = \alpha^{-1}$. Then*

$$\limsup_{t\to 0} \frac{X_t - \mathtt{J}_t}{t^{1/\alpha}(\log|\log t|)^{1-1/\alpha}} = \alpha(\alpha-1)^{-\frac{\alpha-1}{\alpha}}, \qquad \mathbb{P}_0^{\uparrow} - a.s.$$

The same law of the iterated logarithm holds, as $t \to \infty$, under $\mathbb{P}_x^{\uparrow}$, for any $x > 0$.

Proof We give the proof only for the setting that $t \to 0$, with the proof for $t \to \infty$ having a similar structure. The details of the latter are left to the reader.

First, we observe that $X_t - \mathtt{J}_t \le X_t$, for every $t \ge 0$. Hence it is clear from Theorem 10.17 that

$$\limsup_{t\to 0} \frac{X_t - \mathtt{J}_t}{t^{1/\alpha}(\log|\log t|)^{1-1/\alpha}} \le \alpha(\alpha-1)^{-\frac{\alpha-1}{\alpha}}, \qquad \mathbb{P}_0^{\uparrow} - \text{a.s.}$$

For the lower bound, we introduce

$$\Lambda(t) = \alpha(\alpha-1)^{-\frac{\alpha-1}{\alpha}} t^{1/\alpha}(\log|\log t|)^{1-1/\alpha}, \qquad 0 < t < \mathrm{e}^{-1}.$$

We also fix $\epsilon \in (0, 1/2)$ and define the stopping times

$$R_n = \inf\left\{s \ge 1/n \colon \frac{X_s}{\Lambda(t)} \ge (1-\epsilon)\right\}, \qquad n \ge 1.$$

From the above definition, it is clear that $R_n \ge 1/n$ and that R_n goes to 0 as n goes to ∞, $\mathbb{P}_0^{\uparrow}$-a.s. Moreover, spectral negativity implies that $X_{R_n} = \Lambda(R_n)(1-\epsilon)$ on $\{R_n < \infty\}$. From Theorem 10.17, we deduce that R_n is finite, a.s.

By applying the strong Markov property, the Lamperti transform (see Theorem 5.14) and using that $(X, \mathbb{P}_0^\uparrow)$ has no positive jumps, we deduce

$$\mathbb{P}_0^\uparrow\left(\frac{X_{R_n} - \mathtt{J}_{R_n}}{\Lambda(R_n)} \geq (1-2\epsilon)\right) = \mathbb{P}_0^\uparrow\left(\mathtt{J}_{R_n} \leq \frac{\epsilon X_{R_n}}{1-\epsilon}\right)$$
$$= \mathbb{E}_0^\uparrow\left[\mathbb{P}_{X_{R_n}}^\uparrow\left(\mathtt{J}_0 \leq \frac{\epsilon X_0}{1-\epsilon}\right)\right]$$
$$= \mathbf{P}^\uparrow\left(\inf_{t\geq 0} \xi_t^\uparrow \leq \log\left(\frac{\epsilon}{1-\epsilon}\right)\right), \qquad (10.38)$$

where we recall that $(\xi^\uparrow, \mathbf{P}^\uparrow)$ is the Lévy process underlying the Lamperti transform of $(X, \mathbb{P}^\uparrow)$. The right-hand side of (10.38) can be computed explicitly. Indeed, from Theorem 4.6 (iii), the descending ladder height $\hat{H}^\uparrow$ associated to $\xi^\uparrow$ is a β-subordinator with parameters $(1, 1, \alpha - 1)$ which is killed at the time $\mathbf{e}_q$, which is an independent and exponentially distributed random variable with parameter $q = \Gamma(\alpha)$. Thus from Lemma 2.26 and Corollary 4.3, we obtain

$$\mathbf{P}^\uparrow\left(\inf_{t\geq 0} \xi_t^\uparrow \leq \log\left(\frac{\epsilon}{1-\epsilon}\right)\right) = \mathbf{P}^\uparrow\left(\hat{H}_{\mathbf{e}_q-}^\uparrow \geq -\log\left(\frac{\epsilon}{1-\epsilon}\right)\right) = 1 - \left(\frac{1-2\epsilon}{1-\epsilon}\right)^{\alpha-1} =: c_\epsilon,$$

which is strictly positive. Since $R_n \geq 1/n$, we obtain

$$\mathbb{P}_0^\uparrow\left(\frac{X_{R_p} - \mathtt{J}_{R_p}}{\Lambda(R_p)} \geq (1-2\epsilon), \text{ for some } p \geq n\right) \geq \mathbb{P}_0^\uparrow\left(\frac{X_{R_n} - \mathtt{J}_{R_n}}{\Lambda(R_n)} \geq (1-2\epsilon)\right) > c_\epsilon,$$

where the right-hand side is uniformly bounded away from zero for each fixed ϵ. Since R_n converges $\mathbb{P}_0^\uparrow$-a.s. to 0, as n goes to ∞, we have, for all $\epsilon \in (0, 1/2)$, that the left-hand side above decreases to the left-hand side below as we take limsup on the right-hand side, that is,

$$\mathbb{P}_0^\uparrow\left(\frac{X_t - \mathtt{J}_t}{\Lambda(t)} \geq (1-2\epsilon) \text{ i.o. as } t \to 0\right) \geq \limsup_{n\to\infty} \mathbb{P}_0^\uparrow\left(\frac{X_{R_n} - \mathtt{J}_{R_n}}{\Lambda(R_n)} \geq (1-2\epsilon)\right) > 0.$$

The event of the left-hand side is in the sigma-field $\cap_{t>0}\sigma\{X_s : s \leq t\}$ which is trivial under $\mathbb{P}_0^\uparrow$. Therefore

$$\limsup_{t\to 0} \frac{X_t - \mathtt{J}_t}{\Lambda(t)} \geq 1 - 2\epsilon, \qquad \mathbb{P}_0^\uparrow - \text{a.s.}$$

Since we may take ϵ as close to 0 as we like, we deduce the desired lower bound. □

10.4 Stable processes conditioned to limit to 0 from above

In this section, we focus on the asymptotic behaviour near absorption of stable processes conditioned to limit to 0 from above. Recall from Section 5.5

that $(X, \mathbb{P}_x^{\downarrow})$, the stable process conditioned to limit to 0 from above starting from $x > 0$, is defined by (5.32) or equivalently by the change of measure (5.34).

According to Theorem 5.15, the process $(X, \mathbb{P}_x^{\downarrow})$ is a non-conservative positive self-similar Markov process with self-similar index α, whose underlying Lévy process, $(\xi^{\downarrow}, \mathbf{P}^{\downarrow})$, belongs to the class of hypergeometric processes with parameters $(\beta, \gamma, \hat{\beta}, \hat{\gamma}) = (0, \alpha\rho, 0, \alpha\hat{\rho}) \in \mathcal{H}_4$. Moreover, its dual $-\xi^{\downarrow}$ is a hypergeometric process with parameters $(1, \alpha\hat{\rho}, 1, \alpha\rho)$. We note that the latter process is associated to the dual of the stable process conditioned to stay positive, that is, $(X, \hat{\mathbb{P}}_x^{\uparrow})$, for $x \geq 0$. Since

$$0 < -\mathbf{E}^{\downarrow}[\xi_1^{\downarrow}] < \infty,$$

we have from Proposition 5.6 that $(X, \mathbb{P}_x^{\downarrow})$ is equal in law to the process $\tilde{X} = (X_{(\mathtt{D}_1 - t)-}, 0 \leq t \leq \mathtt{D}_1)$, under $\hat{\mathbb{P}}_0^{\uparrow}(\cdot|\Gamma = z)$, for $z \in (0, 1]$, where $\mathtt{D}_1$ was defined in (10.3) and $\Gamma := X_{\mathtt{D}_1 -}$.

We recall that we can state the previous time-reversal property as follows (see for instance the comment after Proposition 5.6). Let

$$\tau^{\{0\}} = \inf\{s \geq 0 \colon X_t = 0\},$$

then, for any $z \in (0, 1]$, the returned process $(X_{(\tau^{\{0\}} - t)-}, 0 \leq t \leq \tau^{\{0\}})$ under $\mathbb{P}_z^{\downarrow}$ has the same law as $(X_t, 0 \leq t < \mathtt{D}_1)$ under $\hat{\mathbb{P}}_0^{\uparrow}(\cdot \,|\, \Gamma = z)$. The previous observation and Theorem 10.12 allow us to obtain the lower envelope of the stable processes conditioned to limit to 0 from above near absorption.

Theorem 10.19 *The lower envelope of the stable processes conditioned to limit to 0 from above near absorption is as follows. Let f be an increasing function, such that either*

$$\lim_{t \to 0} \frac{f(t)^{\alpha}}{t} = 0 \qquad or \qquad \liminf_{t \to 0} \frac{f(t)^{\alpha}}{t} > 0,$$

then, for any $x > 0$,

$$\mathbb{P}_x^{\downarrow}\Big(X_{(\tau^{\{0\}} - t)-} < f(t) \text{ i.o. as } t \to 0\Big) = 0 \text{ or } 1,$$

accordingly as

$$\int_{0+} \frac{f(t)}{t^{\frac{1}{\alpha}+1}} \mathrm{d}t \quad \text{is finite or infinite.}$$

For the upper envelope of $(X, \mathbb{P}^{\downarrow})$ at its absorption time, we first consider the case when the process has negative jumps. Recall that $\underline{X} = (\underline{X}_t, t \geq 0)$ denotes the running infimum process of X, that is,

$$\underline{X}_t = \inf_{s \leq t} X_s, \qquad t \geq 0.$$

Note that, for example, when $z \in (0,1]$, the law of $\underline{X}_{(\tau^{\{0\}}-t)-}$ under $\mathbb{P}_z^{\downarrow}$ is equal to that of $\mathtt{J}_t$ under $\hat{\mathbb{P}}^{\uparrow}(\cdot|\Gamma = z)$. Time reversal of $(X, \mathbb{P}_x^{\downarrow})$ at $\tau^{\{0\}}$ and Theorems 10.14, 10.15, thus gives us the following result.

Theorem 10.20 *Let us assume that $(\alpha, \hat{\rho}) \in \mathcal{A}^+$ and $\alpha \notin \mathbb{Q}$. The upper envelopes of the stable processes conditioned to limit to 0 from above and its past infimum near absorption are as follows. Let $f \in C_0$ such that either*

$$\lim_{t\to 0} \frac{t}{f(t)^{\alpha}} = 0 \qquad or \qquad \liminf_{t\to 0} \frac{t}{f(t)^{\alpha}} > 0,$$

then, for any $x > 0$,

$$\mathbb{P}_x^{\downarrow}\Big(X_{(\tau^{\{0\}}-t)-} > f(t) \text{ i.o. as } t \to 0\Big) = \mathbb{P}_x^{\downarrow}\Big(\underline{X}_{(\tau^{\{0\}}-t)-} > f(t) \text{ i.o. as } t \to 0\Big) = 0 \ \ or \ \ 1,$$

accordingly as

$$\int_{0+} \frac{t^{\hat{\rho}-1}}{f(t)^{\alpha\hat{\rho}}} \mathrm{d}t \quad \text{is finite or infinite.}$$

Finally, we consider the case when the stable processes conditioned to limit to 0 from above has no positive jumps, that is, when $\alpha \in (1,2)$ and $\rho = 1-\alpha^{-1}$. As we will see below, this assumption allows us to obtain three laws of the iterated logarithm for the upper envelope of $(X, \mathbb{P}_x^{\downarrow})$, its past infimum and the reflected process at its past infimum. The following result is a direct consequence of the time reversal property of $(X, \mathbb{P}_x^{\downarrow})$ at $\tau^{\{0\}}$, described below Lemma 10.16, together with the statement of Theorems 10.17 and 10.18.

Theorem 10.21 *Assume that $\alpha \in (1,2)$ and $\rho = 1 - \alpha^{-1}$. Then the upper envelopes near absorption of the stable processes conditioned to limit to 0 from above, its past infimum and the reflected process at its past infimum are described by the following laws of the iterated logarithm,*

$$\limsup_{t\to 0} \frac{X_{(\tau^{\{0\}}-t)-}}{t^{1/\alpha}(\log|\log t|)^{1-1/\alpha}} = \alpha(\alpha-1)^{-\frac{\alpha-1}{\alpha}}, \qquad \mathbb{P}_x^{\downarrow} - a.s.,$$

$$\limsup_{t\to 0} \frac{\underline{X}_{(\tau^{\{0\}}-t)-}}{t^{1/\alpha}(\log|\log t|)^{1-1/\alpha}} = \alpha(\alpha-1)^{-\frac{\alpha-1}{\alpha}}, \qquad \mathbb{P}_x^{\downarrow} - a.s.$$

and

$$\limsup_{t\to 0} \frac{(X-\underline{X})_{(\tau^{\{0\}}-t)-}}{t^{1/\alpha}(\log|\log t|)^{1-1/\alpha}} = \alpha(\alpha-1)^{-\frac{\alpha-1}{\alpha}}, \qquad \mathbb{P}_x^{\downarrow} - a.s.,$$

for all $x > 0$.

10.5 Censored stable processes

In this section, we focus on the asymptotic behaviour of censored stable processes. We only treat the case $\alpha \in (0,1)$. The case that $\alpha \in [1,2)$ can also be treated, however, it requires a level of analysis which lies beyond the space available to us here; we give more commentary below on the associated difficulties.

Recall from Section 5.6 that the censored stable process Z with probabilities $\overrightarrow{\mathbb{P}}:= (\overrightarrow{\mathbb{P}}_x, x > 0)$, is defined by erasing the negative components of the space-time trajectory of the stable process, shunting together the remaining positive sections of path and then killing the resulting process at the first hitting time to the state 0. According to Theorem 5.19, the process $(Z, \overrightarrow{\mathbb{P}})$ is a positive self-similar Markov process with self-similar index α, whose underlying Lévy process $(\overrightarrow{\xi}, \overrightarrow{\mathbf{P}})$ belongs to the class of hypergeometric processes with parameters $(\beta, \gamma, \hat{\beta}, \hat{\gamma}) = (1, \alpha\rho, 1-\alpha, \alpha\hat{\rho})$. We also recall from the discussion below Theorem 5.19 that the censored stable process is conservative if and only if $\alpha \in (0,1]$, otherwise it hits 0 continuously.

In the conservative case, the hypergeometric Lévy process $\overrightarrow{\xi}$ belongs to the class $\mathcal{H}_1$ and, moreover, its dual $-\overrightarrow{\xi}$ is a hypergeometric Lévy process with parameters $(\alpha, \alpha\hat{\rho}, 0, \alpha\rho)$, which belongs to the class $\mathcal{H}_4$ if and only if $\alpha \in (0,1)$. When $\alpha = 1$, the process $\overrightarrow{\xi}$ oscillates and implicitly its associated exponential functional does not converge almost surely. This is one of the difficulties that prevents us from considering this case as techniques we have developed in Chapter 9 do not apply.

In the non-conservative case, that is, $\alpha \in (1,2)$, the hypergeometric Lévy process $\overrightarrow{\xi}$ belongs to the class $\mathcal{H}_2 \setminus \{\hat{\beta} = 0\}$. More precisely, it drifts to $-\infty$ and its exponential functional is well defined, but the behaviour of its tail distribution is not covered in the analysis in this text. That said, the estimates of the tail behaviour will follow from the Mellin transform of its exponential functional which can be deduced from Theorem 4.13 (ii). Such estimates, together with the results in Chapter 9, will allow us to deduce the lower and upper envelopes of the censored stable process $(Z, \overrightarrow{\mathbb{P}})$ at its absorption time $T_0 = \inf\{t > 0 \colon Z_t = 0\}$. In the interests of brevity, however, we leave the details to the interested reader.

Let us thus focus on the case that $\alpha \in (0,1)$. By differentiating the characteristic exponent of $\overrightarrow{\xi}$ and evaluating at 0, it is clear that

$$0 < \overrightarrow{\mathbf{E}}\,[\overrightarrow{\xi}_1] < \infty.$$

Hence, the conditions of Theorem 5.3 are fulfilled, and it makes sense to talk about the censored stable process issued from the origin, that is, $\overrightarrow{\mathbb{P}}_0$ is well defined. Moreover, in the sense of weak convergence on the Skorokhod space (see Section A.10 in the Appendix), $\overrightarrow{\mathbb{P}}_x$ converges to $\overrightarrow{\mathbb{P}}_0$, as $x \to 0$. In other words, we may use the results of Chapter 9 to describe the lower and upper envelope of $(Z, \overrightarrow{\mathbb{P}}_0)$ at 0 and at ∞.

We first describe the lower envelope of $(Z, \overrightarrow{\mathbb{P}}_0)$ at 0 and at ∞. Let us define

$$\hat{I}^{\rightarrow}_{\infty} = \int_0^{\infty} e^{-\alpha \overrightarrow{\xi}_s} ds.$$

From Proposition 4.27, we deduce the upper tail behaviour of the distribution of the exponential functional $\hat{I}^{\rightarrow}_{\infty}$.

Lemma 10.22 *For $\alpha \in (0,1)$ and $\rho \in (0,1)$, we have*

$$\overrightarrow{F}(t) := \overrightarrow{\mathbf{P}}\left(\hat{I}^{\rightarrow}_{\infty} > t\right) \sim \frac{\alpha\Gamma\left(\frac{1-\alpha}{\alpha}\right)}{\Gamma(1-\alpha)} \frac{\sin(\pi\rho)}{\sin(\pi\alpha\rho)} t^{-\frac{1-\alpha}{\alpha}}, \qquad as \quad t \to \infty. \tag{10.39}$$

Proof The proof is similar to that of Lemma 10.11. Recall from the above discussion that $-\overrightarrow{\xi}$ is a hypergeometric Lévy process with parameters $(\beta, \gamma, \hat{\beta}, \hat{\gamma})$ equal to $(\alpha, \alpha\hat{\rho}, 0, \alpha\rho)$ which places it in the class $\mathcal{H}_1 \backslash \{\beta = 1\}$. Next, we appeal to Proposition 4.27, noting that $\hat{\theta}_{\chi} = (1-\alpha)/\alpha$ and $\hat{\theta} = 1 - \alpha$, to give us

$$\overrightarrow{\mathbf{P}}\left(\hat{I}^{\rightarrow}_{\infty} > t\right) \sim \frac{\overrightarrow{\mathtt{M}}\left(\frac{1-\alpha}{\alpha}\right)}{\hat{\psi}^{\rightarrow\prime}(1-\alpha)} t^{-\frac{1-\alpha}{\alpha}}, \qquad t \to \infty, \tag{10.40}$$

where $\overrightarrow{\mathtt{M}}$ denotes the Mellin transform of $\hat{I}^{\rightarrow}_{\infty}$ and $\hat{\psi}^{\rightarrow}(z)$ is the Laplace transform of $-\overrightarrow{\xi}$. Moreover, $\overrightarrow{\mathtt{M}}\,((1-\alpha)/\alpha)$ and $\hat{\psi}^{\rightarrow\prime}(1-\alpha)$ can be computed explicitly. Indeed, from the identities in Theorem 4.13 (i) and 4.37, appealing to some straightforward manipulation using the quasi-periodic properties of double gamma functions (see Appendix A.4) as well as standard properties of gamma functions (see Appendix A.3), we find

$$\overrightarrow{\mathtt{M}}\left(\frac{1-\alpha}{\alpha}\right) = -\Gamma\left(\frac{1-\alpha}{\alpha}\right) \frac{\pi^2}{\Gamma(-\alpha)\Gamma(1-\alpha)\sin(\pi\alpha\rho)\sin(\pi\alpha\hat{\rho})\Gamma(\rho)\Gamma(\hat{\rho})}$$

and

$$\hat{\psi}^{\rightarrow\prime}(1-\alpha) = \frac{\pi}{\Gamma(1-\alpha)\sin(\pi\alpha\hat{\rho})}.$$

The proof follows by combining and tidying up the expressions above using the reflection and recursion formulae for gamma functions. □

For $q \geq 0$, let us introduce $\overrightarrow{T}^{-}(-q) := \{t \geq 0 : -\overrightarrow{\xi}_t \leq -q\}$,

$$\hat{I}^{\rightarrow}_{\overrightarrow{T}^{-}(-q)} := \int_0^{\overrightarrow{T}^{-}(-q)} e^{-\alpha \overrightarrow{\xi}_s} ds \quad \text{and} \quad \overrightarrow{F}_q(t) := \overrightarrow{\mathbf{P}}\left(\hat{I}^{\rightarrow}_{\overrightarrow{T}^{-}(-q)} > t\right).$$

From Lemmas 10.8 and 10.22, we have that $\vec{F}_q\ (t) \asymp \vec{F}\ (t)$ for t large. This equivalence together with Theorems 9.7 and 9.8 imply the following integral tests for lower envelopes.

Theorem 10.23 *Let $\alpha \in (0,1)$. The lower envelope of the censored stable process at 0 and at ∞ is as follows.*

(i) Let f be an increasing function, such that either

$$\lim_{t\to 0} \frac{f(t)^\alpha}{t} = 0 \qquad or \qquad \liminf_{t\to 0} \frac{f(t)^\alpha}{t} > 0,$$

then,

$$\vec{\mathbb{P}}_0\left(Z_t < f(t) \text{ i.o. as } t \to 0\right) = 0 \ \text{ or } \ 1,$$

accordingly as

$$\int_{0+} \frac{f(t)^{1-\alpha}}{t^{\frac{1}{\alpha}}} \mathrm{d}t \quad \text{is finite or infinite.}$$

ii) Let g be an increasing function, such that either

$$\lim_{t\to \infty} \frac{g(t)^\alpha}{t} = 0 \qquad or \qquad \liminf_{t\to \infty} \frac{g(t)^\alpha}{t} > 0,$$

then, for all $x \geq 0$,

$$\vec{\mathbb{P}}_x\left(Z_t < g(t) \text{ i.o. as } t \to \infty\right) = 0 \text{ or } \ 1,$$

accordingly as

$$\int^{\infty} \frac{g(t)^{1-\alpha}}{t^{\frac{1}{\alpha}}} \mathrm{d}t \quad \text{is finite or infinite.}$$

Next, we proceed to describe the upper envelope of the censored stable process at 0 and at ∞ when $\alpha \in (0,1)$. Similarly to previous sections, we first describe the upper envelope of its future infimum $\mathtt{J} = (\mathtt{J}_t \geq 0)$ at 0 and at ∞ and then we compare them with the upper envelopes of the original process.

Recall that $\mathtt{D}_1 = \sup\{t \geq 0 \colon Z_t \leq 1\}$, denotes the last passage time of Z below 1 and consider the random variable $\vec{\Upsilon}$ under $\vec{\mathbf{P}}$, which is independent of $\vec{I}_\infty$ having the same law as $Z_{\mathtt{D}_1-}$, under $\vec{\mathbb{P}}_0$.

The following result gives the behaviour for the lower tail distribution of $(\vec{\Upsilon})^\alpha \vec{I}_\infty$ near 0, with the restriction that $\alpha \notin \mathbb{Q}$. Its proof follows directly from the asymptotic expansion of $\tilde{p}(x)$, as $x \to 0$, given in Theorem 10.5, when the parameters $(\beta, \gamma, \hat{\beta}, \hat{\gamma})$ take the specific values $(\alpha, \alpha\hat{\rho}, 0, \alpha\rho)$.

Lemma 10.24 *Let us assume that $\alpha \notin \mathbb{Q}$. For $\alpha \in (0,1)$ and $\rho \in (0,1)$, we have*

$$\overline{F}^{\leadsto}(t) := \vec{\mathbf{P}}\left((\vec{\Upsilon})^{\alpha}\hat{I}_{\infty}^{\leadsto} < t\right) \sim \frac{\Gamma(1-\alpha)\,\vec{\mathtt{M}}_{\tilde{\Upsilon}}\,(2-\rho)}{\Gamma(1-\alpha\hat{\rho}-\alpha)\Gamma(1-\alpha\rho)}t^{\rho}, \qquad as \quad t \to 0, \tag{10.41}$$

where $\vec{\mathtt{M}}_{\tilde{\Upsilon}}$ denotes the Mellin transform of $(\vec{\Upsilon})^{\alpha}\hat{I}_{\infty}^{\leadsto}$.

The tail behaviour of $\overline{F}^{\leadsto}(t)$ and Proposition 10.9 imply that

$$G^{\leadsto}(t) := \vec{\mathbb{P}}_0\left(S_1 < t\right) \asymp \overline{F}^{\leadsto}(t), \qquad \text{as} \quad t \to 0,$$

where $S_1 = \inf\{t \geq 0 \colon Z_t > 1\}$. Therefore from Theorems 9.7 and 9.8, and Propositions 9.14 and 9.15, we deduce the following integral tests for the upper envelope.

Theorem 10.25 *Let us assume that $\alpha \in (0,1)$ and $\alpha \notin \mathbb{Q}$. The upper envelope of the censored stable process and its future infimum at 0 and at ∞ satisfies:*

(i) Let $f \in C_0$ such that either

$$\lim_{t\to 0}\frac{t}{f(t)^{\alpha}} = 0 \qquad or \qquad \liminf_{t\to 0}\frac{t}{f(t)^{\alpha}} > 0,$$

then,

$$\vec{\mathbb{P}}_0\left(Z_t > f(t) \text{ i.o. as } t \to 0\right) = \vec{\mathbb{P}}_0\left(\mathtt{J}_t > f(t) \text{ i.o. as } t \to 0\right) = 0 \quad or \quad 1,$$

accordingly as

$$\int_{0+}\frac{t^{\rho-1}}{f(t)^{\alpha\rho}}\mathrm{d}t \quad \textit{is finite or infinite.}$$

ii) Let $g \in C_\infty$ such that either

$$\lim_{t\to \infty}\frac{t}{g(t)^{\alpha}} = 0 \qquad or \qquad \liminf_{t\to \infty}\frac{t}{g(t)^{\alpha}} > 0,$$

then, for all $x \geq 0$,

$$\vec{\mathbb{P}}_x\left(Z_t > g(t) \text{ i.o. as } t \to \infty\right) = \vec{\mathbb{P}}_x\left(\mathtt{J}_t > g(t) \text{ i.o. as } t \to \infty\right) = 0 \quad or \quad 1,$$

accordingly as

$$\int^{\infty}\frac{t^{\rho-1}}{g(t)^{\alpha\rho}}\mathrm{d}t \quad \textit{is finite or infinite.}$$

10.6 Isotropic stable processes

We conclude this chapter by studying the lower and upper radial envelopes for the isotropic d-dimensional stable process, denoted here by $(X, \mathbb{P})$. Recall from Section 5.7 that, for any $x > 0$, the radial part of a d-dimensional stable process defined by

$$Z_t = |X_t| \mathbf{1}_{(t<\tau^{\{0\}})}, \qquad t \geq 0,$$

where $\tau^{\{0\}} = \inf\{t > 0 \colon |X_t| = 0\}$. In other words, the radial process is killed and absorbed at its cemetery state whenever X hits 0 for the first time. According to Theorem 5.21, the radial process is a positive self-similar Markov process with self-similar index α, whose underlying Lévy process $(\xi, \mathbf{P})$ has characteristic exponent given by (5.50) such that 2ξ is a hypergeometric process with parameters $(\beta, \gamma, \hat{\beta}, \hat{\gamma}) = (1, \alpha/2, (d-\alpha)/2, \alpha/2)$. We also recall from the discussion below in Theorem 5.21 that the radial part of an isotropic stable process hits 0 continuously if and only if $d = 1$ and $\alpha \in (1, 2)$, otherwise it is conservative.

We first treat the conservative case, i.e when $d \geq \alpha$. The hypergeometric Lévy process 2ξ belongs to the class $\mathcal{H}_1$ and, moreover, its dual -2ξ is an hypergeometric Lévy process with parameters $(\beta, \gamma, \hat{\beta}, \hat{\gamma}) = (1 - (d - \alpha)/2, \alpha/2, 0, \alpha/2)$, which belongs to the class $\mathcal{H}_4$, more precisely, to $\mathcal{H}_1 \backslash \{\beta = 1\}$, if and only if $d > \alpha$. When $d = \alpha = 1$, the process ξ oscillates and its associated exponential functional does not converge almost surely. The techniques we have developed in Chapter 9 therefore cannot help us in the setting $d = \alpha = 1$. Let us assume $d > \alpha$. By differentiating the characteristic exponent of ξ (cf. (5.50)) and evaluating at 0, it is clear that

$$0 < \mathbf{E}\left[\xi_1\right] < \infty,$$

which is consistent with the fact that $\lim_{t\to\infty} |X_t| = \infty$ almost surely. Note, in this setting, we may more simply identify $Z_t = |X_t|$, $t \geq 0$. The conditions of Theorem 5.3 are fulfilled, hence the process $(|X|, \mathbb{P}_x)$ converges towards $(|X|, \mathbb{P}_0)$, as x tends to 0, in the sense of weak convergence on the Skorokhod space (see Section A.10 in the Appendix); this is a completely obvious statement anyway. We can accordingly use the results of Chapter 9 to describe the lower and upper envelope of $(|X|, \mathbb{P}_0)$ at 0 and at ∞.

We first describe the lower envelope of $(|X|, \mathbb{P}_0)$ at 0 and at ∞, for $d > \alpha$. Let us define

$$\hat{I}_\infty := \int_0^\infty \mathrm{e}^{-\alpha\xi_s} \mathrm{d}s = \int_0^\infty \mathrm{e}^{-\frac{\alpha}{2}(2\xi_s)} \mathrm{d}s. \tag{10.42}$$

From Proposition 4.27, we deduce the upper tail behaviour of the distribution of the exponential functional $\hat{I}_\infty$.

Lemma 10.26 *For $d > \alpha$, we have*

$$F(t) := \mathbf{P}\left(\hat{I}_\infty > t\right) \sim \frac{\Gamma\left(\frac{\alpha}{2}\right)\Gamma\left(\frac{d-\alpha}{\alpha}\right)}{\Gamma\left(\frac{d-\alpha}{2}\right)\Gamma\left(\frac{d}{2}\right)} t^{-\frac{d-\alpha}{\alpha}}, \qquad as \quad t \to \infty. \tag{10.43}$$

Proof The proof is similar to those of Lemmas 10.11 and 10.22. Given the properties of -2ξ listed above, taking account of the second equality in (10.42), we may appeal to Proposition 4.27 with $\delta = \alpha/2$, $\hat{\theta}_\chi = (d-\alpha)/\alpha$ and $\hat{\theta} = (d-\alpha)/2$, to write

$$\mathbf{P}\left(\hat{I}_\infty > t\right) \sim \frac{\hat{\mathtt{M}}\left(\frac{d-\alpha}{\alpha}\right)}{\hat{\psi}'\left(\frac{d-\alpha}{2}\right)} t^{-\frac{d-\alpha}{\alpha}}, \qquad t \to \infty, \tag{10.44}$$

where $\hat{\mathtt{M}}$ denotes the Mellin transform of $\hat{I}_\infty$ and $\hat{\psi}(z)$ is the Laplace transform of -2ξ. Moreover, in this case $\hat{\mathtt{M}}((d-\alpha)/\alpha)$ and $\hat{\psi}'((d-\alpha)/2)$ can be computed explicitly. Indeed from the identities in Corollary 4.16 (i) and 4.37, we find

$$\hat{\mathtt{M}}\left(\frac{d-\alpha}{\alpha}\right) = \Gamma\left(\frac{d-\alpha}{\alpha}\right)\frac{\Gamma^2\left(\frac{\alpha}{2}\right)}{\Gamma^2\left(\frac{d-\alpha}{2}\right)} \qquad \text{and} \qquad \hat{\psi}'\left(\frac{d-\alpha}{2}\right) = \frac{\Gamma\left(\frac{\alpha}{2}\right)\Gamma\left(\frac{d}{2}\right)}{\Gamma\left(\frac{d-\alpha}{2}\right)},$$

thus concluding the proof. □

For $q \geq 0$, let us introduce $\hat{T}^-(-q) := \{t \geq 0 \colon \xi_t \leq -q\}$,

$$\hat{I}_{\hat{T}^-(-q)} := \int_0^{\hat{T}^-(-q)} \mathrm{e}^{-\alpha\xi_s}\mathrm{d}s \quad \text{and} \quad F_q(t) := \mathbf{P}\left(\hat{I}_{\hat{T}^-(-q)} > t\right).$$

From Lemmas 10.26 and 10.8, we have that $F_q(t) \asymp F(t)$ for t large. Together with Theorems 9.7 and 9.8, we obtain the following integral tests for the lower envelopes of $|X|$.

Theorem 10.27 *Let $d>\alpha$. The lower envelope of the radial isotropic d-dimensional stable process at 0 and at ∞ is as follows.*

(i) Let f be an increasing function, such that either

$$\lim_{t\to 0}\frac{f(t)^\alpha}{t} = 0 \qquad or \qquad \liminf_{t\to 0}\frac{f(t)^\alpha}{t} > 0,$$

then,

$$\mathbb{P}_0\left(|X_t| < f(t) \text{ i.o. as } t \to 0\right) = 0 \ \text{ or } \ 1,$$

accordingly as

$$\int_{0+}\frac{f(t)^{d-\alpha}}{t^{\frac{d}{\alpha}}}\mathrm{d}t \quad \text{is finite or infinite.}$$

ii) Let g be an increasing function, such that either

$$\lim_{t\to\infty}\frac{g(t)^\alpha}{t}=0 \qquad or \qquad \liminf_{t\to\infty}\frac{g(t)^\alpha}{t}>0,$$

then, for all $x \geq 0$,

$$\mathbb{P}_x\Big(|X_t|<g(t)\ i.o.\ as\ t\to\infty\Big)=\ 0\ or\ \ 1\,,$$

accordingly as

$$\int^\infty \frac{g(t)^{d-\alpha}}{t^{\frac{d}{\alpha}}}\mathrm{d}t \quad \textit{is finite or infinite.}$$

Remaining in the setting that $\alpha < d$, we proceed to describe the upper envelope as $t \to 0$ and $t \to \infty$, again, first taking account of the upper envelope of $\mathtt{J} = (\mathtt{J}_t \geq 0)$, where $\mathtt{J}_t = \inf_{s\geq t}|X_s|$, $t \geq 0$. Recall that $\mathtt{D}_1 = \sup\{t \geq 0\colon |X_t| \leq 1\}$, denotes the last passage time of $|X|$ below 1; and consider the random variable Υ which is independent of $\hat{I}_\infty$ and has the same law, under $\mathbf{P}$, as $|X_{\mathtt{D}_1-}|$, under $\mathbb{P}_0$. We recall that Υ takes values on $[0, 1]$.

The following result gives the behaviour for the lower tail distribution of $\Upsilon^\alpha\hat{I}_\infty$ near 0. Its proof follows directly from the asymptotic expansion of $\tilde{p}(x)$, for x close to 0, given in Corollary 10.6, when the parameters $(\beta, ,\hat{\beta}, \hat{\gamma})$ of the hypergeometric process -2ξ take the specific values $(1-(d-\alpha)/2, \alpha/2, 0, \alpha/2)$. We also recall that $\delta = \alpha/2$ in this case.

Lemma 10.28 *For $d > \alpha$, we have*

$$\overline{F}(t) := \mathbf{P}\left(\Upsilon^\alpha\hat{I}_\infty < t\right) \sim \frac{\Gamma\left(\frac{d}{2}\right)}{\Gamma\left(\frac{d-\alpha}{2}\right)\Gamma\left(1-\frac{\alpha}{2}\right)}t, \qquad as \quad t\to 0. \tag{10.45}$$

Lemma 10.28 and Proposition 10.9 imply that

$$G(t) := \mathbb{P}_0\Big(S_1 < t\Big) \asymp \overline{F}(t), \qquad \text{as} \quad t\to 0,$$

where $S_1 = \inf\{t \geq 0\colon |X_t| > 1\}$. Therefore from Theorems 9.7, 9.8 and Propositions 9.14, 9.15, we deduce the following integral test.

Theorem 10.29 *Let us assume that $d > \alpha$. The upper envelope of the radial isotropic d-dimensional stable process and its future infimum at 0 and at ∞ is as follows.*

(i) Let $f \in \mathcal{C}_0$ such that either

$$\lim_{t\to 0}\frac{t}{f(t)^\alpha}=0 \qquad or \qquad \liminf_{t\to 0}\frac{t}{f(t)^\alpha}>0,$$

then,

$$\mathbb{P}_0\Big(|X_t| > f(t) \text{ i.o. as } t \to 0\Big) = \mathbb{P}_0\Big(\mathtt{J}_t > f(t) \text{ i.o. as } t \to 0\Big) = 0 \ \text{ or } \ 1,$$

accordingly as

$$\int_{0+} \frac{\mathrm{d}t}{f(t)^\alpha} \quad \text{is finite or infinite.}$$

(ii) Let $g \in C_\infty$ such that either

$$\lim_{t\to\infty} \frac{t}{g(t)^\alpha} = 0 \qquad \text{or} \qquad \liminf_{t\to\infty} \frac{t}{g(t)^\alpha} > 0,$$

then, for all $x \geq 0$,

$$\mathbb{P}_x\Big(|X_t| > g(t) \text{ i.o. as } t \to \infty\Big) = \mathbb{P}_x\Big(\mathtt{J}_t > g(t) \text{ i.o. as } t \to \infty\Big) = 0 \ \text{ or } \ 1,$$

accordingly as

$$\int^{\infty} \frac{\mathrm{d}t}{g(t)^\alpha} \quad \text{is finite or infinite.}$$

Now, we consider the remaining case, that is, when $d = 1$ and $\alpha \in (1, 2)$. In this case the process 2ξ is a hypergeometric process with parameters $(\beta, \gamma, \hat\beta, \hat\gamma) = (1, \alpha/2, (1-\alpha)/2, \alpha/2) \in \mathcal{H}_4$, Moreover its dual -2ξ is a hypergeometric process whose parameters $(\beta, \gamma, \hat\beta, \hat\gamma)$ are equal to $((1+\alpha)/2, \alpha/2, 0, \alpha/2)$. Moreover,

$$0 < -\mathbf{E}\left[\xi_1\right] < \infty. \tag{10.46}$$

Let us then consider the dual of the radial isotropic stable process $|X_t|\mathbf{1}_{(t<\tau^{\{0\}})}$, $t \geq 0$, with $\alpha \in (1, 2)$ (see Proposition 5.5), denoted here by $(Z, \hat P_x)$. The latter is a self-similar Markov process with index $\alpha > 0$, whose Lamperti representation is given by

$$Z_t = x \exp\left\{\hat\xi_{\hat\varphi(tx^{-\alpha})}\right\}, \qquad t \geq 0,$$

where $\hat\varphi$ is the right-continuous inverse of

$$\hat I_t = \int_0^t \mathrm{e}^{-\alpha\hat\xi_u} \mathrm{d}u, \qquad t \geq 0.$$

Since $\hat\xi$ has a positive finite mean, see (10.46), the conditions of Theorem 5.3 are fulfilled and the process $(Z, \hat P_x)$ converges towards $(Z, \hat P_0)$, as x goes to 0, in the sense of weak convergence on the Skorokhod space (see Section A.10 in the Appendix). The time reversal property of Proposition 5.6 tells us that for any $x > 0$ and $z \in (0, x]$, the returned process $(|X_{(\tau^{\{0\}}-t)-}|, 0 \leq t \leq \tau^{\{0\}})$, under $\mathbb{P}_z$, has the same law as $(Z_t, 0 \leq t < \mathtt{D}_x)$, under $\hat P_0(\cdot \,|\, \Gamma = z)$.

Let us define

$$I_\infty = \int_0^\infty e^{\alpha\xi_u}du = \int_0^\infty e^{\frac{\alpha}{2}(2\xi_u)}du,$$

thus from Proposition 4.27, we deduce the upper tail behaviour of the distribution of the exponential functional I_∞.

Lemma 10.30 *For $d = 1$ and $\alpha \in (1, 2)$, we have*

$$F(t) := \mathbf{P}(I_\infty > t) \sim \frac{\alpha}{2(\alpha-1)\sqrt{\pi}} \frac{\sin\left(\frac{\pi}{\alpha}\right)\Gamma\left(\frac{3-\alpha}{2}\right)}{\Gamma\left(\frac{\alpha+1}{\alpha}\right)\Gamma\left(\frac{\alpha}{2}\right)} t^{-\frac{\alpha-1}{\alpha}}, \qquad as \quad t \to \infty. \tag{10.47}$$

Proof We proceed similarly as in Lemma 10.26. First, we recall from the above discussion that 2ξ is a hypergeometric Lévy process with parameters $(\beta, \gamma, \hat{\beta}, \hat{\gamma}) = (1, \alpha/2, (1-\alpha)/2, \alpha/2)$ which belongs to the class $\mathcal{H}_4$ defined in (4.35) (more precisely, it belongs to $\mathcal{H}_2 \cap \{\eta - \hat{\gamma} > 0\} \setminus \{\beta = 1\}$). Next, we appeal to Proposition 4.27, taking $\delta = \alpha/2$, $\hat{\theta}_X = (\alpha-1)/\alpha$ and $\hat{\theta} = (\alpha-1)/2$, to write

$$\mathbf{P}(I_\infty > t) \sim \frac{\mathtt{M}\left(\frac{\alpha-1}{\alpha}\right)}{\psi'\left(\frac{\alpha-1}{2}\right)} t^{-\frac{\alpha-1}{\alpha}}, \qquad as \qquad t \to \infty, \tag{10.48}$$

where $\mathtt{M}$ denotes the Mellin transform of I_∞ and $\psi(z)$ is the Laplace transform of 2ξ. Moreover, in this case $\mathtt{M}((\alpha-1)/\alpha)$ and $\psi'((\alpha-1)/2)$ can be computed explicitly. Indeed from the identities in Corollary 4.16 (ii) and 4.37, we find

$$\mathtt{M}\left(\frac{\alpha-1}{\alpha}\right) = \alpha \frac{\sin\left(\frac{\pi}{\alpha}\right)}{4\Gamma\left(\frac{\alpha+1}{\alpha}\right)} \qquad \text{and} \qquad \psi'\left(\frac{d-\alpha}{2}\right) = \frac{\alpha-1}{2} \frac{\sqrt{\pi}\Gamma\left(\frac{\alpha}{2}\right)}{\Gamma\left(\frac{3-\alpha}{2}\right)},$$

thus concluding the proof. □

For $q \geq 0$, we also introduce $T^-(-q) := \{t \geq 0 : \xi_t \leq -q\}$,

$$I_{T^-(-q)} := \int_0^{T^-(-q)} e^{\alpha\xi_s}ds \quad \text{and} \quad F_q(t) := \mathbf{P}\left(I_{T^-(-q)} > t\right).$$

From Lemmas 10.30 and 10.8, we have that $F_q(t) \asymp F(t)$ for t large. Thus, the previous observations and Theorems 9.7 and 9.8 imply the following integral test for the lower envelope of $(Z, \hat{P}_z)$, or equivalently, the radial isotropic stable process reversed from its first visit to the origin.

Theorem 10.31 *Let $d = 1$ and $\alpha \in (1, 2)$. The lower envelope of the radial isotropic stable processes near absorption is as follows. Let f be an increasing function, such that either*

$$\lim_{t\to 0}\frac{f(t)^\alpha}{t}=0 \qquad or \qquad \liminf_{t\to 0}\frac{f(t)^\alpha}{t}>0,$$

then, for any $x>0$,

$$\mathbb{P}_x\Big(|X_{(\tau^{\{0\}}-t)-}|<f(t) \text{ i.o. as } t\to 0\Big)=0 \ \ or \ \ 1,$$

accordingly as

$$\int_{0+}\frac{f(t)^{\alpha-1}}{t^{\frac{2\alpha-1}{\alpha}}}\mathrm{d}t \quad \text{is finite or infinite.}$$

Finally, we describe the upper envelope near absorption at the origin of the radial isotropic stable process with $\alpha\in(1,2)$. Let us introduce its past infimum here denoted by $(\mathtt{m}_t, t\geq 0)$ where

$$\mathtt{m}_t=\inf_{s\leq t}|X_s|\mathbf{1}_{(s<\tau^{\{0\}})}, \qquad t\geq 0.$$

We also consider the random variable $\hat{\Upsilon}$ which is independent of I_∞ and has the same law, under $\mathbf{P}$, as Z_{D_1-}, under $\hat{P}_0$. We recall that Υ takes values on $[0,1]$.

The following result gives the behaviour for the lower tail distribution of $\hat{\Upsilon}^\alpha I_\infty$ near 0. As usual, its proof follows directly from the asymptotic expansion of $\tilde{p}(x)$, for x close to 0, given in Proposition 10.7, when the parameters $(\beta,\gamma,\hat{\beta},\hat{\gamma})$ of the hypergeometric process 2ξ take the specific values $(1,\alpha/2,(1-\alpha)/2,\alpha/2)$. We also recall that $\delta=\alpha/2$.

Lemma 10.32 *For* $d=1$ *and* $\alpha\in(1,2)$, *we have*

$$\overline{F}(t):=\mathbf{P}\left(\hat{\Upsilon}^\alpha I_\infty<t\right)\sim\frac{\alpha}{2\sqrt{\pi}}\frac{\sin\left(\frac{\alpha-1}{\alpha}\pi\right)\Gamma\left(\frac{3-\alpha}{2}\right)}{\Gamma\left(\frac{1}{\alpha}\right)\Gamma\left(\frac{2-\alpha}{2}\right)}t^{\frac{1}{\alpha}}, \qquad as \quad t\to 0. \quad (10.49)$$

Lemma 10.32 and Proposition 10.9 imply that

$$G(t):=\hat{P}_0\big(S_1<t\big)\asymp\overline{F}(t), \qquad \text{as} \quad t\to 0,$$

where $S_1=\inf\{t\geq 0\colon Z_t>1\}$. Therefore from Theorems 9.7 and 9.8, and Propositions 9.14 and 9.15, together with the time reversal property of the radial isotropic stable process from the hitting time $\tau^{\{0\}}$, described above, we get the following result.

Theorem 10.33 *Let us assume that* $d=1$ *and* $\alpha\in(1,2)$. *The upper envelopes of the radial isotropic stable process and its past infimum near absorption are as follows. Let* $f\in C_0$ *such that either*

$$\lim_{t\to 0}\frac{t}{f(t)^\alpha}=0 \qquad or \qquad \liminf_{t\to 0}\frac{t}{f(t)^\alpha}>0,$$

then, for any $x > 0$,

$$\mathbb{P}_x\Big(|X_{(\tau^{\{0\}}-t)^-}| > f(t) \text{ i.o. as } t \to 0\Big) = \mathbb{P}_x\Big(\mathfrak{m}_{(\tau^{\{0\}}-t)^-} > f(t) \text{ i.o. as } t \to 0\Big) = 0 \quad or \quad 1,$$

accordingly as

$$\int_{0+} \frac{t^{\frac{1-\alpha}{\alpha}}}{f(t)} \mathrm{d}t \quad \text{is finite or infinite.}$$

10.7 Comments

There are few integral tests for lower and upper envelopes for path transformations of stable processes. The earliest work in this area is Khintchine [105], who considered the upper envelope of the radial part of one-dimensional stable processes. This corresponds to Theorem 10.29 in the case when $\alpha < 1$. Monrad and Silverstein [152] were the first to study the lower and upper envelopes for the stable processes conditioned to stay positive, as $t \to 0$. In particular, they obtained a law of the iterated logarithm in the spectrally negative case. The results in Monrad and Silverstein [152] are stated in terms of the stable process immediately proceeding a local minimum. Although unknown at the time that Monrad and Silverstein [152] was published, the sample paths are the same as those of the stable process conditioned to stay positive, as $t \to 0$. Fourati [72] described the lower and upper envelopes for the stable process conditioned to stay positive and its future infimum, as $t \to 0$ and $t \to \infty$, under the assumption the process has two-sided jumps. Theorems 10.12, 10.14 and 10.15 are based on these results. One negative aspect of the method we pursue here is the need for $\alpha \notin \mathbb{Q}$ for some of the upper envelopes. This restriction does not appear in the work of Monrad and Silverstein [152] or Fourati [72]. The laws of the iterated logarithm of the spectrally positive stable processes conditioned to stay positive, its future infimum and its reflection at its future infimum, as $t \to 0$ and $t \to \infty$ in Theorems 10.17 and 10.18 are taken from Pardo [160]. Theorem 10.27 is taken from Takeuchi [205, 206] where the lower envelope of isotropic stable processes were studied, as $t \to 0$ and $t \to \infty$, respectively. The rest of the results presented here appear to be a new contribution to the literature.

11

Markov additive and self-similar Markov processes

We would like to understand self-similar Markov processes that explore the real line (resp. $\mathbb{R}^d$). That is to say, we are interested in the class of stochastic processes that respect Definition 5.1, albeit the state-space is taken as $\mathbb{R}$ (resp. $\mathbb{R}^d$) in place of $[0, \infty)$.

Like the case of pssMp, it is possible to describe so-called *real self-similar Markov processes*, or *rssMp* for short, via a space-time transformation to another family of stochastic processes. Whereas pssMp are connected to Lévy processes via the Lamperti space-time transformation, rssMp turn out to be connected to a family of stochastic processes whose dynamics are those of a Lévy process with characteristics that change each time an auxiliary and independent Markov chain changes state. Such a process is known as a Markov modulated (Lévy) process or *Markov additive process* (*MAP* for short). The picture for $\mathbb{R}^d$-valued self-similar Markov processes (or just *ssMp* for short) is a further generalisation of this representation in which the Markov chain is replaced by a general Markov process.

As with Chapter 5, our interest in ssMp comes about through their relationship with stable processes in all dimensions and their path transformations. We discuss this relationship in the next chapter. However, in preparation for that, we spend most of this chapter discussing how MAPs are naturally connected to ssMps. We also address some of the many intrinsic properties of MAPs that will be of future use.

11.1 MAPs and the Lamperti–Kiu transform

As alluded to above, a *real self-similar Markov process* with *self-similarity index* $\alpha > 0$ is a regular Feller process, $Z = (Z_t, t \geq 0)$, on $\mathbb{R}\backslash\{0\}$ such that the origin is a cemetery state, which has the property that its probability laws P_x, $x \in \mathbb{R}$ satisfy the *scaling property* that for all $x \in \mathbb{R} \setminus \{0\}$ and $c > 0$,

$$\text{the law of } (cZ_{tc^{-\alpha}}, t \geq 0) \text{ under } P_x \text{ is } P_{cx}. \tag{11.1}$$

The reader may like to visit Section A.11 in the Appendix for a reminder of what a Feller process is.

In the spirit of the Lamperti transform of the previous chapter, we are able to identify each rssMp with a so-called Markov additive process via a transformation of space and time, known as the *Lamperti–Kiu representation*. We shall shortly describe this transformation in detail. However, first we must make clear what we mean by a Markov additive process.

Definition 11.1 Let E be a finite state space such that $|E| = N$. A regular Feller process, $(\xi, J) = ((\xi_t, J_t), t \geq 0)$ on $\mathbb{R} \times E$ with probabilities $\mathbf{P}_{x,i}$, $x \in \mathbb{R}$, $i \in E$, and cemetery state $(-\infty, \dagger)$, which is always visited simultaneously by the pair (ξ, J), is called a *(killed) Markov additive process* if the pair (ξ, J) is such that for any $i, j \in E$, $s, t \geq 0$ and bounded and measurable $f \colon \mathbb{R} \times E \to \mathbb{R}$,

$$\mathbf{E}_{x,i}[f(\xi_{t+s} - \xi_t, J_{t+s})\mathbf{1}_{(t+s<\varsigma)}|\mathcal{G}_t] = \mathbf{E}_{0,j}[f(\xi_s, J_s)\mathbf{1}_{(s<\varsigma)}] \tag{11.2}$$

on the event $\{J_t = j, t < \varsigma\}$, where $\varsigma = \inf\{t > 0 \colon J_t = \dagger\}$ and $(\mathcal{G}_t, t \geq 0)$ is the filtration generated by (ξ, J) with natural enlargement. The process J is thus a Markov chain on E and is called the *modulator* of ξ, whereas the latter is called the *ordinator*.

If μ is a probability distribution on E, we write $\mathbf{P}_{x,\mu} = \sum_{i\in E} \mu_i \mathbf{P}_{x,i}$. We adopt a similar convention for expectations.

The following proposition gives a characterisation of MAPs in terms of a mixture of Lévy processes, a Markov chain and a family of additional jump distributions.

Proposition 11.2 *The pair (ξ, J) is a Markov additive process if and only if, for each $i, j \in E$, there exist a sequence of i.i.d. Lévy processes $(\xi^{i,n})_{n\geq 0}$ and a sequence of i.i.d. random variables $(\Delta^n_{i,j})_{n\geq 0}$, independent of the chain J, such that if $\sigma_0 = 0$ and $(\sigma_n)_{n\geq 1}$ are the jump times of J prior to ς, the process ξ has the representation*

$$\xi_t = \mathbf{1}_{(n>0)}(\xi_{\sigma_n -} + \Delta^n_{J(\sigma_n -), J(\sigma_n)}) + \xi^{J(\sigma_n),n}_{t-\sigma_n}, \qquad t \in [\sigma_n, \sigma_{n+1}),\ n \geq 0,$$

and $\xi_\varsigma = -\infty$.

We are now ready to describe the connection between MAPs and real positive self-similar Markov processes which are killed at the origin. The next theorem generalises its counterpart for positive self-similar Markov processes, namely Theorem 5.2.

Theorem 11.3 (Lamperti–Kiu transform) *Fix $\alpha > 0$. The process Z is a rssMp with index α if and only if there exists a (killed) MAP, (ξ, J), on $\mathbb{R} \times \{-1, 1\}$ such that*

$$Z_t = e^{\xi_{\varphi(t)}} J_{\varphi(t)}, \qquad 0 \leq t < I_\varsigma,$$

where

$$\varphi(t) = \inf\left\{ s > 0 \colon \int_0^s e^{\alpha \xi_u} \, du > t \right\}, \qquad 0 \leq t < I_\varsigma, \tag{11.3}$$

and $I_\varsigma := \int_0^\varsigma e^{\alpha \xi_s} ds$ is the lifetime of Z until absorption at the origin, which acts as a cemetery state. Recall that ς is given in Definition 11.1. Here, we interpret $\exp\{-\infty\} := 0$ and $\inf \emptyset := \infty$.

Intuitively speaking, the relationship of the MAP (ξ, J) to the rssMp Z is that, up to a time change, J dictates the sign of Z, whereas $\exp\{\xi\}$ dictates the radial distance of Z from the origin.

By comparing Definition 5.1 with that of the definition in (11.1), we see that any pssMp is a rssMp. Indeed, the chain J is such that it never leaves the state $+1$, unless to visit the cemetery state † (corresponding to exponential killing). We consider the former to be a degenerate case of the latter. It turns out that there are other 'degenerate' cases in which a rssMp can change sign at most once.

In the forthcoming discussion, we want to rule out these and other cases. Said another way, we shall henceforth only consider rssMp which have the property that

$$P_x(\exists t > 0 \colon Z_t Z_{t-} < 0) = 1, \qquad \text{for all} \quad x \neq 0. \tag{11.4}$$

The forthcoming Remark 11.8 will help shed some light on other implications of this assumption.

11.2 Distributional and path properties of MAPs

The Lamperti–Kiu transform for rssMp resembles the Lamperti transform for pssMp even more closely when one considers how mathematically close MAPs are to Lévy processes. Let us spend a little time in this section dwelling on this fact. This will also be of use shortly when we look at some explicit examples of the Lampert–Kiu transform. We will assume throughout this section that (ξ, J) is back in the setting of Definition 11.1.

For each $i \in E$, it will be convenient to define, on the same probability space, ξ^i as a Lévy process, which does not have monotone paths, whose law is common to the processes $\xi^{i,n}$, $n \geq 1$, which appear in the definition of Proposition 11.2. Similarly, for each $i, j \in E$, define $\Delta_{i,j}$ to be a random variable whose law is common to the variables $\Delta^n_{i,j}$.

Henceforth, we confine ourselves to the setting that J is an ergodic Markov chain. Let the state space E be the finite set $\{1, \ldots, N\}$, for some $N \in \mathbb{N}$. Denote the transition rate matrix of the chain J by $\boldsymbol{Q} = (q_{i,j})_{i,j \in E}$. For each $i \in E$, the characteristic exponent of the Lévy process ξ^i will be written Ψ_i. For each pair of $i, j \in E$, define the Fourier transform $G_{i,j}(z) = \mathbf{E}[\mathrm{e}^{\mathrm{i}z\Delta_{i,j}}]$ of the jump distribution $\Delta_{i,j}$. Write $\boldsymbol{G}(z)$ for the $N \times N$ matrix whose (i, j)th element is $G_{i,j}(z)$. We will adopt the convention that $\Delta_{i,j} = 0$ if $q_{i,j} = 0$, $i \neq j$, and also set $\Delta_{ii} = 0$ for each $i \in E$.

Thanks to Proposition 11.2, we can use the components in the previous paragraph to write down an analogue of the characteristic exponent of a Lévy process. Define the matrix-valued function

$$\boldsymbol{\Psi}(z) = \mathrm{diag}(-\Psi_1(z), \ldots, -\Psi_N(z)) + \boldsymbol{Q} \circ \boldsymbol{G}(z), \tag{11.5}$$

for all $z \in \mathbb{R}$, where $\circ$ indicates elementwise multiplication, also called Hadamard multiplication. It is then known that

$$\mathbf{E}_{0,i}\left[\mathrm{e}^{\mathrm{i}z\xi_t}; J_t = j\right] = (\mathrm{e}^{\boldsymbol{\Psi}(z)t})_{i,j}, \qquad i, j \in E, t \geq 0, \tag{11.6}$$

for all $z \in \mathbb{R}$. Accordingly, $\boldsymbol{\Psi}$ is called the *(characteristic) matrix exponent* of the MAP (ξ, J).

One aspect of the theory of MAPs which differs slightly from that of Lévy processes concerns *duality*. Recall from Section 2.11 that, for a Lévy process, the dual process describes the time-reversed process over a finite time horizon. Thanks to stationary and independent increments, this turns out to be nothing more than the negative of the Lévy process. The situation for MAPs is a little more involved.

First note that, thanks to irreducibility, the Markov chain J necessarily has a stationary distribution. We denote it by the vector $\boldsymbol{\pi} = (\pi_1, \ldots, \pi_N)$. Each MAP has a dual process, which is also a MAP. Its associated probabilities, $\hat{\mathbf{P}}_{x,i}$, $x \in \mathbb{R}$, $i \in E$, are determined by the dual characteristic matrix exponent, given by

$$\hat{\boldsymbol{\Psi}}(z) := \mathrm{diag}(- \Psi_1(-z), \ldots, -\Psi_N(-z)) + \hat{\boldsymbol{Q}} \circ \boldsymbol{G}(-z)^{\mathrm{T}},$$

and $\hat{\boldsymbol{Q}}$ is the intensity matrix of the modulating Markov chain on E with entries given by

$$\hat{q}_{i,j} = \frac{\pi_j}{\pi_i} q_{j,i}, \qquad i, j \in E.$$

Note that the latter can also be written $\hat{Q} = \boldsymbol{\Delta}_\pi^{-1} Q^{\mathrm{T}} \boldsymbol{\Delta}_\pi$, where $\boldsymbol{\Delta}_\pi = \mathrm{diag}(\pi)$, the matrix with diagonal entries given by π and zeros everywhere else. The matrix $\hat{Q}$ is the intensity matrix of the time-reversed Markov chain J. Hence, when it exists,

$$\hat{\boldsymbol{\Psi}}(z) = \boldsymbol{\Delta}_\pi^{-1} \boldsymbol{\Psi}(-z)^{\mathrm{T}} \boldsymbol{\Delta}_\pi, \tag{11.7}$$

showing that

$$\pi_i \hat{\mathbf{E}}_{0,i}\left[\mathrm{e}^{\mathrm{i}z\xi_t}, J_t = j\right] = \pi_j \mathbf{E}_{0,j}\left[\mathrm{e}^{-\mathrm{i}z\xi_t}, J_t = i\right]. \tag{11.8}$$

At the level of processes, one can understand (11.8) as saying the following.

Lemma 11.4 *The time-reversed process $((\xi_{(t-s)-} - \xi_t, J_{(t-s)-}), s \le t)$ under $\mathbf{P}_{0,\pi}$ is equal in law to $((\xi_s, J_s), s \le t)$ under $\hat{\mathbf{P}}_{0,\pi}$.*

One feature of MAPs that will become of concern later on in this book pertains to the difference between the MAP corresponding to a matrix exponent $\boldsymbol{\Psi}$ and another MAP whose matrix exponent can be written $\boldsymbol{\Delta}_{\boldsymbol{a}}\boldsymbol{\Psi}$, where $\boldsymbol{\Delta}_{\boldsymbol{a}}$ is a diagonal matrix with strictly positive entries $a(i) > 0$, $i \in E$. Quite simply, this transformation in $\boldsymbol{\Psi}$ corresponds to a simple time change in (ξ, J). Specifically, whilst in state $i \in E$, time runs at the new speed $a(i)t$. More precisely, if (ξ', J') is the MAP corresponding to $\boldsymbol{\Delta}_{\boldsymbol{a}}\boldsymbol{\Psi}$ then

$$(\xi'_t, J'_t) = \left(\xi_{\int_0^t a(J_s)\mathrm{d}s}, J_{\int_0^t a(J_s)\mathrm{d}s}\right), \qquad t \ge 0.$$

Suppose now that we take $E = \{1, -1\}$. It is also interesting to ask how the rssMp associated to the MAP (ξ', J'), say Z', relates to the one associated to (ξ, J), previously denoted by Z. Suppose, without loss of generality, that $Z_0 > 0$. By considering the Lamperti representation of Z' until it first crosses below the origin, say $T' = \inf\{t > 0 \colon Z'_t < 0\}$, we note that, on $\{t < T'\}$,

$$\begin{aligned}
\varphi'(t) &= \inf\left\{s > 0\colon \int_0^s \mathrm{e}^{\alpha\xi'_u}\mathrm{d}u > t\right\}\\
&= \inf\left\{s > 0\colon \int_0^s \mathrm{e}^{\alpha\xi_{a(1)u}}\mathrm{d}u > t\right\}\\
&= a(1)^{-1}\inf\left\{a(1)s > 0\colon \int_0^{a(1)s} \mathrm{e}^{\alpha\xi_r}\mathrm{d}r > a(1)t\right\}\\
&= a(1)^{-1}\varphi(a(1)t).
\end{aligned}$$

Hence, as $J'_{\varphi'(t)} = J_{\varphi(a(1)t)} = 1$ (neither J' nor J have left their initial state on $\{t < T'\}$), we have that, on $\{t < T\}$,

$$\mathrm{e}^{\xi'_{\varphi'(t)}} J'_{\varphi'(t)} = \mathrm{e}^{\xi_{\varphi(a(1)t)}} J_{\varphi(a(1)t)}.$$

Now appealing to the Markov property we can extrapolate this argument to positive and negative segments of the path of Z' and deduce that

$$Z'_t = Z_{\int_0^t a(\mathrm{sgn}(Z_s))\mathrm{d}s}, \qquad t \geq 0.$$

As is the case with the characteristic exponent of a Lévy process, the characteristic matrix exponent $\boldsymbol{\Psi}(z)$ may be extended as an analytic function on $\mathbb{C}$ to a larger domain than $\mathbb{R}$, depending on where the moments of ξ are well defined.

Proposition 11.5 *Suppose that $z \in \mathbb{C}$ is such that $\boldsymbol{F}(z) := \boldsymbol{\Psi}(-\mathrm{i}z)$ is defined. Then, the matrix $\boldsymbol{F}(z)$ has a real simple eigenvalue $\chi(z)$, which is larger than the real part of all its other eigenvalues. Furthermore, the corresponding right-eigenvector $\boldsymbol{v}(z) = (v_1(z), \ldots, v_N(z))$ has strictly positive entries and may be normalised such that*

$$\boldsymbol{\pi} \cdot \boldsymbol{v}(z) = 1. \tag{11.9}$$

In the spirit of Section 2.8, it will also be important for us to understand how one may establish Esscher-type changes of measure for MAPs.

Proposition 11.6 *For $x \in \mathbb{R}$ and $i \in E$, define*

$$M_t(x, i) := \mathrm{e}^{\gamma(\xi_t - x) - \chi(\gamma)t} \frac{v_{J_t}(\gamma)}{v_i(\gamma)}, \qquad t \geq 0, \tag{11.10}$$

for some $\gamma \in \mathbb{R}$ with $\chi(\gamma)$ well defined. Then, $(M_t(x, i),\ t \geq 0)$, is a unit-mean martingale with respect to $(\mathcal{G}_t, t \geq 0)$. Moreover, under the change of measure

$$\left.\frac{\mathrm{d}\mathbf{P}^{\gamma}_{x,i}}{\mathrm{d}\mathbf{P}_{x,i}}\right|_{\mathcal{G}_t} = M_t(x, i), \qquad t \geq 0,$$

the process (ξ, J) remains in the class of MAPs and, where defined, its matrix characteristic exponent is given by

$$\boldsymbol{\Psi}_{\gamma}(z) = \boldsymbol{\Delta}_{v}(\gamma)^{-1} \boldsymbol{\Psi}(z - \mathrm{i}\gamma) \boldsymbol{\Delta}_{v}(\gamma) - \chi(\gamma)\mathbf{I}. \tag{11.11}$$

Here, $\mathbf{I}$ is the identity matrix and $\boldsymbol{\Delta}_{v}(\gamma) = \mathrm{diag}(\boldsymbol{v}(\gamma))$.

Just as is the case with the Esscher transform for Lévy processes, a primary effect of the exponential change of measure is to alter the long-term drift of the process. This is stipulated by the strong law of large numbers and the behaviour of the leading eigenvalue χ as a function of γ.

Proposition 11.7 *Suppose that χ is defined in some open interval D of $\mathbb{R}$, then, it is smooth and convex on D.*

Note that, since $\boldsymbol{\Psi}(0) = \boldsymbol{Q}$, it is always the case that $\chi(0) = 0$ and $\boldsymbol{v}(0) = (1, \ldots, 1)$. Hence, for D as in the previous proposition, we must necessarily have $0 \in D$, in which case $\chi'(0)$ is well defined and finite. When this happens, the strong law of large numbers takes the form of the almost sure limit

$$\lim_{t\to\infty} \frac{\xi_t}{t} = \chi'(0), \tag{11.12}$$

and we call $\chi'(0)$ the *drift* of the MAP.

When $\gamma \in D$ is a non-zero root of χ, convexity dictates that, when $\chi'(0) < 0$, we must have that $\gamma > 0$ and $\chi'(\gamma) > 0$. Conversely, when $\chi'(0) > 0$, we must have that $\gamma < 0$ and $\chi'(\gamma) < 0$. If $\chi'(0) = 0$, then no such root γ exists. A natural consequence of the change of measure in Proposition 11.6 is that under $\mathbf{P}^{\gamma}_{i,x}$, the MAP (ξ, J) acquires a new drift, which, by inspection, must be equal to $\chi'(\gamma)$. It follows that, when $\gamma < 0$, the drift of (ξ, J) switches from a positive to a negative value and when $\gamma > 0$, the drift switches from negative to positive.

Remark 11.8 Recall from the Lamperti–Kiu transform in Theorem 11.3 that the radial component of a rssMp, Z, is controlled by ξ of the underlying MAP. In the presence of the assumption (11.4), it is the long term behaviour of ξ that dictates whether Z limits continuously to the origin or not. In particular, if $\zeta = \inf\{t > 0\colon Z_t = 0\}$, then $Z_{\zeta-} = 0$ when $\zeta < \infty$. As such, the existence of a strong law of large numbers of the form (11.12) for the underlying MAP can be instructive.

11.3 Excursion theory for MAPs

Just as is the case with Lévy processes, the exponents of MAPs are also known to have a Wiener–Hopf factorisation. However, in the MAP setting, the two factors correspond to the matrix exponent of the ascending (resp. descending) ladder processes. These are themselves MAPs with trajectories whose range agrees with the range and state of the modulating chain at times of new maxima (resp. minima). In order to deal with the Wiener–Hopf factorisation for MAPs in the next section, we first spend time in this section developing a better understanding of what we mean by the aforementioned ascending ladder MAP.

Let $\mathrm{R}^{(x)}_t = (x \vee \overline{\xi}_t) - \xi_t$, $t \geq 0$, where $\overline{\xi}_t = \sup_{s\leq t} \xi_s$. As a pair, the process $(\mathrm{R}^{(x)}, J)$ is a strong Markov process. To see why, let us momentarily write $J^{(i)}$ in place of J to indicate the initial value of J. Suppose that T is any stopping time with respect to the natural filtration of (ξ, J), then, on $\{T < \infty\}$,

$$\left((\mathrm{R}^{(x)}_{T+t}, J^{(i)}_{T+t}\right) \stackrel{d}{=} \left(\left(x \vee \sup_{s\le T}\xi_s \vee (\sup_{s\le t}\tilde{\xi}_s + \xi_T)\right) - \xi_T - \tilde{\xi}_t\,,\ \tilde{J}_t^{(J_T^{(i)})}\right)$$
$$= \left(\left(\mathrm{R}^{(x)}_T \vee \sup_{s\le t}\tilde{\xi}_s\right) - \tilde{\xi}_t\,,\ \tilde{J}_t^{(J_T^{(i)})}\right)$$
$$= \left(\tilde{Y}_t^{(\mathrm{R}_T^x)},\ \tilde{J}_t^{(J_T^{(i)})}\right), \qquad t \ge 0,$$

where, for $i \in E$, $\tilde{J}^{(i)}$ is an independent copy of $J^{(i)}$, $\tilde{\xi}$ is an independent copy of ξ and, accordingly, for $x, t \ge 0$, $\tilde{Y}^{(x)} = (x \vee \sup_{s\le t}\tilde{\xi}_s) - \tilde{\xi}_t$. It is clear that the strong Markov property now follows.

For convenience, write R in place of $\mathrm{R}^{(0)}$. Since (R, J) is a strong Markov process, by the general theory there exists a local time at the point $(0, i)$, which we henceforth denote by $(L_t^{(i)}, t \ge 0)$. Now consider the process

$$L_t := \sum_{i\in E} L_t^{(i)}, \qquad t \ge 0,$$

where each of the local time processes $L^{(i)}$ may be chosen up to an arbitrary scaling constant.

Note that for a fixed i, the standard theory of local times gives us that the inverse local time process, $(L^{(i)})^{-1}$, is a subordinator. For each $i \ne j$ in E, the points of increase of $L^{(i)}$ and $L^{(j)}$ are disjoint. It therefore follows that $(L^{-1}, H^+, J^+) := ((L_t^{-1}, H_t^+, J_t^+), t \ge 0)$ is a (killed if $L_\infty < \infty$) bivariate Markov additive subordinator (that is to say a MAP whose first two components are ordinates with non-decreasing paths and the third component is the modulator), where

$$H_t^+ := \xi_{L_t^{-1}} \text{ and } J_t^+ := J_{L_t^{-1}}, \qquad \text{if } L_t^{-1} < \infty, \tag{11.13}$$

and $H_t^+ := \infty$ and $J_t^+ := \dagger$ otherwise. Note that the rate at which the process (L^{-1}, H^+, J^+) is killed (and sent to the state $(\infty, \infty, \dagger)$) depends on the state of the chain J^+ when killing occurs. The MAP (L^{-1}, H^+, J^+) has the important role of characterising the times $\{t \ge 0 \colon \overline{\xi}_t = \xi_t\}$, as well as the corresponding positions of ξ and the state of the chain J at these times.

Next define the successive excursions of ξ from its running maximum,

$$(\epsilon_t, J_t^\epsilon) = ((\epsilon_t(s), J_t^\epsilon(s)),\ 0 \le s \le \Delta L_t^{-1})$$
$$:= ((\xi_{L_{t-}^{-1}+s} - \xi_{L_{t-}^{-1}}, J_{L_{t-}^{-1}+s}),\ 0 \le s \le \Delta L_t^{-1}),$$

if $\Delta L_t^{-1} := L_t^{-1} - L_{t-}^{-1} > 0$. To be clear, where J_t^+ is the value of the modulation process J on entry to the excursion ϵ_t and $J_t^\epsilon(\cdot)$ is the evolution of the modulation process $J(\cdot)$ during the excursion ϵ_t. In particular, J^+ and J_t^ϵ should not be confused and are only related by the equality $J_t^\epsilon(0) = J_{t-}^+$.

Let $\overline{\mathcal{U}}(\mathbb{R} \times E)$ be the space of $\mathbb{R} \times E$-valued paths taking the form $((\epsilon(s), J^{\epsilon}(s))\colon s \leq \zeta)$, where ζ is the path lifetime, which are right-continuous with left limits and for which $\epsilon(s)$ is strictly negative-valued for $s \in (0, \zeta)$ with $\epsilon(\zeta) \geq 0$. The process

$$\left((\epsilon, J^{\epsilon}), J^{+}\right) := \left(((\epsilon_t, J^{\epsilon}_t), J^{+}_t), t \geq 0 \text{ and } \Delta L^{-1}_t > 0\right) \tag{11.14}$$

is a Markov additive Poisson point process (see Section A.13 of the Appendix) with family of excursion measures n_i, $i \in E$, on the Skorokhod space of $\mathbb{R} \times E$ (see Section A.10 of the Appendix), which are concentrated on $\overline{\mathcal{U}}(\mathbb{R} \times E)$. Moreover, this Markov additive Poisson point process is killed at time L_∞.

As a bivariate Markov additive subordinator, the process (L^{-1}, H^{+}, J^{+}) has a matrix Laplace exponent given by

$$\mathbf{E}_{0,i}[\mathrm{e}^{-\gamma L^{-1}_t - \lambda H^{+}_t}, J^{+}_t = j] = (\mathrm{e}^{-\boldsymbol{\kappa}(\gamma,\lambda)t})_{i,j}, \qquad \gamma, \lambda \geq 0,$$

where we are using $\exp(-\infty) := 0$ rather than stipulating that $\{t < L_\infty\}$ and the matrix $\boldsymbol{\kappa}(\gamma, \lambda)$ has the structure

$$\boldsymbol{\kappa}(\gamma, \lambda) = \operatorname{diag}(\Phi_1(\gamma, \lambda), \ldots, \Phi_N(\gamma, \lambda)) - \boldsymbol{\Lambda} \circ \boldsymbol{K}(\gamma, \lambda), \qquad \gamma, \lambda \geq 0, \tag{11.15}$$

such that, for $i \in E$, $\Phi_i(\gamma, \lambda)$ is the subordinator exponent that describes the movement of (L^{-1}, H^{+}) when the modulating chain J^{+} is in state i. Moreover, $\boldsymbol{\Lambda}$ is the intensity of J^{+} and the matrix $\boldsymbol{K}(\gamma, \lambda) = (\boldsymbol{K}(\gamma, \lambda))_{i,j}$ is such that, for $i \neq j$ in E, its (i, j)th entry is the Laplace transform of $F^{+}_{i,j}(\mathrm{d}y, \mathrm{d}x)$, the joint distribution of the additional jump incurred by (L^{-1}, H^{+}) when the modulating chain changes state from i to j. The diagonal elements of $\boldsymbol{K}(\gamma, \lambda)$ are set to unity. Later on, we will abuse our notation and write, for $i \in E$, $\Phi_i(\lambda)$ in place of $\Phi_i(0, \lambda)$, $\boldsymbol{\kappa}(\lambda)$ in place of $\boldsymbol{\kappa}(0, \lambda)$ and $F^{+}_{i,j}(\mathrm{d}x)$ in place of $F^{+}_{i,j}(\mathbb{R}^{+}, \mathrm{d}x)$.

For $i \in E$, we can now identify the exponents

$$\Phi_i(\gamma, \lambda) = \mathrm{n}_i(\zeta = \infty) + \mathrm{a}_i\gamma + \mathrm{b}_i\lambda + \int_0^\infty \int_0^\infty (1 - \mathrm{e}^{-\gamma x - \lambda y})\,\mathrm{n}_i(\zeta \in \mathrm{d}x, \epsilon(\zeta) \in \mathrm{d}y, J^{\epsilon}(\zeta) = i), \tag{11.16}$$

for $\gamma, \lambda \geq 0$, where $\mathrm{a}_i, \mathrm{b}_i \geq 0$ and $\zeta = \inf\{s > 0\colon \epsilon(s) > 0\}$ the excursion length. Note in particular that the matrix

$$\operatorname{diag}(\mathrm{n}_1(\zeta = \infty), \ldots, \mathrm{n}_N(\zeta = \infty))$$

encodes the respective killing rates of (L^{-1}, H^{+}) when J^{+} is in each state of E.

On a final note, since the local times $L^{(i)}$ may be chosen up to an arbitrary scaling constant, the matrix exponent $\boldsymbol{\kappa}(\gamma, \lambda)$ can only be defined up to pre-multiplication of a diagonal matrix with strictly positive entries (henceforth referred to as a *strictly positive diagonal matrix*).

For convenience, we assume that ξ is a non-lattice. This is the case, for example, if the jump measures associated to ξ^i, $i \in E$, and $F^+_{i,j}(\mathrm{d}x)$, $i \neq j, i, j \in E$, are diffuse on $(0, \infty)$. Now suppose that $\mathbf{e}_q$ is an independent exponentially distributed random variable with rate $q > 0$. Consider an adapted version of the Markov additive point process of excursions described above in which each excursion $(\epsilon_t, J^\epsilon_t)$ is marked with an independent copy of $\mathbf{e}_q$, denoted by $\mathbf{e}^{(t)}_q$, for $t \geq 0$. Let $\overline{m}_t = \sup\{s \leq t : \overline{\xi}_s = \xi_s\}$. Poisson thinning dictates that the pair $(\overline{m}_{\mathbf{e}_q}, \overline{\xi}_{\mathbf{e}_q})$ is equal in law to the position of the process (L^{-1}, H^+) when conditioned on $\{\Delta L^{-1}_t < \mathbf{e}^{(t)}_q$, for all $t \geq 0\}$ and stopped with rate matrix

$$
\begin{aligned}
&\mathrm{diag}(\mathrm{a}_1 q + \mathrm{n}_1(\zeta > \mathbf{e}_q), \ldots, \mathrm{a}_N q + \mathrm{n}_N(\zeta > \mathbf{e}_q))\\
&\quad = \mathrm{diag}(\mathrm{a}_1 q + \mathrm{n}_1(1 - \mathrm{e}^{-q\zeta}), \ldots, \mathrm{a}_N q + \mathrm{n}_N(1 - \mathrm{e}^{-q\zeta}))\\
&\quad = \mathrm{diag}(\Phi_1(q, 0), \ldots, \Phi_N(q, 0)).
\end{aligned}
$$

In particular, the conditioned process is stopped at a random time θ_q with the property that

$$
\mathbf{P}_{0,i}(\theta_q > t \,|\, \sigma(J^+_s : s \leq t)) = \exp\Big(-\int_0^t \Phi_{J^+_s}(q, 0)\, ds\Big).
$$

The aforementioned conditioned process has matrix exponent which can be derived from the matrix exponent $\boldsymbol{\kappa}(\gamma, \lambda)$. Indeed, whereas in $\boldsymbol{\kappa}(\gamma, \lambda)$ the subordinator exponents in each of the diagonal entries are represented as $\Phi_i(\gamma, \lambda)$, in the conditioned process, this is replaced by

$$
\begin{aligned}
&\mathrm{n}_i(\zeta = \infty) + \mathrm{a}_i\gamma + \mathrm{b}_i\lambda\\
&\quad + \int_0^\infty \int_0^\infty (1 - \mathrm{e}^{-\gamma x - \lambda y})\mathrm{e}^{-qx}\mathrm{n}_i(\zeta \in \mathrm{d}x, \epsilon(\zeta) \in \mathrm{d}y, J^\epsilon(\zeta) = i), \qquad (11.17)
\end{aligned}
$$

for $\gamma, \lambda \geq 0$, which is also equal to $\Phi_i(q+\gamma, \lambda) - \Phi_i(q, 0)$. Hence the conditioned process has matrix exponent given by

$$
\begin{aligned}
\tilde{\boldsymbol{\kappa}}(\gamma, \lambda) := \;&\mathrm{diag}(\Phi_1(q + \gamma, \lambda) - \Phi_1(q, 0), \ldots, \Phi_N(q + \gamma, \lambda) - \Phi_N(q, 0))\\
&- \boldsymbol{\Lambda} \circ \boldsymbol{K}(\gamma, \lambda), \qquad (11.18)
\end{aligned}
$$

for $\gamma, \lambda \geq 0$.

11.4 Matrix Wiener–Hopf factorisation

Now that it is understood what the ascending ladder MAP means, we are ready to state and prove the promised Wiener–Hopf factorisation for MAPs. Write $\hat{\boldsymbol{\kappa}}(\gamma, \lambda)$ for the matrix exponent of the ascending ladder MAP to the dual of (ξ, J). That is to say, $\hat{\boldsymbol{\kappa}}(\gamma, \lambda)$ is to $\hat{\boldsymbol{\Psi}}(z)$ what $\boldsymbol{\kappa}(\gamma, \lambda)$ is to $\boldsymbol{\Psi}(z)$.

Theorem 11.9 *For $z \in \mathbb{R}\backslash\{0\}$ and $\gamma \geq 0$, up to the identification of $\boldsymbol{\kappa}$ (or equivalently $\hat{\boldsymbol{\kappa}}$) by a pre-multiplicative strictly positive diagonal matrix,*

$$\gamma \boldsymbol{I} - \boldsymbol{\Psi}(\mathrm{i}z) = \boldsymbol{\Delta}_\pi^{-1}[\hat{\boldsymbol{\kappa}}(\gamma, \mathrm{i}z)^{\mathrm{T}}]\boldsymbol{\Delta}_\pi \boldsymbol{\kappa}(\gamma, -\mathrm{i}z).$$

The Wiener–Hopf factorisation, both here in the MAP setting as well as in the original setting of Lévy processes, is a natural consequence of a well-established method of splitting trajectories at their maximum over an independent and exponentially distributed time. Let us therefore consider some of the distributional properties of the MAP associated with this kind of path splitting.

For short, we will write $J_{\overline{m}_{\mathbf{e}_q}} = J_{L^{-1}_{L_{\mathbf{e}_q}}-}$. Intuitively speaking, $J_{\overline{m}_{\mathbf{e}_q}}$ takes the value of J that couples with the position $\xi_{\overline{m}_{\mathbf{e}_q}}$ if $\xi_{\overline{m}_{\mathbf{e}_q}} = \overline{\xi}_{\mathbf{e}_q}$ and otherwise it takes the value of J that couples with the position $\xi_{\overline{m}_{\mathbf{e}_q}-}$ if $\xi_{\overline{m}_{\mathbf{e}_q}-} = \overline{\xi}_{\mathbf{e}_q}$. The latter occurs if ξ jumps away from its supremum at the moment that the supremum is attained. Indeed such an occurrence can happen in concurrence with a transition in the chain J away from its current state, in which case the spatial component of the incoming excursion from the maximum would start by a jump and an instantaneous switch in the state of J^ϵ.

Taking account of the exponents (11.16) and (11.17), and in particular that the ordinate L^{-1} has both a drift component and jump component, the pair $(\overline{\xi}_{\mathbf{e}_q}, \overline{m}_{\mathbf{e}_q})$ has matrix Laplace transform given by

$$\begin{aligned}
&\mathbf{E}_{0,i}\left[\mathrm{e}^{-\gamma\overline{m}_{\mathbf{e}_q}-\lambda\overline{\xi}_{\mathbf{e}_q}}, J_{\overline{m}_{\mathbf{e}_q}} = j\right] \\
&= \mathbf{E}_{0,i}\left[\sum_{t\geq 0}\mathrm{e}^{-\gamma L^{-1}_{t-}-\lambda H^+_{t-}}\mathbf{1}_{(J^+_{t-}=j)}\mathbf{1}_{\{\Delta L^{-1}_s<\mathbf{e}^{(s)}_q,\, s<t\}}\mathbf{1}_{(\Delta L^{-1}_t\geq \mathbf{e}^{(t)}_q)}\right] \\
&\quad + \mathbf{E}_{0,i}\left[\int_0^\infty \mathrm{e}^{-\gamma L^{-1}_{t-}-\lambda H^+_{t-}}\mathbf{1}_{(J^+_{t-}=j)}\mathbf{1}_{\{\Delta L^{-1}_s<\mathbf{e}^{(s)}_q,\, s<t\}}\mathrm{a}_j\mathrm{d}t\right] \\
&= \left(\mathrm{a}_j q + \mathrm{n}_j(1-\mathrm{e}^{-q\zeta})\right)\mathbf{E}_{0,i}\left[\int_0^\infty \mathrm{e}^{-\gamma L^{-1}_t-\lambda H^+_t}\mathbf{1}_{(J^+_t=j)}\mathbf{1}_{(t<\theta_q)}\mathrm{d}t\right] \\
&= \Phi_j(q,0)\int_0^\infty \mathbf{E}_{0,i}\left[\mathrm{e}^{-\gamma L^{-1}_t-\lambda H^+_t}\mathbf{1}_{(J^+_t=j)}\mathbf{1}_{(t<\theta_q)}\right]\mathrm{d}t, \qquad (11.19)
\end{aligned}$$

for $\gamma, \lambda \geq 0$, where we are using the definition $\exp(-\infty) = 0$ rather than including an indicator of the event $\{t < L_\infty\}$ in the expectations. Note that the final expectation above can be written in terms of the matrix Laplace exponent of $\tilde{\boldsymbol{\kappa}}(\gamma, \lambda)$ with an added potential corresponding to $\mathrm{diag}(\Phi_1(q,0), \ldots, \Phi_N(q,0))$, that is,

$$\boldsymbol{\kappa}(q+\gamma, \lambda) = \mathrm{diag}(\Phi_1(q+\gamma,\lambda), \ldots, \Phi_N(q+\gamma,\lambda)) - \boldsymbol{\Lambda}\circ\boldsymbol{K}(\gamma,\lambda), \qquad \gamma, \lambda \geq 0.$$

Indeed, one has,

$$\mathbf{E}_{0,i}\left[e^{-\gamma L_t^{-1}-\lambda H_t^+}\mathbf{1}_{(J_t^+=j)}\mathbf{1}_{(t<\theta_q)}\right] = [e^{-\boldsymbol{\kappa}(q+\gamma,\lambda)t}]_{i,j}, \qquad i,j\in E.$$

Continuing the computation in (11.19), we now have the following result.

Theorem 11.10 *For $i, j \in E$, $\gamma, \lambda \geq 0$ and $q > 0$,*

$$\mathbf{E}_{0,i}[e^{-\gamma\overline{m}_{\mathbf{e}_q}-\lambda\overline{\xi}_{\mathbf{e}_q}}, J_{\overline{m}_{\mathbf{e}_q}} = j] = \Phi_j(q,0)[\boldsymbol{\kappa}(q+\gamma,\lambda)^{-1}]_{i,j}. \tag{11.20}$$

We can go a little further in our analysis. With the help of Poisson thinning, we have, on the event $\{J^+_{\theta_q} = j\}$, that the excursion $\epsilon_{J^+_{\theta_q}}$ is independent of

$$((L_t^{-1}, H_t^+, J_t^+), t < \theta_q).$$

In particular, on $\{J^+_{\theta_q} = j\}$, we have that $(\overline{m}_{\mathbf{e}_q}, \overline{\xi}_{\mathbf{e}_q})$ is independent of $(\mathbf{e}_q - \overline{m}_{\mathbf{e}_q}, \xi_{\mathbf{e}_q} - \overline{\xi}_{\mathbf{e}_q})$. Moreover, duality allows us to conclude that, when the modulating chain is sampled under its stationary distribution, on the event $\{J^+_{\theta_q} = j, J_{\mathbf{e}_q} = k\} = \{J_{\overline{m}_{\mathbf{e}_q}} = j, J_{\mathbf{e}_q} = k\}$ the pair $(\mathbf{e}_q - \overline{m}_{\mathbf{e}_q}, \overline{\xi}_{\mathbf{e}_q} - \xi_{\mathbf{e}_q})$ is equal in law to the pair $(\overline{\hat{m}}_{\mathbf{e}_q}, \overline{\hat{\xi}}_{\mathbf{e}_q})$ on $\{\hat{J}_0 = k, \hat{J}_{\overline{\hat{m}}_{\mathbf{e}_q}} = j\}$, where $(\hat{\xi}, \hat{J})$ is equal in law to the dual of (ξ, J) in the sense of Lemma 11.4, $\overline{\hat{\xi}}_t = \sup_{s\leq t}\hat{\xi}_s$ and $\overline{\hat{m}}_t = \sup\{s \leq t : \overline{\hat{\xi}}_s = \hat{\xi}_s\}$.

From the previous discussion, we may now deduce, for example, that, for $i, j, k \in E$ and $\gamma_1, \gamma_2, \lambda_1, \lambda_2 \geq 0$,

$$\begin{aligned}
&\mathbf{E}_{0,i}\left[e^{-\gamma_1\overline{m}_{\mathbf{e}_q}-\lambda_1\overline{\xi}_{\mathbf{e}_q}}e^{-\gamma_2(\mathbf{e}_q-\overline{m}_{\mathbf{e}_q})-\lambda_2(\overline{\xi}_{\mathbf{e}_q}-\xi_{\mathbf{e}_q})}, J_{\overline{m}_{\mathbf{e}_q}} = j, J_{\mathbf{e}_q} = k\right]\\
&= \mathbf{E}_{0,i}\left[e^{-\gamma_1\overline{m}_{\mathbf{e}_q}-\lambda_1\overline{\xi}_{\mathbf{e}_q}}, J_{\overline{m}_{\mathbf{e}_q}} = j\right]\frac{\pi_k}{\pi_j}\hat{\mathbf{E}}_{0,k}\left[e^{-\gamma_2\overline{m}_{\mathbf{e}_q}-\lambda_2\overline{\xi}_{\mathbf{e}_q}}, J_{\overline{m}_{\mathbf{e}_q}} = j\right].
\end{aligned} \tag{11.21}$$

We can also use the ideas above to prove the following technical lemma which will be used in the proof of the MAP Wiener–Hopf factorisation. Using familiar notation, $\hat{\Phi}_j(\gamma, \lambda)$, $j \in E$, is to $\hat{\boldsymbol{\kappa}}(\gamma, \lambda)$ what $\Phi_j(\gamma, \lambda)$, $j \in E$, is to $\boldsymbol{\kappa}(\gamma, \lambda)$.

Lemma 11.11 *For each $j \in E$,*

$$c_j := \lim_{q\downarrow 0}\frac{\Phi_j(q,0)\hat{\Phi}_j(q,0)}{q} \tag{11.22}$$

exists in $[0, \infty)$ and there exists

$$c := \sum_{j\in E}\lim_{q\downarrow 0}\frac{\Phi_j(q,0)\hat{\Phi}_j(q,0)}{q} \in (0, \infty).$$

Proof On the one hand, for all $i, k \in E$ and $\gamma > 0$,

$$\mathbf{E}_{0,i}\left[\mathrm{e}^{-\gamma \mathbf{e}_q}, J_{\mathbf{e}_q} = k\right] = \left[\int_0^\infty q\mathrm{e}^{-(\gamma+q)t}\mathrm{e}^{\boldsymbol{Q}t}\mathrm{d}t\right]_{i,k} = q\left[((q+\gamma)\boldsymbol{I} - \boldsymbol{Q})^{-1}\right]_{i,k}.$$

On the other hand, from (11.21) and (11.20), for all $i, k \in E$ and $\gamma > 0$,

$$\begin{aligned}
&\mathbf{E}_{0,i}[\mathrm{e}^{-\gamma \mathbf{e}_q}, J_{\mathbf{e}_q} = k] \\
&= \sum_{j\in E} \mathbf{E}_{0,i}[\mathrm{e}^{-\gamma(\overline{m}_{\mathbf{e}_q} + \mathbf{e}_q - \overline{m}_{\mathbf{e}_q})}, J_{\overline{m}_{\mathbf{e}_q}} = j,\ J_{\mathbf{e}_q} = k] \\
&= \sum_{j\in E} \Phi_j(q,0)[\boldsymbol{\kappa}(q+\gamma,0)^{-1}]_{i,j}\hat{\Phi}_j(q,0)[\hat{\boldsymbol{\kappa}}(q+\gamma,0)^{-1}]_{k,j}\frac{\pi_k}{\pi_j}.
\end{aligned}$$

Taking limits as $q \downarrow 0$, it follows from continuity that

$$\left[(\gamma\boldsymbol{I} - \boldsymbol{Q})^{-1}\right]_{i,k} = \sum_{j\in E} \lim_{q\downarrow 0} \frac{\Phi_j(q,0)\hat{\Phi}_j(q,0)}{q}[\boldsymbol{\kappa}(\gamma,0)^{-1}]_{i,j}[\hat{\boldsymbol{\kappa}}(\gamma,0)^{-1}]_{k,j}\frac{\pi_k}{\pi_j},$$

where the limit on the right-hand side exists because the limit exits on the left-hand side. The statement of the theorem now follows. □

We are now ready to prove Theorem 11.9 as promised earlier.

Proof of Theorem 11.9 We start by sampling ξ over an independent and exponentially distributed time horizon denoted, as usual, by $\mathbf{e}_q$. By splitting at the maximum, applying duality and appealing to the identity (11.20), we have for $\gamma \geq 0$,

$$\begin{aligned}
&\mathbf{E}_{0,i}[\mathrm{e}^{-\gamma \mathbf{e}_q + \mathrm{i}z\xi_{\mathbf{e}_q}}, J_{\mathbf{e}_q} = j] \\
&= \sum_{k\in E} \mathbf{E}_{0,i}[\mathrm{e}^{-\gamma(\mathbf{e}_q - \overline{m}_{\mathbf{e}_q} + \overline{m}_{\mathbf{e}_q}) + \mathrm{i}z\overline{\xi}_{\mathbf{e}_q}}\mathrm{e}^{\mathrm{i}z(\xi_{\mathbf{e}_q} - \overline{\xi}_{\mathbf{e}_q})}, J_{\overline{m}_{\mathbf{e}_q}} = k,\ J_{\mathbf{e}_q} = j] \\
&= \sum_{k\in E} \mathbf{E}_{0,i}[\mathrm{e}^{-\gamma\overline{m}_{\mathbf{e}_q} + \mathrm{i}z\overline{\xi}_{\mathbf{e}_q}}, J_{\overline{m}_{\mathbf{e}_q}} = k]\frac{\pi_j}{\pi_k}\hat{\mathbf{E}}^{0,j}[\mathrm{e}^{-\gamma\overline{m}_{\mathbf{e}_q} - \mathrm{i}z\overline{\xi}_{\mathbf{e}_q}}, J_{\overline{m}_{\mathbf{e}_q}} = k] \\
&= \sum_{k\in E} \Phi_k(q,0)[\boldsymbol{\kappa}(q+\gamma,-\mathrm{i}z)^{-1}]_{i,k}\frac{\pi_j}{\pi_k}\hat{\Phi}_k(q,0)[\hat{\boldsymbol{\kappa}}(q+\gamma,\mathrm{i}z)^{-1}]_{j,k}.
\end{aligned}$$

Noting that we can write the left-hand side above as $q[((q+\gamma)\boldsymbol{I} - \boldsymbol{\Psi}(\mathrm{i}z))^{-1}]_{i,j}$, we can divide by q and take limits as $q \downarrow 0$ to find that

$$[(\gamma\boldsymbol{I} - \boldsymbol{\Psi}(\mathrm{i}z))^{-1}]_{i,j} = \sum_{k\in E} c_k[\boldsymbol{\kappa}(\gamma,-\mathrm{i}z)^{-1}]_{i,k}\frac{\pi_j}{\pi_k}[[\hat{\boldsymbol{\kappa}}(\gamma,\mathrm{i}z)^{\mathrm{T}}]^{-1}]_{k,j},$$

where we recall that the constants c_k, $k \in E$ were introduced in (11.22). In matrix form, the above equality can be rewritten as

$$(\gamma \boldsymbol{I} - \boldsymbol{\Psi}(\mathrm{i}z))^{-1} = \boldsymbol{\kappa}(\gamma, -\mathrm{i}z)^{-1} \boldsymbol{\Delta}_{c/\pi} [\hat{\boldsymbol{\kappa}}(\gamma, \mathrm{i}z)^{\mathrm{T}}]^{-1} \boldsymbol{\Delta}_{\pi}, \tag{11.23}$$

where $\boldsymbol{\Delta}_{c/\pi} = \mathrm{diag}(c_1/\pi_1, \ldots, c_N/\pi_N)$. With the exception of possibly $\boldsymbol{\Delta}_{c/\pi}$ (on account of the fact that some of the constants c_k may be zero), all of the matrices in (11.23) are invertible. Hence the matrix $\boldsymbol{\Delta}_{c/\pi}$ must also be invertible and, on account of its diagonal form, it necessarily follows that $c_k > 0$. The proof is now completed by inverting the matrices on both left- and right-hand sides of (11.23) and noting that, without loss of generality, the constants c_k may be taken as unity by choosing an appropriate normalisation of local times (which, in turn, means that, for each $j \in E$, the equality in (11.22) can be determined up to a multiplicative constant). This is equivalent to the statement that $\boldsymbol{\kappa}$, or equivalently $\hat{\boldsymbol{\kappa}}$, is identified up to a pre-multiplicative strictly positive diagonal matrix. □

We may think of the factorisation in Theorem 11.9 as a space-time factorisation (in the sense that one captures information about the space-time point of the maximum over an independent and exponentially distributed random time). A purely spatial matrix Wiener–Hopf factorisation occurs by setting $\gamma = 0$. In that case, we have that, for $\theta \in \mathbb{R}$, up to the pre-multiplication of $\boldsymbol{\kappa}$ (or $\hat{\boldsymbol{\kappa}}$) by a strictly positive diagonal matrix,

$$-\boldsymbol{\Psi}(\mathrm{i}\theta) = \boldsymbol{\Delta}_{\pi}^{-1} \hat{\boldsymbol{\kappa}}(\mathrm{i}\theta)^{\mathrm{T}} \boldsymbol{\Delta}_{\pi} \boldsymbol{\kappa}(-\mathrm{i}\theta), \qquad \theta \in \mathbb{R}, \tag{11.24}$$

where we recall the abuse of notation, $\boldsymbol{\kappa}(-\mathrm{i}\theta) = \boldsymbol{\kappa}(0, -\mathrm{i}\theta)$ and $\hat{\boldsymbol{\kappa}}(\mathrm{i}\theta) = \hat{\boldsymbol{\kappa}}(0, \mathrm{i}\theta)$.

This is consistent with the Wiener–Hopf factorisation for Lévy processes (2.31) as, in that setting, the dual process is its negative; hence the ascending ladder height process of the dual is equal to the descending ladder height process. In the Lévy setting the factorisation is valid up to a meaningless multiplicative constant. This is due to the fact that the ladder height exponents appearing in the factorisation can only be found up to a multiplicative constants, corresponding to an arbitrary linear scaling in local time (which does not affect the range of the ladder height processes themselves).

An equivalent way of stating (11.24) is to say that the Wiener–Hopf factorisation in (11.24) can only be identified up to pre-multiplication of $\boldsymbol{\Psi}$ by a strictly positive diagonal matrix. To see why this is the case, suppose, on the right-hand side of (11.24), we write $\boldsymbol{\kappa} = \boldsymbol{\Delta}_a \boldsymbol{\kappa}'$ and $\hat{\boldsymbol{\kappa}} = \boldsymbol{\Delta}_b \hat{\boldsymbol{\kappa}}'$, where $\boldsymbol{\Delta}_a$ and $\boldsymbol{\Delta}_b$ are two strictly positive diagonal matrices. Let $\boldsymbol{\pi}'$ be the vector with entries $a_i \pi_i b_i / \sum_{j \in E} a_j \pi_j b_j$, for $i \in E$. Then we have

$$-\boldsymbol{\Psi}(\mathrm{i}\theta) = \boldsymbol{\Delta}_{\pi}^{-1}\hat{\boldsymbol{\kappa}}'(\mathrm{i}\theta)^{\mathrm{T}}\boldsymbol{\Delta}_b\boldsymbol{\Delta}_{\pi}\boldsymbol{\Delta}_a\boldsymbol{\kappa}'(-\mathrm{i}\theta)$$
$$= (\boldsymbol{\Delta}_a\boldsymbol{\Delta}_b)\boldsymbol{\Delta}_{\pi'}^{-1}\hat{\boldsymbol{\kappa}}'(\mathrm{i}\theta)^{\mathrm{T}}\boldsymbol{\Delta}_{\pi'}\boldsymbol{\kappa}'(-\mathrm{i}\theta), \tag{11.25}$$

for $\theta \in \mathbb{R}$. Note that

$$(\boldsymbol{\pi}')^{\mathrm{T}}[(\boldsymbol{\Delta}_a\boldsymbol{\Delta}_b)^{-1}\boldsymbol{\Psi}(0)] = \frac{\boldsymbol{\pi}^{\mathrm{T}}\boldsymbol{\Psi}(0)}{\sum_{j\in E} a_j\pi_j b_j} = 0,$$

showing that $\boldsymbol{\pi}'$ is the stationary distribution of the modulating chain that corresponds to the matrix exponent $\boldsymbol{\Delta}_{ab}^{-1}\boldsymbol{\Psi}$, where $\boldsymbol{\Delta}_{ab}^{-1} := \boldsymbol{\Delta}_a^{-1}\boldsymbol{\Delta}_b^{-1}$. We now see that (11.25) identifies the factorisation of $\boldsymbol{\Delta}_{ab}^{-1}\boldsymbol{\Psi}$. Recall from the discussion preceding Proposition 11.5, premultiplication of $\boldsymbol{\Psi}$ by a diagonal matrix corresponds to a linear time change, but this does not affect the spatial range of the MAP.

11.5 Self-similar Markov processes in $\mathbb{R}^d$

The notion of a self-similar Markov process (ssMp) in higher dimensions is defined in the same way as in the one-dimensional setting, when (11.1) it is taken as the key defining property, albeit that, now, the process is $\mathbb{R}^d$-valued. The identification of all $\mathbb{R}^d$-valued self-similar Markov processes as a space-time change of a Markov additive process also carries through, providing we understand the notion of a MAP in the appropriate way in higher dimensions.

Definition 11.12 Suppose that E, with annexed cemetery state $(-\infty, \dagger)$, is metrisable to make a locally compact, complete and separable metric space. A regular Feller process $(\xi, \Theta) = ((\xi_t, \Theta_t), t \geq 0)$ on $\mathbb{R} \times E$ with probabilities $\mathbf{P}_{x,\theta}$, $x \in \mathbb{R}$, $\theta \in E$, is called a *Markov additive process* (MAP) such that, for every bounded measurable function $f\colon \mathbb{R}\times E \to \mathbb{R}$, $t, s \geq 0$ and $(x, \theta) \in \mathbb{R}\times E$, on $\{\Theta_t = \phi, t < \varsigma\}$,

$$\mathbf{E}_{x,\theta}[f(\xi_{t+s} - \xi_t, \Theta_{t+s})\mathbf{1}_{(t+s<\varsigma)}|\mathcal{G}_t] = \mathbf{E}_{0,\phi}[f(\xi_s, \Theta_s)\mathbf{1}_{(s<\varsigma)}],$$

where $\varsigma = \inf\{t > 0\colon \Theta_t = \dagger\}$, $\xi_\varsigma = -\infty$ and with, a slight abuse of notation, $(\mathcal{G}_t, t \geq 0)$ is again taken as the filtration generated by the *MAP* with natural enlargement. Similarly to Definition 11.1, Θ alone is a regular Feller processes and referred to as the modulator, whilst ξ is called the ordinate.

In one dimension the role of Θ was played by the Markov chain J on $E = \{-1, 1\}$. This choice of J feeds into the positive or negative positioning of a self-similar Markov process through the Lamperti–Kiu transform with ξ helping to describe the radial distance from the origin. In higher dimensions we will still use ξ to help describe a radial distance from the origin and, by taking $E = \mathbb{S}^{d-1} := \{x \in \mathbb{R}^d : |x| = 1\}$, the process Θ will help describe spatial orientation in the spirit of a generalised polar decomposition. The following theorem is the higher-dimensional analogue of Theorem 11.3 and is attributed to Kiu [111] with additional clarification from [1], building on the original work of Lamperti [139]. As with Theorem 11.3, we omit its proof.

Theorem 11.13 (Generalised Lamperti–Kiu transform) *Fix $\alpha > 0$. The process Z is a ssMp with index α if and only if there exists a (killed) MAP, (ξ, Θ) on $\mathbb{R} \times \mathbb{S}^{d-1}$ such that*

$$Z_t := e^{\xi_{\varphi(t)}} \Theta_{\varphi(t)}, \qquad t \leq I_\varsigma, \tag{11.26}$$

where

$$\varphi(t) = \inf\left\{ s > 0 \colon \int_0^s e^{\alpha \xi_u} \, du > t \right\}, \qquad t \leq I_\varsigma,$$

and $I_\varsigma = \int_0^\varsigma e^{\alpha \xi_s} ds$ is the lifetime of Z until absorption at the origin, which acts as a cemetery state. Here, we interpret $\exp\{-\infty\} \times \dagger := 0$ and $\inf \emptyset := \infty$.

Note that, in the representation (11.26), the lifetime of the process is given by

$$\zeta = \inf\{t > 0 \colon Z_t = 0\}$$

and satisfies $\zeta = I_\varsigma$.

For each $x \in \mathbb{R}^d \backslash \{0\}$, the *skew product decomposition* (for $d \geq 2$), is the unique representation

$$x = (|x|, \arg(x)), \tag{11.27}$$

where $\arg(x) = x/|x|$ is a vector on $\mathbb{S}^{d-1}$, the d-dimensional sphere of unit radius. Conversely, any $x \in \mathbb{R} \times \mathbb{S}^{d-1}$ taking the form (11.27) belongs to $\mathbb{R}^d$. The representation (11.26), therefore, gives us a d-dimensional skew product decomposition of self-similar Markov processes.

In previous sections, for MAPs whose modulator state space is finite, we gave a very detailed analysis of how an excursion theory and some of its ramifications can be developed much in the spirit of the analogous theory for Lévy processes. The analogue for MAPs which conform to the greater generality of Definition 11.12 is a much more difficult area to discuss and beyond the scope of this text.

In this text, the use of such general MAPs will only occur in the setting of d-dimensional stable processes. As such, a result that will be of particular interest to us pertains to isotropy. In particular, we are interested in how the property of isotropy on Z is manifested in the underlying MAP, (ξ, Θ).

Recall that a measure μ on $\mathbb{R}^d$ is isotropic if for $B \in \mathcal{B}(\mathbb{R}^d)$, $\mu(B) = \mu(U^{-1}B)$ for every orthogonal d-dimensional matrix U. In this spirit, we can thus define an isotropic ssMp, $Z = (Z_t, t \geq 0)$ to have the property that, for every orthogonal d-dimensional matrix U and $x \in \mathbb{R}^d$, the law of $(U^{-1}Z, P_x)$ is equal to that of $(Z, P_{U^{-1}x})$.

Theorem 11.14 *Suppose Z is a ssMp, with underlying MAP (ξ, Θ). Then Z is isotropic if and only if $((\xi, U^{-1}\Theta), \mathbf{P}_{x,\theta})$ is equal in law to $((\xi, \Theta), \mathbf{P}_{x,U^{-1}\theta})$, for every orthogonal d-dimensional matrix U and $x \in \mathbb{R}$, $\theta \in \mathbb{S}^{d-1}$.*

Proof Suppose first that Z is an isotropic ssMp. On the event $\{t < \zeta\}$, since

$$\int_0^{\varphi(t)} \mathrm{e}^{\alpha\xi_u} \mathrm{d}u = t \quad \text{and hence} \quad \frac{\mathrm{d}\varphi(t)}{\mathrm{d}t} = \mathrm{e}^{-\alpha\xi_{\varphi(t)}} = |Z_t|^{-\alpha},$$

we have that

$$\varphi(t) = \int_0^t |Z_u|^{-\alpha} \mathrm{d}u. \tag{11.28}$$

Let us introduce its right continuous inverse, on $\{t < \varsigma\}$, as follows

$$A(t) = \inf\left\{s > 0\colon \int_0^s |Z_u|^{-\alpha} \mathrm{d}u > t\right\}. \tag{11.29}$$

Hence, we see that, on $\{t < \varsigma\}$,

$$(\xi_t, \Theta_t) = (\log|Z_{A(t)}|, \arg(Z_{A(t)})), \qquad t \geq 0, \tag{11.30}$$

where the random times $A(t)$ are stopping times in the filtration generated by Z with natural enlargement.

Now suppose that U is any orthogonal d-dimensional matrix. Since Z is isotropic, we see from (11.29) and (11.30) that, for $x \in \mathbb{R}^d\backslash\{0\}$,

$$\begin{aligned} ((\xi, U^{-1}\Theta), \mathbf{P}_{\log|x|,\arg(x)}) &= ((\log|Z_{A(t)}|, U^{-1}\arg(Z_{A(t)})), P_x) \\ &\overset{(d)}{=} ((\log|Z_{A(t)}|, \arg(Z_{A(t)})), P_{U^{-1}x}) \\ &= ((\xi, \Theta), \mathbf{P}_{\log|x|, U^{-1}\arg(x)}) \end{aligned} \tag{11.31}$$

and the 'only if' direction is proved.

For the converse statement, suppose that the left-hand side and right-hand side in (11.31) are equal in distribution for all orthogonal d-dimensional matrices U and $x \in \mathbb{R}^d\backslash\{0\}$. Setting $Z' = U^{-1}Z$ and letting A' play the role of (11.29) but for Z', since $|Z'| = |Z|$, and $\arg(Z') = U^{-1}\arg(Z)$, we have

$$
\begin{aligned}
((\log|Z'_{A'(t)}|, \arg(Z'_{A'(t)})), P_x) &= ((\log|Z_{A(t)}|, U^{-1}\arg(Z_{A(t)})), P_x)\\
&= ((\xi, U^{-1}\Theta), \mathbf{P}_{\log|x|,\arg(x)})\\
&\stackrel{(d)}{=} ((\xi, \Theta), \mathbf{P}_{\log|x|, U^{-1}\arg(x)})\\
&= ((\log|Z_{A(t)}|, \arg(Z_{A(t)})), P_{U^{-1}x}). \qquad (11.32)
\end{aligned}
$$

This concludes the 'if' part of the proof. □

Corollary 11.15 *If Z is an isotropic ssMp, then $|Z|$ is equal in law to a pssMp and hence ξ is a Lévy process.*

Proof It suffices to verify that $|Z|$ is a Markov process with the required scaling property whose semigroup has the Feller property given in Definition A.14.

The scaling property of $|Z|$ follows directly from that of Z. Moreover we have, for bounded measurable $g\colon [0,\infty) \to \mathbb{R}$ and $s, t \geq 0$, on $\{t < \zeta\}$,

$$
\begin{aligned}
E_x[g(|Z_{t+s}|)\mathbf{1}_{(t+s<\zeta)} \,|\, \sigma(Z_u, u \leq t)] &= \mathbf{E}_{y,\theta}[g(\mathrm{e}^{\xi_{\varphi(s)}})\mathbf{1}_{(s<\zeta)}]_{y=\log|Z_t|,\,\theta=\arg(Z_t)}\\
&\stackrel{(d)}{=} \mathbf{E}_{y,1}[g(\mathrm{e}^{\xi_{\varphi(s)}})\mathbf{1}_{(s<\zeta)}]_{y=\log|Z_t|}\\
&= E_z[g(|Z_s|)\mathbf{1}_{(s<\zeta)}]_{z=|Z_t|1},
\end{aligned}
$$

where $1 = (1, 0, \ldots, 0) \in \mathbb{R}^d$ is the 'North Pole' on $\mathbb{S}^{d-1}$ and we have used Theorem 11.14. This ensures the Markov property.

To verify the Feller property, we note that, for $x \in \mathbb{R}^d\backslash\{0\}$, (Z, P_x) is equal in law to

$$
Z_t^{(x)} := |x|\mathrm{e}^{\xi_{\varphi(|x|^{-\alpha}t)}}\Theta_{\varphi(|x|^{-\alpha}t)}, \qquad t < |x|^\alpha \int_0^\varsigma \mathrm{e}^{\alpha\xi_u}\mathrm{d}u,
$$

under $\mathbf{P}_{0,\arg(x)}$. Hence for all continuous $g\colon [0,\infty) \to \mathbb{R}$ vanishing at ∞,

$$
E_x[g(|Z_t|)\mathbf{1}_{(t<\zeta)}] = \mathbf{E}_{0,\arg(x)}\left[g(|x|\mathrm{e}^{\xi_{\varphi(|x|^{-\alpha}t)}})\mathbf{1}_{(t<\zeta)}\right] = \mathbf{E}_{0,1}\left[g(|x|\mathrm{e}^{\xi_{\varphi(|x|^{-\alpha}t)}})\mathbf{1}_{(s<\zeta)}\right],
$$

where we have again used Theorem 11.14. The remaining conditions that will identify $|Z|$ as a regular Feller process can now be easily verified using the fact that such properties hold for Z as a ssMp. □

11.6 Starting at the origin

In one dimension, the representation (11.3) is not meaningful when $x = 0$. As with the setting of pssMp, one is equally concerned with the question as to whether 0 may be considered as part of the state space for a rssMp. That is to say, the question as to whether there exists a notion of P_0, which is consistent with the family $(P_x, x \neq 0)$ in an appropriate sense.

Again we divide the discussion into the setting of conservative and non-conservative rssMp. What is known in these two settings mirrors in a very natural way Theorems 5.3 and 5.7 for pssMp.

Theorem 11.16 *Assume that Z is a conservative real self-similar Markov process. Moreover, suppose that the MAP $((\xi, J), \mathbf{P})$, associated with Z through the Lamperti–Kiu transform, is such that ξ is not concentrated on a lattice and its ascending ladder height process H^+ satisfies $\mathbf{E}_{0,\pi}[H_1^+] < \infty$. Then $P_0 := \lim_{x\downarrow 0} P_x$ exists, in the sense convergence on the Skorokhod space, under which Z leaves the origin continuously. Conversely, if $\mathbf{E}_{0,\pi}[H_1^+] = \infty$, then this limit does not exist. Under the additional assumption that $\mathbf{E}_{0,\pi}[\xi_1] > 0$, for any positive measurable function f and $t > 0$,*

$$E_0[f(Z_t)] = \frac{1}{\alpha \hat{\mathbf{E}}_{0,\pi}[|\xi_1|]} \sum_{i=\pm 1} \pi_i \hat{\mathbf{E}}_{0,i}\left[\frac{1}{I_\infty} f\left(i\,(t/I_\infty)^{1/\alpha}\right)\right], \tag{11.33}$$

where $I_\infty = \int_0^\infty \exp\{\alpha \xi_s\} \mathrm{d}s$.

Similarly, with the same notion of recurrent extension described in Theorem 5.7, we have the following result giving a complete characterisation of when P_0 can be defined as such in the non-conservative setting.

Theorem 11.17 *Assume that X is a non-conservative real self-similar Markov process with index of self-similarity given by $\alpha > 0$. Suppose that $((\xi, J), \mathbf{P})$ is the (killed) MAP associated with X through the Lamperti–Kiu transform. Then there exists a unique recurrent extension of X which leaves 0 continuously if and only if there exists a Cramér number $\beta \in (0, \alpha)$ such*

$$\sum_{j=\pm 1} \mathbf{E}_{0,i}[\mathrm{e}^{\beta \xi_1}; J_1 = j] v_j = v_i, \qquad i = \pm 1,$$

or equivalently $\chi(\beta) = 0$.

The same questions may be asked in dimensions $d \geq 2$. However, significantly less is known at the time of writing in terms of the characterisation of entrance laws, and there are still many open questions in that direction.

11.7 Comments

The structure of real self-similar Markov processes has been investigated by Chybiryakov [51] and Kiu [111] in the symmetric case and by Chaumont et al. [47] in the one-dimensional case. In Section 11.1 we summarise the representation of self-similar Markov processes in one dimension based on Chaumont et al. [47], albeit with an interpretation of their results in terms of a two-state Markov additive process.

The convenience of MAPs to describe the Lamperti–Kiu transform was first noticed in Kuznetsov et al. [120]. The representation of MAPs in Proposition 11.2 is classical and has been previously explored in the context of queuing and ruin theory; see, for example, Asmussen [7] and Asmussen and Albrecher [8]. Our presentation Section 11.2 also makes use of observations taken from Kuznetsov et al. [120] and Kyprianou [124].

The notion of a Wiener–Hopf factorisation for Markov additive processes is well known and it dates back to early studies of queuing theory and inventory models. In this respect, the book of Prabhu [174], as well as Asmussen's book on queueing theory [7], focus predominantly on the setting of Markov additive random walks. Kaspi [102] has explored the continuous case using excursion theory and first showed a version of Theorem 11.9. Sections 11.3 and 11.4 are largely based on the excursion theoretic treatment of MAPs found in the Appendix of Dereich et al. [57].

Section 11.5 introduces general MAPs, much of which can be found in the work of Çinlar [52, 53, 54], and how they fit into a generalised Lamperti–Kiu transformation as found in Kiu [111], Alili et al. [1] and Kyprianou et al. [133]. The isotropic properties of MAPs in relation to their associated ssMp is taken from Kyprianou [125].

Finally, the inclusion of 0 to the state space in the definition of a rssMp in the conservative case was resolved by Dereich et al. [57]. The representation of the entrance law (11.33) has not been developed in the literature, with the exception of the special setting of a stable process conditioned to avoid the origin (cf. Kyprianou et al. [129]). Nonetheless, it is not difficult to derive by copying ideas from the proof of the analogous result in Theorem 5.3, for example, using ideas from Bertoin and Yor [27]. Theorem 11.17 was derived in Pardo et al. [161]. As alluded to at the end of Section 11.6, little is known about entrance laws in the general setting of d-dimensional ssMp. The most recent contribution in this respect is due to Kyprianou et al. [133].

12

Stable processes as self-similar Markov processes

In the previous chapter we have shown how self-similar Markov processes may be expressed in terms of MAPs via the Lamperti–Kiu transform. Moreover, when reflecting on the fact that the Lamperti–Kiu transform is to a ssMp what the Lamperti transform is to a pssMp, we have also shown how MAPs play an analogous role to Lévy processes. Ultimately, we are interested in understanding the structure of the MAPs that underlie stable processes, which are of course ssMp.

This is precisely what we do in this chapter. Indeed, we look a little further beyond the relationship of stable processes to MAPs and consider a special family of self-similar Markov processes that are related to stable processes through a Doob h-transform. The latter turns out to be the result of conditioning a stable process to either avoid or limit continuously to the origin.

The highlight of this chapter, however, is the Riesz–Bogdan–Żak transform. This uses the relationship between stable processes and their underlying MAPs to furnish us with a remarkable connection between space-time transformed stable processes and the conditioned stable processes considered in this chapter. The Riesz–Bogdan–Żak transform turns out to be an incredibly robust tool for developing fluctuation identities. In this respect, we will use it extensively in Chapters 14 and 15. However, as a first example of its applicability, we return to Theorem 6.12 in this chapter and provide its missing proof using the Riesz–Bogdan–Żak transform.

12.1 Stable processes and their h-transforms as ssMp

In one dimension, the most obvious example of a rssMp, which is not a pssMp, is a two-sided jumping stable process killed on hitting the origin. Recall that $(X, \mathbb{P}_x)$ denotes the stable process starting from $x \in \mathbb{R}$. The qualification of hitting the origin is an issue if and only if $\alpha \in (1,2)$ as otherwise, the stable process almost surely never hits the origin. Nonetheless we consider both regimes in this section. We name the underlying process that emerges through the Lamperti–Kiu transform a *Lamperti-stable MAP*. For this fundamental example, we can compute the associated characteristic matrix exponent explicitly. Recall that the state space of the underlying modulating chain in the Lamperti-stable MAP is $\{-1,1\}$. Accordingly, we henceforth arrange any matrix $\boldsymbol{A}$ pertaining to this MAP with the ordering

$$\begin{pmatrix} A_{1,1} & A_{1,-1} \\ A_{-1,1} & A_{-1,-1} \end{pmatrix}.$$

Lemma 12.1 *Suppose that X is a stable process with two-sided jumps. The characteristic matrix exponent of the Lamperti-stable MAP is given by*

$$\boldsymbol{\Psi}(z) = \begin{bmatrix} -\dfrac{\Gamma(\alpha - \mathrm{i}z)\Gamma(1+\mathrm{i}z)}{\Gamma(\alpha\hat{\rho} - \mathrm{i}z)\Gamma(1-\alpha\hat{\rho}+\mathrm{i}z)} & \dfrac{\Gamma(\alpha - \mathrm{i}z)\Gamma(1+\mathrm{i}z)}{\Gamma(\alpha\hat{\rho})\Gamma(1-\alpha\hat{\rho})} \\ \\ \dfrac{\Gamma(\alpha - \mathrm{i}z)\Gamma(1+\mathrm{i}z)}{\Gamma(\alpha\rho)\Gamma(1-\alpha\rho)} & -\dfrac{\Gamma(\alpha - \mathrm{i}z)\Gamma(1+\mathrm{i}z)}{\Gamma(\alpha\rho - \mathrm{i}z)\Gamma(1-\alpha\rho+\mathrm{i}z)} \end{bmatrix}, \quad (12.1)$$

for $z \in \mathbb{R}$. Moreover, the relation (12.1) can be analytically extended in $\mathbb{C}$ so that $\mathrm{Re}(\mathrm{i}z) \in (-1,\alpha)$.

Proof In (11.5) it suffices only to compute $\Psi_1(z)$, $q_{1,-1}$ and $G_{1,-1}(z)$. Indeed, this follows by virtue of the fact that $-X$ has the same law as X, albeit with the roles of ρ and $\hat{\rho}$ interchanged and so, for example, $\Psi_{-1}(z) = \Psi_1(z)|_{\rho\leftrightarrow\hat{\rho}}$.

We also note that until X first enters $(-\infty,0)$, its trajectory agrees with that of the pssMp given in (5.12). It follows immediately that

$$\Psi_1(z) = \Psi^*(z) - \Psi^*(0), \qquad z \in \mathbb{R},$$

where Ψ^* was defined in (5.16), and that $q_{1,-1} = \Psi^*(0)$. Since $q_{1,1} = -q_{1,-1}$, we have that

$$[\boldsymbol{\Psi}(z)]_{1,1} = -\Psi_1(z) - q_{1,1} = -\Psi^*(z), \qquad z \in \mathbb{R}.$$

It remains to identify $[\boldsymbol{\Psi}(z)]_{1,-1}$. To this end, we can appeal to the Lamperti–Kiu transform to identify the distributional equality

$$\mathrm{e}^{\Delta_{1,-1}} = \frac{|X_{\tau_0^-}|}{X_{\tau_0^- -}},$$

where $\tau_0^- = \inf\{t > 0 \colon X_t < 0\}$. This has the implication that

$$G_{1,-1}(z) = \mathbb{E}_x\left[\left(-\frac{X_{\tau_0^-}}{X_{\tau_0^- -}}\right)^{\mathrm{i}z}\right], \qquad z \in \mathbb{R},$$

which is convenient as this transform was computed in (5.43). We now have that

$$[\boldsymbol{\Psi}(z)]_{1,-1} = q_{1,-1}G_{1,-1}(z) = \Psi^*(0)\frac{\Gamma(\alpha - \mathrm{i}z)\Gamma(1+\mathrm{i}z)}{\Gamma(\alpha)} = \frac{\Gamma(\alpha - \mathrm{i}z)\Gamma(1+\mathrm{i}z)}{\Gamma(\alpha\hat{\rho})\Gamma(1-\alpha\hat{\rho})},$$

for $z \in \mathbb{R}$, as required.

The analytic extension of (12.1) is now a straightforward procedure. □

Without checking the value of $\chi'(0)$, we are able to deduce the long term behaviour of the Lamperti-stable MAP from the transience/recurrence properties of the stable process.

We know that, when $\alpha \in (1,2)$, the stable process is recurrent and $\mathbb{P}_x(\tau^{\{0\}} < \infty) = 1$, for all $x \neq 0$. In that case, the Lamperti–Kiu representation dictates that

$$\lim_{t\to\infty} \xi_t = -\infty.$$

When $\alpha \in (0,1)$, we also know that the stable process is transient and never hits the origin so that $\mathbb{P}_x(\lim_{t\to\infty} |X_t| = \infty) = 1$, for all $x \in \mathbb{R}$. Once again, the Lamperti–Kiu transform tells us that

$$\lim_{t\to\infty} \xi_t = \infty.$$

Finally, when $\alpha = 1$, we know that $\limsup_{t\to\infty} |X_t| = \infty$ and $\liminf_{t\to\infty} |X_t| = 0$ and X never hits the origin. As such, the Lamperti–Kiu representation forces us to conclude that ξ oscillates.

There is a second example of a rssMp that we can describe to the same degree of detail as stable processes. This comes about by a change of measure, which corresponds to a Doob h-transform to the semigroup of a two-sided jumping stable process killed on first hitting the origin if $\alpha \in (1,2)$.

Proposition 12.2 *Suppose that $(X, \mathbb{P}_x)$, $x \in \mathbb{R}$, is a one-dimensional stable process with two-sided jumps. Recall that $(\mathcal{F}_t, t \geq 0)$ is the filtration generated by X, which is naturally enlarged (see Remark A.13 in the Appendix) and that we use ζ for the lifetime of a killed process. Consider the change of measure*

$$\left.\frac{\mathrm{d}\mathbb{P}^\circ_x}{\mathrm{d}\mathbb{P}_x}\right|_{\mathcal{F}_t} = \frac{h(X_t)}{h(x)}\mathbf{1}_{(t<\zeta)}, \tag{12.2}$$

where

$$h(x) = \left(\sin(\pi\alpha\hat{\rho})\mathbf{1}_{(x\geq 0)} + \sin(\pi\alpha\rho)\mathbf{1}_{(x<0)}\right)|x|^{\alpha-1} \tag{12.3}$$

and $\zeta = \inf\{t > 0 \colon X_t = 0\}$. Then $(X, \mathbb{P}^\circ_x)$, $x \in \mathbb{R}\backslash\{0\}$, is a rssMp with matrix exponent given by

$$\boldsymbol{\Psi}^{\circ}(z) = \begin{bmatrix} -\dfrac{\Gamma(1-\mathrm{i}z)\Gamma(\alpha+\mathrm{i}z)}{\Gamma(1-\alpha\rho-\mathrm{i}z)\Gamma(\alpha\rho+\mathrm{i}z)} & \dfrac{\Gamma(1-\mathrm{i}z)\Gamma(\alpha+\mathrm{i}z)}{\Gamma(\alpha\rho)\Gamma(1-\alpha\rho)} \\ \\ \dfrac{\Gamma(1-\mathrm{i}z)\Gamma(\alpha+\mathrm{i}z)}{\Gamma(\alpha\hat{\rho})\Gamma(1-\alpha\hat{\rho})} & -\dfrac{\Gamma(1-\mathrm{i}z)\Gamma(\alpha+\mathrm{i}z)}{\Gamma(1-\alpha\hat{\rho}-\mathrm{i}z)\Gamma(\alpha\hat{\rho}+\mathrm{i}z)} \end{bmatrix}, \tag{12.4}$$

for $\mathrm{Re}(\mathrm{i}z) \in (-\alpha, 1)$.

Proof The basic idea of the proof is to show that the change of measure (12.2) is induced by a change of measure on the MAP underlying the stable process. A straightforward computation using the reflection formula for gamma functions shows that, for $\mathrm{Re}(\mathrm{i}z) \in (-1, \alpha)$,

$$\begin{aligned}\det\boldsymbol{\Psi}(z) = &\frac{\Gamma(\alpha-\mathrm{i}z)^2\Gamma(1+\mathrm{i}z)^2}{\pi^2} \\ &\times \{\sin(\pi(\alpha\rho-\mathrm{i}z))\sin(\pi(\alpha\hat{\rho}-\mathrm{i}z)) - \sin(\pi\alpha\rho)\sin(\pi\alpha\hat{\rho})\}.\end{aligned}$$

We see that $\det\boldsymbol{\Psi}(z) = 0$ has a root at $\mathrm{i}z = \alpha - 1$. Indeed, one verifies

$$\begin{aligned}\sin(\pi(\alpha\rho-\alpha+1))\sin(\pi(\alpha\hat{\rho}-\alpha+1)) &= \sin(\pi(-\alpha\hat{\rho}+1))\sin(\pi(-\alpha\rho+1)) \\ &= \sin(\pi\alpha\rho)\sin(\pi\alpha\hat{\rho}).\end{aligned}$$

In turn, this implies that $\chi(\alpha-1) = 0$. Note also that, for this particular value of γ, we have that $\gamma \in (-1, \alpha)$ and hence $\mathrm{Re}(\mathrm{i}z) \in (-1, \alpha)$ when $z = -\mathrm{i}\gamma$.

One also easily checks that

$$\boldsymbol{v}(\alpha-1) \propto \begin{bmatrix} \sin(\pi\alpha\hat{\rho}) \\ \sin(\pi\alpha\rho) \end{bmatrix}.$$

We claim that with $\gamma = \alpha - 1$, the exponential change of measure (11.10) corresponds precisely to (12.2) when (ξ, J) is the MAP underlying the stable process. To see this, first note that the time change $\varphi(t)$ is a stopping time and so we consider the change of measure (11.10) at this stopping time. In this respect, thanks to the Lamperti–Kiu transform, we use $\exp\{\xi_{\varphi(t)}\} = |X_t|$, $J_{\varphi(t)} = \mathrm{sgn}(X_t)$ and the ratio of constants, coming from (12.3), as they appear in the expression for (12.2) matches the term $v_{J_{\varphi(t)}}(\alpha-1)/v_{J_0}(\alpha-1)$.

It is now a straightforward exercise to check from (11.11) that the MAP associated to the process $(X, \mathbb{P}^{\circ}_x)$, $x \in \mathbb{R}\backslash\{0\}$, that is to say $\boldsymbol{\Psi}_{\alpha-1}(z)$ agrees with $\boldsymbol{\Psi}^{\circ}(z)$, for $\mathrm{Re}(\mathrm{i}z) \in (-\alpha, 1)$. The reader will note that in doing this, it will be useful to use the reflection formula for the gamma function in dealing with the terms coming from $\boldsymbol{\Delta}_{\boldsymbol{v}}(\alpha-1)$ in the expression for $\boldsymbol{\Psi}_{\alpha-1}(z)$ coming from (11.11). □

It is an interesting exercise to verify directly the scaling property of $(X, \mathbb{P}^\circ_x)$, $x \in \mathbb{R}\backslash\{0\}$. This can be done much in the spirit of the computation in (5.30), noting that the stopping time $\tau^{\{0\}}$ scales in a similar way to (5.38) from the scaling of X. Indeed, if, for each $c > 0$, we let $X^c_t = cX_{c^{-\alpha}t}$, $t \geq 0$, and write $\tau^{\{0\}}_c = \inf\{t > 0\colon X^c_t = 0\}$ (reserving $\tau^{\{0\}}$ in place of $\tau^{\{0\}}_1$), then we have

$$\tau^{\{0\}}_c = c^\alpha \inf\{c^{-\alpha}t > 0\colon cX_{c^{-\alpha}t} = 0\} = c^\alpha \tau^{\{0\}}.$$

Together with the fact that, for $c > 0$, $h(cx) = c^{\alpha-1}h(x)$, we have, for bounded measurable f, $x \in \mathbb{R}\backslash\{0\}$ and $t \geq 0$,

$$\begin{aligned}
\mathbb{E}^\circ_x[f(X^c_s\colon s \leq t), t < \zeta] &= \mathbb{E}_x\left[\frac{c^{\alpha-1}h(X_{c^{-\alpha}t})}{c^{\alpha-1}h(x)} f(cX_{c^{-\alpha}s}\colon s \leq t)\mathbf{1}_{(c^{-\alpha}t<\tau^{\{0\}})}\right] \\
&= \mathbb{E}_x\left[\frac{h(X^c_t)}{h(cx)} f(X^c_s, s \leq t)\mathbf{1}_{(t<\tau^{\{0\}}_c)}\right] \\
&= \mathbb{E}^\circ_{cx}\left[f(X_s, s \leq t)\right].
\end{aligned} \tag{12.5}$$

In other words, the law of $(X^c, \mathbb{P}^\circ_x)$ agrees with $(X, \mathbb{P}^\circ_{cx})$ for $x \in \mathbb{R}\backslash\{0\}$.

Intuitively speaking, when $\alpha \in (0, 1)$, the change of measure (12.2) rewards paths that visit close neighbourhoods of the origin and penalises paths that wander large distances away from the origin. Conversely, when $\alpha \in (1, 2)$, the change of measure does the opposite. It penalises those paths that approach the origin and rewards those that stray away from the origin. As we shall shortly see in Section 12.2, for $\alpha \in (0, 1)$, the change of measure (12.2) corresponds to conditioning the stable process, in an appropriate sense, to limit continuously to the origin and, when $\alpha \in (1, 2)$, it corresponds to conditioning the process to avoid the origin.

Let us now turn to higher dimensions. The most prominent example of a d-dimensional ssMp that will be of use to us is of course the isotropic stable process in $\mathbb{R}^d$ itself. Recall that $(X, \mathbb{P}_x)$, denotes the isotropic d-dimensional stable process starting from $x \in \mathbb{R}^d$, whose characteristic exponent is of the form

$$\Psi(z) = |z|^\alpha, \qquad z \in \mathbb{R}^d,$$

and, accordingly, from (1.47), the underlying Lévy measure satisfies

$$\Pi(B) = c(\alpha)\int_{\mathbb{S}^{d-1}} \sigma_1(\mathrm{d}\phi)\int_{(0,\infty)} \mathbf{1}_B(r\phi)\frac{\mathrm{d}r}{r^{\alpha+1}} = c(\alpha)\int_B \frac{1}{|z|^{\alpha+d}}\mathrm{d}z,$$

for $B \in \mathcal{B}(\mathbb{R}^d)$, where $\sigma_1(\mathrm{d}\phi)$ is the surface measure on $\mathbb{S}^{d-1}$ normalised to have unit mass and

$$c(\alpha) = 2^{\alpha-1}\pi^{-d}\frac{\Gamma((d+\alpha)/2)\Gamma(d/2)}{|\Gamma(-\alpha/2)|}.$$

The description of the underlying MAP is somewhat less straightforward to characterise. We know, however, that the stable process is a pure jump process and accordingly the underlying MAP must be too. The theorem below uses the compensation formula for the jumps of the stable process as a way of capturing the jump dynamics of the MAP. We will use the usual notation in the stable setting. That is, (ξ, Θ) with probabilities $\mathbf{P}_{x,\theta}$, $x \in \mathbb{R}$, $\theta \in \mathbb{S}^d$, is the MAP underlying the stable process. We will work with the increments $\Delta\xi_t = \xi_t - \xi_{t-} \in \mathbb{R}$, $t \geq 0$, and recall that 1 is the 'North Pole'.

Theorem 12.3 *Suppose that f is a positive, bounded, measurable function on $[0,\infty) \times \mathbb{R} \times \mathbb{R} \times \mathbb{S}^{d-1} \times \mathbb{S}^{d-1}$ such that $f(\cdot,\cdot,0,\cdot,\cdot) = 0$, then, for all $\theta \in \mathbb{S}^{d-1}$,*

$$\mathbf{E}_{0,\theta}\left[\sum_{s>0} f(s, \xi_{s-}, \Delta\xi_s, \Theta_{s-}, \Theta_s)\right]$$
$$= \int_0^\infty \int_{\mathbb{R}} \int_{\mathbb{S}^{d-1}} \int_{\mathbb{S}^{d-1}} \int_{\mathbb{R}} V_\theta(\mathrm{d}s, \mathrm{d}x, \mathrm{d}\vartheta)\sigma_1(\mathrm{d}\phi)\mathrm{d}y \frac{c(\alpha)\mathrm{e}^{yd}}{|\mathrm{e}^y\phi - \vartheta|^{\alpha+d}} f(s, x, y, \vartheta, \phi), \tag{12.6}$$

where

$$V_\theta(\mathrm{d}s, dx, \mathrm{d}\vartheta) = \mathbf{P}_{0,\theta}(\xi_s \in \mathrm{d}x, \Theta_s \in \mathrm{d}\vartheta)\mathrm{d}s, \qquad x \in \mathbb{R}, \vartheta \in \mathbb{S}^{d-1}, s \geq 0,$$

is the space-time potential of (ξ, Θ), $\sigma_1(d\phi)$ is the surface measure on $\mathbb{S}^{d-1}$ normalised to have unit mass and $c(\alpha) = 2^{\alpha-1}\pi^{-d}\Gamma((d+\alpha)/2)\Gamma(d/2)/\left|\Gamma(-\alpha/2)\right|$.

Proof According to the generalised Lamperti–Kiu transformation (11.26), we have

$$\xi_t = \log(|X_{A(t)}|/|X_0|), \qquad \Theta_t = \frac{X_{A(t)}}{|X_{A(t)}|}, \qquad t \geq 0,$$

where $A(t) = \inf\{s > 0 \colon \int_0^s |X_u|^{-\alpha}\mathrm{d}u > t\}$; see also (11.29). Let f be given as in the statement of the theorem. Writing the left-hand side of (12.6) in terms of the stable process, we have for all $\theta \in \mathbb{S}^{d-1}$,

$$\mathbf{E}_{0,\theta}\left[\sum_{s>0} f(s, \xi_{s-}, \Delta\xi_s, \Theta_{s-}, \Theta_s)\right]$$
$$= \mathbb{E}_\theta\left[\sum_{s>0} f\left(\int_0^s |X_u|^{-\alpha}\mathrm{d}u, \log|X_{s-}|, \log\left(|X_s|/|X_{s-}|\right), \mathrm{Arg}(X_{s-}), \mathrm{Arg}(X_s)\right)\right].$$

Next note that, for $t \geq 0$,

$$\frac{|X_s|}{|X_{s-}|} = \left|\frac{X_{s-}}{|X_{s-}|} + \frac{\Delta X_s}{|X_{s-}|}\right| = \left|\mathrm{Arg}(X_{s-}) + \frac{\Delta X_s}{|X_{s-}|}\right|$$

and

$$\mathrm{Arg}(X_s) = \frac{X_s}{|X_s|} = \frac{\dfrac{X_{s-}}{|X_{s-}|} + \dfrac{\Delta X_s}{|X_{s-}|}}{\left|\dfrac{X_{s-}}{|X_{s-}|} + \dfrac{\Delta X_s}{|X_{s-}|}\right|} = \frac{\mathrm{Arg}(X_{s-}) + \dfrac{\Delta X_s}{|X_{s-}|}}{\left|\mathrm{Arg}(X_{s-}) + \dfrac{\Delta X_s}{|X_{s-}|}\right|}.$$

The compensation formula for the Poisson random measure of jumps of X now tells us that

$$\begin{aligned}
&\mathbf{E}_{0,\theta}\left[\sum_{s>0} f(s, \xi_{s-}, \Delta\xi_s, \Theta_{s-}, \Theta_s)\right] \\
&= \mathbb{E}_\theta\Bigg[\int_0^\infty \mathrm{d}s \int_{\mathbb{S}^{d-1}} \sigma_1(\mathrm{d}\phi) \int_0^\infty \mathrm{d}r \frac{c(\alpha)}{r^{1+\alpha}} \\
&\qquad f\Bigg(\int_0^s |X_u|^{-\alpha}\mathrm{d}u, \log|X_{s-}|, \log\left|\mathrm{Arg}(X_{s-}) + \frac{r\phi}{|X_{s-}|}\right|, \\
&\qquad\qquad \mathrm{Arg}(X_{s-}), \frac{\mathrm{Arg}(X_{s-}) + \frac{r\phi}{|X_{s-}|}}{\left|\mathrm{Arg}(X_{s-}) + \frac{r\phi}{|X_{s-}|}\right|}\Bigg)\Bigg] \\
&= \mathbf{E}_{0,\theta}\left[\int_0^\infty \mathrm{d}v \int_{\mathbb{S}^{d-1}} \sigma_1(\mathrm{d}\phi) \int_0^\infty \mathrm{d}u \frac{c(\alpha)}{u^{1+\alpha}} f\left(v, \xi_v, \log|\Theta_v + u\phi|, \Theta_v, \frac{\Theta_v + u\phi}{|\Theta_v + u\phi|}\right)\right] \\
&= \mathbf{E}_{0,\theta}\left[\int_0^\infty \mathrm{d}v \int_{\mathbb{R}^d} \mathrm{d}z \frac{\tilde{c}(\alpha)}{|z|^{\alpha+d}} f\left(v, \xi_v, \log|\Theta_v + z|, \Theta_v, \frac{\Theta_v + z}{|\Theta_v + z|}\right)\right] \\
&= \mathbf{E}_{0,\theta}\left[\int_0^\infty \mathrm{d}v \int_{\mathbb{R}^d} \mathrm{d}w \frac{\tilde{c}(\alpha)}{|w - \Theta_v|^{\alpha+d}} f\left(v, \xi_v, \log|w|, \Theta_v, \frac{w}{|w|}\right)\right],
\end{aligned} \tag{12.7}$$

where in the second equality, we first make the change of variables $u = r/|X_{s-}|$ and then $v = \int_0^s |X_u|^{-\alpha}\mathrm{d}u$ and, in the third equality, we convert to Cartesian coordinates with $\tilde{c}(\alpha) = 2^{\alpha-1}\pi^{-d}\Gamma((d+\alpha)/2)\Gamma(d/2)/\left|\Gamma(-\alpha/2)\right|$. Converting the right-hand side above back to skew product variables, we thus get

$$\begin{aligned}
&\mathbf{E}_{0,\theta}\left[\sum_{s>0} f(s, \xi_{s-}, \Delta\xi_s, \Theta_{s-}, \Theta_s)\right] \\
&= \mathbf{E}_{0,\theta}\left[\int_0^\infty \mathrm{d}v \int_{\mathbb{S}^{d-1}} \sigma_1(\mathrm{d}\phi) \int_0^\infty \mathrm{d}r \frac{c(\alpha) r^{d-1}}{|r\phi - \Theta_v|^{\alpha+d}} f(v, \xi_v, \log r, \Theta_v, \phi)\right] \\
&= \mathbf{E}_{0,\theta}\left[\int_0^\infty \mathrm{d}v \int_{\mathbb{S}^{d-1}} \sigma_1(\mathrm{d}\phi) \int_{\mathbb{R}} \mathrm{d}y \frac{c(\alpha)\mathrm{e}^{yd}}{|\mathrm{e}^y\phi - \Theta_v|^{\alpha+d}} f(v, \xi_v, y, \Theta_v, \phi)\right] \\
&= \int_0^\infty \int_{\mathbb{R}} \int_{\mathbb{S}^{d-1}} \int_{\mathbb{S}^{d-1}} \int_{\mathbb{R}} V_\theta(\mathrm{d}v, \mathrm{d}x, \mathrm{d}\vartheta)\sigma_1(\mathrm{d}\phi)\mathrm{d}y \frac{c(\alpha)\mathrm{e}^{yd}}{|\mathrm{e}^y\phi - \vartheta|^{\alpha+d}} f(v, x, y, \vartheta, \phi),
\end{aligned} \tag{12.8}$$

as required. □

The radial component of an isotropic d-dimensional stable process, which can be singled out by Corollary 11.15, has already been studied in Theorem 5.21.

The second example of a d-dimensional ssMp takes inspiration from Proposition 12.2.

Proposition 12.4 *In the spirit of* (12.2) *we define, for an isotropic d-dimensional stable process,* $(X, \mathbb{P}_x)$, $x \in \mathbb{R}^d\backslash\{0\}$,

$$\left.\frac{\mathrm{d}\mathbb{P}^\circ_x}{\mathrm{d}\mathbb{P}_x}\right|_{\mathcal{F}_t} = \frac{|X_t|^{\alpha-d}}{|x|^{\alpha-d}}\mathbf{1}_{(t<\zeta)}, \qquad t \geq 0, \tag{12.9}$$

where $(\mathcal{F}_t, t \geq 0)$ *is the usual naturally enlarged filtration for X (cf. Remark A.13 in the Appendix) and* $\zeta = \inf\{t > 0\colon |X_t| = 0\}$. *For* $d \geq 2$, *the process* $(X, \mathbb{P}^\circ)$ *is a ssMp. Moreover,* $(|X|, \mathbb{P}^\circ)$ *is a pssMp with underlying Lévy process that has characteristic exponent*

$$\Psi^\circ(z) = 2^\alpha \frac{\Gamma(\frac{1}{2}(-\mathrm{i}z + d))}{\Gamma(-\frac{1}{2}(\mathrm{i}z + \alpha - d))} \frac{\Gamma(\frac{1}{2}(\mathrm{i}z + \alpha))}{\Gamma(\frac{1}{2}\mathrm{i}z)}, \qquad z \in \mathbb{R}. \tag{12.10}$$

Proof The proof uses similar reasoning to the proof of Proposition 12.2. Recalling that Ψ in (5.50) is the characteristic exponent of the Lévy process ξ which underlies the radial component of a stable process in d-dimensions, we easily verify that $\Psi(-\mathrm{i}(\alpha - d)) = 0$. It follows that

$$\exp\{(\alpha - d)\xi_t\}, \qquad t \geq 0, \tag{12.11}$$

is a martingale. Moreover, under the change of measure induced by this martingale, ξ remains in the class of Lévy processes, but now with characteristic exponent $\Psi^\circ(z) = \Psi(z - \mathrm{i}(\alpha - d))$, $z \in \mathbb{R}$. Noting that $\varphi(t)$ is a stopping time in the filtration of ξ, we see that (12.9) also represents the aforesaid change of measure.

It is now clear that at least $(|X|, \mathbb{P}^\circ)$ is pssMp with Ψ° characterising its underlying Lévy process. However, technically speaking, we don't know if the underlying MAP (ξ, Θ) remains in the family of MAPs under the change of measure (12.11). Hence, it is not necessarily clear that $(X, \mathbb{P}^\circ)$ is a ssMp. This can be resolved by appealing to calculations similar to those in (12.5), which tell us that this is the case, moreover, the Feller property is easily seen to be preserved by the change of measure (12.9). □

We can again note that, for $d \geq 2$, the change of measure (12.9) rewards paths that remain close to the origin and penalises those that stray far from the origin. Recalling that stable processes are transient in dimension two and above, we will see in the next section that, just as in Proposition 12.2 when

$\alpha < d = 1$, $\mathbb{P}^\circ_x$, $x \in \mathbb{R}^d\backslash\{0\}$, corresponds to conditioning the stable process to limit continuously to the origin.

We complete this section by noting that our choice of notation $\mathbb{P}^\circ_x$, $x \in \mathbb{R}^d\backslash\{0\}$, is consistent across Propositions 12.2 and 12.4. Indeed, if we understand $d = 1$ in the latter proposition (in particular, X is taken to be a one-dimensional symmetric stable process), then it agrees with the conclusion of the former.

12.2 Stable processes conditioned to avoid or hit 0

In this section, as alluded to previously, we demonstrate that the change of measure that leads to $\mathbb{P}^\circ_x$, $x \subset \mathbb{R}^d\backslash\{0\}$, defined by (12.2) for $d = 1$ and by (12.9) for $d \geq 2$, corresponds to conditioning the path of a stable process. We exclude the case $\alpha = d = 1$ from the discussion as the density in the change of measure in (12.2) is equal to unity.

In one dimension, we shall implicitly understand that there are two-sided jumps (i.e. $0 < \alpha\rho, \alpha\hat{\rho} < 1$). Recall that such stable processes cannot creep and have the property that, when $\alpha \in (1, 2)$, $\mathbb{P}_x(\tau^{\{0\}} < \infty) = 1$, for all $x \neq 0$, and hence the process is recurrent. Otherwise, when $\alpha \in (0, 1]$, $\mathbb{P}_x(\tau^{\{0\}} = \infty) = 1$ and the process is transient. Moreover, in higher dimensions, we also recall that the isotropic stable process cannot hit points and is transient.

Theorem 12.5 *Suppose that X is a d-dimensional stable process with index $\alpha \in (0, 2)$, where we understand it to have two-sided jumps in the case $d = 1$ and to be isotropic in the case $d \geq 2$. Define ζ for the lifetime of killed trajectories on the Skorokhod space (see Section A.10 in the Appendix), for any Borel set D, let $\tau^D := \inf\{s \geq 0 \colon X_s \in D\}$ and define $B_a = \{x \colon |x| < a\}$.*

(i) If $d > \alpha$, then, for all $A \in \mathcal{F}_t$, $t > 0$,

$$\mathbb{P}^\circ_x(A,\ t < \zeta) = \lim_{a\to 0} \mathbb{P}_x(A \cap \{t < \tau^{B_a}\} \,|\, \tau^{B_a} < \infty), \tag{12:12}$$

so that the process $(X, \mathbb{P}^\circ_x)$, $x \neq 0$, is non-conservative.

(ii) If $d < \alpha$, then, for all $A \in \mathcal{F}_t$, $t \geq 0$,

$$\mathbb{P}^\circ_x(A) = \lim_{a\to\infty} \mathbb{P}_x(A \cap \{t < \tau^{B_a^c}\} \,|\, \tau^{B_a^c} < \tau^{\{0\}}), \tag{12.13}$$

so that the process $(X, \mathbb{P}^\circ_x)$, $x \neq 0$, is conservative.

Remark 12.6 We note that the first part of the above theorem corresponds to conditioning the process to hit 0 continuously, whilst the second part corresponds to conditioning it to avoid the origin. Accordingly, we refer to $(X, \mathbb{P}^\circ)$ in these terms.

Proof of Theorem 12.5 Recalling that $\tau^{\{0\}} = \infty$ almost surely for $d > \alpha$, a simple application of Bayes formula and the Markov property tells us that

$$\mathbb{P}_x(A \cap \{t < \tau^{B_a}\} \,|\, \tau^{B_a} < \tau^{\{0\}}) = \mathbb{E}_x\left[\mathbf{1}_{(A\cap\{t<\tau^{B_a}\})} \frac{\mathbb{P}_{X_t}(\tau^{B_a} < \tau^{\{0\}})}{\mathbb{P}_x(\tau^{B_a} < \tau^{\{0\}})}\right]. \tag{12.14}$$

When $1 = d > \alpha$, Corollary 6.11 gives us that, for example, for $x > a$,

$$\mathbb{P}_x(\tau^{B_a} < \infty) = \frac{\Gamma(1-\alpha\rho)}{\Gamma(\alpha\hat{\rho})\Gamma(1-\alpha)} \int_{\frac{x-a}{x+a}}^{1} t^{\alpha\hat{\rho}-1}(1-t)^{-\alpha}\,\mathrm{d}t.$$

One may deduce that, for $x, y \neq 0$,

$$\sup_{|a|<1} \frac{\mathbb{P}_y(\tau^{B_a} < \infty)}{\mathbb{P}_x(\tau^{B_a} < \infty)} < C\left|\frac{y}{x}\right|^{\alpha-1} \quad \text{and} \quad \lim_{a\to 0} \frac{\mathbb{P}_y(\tau^{B_a} < \infty)}{\mathbb{P}_x(\tau^{B_a} < \infty)} = \frac{h(y)}{h(x)}, \tag{12.15}$$

where h was given in (12.3). Noting that the role of y is played by X_t in (12.15), the proof of (12.12) is completed using dominated convergence in (12.14), with the help of (3.17) and the remarks on moments in Section 2.18.

In the case that $d \geq 2(> \alpha)$, we note that, for $|x| > a$,

$$\mathbb{P}_x(\tau^{B_a} < \infty) = \mathbf{P}_{0,\mathbf{1}}(\tau^{-}_{\log(a/|x|)} < \infty),$$

where $\tau^{-}_{\log(a/|x|)} = \inf\{t > 0 \colon \xi_t < \log(a/|x|)\}$ and $\mathbf{1} = (1, 0, \ldots, 0)$ is the 'North Pole' on $\mathbb{S}^{d-1}$ and ξ is the Lévy process that underlies $|X|$ via the Lamperti transform (cf. Section 5.7). Recalling that ξ is a hypergeometric Lévy process with exponent (5.50), momentarily assuming its descending ladder renewal measure has a density on $(0, \infty)$, say $\hat{u}$, we note that it must satisfy

$$\int_0^\infty \mathrm{e}^{-\lambda y}\hat{u}(y/2)\mathrm{d}y = 2\frac{\Gamma(\lambda + \frac{1}{2}(d-\alpha))}{\Gamma(\lambda + \frac{1}{2}d)}, \qquad \lambda \geq 0.$$

Now appealing to Lemma 2.26 and Corollary 4.3, we deduce that

$$\mathbb{P}_x(\tau^{B_a} < \infty) \propto \int_{-\log(a/|x|)}^\infty \mathrm{e}^{-(d-\alpha)z}(1-\mathrm{e}^{-2z})^{\frac{1}{2}\alpha-1}\mathrm{d}z = \int_0^{a/|x|} t^{(d-\alpha)-1}(1-t^2)^{\frac{1}{2}\alpha-1}\mathrm{d}t.$$

Reasoning as above,

$$\sup_{|a|<1} \frac{\mathbb{P}_y(\tau^{B_a} < \infty)}{\mathbb{P}_x(\tau^{B_a} < \infty)} < C\left|\frac{y}{x}\right|^{\alpha-d} \quad \text{and} \quad \lim_{a\to 0} \frac{\mathbb{P}_y(\tau^{B_a} < \infty)}{\mathbb{P}_x(\tau^{B_a} < \infty)} = \left|\frac{y}{x}\right|^{\alpha-d},$$

and, again, the proof is completed using dominated convergence, with the help of (3.17).

Now recall from Theorem 6.7, when $\alpha > d(= 1)$, after translation and scaling, also using the integral expression for the Beta function (A.18) in the Appendix, we have that, for example, for $x \in (0, a)$,

$$\mathbb{P}_x(\tau^{B_a^c} < \tau^{\{0\}}) = (\alpha - 1)\left(\frac{x}{a}\right)^{\alpha-1} \int_1^{a/x} (t-1)^{\alpha\rho-1}(t+1)^{\alpha\hat{\rho}-1}\mathrm{d}t.$$

As above, we can easily check that, for $x, y \neq 0$,

$$\sup_{|a|>1} \frac{\mathbb{P}_y(\tau^{B_a^c} < \tau^{\{0\}})}{\mathbb{P}_x(\tau^{B_a^c} < \tau^{\{0\}})} < C\left|\frac{y}{x}\right|^{\alpha-1} \quad \text{and} \quad \lim_{a\to\infty} \frac{\mathbb{P}_y(\tau^{B_a^c} < \tau^{\{0\}})}{\mathbb{P}_x(\tau^{B_a^c} < \tau^{\{0\}})} = \frac{h(y)}{h(x)}.$$

The proof can now be completed as in the other two cases above. □

12.3 One-dimensional Riesz–Bogdan–Żak transform

The changes of measure, (12.2) in one dimension and (12.9) in higher dimension, also play an important role in a remarkable space-time path transformation: the Riesz–Bogdan–Żak transform. Later on in this text, we will use it to analyse a number of complicated path functionals of stable processes in dimensions one and two.

In this section, we introduce the Riesz–Bogdan–Żak transform in one dimension and in the next section, we revisit the case of first entry into a bounded interval to demonstrate usefulness by reproving the first entry distribution in Theorem 6.9 and offering the promised proof of Theorem 6.12 for the resolvent of the stable process until first entry into a strip. Moreover, we also use it to prove Theorem 6.18, giving the density of the potential of a one-dimensional stable process killed on hitting the origin, when $\alpha \in (1, 2)$. The final section of this chapter will give the Riesz–Bogdan–Żak transform in higher dimensions.

Theorem 12.7 (The one-dimensional Riesz–Bogdan–Żak transform) *Suppose that X is a one-dimensional stable process with two-sided jumps. Define*

$$\eta(t) = \inf\left\{s > 0 : \int_0^s |X_u|^{-2\alpha}\mathrm{d}u > t\right\}, \qquad t \geq 0. \tag{12.16}$$

Then, for all $x \in \mathbb{R}\backslash\{0\}$, $(-1/X_{\eta(t)}, \eta(t) < \tau^{\{0\}})$ under $\mathbb{P}_x$ is equal in law to $(X, \mathbb{P}^\circ_{-1/x})$.

Proof First note that, if X is an (α, ρ) stable process, then $-X$ is an $(\alpha, \hat{\rho})$ stable process. Next, we show that $(-1/X_{\eta(t)}, \eta(t) < \tau^{\{0\}})$ is a rssMp with index α by analysing its Lamperti–Kiu decomposition.

To this end, note that, if $(\hat{\xi}, \hat{J})$ is the MAP that underlies $\hat{X} := -X$, then its matrix exponent, say $\hat{\boldsymbol{\Psi}}(z)$, is equal to (12.1) with the roles of ρ and $\hat{\rho}$ interchanged. As $\hat{X}$ is a rssMp, we have

$$\hat{X}_t = \mathrm{e}^{\hat{\xi}_{\varphi(t)}} \hat{J}_{\varphi(t)}, \qquad t < \tau^{\{0\}},$$

where

$$\int_0^{\varphi(t)} \mathrm{e}^{\alpha \hat{\xi}_s} \mathrm{d}s = t.$$

Noting that

$$\int_0^{\eta(t)} \mathrm{e}^{-2\alpha \hat{\xi}_{\varphi(u)}} \mathrm{d}u = t, \qquad \eta(t) < \tau^{\{0\}},$$

a straightforward differentiation of the last two integrals shows that, respectively,

$$\frac{\mathrm{d}\varphi(t)}{\mathrm{d}t} = \mathrm{e}^{-\alpha \hat{\xi}_{\varphi(t)}} \text{ and } \frac{\mathrm{d}\eta(t)}{\mathrm{d}t} = \mathrm{e}^{2\alpha \hat{\xi}_{\varphi\circ\eta(t)}}, \qquad \eta(t) < \tau^{\{0\}}.$$

The chain rule now tells us that

$$\frac{\mathrm{d}(\varphi \circ \eta)(t)}{\mathrm{d}t} = \left.\frac{\mathrm{d}\varphi(s)}{\mathrm{d}s}\right|_{s=\eta(t)} \frac{\mathrm{d}\eta(t)}{\mathrm{d}t} = \mathrm{e}^{\alpha \hat{\xi}_{\varphi\circ\eta(t)}}, \tag{12.17}$$

and hence,

$$\int_0^{\varphi\circ\eta(t)} \mathrm{e}^{-\alpha \hat{\xi}_u} \mathrm{d}u = t, \qquad \eta(t) < \tau^{\{0\}}.$$

The qualification that $\eta(t) < \tau^{\{0\}}$ only matters when $\alpha \in (1,2)$. In that case, the fact that $\mathbb{P}_x(\tau^{\{0\}} < \infty) = 1$ for all $x \in \mathbb{R}$ implies that $\lim_{t\to\infty} \hat{\xi}_t = -\infty$ almost surely. As a consequence, it follows that $\int_0^\infty \mathrm{e}^{-\alpha \hat{\xi}_u} \mathrm{d}u = \infty$ and hence $\lim_{t\to\infty} \varphi \circ \eta(t) = \infty$. That is to say, we have $\lim_{t\to\infty} \eta(t) = \tau^{\{0\}}$. Noting that $1/\hat{J}_s = \hat{J}_s$, $s \geq 0$, it now follows that

$$\frac{1}{\hat{X}_{\eta(t)}} = \mathrm{e}^{-\hat{\xi}_{\varphi\circ\eta(t)}} \hat{J}_{\varphi\circ\eta(t)}, \qquad t < \tau^{\{0\}},$$

is the representation of a rssMp whose underlying MAP has matrix exponent given by $\hat{\boldsymbol{\Psi}}(-z)$, whenever it is well defined. Recalling the definition of $\hat{\boldsymbol{\Psi}}(z)$, we see that the MAP that underlies $(-1/X_{\eta(t)}, t \geq 0)$ via the Lamperti–Kiu transform is identically equal in law to the MAP with matrix exponent given by $\boldsymbol{\Psi}^\circ$ given in (12.4). □

12.4 First entrance into a bounded interval revisited

As promised, now that we have the Riesz–Bogdan–Żak transformation in hand for one-dimensional stable processes, we can return to the problem of describing the first entry of a stable process into a strip. In particular, we can finally give the proof of Theorem 6.12. Throughout this section, we will assume that X has two-sided jumps; that is to say $0 < \alpha\rho, \alpha\hat{\rho} < 1$.

Let us start our discussion by returning to Theorem 6.9. By scaling and spatial homogeneity, we can capture the analogue of the conclusion of Theorem 6.9 in one formula for first entry into the interval $(-1, 1)$,

$$\begin{aligned}
&\mathbb{P}_x(X_{\tau^{(-1,1)}} \in \mathrm{d}y,\ \tau^{(-1,1)} < \infty) \\
&= \frac{\sin(\pi\alpha\hat{\rho})}{\pi}(1+y)^{-\alpha\rho}(1-y)^{-\alpha\hat{\rho}}\Bigg((x-1)^{\alpha\hat{\rho}}(1+x)^{\alpha\rho}(x-y)^{-1} \\
&\qquad -(\alpha-1)_+\int_1^x (t-1)^{\alpha\hat{\rho}-1}(t+1)^{\alpha\rho-1}\,\mathrm{d}t\Bigg)\mathrm{d}y, \qquad (12.18)
\end{aligned}$$

for $y \in (-1, 1)$, where $(\alpha-1)_+ = \max\{\alpha - 1, 0\}$ and

$$\tau^{(-1,1)} = \inf\{t > 0\colon X_t \in (-1, 1)\}.$$

We proved this result by converting the left-hand side into a statement about a first passage problem of $\vec{\xi}$ and taking advantage of the fact that the Wiener–Hopf factorisation of $\vec{\xi}$ is mathematically tractable. In the hands of the Riesz–Bogdan–Żak transformation, however, the right-hand side of (12.18) can be derived directly from the conclusion of Corollary 6.3 (resp. Corollary 6.8) when $\alpha \in (0, 1]$ (resp. when $\alpha \in (1, 2)$).

Consider for example the case that $\alpha \in (0, 1]$. Noting that the stable process cannot hit points, Theorem 12.7 tells us that the range of $-1/X$, agrees, up to a Doob h-transform, with the range of X. In particular, the time change that is involved in the Riesz–Bogdan–Żak transform is of no consequence when considering first passage distributions. Therefore, we have that for bounded measurable functions f on $(-1, 1)$,

$$\begin{aligned}
&\int_{(-1,1)} \frac{h(y)}{h(x)} f(y)\mathbb{P}_x(X_{\tau^{(-1,1)}} \in \mathrm{d}y,\ \tau^{(-1,1)} < \infty) \\
&\qquad = \int_{(-1,1)^c} f(1/y)\hat{\mathbb{P}}_{1/x}(X_{\tau_1^+ \wedge \tau_{-1}^-} \in \mathrm{d}y), \qquad (12.19)
\end{aligned}$$

where $\hat{\mathbb{P}}_z$, $z \in \mathbb{R}$, is the law of $-X$ issued with $-X_0 = z$,

$$\tau_1^+ = \inf\{t > 0\colon X_t > 1\}, \qquad \tau_{-1}^- = \inf\{t > 0\colon X_t < -1\},$$

and h is given in (12.3). Spatial homogeneity in Corollary 6.3 allows us to deduce that, for $z > 1$ and $x \in (-1, 1)$,

$$\begin{aligned}&\mathbb{P}_x(X_{\tau_1^+} \in \mathrm{d}z; \tau_1^+ < \tau_{-1}^-)\\&= \frac{\sin(\pi\alpha\rho)}{\pi}(1-x)^{\alpha\rho}(1+x)^{\alpha\hat{\rho}}(z-1)^{-\alpha\rho}(z+1)^{-\alpha\hat{\rho}}(z-x)^{-1}\mathrm{d}z.\end{aligned}$$

If we now additionally suppose that $\operatorname{supp} f \subset (0, 1)$, then for $x > 1$,

$$\begin{aligned}&\int_{(0,1)} \frac{y^{\alpha-1}}{x^{\alpha-1}} f(y)\mathbb{P}_x(X_{\tau^{(-1,1)}} \in \mathrm{d}y,\ \tau^{(-1,1)} < \infty)\\&= \frac{\sin(\pi\alpha\hat{\rho})}{\pi}\left(1-\frac{1}{x}\right)^{\alpha\rho}\left(1+\frac{1}{x}\right)^{\alpha\hat{\rho}}\int_1^\infty f(1/z)(z-1)^{-\alpha\rho}(z+1)^{-\alpha\hat{\rho}}\left(z-\frac{1}{x}\right)^{-1}\mathrm{d}z\\&= \frac{\sin(\pi\alpha\hat{\rho})}{\pi}(x-1)^{\alpha\rho}(x+1)^{\alpha\hat{\rho}}\int_0^1 \frac{y^{\alpha-1}}{x^{\alpha-1}} f(y)(1-y)^{-\alpha\rho}(1+y)^{-\alpha\hat{\rho}}(x-y)^{-1}\,\mathrm{d}y.\end{aligned}$$

Comparing the left- and right-hand sides above, we recover the identity for (12.18) when $y \in (0, 1)$ and $\alpha \in (0, 1]$. Taking care to note that $(-1, 0)$ maps to $(-\infty, -1)$ under the inversion $x \mapsto 1/x$, a similar computation can be used to verify (12.18) for $y \in (-1, 0)$ and $\alpha \in (0, 1]$. Moreover, it should now be clear that this method works equally well when $\alpha \in (1, 2)$, albeit that we need to replace $\hat{\mathbb{P}}_{1/x}(X_{\tau_1^+\wedge\tau_{-1}^-} \in \mathrm{d}y)$ in (12.19) by $\hat{\mathbb{P}}_x(X_{\tau_1^+} \in \mathrm{d}y, \tau_1^+ < \tau_{-1}^- \wedge \tau^{\{0\}})$ and then appeal to Corollary 6.8 in place of Corollary 6.3. As an exercise, the reader is now encouraged to verify these computations, but also to apply this method to derive Corollary 6.11 from Corollary 6.1.

The technique of spatial inversion via the Riesz–Bogdan–Żak transformation, can also be used to deal with potential measures, albeit that we must be careful to take account of the time change that comes with it. This is precisely what we will do in the proof of Theorem 6.12 which we address now.

Proof of Theorem 6.12 Let us write

$$U_\circ^{(-1,1)}(x, \mathrm{d}y) = \int_0^\infty \mathbb{P}_x^\circ(X_t \in \mathrm{d}y,\ t < \tau_1^+ \wedge \tau_{-1}^-)\,\mathrm{d}t, \qquad |x|, |y| < 1,$$

where we recall that the process $(X, \mathbb{P}_x^\circ)$, $x \in \mathbb{R}\backslash\{0\}$ is the result of the change of measure (12.2) appearing in the Riesz–Bogdan–Żak transform, Theorem 12.7. Let us preemptively assume that $U_\circ^{(-1,1)}(x, \mathrm{d}y)$ has a density with respect to Lebesgue measure, written $u_\circ^{(-1,1)}(x, y)$, for $|x|, |y| < 1$.

On the one hand, we have, for $|x|, |y| < 1$,

$$u_\circ^{(-1,1)}(x, y) = \frac{h(y)}{h(x)} u_{\{0\}}^{(-1,1)}(x, y), \tag{12.20}$$

where h was given in (12.3) and $u^{(-1,1)}_{\{0\}}(x,y)$ is the assumed density of

$$U^{(-1,1)}_{\{0\}}(x,\mathrm{d}y) = \int_0^\infty \mathbb{P}_x(X_t \in \mathrm{d}y,\, t < \tau^{\{0\}} \wedge \tau_1^+ \wedge \tau_{-1}^-).$$

By path counting and the strong Markov Property, we have that

$$U^{(-1,1)}(x,\mathrm{d}y) = U^{(-1,1)}_{\{0\}}(x,\mathrm{d}y) + \mathbb{P}_x(\tau^{\{0\}} < \tau_1^+ \wedge \tau_{-1}^-)U^{(-1,1)}(0,\mathrm{d}y), \qquad (12.21)$$

for $|x|, |y| < 1$, where we interpret the second term on the right-hand side as zero if $\alpha \in (0,1]$. Note that the existence of the density in Theorem 6.4 together with the above equality ensure that the densities $u^{(-1,1)}_\circ$ and $u^{(-1,1)}_{\{0\}}$ both exist. Combining (12.20) and (12.21),we thus have

$$u^{(-1,1)}_\circ(x,y) = \frac{h(y)}{h(x)}\left(u^{(-1,1)}(x,y) - \mathbb{P}_x(\tau^{\{0\}} < \tau_1^+ \wedge \tau_{-1}^-)u^{(-1,1)}(0,y)\right). \qquad (12.22)$$

The Riesz–Bogdan–Żak transform, $\eta(t)$, satisfies (12.16) and hence differentiating the identity $\int_0^{\eta(t)} |X_u|^{-2\alpha}\mathrm{d}u = t$, we get

$$|X_{\eta(t)}|^{-2\alpha}\mathrm{d}\eta(t) = \mathrm{d}t.$$

Let us introduce the following transformation $K\colon \mathbb{R}\backslash\{0\} \to \mathbb{R}\backslash\{0\}$, by

$$K(x) = \frac{x}{|x|^2} = \frac{1}{x}, \qquad x \in \mathbb{R}\backslash\{0\}.$$

(Our choice notation here is arguably excessive, however, it will be convenient for later reflection when considering the functionality of the Riesz–Bogdan–Żak transformation in higher dimensions; see Section 12.6.) It follows that, for bounded and measurable functions f,

$$\begin{aligned}
&\int_{(-1,1)} f(y)u^{(-1,1)}_\circ(x,y)\,\mathrm{d}y \\
&= \mathbb{E}_{-K(x)}\left[\int_0^\infty f(-K(X_s))|X_s|^{-2\alpha}\mathbf{1}_{(s<\tau^{(-1,1)})}\mathrm{d}s\right] \\
&= \int_{(-1,1)^c} f(-K(z))|z|^{-2\alpha}u^{(-1,1)^c}(-K(x),z)\,\mathrm{d}z \\
&= \int_{(-1,1)} f(y)|y|^{2\alpha}u^{(-1,1)^c}(-K(x),-K(y))|y|^{-2}\,\mathrm{d}y, \qquad (12.23)
\end{aligned}$$

where the density $u^{(-1,1)^c}$ is ensured by the density in the integral on the left-hand side above and we have used the easily proved fact that $\mathrm{d}(K(z)) = |z|^{-2}\mathrm{d}z$. Putting (12.23) and (12.22) together, noting that $K(K(x)) = K(x)$, we conclude that, for $|x|, |y| > 1$,

$$
\begin{aligned}
&u^{(-1,1)^c}(-x,-y)\\
&= u^{(-1,1)^c}(x,y)|_{\rho\leftrightarrow\hat\rho}\\
&= |y|^{2\alpha-2}\frac{h(K(y))}{h(K(x))}\left(u^{(-1,1)}(K(x),K(y)) - \mathbb{P}_{K(x)}(\tau^{\{0\}} < \tau_1^+ \wedge \tau_{-1}^-)u^{(-1,1)}(0,K(y))\right),
\end{aligned}
\tag{12.24}
$$

where $\rho \leftrightarrow \hat\rho$ is understood to mean that the roles of ρ and $\hat\rho$ are interchanged.

Let us now focus, for example, on the setting $y > x > 1$. Appealing to spatial homogeneity and scaling, with the help of Theorems 6.7 and 6.4, we can develop (12.24) and get

$$
\begin{aligned}
&u^{(-1,1)^c}(x,y)\\
&= \frac{2^{1-\alpha}}{\Gamma(\alpha\rho)\Gamma(\alpha\hat\rho)}\Bigg(|y-x|^{\alpha-1}\int_1^{\left|\frac{1-xy}{y-x}\right|}(s+1)^{\alpha\rho-1}(s-1)^{\alpha\hat\rho-1}\,\mathrm{d}s\\
&\qquad -(\alpha-1)_+\int_1^x (s+1)^{\alpha\rho-1}(s-1)^{\alpha\hat\rho-1}\,\mathrm{d}s\int_1^y (s+1)^{\alpha\hat\rho-1}(s-1)^{\alpha\rho-1}\,\mathrm{d}s\Bigg),
\end{aligned}
$$

where we have used again that $|K(x) - K(y)| = |x-y|/|x||y|$, and, in particular, that $|1 - K(x)K(y)| = |1-xy|/|x||y|$.

With some additional minor computations, the remaining cases follow similarly giving a complete description of $u^{(-1,1)^c}(x,y)$, for all $|x|, |y| > 1$. The details are left to the reader.

Finally, to recover the precise statement of Theorem 6.12, we need to transform the identity for the symmetric interval $(-1,1)$ to the interval $(0,a)$. This is a standard procedure which uses the scaling and spatial homogeneity of the stable process.

Indeed, for $a > 0$, we have that $(X_t, t \geq 0)$ under $\mathbb{P}_x$ is equal in law to $(aX_{(a/2)^{-\alpha}t}/2, t \geq 0)$ under $\mathbb{P}_{2x/a}$. Hence, for bounded, measurable functions f on $\mathbb{R}$ and $|x| > 1$,

$$
\int_{\mathbb{R}} f(y)u^{(0,a)^c}(x,y)\mathrm{d}y = \left(\frac{a}{2}\right)^{\alpha}\mathbb{E}_{2x/a}\left[\int_0^{\tau^{(0,2)^c}} f(aX_s/2)\mathrm{d}s\right].
$$

Now, invoking spatial homogeneity on the right-hand side above, we get

$$
\begin{aligned}
\int_{\mathbb{R}} f(y)u^{(0,a)^c}(x,y)\mathrm{d}y &= \left(\frac{a}{2}\right)^{\alpha}\mathbb{E}_{\frac{2x}{a}-1}\left[\int_0^{\tau^{(-1,1)^c}} f(a(X_s+1)/2)\mathrm{d}s\right]\\
&= \left(\frac{a}{2}\right)^{\alpha}\int_{\mathbb{R}} f(a(z+1)/2)u^{(-1,1)^c}\left(\frac{2x}{a}-1, z\right)\mathrm{d}z\\
&= \left(\frac{a}{2}\right)^{\alpha-1}\int_{\mathbb{R}} f(y)u^{(-1,1)^c}\left(\frac{2x}{a}-1, \frac{2y}{a}-1\right)\mathrm{d}z.
\end{aligned}
$$

Thus, we have the relation

$$u^{(0,a)^c}(x,y) = \left(\frac{a}{2}\right)^{\alpha-1} u^{(-1,1)^c}\left(\frac{2x}{a} - 1, \frac{2y}{a} - 1\right),$$

for $x, y \notin [0, a]$. □

12.5 First hitting of a point revisited

We return to the identity for the resolvent of a stable process killed on hitting the origin in Theorem 6.18 and give the promised proof. As in the previous section, the Riesz–Bogdan–Żak transformation will play an instrumental role in its proof. Recall that necessarily we need $\alpha \in (1, 2)$, in which case points are hit with probability one. Moreover, the statement of Theorem 6.18 excludes the setting of spectrally one-sided processes, hence $0 < \alpha\hat{\rho}, \alpha\rho < 1$.

Proof of Theorem 6.18 Suppose that f is a bounded measurable function in $\mathbb{R}$. We are interested in the resolvent measure $U^{\{0\}}(x, \mathrm{d}y)$ which satisfies

$$\int_{\mathbb{R}} f(y) U^{\{0\}}(x, \mathrm{d}y) = \mathbb{E}_x\left[\int_0^\infty f(X_t)\mathbf{1}_{(t<\tau^{\{0\}})}\,\mathrm{d}t\right].$$

Recalling the change of measure (12.2), we have that

$$\int_{\mathbb{R}} f(y)\frac{h(y)}{h(x)} U^{\{0\}}(x, \mathrm{d}y) = \mathbb{E}^\circ_x\left[\int_0^\infty f(X_t)\,\mathrm{d}t\right].$$

Let us momentarily focus our attention on the setting that $x, y > 0$ in $U^{\{0\}}(x, \mathrm{d}y)$. In that case, we can write

$$\int_{[0,\infty)} f(y)\frac{h(y)}{h(x)} U^{\{0\}}(x, \mathrm{d}y) = \mathbb{E}^\circ_x\left[\int_0^\infty f(X_t)\mathbf{1}_{(X_t>0)}\,\mathrm{d}t\right] = \mathbb{E}^\circ_x\left[\int_0^\infty f(\vec{Z}^\circ_t)\,\mathrm{d}t\right],$$

where $\vec{Z}^\circ = (\vec{Z}^\circ_t, t \geq 0)$ is the pssMp which is derived by censoring away the negative sections of the path of $(X, \mathbb{P}^\circ_x)$, $x \in \mathbb{R}$ in the spirit of what we have already seen for stable processes, cf. Section 5.6. Suppose that we denote the Lévy process that underlies Z by $\vec{\xi}^\circ$, with probabilities $\vec{\mathbf{P}}^\circ_x$, $x \in \mathbb{R}$. Taking account of the time change in the Lamperti transform (5.4), we thus have on the one hand that

$$\begin{aligned}\int_{[0,\infty)} f(y)\frac{h(y)}{h(x)} U^{\{0\}}(x, \mathrm{d}y) &= \vec{\mathbf{P}}^\circ_{\log x}\left[\int_0^\infty f(\mathrm{e}^{\vec{\xi}^\circ_t})\mathrm{e}^{\alpha\vec{\xi}^\circ_t}\mathrm{d}t\right] \\ &= \int_{\mathbb{R}} f(\mathrm{e}^z)\mathrm{e}^{\alpha z}\,\vec{u}^\circ(z - (\log x))\,\mathrm{d}z \\ &= \int_{\mathbb{R}} f(y)y^{\alpha-1}\,\vec{u}^\circ(\log(y/x))\,\mathrm{d}y, \qquad (12.25)\end{aligned}$$

where we have pre-emptively assumed that $\overset{\rightsquigarrow}{\xi}^\circ$ has a resolvent density, which we have denoted by $\overset{\rightsquigarrow}{u}^\circ$. This is a reasonable assumption because of the following heuristics.

The time change given in (5.35) that corresponds to censoring the stable process presents a series of stopping times. As such, incorporating path censoring with (12.2) we get a description of the change of measure between the censored process $\overset{\rightsquigarrow}{Z}^\circ$ and the censored stable process discussed in (5.6). In effect, this is tantamount to a Doob h-transform between the two positive-valued processes with h function taking the form $h(x) = x^{\alpha-1}$, $x \geq 0$. As we have seen in Sections 5.4 and 5.32, this Doob h-transform also plays out as an Esscher transform between the underlying Lévy processes $\overset{\rightsquigarrow}{\xi}$ (for the censored stable process) and $\overset{\rightsquigarrow}{\xi}^\circ$ (for $\overset{\rightsquigarrow}{Z}^\circ$). In particular, we have

$$\left.\frac{\mathrm{d}\overset{\rightsquigarrow}{\mathbf{P}}^\circ}{\mathrm{d}\overset{\rightsquigarrow}{\mathbf{P}}}\right|_{\sigma(\overset{\rightsquigarrow}{\xi}_s:\, s\leq t)} = \mathrm{e}^{(\alpha-1)\overset{\rightsquigarrow}{\xi}_t}, \qquad t \geq 0.$$

It is thus straightforward to show that

$$\overset{\rightsquigarrow}{u}^\circ(x) = \overset{\rightsquigarrow}{u}(x)\mathrm{e}^{(\alpha-1)x}, \qquad x \in \mathbb{R}. \tag{12.26}$$

Consolidating (12.25) and (12.26), noting in particular from (12.2) that $h(x) = s(-x)|x|^{\alpha-1}$, where $s(x) = \sin(\pi\alpha\rho)\mathbf{1}_{(x\geq 0)} + \sin(\pi\alpha\hat{\rho})\mathbf{1}_{(x<0)}$, we thus conclude that

$$U^{\{0\}}(x, \mathrm{d}y) = y^{\alpha-1}\,\overset{\rightsquigarrow}{u}(\log(y/x))\,\mathrm{d}y, \qquad x, y > 0, \tag{12.27}$$

where we recall that $\overset{\rightsquigarrow}{u}$ has been computed explicitly in Theorem 6.15.

Bringing across the specific form of $\overset{\rightsquigarrow}{u}$ from Theorem 6.15, we can now read off that, for $x, y > 0$, $U^{\{0\}}(x, \mathrm{d}y)$ is absolutely continuous with density, $u^{\{0\}}(x, y)$, taking the form

$$u^{\{0\}}(x, y)$$
$$= -\frac{1}{\pi^2}\Gamma(1-\alpha)\begin{cases} \sin(\pi\alpha\rho)y^{\alpha-1} - \sin(\pi\alpha\rho)(y-x)^{\alpha-1} + \sin(\pi\alpha\hat{\rho})x^{\alpha-1}, & y > x, \\ \sin(\pi\alpha\rho)y^{\alpha-1} + \sin(\pi\alpha\hat{\rho})x^{\alpha-1} - \sin(\pi\alpha\hat{\rho})(x-y)^{\alpha-1}, & y < x, \end{cases}$$

which is consistent with the statement of the theorem.

Note in particular that, for $y > 0$,

$$u^{\{0\}}(y, y) := \lim_{x\to y} u^{\{0\}}(x, y) = -\frac{1}{\pi^2}\Gamma(1-\alpha)y^{\alpha-1}(\sin(\pi\alpha\rho) + \sin(\pi\alpha\hat{\rho})).$$

We can use this limit to deal with the case that $x < 0 < y$. Indeed, recall the identity (6.21) which tells us in particular that

$$U^{\{0\}}(x, \mathrm{d}y) = \mathbb{P}_x(\tau^{\{y\}} < \tau^{\{0\}})u^{\{0\}}(y, y)\mathrm{d}y, \qquad y > 0.$$

Hence, recalling the expression for $\mathbb{P}_x(\tau^{\{y\}} < \tau^{\{0\}})$ in Theorem 6.16, we recover the required identity in the regime that $x < 0 < y$.

By working instead with $-X$ (or equivalently censoring out the positive parts of the path of $(X, \mathbb{P}^\circ_x)$, $x \in \mathbb{R}$) we easily conclude that the same identities hold when $x, y < 0$ and $x > 0 > y$, simply by interchanging the roles of ρ and $\hat{\rho}$. □

12.6 Riesz–Bogdan–Żak transformation in dimension $d \geq 2$

We complete this chapter, as promised, by offering the d-dimensional version of the Riesz–Bogdan–Żak transformation for higher-dimensional, albeit isotropic, stable processes. Recall that the transformation $K\colon \mathbb{R}^d \to \mathbb{R}^d$, but now for dimension d, is defined as follows,

$$K(x) = \frac{x}{|x|^2}, \qquad x \in \mathbb{R}^d.$$

This transformation inverts space through the unit sphere $\{x \in \mathbb{R}^d : |x| = 1\}$ and accordingly, it is not surprising that $K(K(x)) = x$. To see how the K-transform maps $\mathbb{R}^d$ into itself, write $x \in \mathbb{R}^d$ in skew product form $x = (|x|, \text{Arg}(x))$, and note that

$$K(x) = (|x|^{-1}, \text{Arg}(x)), \qquad x \in \mathbb{R}^d,$$

showing that the K-transform 'radially inverts' elements of $\mathbb{R}^d$ through $\mathbb{S}^{d-1}$.

Theorem 12.8 (d-dimensional Riesz–Bogdan–Żak Transform, $d \geq 2$) *Suppose that X is a d-dimensional isotropic stable process with $d \geq 2$. Define*

$$\eta(t) = \inf\left\{s > 0\colon \int_0^s |X_u|^{-2\alpha} \mathrm{d}u > t\right\}, \qquad t \geq 0.$$

Then, for all $x \in \mathbb{R}^d \backslash \{0\}$, $(K(X_{\eta(t)}), t \geq 0)$ under $\mathbb{P}_x$ is equal in law to $(X, \mathbb{P}^\circ_{K(x)})$.

Proof As with the proof of Theorem 12.3, it is straightforward to check that $(KX_{\eta(t)}, t \geq 0)$ is a ssMp. Indeed, in skew product form,

$$KX_{\eta(t)} = \mathrm{e}^{-\xi_{\varphi\circ\eta(t)}} \Theta_{\varphi\circ\eta(t)}, \qquad t \geq 0,$$

and, just as in the computation (12.17), one easily verifies again that

$$\varphi \circ \eta(t) = \inf\left\{s > 0\colon \int_0^s \mathrm{e}^{-\alpha\xi_u} \mathrm{d}u > t\right\}.$$

It is thus clear that $(KX_{\eta(t)}, t \geq 0)$ is a ssMp with underlying MAP equal to $(-\xi, \Theta)$. To complete the proof, it therefore suffices to check that $(-\xi, \Theta)$ is also the MAP which underlies the ssMp $(X, \mathbb{P}^\circ_x)$, $x \in \mathbb{R}^d \backslash \{0\}$.

To this end, we note that $(X, \mathbb{P}^\circ_x)$, $x \in \mathbb{R}^d\backslash\{0\}$, is a pure jump process and hence entirely characterised by its jump rate. To understand why at a heuristic level, note that, as a Feller process, it is in possession of an infinitesimal generator, say $\mathcal{L}^\circ$. Indeed, standard theory tells us that

$$\mathcal{L}^\circ f(x) = \lim_{t\downarrow 0} \frac{\mathbb{E}^\circ_x[f(X_t)] - f(x)}{t} = \lim_{t\downarrow 0} \frac{\mathbb{E}_x[|X_t|^{\alpha-d} f(X_t)] - |x|^{\alpha-d} f(x)}{|x|^{\alpha-d} t}, \quad (12.28)$$

for twice continuously differentiable and compactly supported functions f, where $x \in \mathbb{R}^d\backslash\{0\}$. That is to say

$$\mathcal{L}^\circ f(x) = \frac{1}{h(x)} \mathcal{L}(hf)(x), \quad (12.29)$$

where $h(x) = |x|^{\alpha-d}$ and $\mathcal{L}$ is the infinitesimal generator of the stable process, which has action

$$\mathcal{L}f(x) = \mathrm{a} \cdot \nabla f(x) + \int_{\mathbb{R}^d} [f(x+y) - f(x) - \mathbf{1}_{(|y|\leq 1)} y \cdot \nabla f(x)] \Pi(\mathrm{d}y), \qquad x \in \mathbb{R}^d,$$

for twice continuously differentiable and compactly supported functions f. The vector $\mathrm{a} \in \mathbb{R}^d$ depends on $\alpha \in (0,2)$, and its value is prescribed by Remark 1.25. Straightforward algebra, appealing to the fact that $\mathcal{L}h = 0$ shows that, for twice continuously differentiable and compactly supported functions, f, the infinitesimal generator of the conditioned process (12.29) takes the form

$$\mathcal{L}^\circ f(x) = \mathrm{a} \cdot \nabla f(x) + \int_{\mathbb{R}^d} [f(x+y) - f(x) - \mathbf{1}_{(|y|\leq 1)} y \cdot \nabla f(x)] \frac{h(x+y)}{h(x)} \Pi(\mathrm{d}y), \quad (12.30)$$

for $|x| > 0$, where Π is the stable Lévy measure given in (3.23). The integral component in $\mathcal{L}^\circ$ tells us that the instantaneous rate at which jumps arrive for the conditioned process, when positioned at x, is given by

$$\Pi^\circ(x, B) := \int_B \frac{h(x+y)}{h(x)} \Pi(\mathrm{d}y),$$

where $|x| > 0$ and B is Borel in $\mathbb{R}^d$.

Since its jump rates entirely characterise $(X, \mathbb{P}^\circ_x)$, $x \in \mathbb{R}^d\backslash\{0\}$, to draw the conclusion we are working towards, we need to study the probabilities $\mathbf{P}^\circ_{y,\vartheta}$, $y \in \mathbb{R}$ and $\vartheta \in \mathbb{S}^{d-1}$, for the underlying MAP via an identity similar to (12.6). To this end, suppose that f is a positive, bounded, measurable and function on $[0,\infty) \times \mathbb{R} \times \mathbb{R} \times \mathbb{S}^{d-1} \times \mathbb{S}^{d-1}$ such that $f(\cdot, \cdot, 0, \cdot, \cdot) = 0$. Write

$$\mathbf{E}^{\circ}_{0,\theta}\left[\sum_{s>0} f(s,\xi_{s-},\Delta\xi_s,\Theta_{s-},\Theta_s)\right]$$
$$= \lim_{t\to\infty} \mathbf{E}_{0,\theta}\left[\mathcal{M}_t \sum_{0<s\le t} f(s,\xi_{s-},\Delta\xi_s,\Theta_{s-},\Theta_s)\right], \qquad (12.31)$$

where $\mathcal{M}_t = \exp\{(\alpha-d)\xi_t\}$, $t\ge 0$, is the martingale density corresponding to (12.9) and the limit is justified by monotone convergence. Suppose we write Σ_t for the sum term in the final expectation above. The semi-martingale change of variable formula tells us that

$$\mathcal{M}_t\Sigma_t = \mathcal{M}_0(\theta)\Sigma_0 + \int_0^t \Sigma_{s-}\mathrm{d}\mathcal{M}_s + \int_0^t \mathcal{M}_{s-}\mathrm{d}\Sigma_s + [\mathcal{M},\Sigma]_t, \qquad t\ge 0,$$

where $[\mathcal{M},\Sigma]_t$ is the quadratic co-variation term. On account of the fact that $(\Sigma_t, t\ge 0)$, has bounded variation, the latter term takes the form $[\mathcal{M},\Sigma]_t = \sum_{s\le t}\Delta M_t\Delta\Sigma_t$. As a consequence,

$$\mathcal{M}_t\Sigma_t = \mathcal{M}_0(\theta)\Sigma_0 + \int_0^t \Sigma_{s-}\mathrm{d}\mathcal{M}_s + \int_0^t \mathcal{M}_s\mathrm{d}\Sigma_s, \qquad t\ge 0. \qquad (12.32)$$

Moreover, after taking expectations and then taking limits as $t\to\infty$, with the help of (12.31) and monotone convergence, as the first in integral in (12.32) is a martingale and $\Sigma_0 = 0$, the only surviving terms give us

$$\mathbf{E}^{\circ}_{0,\theta}\left[\sum_{s>0} f(s,,\xi_{s-},\Delta\xi_s,\Theta_{s-},\Theta_s)\right]$$
$$= \mathbf{E}_{0,\theta}\left[\sum_{s>0} \mathrm{e}^{(\alpha-d)(\xi_{s-}+\Delta\xi_s)} f(s,,\xi_{s-},\Delta\xi_s,\Theta_{s-},\Theta_s)\right]$$
$$= \mathbb{E}_\theta\Bigg[\int_0^\infty \mathrm{d}s \int_{\mathbb{S}^{d-1}} \sigma_1(\mathrm{d}\phi)\int_0^\infty \frac{c(\alpha)\mathrm{d}r}{r^{1+\alpha}}|X_{s-}|^{\alpha-d}\left|\mathrm{Arg}(X_{s-}) + \frac{r\phi}{|X_{s-}|^{\alpha-d}}\right|^{\alpha-d}$$
$$f\Bigg(\int_0^s |X_u|^{-2\alpha}\mathrm{d}u, \log|X_{s-}|, \log\left|\mathrm{Arg}(X_{s-}) + \frac{r\phi}{|X_{s-}|}\right|,$$
$$\mathrm{Arg}(X_{s-}), \frac{\mathrm{Arg}(X_{s-}) + \frac{r\phi}{|X_{s-}|}}{\left|\mathrm{Arg}(X_{s-}) + \frac{r\phi}{|X_{s-}|}\right|}\Bigg)\Bigg],$$

where the second equality proceeds in a similar fashion to the calculation in (12.7). Now, picking up from the second equality of (12.8) with $f(\cdot,\xi,\Delta,\cdot,\cdot)$ replaced by $\exp((\alpha-d)(\xi+\Delta))f(\cdot,\xi,\Delta,\cdot,\cdot)$, we get

$$
\begin{aligned}
&\mathbf{E}^{\circ}_{0,\theta}\left[\sum_{s>0} f(s, \xi_{s-}, \Delta\xi_s, \Theta_{s-}, \Theta_s)\right] \\
&= \mathbf{E}_{0,\theta}\left[\int_0^\infty \mathrm{d}v \int_{\mathbb{S}^{d-1}} \sigma_1(\mathrm{d}\phi) \int_0^\infty \mathrm{d}r \frac{c(\alpha)\mathrm{e}^{(\alpha-d)\xi_v}\,|\Theta_v + r\phi|^{\alpha-d}}{r^{1+\alpha}}\right. \\
&\qquad\qquad \left. \times f\left(v, \xi_v, \log|\Theta_v + r\phi|, \Theta_v, \frac{\Theta_v + r\phi}{|\Theta_v + r\phi|}\right)\right] \\
&= \mathbf{E}_{0,\theta}\left[\int_0^\infty \mathrm{d}v \int_{\mathbb{R}^d} \mathrm{d}z \frac{c(\alpha)\mathrm{e}^{(\alpha-d)\xi_v}\,|\Theta_v + z|^{\alpha-d}}{|z|^{\alpha+d}} f\left(v, \xi_v, \log|\Theta_v + z|, \Theta_v, \frac{\Theta_v + z}{|\Theta_v + z|}\right)\right] \\
&= \mathbf{E}_{0,\theta}\left[\int_0^\infty \mathrm{d}v \int_{\mathbb{S}^{d-1}} \sigma_1(\mathrm{d}\phi) \int_0^\infty \mathrm{d}r \frac{c(\alpha) r^{\alpha-1}}{|r\phi - \Theta_v|^{\alpha+d}} \mathrm{e}^{(\alpha-d)\xi_v} f(v, \xi_v, \log r, \Theta_v, \phi)\right],
\end{aligned}
$$

where we have converted to Cartesian coordinates in the second equality and back to skew product variables in the final equality after an easy change of variables. We may now proceed using the resolvent $V_\theta(\mathrm{d}s, \mathrm{d}x, \mathrm{d}\vartheta)$ to get

$$
\begin{aligned}
&\mathbf{E}^{\circ}_{0,\theta}\left[\sum_{s>0} f(s, \xi_{s-}, \Delta\xi_s, \Theta_{s-}, \Theta_s)\right] \\
&= \int_0^\infty \int_{\mathbb{R}} \int_{\mathbb{S}^{d-1}} \int_{\mathbb{S}^{d-1}} \int_{\mathbb{R}} V_\theta(\mathrm{d}v, \mathrm{d}x, \mathrm{d}\vartheta) \sigma_1(\mathrm{d}\phi) \mathrm{d}y \frac{c(\alpha)\mathrm{e}^{(\alpha-d)x}\mathrm{e}^{y\alpha}}{|\mathrm{e}^y\phi - \vartheta|^{\alpha+d}} f(v, x, y, \vartheta, \phi) \\
&= \int_0^\infty \int_{\mathbb{R}} \int_{\mathbb{S}^{d-1}} \int_{\mathbb{S}^{d-1}} \int_{\mathbb{R}} V_\theta(\mathrm{d}v, \mathrm{d}x, \mathrm{d}\vartheta) \sigma_1(\mathrm{d}\phi) \mathrm{d}w \frac{c(\alpha)\mathrm{e}^{(\alpha-d)x}\mathrm{e}^{wd}}{|\phi - \mathrm{e}^w\vartheta|^{\alpha+d}} f(v, x, -w, \vartheta, \phi) \\
&= \int_0^\infty \int_{\mathbb{R}} \int_{\mathbb{S}^{d-1}} \int_{\mathbb{S}^{d-1}} \int_{\mathbb{R}} V_\theta(\mathrm{d}v, \mathrm{d}x, \mathrm{d}\vartheta) \sigma_1(\mathrm{d}\phi) \mathrm{d}w \frac{c(\alpha)\mathrm{e}^{(\alpha-d)x}\mathrm{e}^{wd}}{|\mathrm{e}^w\phi - \vartheta|^{\alpha+d}} f(v, x, -w, \vartheta, \phi).
\end{aligned}
$$

In the penultimate equality, we simply change variables $y = -w$, and in the final equality, we note that $|\phi - \mathrm{e}^w\vartheta|^2 = |\mathrm{e}^w\phi - \vartheta|^2$ on account of the fact that

$$
(\phi - \mathrm{e}^w\vartheta)\cdot(\phi - \mathrm{e}^w\vartheta) = 1 - 2\mathrm{e}^w\vartheta\cdot\phi + \mathrm{e}^{2w} = (\mathrm{e}^w\phi - \vartheta)\cdot(\mathrm{e}^w\phi - \vartheta).
$$

In conclusion, we have

$$
\begin{aligned}
&\mathbf{E}^{\circ}_{0,\theta}\left[\sum_{s>0} f(s, \xi_{s-}, \Delta\xi_s, \Theta_{s-}, \Theta_s)\right] \\
&= \int_0^\infty \int_{\mathbb{R}} \int_{\mathbb{S}^{d-1}} \int_{\mathbb{S}^{d-1}} \int_{\mathbb{R}} V^{\circ}_\theta(\mathrm{d}v, \mathrm{d}x, \mathrm{d}\vartheta) \sigma_1(\mathrm{d}\phi) \mathrm{d}w \frac{c(\alpha)\mathrm{e}^{wd}}{|\mathrm{e}^w\phi - \vartheta|^{\alpha+d}} f(v, x, -w, \vartheta, \phi),
\end{aligned}
\tag{12.33}
$$

where for $s > 0$, $x \in \mathbb{R}$, $\vartheta \in \mathbb{S}^{d-1}$,

$$
V^{\circ}_\theta(\mathrm{d}s, \mathrm{d}x, \mathrm{d}\vartheta) = \mathbf{P}^{\circ}_{0,\theta}(\xi_s \in \mathrm{d}x, \Theta_s \in \mathrm{d}\vartheta)\mathrm{d}s = V_\theta(\mathrm{d}s, \mathrm{d}x, \mathrm{d}\vartheta)\mathrm{e}^{(\alpha-d)x}
$$

is the space-time potential of (ξ, Θ).

Comparing the right-hand side of (12.33) above with that of (12.6), it now becomes clear that the jump structure of (ξ, Θ) under $\mathbf{P}^{\circ}_{x,\theta}$, $x \in \mathbb{R}$, $\theta \in \mathbb{S}^{d-1}$, is precisely that of $(-\xi, \Theta)$ under $\mathbf{P}_{x,\theta}$, $x \in \mathbb{R}$, $\theta \in \mathbb{S}^{d-1}$.

In conclusion, this is now sufficient to deduce that $(X, \mathbb{P}^{\circ}_{Kx})$, $|x| > 0$, is equal in law to $(KX_{\eta(t)}, t \geq 0)$ under $\mathbb{P}_x$, as both are self-similar Markov processes with the same underlying MAP. □

Reviewing the proofs of the previous Theorems 12.8 and 12.7, we also have the following corollary at no extra cost.

Corollary 12.9 *When $d = 1$, the process (ξ°, J°) is equal in law to $(-\xi, J)$ and, when $d \geq 2$, the process $(\xi^{\circ}, \Theta^{\circ})$ is equal in law to $(-\xi, \Theta)$.*

12.7 Radial asymptotics for $d \geq 2$

Recall from Corollary 11.15 that $|X|$ under both $\mathbb{P}$ and $\mathbb{P}^{\circ}$ are positive self-similar Markov processes. The Reisz–Bogdan–Żak transform, in particular Corollary 12.9, tells us that the underlying Lévy processes associated via the Lamperti transform are ξ and $-\xi$ respectively. In this respect, two processes $(|X|, \mathbb{P})$ and $(|X|, \mathbb{P}^{\circ})$ are dual to one another in the sense described in Section 5.2. Indeed, Proposition 5.6 tells us that the process $(|X_{(\mathtt{D}_x - t)-}|, 0 \leq t \leq \mathtt{D}_x)$, under $\mathbb{P}_0(\cdot \,|\, |X_{\mathtt{D}_x-}| = z)$, for $z \leq x$ in the support set of the law of $|X_{\mathtt{D}_x-}|$, is equal in law to $(|X|, \mathbb{P}^{\circ}_z)$, where $\mathtt{D}_x = \sup\{t > 0 \colon |X_t| \leq x\}$. As remarked below Proposition 5.6, there is a second way of seeing this duality. Let $\tau^{\{0\}} = \inf\{t > 0 \colon |X_t| = 0\}$. Then $(|X_{(\tau^{\{0\}}-t)-}|, 0 \leq t \leq \tau^{\{0\}})$ under $\mathbb{P}^{\circ}_z$ has the same law as $(|X_t|, 0 \leq t < \mathtt{D}_x)$, under $\mathbb{P}_0(\cdot \,|\, |X_{\mathtt{D}_x-}| = z)$, for any $z \leq x$. In particular, this means that we can read out of Theorem 10.27 and Theorem 10.29 the result given below, for which we recall that $\mathtt{J}_t = \inf_{s \geq t} |X_s|$, $t \geq 0$.

Theorem 12.10 *Let $d > \alpha$ and f be an increasing function such that either*

$$\lim_{t \to 0} \frac{t}{f(t)^{\alpha}} = 0 \qquad \text{or} \qquad \liminf_{t \to 0} \frac{t}{f(t)^{\alpha}} > 0$$

holds. The lower and upper envelope of the isotropic d-dimensional stable process conditioned to continuously absorb at 0 and at ∞ are characterised by the following integral tests.

(i) For all $|x| > 0$,

$$\mathbb{P}^{\circ}_x(|X_{(\tau^{\{0\}}-t)-}| < f(t) \text{ i.o. as } t \to 0) = 0 \text{ or } 1,$$

accordingly as

$$\int_{0+} \frac{f(t)^{d-\alpha}}{t^{\frac{d}{\alpha}}} \mathrm{d}t \quad \textit{is finite or infinite.}$$

(ii) Suppose additionally that $f \in C_0$ *(cf. Definition 9.10), then, for all* $|x| > 0$,

$$\mathbb{P}^\circ_x(|X_{(\tau^{\{0\}}-t)-}| > f(t) \text{ i.o. as } t \to 0)$$
$$= \mathbb{P}^\circ_x(\mathtt{J}_{(\tau^{\{0\}}-t)-} > f(t) \text{ i.o. as } t \to 0) = 0 \ \text{ or } \ 1,$$

accordingly as

$$\int_{0+} \frac{\mathrm{d}t}{f(t)^\alpha} \quad \textit{is finite or infinite.}$$

12.8 Comments

The identification of the MAP underlying the stable process when seen as a rssMp is taken from Chaumont et al. [47] and Kuznetsov et al. [120]. Theorem 12.3 generalises a similar result for planar stable processes in Bertoin and Werner [25] and is taken from Kyprianou [125]. Section 12.2 is based on computations found in Kyprianou et al. [135].

The Riesz–Bogdan–Żak transform for isotropic stable process in Theorem 12.8 was first proved by Bogdan and Żak [40], inspired by calculations in the classical works of Riesz [177, 178]. The result in one dimension allowing for non-symmetric stable processes, Theorem 12.7, was proved in Kyprianou [124].

Whilst Theorem 6.12 was first proved in Profeta and Simon [175], the proof we give here is different, relying on the Riesz–Bogran–Żak transform, and first appeared in Kyprianou [125]. The proof of Theorem 6.18 given in this chapter is also very recent and is taken from the same reference.

A more general perspective of Riesz–Bogdan–Żak-type transformations for general self-similar Markov processes was recently given in Alili et al. [1]. The proof we give here is taken from Kyprianou [125]. As part of the proof, we have used the integration by parts formula for general semi-martingale calculus, for which, one can refer to, for example, p. 86 of Protter [176].

Theorem 12.10 is a new result which combines the new results in Chapter 10 with the Riesz–Bogdan–Żak transform, in particular the conclusion of Corollary 12.9.

13

Radial reflection and the deep factorisation

We have seen that the Wiener–Hopf factorisation for one-dimensional stable processes has provided some key results, for example Theorem 3.4, based on which many other identities can be developed. In this section, we consider the so-called *deep factorisation* of one-dimensional stable processes which have both positive and negative jumps. That is to say, we consider the factorisation of the matrix exponent of the Lamperti-stable MAP. Our main motivation for doing this is that the factorisation contains information about the stable process reflected in its radial maximum. In particular, by examining the deep factorisation, we are able to derive an exact formula for the limiting distribution of the radially reflected stable process when $\alpha \in (0, 1)$ and $\rho \in (0, 1)$, that is, no monotone paths.

13.1 Radially reflected stable processes when $\alpha \in (0, 1)$

Let us start by introducing the radially reflected one-dimensional stable processes

$$R_t^{(x,m)} = \frac{X_t^{(x)}}{M_t^{(x,m)}}, \qquad t \geq 0,$$

where, as usual $X^{(x)}$ is a stable process with its point of issue indicated as $x \in \mathbb{R}$ and for $m > 0$, $M_t^{(x,m)} = \sup_{s \leq t} |X_s^{(x)}| \vee m$, $t \geq 0$. In particular, $R_0 = x/(x \vee m)$. Just as we often omit the point of issue of X, we will similarly write R without indicating its point of issue. As with many path transformations we have considered, we assume that X has two-sided jumps.

Lemma 13.1 *Suppose that X has two-sided jumps. The pair of processes (X, R) are Feller.*

Proof Suppose that τ is a stopping time with respect to $(\mathcal{F}_t, t \geq 0)$, the filtration generated X with natural enlargement. It is easy to verify, using the scaling and Markov properties, that for $s \geq 0$, on $\{\tau < \infty\}$,

$$M^{(x,m)}_{\tau+s} = \tilde{M}^{(x',m')}_s, \qquad X^{(x)}_{\tau+s} = \tilde{X}^{(x')}_s,$$

where, $m' = M^{(x,m)}_\tau$ and $x' = X^{(x)}_\tau$ and for constant (x', m'), $(\tilde{M}^{(x',m')}, \tilde{X}^{(x')})$ is an independent copy of $(M^{(x',m')}, X^{(x')})$. We thus have that, on $\{\tau < \infty\}$, for bounded, measurable $f : \mathbb{R} \times (0,\infty) \to [0,\infty)$,

$$\mathbb{E}[f(X^{(x)}_{\tau+s}, R^{(x,m)}_{\tau+s})|\mathcal{F}_\tau] = g_s\left(X^{(x)}_\tau, X^{(x)}_\tau / R^{(x,m)}_\tau\right), \qquad s \geq 0,$$

where $g_s(x,m) = \mathbb{E}[f(X^{(x)}_s, R^{(x,m)}_s)]$. It follows that the pair (R, X) is a strong Markov process.

The remaining aspects of the Feller property can easily be derived appealing to the continuity properties of $X^{(x)}$ and $M^{(x,m)}$ and dominated convergence. □

Whilst the process R is not necessarily Markovian alone, one might expect the existence of a limiting distribution, in the sense of the convergence of the law of R_t on $[-1,1]$ as $t \to \infty$. The following result illustrates this prediction in explicit detail. Its proof becomes the main motivation to examine the matrix factorisation, also called the *deep factorisation*, of the underlying MAP of the associated stable process.

Theorem 13.2 *Suppose that $\alpha, \rho \in (0,1)$. Let $x \in (-1,1)$, then under $\mathbb{P}_x$, R has a limiting distribution μ, concentrated on $[-1,1]$, given by*

$$\frac{d\mu(y)}{dy} = 2^{-\alpha}\frac{\Gamma(\alpha)}{\Gamma(\alpha\rho)\Gamma(\alpha\hat{\rho})}\left[(1-y)^{\alpha\hat{\rho}-1}(1+y)^{\alpha\rho} + (1-y)^{\alpha\hat{\rho}}(1+y)^{\alpha\rho-1}\right],$$

for $y \in [-1,1]$.

To see the heuristic connection between Theorem 13.2 and the deep factorisation, we note from the Lamperti–Kiu transform in Theorem 11.3 that

$$\left(\frac{X_t}{M_t}, \operatorname{sgn}(X_t)\right) = \left(e^{\xi_{\varphi(t)} - \bar{\xi}_{\varphi(t)}}, J_{\varphi(t)}\right), \qquad t \geq 0, \tag{13.1}$$

where $\varphi(t) = \inf\{s > 0 : \int_0^s \exp(\xi_u)du > t\}$ and (ξ, J) is the Lamperti-stable MAP with probabilities $\mathbf{P}_{x,i}$, $x \in \mathbb{R}$, $i \in \{1,-1\}$. When $\alpha \in (0,1)$, we know that $\lim_{t\to\infty} |X_t| = \infty$ and hence $\lim_{t\to\infty} \xi_t = \infty$. Were it not for the time change $\varphi(t)$ in (13.1), we can see that, the limit we are after would lead us to

$$\lim_{t\to\infty} \mathbf{E}_{0,1}\left[f(e^{-(\bar{\xi}_t - \xi_t)})\right] = \lim_{q\downarrow 0} \mathbf{E}_{0,1}\left[f(e^{-(\bar{\xi}_{\mathbf{e}_q} - \xi_{\mathbf{e}_q})})\right], \tag{13.2}$$

for all bounded, continuous functions f, as soon as the limit exists (on either side of the equality), where $\mathbf{e}_q$ is an independent and exponentially distributed random variable. Rather than being the ordinate of a MAP, let us momentarily suppose we are dealing with the case that ξ is a Lévy process (which, in fact, is the case when X is symmetric). The existence of a non-trivial limit in (13.2) is a classical question from the theory of Lévy processes for which it is needed that $\lim_{t\to\infty} \xi_t = \infty$; see the discussion in Section 2.15. This helps explain the requirement that $\alpha \in (0,1)$ in the statement of Theorem 13.2. Indeed, in the symmetric setting, the process ξ is a Lévy process (ignoring the role of J), and it is known in that setting that a stationary distribution of $\overline{\xi} - \xi$ exists if and only if $\lim_{t\to\infty} \xi_t = \infty$.

The above heuristic aside, the stationary distribution that we are after must also additionally take account of the time change $\varphi(t)$ in the limit. Nonetheless, the limiting expectation on the right-hand side of (13.2) can, in principle, be accessed by the kind of computations that were dealt with in Section 11.4, when splitting a MAP at its maximum; see for example (11.21). As explained in that section, splitting at the maximum is what lies at the heart of the matrix Wiener–Hopf factorisation for MAPs. Indeed, from (11.23), it would appear that the distributional information we are after is hidden in the inverse matrix $\boldsymbol{\kappa}(-\mathrm{i}z)^{-1}$, where $\boldsymbol{\kappa}$ is the matrix Laplace exponent of the ascending ladder height MAP of ξ given by (11.15).

With this heuristic in mind and with a view to proving Theorem 13.2 we thus turn our attention to an appropriate development of the deep factorisation of the stable process.

13.2 Deep inverse factorisation of the stable process

Recall that the Lamperti-stable MAP has matrix exponent, in the sense of (11.6), given by

$$\boldsymbol{\Psi}(z) = \begin{bmatrix} -\dfrac{\Gamma(\alpha - \mathrm{i}z)\Gamma(1+\mathrm{i}z)}{\Gamma(\alpha\hat{\rho} - \mathrm{i}z)\Gamma(1-\alpha\hat{\rho}+\mathrm{i}z)} & \dfrac{\Gamma(\alpha - \mathrm{i}z)\Gamma(1+\mathrm{i}z)}{\Gamma(\alpha\hat{\rho})\Gamma(1-\alpha\hat{\rho})} \\ \\ \dfrac{\Gamma(\alpha - \mathrm{i}z)\Gamma(1+\mathrm{i}z)}{\Gamma(\alpha\rho)\Gamma(1-\alpha\rho)} & -\dfrac{\Gamma(\alpha - \mathrm{i}z)\Gamma(1+\mathrm{i}z)}{\Gamma(\alpha\rho - \mathrm{i}z)\Gamma(1-\alpha\rho+\mathrm{i}z)} \end{bmatrix}, \quad (13.3)$$

for $\mathrm{Re}(\mathrm{i}z) \in (-1, \alpha)$. Moreover, from (11.24), the Wiener–Hopf factorisation takes the form

$$-\boldsymbol{\Psi}(\mathrm{i}\theta) = \boldsymbol{\Delta}_\pi^{-1}\hat{\boldsymbol{\kappa}}(\mathrm{i}\theta)^{\mathrm{T}}\boldsymbol{\Delta}_\pi\boldsymbol{\kappa}(-\mathrm{i}\theta), \qquad \theta \in \mathbb{R}, \quad (13.4)$$

up to pre-multiplication of $\boldsymbol{\Psi}$ by a strictly positive diagonal matrix. In (13.4), we recall that $\boldsymbol{\kappa}$ is the matrix exponent of the ascending ladder MAP, with $\hat{\boldsymbol{\kappa}}$ being that of the dual ascending ladder MAP, and $\boldsymbol{\Delta}_\pi$ is the diagonal matrix constructed from the stationary distribution of the underlying switching chain, denoted by $\boldsymbol{\pi}$.

In the setting of the stable process (13.4) is called the *deep factorisation* (to distinguish from the actual factorisation of the characteristic exponent of the stable process as a Lévy process). Although the form of (13.3) is relatively appealing and we can compute directly the quantity $\boldsymbol{\pi}$, it turns out to be rather difficult to directly identify the individual exponents $\boldsymbol{\kappa}$ and $\hat{\boldsymbol{\kappa}}$ from an explicit factorisation of (13.3).

There is an argument, however, which points towards one seeking instead matrix inverse $\boldsymbol{\kappa}^{-1}$. Indeed, we can take inspiration by looking all the way back to Theorem 2.22. In the notation given there, recalling the identity (2.33), one may otherwise read the identity (2.34) of Theorem 2.33 in its inverse form,

$$\frac{1}{\Psi_p(z)} = \frac{1}{\kappa_p(-\mathrm{i}z)} \times \frac{1}{\hat{\kappa}_p(\mathrm{i}z)}, \qquad z \in \mathbb{R}.$$

Note, an inverse factorisation holds in the MAP case by a similar direct inversion of the statement in Theorem 11.9. The inverse factorisation points towards potentials. Indeed $1/\Psi_p(z)$, $z \in \mathbb{R}$, is nothing more than the Fourier transform of the p-potential measure of the underlying Lévy process, Y, meaning

$$\int_0^\infty \mathrm{e}^{-qt}\mathrm{P}(Y_t \in \mathrm{d}x)\mathrm{d}t, \qquad x \in \mathbb{R}.$$

Similarly, $1/\kappa_p(\lambda)$, $\lambda \geq 0$ is the Laplace transform of the p-potential of the ascending ladder height process, H.

We therefore adopt a method which allows us to compute the two inverse matrix exponents $\boldsymbol{\kappa}^{-1}$ and $\hat{\boldsymbol{\kappa}}^{-1}$ directly, without the need to directly unravel the matrix $\boldsymbol{\Psi}$. Our method is based around extracting the information we need directly from fluctuation theory of the underlying MAP. It also allows us to obtain the potential density matrices for the ascending ladder MAPs. That is to say, the matrices $\boldsymbol{u}(x)$ and $\hat{\boldsymbol{u}}(x)$, $x \geq 0$, whose Laplace transforms are precisely $\boldsymbol{\kappa}^{-1}$ and $\hat{\boldsymbol{\kappa}}^{-1}$. These will turn out to be of importance in addressing the proof of Theorem 13.2.

To this end, for $a, b, c \in \mathbb{R}$, define

$$\Xi(a,b,c) := \int_0^1 u^a(1-u)^b(1+u)^c\mathrm{d}u, \tag{13.5}$$

whenever the integral can be performed. Note that the function Ξ can also be written in terms of the ${}_2F_1$ hypergeometric function, specifically

$$\Xi(a,b,c) = \frac{\Gamma(a+1)\Gamma(b+1)}{\Gamma(a+b+2)}\,{}_2F_1\left(-c, a+1, a+b+2, -1\right),$$

providing $a, b > -1$. See (A.31) in the Appendix.

Theorem 13.3 *Suppose that X is an α-stable process with two-sided jumps. Then we have that, up to post-multiplication by a strictly positive diagonal matrix, the factors* $\boldsymbol{\kappa}^{-1}$ *and* $\hat{\boldsymbol{\kappa}}^{-1}$ *are given by the following relations.*

(i) When $\alpha \in (0,1)$ *and* $\rho \in (0,1)$*, we have, for all* $\lambda \geq 0$,

$$\boldsymbol{\kappa}^{-1}(\lambda) = \begin{bmatrix} \Xi(\lambda-1, \alpha\rho-1, \alpha\hat{\rho}) & \Xi(\lambda-1, \alpha\rho, \alpha\hat{\rho}-1) \\ \Xi(\lambda-1, \alpha\hat{\rho}, \alpha\rho-1) & \Xi(\lambda-1, \alpha\hat{\rho}-1, \alpha\rho) \end{bmatrix}$$

and

$$\begin{bmatrix} \frac{\Gamma(\alpha\hat{\rho})}{\Gamma(1-\alpha\rho)} & 0 \\ 0 & \frac{\Gamma(\alpha\rho)}{\Gamma(1-\alpha\hat{\rho})} \end{bmatrix} \hat{\boldsymbol{\kappa}}^{-1}(\lambda) = \begin{bmatrix} \Xi(\lambda-\alpha, \alpha\hat{\rho}-1, \alpha\rho) & \Xi(\lambda-\alpha, \alpha\hat{\rho}, \alpha\rho-1) \\ \Xi(\lambda-\alpha, \alpha\rho, \alpha\hat{\rho}-1) & \Xi(\lambda-\alpha, \alpha\rho-1, \alpha\hat{\rho}) \end{bmatrix}.$$

(ii) When $\alpha = 1$*, we have, for all* $\lambda \geq 0$,

$$\boldsymbol{\kappa}^{-1}(\lambda) = \hat{\boldsymbol{\kappa}}^{-1}(\lambda) = \begin{bmatrix} \Xi(\lambda-1, -1/2, 1/2) & \Xi(\lambda-1, 1/2, -1/2) \\ \Xi(\lambda-1, 1/2, -1/2) & \Xi(\lambda-1, -1/2, 1/2) \end{bmatrix}.$$

(iii) When $\alpha \in (1,2)$ *and* $0 < \alpha\rho, \alpha\hat{\rho} < 1$*, we have, for all* $\lambda \geq 0$,

$$\begin{aligned}\boldsymbol{\kappa}^{-1}(\lambda) &= \begin{bmatrix} \Xi(\lambda-1, \alpha\rho-1, \alpha\hat{\rho}) & \Xi(\lambda-1, \alpha\rho, \alpha\hat{\rho}-1) \\ \Xi(\lambda-1, \alpha\hat{\rho}, \alpha\rho-1) & \Xi(\lambda-1, \alpha\hat{\rho}-1, \alpha\rho) \end{bmatrix} \\ &\quad - \frac{(\alpha-1)}{(\lambda+\alpha-1)} \begin{bmatrix} \Xi(\lambda-1, \alpha\rho-1, \alpha\hat{\rho}-1) & \Xi(\lambda-1, \alpha\rho-1, \alpha\hat{\rho}-1) \\ \Xi(\lambda-1, \alpha\hat{\rho}-1, \alpha\rho-1) & \Xi(\lambda-1, \alpha\hat{\rho}-1, \alpha\rho-1) \end{bmatrix}\end{aligned}$$

and

$$\begin{aligned}\begin{bmatrix} \frac{\Gamma(\alpha\hat{\rho})}{\Gamma(1-\alpha\rho)} & 0 \\ 0 & \frac{\Gamma(\alpha\rho)}{\Gamma(1-\alpha\hat{\rho})} \end{bmatrix} \hat{\boldsymbol{\kappa}}^{-1}(\lambda) &= \begin{bmatrix} \Xi(\lambda-\alpha, \alpha\hat{\rho}-1, \alpha\rho) & \Xi(\lambda-\alpha, \alpha\hat{\rho}, \alpha\rho-1) \\ \Xi(\lambda-\alpha, \alpha\rho, \alpha\hat{\rho}-1) & \Xi(\lambda-\alpha, \alpha\rho-1, \alpha\hat{\rho}) \end{bmatrix} \\ &\quad - \frac{(\alpha-1)}{\lambda} \begin{bmatrix} \Xi(\lambda-\alpha, \alpha\hat{\rho}-1, \alpha\rho-1) & \Xi(\lambda-\alpha, \alpha\hat{\rho}-1, \alpha\rho-1) \\ \Xi(\lambda-\alpha, \alpha\rho-1, \alpha\hat{\rho}-1) & \Xi(\lambda-\alpha, \alpha\rho-1, \alpha\hat{\rho}-1) \end{bmatrix}.\end{aligned}$$

13.3 Ladder MAP matrix potentials

As alluded to above, exploring the inverse ascending ladder MAP matrix exponents is equivalent to exploring their associated potential measures. For $x \geq 0$ and $i, j \in \{-1, 1\}$, we write

$$U_{i,j}(x) = \int_0^\infty \mathbf{P}_{0,i}(H_t^+ \leq x,\ J_t^+ = j,\ t < \varsigma)\,\mathrm{d}t.$$

Recall the process H^+ has a cemetery state $+\infty$ and ς is the process lifetime, which may otherwise be identified as L_∞, where L is the local time of $\overline{\xi} - \xi$ at 0; see the discussion in Section 11.3. Assuming that $U_{i,j}(x)$ is absolutely continuous on $[0, \infty)$, for $i, j \in \{-1, 1\}$, if we let $\boldsymbol{u}_{i,j}(x)$ denote its density, then $\boldsymbol{U}$ is the unique matrix valued function such that, for $\lambda \geq 0$,

$$\int_0^\infty \mathrm{e}^{-\lambda x}\boldsymbol{u}_{i,j}(x)\mathrm{d}x = \int_0^\infty \mathbf{E}_{0,i}[\mathrm{e}^{-\lambda H_t^+},\ J_t^+ = j]\mathrm{d}t = [\boldsymbol{\kappa}^{-1}(\lambda)]_{i,j}.$$

Similarly, we can define the matrix $\hat{\boldsymbol{U}}(x)$ for the dual MAP and, when it exists, we will write $\hat{\boldsymbol{u}}$ for its density, in which case, it necessarily satisfies

$$\int_0^\infty \mathrm{e}^{-\lambda x}\hat{\boldsymbol{u}}_{i,j}(x)\mathrm{d}x = [\hat{\boldsymbol{\kappa}}^{-1}(\lambda)]_{i,j}, \qquad \text{for each } i, j = \pm 1, \lambda \geq 0.$$

Using the previous two matrix Laplace transforms, the following result has Theorem 13.3 as its corollary. (Indeed, we leave the very straightforward verification of this fact as an exercise for the reader). In order to state the next result, we need to introduce the notation

$$\chi(a, b, c; x) = x^a(1 - x)^b(1 + x)^c, \qquad x \geq 0,$$

and $a, b, c \in \mathbb{R}$, so that, when the integral can be performed,

$$\int_0^\infty \mathrm{e}^{-\lambda x}\chi(a, b, c; \mathrm{e}^{-x})\mathrm{d}x = \Xi(\lambda + a - 1, b, c), \qquad \lambda > 0. \tag{13.6}$$

Theorem 13.4 *Under the conditions of Theorem 13.3, up to post-multiplication by a strictly positive diagonal matrix, the ascending ladder potential $\boldsymbol{U}$ has a density with respect to Lebesgue measure, given by the following relations.*

(i) When $\alpha \in (0, 1)$, we have, for $x \geq 0$,

$$\boldsymbol{u}(x) = \begin{bmatrix} \chi(0, \alpha\rho - 1, \alpha\hat{\rho}; \mathrm{e}^{-x}) & \chi(0, \alpha\rho, \alpha\hat{\rho} - 1; \mathrm{e}^{-x}) \\ \\ \chi(0, \alpha\hat{\rho}, \alpha\rho - 1; \mathrm{e}^{-x}) & \chi(0, \alpha\hat{\rho} - 1, \alpha\rho; \mathrm{e}^{-x}) \end{bmatrix}$$

and

$$\begin{bmatrix} \frac{\Gamma(\alpha\hat\rho)}{\Gamma(1-\alpha\rho)} & 0 \\ 0 & \frac{\Gamma(\alpha\rho)}{\Gamma(1-\alpha\hat\rho)} \end{bmatrix} \hat{\boldsymbol{u}}(x) = \begin{bmatrix} \chi(1-\alpha, \alpha\hat\rho - 1, \alpha\rho; \mathrm{e}^{-x}) & \chi(1-\alpha, \alpha\hat\rho, \alpha\rho - 1; \mathrm{e}^{-x}) \\ \chi(1-\alpha, \alpha\rho, \alpha\hat\rho - 1; \mathrm{e}^{-x}) & \chi(1-\alpha, \alpha\rho - 1, \alpha\hat\rho; \mathrm{e}^{-x}) \end{bmatrix}.$$

(ii) When $\alpha = 1$, we have for $x \geq 0$, $\boldsymbol{u}(x) = \hat{\boldsymbol{u}}(x)$ and

$$\boldsymbol{u}(x) = \begin{bmatrix} \chi(0, -1/2, 1/2; \mathrm{e}^{-x}) & \chi(0, 1/2, -1/2; \mathrm{e}^{-x}) \\ \chi(0, 1/2, -1/2; \mathrm{e}^{-x}) & \chi(0, -1/2, 1/2; \mathrm{e}^{-x}) \end{bmatrix}.$$

(iii) When $\alpha \in (1, 2)$, we have, for $x \geq 0$,

$$\boldsymbol{u}(x) = \begin{bmatrix} \chi(0, \alpha\rho - 1, \alpha\hat\rho; \mathrm{e}^{-x}) & \chi(0, \alpha\rho, \alpha\hat\rho - 1; \mathrm{e}^{-x}) \\ \chi(0, \alpha\hat\rho, \alpha\rho - 1; \mathrm{e}^{-x}) & \chi(0, \alpha\hat\rho - 1, \alpha\rho; \mathrm{e}^{-x}) \end{bmatrix}$$
$$- (\alpha - 1)\mathrm{e}^{-(\alpha-1)x} \int_{\mathrm{e}^{-x}}^{1} \mathrm{d}u \begin{bmatrix} \chi(-\alpha, \alpha\rho - 1, \alpha\hat\rho - 1; u) & \chi(-\alpha, \alpha\rho - 1, \alpha\hat\rho - 1; u) \\ \chi(-\alpha, \alpha\hat\rho - 1, \alpha\rho - 1; u) & \chi(-\alpha, \alpha\hat\rho - 1, \alpha\rho - 1; u) \end{bmatrix}$$

and

$$\begin{bmatrix} \frac{\Gamma(\alpha\hat\rho)}{\Gamma(1-\alpha\rho)} & 0 \\ 0 & \frac{\Gamma(\alpha\rho)}{\Gamma(1-\alpha\hat\rho)} \end{bmatrix} \hat{\boldsymbol{u}}(x) = \begin{bmatrix} \chi(\alpha - 1, \alpha\hat\rho - 1, \alpha\rho; \mathrm{e}^{-x}) & \chi(\alpha - 1, \alpha\hat\rho, \alpha\rho - 1; \mathrm{e}^{-x}) \\ \chi(\alpha - 1, \alpha\rho, \alpha\hat\rho - 1; \mathrm{e}^{-x}) & \chi(\alpha - 1, \alpha\rho - 1, \alpha\hat\rho; \mathrm{e}^{-x}) \end{bmatrix}$$
$$- (\alpha - 1) \int_{\mathrm{e}^{-x}}^{1} \mathrm{d}u \begin{bmatrix} \chi(-\alpha, \alpha\rho - 1, \alpha\hat\rho - 1; u) & \chi(-\alpha, \alpha\rho - 1, \alpha\hat\rho - 1; u) \\ \chi(-\alpha, \alpha\hat\rho - 1, \alpha\rho - 1; u) & \chi(-\alpha, \alpha\hat\rho - 1, \alpha\rho - 1; u) \end{bmatrix}.$$

Just as we can construct an ascending ladder MAP (H^+, J^+), we can similarly construct the descending ladder MAP, henceforth referred to as (H^-, J^-), and accordingly we can talk about the associated descending ladder potential matrix $\boldsymbol{U}^-$. On our way to proving Theorem 13.4, we will see that $x \mapsto U^-_{i,j}(x)$, $x \geq 0$, is differentiable, and we will pre-emptively denote by $\boldsymbol{u}^-$ the derivative matrix of $\boldsymbol{U}^-$. (Note, *a priori*, as an increasing function it is Lebesgue almost everywhere differentiable.) Denote by $\boldsymbol{\kappa}^-$ the MAP–exponent of (H^-, J^-). Our means of accessing $\hat{\boldsymbol{\kappa}}$ will be via $\boldsymbol{\kappa}^-$.

Lemma 13.5 *Up to pre-multiplying by a strictly positive diagonal matrix,*

$$\boldsymbol{\kappa}^-(\lambda) = \hat{\boldsymbol{\kappa}}(\lambda), \qquad \lambda \geq 0. \tag{13.7}$$

Proof By considering $\boldsymbol{\Psi}(0) = \boldsymbol{Q}$ and $\boldsymbol{\pi}^{\mathrm{T}}\boldsymbol{Q} = \mathbf{0}$, we can derive

$$\boldsymbol{\pi} \propto \begin{bmatrix} \sin(\pi\alpha\rho) \\ \sin(\pi\alpha\hat{\rho}) \end{bmatrix}. \tag{13.8}$$

Using this expression for $\boldsymbol{\pi}$, the identity $\hat{\boldsymbol{\Psi}}(z) = \boldsymbol{\Delta}_\pi^{-1}\boldsymbol{\Psi}(-z)^{\mathrm{T}}\boldsymbol{\Delta}_\pi$ from (11.7) and (13.3), a straightforward computation gives us the dual Matrix exponent

$$\hat{\boldsymbol{\Psi}}(z) = \begin{bmatrix} -\dfrac{\Gamma(\alpha + \mathrm{i}z)\Gamma(1 - \mathrm{i}z)}{\Gamma(\alpha\hat{\rho} + \mathrm{i}z)\Gamma(1 - \alpha\hat{\rho} - \mathrm{i}z)} & \dfrac{\Gamma(\alpha + \mathrm{i}z)\Gamma(1 - \mathrm{i}z)}{\Gamma(\alpha\hat{\rho})\Gamma(1 - \alpha\hat{\rho})} \\ \\ \dfrac{\Gamma(\alpha + \mathrm{i}z)\Gamma(1 - \mathrm{i}z)}{\Gamma(\alpha\rho)\Gamma(1 - \alpha\rho)} & -\dfrac{\Gamma(\alpha + \mathrm{i}z)\Gamma(1 - \mathrm{i}z)}{\Gamma(\alpha\rho + \mathrm{i}z)\Gamma(1 - \alpha\rho - \mathrm{i}z)} \end{bmatrix}, \tag{13.9}$$

which is well defined for $\mathrm{Re}(\mathrm{i}z) \in (-\alpha, 1)$. It is not hard to check, by inspecting (13.9) and (13.3), that the simpler relation $\boldsymbol{\Psi}(-z) = \hat{\boldsymbol{\Psi}}(z)$ also holds, for $z \in \mathbb{R}$. As a consequence, the MAP $(-\xi, J)$ is equal in law to $(\hat{\xi}, \hat{J})$. Since $\boldsymbol{\kappa}^-$ and $\hat{\boldsymbol{\kappa}}$ are the matrix Laplace exponent of the ascending ladder height processes of the MAPs $(-\xi, J)$ and $(\hat{\xi}, \hat{J})$, respectively, it follows that $\boldsymbol{\kappa}^-(\lambda) = \hat{\boldsymbol{\kappa}}(\lambda)$ as required. □

Note that (13.7) also implies that, when it exists, up to post-multiplication by a strictly positive matrix,

$$\hat{\boldsymbol{u}}(x) = \boldsymbol{u}^-(x), \qquad x \geq 0. \tag{13.10}$$

Recalling (13.3), we can also verify by inspection that

$$\hat{\boldsymbol{\Psi}}(z) = \boldsymbol{\Delta}_{\boldsymbol{\pi}}^{-1}\ \boldsymbol{\Psi}(z - \mathrm{i}(\alpha - 1))|_{\rho\leftrightarrow\hat{\rho}}\,\boldsymbol{\Delta}_{\boldsymbol{\pi}}, \qquad z \in \mathbb{R}, \tag{13.11}$$

where $\rho \leftrightarrow \hat{\rho}$ means that the roles of ρ and $\hat{\rho}$ are exchanged. Although complicated, this identity may also be verified by comparing the relationships between $\hat{\boldsymbol{\Psi}}(z)$, $\boldsymbol{\Psi}_{\alpha-1}(z)$ and $\boldsymbol{\Psi}^\circ(z)$ in the proofs of Theorems 12.2 and 12.7.

Writing out the Wiener–Hopf factorisation (11.24) in place of $\hat{\boldsymbol{\Psi}}(z)$ and $\boldsymbol{\Psi}(z - \mathrm{i}(\alpha - 1))$ in (13.11), we see that

$$\hat{\boldsymbol{\kappa}}(\lambda) = \boldsymbol{\Delta}_{\boldsymbol{\pi}}^{-1}\ \boldsymbol{\kappa}(\lambda + 1 - \alpha)|_{\rho\leftrightarrow\hat{\rho}}\,\boldsymbol{\Delta}_{\boldsymbol{\pi}}, \qquad \lambda \geq 0, \tag{13.12}$$

or equivalently,

$$\boldsymbol{\kappa}(\lambda)^{-1} = [\boldsymbol{\Delta}_{\boldsymbol{\pi}}\ \hat{\boldsymbol{\kappa}}(\lambda + \alpha - 1)^{-1}\boldsymbol{\Delta}_{\boldsymbol{\pi}}^{-1}]\big|_{\rho\leftrightarrow\hat{\rho}}, \qquad \lambda \geq 0. \tag{13.13}$$

Hence, referring back to (13.10), we have

$$\boldsymbol{u}(x) = [\mathrm{e}^{(1-\alpha)x}\boldsymbol{\Delta}_{\boldsymbol{\pi}}\ \boldsymbol{u}^-(x)\boldsymbol{\Delta}_{\boldsymbol{\pi}}^{-1}]\big|_{\rho\leftrightarrow\hat{\rho}}, \qquad x \geq 0. \tag{13.14}$$

When combining with (13.10) and (13.14), Lemma 13.6 below proves Theorem 13.4 (i). By taking Laplace transforms using (13.6), Lemma 13.6 also gives us Theorem 13.3 (i).

Lemma 13.6 *Suppose that* $\alpha, \rho \in (0,1)$. *Then for all* $x \geq 0$, *up to post-multiplication by a strictly positive diagonal matrix,*

$$\begin{bmatrix} \frac{\Gamma(\alpha\hat{\rho})}{\Gamma(1-\alpha\rho)} & 0 \\ 0 & \frac{\Gamma(\alpha\rho)}{\Gamma(1-\alpha\hat{\rho})} \end{bmatrix} \boldsymbol{u}^-(x)$$
$$= \begin{bmatrix} \chi(1-\alpha, \alpha\hat{\rho}-1, \alpha\rho; \mathrm{e}^{-x}) & \chi(1-\alpha, \alpha\hat{\rho}, \alpha\rho-1; \mathrm{e}^{-x}) \\ \\ \chi(1-\alpha, \alpha\rho, \alpha\hat{\rho}-1; \mathrm{e}^{-x}) & \chi(1-\alpha, \alpha\rho-1, \alpha\hat{\rho}; \mathrm{e}^{-x}) \end{bmatrix}.$$

Proof Let $T_0^- := \inf\{t \geq 0 : \xi_t < 0\}$ and recall from Section 6.4 that $\underline{m}$ is such that $|X_t| \geq |X_{\underline{m}}|$ for all $t \geq 0$. Then we claim that, for $i, j = \{-1, 1\}$ and $y > 0$,

$$\mathbf{P}_{y,i}(T_0^- = \infty,\ J_{\varphi(\underline{m})} = j) = U_{i,j}^-(y), \tag{13.15}$$

where we recall that φ is the time change in the Lamperti–Kiu transform 11.3. To see why, suppose that we write η_i^-, $i = \pm 1$ for the rates at which the descending ladder MAP (H^-, J^-) is killed. Then, appealing to the compensation formula for Markov additive Poisson point processes (see (A.47) in the Appendix), we have

$$\begin{aligned}
&\mathbf{P}_{x,i}(T_0^- = \infty, J_{\varphi(\underline{m})} = j) \\
&\quad = \mathbf{E}_{0,i}\left[\mathbf{E}_{0,i}\Big[\sum_{t\geq 0} \mathbf{1}_{(H_{t-}^- \leq x,\, J_t^- = j)} \Big| \sigma(J_u^- : u \geq 0)\Big]\right] \\
&\quad = \mathbf{E}_{0,i}\Big[\int_0^\infty \mathbf{1}_{(H_{t-}^- \leq x,\, J_{t-}^- = j)}\, \mathrm{d}t\Big]\eta_j^- \\
&\quad = U_{i,j}^-(x)\eta_j^-.
\end{aligned}$$

Since we know that the matrix potential $U^-(x)$ is unique up to post-multiplication by a strictly positive diagonal matrix, we may absorb the constant η_j^- into the definition of $U_{i,j}^-(x)$.

Now note that the event $\{T_0^- = \infty, J_{\varphi(\underline{m})} = 1\}$ occurs if and only if $\{\tau^{(-1,1)} = \infty\}$ and furthermore the point at which X is closest to the origin is positive, that is, $X_{\underline{m}} > 0$. Thus $\{T_0^- = \infty, J_{\varphi(\underline{m})} = 1\}$ occurs if and only if $\{X_{\underline{m}} > 1\}$ does. Using Theorem 6.13 and, specifically, (6.11), we have that

$$
\begin{aligned}
U^-_{1,1}(x) &= \mathbf{P}_{x,i}(T_0^- = \infty,\ J_{\varphi(\underline{m})} = 1) \\
&= \mathbb{P}_{e^x}(X_{\underline{m}} > 1) \\
&= \frac{1}{2}\int_1^{e^x} (e^x + z) G'\left(\frac{e^x}{z}\right) z^{-2} dz \\
&= \frac{1}{2}\int_1^{e^x} (1 + 1/u) G'(u) du,
\end{aligned}
\tag{13.16}
$$

where, for $x > 1$,

$$
G(x) = \frac{\Gamma(1-\alpha\rho)}{\Gamma(\alpha\hat{\rho})\Gamma(1-\alpha)} \int_0^{(x-1)/(x+1)} t^{\alpha\hat{\rho}-1}(1-t)^{-\alpha}\, dt
$$

and in the final equality we have used the substitution $u = e^x/z$. Differentiating the above equation, we get that

$$
\begin{aligned}
u^-_{1,1}(x) &= \frac{1}{2}(e^x + 1) G'(e^x) \\
&= \frac{\Gamma(1-\alpha\rho)}{2^\alpha \Gamma(1-\alpha)\Gamma(\alpha\hat{\rho})} (e^x - 1)^{\alpha\hat{\rho}-1}(e^x + 1)^{\alpha\rho}.
\end{aligned}
\tag{13.17}
$$

Similarly, considering the event $\{T_0^- = \infty,\ J_{\varphi(\underline{m})} = -1\}$, we get that

$$
u^-_{1,-1}(x) = \frac{1}{2}(e^x - 1) G'(e^x) = \frac{\Gamma(1-\alpha\rho)}{2^\alpha \Gamma(1-\alpha)\Gamma(\alpha\hat{\rho})} (e^x - 1)^{\alpha\hat{\rho}}(e^x + 1)^{\alpha\rho-1}.
$$

To derive the row $u^-_{-1,j}$ we can use $-X$ in the computations above, which implies that $u^-_{-1,j}$ is the same as $u^-_{1,-j}$, albeit that the roles of ρ and $\hat{\rho}$ are exchanged. This concludes the proof of the lemma. □

The reader will note at this point that the expression for $u(x)$ in the statement of Theorem 13.4 (i) is not precisely what one obtains by performing the calculation in (13.14). In fact, the matrix that emerges takes the form

$$
\begin{bmatrix}
\frac{\Gamma(1-\alpha\hat{\rho})}{\Gamma(\alpha\rho)} \chi(0, \alpha\rho - 1, \alpha\hat{\rho}; e^{-x}) & \frac{\Gamma(1-\alpha\rho)}{\Gamma(\alpha\hat{\rho})} \chi(0, \alpha\rho, \alpha\hat{\rho} - 1; e^{-x}) \\
\\
\frac{\Gamma(1-\alpha\hat{\rho})}{\Gamma(\alpha\rho)} \chi(0, \alpha\hat{\rho}, \alpha\rho - 1; e^{-x}) & \frac{\Gamma(1-\alpha\rho)}{\Gamma(\alpha\hat{\rho})} \chi(0, \alpha\hat{\rho} - 1, \alpha\rho; e^{-x})
\end{bmatrix}
$$

and this is equal to the expression for $u(x)$ in the statement of Theorem 13.4 (i) post multiplied by the matrix

$$
\begin{bmatrix}
\frac{\Gamma(1-\alpha\hat{\rho})}{\Gamma(\alpha\rho)} & 0 \\
0 & \frac{\Gamma(1-\alpha\rho)}{\Gamma(\alpha\hat{\rho})}
\end{bmatrix}.
$$

Following the calculations above, we establish the analogue of Lemma 13.6 for the case that $\alpha \in (1, 2)$, and hence the proof of Theorem 13.4 (iii). Since

$\alpha \in (1,2)$ we have that $\tau^{\{0\}} := \inf\{t \geq 0 : X_t = 0\} < \infty$ and $X_{\tau^{\{0\}}-} = 0$ almost surely. Hence, for the MAP (ξ, J), it is the case that ξ drifts to $-\infty$. Recall that $\overline{m}$ is the unique time on $[0, \tau^{\{0\}}]$ such that

$$|X_{\overline{m}}| \geq |X_t|, \qquad \text{for all } t < \tau^{\{0\}}.$$

We are interested in the quantity $X_{\overline{m}}$ as it allows us access to $\boldsymbol{u}(x)$, $x \geq 0$.

Proof of Theorem 13.4 (iii). Let $T_0^+ := \inf\{t \geq 0 : \xi_t > 0\}$ then similarly to (13.15) we can easily derive the analogous identity

$$\mathbf{P}_{-y,i}(T_0^+ = \infty,\ J_{\varphi(\overline{m})} = j) = \boldsymbol{U}_{i,j}(y),$$

for $i, j = \{-1, 1\}$ and $y > 0$. Similarly to the equality (13.16), we may appeal to Theorem 6.14 to get that

$$\begin{aligned}
&\boldsymbol{u}_{1,1}(x)\\
&= \frac{\mathrm{d}}{\mathrm{d}x}\mathbb{P}_{\mathrm{e}^{-x}}(X_{\overline{m}} \in (\mathrm{e}^{-x}, 1))\\
&= \frac{\alpha-1}{2}\frac{\mathrm{d}}{\mathrm{d}x}\int_{\mathrm{e}^{-x}}^{1} \mathrm{d}z\frac{1}{z^{\alpha}}\mathrm{e}^{(2-\alpha)x}\left((\mathrm{e}^{-x}+z)\overline{G}'\left(\frac{z}{\mathrm{e}^{-x}}\right) - (\alpha-1)\mathrm{e}^{-x}\overline{G}\left(\frac{z}{\mathrm{e}^{-x}}\right)\right)\\
&= \frac{\alpha-1}{2}\frac{\mathrm{d}}{\mathrm{d}x}\int_{1}^{\mathrm{e}^{x}} \mathrm{d}u\,\frac{1}{u^{\alpha}}\left((1+u)\overline{G}'(u) - (\alpha-1)\overline{G}(u)\right)\\
&= \frac{\alpha-1}{2}\mathrm{e}^{-(\alpha-1)x}\left((1+\mathrm{e}^{x})\overline{G}'(\mathrm{e}^{x}) - (\alpha-1)\overline{G}(\mathrm{e}^{x})\right)\\
&= \frac{\alpha-1}{2}\mathrm{e}^{-(\alpha-1)x}\left((\mathrm{e}^{x}-1)^{\alpha\rho-1}(\mathrm{e}^{x}+1)^{\alpha\hat{\rho}} - (\alpha-1)\overline{G}(\mathrm{e}^{x})\right),
\end{aligned} \tag{13.18}$$

where

$$\overline{G}(z) = \int_1^z (t-1)^{\alpha\rho-1}(t+1)^{\alpha\hat{\rho}-1}\mathrm{d}t = \int_{1/z}^{1} \chi(-\alpha, \alpha\rho-1, \alpha\hat{\rho}-1; u)\mathrm{d}u$$

and in the third equality we have used the substitution $u = z/\mathrm{e}^{-x}$. Similar proofs give $\boldsymbol{u}_{i,j}^{-1}(x)$ for the remaining i, j, and this gives us the proof of the first part of Theorem 13.4 (iii).

As alluded to earlier, the proof of the first part of Theorem 13.3 (iii) will follow by taking Laplace transforms. Indeed, a straightforward application of Fubini's Theorem and a change of variable shows that, for $\gamma > \alpha - 1$,

$$\int_0^\infty \mathrm{e}^{-\gamma x}\overline{G}(\mathrm{e}^{x})\mathrm{d}x = \frac{1}{\gamma}\Xi(\gamma-\alpha, \alpha\rho-1, \alpha\hat{\rho}-1), \tag{13.19}$$

where we recall Ξ was defined in (13.6). Moreover, using the substitution $u = \mathrm{e}^{-x}$, one may also easily derive that

$$\int_0^\infty \mathrm{e}^{-(\lambda+\alpha-1)x}(\mathrm{e}^{x}-1)^{\alpha\rho-1}(\mathrm{e}^{x}+1)^{\alpha\hat{\rho}}\mathrm{d}x = \Xi(\lambda-1, \alpha\rho-1, \alpha\hat{\rho}).$$

Integrating (13.18) and using the previous two integral identities, we get that, for $\lambda \geq 0$,

$$\kappa^{-1}_{1,1}(\lambda) = \frac{\alpha - 1}{2}\Xi(\lambda - 1, \alpha\rho - 1, \alpha\hat{\rho}) - \frac{(\alpha-1)^2}{2(\lambda + \alpha - 1)}\Xi(\lambda - 1, \alpha\rho - 1, \alpha\hat{\rho} - 1).$$

Laplace transforming $u^{-1}_{i,j}(x)$ for the remaining i, j similarly completes the proof of Theorem 13.3 (iii).

Next we can appeal to (13.14) and (13.10) and note that, up to post-multiplication by a strictly positive diagonal matrix,

$$\hat{u}(x) = \mathrm{e}^{(\alpha-1)x}\boldsymbol{\Delta}_\pi^{-1}\, u(x)|_{\rho\leftrightarrow\hat{\rho}}\,\boldsymbol{\Delta}_\pi, \qquad x \geq 0.$$

In fact, the normalisation corresponding to the statement of the theorem means that we choose to work with the definition

$$\hat{u}(x) = \mathrm{e}^{(\alpha-1)x}\boldsymbol{\Delta}_\pi^{-1}\, u(x)|_{\rho\leftrightarrow\hat{\rho}}\,\boldsymbol{\Delta}_\pi \begin{bmatrix} \frac{\Gamma(1-\alpha\rho)}{\Gamma(\alpha\hat{\rho})} & 0 \\ 0 & \frac{\Gamma(1-\alpha\hat{\rho})}{\Gamma(\alpha\rho)} \end{bmatrix}, \qquad x \geq 0. \tag{13.20}$$

Appealing to the matrix algebra

$$\boldsymbol{\Delta}_\pi^{-1} M \boldsymbol{\Delta}_\pi \begin{bmatrix} \frac{\Gamma(1-\alpha\rho)}{\Gamma(\alpha\hat{\rho})} & 0 \\ 0 & \frac{\Gamma(1-\alpha\hat{\rho})}{\Gamma(\alpha\rho)} \end{bmatrix} = \begin{bmatrix} \frac{\Gamma(1-\alpha\rho)}{\Gamma(\alpha\hat{\rho})} & 0 \\ 0 & \frac{\Gamma(1-\alpha\hat{\rho})}{\Gamma(\alpha\rho)} \end{bmatrix} M, \tag{13.21}$$

where M is any 2×2 matrix, we also get the relation for the second part of Theorem 13.4 (iii).

Similarly, by observing the relation (13.12) and post multiplying by the same matrix as in (13.20), we get the $\hat{\kappa}^{-1}(\lambda)$ in Theorem 13.3 (iii).

Finally, we deal with the case that $\alpha = 1$. Recall that the process X is a Cauchy process, which has the property that

$$\limsup_{t\to\infty} |X_t| = \infty \qquad \text{and} \qquad \liminf_{t\to\infty} |X_t| = 0.$$

This means that the map (ξ, J) oscillates, and hence the global minimum and maximum both do not exist so that the previous methods cannot be used. Instead, we focus on a two-sided exit problem as an alternative approach. Note, the method we are about to describe also works for the other cases of α, however it is lengthy. □

Proof of Theorem 13.4 (ii) We start by referring back to the conclusion of Theorem 6.2, which tells us in particular that, for $z \in (0, 1)$, $0 \leq v \leq u \leq 1$ and $y \geq 0$,

$$\begin{aligned}&\mathbb{P}_z(1 - \overline{X}_{\tau_1^+ -} \in \mathrm{d}u,\ 1 - X_{\tau_1^+ -} \in \mathrm{d}v,\ X_{\tau_1^+} - 1 \in \mathrm{d}y,\ \tau_1^+ < \tau_0^-)\\ &\quad = \frac{1}{\pi}\frac{z^{1/2}(1-v)^{1/2}}{(1-u-z)^{1/2}(v-u)^{1/2}(1-u)(v+y)^2}\mathrm{d}u\,\mathrm{d}v\,\mathrm{d}y,\end{aligned}$$

where $\tau_0^- := \inf\{t \geq 0 : X_t < 0\}$. We wish to integrate v out of the above equation. To do this, we make the otherwise subtle observation that

$$\begin{aligned}
&\int_u^1 \mathrm{d}v\,(1-v)^{1/2}(v-u)^{-1/2}(v+y)^{-2} \\
&= (u+y)^{-2}(1-u)\int_0^1 \mathrm{d}z\,(1-z)^{1/2}z^{-1/2}\left(1+z\frac{1-u}{u+y}\right)^{-2} \\
&= (u+y)^{-2}(1-u)\frac{\pi}{2}\,{}_2F_1\left(2,1/2,2,-\frac{1-u}{u+y}\right) \\
&= \frac{\pi}{2}(u+y)^{-3/2}(1-u)(1+y)^{-1/2},
\end{aligned}$$

where in the first equality we have used the substitution $z = (v-u)/(1-u)$. In the second equality we have used the integral representation of the ${}_2F_1$ function and the final equality follows from the Euler–transformation (see the Appendix for further clarification of these identities). Hence, for $z \leq 1$, $u \in [0, 1-z)$ and $y \geq 0$,

$$\begin{aligned}
&\mathbb{P}_z(X_{\tau_1^+} - 1 \in \mathrm{d}y,\ 1 - \overline{X}_{\tau_1^+-} \in \mathrm{d}u,\ \tau_1^+ < \tau_0^-) \\
&= \frac{1}{2}\frac{z^{1/2}}{(1-u-z)^{1/2}(1+y)^{1/2}(u+y)^{3/2}}\mathrm{d}u\mathrm{d}y. \qquad (13.22)
\end{aligned}$$

Next, we have that for $z \in (-1, 1)$, $u \in [0, (1-z) \vee 1)$ and $y \geq 0$,

$$\begin{aligned}
&\mathbb{P}_z(1 - \overline{X}_{\tau_1^+-} \in \mathrm{d}u, X_{\tau_1^+} - 1 \in \mathrm{d}y, \overline{X}_{\tau_1^+-} > -\underline{X}_{\tau_1^+-}, \tau_1^+ < \tau_{-1}^-) \\
&= \frac{\partial}{\partial v}\frac{\partial}{\partial y}\mathbb{P}_z(1 - \overline{X}_{\tau_1^+-} \leq v, X_{\tau_1^+} - 1 \leq y, \tau_1^+ < \tau_{u-1}^-)|_{v=u}\,\mathrm{d}u\,\mathrm{d}y \\
&= \frac{\partial}{\partial v}\frac{\partial}{\partial y}\mathbb{P}_{\frac{z+1-u}{2-u}}\left(1 - \overline{X}_{\tau_1^+-} \leq \frac{v}{2-u}, X_{\tau_1^+} - 1 \leq \frac{y}{2-u}, \tau_1^+ < \tau_0^-\right)|_{v=u}\,\mathrm{d}u\,\mathrm{d}y \\
&= \frac{1}{2}(2-u)^{-2}\frac{\left(\frac{z+1-u}{2-u}\right)^{1/2}}{\left(1-\frac{u}{2-u}-\frac{z+1-u}{2-u}\right)^{1/2}\left(1+\frac{y}{2-u}\right)^{1/2}\left(\frac{u+y}{2-u}\right)^{3/2}}\mathrm{d}u\,\mathrm{d}y \\
&= \frac{(1-u+z)^{1/2}}{(1-u-z)^{1/2}}\frac{1}{(2-u+y)^{1/2}(u+y)^{3/2}}\mathrm{d}u\,\mathrm{d}y, \qquad (13.23)
\end{aligned}$$

where in the first equality we have used that the event $\{\overline{X}_{\tau_1^+-} > -\underline{X}_{\tau_1^+-}, 1-\overline{X}_{\tau_1^+-} \in \mathrm{d}u\}$ constrains $\underline{X}$ and is equivalent to $\{\tau_1^+ < \tau_{u-1}^-, 1 - \overline{X}_{\tau_1^+-} \in \mathrm{d}u\}$. In the second equality we have used the scaling property of X and in the third equality we have used (13.22).

For each $x \geq 0$ and $j = \pm 1$,

$$\begin{aligned}
&\frac{\partial}{\partial u}\frac{\partial}{\partial y}\mathbf{P}_{0,j}(x - H^+_{T_x-} \leq u; H^+_{T_x} - x \leq y; J^+_{T_x-} = 1; J^+_{T_x} = 1) \\
&= \frac{\partial}{\partial u}\frac{\partial}{\partial y}\mathbb{P}_j(\overline{X}_{\tau^+_{e^x}-} \geq e^{x-u}; X_{\tau^+_{e^x}} \leq e^{y+x}; X_{\tau^+_{e^x}-} > 0; \tau^+_{e^x} < \tau^-_{-e^x}) \\
&= \frac{\partial}{\partial u}\frac{\partial}{\partial y}\mathbb{P}_{je^{-x}}(\overline{X}_{\tau^+_1-} \geq e^{-u}; X_{\tau^+_1} \leq e^{y}; X_{\tau^+_1-} > 0; \tau^+_1 < \tau^-_{-1}) \\
&= e^{y-u}\frac{(e^{-u} + je^{-x})^{1/2}}{(e^{-u} - je^{-x})^{1/2}}\frac{1}{(e^y + e^{-u})^{1/2}(e^y - e^{-u})^{3/2}}, \qquad (13.24)
\end{aligned}$$

where in the first equality we have related the associated events of the two-sided exit problem to the first passage problem of the radial ascending ladder MAP (H^+, J^+) and second equality we have used the scaling property of X and in the final equality we applied (13.23).

In addition, appealing to the Markov additive Poisson point process structure of jumps for (H^+, J^+), for any $x > 0$, $u \in (0, x)$ and $i = \pm 1$,

$$\begin{aligned}
&\mathbf{P}_{0,i}(x - H^+_{T_x-} \in du; J^+_{T_x-} = 1; J^+_{T_x} = 1) \\
&= \mathbf{E}_{0,i}\left[\sum_{t>0} \mathbf{1}_{(x-H^+_{t-}\in du,\, J^+_{t-}=1)}\mathbf{1}_{(H^+_{t-}+\Delta H^+_t > x,\, J^+_t=1)}\right] \\
&= \boldsymbol{u}_{i,1}(x-u)\mathbf{n}_1(-\epsilon(\zeta) \in [u,\infty))\,du, \qquad (13.25)
\end{aligned}$$

where $\Delta H^+_t = H^+_t - H^+_{t-}$ and the excursion measure $\mathbf{n}_1$ with canonical excursion $\epsilon : [0, \zeta] \to \mathbb{R}$ were introduced in Section 11.3.

Equation (13.24) together with (13.25) gives that for $x \geq 0$,

$$\frac{\boldsymbol{u}_{1,1}(x-u)}{\boldsymbol{u}_{-1,1}(x-u)} = \frac{\mathbf{P}_{0,1}(x - H^+_{T_x-} \in du; J^+_{T_x-} = 1; J^+_{T_x} = 1)/d}{\mathbf{P}_{0,-1}(x - H^+_{T_x-} \in du; J^+_{T_x-} = 1; J^+_{T_x} = 1)/du} = \frac{1 + e^{-(x-u)}}{1 - e^{-(x-u)}}. \qquad (13.26)$$

Next we claim that for any $x \geq 0$,

$$\sum_{i=\pm 1} \boldsymbol{u}_{1,i}(x) = (1 - e^{-x})^{-1/2}(1 + e^{-x})^{1/2} + (1 - e^{-x})^{1/2}(1 + e^{-x})^{-1/2}. \qquad (13.27)$$

To see why, we refer back to Section 5.7 in which it was shown that $(|X_t| : t \geq 0)$ is a positive self-similar Markov process with index α. The sum on the left-hand side of (13.27) is precisely the potential of the ascending ladder height process of the Lévy process which underlies the Lamperti transform of $|X|$. We can verify that the potential of the ascending ladder height process of this Lévy process has the form given by the right-hand side of (13.27) by inspecting the Wiener–Hopf factorisation in (5.50). Specifically, the aforesaid ascending ladder process has Laplace exponent which is proportional to $\Gamma((\lambda+1)/2)/\Gamma(\lambda/2)$, $\lambda \geq 0$. Then the identity in (13.27) can be verified by checking that, up to a multiplicative constant, its Laplace transform agrees with $[\Gamma((\lambda+1)/2)/\Gamma(\lambda/2)]^{-1}$, $\lambda \geq 0$.

Now we can finish the proof. Notice first that the Cauchy process is symmetric, thus $\boldsymbol{u}_{i,j} = \boldsymbol{u}_{-i,-j}$ for each $i, j \in \{1, -1\}$. Thus from (13.27) we get

$$\sum_{i=\pm 1} \boldsymbol{u}_{i,1}(x) = (1 - \mathrm{e}^{-x})^{-1/2}(1 + \mathrm{e}^{-x})^{1/2} + (1 - \mathrm{e}^{-x})^{1/2}(1 + \mathrm{e}^{-x})^{-1/2}. \quad (13.28)$$

Solving the simultaneous equations (13.26) and (13.28) together with the fact $\boldsymbol{u}_{i,j} = \boldsymbol{u}_{-i,-j}$ gives the result for $\boldsymbol{u}$. Finally we note from (13.12) that $\hat{\boldsymbol{u}} = \boldsymbol{u}$. This finishes the proof. □

13.4 Stationary limit of the radially reflected process

Proof of Theorem 13.2 Recall that (R, X) is a Markov process. For $x \in [-1, 1]$ and $j = \pm 1$, when it exists, define

$$\mu_j(dy) := \lim_{t\to\infty} \mathbb{P}_x(|R_t| \in dy;\ \mathrm{sgn}(R_t) = j), \qquad y \in [0, 1].$$

Notice that the stationary distribution μ is given by

$$\mu(A) = \mu_1(A \cap [0, 1]) + \mu_{-1}(-A \cap [0, 1]), \quad (13.29)$$

for Borel A in $[-1, 1]$, and hence it suffices to establish an identity for μ_j. Here we are pre-emptively assuming that each of the two measures on the right-hand side is absolutely continuous with respect to Lebesgue measure and so there is no 'double counting' at zero in (13.29).

Recall from the discussion following Theorem 13.2 that our strategy involves the computation of stationary distribution of the pair $(\bar{\xi} - \xi, J)$. We thus make our first steps in this direction. For $i, j = \pm 1$,

$$\mathbf{E}_{0,i}\left[\mathrm{e}^{-\lambda(\bar{\xi}_{\mathbf{e}_q} - \xi_{\mathbf{e}_q})};\ J_{\mathbf{e}_q} = j\right] = \sum_{k=\pm 1} \mathbf{E}_{0,i}\left[\mathrm{e}^{-\lambda(\bar{\xi}_{\mathbf{e}_q} - \xi_{\mathbf{e}_q})};\ J_{\overline{m}_{\mathbf{e}_q}} = k,\ J_{\mathbf{e}_q} = j\right],$$

where $\mathbf{e}_q$ is an independent and exponentially distributed random variable with rate q and $\overline{m}_{\mathbf{e}_q}$ is the unique time at which ξ obtains its maximum on the time interval $[0, \mathbf{e}_q]$. In the spirit of the computation in (11.21), we can appeal to duality to show that

$$\begin{aligned}
\int_0^1 y^\lambda \tilde{\mu}_j(dy) &:= \lim_{q\downarrow 0} \mathbf{E}_{0,i}\left[\mathrm{e}^{-\lambda(\overline{\xi}_{\mathbf{e}_q}-\xi_{\mathbf{e}_q})}; J_{\mathbf{e}_q} = j\right] \\
&= \lim_{q\downarrow 0} \sum_{k=\pm 1} \mathbf{E}_{0,i}\left[\mathrm{e}^{-\lambda(\overline{\xi}_{\mathbf{e}_q}-\xi_{\mathbf{e}_q})}; J_{\overline{m}_{\mathbf{e}_q}} = k,\ J_{\mathbf{e}_q} = j\right] \\
&= \lim_{q\downarrow 0} \sum_{k=\pm 1} \mathbf{P}_{0,i}(J_{\overline{m}_{\mathbf{e}_q}} = k)\hat{\mathbf{E}}_{0,j}\left[\mathrm{e}^{-\lambda\overline{\xi}_{\mathbf{e}_q}}; J_{\overline{m}_{\mathbf{e}_q}} = k\right]\frac{\pi_j}{\pi_k} \\
&= \sum_{k=\pm 1} \hat{\Phi}_k(0,0)[\hat{\boldsymbol{\kappa}}(\lambda)^{-1}]_{j,k}\frac{\pi_j}{\pi_k},
\end{aligned}$$

where, in the fourth equality, we have used the conclusion of Theorem 11.10 and we have also used the fact that, $\overline{m}_{\mathbf{e}_q}$ converges to $+\infty$ almost surely under $\mathbf{P}_{0,i}$ as $q \to \infty$ on account of the fact that $\limsup_{t\to\infty} |X_t| = \infty$.

Since $[\hat{\boldsymbol{\kappa}}(\lambda)^{-1}]_{j,k}$ is the Laplace transform of $\hat{\boldsymbol{u}}_{j,k}$, and since the latter is defined up to post multiplication by a strictly positive diagonal matrix, it now follows that, there exist constants c_k, for $k = \pm 1$, such that

$$\left.\frac{\mathrm{d}\tilde{\mu}_j(y)}{\mathrm{d}y}\right|_{y=\mathrm{e}^{-x}} = \pi_j \sum_{k=\pm 1} \hat{\boldsymbol{u}}_{j,k}(x)c_k, \qquad x \geq 0.$$

Said another way,

$$\tilde{\mu}_j(\mathrm{d}y) = \frac{\pi_j}{y} \sum_{k=\pm 1} \hat{\boldsymbol{u}}_{j,k}(-\log y)c_k \mathrm{d}y, \qquad y \in [0,1].$$

The constants c_k, $k = \pm 1$, can be found by noting that, for $j = \pm 1$, $\mu_j([0,1]) = \pi_j$ and hence, for $j = \pm 1$,

$$c_1\left(\int_0^\infty \hat{\boldsymbol{u}}_{j,1}(x)\mathrm{d}x\right) + c_{-1}\left(\int_0^\infty \hat{\boldsymbol{u}}_{j,-1}(x)\mathrm{d}x\right) = 1. \tag{13.30}$$

With the help of the integral representation of the hypergeometric function ${}_2F_1$ given in (A.31) as well as the linear combination identity for hypergeometric functions in (A.32) of the Appendix, we have from the expression in Theorem 13.4 (i) that

$$\int_0^\infty [\hat{\boldsymbol{u}}_{1,1}(x) - \hat{\boldsymbol{u}}_{-1,1}(x)]\mathrm{d}x$$
$$= \frac{\Gamma(1-\alpha\rho)}{\Gamma(\alpha\hat{\rho})}\int_0^1 u^{-\alpha}(1-u)^{\alpha\hat{\rho}-1}(1+u)^{\alpha\rho}\mathrm{d}u$$
$$- \frac{\Gamma(1-\alpha\hat{\rho})}{\Gamma(\alpha\rho)}\int_0^1 u^{-\alpha}(1-u)^{\alpha\rho}(1+u)^{\alpha\hat{\rho}-1}\mathrm{d}u$$
$$= \frac{\Gamma(1-\alpha\rho)}{\Gamma(\alpha\hat{\rho})}B(1-\alpha,\alpha\hat{\rho})\ {}_2F_1(-\alpha\rho, 1-\alpha, 1-\alpha\rho; -1)$$
$$- \frac{\Gamma(1-\alpha\hat{\rho})}{\Gamma(\alpha\rho)}B(1-\alpha,\alpha\rho+1)\ {}_2F_1(1-\alpha\hat{\rho}, 1-\alpha, 2-\alpha\hat{\rho}; -1)$$
$$= \Gamma(1-\alpha\rho)\Gamma(1-\alpha\hat{\rho}).$$

Now subtracting (13.30) with $j = -1$ from the same equation with $j = 1$, it appears that

$$\Gamma(1-\alpha\rho)\Gamma(1-\alpha\hat{\rho})(c_1 - c_{-1}) = 0,$$

which is to say, $c_1 = c_{-1}$.

In order to evaluate either of these constants, we appeal to the definition of the Beta function (A.18) to compute

$$\int_0^\infty [\hat{\boldsymbol{u}}_{1,1}(x) + \hat{\boldsymbol{u}}_{1,-1}(x)]\mathrm{d}x$$
$$= \frac{\Gamma(1-\alpha\rho)}{\Gamma(\alpha\hat{\rho})}\int_0^1 \left(u^{-\alpha}(1-u)^{\alpha\hat{\rho}-1}(1+u)^{\alpha\rho} + u^{-\alpha}(1-u)^{\alpha\hat{\rho}}(1+u)^{\alpha\rho-1}\right)\mathrm{d}u$$
$$= 2\frac{\Gamma(1-\alpha\rho)}{\Gamma(\alpha\hat{\rho})}\int_0^1 u^{-\alpha}(1-u)^{\alpha\hat{\rho}-1}(1+u)^{\alpha\rho-1}\mathrm{d}u$$
$$= 2^\alpha\frac{\Gamma(1-\alpha\rho)}{\Gamma(\alpha\hat{\rho})}\int_0^1 v^{\alpha\hat{\rho}-1}(1-v)^{-\alpha}\mathrm{d}v$$
$$= 2^\alpha\Gamma(1-\alpha),$$

where in the third equality, we have made the substitution $v = (1-u)/(1+u)$. It now follows from (13.30) that

$$c_1 = c_{-1} = \frac{1}{2^\alpha\Gamma(1-\alpha)}$$

and hence, for example, on $y \in [0,1]$,

$$\tilde{\mu}(dy) = \frac{2^{-\alpha}\pi}{\Gamma(\alpha\rho)\Gamma(\alpha\hat{\rho})\Gamma(1-\alpha)[\sin(\pi\alpha\rho)+\sin(\pi\alpha\hat{\rho})]}$$
$$\times\left\{y^{-\alpha}(1-y)^{\alpha\hat{\rho}-1}(1+y)^{\alpha\rho} + y^{-\alpha}(1-y)^{\alpha\hat{\rho}}(1+y)^{\alpha\rho-1}\right\}.$$

What is important about our formula for $\tilde{\mu}$ is that it agrees with the normalised stationary occupation measure of $(\exp(-(\overline{\xi}-\xi)), J)$, which is consistent with what general theory would predict, on the assumption that its limiting distribution exists. This leads to the conclusion that the time-changed analogue that we are interested in $(\exp(-(\overline{\xi}_\varphi - \xi_\varphi)), J_\varphi)$ also has a limiting distribution which is equal to its normalised stationary occupation measure (see remarks in the Comments section at the end of this chapter).

From the definition of the time change φ in the Lamperti–Kiu representation of Theorem 11.3, we know that

$$\mathrm{d}\varphi(t) = \mathrm{e}^{\alpha\xi_{\varphi(t)}}\mathrm{d}t, \qquad t > 0,$$

see, for example, the calculation in the proof of Theorem 11.14. In particular, for t sufficiently large so that $M_t > 1$,

$$f(R_t)\,\mathrm{d}t = f\left(\mathrm{e}^{-(\overline{\xi}_s-\xi_s)}J_s\right)\mathrm{e}^{\alpha\xi_s}\mathrm{d}s, \tag{13.31}$$

where $s = \varphi(t)$. Acknowledging that μ is proportional to its stationary occupation measure, it follows that, for $j = \pm 1$ and some constant, $K > 0$,

$$\mu_j(dy) = Ky^\alpha\tilde{\mu}_j(\mathrm{d}y), \qquad y \in [-1, 1]. \tag{13.32}$$

Note that

$$\begin{aligned}
1 &= \int_{[-1,1]} \mu(\mathrm{d}y) \\
&= K\frac{2^{1-\alpha}\pi}{\Gamma(\alpha\rho)\Gamma(\alpha\hat{\rho})\Gamma(1-\alpha)[\sin(\pi\alpha\rho)+\sin(\pi\alpha\hat{\rho})]} \\
&\qquad \times\left\{\int_0^1 (1-y)^{\alpha\hat{\rho}-1}(1+y)^{\alpha\rho-1}\mathrm{d}y + \int_0^1 (1-y)^{\alpha\rho-1}(1+y)^{\alpha\hat{\rho}-1}\mathrm{d}y\right\}.
\end{aligned}$$

Appealing to the hypergeometric identity (A.32) in the Appendix, the curly brackets is equal to

$$\begin{aligned}
&\frac{1}{\alpha\hat{\rho}}\,{}_2F_1(1-\alpha\rho, 1, 1+\alpha\hat{\rho}; -1) + \frac{1}{\alpha\rho}\,{}_2F_1(1-\alpha\hat{\rho}, 1, 1+\alpha\rho; -1) \\
&= \frac{1}{\alpha\hat{\rho}}\frac{1}{\alpha\rho}(\alpha\rho\,{}_2F_1(1-\alpha\rho, 1, 1+\alpha\hat{\rho}; -1) + \alpha\hat{\rho}\,{}_2F_1(1-\alpha\hat{\rho}, 1, 1+\alpha\rho; -1)) \\
&= 2^{\alpha-1}\frac{\Gamma(\alpha\rho)\Gamma(\alpha\hat{\rho})}{\Gamma(\alpha)}
\end{aligned}$$

and hence

$$K = \frac{[\sin(\pi\alpha\rho)+\sin(\pi\alpha\hat{\rho})]}{\sin(\alpha\pi)}.$$

In conclusion, we have that $d\mu(y)/dy$ is equal to

$$\frac{2^{-\alpha}\Gamma(\alpha)}{\Gamma(\alpha\rho)\Gamma(\alpha\hat{\rho})}\begin{cases}(1-y)^{\alpha\hat{\rho}-1}(1+y)^{\alpha\rho}+(1-y)^{\alpha\hat{\rho}}(1+y)^{\alpha\rho-1} & \text{if } y\in[0,1],\\ \\ (1-|y|)^{\alpha\rho}(1+|y|)^{\alpha\hat{\rho}-1}+(1-|y|)^{\alpha\rho-1}(1+|y|)^{\alpha\hat{\rho}} & \text{if } y\in[-1,0),\end{cases}$$

as required. □

13.5 Deep factorisation of the stable process

For stable processes with two-sided jumps, Theorem 13.3 gives the explicit elements of the factorisation of (13.3) in an inverse form as predicted by the relation

$$-\boldsymbol{\Psi}(\mathrm{i}\theta)=\boldsymbol{\Delta}_{\pi}^{-1}\hat{\boldsymbol{\kappa}}(\mathrm{i}\theta)^{\mathrm{T}}\boldsymbol{\Delta}_{\pi}\boldsymbol{\kappa}(-\mathrm{i}\theta),\qquad \theta\in\mathbb{R}. \tag{13.33}$$

That is to say, Theorem 13.3 identifies the elements of the factorisation

$$-\boldsymbol{\Psi}(\mathrm{i}\theta)^{-1}=\boldsymbol{\kappa}(-\mathrm{i}\theta)^{-1}\boldsymbol{\Delta}_{\pi}^{-1}[\hat{\boldsymbol{\kappa}}(\mathrm{i}\theta)^{-1}]^{\mathrm{T}}\boldsymbol{\Delta}_{\pi},\qquad \theta\in\mathbb{R}. \tag{13.34}$$

It remains to ask whether the matrix exponents $\boldsymbol{\kappa}$ and $\hat{\boldsymbol{\kappa}}$ can be identified in explicit form.

Recall that we indicated a direct identification of these factors from the left-hand side of (13.33), that is, the matrix (13.3), is deceptively difficult. It turns out that the factors $\boldsymbol{\kappa}$ and $\hat{\boldsymbol{\kappa}}$ can nonetheless be found directly, albeit using a rather complicated method, which remains beyond the scope of this book.

It is necessary to split the presentation of the deep factorisation into two cases according to whether $\alpha\in(1,2)$ or $\alpha\in(0,1]$, respectively. As before, we exclude the case of one-sided jumps. We recall that this boils down to whether X can hit the origin or not. Note that it suffices to give a description of $\boldsymbol{\kappa}$ on account of the relation (13.12).

To state our main results, we first need to introduce some notation. The family of subordinator exponents (also known as Bernstein functions)

$$\kappa_{q,p}(\lambda):=\int_0^\infty(1-\mathrm{e}^{-\lambda x})\frac{(q\vee p)-1}{(1-\mathrm{e}^{-x})^q(1+\mathrm{e}^{-x})^p}\mathrm{e}^{-\alpha x}\mathrm{d}x,\qquad \lambda\geq 0, \tag{13.35}$$

will be of use to us, where $q,p\in\{\alpha\rho,\alpha\rho+1,\alpha\hat{\rho},\alpha\hat{\rho}+1\}$ such that $q+p=\alpha+1$. It is easy to verify that the above expression is indeed the Laplace exponent of a subordinator as the associated Lévy density behaves like either $x^{-\alpha\rho-1}$ or $x^{-\alpha\hat{\rho}-1}$ as $x\downarrow 0$ and like $\mathrm{e}^{-\alpha x}$ as $x\uparrow\infty$. With this information, it is also straightforward to verify that the mean value $\kappa'_{q,p}(0+)$ is finite.

Theorem 13.7 *When $\alpha, \rho \in (0,1)$, or $\alpha = 1$ and $\rho = 1/2$, up to pre-multiplication by a strictly positive diagonal matrix, the ascending ladder MAP exponent is given by*

$$\boldsymbol{\kappa}(\lambda) = \begin{bmatrix} \kappa_{\alpha\rho+1,\alpha\hat{\rho}}(\lambda) + \dfrac{\sin(\pi\alpha\hat{\rho})}{\sin(\pi\alpha\rho)}\kappa'_{\alpha\hat{\rho},\alpha\rho+1}(0+) & -\dfrac{\sin(\pi\alpha\hat{\rho})}{\sin(\pi\alpha\rho)}\dfrac{\kappa_{\alpha\hat{\rho},\alpha\rho+1}(\lambda)}{\lambda} \\ -\dfrac{\sin(\pi\alpha\rho)}{\sin(\pi\alpha\hat{\rho})}\dfrac{\kappa_{\alpha\rho,\alpha\hat{\rho}+1}(\lambda)}{\lambda} & \kappa_{\alpha\hat{\rho}+1,\alpha\rho}(\lambda) + \dfrac{\sin(\pi\alpha\rho)}{\sin(\pi\alpha\hat{\rho})}\kappa'_{\alpha\rho,\alpha\hat{\rho}+1}(0+) \end{bmatrix},$$

for $\lambda \geq 0$.

The next theorem deals with the case that $\alpha \in (1,2)$. For this we need to introduce another family of Bernstein functions. Define

$$\phi_{q,p}(\lambda) = \int_0^\infty (1 - e^{-\lambda u})\left\{\frac{(q \vee p) - 1}{(1 - e^{-u})^q(1 + e^{-u})^p} - \frac{(\alpha - 1)}{2(1 - e^{-u})^q(1 + e^{-u})^p}\right\} e^{-u} du,$$

for $\lambda \geq 0$, $q, p \in \{\alpha\rho, \alpha\rho + 1, \alpha\hat{\rho}, \alpha\hat{\rho} + 1\}$ such that $q + p = \alpha + 1$. Note, again, that the density in curly brackets can be verified to be positive in all cases and is a Bernstein function since, as before, the associated Lévy density behaves like either $x^{-\alpha\rho-1}$ or $x^{-\alpha\hat{\rho}-1}$ as $x \downarrow 0$ and like e^{-x} as $x \uparrow \infty$. Once again, it is also subsequently straightforward to verify that the mean value $\phi'_{q,p}(0+)$ is finite.

Theorem 13.8 *When $\alpha \in (1,2)$ and $0 < \alpha\rho, \alpha\hat{\rho} < 1$, up to pre-multiplication by a strictly positive diagonal matrix, the ascending ladder MAP exponent is given by*

$$\boldsymbol{\kappa}(\lambda) = \begin{bmatrix} \begin{matrix}\sin(\pi\alpha\rho)\phi_{\alpha\rho+1,\alpha\hat{\rho}}(\lambda + \alpha - 1) \\ + \sin(\pi\alpha\rho)\phi'_{\alpha\hat{\rho},\alpha\rho+1}(0+)\end{matrix} & -\sin(\pi\alpha\hat{\rho})\dfrac{\phi_{\alpha\hat{\rho},\alpha\rho+1}(\lambda + \alpha - 1)}{\lambda + \alpha - 1} \\ -\sin(\pi\alpha\rho)\dfrac{\phi_{\alpha\rho,\alpha\hat{\rho}+1}(\lambda + \alpha - 1)}{\lambda + \alpha - 1} & \begin{matrix}\sin(\pi\alpha\hat{\rho})\phi_{\alpha\hat{\rho}+1,\alpha\rho}(\lambda + \alpha - 1) \\ + \sin(\pi\alpha\hat{\rho})\phi'_{\alpha\rho,\alpha\hat{\rho}+1}(0+)\end{matrix} \end{bmatrix},$$

for $\lambda \geq 0$.

13.6 Comments

No explicit examples of MAP Wiener–Hopf factorisations were known to exist prior to the introduction of the deep factorisation in Kyprianou [124] who proved Theorems 13.7 and 13.8. The proofs of these two theorems identify the matrix entries by considering the asymptotic upward (resp. downward) overshoots of the associated MAP. In turn, this is equivalent to understanding the asymptotic behaviour of first exit out of (resp. first entrance into) intervals of the stable process. This approach relies heavily on an application of the Markov Additive Renewal Theorem. The latter has been explored in the past by Kesten [104], Lalley [138] and Alsmeyer [3, 4].

In a follow up paper, Kyprianou et al. [132] addressed the inverse deep factorisation. The presentation in Sections 13.1, 13.2, 13.3 and 13.4 are lifted directly from there. It is worthy of note that the argument using (13.31), which leads to the claim (13.32), can be formally justified using the use of Revuz measures, see, for example, at the bottom of p.240 in Walsh [212].

Further developments with regard to the deep factorisation in the setting of isotropic d-dimensional stable processes were also pursued in Kyprianou et al. [134]. Details of that work will appear in the forthcoming chapters.

14

Spatial fluctuations and the unit sphere

In this chapter, we restrict ourselves to the setting of the isotropic stable processes in dimension $d \geq 2$ with index $\alpha \in (0, 2)$, denoted, as always, by $X = (X_t, t \geq 0)$, with probabilities $\mathbb{P}_x$, $x \in \mathbb{R}^d$. We are mostly concerned with the setting of first passage problems in relation to the interior and exterior of the unit ball and the surface of the unit sphere. These are geometrical domains that we can take natural advantage of in our analysis by appealing to isotropy, classical techniques from potential analysis that involve inversions of space through spheres and the Riesz–Bogdan–Żak transform. Finally, we also devote some attention to the setting of exit problems from general domains, appealing to a relatively straightforward numerical method, known as the walk-on-spheres algorithm. This is based on the preceding results for first passage problem for the unit ball.

14.1 Sphere inversions

Before handling any of the promised exit problems, let us start in this and the next section by considering two remarkably simple but effective transformations which invert space through a given sphere in $\mathbb{R}^d$. Later on in the chapter, we will be particularly interested in how these spatial transformations can be used to manipulate integrals (traditionally known as Riesz potentials) of the kind

$$U\mu(x) = \int_D |x - y|^{\alpha-d} \mu(\mathrm{d}y), \qquad x \in \mathbb{R}^d, \tag{14.1}$$

where μ will take the form of various finite measures on $D \subseteq \mathbb{R}^d$ and $\alpha \in (0, 2)$.

Fix a point $b \in \mathbb{R}^d$ and a value $r > 0$. A homeomorphism of $\mathbb{R}^d \backslash \{b\}$ defined by

$$x^* = b + \frac{r^2}{|x-b|^2}(x-b), \tag{14.2}$$

is called an *inversion through the sphere* $\mathbb{S}^{d-1}(b,r) := \{x \in \mathbb{R}^d : |x-b| = r\}$. Recall that we have already reserved the special notation $\mathbb{S}^{d-1}$ to mean $\mathbb{S}^{d-1}(0,1)$. Amongst the many properties of this inversion, the most important is that the exterior of $\mathbb{S}^{d-1}(b,r)$ maps to its interior and vice versa; see Figure 14.1.

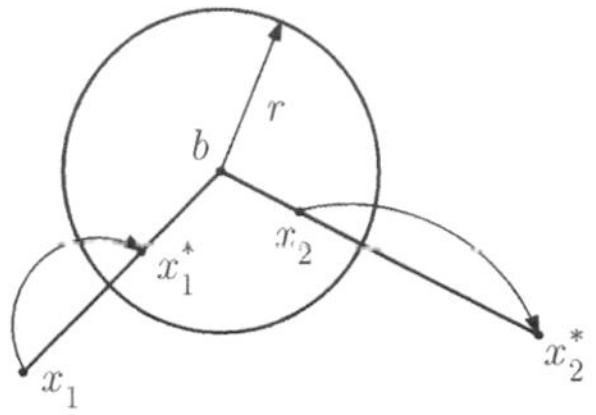

Figure 14.1 Inversion relative to the sphere $\mathbb{S}^{d-1}(b,r)$

Straightforward algebra also tells us that

$$r^2 = |x^* - b||x-b|, \tag{14.3}$$

which, in turn, also gives us that $(x^*)^* = x$, for $x \in \mathbb{R}^d \backslash \{b\}$ and, in particular,

$$x = b + \frac{r^2}{|x^*-b|^2}(x^*-b). \tag{14.4}$$

Moreover, straightforward algebra using (14.2) and (14.4) gives us, for $x, y \in \mathbb{R}\backslash\{b\}$,

$$|x^* - y^*| = \sqrt{(x^*-y^*)\cdot(x^*-y^*)} = \frac{r^2|x-y|}{|x-b||y-b|}. \tag{14.5}$$

Another very important fact about inversion through the sphere $\mathbb{S}^{d-1}(b,r)$ is that a sphere which does not pass through or encircle b will always map to another sphere. To see why, suppose that we consider the image of any sphere $\mathbb{S}^{d-1}(c,R)$, for $c \in \mathbb{R}^d$ and $R > 0$, for which $|c-b| > R$, and denote its image under inversion through $\mathbb{S}^{d-1}(b,r)$ by $\mathbb{S}_*^{d-1}(c,R)$. We can write $\mathbb{S}^{d-1}(c,R) = \{x \in \mathbb{R}^d : |(x-b)-(c-b)|^2 = R^2\}$, which can otherwise be written as $x \in \mathbb{R}^d$ such that

$$|x-b|^2 - 2(x-b)\cdot(c-b) + |c-b|^2 = R^2.$$

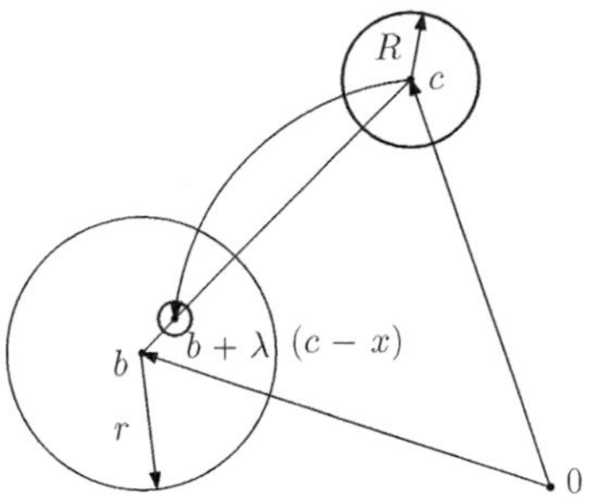

Figure 14.2 The sphere $\mathbb{S}^{d-1}(c,R)$ maps to the sphere $\mathbb{S}_*^{d-1}(c,R)$ under inversion through $\mathbb{S}^{d-1}(b,r)$

From (14.3) and (14.4), after a little algebra, for $x \in \mathbb{S}^{d-1}(c,R)$,

$$|x^* - b|^2 - \lambda(x^* - b)\cdot(c-b) + \lambda^2|c-b|^2 = \eta^2,$$

where

$$\lambda = r^2/(|c-b|^2 - R^2) \quad \text{and} \quad \eta^2 = r^4R^2/(|c-b|^2 - R^2)^2.$$

That is to say, $\mathbb{S}_d^*(c,R) = \{x^* \in \mathbb{R}^d : |(x^* - (b + \lambda(c-b))|^2 = \eta^2\}$ so that $\mathbb{S}_d^*(c,R)$ is mapped to another sphere.

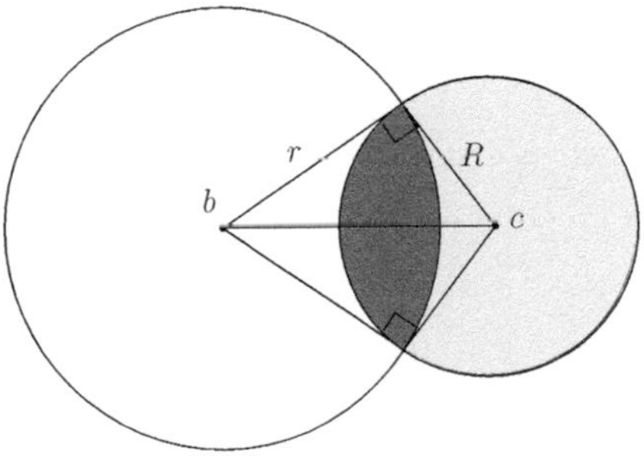

Figure 14.3 The sphere $\mathbb{S}^{d-1}(c,R)$ maps to itself under inversion through $\mathbb{S}^{d-1}(b,r)$ provided the former is orthogonal to the latter, which is equivalent to $r^2+R^2 = |c-b|^2$. In particular, the area contained in the right-most shaded segment is mapped to the area in the left-most shaded segment and vice versa.

We note in particular that $\mathbb{S}_*^{d-1}(c,R) = \mathbb{S}^{d-1}(c,R)$ if and only if $\lambda = 1$, in other words, $r^2 + R^2 = |c-b|^2$. This is equivalent to requiring that the spheres $\mathbb{S}^{d-1}(c,R)$ and $\mathbb{S}^{d-1}(b,r)$ are orthogonal, and therefore necessarily overlapping.

What is additionally interesting about this choice of $\mathbb{S}^{d-1}(c,R)$ is that its interior maps to its interior and its exterior to its exterior.

14.2 Sphere inversions with reflection

A variant of the transformation (14.2) takes the form

$$x^{\diamond} = b - \frac{r^2}{|x-b|^2}(x-b), \tag{14.6}$$

for a fixed $b \in \mathbb{R}^d$ and $x \in \mathbb{R}^d$, which similarly has the self-inverse property (14.4). It is also quite straightforward to show that

$$r^2 = |x^{\diamond} - b||x-b|, \tag{14.7}$$

and

$$|x^{\diamond} - y^{\diamond}| = \frac{r^2|x-y|}{|x-b||y-b|} \tag{14.8}$$

still hold in the spirit of (14.3) and (14.5), respectively.

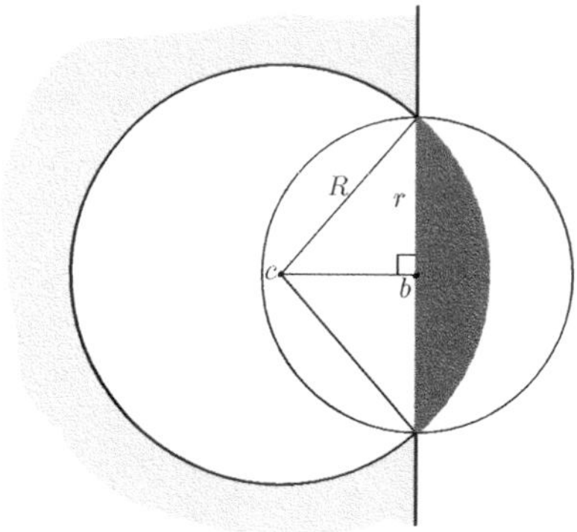

Figure 14.4 The sphere $\mathbb{S}^{d-1}(c,R)$ maps to itself through $\mathbb{S}^{d-1}(b,r)$ via (14.6) providing $|c-b|^2 + r^2 = R^2$. However, this time, the exterior of the sphere $\mathbb{S}^{d-1}(c,R)$ maps to the interior of the sphere $\mathbb{S}^{d-1}(c,R)$ and vice versa. For example, the region in the exterior of $\mathbb{S}^{d-1}(c,R)$ contained in the left-most shaded region maps to the portion of the interior of $\mathbb{S}^{d-1}(c,R)$ contained by the right-most shaded region.

Intuitively speaking, $x^{\diamond}$ performs the same sphere inversion as x^*, albeit with the additional feature that there is pointwise reflection about b. As such, any sphere $\mathbb{S}^{d-1}(c,R)$ will map to another sphere, say $\mathbb{S}^{d-1}_{\diamond}(c,R)$ so long as

$|c-b| < R$. We are again interested in choices of c and R such that $\mathbb{S}_\diamond^{d-1}(c,R) = \mathbb{S}^{d-1}(c,R)$. This turns out to be possible so long as $R^2 = r^2 + |c-b|^2$. Moreover, in that case, the interior of $\mathbb{S}^{d-1}(c,R)$ maps to its exterior and its exterior to its interior; see Figure 14.4.

To see how this is possible, we need to prove a new identity for $x^\diamond$. We claim that

$$|x^\diamond - c|^2 - R^2 = \frac{|x^\diamond - b|^2}{r^2}(R^2 - |x-c|^2), \qquad x \in \mathbb{R}^d. \tag{14.9}$$

Indeed, recalling that $|x^\diamond - b||x-b| = r^2$, we can write

$$x = b + \frac{(x-b)}{|x-b|}|x-b| \quad \text{and} \quad x^\diamond = b - |x^\diamond - b|\frac{(x-b)}{|x-b|}.$$

Hence, as $|b-c|^2 + r^2 = R^2$ and using again that $|x^\diamond - b||x-b| = r^2$, we have

$$\begin{aligned}
|x^\diamond - c|^2 - R^2 &= |(x^\diamond - b) + (b-c)|^2 - R^2 \\
&= |x^\diamond - b|^2 - 2|x^\diamond - b|\frac{(x-b)\cdot(b-c)}{|x-b|} - r^2 \\
&= \frac{|x^\diamond - b|^2}{r^2}\left(r^2 - 2(x-b)\cdot(b-c) - |x-b|^2\right) \\
&= \frac{|x^\diamond - b|^2}{r^2}\left(R^2 - |b-c|^2 - 2|x-b|\frac{(x-b)}{|x-b|}\cdot(b-c) - |x-b|^2\right) \\
&= \frac{|x^\diamond - b|^2}{r^2}\left(R^2 - |x-c|^2\right),
\end{aligned} \tag{14.10}$$

which proves (14.9).

It is now immediately apparent that $|x^\diamond - c|^2 < R^2$ if and only if $|x-c|^2 > R^2$, and $|x^\diamond - c|^2 = R^2$ if and only if $|x-c|^2 = R^2$ which is to say that $\mathbb{S}_\diamond^{d-1}(c,R) = \mathbb{S}^{d-1}(c,R)$ and that the interior of $\mathbb{S}^{d-1}(c,R)$ maps to its exterior and its exterior maps to its interior as claimed.

14.3 First hitting of a sphere

Let us turn to the problem of understanding the distribution of the position of X on first hitting of the sphere $\mathbb{S}^{d-1}(0,a) = \{x \in \mathbb{R}^d : |x| = a\}$. To this end, we introduce the notation

$$\tau_a^\odot = \inf\{t > 0 : |X_t| = a\},$$

for $a > 0$. Moreover, we will write $\sigma_a(\mathrm{d}z)$ for the surface measure on $\mathbb{S}^{d-1}(0,a)$, normalised to have unit total mass.

Theorem 14.1 *Define the function*

$$h_a^{\odot}(x,y) = \frac{\Gamma\left(\frac{\alpha+d}{2}-1\right)\Gamma\left(\frac{\alpha}{2}\right)}{\Gamma\left(\frac{d}{2}\right)\Gamma(\alpha-1)}\frac{\left|\|x\|^2-a^2\right|^{\alpha-1}a^{d-\alpha}}{|x-y|^{\alpha+d-2}},$$

for $|x| \neq a$, $|y| = a$ and $a > 0$. Then, if $\alpha \in (1,2)$,

$$\mathbb{P}_x(X_{\tau_a^{\odot}} \in \mathrm{d}y) = h_a^{\odot}(x,y)\sigma_a(\mathrm{d}y)\mathbf{1}_{(|x|\neq a)} + \delta_x(\mathrm{d}y)\mathbf{1}_{(|x|=a)},$$

for $|y| = a$.

Before proving this theorem, we need to address a number of preliminary results. Our first such result lays out a unique characterisation of the desired hitting distribution. This serves as our principal mechanism for finding it.

Theorem 14.2 *For $|x|, a > 0$, if $\alpha \in (1,2)$, then*

$$\mathbb{P}_x(\tau_a^{\odot} < \infty)$$
$$= \frac{\Gamma\left(\frac{\alpha+d}{2}-1\right)\Gamma\left(\frac{\alpha}{2}\right)}{\Gamma\left(\frac{d}{2}\right)\Gamma(\alpha-1)}\begin{cases} {}_2F_1((d-\alpha)/2, 1-\alpha/2, d/2; |x|^2/a^2), & a > |x|, \\ \left(\frac{|x|}{a}\right)^{\alpha-d} {}_2F_1((d-\alpha)/2, 1-\alpha/2, d/2; a^2/|x|^2), & a \leq |x|. \end{cases}$$

Otherwise, if $\alpha \in (0,1]$, then $\mathbb{P}_x(\tau_a^{\odot} = \infty) = 1$, for all $|x| \neq a > 0$.

Proof From Section 5.7, we know that $|X|$ is a positive self-similar Markov process. Denote the underlying Lévy processes associated through the Lamperti transform by ξ with probabilities $\mathbf{P}_x$, $x \in \mathbb{R}$.

$$\mathbb{P}_x(\tau_a^{\odot} < \infty) = \mathbf{P}_{\log|x|}(\tau^{\{\log a\}} < \infty) = \mathbf{P}_0(\tau^{\{\log(a/|x|)\}} < \infty),$$

where $\tau^{\{z\}} = \inf\{t > 0 : \xi_t = z\}$, $z \in \mathbb{R}$. From this observation, we note the ability of X to hit the sphere $\mathbb{S}^{d-1}(0,a)$ with positive probability, boils down to the ability of ξ to hit points with positive probability. In this respect, Theorem 2.18 tells us that a necessary and sufficient condition is the integrability of $(1+\Psi(z))^{-1}$ in (2.28).

The characteristic exponent of ξ is given by Theorem 5.21, and in particular, appealing to (A.16),

$$\frac{1}{\Psi(z)} = \frac{1}{2^{\alpha}}\frac{\Gamma(-\frac{1}{2}\mathrm{i}z)}{\Gamma(\frac{1}{2}(-\mathrm{i}z+\alpha))}\frac{\Gamma(\frac{1}{2}(\mathrm{i}z+d-\alpha))}{\Gamma(\frac{1}{2}(\mathrm{i}z+d))} \sim z^{-\alpha} \tag{14.11}$$

uniformly on $\mathbb{R}$, as $|z| \to \infty$. We thus conclude from (2.28) that each sphere $\mathbb{S}^{d-1}(0,a)$ can be reached with positive probability from any x with $|x| \neq a$ if and only if $\alpha \in (1,2)$. Moreover, when $\alpha \in (1,2)$, Lemma 2.19 gives us the identity

$$\mathbb{P}_x(\tau_a^{\odot} < \infty) = \frac{u_\xi(\log(a/|x|))}{u_\xi(0)}, \tag{14.12}$$

where, up to a multiplicative constant, the potential density u_ξ can be computed via a Laplace inversion in the spirit of the computations completed in the proof of Theorem 6.15.

To this end, we note (again from Theorem 5.21) that the Laplace exponent $-\Psi(-\mathrm{i}z)$ of ξ is well defined for $\mathrm{Re}(z) \in (-d, \alpha)$ with roots at 0 and $\alpha - d$. The transience of X, for $d \geq 2$, ensures that $\mathbf{E}[\xi_1] > 0$ and hence, as $-\Psi(-\mathrm{i}z)$ is convex for real z, we easily deduce that $\mathrm{Re}(\Psi(-\mathrm{i}z)) > 0$, for $\mathrm{Re}(z) \in (\alpha - d, 0)$. In particular, it follows that the Laplace transform of u_ξ is well defined for $\mathrm{Re}(z) \in (\alpha - d, 0)$ as

$$\int_{\mathbb{R}} \mathrm{e}^{zx} u_\xi(x)\mathrm{d}x = \int_0^\infty \mathrm{e}^{-\Psi(-\mathrm{i}z)t}\mathrm{d}t = \frac{1}{\Psi(-\mathrm{i}z)}.$$

We may thus compute u_ξ as a Laplace inversion in the form

$$u_\xi(x) = \frac{1}{2\pi\mathrm{i}} \int_{c+\mathrm{i}\mathbb{R}} \frac{\mathrm{e}^{-zx}}{\Psi(-\mathrm{i}z)}\mathrm{d}z, \qquad x \in \mathbb{R},$$

providing $c \in (\alpha - d, 0)$.

As we have seen in other cases, this integral can be computed using relatively straightforward residue calculus. Indeed, from (14.11) we note that $1/\Psi(-\mathrm{i}z)$ has simple poles at $\{2n, n \geq 0\}$, and $\{-2n - (d-\alpha) \colon n \geq 0\}$.

We can construct a contour integral, $\gamma_R = \{c + \mathrm{i}x \colon |x| \leq R\} \cup \{c + R\mathrm{e}^{\mathrm{i}\theta} : \theta \in (\pi/2, 3\pi/2)\}$, where $c \in (\alpha - d, 0)$; see Figure 14.5.

Residue calculus now gives us

$$\begin{aligned}\frac{1}{2\pi\mathrm{i}} &\int_{c-\mathrm{i}R}^{c+\mathrm{i}R} \frac{\mathrm{e}^{-zx}}{\Psi(-\mathrm{i}z)}\mathrm{d}z \\ &= -\frac{1}{2\pi\mathrm{i}} \int_{c+R\mathrm{e}^{\mathrm{i}\theta}\colon \theta\in(\pi/2,3\pi/2)} \frac{\mathrm{e}^{-zx}}{\Psi(-\mathrm{i}z)}\mathrm{d}z \\ &\quad + \sum_{0\leq n\leq \lfloor R\rfloor} \mathrm{Res}\left(\frac{\mathrm{e}^{-zx}}{\Psi(-\mathrm{i}z)}; z = -2n - (d-\alpha)\right).\end{aligned} \tag{14.13}$$

Now fix $x \leq 0$. Appealing again to the uniform estimate (A.16), the assumption $x \leq 0$ and the fact that the arc length of $\{c + R\mathrm{e}^{\mathrm{i}\theta} \colon \theta \in (\pi/2, 3\pi/2)\}$ is πR, we have

$$\left|\int_{c+R\mathrm{e}^{\mathrm{i}\theta}\colon \theta\in(\pi/2,3\pi/2)} \frac{\mathrm{e}^{-xz}}{\Psi(-\mathrm{i}z)}\mathrm{d}z\right| \leq CR^{-(\alpha-1)} \to 0,$$

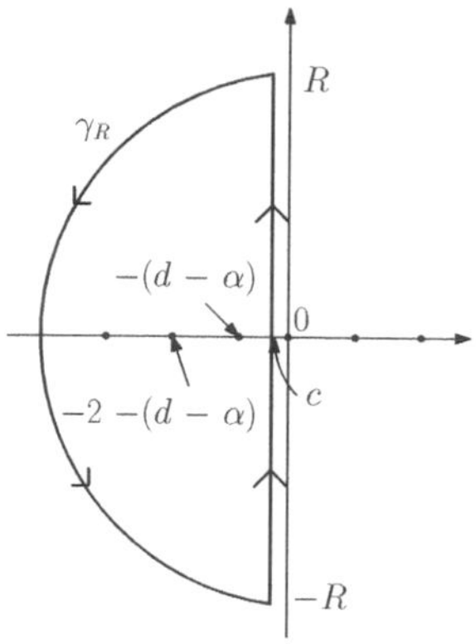

Figure 14.5 The contour integral γ_R

as $R \to \infty$, for some constant $C > 0$. By taking limits in (14.13), we now have

$$u_\xi(x) = \sum_{n\geq 0} \mathrm{Res}\left(\frac{\mathrm{e}^{-zx}}{\Psi(-\mathrm{i}z)}; z = -2n - (d-\alpha)\right). \tag{14.14}$$

Using the residues of the gamma function in (A.11), that is, for $n = 0, 1, \ldots$, $\mathrm{Res}(\Gamma(z); z = -n) = (-1)^n/n!$, we have in (14.14), for $x \leq 0$,

$$\begin{aligned}
u_\xi(x) &= \mathrm{e}^{x(d-\alpha)} \sum_{n=0}^{\infty} (-1)^n \frac{\Gamma(n + (d-\alpha)/2)}{\Gamma(-n+\alpha/2)\Gamma(n+d/2)} \frac{\mathrm{e}^{2nx}}{n!} \\
&= \mathrm{e}^{x(d-\alpha)} \frac{\Gamma((d-\alpha)/2)}{\Gamma(\alpha/2)\Gamma(d/2)} \sum_{0}^{\infty} \frac{((d-\alpha)/2)_n (1-\alpha/2)_n}{(d/2)_n} \frac{\mathrm{e}^{2nx}}{n!} \\
&= \mathrm{e}^{x(d-\alpha)} \frac{\Gamma((d-\alpha)/2)}{\Gamma(\alpha/2)\Gamma(d/2)} {}_2F_1((d-\alpha)/2, 1-\alpha/2, d/2; \mathrm{e}^{2x}),
\end{aligned} \tag{14.15}$$

where $(a)_n = \Gamma(n+a)/\Gamma(a)$ is the Pochhammer symbol and we have used the relation (A.13) and the recursion formula for gamma functions.

Note, in particular, this tells us that

$$\begin{aligned}
u_\xi(0) &= \frac{\Gamma((d-\alpha)/2)}{\Gamma(\alpha/2)\Gamma(d/2)} {}_2F_1((d-\alpha)/2, 1-\alpha/2, d/2; 1) \\
&= \frac{\Gamma((d-\alpha)/2)}{\Gamma(\alpha/2)\Gamma(d/2)} \frac{\Gamma(d/2)\Gamma(\alpha-1)}{\Gamma(\alpha/2)\Gamma((\alpha+d)/2-1)}.
\end{aligned}$$

Feeding (14.15) back in (14.12), we have thus established the asserted identity for $\mathbb{P}_x(\tau_a^{\odot} < \infty)$ when $a \leq |x|$.

To deal with the case $a > |x|$, we can appeal to the Riesz–Bogdan–Żak transform in Theorem 12.8 to help us. Let us momentarily set $a = 1$ and note that, for $|x| < 1$,

$$\mathbb{P}_{x/|x|^2}(\tau_1^{\odot} < \infty) = \mathbb{E}_x\left[\frac{|X_{\tau_1^{\odot}}|^{\alpha-d}}{|x|^{\alpha-d}}\mathbf{1}_{(\tau_1^{\odot}<\infty)}\right] = \frac{1}{|x|^{\alpha-d}}\mathbb{P}_x(\tau_1^{\odot} < \infty) \tag{14.16}$$

and hence,

$$\mathbb{P}_x(\tau_1^{\odot} < \infty) = \frac{\Gamma\left(\frac{\alpha+d}{2}-1\right)\Gamma\left(\frac{\alpha}{2}\right)}{\Gamma\left(\frac{d}{2}\right)\Gamma(\alpha-1)}{}_2F_1((d-\alpha)/2, 1-\alpha/2, d/2; |x|^2).$$

To obtain from this the desired formula for $\mathbb{P}_x(\tau_a^{\odot} < \infty)$, where $|x| < a$, we can appeal to a straightforward scaling argument. We leave the details to the reader. □

Theorem 14.3 *Suppose $\alpha \in (1, 2)$. For all $x \in \mathbb{R}^d$,*

$$\mathbb{P}_x(\tau_1^{\odot} < \infty) = \frac{\Gamma\left(\frac{\alpha+d}{2}-1\right)\Gamma\left(\frac{\alpha}{2}\right)}{\Gamma\left(\frac{d}{2}\right)\Gamma(\alpha-1)}\int_{\mathbb{S}^{d-1}}|z-x|^{\alpha-d}\sigma_1(\mathrm{d}z). \tag{14.17}$$

In particular, for all $x \in \mathbb{S}^{d-1}$,

$$\int_{\mathbb{S}^{d-1}}|z-x|^{\alpha-d}\sigma_1(\mathrm{d}z) = \frac{\Gamma\left(\frac{d}{2}\right)\Gamma(\alpha-1)}{\Gamma\left(\frac{\alpha+d}{2}-1\right)\Gamma\left(\frac{\alpha}{2}\right)}. \tag{14.18}$$

Proof The key to the proof is two integral identities. The first states that, for any $0 < |a| < 1$ and $\nu > 0$,

$$\int_0^{\pi}\frac{(\sin\theta)^{d-2}}{(a^2+2a\cos\theta+1)^{\nu}}\mathrm{d}\theta = \pi^{1/2}\frac{\Gamma\left(\frac{d-1}{2}\right)}{\Gamma\left(\frac{d}{2}\right)}{}_2F_1\left(\nu, \nu-\frac{d}{2}+1, \frac{d}{2}; a^2\right). \tag{14.19}$$

The second pertains to surface integrals over $\mathbb{S}^{d-1}$ and states that, for non-negative and measurable f on $[0, \infty)$,

$$\int_{\mathbb{S}^{d-1}} f(x\cdot z)\sigma_1(\mathrm{d}z) = \pi^{-1/2}\frac{\Gamma\left(\frac{d}{2}\right)}{\Gamma\left(\frac{d-1}{2}\right)}\int_0^{\pi} f(|x|\cos\theta)(\sin\theta)^{d-2}\mathrm{d}\theta. \tag{14.20}$$

Recalling the statement of Theorem 14.2, allowing a to play the role of $-|x|$, setting $r = 1$ and taking $\nu = (d-\alpha)/2$ in (14.19) and

$$f(x\cdot z) = (|x|^2 - 2x\cdot z + 1)^{-(d-\alpha)/2} = |x-z|^{\alpha-d},$$

we can identify, for $|x| < 1$,

$$
\begin{aligned}
\mathbb{P}_x(\tau_1^{\odot} < \infty) &= \frac{\Gamma\left(\frac{\alpha+d}{2} - 1\right)\Gamma\left(\frac{\alpha}{2}\right)}{\Gamma\left(\frac{d}{2}\right)\Gamma(\alpha - 1)} {}_2F_1\left(\frac{d-\alpha}{2}, 1 - \frac{\alpha}{2}, \frac{d}{2}; |x|^2\right) \\
&= \pi^{-1/2}\frac{\Gamma\left(\frac{\alpha+d}{2} - 1\right)\Gamma\left(\frac{\alpha}{2}\right)}{\Gamma\left(\frac{d-1}{2}\right)\Gamma(\alpha - 1)} \int_0^{\pi} \frac{\sin^{d-2}\theta}{(|x|^2 + 2|x|\cos\theta + 1)^{(d-\alpha)/2}} d\theta \\
&= \frac{\Gamma\left(\frac{\alpha+d}{2} - 1\right)\Gamma\left(\frac{\alpha}{2}\right)}{\Gamma\left(\frac{d}{2}\right)\Gamma(\alpha - 1)} \int_{\mathbb{S}^{d-1}} \frac{1}{(|x|^2 - 2x\cdot z + 1)^{(d-\alpha)/2}} \sigma_1(\mathrm{d}z) \\
&= \frac{\Gamma\left(\frac{\alpha+d}{2} - 1\right)\Gamma\left(\frac{\alpha}{2}\right)}{\Gamma\left(\frac{d}{2}\right)\Gamma(\alpha - 1)} \int_{\mathbb{S}^{d-1}} |x - z|^{\alpha-d} \sigma_1(\mathrm{d}z).
\end{aligned}
$$

The case that $|x| > 1$ can be dealt with using the same application of the Riesz–Bogdan–Żak from Theorem 12.8 as in (14.16) so that, recalling that $Kx := x/|x|^2$ is self inverse and $|Kx| = 1/|x|$,

$$
\begin{aligned}
\mathbb{P}_x(\tau_1^{\odot} < \infty) &= |x|^{\alpha-d}\mathbb{P}_{Kx}(\tau_1^{\odot} < \infty) \\
&= \frac{\Gamma\left(\frac{\alpha+d}{2} - 1\right)\Gamma\left(\frac{\alpha}{2}\right)}{\Gamma\left(\frac{d}{2}\right)\Gamma(\alpha - 1)} \int_{\mathbb{S}^{d-1}} |x|^{\alpha-d}|Kx - Kz|^{\alpha-d} \sigma_1(\mathrm{d}z) \\
&= \frac{\Gamma\left(\frac{\alpha+d}{2} - 1\right)\Gamma\left(\frac{\alpha}{2}\right)}{\Gamma\left(\frac{d}{2}\right)\Gamma(\alpha - 1)} \int_{\mathbb{S}^{d-1}} |x - z|^{\alpha-d} \sigma_1(\mathrm{d}z),
\end{aligned}
$$

where we have used that, for $z \in \mathbb{S}^{d-1}$, we have $Kz = z$ and $|Kx - Kz| = |x - z|/(|x||z|) = |x - z|/|x|$.

Finally, the case that $|x| = 1$ is recovered by taking a limit as $|x| \to 1$ in either of the previous cases, which leads to (14.18). □

Lemma 14.4 *Suppose that* $\alpha \in (1, 2)$. *Write* $\mu_x^{\odot}(\mathrm{d}y) = \mathbb{P}_x(X_{\tau_1^{\odot}} \in \mathrm{d}y)$ *on* $\mathbb{S}^{d-1}$, *where* $x \in \mathbb{R}^d \backslash \mathbb{S}^{d-1}$. *Then the measure* $\mu_x^{\odot}$ *is the unique solution to*

$$
|x - y|^{\alpha-d} = \int_{\mathbb{S}^{d-1}} |z - y|^{\alpha-d}\mu(\mathrm{d}z), \qquad y \in \mathbb{S}^{d-1}. \tag{14.21}
$$

Proof We begin by recalling the expression for the resolvent of the stable process in Theorem 3.11 which states that, due to transience,

$$
\int_0^{\infty} \mathbb{P}_x(X_t \in \mathrm{d}y)\mathrm{d}t = C(\alpha)|x - y|^{\alpha-d}\mathrm{d}y, \qquad x, y \in \mathbb{R}^d,
$$

where $C(\alpha)$ is an unimportant constant in the following discussion. Suppose now that we fix an arbitrary $y \in \mathbb{S}^{d-1}$. Then a straightforward application of the strong Markov property tells us that, for $x \in \mathbb{R}^d \backslash \mathbb{S}^{d-1}$,

$$\begin{aligned}\int_0^\infty \mathbb{P}_x(X_t \in \mathrm{d}y)\mathrm{d}t &= \mathbb{E}_x\left[\mathbf{1}_{(\tau_1^\odot<\infty)} \int_0^\infty \mathbb{P}_z(X_t \in \mathrm{d}y)|_{z=X_{\tau_1^\odot}} \mathrm{d}t\right] \\ &= \mathbb{E}_x\left[\mathbf{1}_{(\tau_1^\odot<\infty)} C(\alpha)|X_{\tau_1^\odot} - y|^{\alpha-d}\right]\mathrm{d}y \\ &= C(\alpha)\int_{\mathbb{S}^{d-1}} |z-y|^{\alpha-d}\mathbb{P}_x(X_{\tau_1^\odot} \in \mathrm{d}z)\mathrm{d}y,\end{aligned}$$

which shows that $\mu_x^\odot$ is a solution to (14.21).

Let us now address the issue of uniqueness in (14.33). Suppose that ν is a signed measure on $\mathbb{S}^{d-1}$ which satisfies

$$\int_{\mathbb{S}^{d-1}}\int_{\mathbb{S}^{d-1}} |z-y|^{\alpha-d}\,|\nu|(\mathrm{d}z)\,|\nu|(\mathrm{d}y) < \infty. \tag{14.22}$$

Here, we understand $|\nu| = \nu^+ + \nu^-$, when we represent $\nu = \nu^+ - \nu^-$. We claim that

$$\int_{\mathbb{S}^{d-1}} |z-y|^{\alpha-d}\nu(\mathrm{d}z) = 0 \tag{14.23}$$

implies that the measure $\nu \equiv 0$. The significance of this claim is that it would immediately imply that (14.21) has a unique solution in the class of probability measures by setting $\nu = \mu - \mu_x^\odot$, where $\mu(\mathrm{d}y)$ is any other probability measure supported on $\mathbb{S}^{d-1}$ which solves (14.21). Indeed, with this choice of ν, (14.22) holds on account of the fact that, for $|x| \neq 1$,

$$\int_{\mathbb{S}^{d-1}}\int_{\mathbb{S}^{d-1}} |z-y|^{\alpha-d}\mu_x^\odot(\mathrm{d}z)\mu(\mathrm{d}y) = \int_{\mathbb{S}^{d-1}} |x-y|^{\alpha-d}\mu(\mathrm{d}y) \le (1+|x|)^{\alpha-d},$$

where we have used rotational symmetry in the inequality so that the largest value the integrand can take occurs when $y = -x/|x|$ (note that $\alpha - d < 0$, so we need to minimise $|x-y|$).

Verifying the claim (14.23) is not too difficult. Indeed, if we write $\mathrm{p}_t(z)$ for the transition density of X, that is, $\mathbb{P}_x(X_t \in \mathrm{d}y) = \mathrm{p}_t(y-x)\mathrm{d}y$, then a standard Fourier inversion tells us that, for $x \in \mathbb{R}^d$,

$$\mathrm{p}_t(x) = (2\pi)^{-d}\int_{\mathbb{R}^d} \mathrm{e}^{\mathrm{i}\theta\cdot x}\mathrm{e}^{-|\theta|^\alpha t}\mathrm{d}\theta.$$

Note that

$$\begin{aligned}
C(\alpha) &\int_{\mathbb{S}^{d-1}} \int_{\mathbb{S}^{d-1}} |z-y|^{\alpha-d}\, \nu(\mathrm{d}z)\, \nu(\mathrm{d}y) \\
&= \int_{\mathbb{S}^{d-1}} \int_{\mathbb{S}^{d-1}} \int_0^\infty \mathrm{p}_t(z-y)\, \mathrm{d}t\, \nu(\mathrm{d}z)\, \nu(\mathrm{d}y) \\
&= \int_{\mathbb{S}^{d-1}} \int_{\mathbb{S}^{d-1}} \int_0^\infty (2\pi)^{-d} \int_{\mathbb{R}^d} \mathrm{e}^{\mathrm{i}\theta\cdot(z-y)} \mathrm{e}^{-|\theta|^\alpha t} \mathrm{d}\theta\, \mathrm{d}t\, \nu(\mathrm{d}z)\, \nu(\mathrm{d}y) \\
&= (2\pi)^{-d} \int_0^\infty \mathrm{d}t \int_{\mathbb{R}^d} \mathrm{d}\theta\, \mathrm{e}^{-|\theta|^\alpha t} \left(\int_{\mathbb{S}^{d-1}} \mathrm{e}^{-\mathrm{i}\theta\cdot y} \nu(\mathrm{d}y) \right) \left(\int_{\mathbb{S}^{d-1}} \mathrm{e}^{\mathrm{i}\theta\cdot z} \nu(\mathrm{d}z) \right) \\
&= (2\pi)^{-d} \int_0^\infty \mathrm{d}t \int_{\mathbb{R}^d} \mathrm{d}\theta\, \mathrm{e}^{-|\theta|^\alpha t} |\phi(\theta)|^2,
\end{aligned}$$

where $\phi(\theta) = \int_{\mathbb{S}^{d-1}} \mathrm{e}^{\mathrm{i}\theta\cdot z} \nu(\mathrm{d}z)$. The assumption that $\int_{\mathbb{S}^{d-1}} |z-y|^{\alpha-d} \nu(\mathrm{d}z) = 0$ thus implies that $\phi \equiv 0$, which, in turn, implies $\nu \equiv 0$, as claimed, and hence (14.21) has a unique solution. □

Proof of Theorem 14.1 As usual, a scaling argument means that it is sufficient to deal with the case $a = 1$. First assume that $|x| > 1$. Starting with the equality (14.18), we want to apply the transformation (14.2) through the sphere $\mathbb{S}^{d-1}(x, (|x|^2 - 1)^{1/2})$, remembering that this transformation maps $\mathbb{S}^{d-1}$ to itself. If we write $y = rA(\theta)$, where $r = |y| > 0$ and $A(\theta) = \mathrm{Arg}(y)$ for parameterisation $\theta = (\theta_1, \dots, \theta_{d-1})$, where $\theta_j \in [0, \pi]$ and $\theta_{d-1} \in [0, 2\pi)$ then

$$\mathrm{d}y = r^{d-1} \mathcal{J}(\theta) \mathrm{d}\theta \mathrm{d}r = \frac{2\pi^{d/2}}{\Gamma(d/2)} r^{d-1} \sigma_1(\mathrm{d}\theta) \mathrm{d}r, \tag{14.24}$$

where $\mathcal{J}$ is the part of the Jacobian of y with respect to (r, θ) which depends on θ and $\sigma_1(\mathrm{d}y)$ is the surface measure on $\mathbb{S}^{d-1}$ normalised to have unit mass. Recall the transformation $Kz = z/|z|^2$. Suppose now that we write $z = wA(\theta)$ and set $y = Kz = w^{-1}A(\theta)$, then, if we set $r = w^{-1}$,

$$\mathrm{d}y = r^{d-1} \mathcal{J}(\theta)\Big|_{r=w^{-1}} \mathrm{d}\theta \frac{\mathrm{d}r}{\mathrm{d}w} \mathrm{d}w = w^{-2d} \cdot w^{d-1} \mathcal{J}(\theta) \mathrm{d}w \mathrm{d}\theta = |z|^{-2d} \mathrm{d}z. \tag{14.25}$$

Now taking account of the fact that, transforming through the sphere $\mathbb{S}^{d-1}(x, (|x|^2 - 1)^{1/2})$, that is, $z \mapsto z^* = x + K\tilde{z}$, where $\tilde{z} = (1 - |x|^2)^{-1}(z - x)$, we can work with the differential change of Cartesian variables

$$\mathrm{d}z^* = (|x|^2 - 1)^{2d} |z - x|^{-2d} \prod_{i=1}^d \frac{\mathrm{d}z_i}{(|x|^2 - 1)} = (|x|^2 - 1)^d |z - x|^{-2d} \mathrm{d}z, \tag{14.26}$$

where $z = (z_1, \ldots, z_d)$. In particular, writing $z = rA(\theta)$ and $z^* = r^*A(\theta^*)$, and appealing to (14.24) and (14.5), this tells us that

$$\frac{2\pi^{d/2}}{\Gamma(d/2)}(r^*)^{d-1}\sigma_1(d\theta^*)dr^* = \left\{\frac{2\pi^{d/2}}{\Gamma(d/2)}\frac{(|x|^2-1)^{d-1}}{|z-x|^{2d-2}}r^{d-1}\sigma_1(d\theta)\right\}\left\{\frac{(|x|^2-1)}{|z-x|^2}dr\right\}.$$

In particular, we note that since $|z^*| = 1$ if and only if $|z| = 1$, then we can identify the change of variable in the surface measure σ_1 satisfies

$$\sigma_1(dz^*) = \frac{(|x|^2-1)^{d-1}}{|z-x|^{2d-2}}\sigma_1(dz), \qquad z \in \mathbb{S}^{d-1}.$$

Taking account of (14.3), this can equivalently be written as

$$\frac{1}{|z^*-x|^{d-1}}\sigma_1(dz^*) = \frac{1}{|z-x|^{d-1}}\sigma_1(dz), \qquad z \in \mathbb{S}^{d-1}. \tag{14.27}$$

Returning to (14.18), with the help of (14.5) and (14.3) for the transformation (14.2) in $\mathbb{S}^{d-1}(x, (|x|^2-1)^{1/2})$ and (14.27), this gives us for $x \in \mathbb{R}^d\backslash\mathbb{S}^{d-1}$ and $y \in \mathbb{S}^{d-1}$,

$$\begin{aligned}\frac{\Gamma\left(\frac{d}{2}\right)\Gamma(\alpha-1)}{\Gamma\left(\frac{\alpha+d}{2}-1\right)\Gamma\left(\frac{\alpha}{2}\right)} &= \int_{\mathbb{S}^{d-1}} |z^*-x|^{d-1}|z^*-y^*|^{\alpha-d}\frac{\sigma_1(dz^*)}{|z^*-x|^{d-1}}\\ &= \frac{(|x|^2-1)^{\alpha-1}}{|y-x|^{\alpha-d}}\int_{\mathbb{S}^{d-1}}\frac{|z-y|^{\alpha-d}}{|z-x|^{\alpha+d-2}}\sigma_1(dz),\end{aligned} \tag{14.28}$$

where we recall that the transformation (14.2) maps $\mathbb{S}^{d-1}$ to itself. In other words,

$$|x-y|^{\alpha-d} = \int_{\mathbb{S}^{d-1}} |z-y|^{\alpha-d}\frac{\Gamma\left(\frac{\alpha+d}{2}-1\right)\Gamma\left(\frac{\alpha}{2}\right)}{\Gamma\left(\frac{d}{2}\right)\Gamma(\alpha-1)}\frac{(|x|^2-1)^{\alpha-1}}{|z-x|^{\alpha+d-2}}\sigma_1(dz),$$

which, by the uniqueness given in Lemma 14.4, establishes the statement of the theorem for $|x| > 1$.

Finally, for the case $|x| < 1$, we can appeal to similar reasoning as in the proof of Theorem 14.2 and use the Riesz–Bogdan–Żak transform to establish the identity. The details are left to the reader. □

Remark 14.5 Although we have excluded the setting that X is a Brownian motion, that is, the case $\alpha = 2$, our analysis can be easily adapted to include it. In that case, the conclusion of Theorems 14.1 and 14.2 provide us with the classical Newtonian Poisson potential formula. Indeed, for $|x| < a$,

$$\mathbb{P}_x(\tau_a^{\ominus} < \infty) = 1 = \int_{\mathbb{S}^{d-1}(0,a)}\frac{a^{d-2}(a^2-|x|^2)}{|z-x|^d}\sigma_a(dz). \tag{14.29}$$

Similarly, we can also reproduce the classical conclusion for the case that X is a Brownian motion that, for $|x| > a$,

$$\mathbb{P}_x(\tau_a^\odot < \infty) = \left(\frac{|x|}{a}\right)^{2-d} = \int_{\mathbb{S}^{d-1}(0,a)} \frac{a^{d-2}(|x|^2 - a^2)}{|z - x|^d} \sigma_a(\mathrm{d}z). \qquad (14.30)$$

The identities (14.29) and (14.30) will be of use later.

Next, we turn our attention to deriving the resolvent of the stable process killed on hitting $\mathbb{S}^{d-1}$. We will appeal to the simple principle of path counting. To this end, write the associated resolvent measure as

$$U^\odot(x, \mathrm{d}y) = \int_0^\infty \mathbb{P}_x(X_t \in \mathrm{d}y,\ t < \tau_1^\odot)\mathrm{d}t, \qquad x, y \in \mathbb{R}^d \backslash \mathbb{S}^{d-1}.$$

Theorem 14.6 *Define $Q(x) = \mathbb{P}_x(\tau^\odot < \infty)$, for $x \in \mathbb{R}^d$. Then, for all $x, y \in \mathbb{R}^d \backslash \mathbb{S}^{d-1}$,*

$$U^\odot(x, \mathrm{d}y) = 2^{-\alpha}\pi^{-d/2}\frac{\Gamma((d-\alpha)/2)}{\Gamma(\alpha/2)}|x - y|^{\alpha-d}\left(1 - Q\left(\frac{y}{|y - x|}\left|x - \frac{y}{|y|^2}\right|\right)\right)\mathrm{d}y.$$

Proof Let us preemptively assume that $U^\odot(x, \mathrm{d}y)$ has density with respect to Lebesgue measure, written $u^\odot(x, y)$, $x, y \in \mathbb{R}^d \backslash \mathbb{S}^{d-1}$. As alluded to above, straightforward path counting and the strong Markov property tells us that

$$u^\odot(x, y) = C_\alpha |x - y|^{\alpha-d} - C_\alpha \int_{\mathbb{S}^{d-1}} |z - y|^{\alpha-d} h_1^\odot(x, z)\sigma_1(\mathrm{d}z),$$

where for convenience we have written $C_\alpha = 2^{-\alpha}\pi^{-d/2}\Gamma((d-\alpha)/2)/\Gamma(\alpha/2)$. In order to deal with the integral on the right-hand side above, we need to split our computations into the cases that $|x| < 1$ and $|x| > 1$.

First assume that $|x| > 1$. We appeal to a sphere inversion of the type (14.2) via the sphere $\mathbb{S}^{d-1}(x, (|x|^2 - 1)^{1/2})$ in a manner similar to the computation in (14.28). Indeed, reading only the second equality of (14.28), we see that

$$\begin{aligned} &C_\alpha \int_{\mathbb{S}^{d-1}} |z - y|^{\alpha-d} h_1^\odot(x, z)\sigma_1(\mathrm{d}z) \\ &= C_\alpha \frac{\Gamma\left(\frac{\alpha+d}{2} - 1\right)\Gamma\left(\frac{\alpha}{2}\right)}{\Gamma\left(\frac{d}{2}\right)\Gamma(\alpha - 1)} |x - y|^{\alpha-d} \int_{\mathbb{S}^{d-1}} |z^* - y^*|^{\alpha-d}\sigma_1(\mathrm{d}z^*) \\ &= C_\alpha |x - y|^{\alpha-d} Q(y^*), \end{aligned}$$

where in the second equality above, we have used Theorem 14.3. Recalling that $y^* = x + (|x|^2 - 1)(y - x)/|y - x|^2$, a straightforward piece of algebra tells us that

$$|y^*|^2|y - x|^2 = |y|^2\left|x - \frac{y}{|y|^2}\right|^2$$

and the result follows as soon as we note that isometry implies that $Q(y^*) = Q(|y^*|y/|y|)$.

For the case that $|x| < 1$, we can again appeal to the Riesz–Bogdan–Żak transform in Theorem 12.7. Recall that this transform states that, for $x \neq 0$, $(KX_{\eta(t)}, t \geq 0)$ under $\mathbb{P}_{Kx}$ is equal in law to $(X_t, t \geq 0)$ under $\mathbb{P}^{\circ}_x$, where $\eta(t) = \inf\{s > 0 \colon \int_0^s |X_u|^{-2\alpha} \mathrm{d}u > t\}$. Noting that, since $\int_0^{\eta(t)} |X_u|^{-2\alpha} \mathrm{d}u = t$, if we write $s = \eta(t)$, then

$$|X_s|^{-2\alpha}\mathrm{d}s = \mathrm{d}t, \qquad t > 0,$$

and hence we have that, for $|x| < 1$ and $x \in \mathbb{R}^d \backslash \mathbb{S}^{d-1}$,

$$\begin{aligned}
\int_{\mathbb{R}^d} f(y)\frac{|y|^{\alpha-d}}{|x|^{\alpha-d}} u^{\odot}(x,y)\mathrm{d}y &= \int_0^\infty \mathbb{E}_x\left[\frac{|X_t|^{\alpha-d}}{|x|^{\alpha-d}} f(X_t);\, t < \tau^{\odot}\right]\mathrm{d}t \\
&= \mathbb{E}^{\circ}_x\left[\int_0^\infty f(X_t)\mathbf{1}_{(t<\tau^{\odot})}\mathrm{d}t\right] \\
&= \mathbb{E}_{Kx}\left[\int_0^\infty f(KX_s)\mathbf{1}_{(s<\tau^{\odot})}|X_s|^{-2\alpha}\mathrm{d}s\right] \\
&= \int_{\mathbb{R}^d} f(Ky)|y|^{-2\alpha} u^{\odot}(Kx, y)\mathrm{d}y \\
&= \int_{\mathbb{R}^d} f(z)|z|^{2(\alpha-d)} u^{\odot}(Kx, Kz)\mathrm{d}z,
\end{aligned}$$

where we recall that $Kx = x/|x|^2$ and we have used (14.25) in the final equality. We can now appeal to the expression we have just derived for $u^{\odot}$ previously on account of the fact that $|Kx| = 1/|x| > 1$. Equation (14.5) for the transform K tells us that $|Ky - Kx| = |x - y|/|x||y|$. Hence we have, for $|x| < 1$ and $y \in \mathbb{R}^d \backslash \mathbb{S}^{d-1}$,

$$\begin{aligned}
u^{\odot}(x,y) &= C_\alpha |x|^{\alpha-d}|y|^{\alpha-d}|Kx - Ky|^{\alpha-d}\left(1 - Q\left(\frac{Ky}{|Ky - Kx|}|Kx - KKy|\right)\right) \\
&= C_\alpha |x - y|^{\alpha-d}\left(1 - Q\left(\frac{y}{|y - x|}|x - Ky|\right)\right),
\end{aligned}$$

as required. □

14.4 First entrance and exit of a ball

Let us start by defining the stopping times

$$\tau_a^{\oplus} := \inf\{t > 0 \colon |X_t| < a\} \text{ and } \tau_a^{\ominus} := \inf\{t > 0 \colon |X_t| > a\},$$

for $a > 0$. Recall that X is transient in dimension $d \geq 2$ and hence $\mathbb{P}_x(\tau_a^{\oplus} < \infty) < 1$, for all $|x| \geq a$, and $\mathbb{P}_x(\tau_a^{\ominus} < \infty) = 1$, for all $|x| \leq 1$.

We will first establish the distribution of $X_{\tau_a^\ominus}$ and then use the Riesz–Bogdan–Żak transform to give directly the distribution of $X_{\tau_a^\oplus}$.

Theorem 14.7 *Define the function*

$$g_a(x, y) = \pi^{-(d/2+1)}\Gamma(d/2)\sin(\pi\alpha/2)\frac{\left|a^2 - |x|^2\right|^{\alpha/2}}{\left|a^2 - |y|^2\right|^{\alpha/2}}|x - y|^{-d},$$

for $x, y \in \mathbb{R}^d\backslash\mathbb{S}^{d-1}$, *and* $a > 0$.

(i) Suppose that $|x| < a$, *then*

$$\mathbb{P}_x(X_{\tau_a^\ominus} \in \mathrm{d}y) = g_a(x, y)\mathrm{d}y, \qquad |y| \geq a. \tag{14.31}$$

(ii) Suppose that $|x| > a$, *then*

$$\mathbb{P}_x(X_{\tau_a^\oplus} \in \mathrm{d}y,\ \tau_a^\oplus < \infty) = g_a(x, y)\mathrm{d}y, \qquad |y| \leq a. \tag{14.32}$$

The fact that the Riesz–Bogdan–Żak transform lies behind the relationship between the distributions of $X_{\tau_a^\ominus}$ and $X_{\tau_a^\oplus}$ should reassure the reader that one uses a single function, albeit with symmetries relative to the ball of unit radius, to describe both.

The proof of the above theorem is quite involved, and we need to pass first through an intermediate result, which offers an analogous approach to the use of Lemma 14.4 in proving Theorem 14.1. We omit its proof as it follows an almost identical thread to that of Lemma 14.4.

Lemma 14.8 *Uniquely in the class of probability distributions supported on the exterior of* $\mathbb{S}^{d-1}$, *we have that, for* $|x| < 1$, $\mu_x^\ominus(\mathrm{d}z) := \mathbb{P}_x(X_{\tau_1^\ominus} \in \mathrm{d}z)$ *solves*

$$|x - y|^{\alpha-d} = \int_{|z|\geq 1} |z - y|^{\alpha-d}\mu(\mathrm{d}z), \qquad |y| > 1, \tag{14.33}$$

and, for $|x| > 1$, *again uniquely in the class of probability distributions supported on the interior of* $\mathbb{S}^{d-1}$, $\mu_x^\oplus(\mathrm{d}z) := \mathbb{P}_x(X_{\tau_1^\oplus} \in \mathrm{d}z,\ \tau_1^\oplus < \infty)$ *solves*

$$|x - y|^{\alpha-d} = \int_{|z|\leq 1} |z - y|^{\alpha-d}\mu(\mathrm{d}z), \qquad |y| < 1. \tag{14.34}$$

We now turn our attention to showing that the unique solution to (14.33) and (14.34) are given by $\mu_x^\ominus(\mathrm{d}z) = g_1(x, z)\mathrm{d}z$, $|z| > 1 > |x|$ and $\mu_x^\oplus(\mathrm{d}z) = g_1(x, z)\mathrm{d}z$, $|z| < 1 < |x|$, respectively.

Proof of Theorem 14.7 (i) We only give the proof in the case that $a = 1$. The proof for general $a > 0$ is achieved using the scaling property. For example, for Borel set D and $|x| < 1$,

$$\begin{aligned}\int_D g_a(x,z)\mathrm{d}z &= \mathbb{P}_x(X_{\tau_a^\ominus} \in D) \\ &= \mathbb{P}_{x/a}(X_{\tau_1^\ominus} \in a^{-1}D) \\ &= \int_{a^{-1}D} g_1(x/a,z)\mathrm{d}z \\ &= \int_D g_1(x/a,y/a)a^{-d}\mathrm{d}y.\end{aligned}$$

Hence, $g_a(x,y) = g_1(x/a,y/a)/a^d$.

For the case $a = 1$, the proof is complete as soon as we can verify that

$$|x-y|^{\alpha-d} = c_{\alpha,d}\int_{|z|\geq 1} |z-y|^{\alpha-d}\frac{|1-|x|^2|^{\alpha/2}}{|1-|z|^2|^{\alpha/2}}|x-z|^{-d}\mathrm{d}z, \tag{14.35}$$

for $|y| > 1 > |x|$, where

$$c_{\alpha,d} = \pi^{-(1+d/2)}\,\Gamma(d/2)\,\sin(\pi\alpha/2).$$

Starting with the integral on the right-hand side of (14.35), we will appeal to the transformation (14.6) through the sphere $\mathbb{S}^{d-1}(x,(1-|x|^2)^{1/2})$, noting in particular that

$$|z^\diamond - y^\diamond| = (1-|x|^2)\frac{|z-y|}{|z-x||y-x|} \quad\text{and}\quad |z|^2 - 1 = \frac{|z-x|^2}{1-|x|^2}(1-|z^\diamond|^2), \tag{14.36}$$

where the second identity comes from (14.10). A similar analysis of the differential calculus associated to (14.6) gives us

$$\mathrm{d}z^\diamond = (1-|x|^2)^d|z-x|^{-2d}\mathrm{d}z, \qquad z\in\mathbb{R}^d, \tag{14.37}$$

just as in (14.26).

Now we can use (14.36) and (14.37) to compute, for $|x| < 1 < |y|$,

$$\begin{aligned}&c_{\alpha,d}\int_{|z|\geq 1} |z-y|^{\alpha-d}\frac{|1-|x|^2|^{\alpha/2}}{|1-|z|^2|^{\alpha/2}}|x-z|^{-d}\mathrm{d}z \\ &\qquad = c_{\alpha,d}|y-x|^{\alpha-d}\int_{|z^\diamond|\leq 1}\frac{|z^\diamond - y^\diamond|^{\alpha-d}}{|1-|z^\diamond|^2|^{\alpha/2}}\mathrm{d}z^\diamond.\end{aligned} \tag{14.38}$$

Next, we perform another transformation of the type (14.6), albeit through the sphere $\mathbb{S}^{d-1}(y^\diamond,(1-|y^\diamond|^2)^{1/2})$. In a similar fashion to the calculation that led to the right-hand side of (14.38), using (14.7) and the second equality in (14.36), we obtain

$$c_{\alpha,d}\int_{|z|\geq 1}|z-y|^{\alpha-d}\frac{|1-|x|^2|^{\alpha/2}}{|1-|z|^2|^{\alpha/2}}|x-z|^{-d}\mathrm{d}z$$
$$= c_{\alpha,d}|y-x|^{\alpha-d}\int_{|w|\geq 1}\frac{|1-|y^\circ|^2|^{\alpha/2}}{|1-|w|^2|^{\alpha/2}}|w-y^\circ|^{-d}\mathrm{d}w. \quad (14.39)$$

The question now remains as to whether the integral on the right-hand side of (14.39) is equal to $1/c_{\alpha,d}$. We resolve this issue by recalling that the surface of a sphere of radius r is given by $2\pi^{d/2}r^{d-1}/\Gamma(d/2)$. Moreover, writing, for $|y^\circ|<1$,

$$\int_{|w|\geq 1}\frac{1}{|1-|w|^2|^{\alpha/2}}|w-y^\circ|^{-d}\mathrm{d}w$$
$$= \frac{2\pi^{d/2}}{\Gamma(d/2)}\int_1^\infty r^{d-1}\mathrm{d}r\int_{\mathbb{S}^{d-1}(0,r)}\frac{1}{|1-|z|^2|^{\alpha/2}}|z-y^\circ|^{-d}\sigma_r(\mathrm{d}z)$$
$$= \frac{2\pi^{d/2}}{\Gamma(d/2)}\int_1^\infty \frac{r^{d-1}\mathrm{d}r}{|1-r^2|^{\alpha/2}}\int_{\mathbb{S}^{d-1}(0,r)}|z-y^\circ|^{-d}\sigma_r(\mathrm{d}z), \quad (14.40)$$

where $\sigma_r(\mathrm{d}z)$ is the surface measure on $\mathbb{S}^{d-1}(0,r)$ normalised to have unit total mass. In order to continue, we recall Remark 14.5, in particular the Newtonian Poisson formula (14.29) which

$$\int_{\mathbb{S}^{d-1}(0,r)}\frac{r^{d-2}(r^2-|y^\circ|^2)}{|z-y^\circ|^d}\sigma_r(\mathrm{d}z)=1, \qquad |y^\circ|<1<r. \quad (14.41)$$

The identity (14.41) allows us to continue to develop the right-hand side of (14.40), so that we have

$$\int_{|v|\geq 1}\frac{1}{|1-|w|^2|^{\alpha/2}}|v-y^\circ|^{-d}\mathrm{d}w = \frac{\pi^{d/2}}{\Gamma(d/2)}\int_1^\infty\frac{2r}{(r^2-1)^{\alpha/2}(r^2-|y^\circ|^2)}\mathrm{d}r. \quad (14.42)$$

A further change of variable, first $s=(r^2-1)/(1-|y^\circ|^2)$ and the representation of the beta function in (A.18) gives

$$\int_1^\infty\frac{2r}{(r^2-1)^{\alpha/2}(r^2-|y^\circ|^2)}\mathrm{d}r = \frac{1}{(1-|y^\circ|^2)^{\alpha/2}}\int_0^\infty s^{-\alpha/2}(1+s)^{-1}\mathrm{d}s$$
$$= \frac{1}{(1-|y^\circ|^2)^{\alpha/2}}\Gamma(\alpha/2)\Gamma(1-\alpha/2)$$
$$= \frac{\pi}{\sin(\alpha\pi/2)}\frac{1}{(1-|y^\circ|^2)^{\alpha/2}}.$$

Plugging back into (14.42), and then into (14.39), we end up with

$$\int_{|z|\geq 1} |z-y|^{\alpha-d} \frac{|1-|x|^2|^{\alpha/2}}{|1-|z|^2|^{\alpha/2}} |x-z|^{-d} \mathrm{d}z = \frac{\pi^{1+d/2}}{\Gamma(d/2)\sin(\alpha\pi/2)} = \frac{1}{c_{\alpha,d}}$$

as required.

The identity (14.35) is thus affirmed for all $|x| < 1 < |y|$ and, hence, the first part of Theorem 14.7 is proved. □

Proof of Theorem 14.7 (ii) As with the case of part (i), a standard scaling argument allows us to reduce the proof to the case that $a = 1$. In that case, we can appeal to the Riesz–Bogdan–Żak transform in Theorem 12.8 and note that for Borel set $D \subseteq \{u \colon |u| \leq 1\}$ and $|x| > 1$,

$$\mathbb{P}_x(X_{\tau_1^\oplus} \in D) = \mathbb{P}^\circ_{Kx}(KX_{\tau_1^\ominus} \in D),$$

where we recall that $Kx = x/|x|^2$ and $\mathbb{P}^\circ_x$, $x \neq 0$, is the result of the Doob h-transform in (12.9). It follows that

$$\begin{aligned}
&\mathbb{P}_x(X_{\tau_1^\oplus} \in D)\\
&= \int_{KD} \frac{|y|^{\alpha-d}}{|Kx|^{\alpha-d}} g_1(Kx, y)\mathrm{d}y\\
&= c_{\alpha,d} \int_{KD} |z|^{d-\alpha} |Kx|^{d-\alpha} \frac{|1-|Kx|^2|^{\alpha/2}}{|1-|y|^2|^{\alpha/2}} |Kx-y|^{-d}\mathrm{d}y\\
&= c_{\alpha,d} \int_D |z|^{2d} \frac{|1-|x|^2|^{\alpha/2}}{|1-|z|^2|^{\alpha/2}} |x-z|^{-d}|z|^{-2d}\mathrm{d}z, \qquad (14.43)
\end{aligned}$$

where $KD = \{Kx \colon x \in D\}$ and we have used the change of variable $y = Kz$ together with (14.25) in the final equality. The required identity now follows directly from (14.43). □

Recalling that in dimensions $d \geq 2$, the stable process is transient, it makes sense to compute the probability that the ball of radius a around the origin is never entered. That is to say, to compute the total mass of the measure $\mu^\oplus$. Naturally one can do this by marginalising the distribution given in Theorem 14.7, however, it turns out to be simpler to make use of the Lamperti representation of $|X|$.

Lemma 14.9 *We have for $|x| > a$,*

$$\mathbb{P}_x(\tau_a^\oplus = \infty) = \frac{\Gamma(d/2)}{\Gamma((d-\alpha)/2)\Gamma(\alpha/2)} \int_0^{(|x|^2/a^2)-1} (u+1)^{-d/2} u^{\alpha/2-1}\mathrm{d}u.$$

Proof From Theorem 5.21 we have the Wiener–Hopf factorisation of the characteristic exponent of ξ, the Lévy process appearing in the Lamperti transform of $|X|$. In particular, its descending ladder height process has Laplace exponent given by $\Gamma((\lambda + d - \alpha)/2)/\Gamma((\lambda + d)/2)$, $\lambda \geq 0$. If we denote its descending ladder height potential measure by U^-_ξ, then, from (2.38), we have that

$$\int_{[0,\infty)} \mathrm{e}^{-\lambda x} U^-_\xi(\mathrm{d}x) = \frac{\Gamma((\lambda + d)/2)}{\Gamma((\lambda + d - \alpha)/2)}, \qquad \lambda \geq 0.$$

In light of Corollary 4.3, this transform can be inverted explicitly and, pre-emptively assuming that it has a density, denoted by $u^-_\xi(x)$, $x \geq 0$, we have

$$\int_{[0,\infty)} \mathrm{e}^{-2\lambda x} u^-_\xi(x)\mathrm{d}x = \int_0^\infty \mathrm{e}^{-\lambda x} \frac{1}{\Gamma(\alpha/2)} \mathrm{e}^{-(d-\alpha)x/2}(1 - \mathrm{e}^{-x})^{\alpha/2-1}\mathrm{d}x,$$

so that

$$U^-_\xi(\mathrm{d}x) = \frac{2}{\Gamma(\alpha/2)} \mathrm{e}^{-(d-\alpha)x}(1 - \mathrm{e}^{-2x})^{\alpha/2-1}\mathrm{d}x.$$

Next, with the help of Lemma 2.26 we note that

$$\begin{aligned}\mathbb{P}_x(\tau^\oplus_a < \infty) &= \frac{\Gamma((d - \alpha)/2)}{\Gamma(d/2)\Gamma(\alpha/2)} \int_{\log(|x|/a)}^\infty 2\mathrm{e}^{-(d-\alpha)y}(1 - \mathrm{e}^{-2y})^{\alpha/2-1}\mathrm{d}y \\ &= \frac{\Gamma((d - \alpha)/2)}{\Gamma(d/2)\Gamma(\alpha/2)} \int_{(|x|/a)^2-1}^\infty (u + 1)^{-d/2}u^{\alpha/2-1}\mathrm{d}u\end{aligned}$$

where, in the second equality, we have made the substitution $u = \mathrm{e}^{2y} - 1$. Recalling from (A.18) that

$$\frac{\Gamma(d/2)\Gamma(\alpha/2)}{\Gamma((d - \alpha)/2)} = \int_0^\infty (u + 1)^{-d/2}u^{\alpha/2-1}\mathrm{d}u,$$

the proof is complete. □

The conclusion of Theorem 14.7 also gives us the opportunity to study the potentials

$$U^\ominus_a(x, \mathrm{d}y) = \int_0^\infty \mathbb{P}_x(X_t \in \mathrm{d}y,\ t < \tau^\ominus_a)\mathrm{d}t$$

and

$$U^\oplus_a(x, \mathrm{d}y) = \int_0^\infty \mathbb{P}_x(X_t \in \mathrm{d}y,\ t < \tau^\oplus_a)\mathrm{d}t,$$

for $|x|, |y| < a$ and $|x|, |y| > a$, respectively.

Theorem 14.10 *For $a > 0$, define the function*

$$u_a(x, y) = 2^{-\alpha}\pi^{-d/2}\frac{\Gamma(d/2)}{\Gamma(\alpha/2)^2}|x - y|^{\alpha-d} \int_0^{\zeta_a(x,y)} (u + 1)^{-d/2}u^{\alpha/2-1}\mathrm{d}u, \qquad (14.44)$$

where $\zeta_a(x, y) = (a^2 - |x|^2)(a^2 - |y|^2)/a^2|x - y|^2$.

(i) In the case that $|x|, |y| < a$,

$$U_a^{\ominus}(x, \mathrm{d}y) = u_a(x, y)\mathrm{d}y.$$

(ii) In the case that $|x|, |y| > a$,

$$U_a^{\oplus}(x, \mathrm{d}y) = u_a(x, y)\mathrm{d}y.$$

Proof As with Theorem 14.7, it suffices to consider the case that $a = 1$, thanks to scaling. The method of the proof of the first part boils down to counting paths. More precisely, suppose we momentarily assume density of $U_1^{\ominus}(x, \mathrm{d}y)$ exists, for $|x|, |y| < a$, and write it as $u_1^{\ominus}(x, y)$. We have, by the strong Markov property, that,

$$\kappa_{\alpha,d}|x - y|^{\alpha-d} = u_1^{\ominus}(x, y) + \int_{|z|>1} \kappa_{\alpha,d}|z - y|^{\alpha-d}\mu_x^{\ominus}(\mathrm{d}z),$$

for $|x|, |y| < 1$, where $\kappa_{\alpha,d} = 2^{-\alpha}\pi^{-d/2}\Gamma((d-\alpha)/2)/\Gamma(\alpha/2)$. To complete the proof, it thus suffices to show that

$$\begin{aligned} &|x - y|^{\alpha-d} - \int_{|z|>1} |z - y|^{\alpha-d} g_1(x, z)\mathrm{d}z \\ &= \frac{\Gamma(d/2)}{\Gamma(\alpha/2)\Gamma((\alpha-d)/2)}|x - y|^{\alpha-d} \int_0^{\zeta_1(x,y)} (u+1)^{-d/2}u^{\alpha/2-1}\mathrm{d}u, \end{aligned} \tag{14.45}$$

where we recall that the kernel g_1 was given in Theorem 14.7. To this end, start with the integral on the left-hand side of (14.45). From (14.38) we have already shown that, by performing the transformation (14.6) through the sphere $\mathbb{S}^{d-1}(x, (1 - |x|^2)^{1/2})$,

$$\begin{aligned} &\int_{|z|>1} |z - y|^{\alpha-d} g_1(x, z)\mathrm{d}z \\ &= \pi^{-(d/2+1)}\Gamma(d/2)\sin(\pi\alpha/2)\int_{|z|\geq 1} |z - y|^{\alpha-d}\frac{|1 - |x|^2|^{\alpha/2}}{|1 - |z|^2|^{\alpha/2}}|x - z|^{-d}\mathrm{d}z \\ &= \pi^{-(d/2+1)}\Gamma(d/2)\sin(\pi\alpha/2)|y - x|^{\alpha-d}\int_{|w|\leq 1}\frac{|w - y^{\circ}|^{\alpha-d}}{|1 - |w|^2|^{\alpha/2}}\mathrm{d}w. \end{aligned} \tag{14.46}$$

Next, we want to apply the transformation (14.2) through the sphere $\mathbb{S}^{d-1}(y^{\circ}, (|y^{\circ}|^2 - 1)^{1/2})$, noting that a similar calculation to the one in (14.10) will give us that, if $w^* = y^{\circ} + |w - y^{\circ}|^{-2}(w - y^{\circ})(|y^{\circ}| - 1)$, then

$$|w^*|^2 - 1 = \frac{|w - y^{\circ}|}{|y^{\circ}|^2 - 1}(|w|^2 - 1)$$

and also a similar calculation to the one in (14.26) shows us that

$$\mathrm{d}w^* = (|y^{\circ}|^2 - 1)^d|w - y^{\circ}|^{-2d}\mathrm{d}w.$$

Following the manipulations in (14.40) and (14.42), albeit using (14.30) in place of (14.29), recalling that $|y^\circ| > 1$, we get

$$
\begin{aligned}
&\int_{|w|\le 1} \frac{|w-y^\circ|^{\alpha-d}}{|1-|w|^2|^{\alpha/2}}\mathrm{d}w \\
&= \int_{|w^*|\le 1} \frac{(|y^\circ|^2-1)^{\alpha/2}}{|1-|w^*|^2|^{\alpha/2}}|w^*-y^\circ|^{-d}\mathrm{d}w^* \\
&= \frac{2\pi^{d/2}}{\Gamma(d/2)}(|y^\circ|^2-1)^{\alpha/2}\int_0^1 \frac{r^{d-1}\mathrm{d}r}{(1-r^2)^{\alpha/2}}\int_{\mathbb{S}^{d-1}(0,r)}|z-y^\circ|^{-d}\sigma_r(\mathrm{d}z) \\
&= \frac{\pi^{d/2}}{\Gamma(d/2)}(|y^\circ|^2-1)^{\alpha/2}\int_0^1 \frac{2r}{(1-r^2)^{\alpha/2}(|y^\circ|^2-r^2)}\left(\frac{|y^\circ|}{r}\right)^{2-d}\mathrm{d}r \\
&= \frac{\pi^{d/2}}{\Gamma(d/2)}(|y^\circ|^2-1)^{\alpha/2}|y^\circ|^{2-d}\int_0^1 v^{d/2-1}(1-v)^{-\alpha/2}(|y^\circ|^2-v)^{-1}\mathrm{d}v,
\end{aligned}
\tag{14.47}
$$

where we have made the change of variable $v = r^2$ in the final equality. Note, however, that, with the help of (A.31), (A.33) and (A.35), the integral on the right-hand side of (14.47) satisfies

$$
\begin{aligned}
&\int_0^1 v^{d/2-1}(1-v)^{-\alpha/2}(|y^\circ|^2-v)^{-1}\mathrm{d}v \\
&= |y^\circ|^{-2}\frac{\Gamma(d/2)\Gamma(1-\alpha/2)}{\Gamma(1+(d-\alpha)/2)}{}_2F_1(1,\, d/2,\, 1+(d-\alpha)/2;\, |y^\circ|^{-2}) \\
&= |y^\circ|^{-2}(1-|y^\circ|^{-2})^{-\alpha/2}\frac{\Gamma(d/2)\Gamma(1-\alpha/2)}{\Gamma(1+(d-\alpha)/2)} \\
&\qquad \times {}_2F_1((d-\alpha)/2,\, 1-\alpha/2,\, 1+(d-\alpha)/2;\, |y^\circ|^{-2}) \\
&= \frac{\Gamma(d/2)\Gamma(1-\alpha/2)}{\Gamma((d-\alpha)/2)}|y^\circ|^{d-2}(|y^\circ|^2-1)^{-\alpha/2} \\
&\qquad \times \int_0^{|y^\circ|^{-2}} s^{\frac{(d-\alpha)}{2}-1}(1-s)^{\frac{\alpha}{2}-1}\mathrm{d}s.
\end{aligned}
\tag{14.48}
$$

Now putting (14.48) into (14.47), then into (14.46) and the latter into the left-hand side of (14.45), we get

$$
\begin{aligned}
&|x-y|^{\alpha-d}\left(1-\frac{\Gamma(d/2)}{\Gamma(\alpha/2)\Gamma((d-\alpha)/2)}\int_0^{|y^\circ|^{-2}} s^{\frac{(d-\alpha)}{2}-1}(1-s)^{\frac{\alpha}{2}-1}\mathrm{d}s\right) \\
&= \frac{\Gamma(d/2)}{\Gamma(\alpha/2)\Gamma((d-\alpha)/2)}|x-y|^{\alpha-d}\int_{|y^\circ|^{-2}}^1 s^{\frac{(d-\alpha)}{2}-1}(1-s)^{\frac{\alpha}{2}-1}\mathrm{d}s \\
&= \frac{\Gamma(d/2)}{\Gamma(\alpha/2)\Gamma((d-\alpha)/2)}|x-y|^{\alpha-d}\int_0^{|y^\circ|^2-1}(1+u)^{-\frac{d}{2}}u^{\frac{\alpha}{2}-1}\mathrm{d}u
\end{aligned}
$$

where, in the final equality, we made the change of variables $u = (1 - s)/s$. Recalling from the second equality in (14.36) that $|y^\circ|^2 - 1 = \zeta_1(x, y)$, we finally come to rest at the conclusion that the left-hand side of (14.45) agrees with the required right-hand side, thus completing the proof.

For part (ii) of the theorem we set $a = 1$ as usual, with the general case deduced by scaling. We can use similar reasoning as in the proof of Theorem 14.6 and note from the Riesz–Bogdan–Żak transform in Theorem 12.7 that

$$|X_s|^{-2\alpha}\mathrm{d}s = \mathrm{d}t, \qquad t > 0.$$

Hence, for $|x| > 1$,

$$\begin{aligned}\int_{|y|>1} \frac{|z|^{\alpha-d}}{|x|^{\alpha-d}} u_1^{\oplus}(x, z) f(z)\mathrm{d}z &= \mathbb{E}_x^\circ\left[\int_0^{\tau_1^{\oplus}} f(X_t)\mathrm{d}t\right] \\ &= \mathbb{E}_{Kx}\left[\int_0^{\tau_1^{\ominus}} f(KX_{\eta(t)})\mathrm{d}t\right] \\ &= \mathbb{E}_{Kx}\left[\int_0^{\tau_1^{\ominus}} f(KX_s)|X_s|^{-2\alpha}\mathrm{d}s\right] \\ &= \int_{|y|<1} u_1^{\ominus}(Kx, y) f(Ky)|y|^{-2\alpha}\mathrm{d}y,\end{aligned}$$

where we have pre-emptively assumed that the resolvent associated to (11.19) has a density, which we have denoted by $u_1^{\oplus}(x, y)$. In the integral on the right-hand side above, we can make the change of variables $y = Kz$, which is equivalent to $z = Ky$. Noting that $\mathrm{d}y = \mathrm{d}z/|z|^{2d}$, we get

$$\int_{|y|>1} \frac{|z|^{\alpha-d}}{|x|^{\alpha-d}} u_1^{\oplus}(x, z) f(z)\mathrm{d}z = \int_{|z|>1} u_1^{\ominus}(Kx, Kz) f(z) \frac{|z|^{2\alpha}}{|z|^{2d}}\mathrm{d}z,$$

from which, together with the help of identity (14.44), we can conclude that a density for $U_1^{\oplus}$ does indeed exist and, for $|x|, |z| > 1$,

$$\begin{aligned}u_1^{\oplus}(x, z) &= \frac{|x|^{\alpha-d}}{|z|^{\alpha-d}} u_1^{\ominus}(Kx, Kz) \frac{|z|^{2\alpha}}{|z|^{2d}} \\ &= 2^{-\alpha}\pi^{-d/2} \frac{\Gamma(d/2)}{\Gamma(\alpha/2)^2} \frac{|x|^{\alpha-d}}{|z|^{\alpha-d}} \frac{|z|^{2\alpha}}{|z|^{2d}} |Kx - Kz|^{\alpha-d} \int_0^{\zeta_1(Kx,Kz)} (u+1)^{-d/2} u^{\alpha/2-1}\mathrm{d}u.\end{aligned}$$

Hence, after a little algebra, for $|x|, |z| > 1$,

$$u_1^{\oplus}(x, z) = 2^{-\alpha}\pi^{-d/2} \frac{\Gamma(d/2)}{\Gamma(\alpha/2)^2} |x - z|^{\alpha-d} \int_0^{\zeta_1(x,z)} (u + 1)^{-d/2} u^{\alpha/2-1}\mathrm{d}u,$$

where we have again used the fact that $|Kx - Kz| = |x - z|/|x||z|$ so that

$$\zeta_1(Kx, Kz) = (|x|^2 - 1)(|z|^2 - 1)/|x - z|^2 = \zeta_1(x, z)$$

and the result is proved. □

14.5 Walk-on-spheres and first exit of general domains

In the previous sections, we have provided a very extensive analysis of first exit (and thanks to the Riesz–Bogdan–Żak transform) and first entrance of the stable process into a sphere. But are there other domains D for which exit distributions of the form $\mathbb{P}_x(X_{\tau_D} \in \mathrm{d}y : \tau_D < \infty)$ can be identified explicitly, where $\tau_D = \inf\{t > 0 : X_t \notin D\}$? Despite there being an extensive literature, in particular, in the domain of potential analysis, there are very few scenarios, aside from those already reported on in this and previous chapters, where explicit identities can be obtained.

Classical Monte Carlo simulation is a method to empirically inform the distribution of the first exit position X_{τ_D} by appealing to numerical experiments. Indeed, suppose $\{X^i_{\tau^i_D} : i = 1, \ldots, n\}$ are n independent runs of the path of the stable process until first exiting a given domain D, where $X^i_0 = x$ and $\tau^i_D = \inf\{t > 0 : X^i_t \notin D\}$, for $i = 1, \ldots, n$. Then appealing to the Law of Large Numbers to conclude under appropriate conditions,

$$\frac{1}{n}\sum_{i=1}^{n} f(X^i_{\tau^i_D}) \approx \mathbb{E}_x[f(X_{\tau_D})],$$

for bounded measurable $f : \mathbb{R}^d \mapsto \mathbb{R}^d$, provides a numerical algorithm for simulating the, otherwise deterministic, expectation $\mathbb{E}_x[f(X_{\tau_D})]$. One of the problems here is that it is not clear how to efficiently simulate the path of a stable process. For sure, simulating an entire path until first exiting D will be a costly operation.

It turns out that the problem of first exit of the stable process from the unit ball can be used to form the basis of a very exact algorithm that provides a shortcut to the suggested aforementioned Monte Carlo algorithm. The algorithm is called *walk-on-spheres* and consists of constructing a sequence of exit problems from judiciously sized and positioned balls such that the algorithm ends with the exact distribution of X_{τ_D}. Naturally, there will be some constraints on the type of domain that is permitted, but the method is nonetheless robust. As such we will, by default unless otherwise mentioned, assume that D is an (open) convex domain but not necessarily bounded.

The walk-on-spheres algorithm consists of constructing a sequence of random variables $\chi = (\chi_i, i = 0, \ldots, N)$, where N is random and dependent on the evolution of χ and χ_N is exactly equal in distribution to X_{τ_D}. The algorithm evolves as follows:

1. Set $\chi_0 = x \in D$.
2. Given χ_{n-1} and that the algorithm has not yet ended, define the radius

$$r_n := \sup\{r > 0\colon \mathbb{S}^{d-1}(X_{n-1}, r) \subseteq D\}$$

   and set

$$B_n = \{x \in \mathbb{R}^d\colon |x - \chi_{n-1}| \le r_n\}, \tag{14.49}$$

   the largest ball centred at χ_{n-1} that is contained in the closure of D.
3. Next, given χ_{n-1} and r_n, sample the point χ_n from the distribution of $X_{\tau_{B_n}}$. Equivalently, $\chi_n = \chi_{n-1} + \Delta_n$, where Δ_n is independently sampled using Theorem 14.7 (i), according to the law of $X_{\tau_r^\oplus}$ under $\mathbb{P}_0$ with $r = r_n$.
4. If $\chi_n \notin D$, then $N = n$ and the algorithm stops.
5. Otherwise, go to step 2.

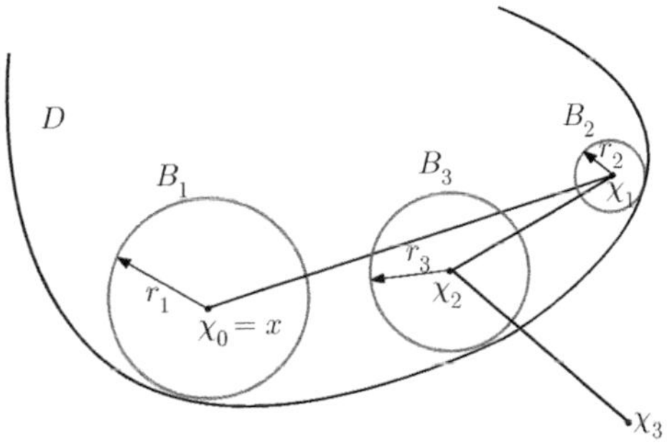

Figure 14.6 Steps of the walk-on-sphere algorithm until exiting the convex domain D in the stable setting. In this realisation, $N = 3$.

The next theorem shows that the walk-on-spheres algorithm comes to an end remarkably fast. Heuristically speaking, it shows that the probability that the algorithm takes more than n steps decays exponentially in n, irrespective of the choice of $\chi_0 \in D$.

Theorem 14.11 *Suppose that D is a convex domain such that D^c has positive d-dimensional Lebesgue measure. For $\chi_0 = x \in D$ the terminal value χ_N is equal in law to X_{τ_D} under $\mathbb{P}_x$. Moreover, there exists a constant $p = p(\alpha, d) > 0$ (independent of x and D) and an integer-valued random variable Γ, defined on the same probability space as $(X, \mathbb{P}_x)$, such that $N \le \Gamma$ almost surely, where*

$$\mathbb{P}_x(\Gamma = k) = (1-p)^{k-1}p, \qquad k \in \mathbb{N}.$$

This implies that

$$\mathbb{P}_x(N \geq k) \leq (1-p)^{k-1}, \qquad k \geq 1. \tag{14.50}$$

It is worth remarking that, although Γ has the same distribution for each $x \in D$, it is not the same random variable for each $x \in D$ as they are constructed on different probability spaces. It is also trivial to note that the walk-on-spheres algorithm offers efficient convergence. Within the assumptions of Theorem 14.11, it is clear that, whenever $\mathbb{E}_x[|f(X_{\tau_D})|] < \infty$,

$$\lim_{n\to\infty} \frac{1}{n} \sum_{i=1}^{n} f(\chi^i_{N^i}) = \mathbb{E}_x[f(\chi_N)] = \mathbb{E}_x[f(X_{\tau_D})] =: u_f(x), \tag{14.51}$$

almost surely, where $(\chi^i_n, n \leq N^i)$, $i \geq 1$ are i.i.d. copies of the walk-on-spheres with $\chi^i_0 = x \in D$, $i \geq 1$. Moreover, with the slightly stronger assumption that $\mathbb{E}_x[f(X_{\tau_D})^2] < \infty$, in the sense of weak convergence,

$$\lim_{n\to\infty} n^{1/2}\left(\frac{1}{n}\sum_{i=1}^{n} f(\chi^i_{N^i}) - u_f(x)\right) \overset{d}{\Longrightarrow} \mathcal{N}(0, \mathrm{Var}(f(\chi_N))), \tag{14.52}$$

where $\mathcal{N}(\mu, \sigma^2)$ is a Gaussian random variable with mean μ and standard deviation σ.

We can provide relatively simple sufficient analytical conditions on f to ensure that $\mathbb{E}_x[|f(X_{\tau_D})|] < \infty$.

Lemma 14.12 *Suppose that D is convex and bounded, and that* $f\colon \mathbb{R}^d \mapsto \mathbb{R}$ *is continuous and satisfies*

$$\int_{\mathbb{R}^d} \frac{|f(z)|}{1+|z|^{d+\alpha}}\,\mathrm{d}z < \infty. \tag{14.53}$$

Then $\mathbb{E}_x[|f(X_{\tau_D})|] < \infty$, *for all* $x \in D$.

Proof On account of the fact that D is bounded, we can define a ball of sufficiently large radius $R > 0$, say

$$\mathbb{B}(x,R) := \{z \in \mathbb{R}^d : |z-x| \leq R\},$$

centred at x, such that D is compactly embedded in $\mathbb{B}(x,R)$ and hence $\tau_D \leq \tau_{\mathbb{B}(x,R)}$ almost surely, irrespective of the initial position of X. Then

$$
\begin{aligned}
&\mathbb{E}_x[|f(X_{\tau_D})|] \\
&= \mathbb{E}_x[|f(X_{\tau_D})|\mathbf{1}_{(\tau_D=\tau_{\mathbb{B}(x,R)})}] + \mathbb{E}_x[|f(X_{\tau_D})|\mathbf{1}_{(\tau_D<\tau_{\mathbb{B}(x,R)})}] \\
&\le \mathbb{E}_x[|f(X_{\tau_{\mathbb{B}(x,R)}})|] + \sup_{y\in\mathbb{B}(x,R)\setminus D} |f(x)| \\
&= \mathbb{E}[|f(x+RX_{\tau_{B(0,1)}})|] + \sup_{y\in\mathbb{B}(x,R)\setminus D} |f(y)| \\
&= \pi^{-(d/2+1)}\,\Gamma(d/2)\,\sin(\pi\alpha/2)\int_{|y|>1} \frac{|f(x+Ry)|}{\left|1-|y|^2\right|^{\alpha/2}|y|^d}\,\mathrm{d}y + \sup_{y\in\mathbb{B}(x,R)\setminus D} |f(y)| \\
&= C\int_{\mathbb{R}^d} \frac{|f(z)|}{1+|z|^{d+\alpha}}\,\mathrm{d}y + \sup_{y\in\mathbb{B}(x,R)\setminus D} |f(y)| < \infty, \qquad (14.54)
\end{aligned}
$$

for some constant $C \in (0,\infty)$, which does not depend on x (this is ensured thanks to the boundedness of D). In the inequality we have used the fact that, on $\{\tau_D < \tau_{\mathbb{B}(x,R)}\}$, we have $X_{\tau_D} \in \mathbb{B}(x,R)\setminus D$, moreover, that, as a continuous function in $\mathbb{R}^d$, f is bounded in $\mathbb{B}(x,R)\setminus D$. In the second equality, we have used spatial homogeneity and the scaling property of stable processes. In the third equality, we have used Theorem 14.7. The fourth equality follows by changing variables $z = x + Ry$ in the integral, appropriately estimating the denominator and using (14.53). □

Proof of Theorem 14.11 The distributional equivalence of the terminal value of the algorithm, χ_N, and the first point of exit from D of the stable process, X_{τ_D}, is a straightforward consequence of the Markov property and stationary and independent increments.

For the remaining part of the proof, we need more notation. For convenience, we shall henceforth write $X^{(x)} = (X^{(x)}_t : t \ge 0)$ to indicate the dependency of X on its initial position $X_0 = x$ (and thus is equivalent to writing $(X, \mathbb{P}_x)$). For any $x = (x_1, \ldots, x_d) \in \mathbb{R}^d$ such that $x_1 > 0$, we have

$$V = \{(z_1, \ldots, z_d) \in \mathbb{R}^d : z_1 > 0\},$$

for the open half-space containing x and denote its boundary

$$\partial V = \{(z_1, \ldots, z_d) \in \mathbb{R}^d : z_1 = 0\}.$$

Finally write $1 = (1, 0, \ldots, 0) \in \mathbb{R}^d$ for the 'North Pole' of $\mathbb{S}^{d-1}$.

Without loss of generality, by appealing to the spatial homogeneity and isotropy of X which allows us to appropriately choose our coordinate system, we may suppose that $x = |x|\,1 \in D$ is such that ∂V is a tangent hyperplane to both D and B_1; recall the latter of these two was defined in (14.49). The scaling property of X ensures that we can write

$$X^{(x)}_s = |x|\hat{X}^{(1)}_{|x|^{-\alpha}s}, \qquad s \ge 0, \tag{14.55}$$

where $\hat{X}^{(x)}$ is equal in law to $X^{(x)}$. Note that

$$\begin{aligned}\tau_{\mathbb{B}(x,|x|)} &= \inf\{t > 0 : X_t^{(x)} \notin \mathbb{B}(x,|x|)\} \\ &= |x|^\alpha \inf\{|x|^{-\alpha}t > 0 : |x|\hat{X}^{(1)}_{|x|^{-\alpha}t} \notin \mathbb{B}(x,|x|)\} \\ &= |x|^\alpha \inf\{u > 0 : \hat{X}^{(1)}_u \notin \mathbb{B}(1,1)\} \\ &=: |x|^\alpha \hat{\tau}_{\mathbb{B}(1,1)}. \end{aligned} \tag{14.56}$$

It follows that

$$X^{(x)}_{\tau_{B_1}} = |x|\hat{X}^{(1)}_{|x|^{-\alpha}|x|^\alpha \hat{\tau}_{\mathbb{B}(1,1)}} \stackrel{(d)}{=} |x|X^{(1)}_{\tau_{\mathbb{B}(1,1)}}. \tag{14.57}$$

A similar computation will show that $X^{(x)}_{\tau_V}$ is equal in distribution to $|x|\, X^{(1)}_{\tau_V}$.

Next, with the same assumptions on the coordinate system that allows us to wrte $x = |x|\,1$, we define the indicator random variables

$$I_D = \mathbf{1}_{(X_{\tau_{B_1}} \notin D)} \quad \text{and} \quad I_V = \mathbf{1}_{(X_{\tau_{B_1}} \notin V)}.$$

Sometimes we will write $I_D(x)$ and $I_V(x)$ to indicate that $X_0 = x$. Note by convexity of D, we have that

$$\mathbb{P}_x(I_D \geq I_V) = 1, \text{ and for all } x \in D. \tag{14.58}$$

Moreover, independently of $x \in D$, from Theorem 14.7 and the scaling in (14.57), we have

$$\begin{aligned}&\mathbb{P}_x(X_{\tau_{B_1}} \notin V) \\ &= \mathbb{P}_1(X_{\tau_{\mathbb{B}(1,1)}} \notin V) \\ &= \frac{\Gamma(d/2)}{\pi^{(d+2)/2}} \sin(\pi\alpha/2) \int_{x_1<-1} \left|1 - |x|^2\right|^{-\alpha/2} |x|^{-d}\, dx \\ &=: p(\alpha, d),\end{aligned}$$

which is a number in (0, 1).

Now suppose we condition on the previous positions of the walk-on-spheres, $\chi_0, \ldots, \chi_{k-1}$ as well as on the event $\{N > k-1\}$. As noted previously, thanks to stationary and independent increments as well as isotropy, we can always choose a coordinate system, or equivalently reorient and scale D in such a way that $\chi_k = |\chi_k|1$. This has the implication that, with the aforesaid conditioning, $\mathbf{1}_{\{N=k\}}$ is independent of $\chi_0, \ldots, \chi_{k-2}$ and equal in law to $I_D(\chi_{k-1})$. On the other hand, the event $I_V(\chi_{k-1})$ is independent of $\chi_0, \ldots, \chi_{k-1}$ and equal in law to a Bernoulli random variable with probability of success $p = p(\alpha, d)$. In particular, the sequence $I_V(\chi_k)$, $k \geq 0$ is a sequence of i.i.d. Bernoulli trials. That is to say, if we define

$$\Gamma = \min\{n \geq 1 : I_V(\chi_k) = 1\},$$

then it is geometrically distributed with parameter p.

As observed in (14.58), we also have that $\mathbb{P}_x(I_D \geq I_V)|_{x=\chi_k} = 1$, $k < N$, that is to say, $\{I_V(\chi_k) = 1\} \subseteq \{I_D(\chi_k) = 1\}$, for $k < N$, and hence

$$\min\{n \geq 1 : I_D(\chi_k) = 1\} \leq \min\{n \geq 1 : I_V(\chi_k) = 1\}$$

almost surely. In other words, we have $N \leq \Gamma$, almost surely, as required and (14.50) follows immediately. □

Thanks to Theorem 14.10 (i), where an expression for

$$U_a^{\ominus}(x, \mathrm{d}y) := \mathbb{E}_x\left[\int_0^{\tau_a^{\ominus}} \mathbf{1}_{(X_t \in \mathrm{d}y)} \mathrm{d}t\right]$$

is given, the walk-on-spheres algorithm can also be used to evaluate expressions, such as

$$u_{f,g}(x) := \mathbb{E}_x[f(X_{\tau_D})] + \mathbb{E}_x\left[\int_0^{\tau_D} g(X_s)\,\mathrm{d}s\right], \qquad x \in D, \tag{14.59}$$

for all f satisfying (14.53) and g such that the second expectation above is finite. A simple way to guarantee this is to ensure that g is uniformly bounded and the domain D is bounded.

Suppose the latter are true, it is straightforward to compare any moment $\mathbb{E}_x\left[\left(\int_0^{\tau_D} g(X_s)\,\mathrm{d}s\right)^n\right]$ with $\mathbb{E}_x[\tau_D^n]$, for $n \in \mathbb{N}$. Moreover, as D is bounded, we can easily compare $\mathbb{E}_x[\tau_D^n]$ with the nth moment of the time it takes X to exit a sufficiently large ball centred at x, say $\mathbb{B}(x, R)$. Then, recalling the jumps of X arrive as a Poisson point process with intensity Π given by (3.23), we known that, after an exponential amount of time with parameter $\Pi(\{x : |x| > R\}) < \infty$, a jump will occur which is sufficient to ensure that X exits $\mathbb{B}(x, R)$, and hence D. It follows that we can stochastically bound τ_D by an exponential random variable. This ensures the existence of all of its moments and hence $\mathbb{E}_x\left[\left(\int_0^{\tau_D} g(X_s)\,\mathrm{d}s\right)^n\right] < \infty$, for all $n \in \mathbb{N}$.

Lemma 14.13 *Suppose that D is convex and bounded, f satisfies* (14.53) *and g is bounded and measurable. Then, for $x \in D$, we have the representation*

$$u_{f,g}(x) = \mathbb{E}_x[f(\chi_N)] + \mathbb{E}_x\left[\sum_{n=0}^{N-1} U_1^{\ominus}[r_n^{\alpha} g(\chi_n + r_n \cdot)](0)\right],$$

where

$$U_a^{\ominus}[g](x) = \int_{|y|<a} g(y) U_a^{\ominus}(x, \mathrm{d}y).$$

As before, this suggests that the walk-on-spheres algorithm provides us a Monte Carlo method for evaluating $u_{f,g}$ via the Strong Law of Large Numbers;

the Central Limit Theorem again gives us the rate of convergence. Indeed, assuming that D is convex and bounded, that f satisfies (14.53) and g is bounded and measurable, we have

$$\lim_{n\to\infty} \frac{1}{n}\sum_{i=1}^{n} \omega^i_{f,g} = u_{f,g}(x),$$

almost surely, where $\omega^i_{f,g}$, $i \geq 1$, are i.i.d. copies of

$$\omega_{f,g} = f(\chi_N) + \sum_{n=0}^{N-1} U_1^{\ominus}[r_n^\alpha g(\chi_n + r_n\cdot)](0).$$

Moreover, when additionally $\int_{\mathbb{R}^d} f(x)^2(1+|x|^{\alpha+d})^{-1}\,\mathrm{d}x < \infty$. then $\mathrm{Var}(\omega_{f,g}) < \infty$ and in the sense of weak convergence,

$$n^{1/2}\left(\frac{1}{n}\sum_{i=1}^{n}\omega^i_{f,g} - u_{f,g}(x)\right) \overset{d}{\Longrightarrow} \mathcal{N}(0, \mathrm{Var}(\omega_{f,g})).$$

Proof of Lemma 14.13 Given the walk-on-spheres $(\chi_n, n \leq N)$ with $\chi_0 = x \in D$, define σ_n jointly with χ_n so that, given χ_{n-1}, (χ_n, σ_n) is equal in law to $(X_{\tau_{B_n}}, \tau_{B_n})$ under $\mathbb{P}_{\chi_{n-1}}$. We can now represent the second expectation on the right-hand side of (14.59) in the form

$$\mathbb{E}_x\left[\sum_{n\geq 0}\mathbf{1}_{(\chi_n\in D)}\int_0^{\sigma_{n+1}} g(\chi_n + X_s^{(n+1)})\,\mathrm{d}s\right], \qquad x \in D, \tag{14.60}$$

where $X^{(n)}$, $n \geq 0$, are independent copies of $(X, \mathbb{P}_0)$. Applying Fubini's Theorem, then conditioning each expectation on $\mathcal{F}_n := \sigma(\chi_k : k \leq n)$ followed by Fubini's Theorem again, we have

$$\begin{aligned}
\mathbb{E}_x\left[\int_0^{\tau_D} g(X_s)\,\mathrm{d}s\right] &= \sum_{n\geq 0}\mathbb{E}_x\left[\mathbf{1}_{(\chi_n\in D)}\,\mathbb{E}_y\left[\int_0^{\tau_{B_r}} g(X_s)\,\mathrm{d}s\right]\Bigg|_{y=\chi_n, r=r_n}\right]\\
&= \sum_{n\geq 0}\mathbb{E}_x\left[\mathbf{1}_{(\chi_n\in D)} U_{r_n}^{\ominus}[g](\chi_n)\right]\\
&= \mathbb{E}_x\left[\sum_{n=0}^{N-1} U_{r_n}^{\ominus}[g](\chi_n)\right].
\end{aligned}$$

The proof is completed once we show that $U_r^{\ominus}[g](x) = r^\alpha U_1^{\ominus}[g(x + r\cdot)](0)$, for $r > 0$, $x \in \mathbb{R}^d$ and bounded, measurable g. To this end, we appeal to spatial homogeneity and familiar computations using the scaling property of stable processes to compute

$$\begin{aligned} U_r^{\ominus}[g](x) &= \mathbb{E}_x\left[\int_0^{\tau_r^{\ominus}} g(X_t)\,\mathrm{d}t\right] \\ &= \mathbb{E}_0\left[\int_0^{\tau_r^{\ominus}} g(x+X_t)\,\mathrm{d}t\right] \\ &= \mathbb{E}_0\left[\int_0^{\tau_1^{\ominus}} r^{\alpha}\, g(x+rX_s)\,\mathrm{d}s\right] \\ &= r^{\alpha}\, U_1^{\ominus}[g(x+r\,\cdot))](0), \end{aligned}$$

which completes the proof. □

14.6 Comments

The sphere inversions discussed in Sections 14.1 and 14.2 are classical techniques from potential analysis of both Brownian motion and stable processes. Of many possible published texts in this field, we refer to the monographs of Port and Stone [173] for the former and Bliedtner and Hansen [36] for the latter. Section 14.3 is based on Port [172], albeit with a modern interpretation of the proofs, taking advantage of the Lamperti representation of stable processes as given in Kyprianou [125]. In the case of Theorem 14.1 we opt for a slightly different representation of the sphere hitting probability, working directly with hypergeometric functions rather than Legendre functions as originally given in Port [172]. The integral identities (14.19) and (14.20), used in the proof of Theorem 14.3, can be found in formula 3.655.2 of Gradshteyn and Ryzhik [82] and Blumenson [37], respectively. Section 14.4 is based on Blumenthal et al. [39], again with a modern interpretation coming from Bliedtner and Hansen [36] and Kyprianou [125].

The basic idea of the 'Walk-on-Spheres' algorithm originates in the setting of Brownian motion from the classical paper of Muller [155]. Unlike the stable setting the Brownian version of the algorithm does not end after a finite number of steps. As the Brownian motion approaches the boundary of the domain D, the sequence of inscribed spheres will typically become arbitrarily small in size (not necessarily in a monotone way). In order to bring the algorithm to an end, Muller suggests that once the walk-on-spheres algorithm comes within $\varepsilon > 0$ of the boundary, then it should end and adopt the closest point on the boundary as its terminal position. Muller [155] and Motoo [154] proved that with the aforementioned stopping on hitting an ε-thickening of the boundary, the mean number of steps in the algorithm is of order $|\log \varepsilon|$.

Since this foundational work, a significant volume of literature has evolved, largely in the numerical analysis domain, addressing variants of this problem as well as new proofs of the original Muller–Motoo results. The literature is far to extensive to give a complete list here, but some key references include [33, 93, 94, 151, 186, 187]. Moreover, applications of this method in the diffusive setting are also quite significant. Examples include turbulent fluids, contaminant transport in fractured rocks, chaotic dynamics and disordered quantum ensembles; see [112, 113, 192]. The theory for walk-on-spheres algorithm in the stable setting that we describe here naturally presents significantly different behaviour on account of the inclusion of jumps. The results discussed in Section 14.5 appeared in Kyprianou et al. [126].

15

Applications of radial excursion theory

In this chapter, we continue our analysis of fluctuations of the isotropic stable processes in dimension $d \geq 2$ with index $\alpha \in (0, 2)$. We consider the notion of excursions from the radial minimum and how this informs the distribution of the point of closest reach to the origin within the range of X. In turn, this allows us to look at the problem of first passage into and out of the unit sphere in greater detail, including more information about the law of the pre-entry (resp. pre-exit) position as well as the point of closest reach prior to first entry (resp. first exit). In the spirit of Theorem 13.2 we also consider the d-dimensional stable process reflected in its radial maximum.

15.1 Radial excursions

One of the principal tools that we will use in our computations is that of radial excursion theory of X from its running minimum, and similarly from its running maximum. Recall that X is a self-similar Markov process whose underlying MAP (ξ, Θ), via the general Lamperti–Kiu decomposition (11.26), has probabilities written $\mathbf{P} = (\mathbf{P}_{x,\theta},\ x \in \mathbb{R}, \theta \in \mathbb{S}^{d-1})$. Moreover, $|X|$ is a positive self-similar Markov process whose underlying Lévy process is ξ.

In the spirit of the discussion in Section 11.3, by considering, say, $\ell = (\ell_t, t \geq 0)$, the local time at 0 of the reflected Lévy process $\xi_t - \underline{\xi}_t$, $t \geq 0$, where $\underline{\xi}_t := \inf_{s \leq t} \xi_s$, $t \geq 0$, we can identify the descending ladder MAP (H^-_t, Θ^-_t), $t \geq 0$, of the process (ξ, Θ) in the obvious way. That is, we write

$$(H^-_t, \Theta^-_t) = (-\xi_{\ell^{-1}_t}, \Theta_{\ell^{-1}_t}), \qquad t < \ell_\infty,$$

and otherwise, we send (H^-_t, Θ^-_t) to a cemetery state, say $(\infty, \dagger)$. It is worth noting that, in the language of killed processes that we have used in previous chapters, we have explicitly identified the killing time as ℓ_∞.

Because ξ alone is a Lévy process, it is also the case that H^-, alone, is a subordinator. Hence, although the local time ℓ pertains to the reflected Lévy process $\xi - \underline{\xi}$, it serves as an adequate choice for the local time of the Markov process $(\xi - \underline{\xi}, \Theta)$ on the set $\{0\} \times \mathbb{S}^{d-1}$.

More precisely, suppose we define

$$\mathsf{g}_t = \sup\{s < t \colon \xi_s = \underline{\xi}_s\} \quad \text{and} \quad \mathsf{d}_t = \inf\{s > t \colon \xi_s = \underline{\xi}_s\},$$

which code the left and right end points of excursions of ξ from its minimum, respectively. The regularity of ξ for $(-\infty, 0)$ and $(0, \infty)$ ensures that both g_t and d_t are well defined for $t \geq 0$, as is $\mathsf{g}_\infty = \lim_{t\to\infty} \mathsf{g}_t$. For all $t > 0$ such that $\mathsf{d}_t > \mathsf{g}_t$ the process

$$(\epsilon_{\mathsf{g}_t}(s), \Theta^{\epsilon}_{\mathsf{g}_t}(s)) := (\xi_{\mathsf{g}_t+s} - \xi_{\mathsf{g}_t}, \Theta_{\mathsf{g}_t+s}), \qquad 0 \leq s \leq \varsigma_{\mathsf{g}_t} := \mathsf{d}_t - \mathsf{g}_t,$$

codes the excursion of $(\xi - \underline{\xi}, \Theta)$ from the set $(0, \mathbb{S}^{d-1})$ which straddles time t. Such excursions live in the space of $\mathcal{U}(\mathbb{R} \times \mathbb{S}^{d-1})$, the space of càdlàg paths $((\epsilon(s), \Theta^{\epsilon}(s)), 0 \leq s \leq \varsigma)$ with lifetime $\varsigma = \inf\{s > 0 \colon \epsilon(s) < 0\} > 0$ such that $(\epsilon(0), \Theta^{\epsilon}(0)) \in \{0\} \times \mathbb{S}^{d-1}$, $(\epsilon(s), \Theta^{\epsilon}(s)) \in (0, \infty) \times \mathbb{S}^{d-1}$, for $0 < s < \varsigma$, and $\epsilon(\varsigma) \in (-\infty, 0)$.

The excursions of $(\xi - \underline{\xi}, \Theta)$ form a Markov additive Poisson point process; see Definition A.19 in the Appendix. In the current setting, we can describe it using Maisonneuve's theory of so-called exit systems.

For $t > 0$, let $R_t = \mathsf{d}_t - t$, and define the set

$$G = \{t > 0 \colon R_{t-} = 0, R_t > 0\} = \{\mathsf{g}_s \colon s \geq 0\}$$

of the left extrema of excursions from 0 for $\xi - \underline{\xi}$. Necessarily G is a countable set of times. Recall that, we used $(\mathcal{F}_t, t \geq 0)$ to denote the filtration generated by X. However, for convenience and precision of what shortly follows, we will insist that $(\mathcal{F}_t, t \geq 0)$ is naturally enlarged (which thus implies it is right-continuous, see Remark A.13 in the Appendix).

By an additive functional $A = (A_t, t \geq 0)$ with respect to $(\mathcal{F}_t, t \geq 0)$, we mean that $A_0 = 0$, $t \mapsto A_t$ is non-decreasing, A is right-continuous, A_t is $\mathcal{F}_t$-measurable and, for $s, t \geq 0$, $A_{t+s} = A_t + \vartheta_t \circ A_s$, where $(\vartheta_t, t \geq 0)$, is the Markov shift operator, with each of these statements holding $\mathbf{P}$-almost surely.

Maisonneuve's theory of exit systems now implies that there exist an *additive functional*, $A = (A_t, t \geq 0)$, supported by the closure of the times $\{t \geq 0 \colon (\xi_t - \underline{\xi}_t, \Theta_t) \in \{0\} \times \mathbb{S}^{d-1}\}$, with $\mathbf{E}_{x,\theta}[\int_0^\infty \mathrm{e}^{-t} \mathrm{d}A_t] < \infty$, for all $x \in \mathbb{R}$, $\theta \in \mathbb{S}^{d-1}$, and a family of *excursion measures*, $(\mathbb{N}_\theta, \theta \in \mathbb{S}^{d-1})$, such that

(i) the map $\theta \mapsto \mathbb{N}_\theta$ is a kernel from $(\mathbb{S}^{d-1}, \mathcal{B}(\mathbb{S}^{d-1}))$ to the Skorokhod space of $\mathbb{R} \times \mathbb{S}^{d-1}$, cf. Section A.10 in the Appendix, which is concentrated on $\underline{\mathcal{U}}(\mathbb{R} \times \mathbb{S}^{d-1})$;

(ii) $\mathbb{N}_\theta(1 - e^{-\varsigma}) < \infty$;

(iii) for $x \neq 0$ and $\theta \in \mathbb{S}^{d-1}$, we have the *exit formula*

$$\mathbf{E}_{x,\theta}\left[\sum_{t\in G} F((\xi_s, \Theta_s), s < g)\, H(\epsilon_t, \Theta^\epsilon_t)\right]$$
$$= \mathbf{E}_{x,\theta}\left[\int_0^\infty F((\xi_s, \Theta_s), s < g)\, \mathbb{N}_{\Theta_t}(H(\epsilon, \Theta^\epsilon))\mathrm{d}A_t\right], \qquad (15.1)$$

where F is continuous and H is measurable on the space of càdlàg paths $\mathcal{D}(\mathbb{R} \times \mathbb{S}^{d-1})$, see Section A.10 in the Appendix;

(iv) under any measure $\mathbb{N}_\theta$ the process $(\epsilon(s)), \Theta^\epsilon(s), s < \varsigma)$ is Markovian with the same semigroup as (ξ, Θ) killed at its first hitting time of $(-\infty, 0] \times \mathbb{S}^{d-1}$.

Let us make a number of remarks concerning (15.1). First, the choice of additive functional Λ is not uniquely defined. A different choice of A will result in a different definition of the excursion measures $(\mathbb{N}_\theta, \theta \in \mathbb{S}^{d-1})$. That said, there is a natural choice we can make here, which is $A = \ell$.

Second, the reader should be careful not to confuse the index t of $(\epsilon_t(s), \Theta^\epsilon_t(s))$ with the index s. The former is the local time to which the excursion is associated, the latter is the time into the excursion. As such,

$$(\epsilon_t, \Theta^\epsilon_t) = ((\epsilon_t(s), \Theta^\epsilon_t(s)), s \leq \varsigma_t)$$

is the entire excursion whose left end point begins at real time ℓ^{-1}_{t-}, where ς_t is the excursion lifetime.

Finally, we also note that thanks to path regularity of ξ for the upper and lower half-line, every excursion indexed by $t \in G$ is an accumulation point of increasing index times t_i in G. That is to say, for any ε such that $t - \varepsilon > 0$, if $\{t_i : i \geq 1\} = G \cap (t - \varepsilon, t)$, then $\sup_{i\geq 1} t_i = t$.

Taking account of the Lamperti–Kiu transform (11.26), it is natural to consider how the excursion of $(\xi - \underline{\xi}, \Theta)$ from $\{0\} \times \mathbb{S}^{d-1}$ translates into what we might refer to as a *radial excursion theory* for the process

$$\Upsilon_t := e^{\xi_t}\Theta_t, \qquad t \geq 0. \qquad (15.2)$$

To see why 'radial excursion theory' is an appropriate terminology here, if we ignore the time change in (11.26), we see that the radial minima of the process Υ agree with the radial minima of the stable process X. Indeed, there is a bijection between excursions of $(\xi - \underline{\xi}, \Theta)$ from $\{0\} \times \mathbb{S}^{d-1}$ and excursions of $(\Upsilon_t / \inf_{s\le t} |\Upsilon_s|, t \ge 0)$, from $\mathbb{S}^{d-1}$, or equivalently excursions of Υ from its running radial infimum. Moreover, we see that, for all $t > 0$ such that $\mathtt{d}_t > \mathtt{g}_t$,

$$\Upsilon_{\mathtt{g}_t+s} = \mathrm{e}^{\xi_{\mathtt{g}_t}} \mathrm{e}^{\epsilon_{\mathtt{g}_t}(s)} \Theta^{\epsilon}_{\mathtt{g}_t}(s) = |\Upsilon_{\mathtt{g}_t}| \mathrm{e}^{\epsilon_{\mathtt{g}_t}(s)} \Theta^{\epsilon}_{\mathtt{g}_t}(s), \qquad s \le \varsigma_{\mathtt{g}_t}.$$

This will be useful to keep in mind in the forthcoming excursion computations.

The exit system (15.1) for the process (15.2) will turn out to be important for developing the main results of this chapter. Let us provide some further motivating discussion in this respect. From Theorem 5.21 we know explicitly the Wiener–Hopf factorisation of the Lévy process ξ. Moreover, referring to the discussion in Sections 4.3 and 4.1, it is not difficult to see that both the ascending and descending ladder height processes of ξ are subordinators with infinite jump activity. This implies the previously mentioned fact that ξ is regular for both the upper and lower half-lines. In turn, this means that when X is started from any point on $\mathbb{S}^{d-1}(0, r)$, it instantaneously visits both the interior and the exterior of $\mathbb{S}^{d-1}(0, r)$ almost surely.

These facts, together with the property that paths are right-continuous with left-limits, ensure that, with

$$\mathtt{G}(t) := \inf\{s \ge 0 \colon |X_s| = \inf_{u\le s} |X_u|\}, \qquad t \ge 0,$$

the quantity $X_{\mathtt{G}(t)}$ is well defined as the point of closest reach to the origin up to time t in the sense that $X_{\mathtt{G}(t)-} = X_{\mathtt{G}(t)}$ and

$$|X_{\mathtt{G}(t)}| = \inf_{s\le t} |X_s|.$$

The process $(\mathtt{G}(t), t \ge 0)$ is increasing and hence

$$\mathtt{G}(\infty) = \lim_{t\to\infty} \mathtt{G}(t)$$

exists almost surely. Moreover, $\mathtt{G}(\infty) = \mathtt{G}(t)$ for all t sufficiently large as $\lim_{t\to\infty} |X_t| = \infty$ thanks to transience of stable processes in high dimension. Hence

$$|X_{\mathtt{G}(\infty)}| = \inf_{s\ge 0} |X_s|.$$

We can now use (15.1) to develop distributional identities for $X_{\mathtt{G}(\infty)}$. We have, for bounded measurable f on $\mathbb{R}^d$,

$$
\begin{aligned}
\mathbb{E}_x[f(X_{\mathsf{G}(\infty)})] &= \mathbf{E}_{\log|x|,\arg(x)}\left[\sum_{t\in G} f(\mathrm{e}^{\xi_t}\Theta_t)\mathbf{1}_{(\varsigma_t=\infty)}\right] \\
&= \mathbf{E}_{\log|x|,\arg(x)}\left[\int_0^\infty f(\mathrm{e}^{\xi_t}\Theta_t)\mathbb{N}_{\Theta_t}(\varsigma=\infty)\mathrm{d}\ell_t\right] \\
&= \mathbf{E}_{\log|x|,\arg(x)}\left[\int_0^{\ell_\infty} f(\mathrm{e}^{-H_t^-}\Theta_t^-)\mathbb{N}_{\Theta_t^-}(\varsigma=\infty)\mathrm{d}t\right] \\
&= \int_{|z|<|x|} U_x^-(\mathrm{d}z)f(z)\mathbb{N}_{\arg(z)}(\varsigma=\infty), \qquad (15.3)
\end{aligned}
$$

where

$$
U_x^-(\mathrm{d}z) := \int_0^\infty \mathbf{P}_{\log|x|,\arg(x)}(\mathrm{e}^{-H_t^-}\Theta_t^- \in \mathrm{d}z,\ t<\ell_\infty)\mathrm{d}t, \qquad |z|\le|x|.
$$

Thanks, again, to the transience of X, which implies that $\lim_{t\to\infty}\xi_t=\infty$, we know that (H^-,Θ^-) experiences killing at a rate that occurs, in principle, in a state-dependent manner, specifically $\mathbb{N}_\theta(\varsigma=\infty)$, $\theta\in\mathbb{S}^{d-1}$. Isotropy allows us to deduce that all such rates take a common value and, thanks to the arbitrary scaling of local time ℓ, we can choose this common value to be unity. Said another way, ℓ_∞ is exponentially distributed with rate 1.

In conclusion, we reach the identity

$$
\mathbb{E}_x[f(X_{\mathsf{G}(\infty)})] = \int_{|z|<|x|} U_x^-(\mathrm{d}z)f(z) \qquad (15.4)
$$

or equivalently, the law of $X_{\mathsf{G}(\infty)}$ under $\mathbb{P}_x$, $x\neq 0$, is nothing more than the quantity $U_x^-(\mathrm{d}z)$, $|z|\le|x|$. From this analysis and the conclusion of Lemma 14.9, we also get another handy identity which will soon be of use. For $r<|x|$, $\mathbb{P}_x(\tau_r^\oplus=\infty)=\mathbb{P}_x(|\mathsf{G}(\infty)|>r)$, where

$$
\tau_r^\oplus = \inf\{t>0\colon |X_t|<r\},
$$

and hence

$$
\begin{aligned}
\mathbb{P}_x(\tau_r^\oplus=\infty) &= \int_{r<|z|<|x|} U_x^-(\mathrm{d}z) \\
&= \frac{\Gamma(d/2)}{\Gamma((d-\alpha)/2)\Gamma(\alpha/2)}\int_0^{(|x|^2/r^2)-1}(u+1)^{-d/2}u^{\alpha/2-1}\mathrm{d}u. \qquad (15.5)
\end{aligned}
$$

Another identity where we gain some insight into the quantity U_x^- is the first passage result in Theorem 14.7. For example, the following identity emerges very quickly from (15.1). Indeed, for bounded measurable functions f,g on $\mathbb{R}^d$, noting that the range of Υ and X are the same,

$$\mathbb{E}_x[g(X_{G(\tau_1^\oplus-)})f(X_{\tau_1^\oplus}); \tau_1^\oplus < \infty]$$
$$= \mathbf{E}_{\log|z|,\arg(x)}\left[\sum_{t\in G} \mathbf{1}_{(e^{\xi_t}>1)}\mathbf{1}_{(e^{\xi_t+\epsilon_t(\varsigma_t)}<1)}g(e^{\xi_t}\Theta_t)f(e^{\xi_t+\epsilon_t(\varsigma_t)}\Theta^{\epsilon_t}(\varsigma_t))\right]$$
$$= \int_{1<|z|<|x|} U_x^-(\mathrm{d}z)\int_{|y||z|<1} \mathbb{N}_{\arg(z)}(e^{\epsilon(\varsigma)}\Theta^{\epsilon}(\varsigma) \in \mathrm{d}y; \varsigma < \infty)g(z)f(|z|y). \quad (15.6)$$

With judicious computations, by restricting the computation above to the case $g \equiv 1$, one might expect to be able to extract an identity for U_x^- from (14.32). For example, developing (15.6) we might write, for $|x| > 1$ and bounded measurable f on $[0, \infty)$,

$$\mathbb{E}_x[f(|X_{\tau_1^\oplus}|); \tau_1^\oplus < \infty]$$
$$= \int_{1<|z|<|x|} U_x^-(\mathrm{d}z)\int_{y>\log|z|} \mathbb{N}_{\arg(z)}(|\epsilon(\varsigma)| \in \mathrm{d}y; \varsigma < \infty)f(|z|e^{-y})$$
$$= \int_{1<|z|<|x|} U_x^-(\mathrm{d}z)\int_{y>\log|z|} \nu(\mathrm{d}y)f(|z|e^{-y}), \quad (15.7)$$

where we have appealed to isotropy to ensure that $\nu(\mathrm{d}y) := \mathbb{N}_{\arg(z)}(|\epsilon(\varsigma)| \in \mathrm{d}y, \varsigma < \infty)$ does not depend on the value of z. Since, for each excursion ϵ, $|\epsilon(\varsigma)|$ on $\{\varsigma < \infty\}$ is the magnitude of the (downwards) increment in the logarithm of the radial minimum, we also note that ν is also the Lévy measure of the subordinator H^-.

Recall from Theorem 5.21 that the Wiener–Hopf factorisation for ξ is known, namely

$$\Psi(\theta) = \frac{\Gamma(\frac{1}{2}(-\mathrm{i}\theta + \alpha))}{\Gamma(-\frac{1}{2}\mathrm{i}\theta)} \times \frac{\Gamma(\frac{1}{2}(\mathrm{i}\theta + d))}{\Gamma(\frac{1}{2}(\mathrm{i}\theta + d - \alpha))}, \qquad \theta \in \mathbb{R}. \quad (15.8)$$

The normalisation of ℓ (i.e. the choice that ℓ_∞ is exponential with rate 1) is equivalent to the requirement that $\Phi^-(0) = 1$, where Φ^- is the Laplace exponent of H^- and hence

$$\Phi^-(\lambda) = \int_{(0,\infty)} (1-e^{-\lambda y})\nu(\mathrm{d}y) = \frac{\Gamma((d-\alpha)/2)\Gamma((\lambda+d)/2)}{\Gamma(d/2)\Gamma((\lambda+d-\alpha)/2)}, \qquad \lambda \geq 0. \quad (15.9)$$

In the spirit of Proposition 4.1, we can invert the identity 15.9 to find ν. For the sake of brevity, the details are left to the reader and we have

$$\nu(\mathrm{d}y) = \frac{\alpha\Gamma((d-\alpha)/2)}{\Gamma(d/2)\Gamma(1-\alpha/2)}(1-e^{-2y})^{-\frac{\alpha}{2}-1}e^{-dy}\mathrm{d}y, \qquad y > 0. \quad (15.10)$$

Despite the fact that the left-hand sides of (15.7) and (15.10) are explicitly available, it seems difficult to back out an expression for the measure U_x^- in a straightforward manner. Nonetheless, with careful approximation, we will do precisely this in the next section.

15.2 The Point of closest reach to the origin

Our objective here is to make the identity (15.4) explicit and, thereby, we will obtain a concrete formula for the measure $U_x^-(\mathrm{d}y)$, $|y| < |x|$, which is stated as a corollary immediately after the main result of this section below.

Theorem 15.1 *The law of the point of closest reach to the origin is given by*

$$\mathbb{P}_x(X_{\mathtt{G}(\infty)} \in \mathrm{d}y) = \pi^{-d/2}\frac{\Gamma(d/2)^2}{\Gamma((d-\alpha)/2)\Gamma(\alpha/2)}\frac{(|x|^2-|y|^2)^{\alpha/2}}{|x-y|^d|y|^\alpha}\mathrm{d}y,$$

for $|y| < |x|$.

Corollary 15.2 *For all* $|x| > 0$,

$$U_x^-(\mathrm{d}y) = \pi^{-d/2}\frac{\Gamma(d/2)^2}{\Gamma((d-\alpha)/2)\Gamma(\alpha/2)}\frac{(|x|^2-|y|^2)^{\alpha/2}}{|y|^\alpha|y-x|^d}, \qquad |y| < |x|. \tag{15.11}$$

Before embarking on the proof of Theorem 15.1, let us introduce some notation. First define, for $x \neq 0$, $|x| > r$, $\delta > 0$ and continuous, positive and bounded f on $\mathbb{R}^d$,

$$\Delta_r^\delta f(x) := \frac{1}{\delta}\mathbb{E}_x\left[f(\arg(X_{\mathtt{G}_\infty})), |X_{\mathtt{G}_\infty}| \in [r-\delta, r]\right]. \tag{15.12}$$

The crux of our proof will be to establish a limit of $\Delta_r^\delta f(x)$ in concrete terms as $\delta \to 0$.

Note that, by conditioning on first entry into the ball of radius r, with the help of the first entrance law (14.32) and (15.4), we have

$$\begin{aligned}
&\Delta_r^\delta f(x)\\
&= \frac{1}{\delta}\int_{|y|\in[r-\delta,r]} \mathbb{P}_x(X_{\tau_r^\oplus} \in \mathrm{d}y;\ \tau_r^\oplus < \infty)\mathbb{E}_y\left[f(\arg(X_{\mathtt{G}_\infty}));\ |X_{\mathtt{G}_\infty}| \in [r-\delta, |y|]\right]\\
&= \frac{1}{\delta}C_{\alpha,d}\int_{|y|\in[r-\delta,r]} \mathrm{d}y\left|\frac{r^2-|x|^2}{r^2-|y|^2}\right|^{\alpha/2}|y-x|^{-d}\mathbb{E}_y\left[f(\arg(X_{\mathtt{G}_\infty}));\ |X_{\mathtt{G}_\infty}| \in [r-\delta, |y|]\right]\\
&= \frac{1}{\delta}C_{\alpha,d}|r^2-|x|^2|^{\alpha/2}\int_{|y|\in(r-\delta,r]} \mathrm{d}y\frac{|y-x|^{-d}}{|r^2-|y|^2|^{\alpha/2}}\int_{r-\delta\le|z|\le|y|} U_y^-(\mathrm{d}z)f(\arg(z)),
\end{aligned} \tag{15.13}$$

where

$$C_{\alpha,d} = \pi^{-(d/2+1)}\Gamma(d/2)\sin\left(\frac{\pi\alpha}{2}\right).$$

Our next objective is to try and replace $\int_{r-\delta\le|z|\le|y|} U_y^-(\mathrm{d}z)f(\arg(z))$ by a simpler object which can be asymptotically estimated in the limit. To this end, we need some technical lemmas.

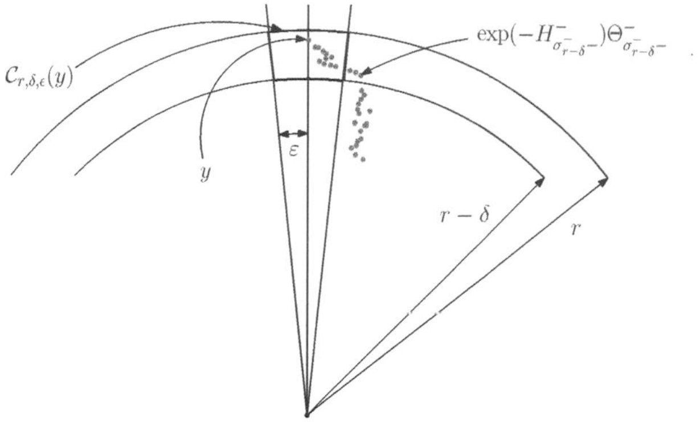

Figure 15.1 The process $\exp(-H^-)\Theta^-$ in relation to the domain $C_{r,\delta,\varepsilon}(y)$

Lemma 15.3 *Suppose that f is a bounded continuous function on $\mathbb{R}^d$ and $r>0$. Then*

$$\lim_{\delta\to 0}\sup_{|y|\in(r-\delta,r]}\left|\frac{\int_{r-\delta\le|z|\le|y|} U_y^-(\mathrm{d}z)f(z)}{\int_{r-\delta\le|z|\le|y|} U_y^-(\mathrm{d}z)} - f(y)\right| = 0.$$

Proof Suppose that $C_{r,\delta,\varepsilon}(y)$ is the geometric region which coincides with the intersection of a cone with axis of symmetry along y with radial extent ε, say C_ε, and the annulus $\{z\in\mathbb{R}^d : r-\delta\le|z|\le r\}$; see Figure 15.1. Choose ε,δ such that

$$\sup_{z\in C_{r,\delta,\varepsilon}(y)}|f(z)-f(y)|<\varepsilon',$$

for some choice of $\varepsilon'\ll 1$. We have

$$\begin{aligned}&\sup_{|y|\in(r-\delta,r]}\left|\frac{\int_{r-\delta\le|z|\le|y|} U_y^-(\mathrm{d}z)f(z)}{\int_{r-\delta\le|z|\le|y|} U_y^-(\mathrm{d}z)} - f(y)\right|\\ &\qquad\le\varepsilon' + 2\|f\|_\infty\sup_{|y|\in(r-\delta,r]}\frac{\int_{r-\delta\le|z|\le|y|} U_y^-(\mathrm{d}z)\mathbf{1}_{(z\notin C_{r,\delta,\varepsilon}(y))}}{\int_{r-\delta\le|z|\le|y|} U_y^-(\mathrm{d}z)}.\end{aligned}\qquad(15.14)$$

In order to deal with the second term on the right-hand side of (15.14), taking inspiration from the computations in (15.6) and (15.7), note that, for $|y| \in (r-\delta, r]$,

$$
\begin{aligned}
&\sup_{|y|\in(r-\delta,r]} \int_{r-\delta\le|z|\le|y|} U_y^-(\mathrm{d}z)\mathbf{1}_{(z\notin C_{r,\delta,\varepsilon}(y))}\nu\left(\log\left(\frac{|z|}{r-\delta}\right),\infty\right)\\
&\quad= \sup_{|y|\in(r-\delta,r]} \mathbb{P}_y(\exp(-H^-_{\sigma^-_{r-\delta}-})\Theta^-_{\sigma^-_{r-\delta}-} \notin C_{r,\delta,\varepsilon}(y),\ \sigma^-_{r-\delta}<\infty)\\
&\quad= \sup_{\beta\in(r-\delta,r]} \mathbb{P}_{\beta 1}(\exp(-H^-_{\sigma^-_{r-\delta}-})\Theta^-_{\sigma^-_{r-\delta}-} \notin C_{r,\delta,\varepsilon}(\beta 1),\ \sigma^-_{r-\delta}<\infty)\\
&\quad\le \sup_{\beta\in(r-\delta,r]} \mathbb{P}_{\beta 1}(\Theta^-_{\sigma^-_{r-\delta}-} \notin C_\varepsilon\cap\mathbb{S}^{d-1},\ \sigma^-_{r-\delta}<\infty)\\
&\quad\le \sup_{\beta\in(r-\delta,r]} \mathbb{P}_{\beta 1}(\upsilon_\varepsilon<\sigma^-_{r-\delta})\\
&\quad\le \mathbb{P}_{r1}(\upsilon_\varepsilon<\sigma^-_{r-\delta}),
\end{aligned}
\tag{15.15}
$$

where $1 = (1, 0, \ldots, 0)$ is the 'North Pole' on $\mathbb{S}^d$,

$$\sigma^-_{r-\delta} = \inf\{t>0\colon \exp(-H^-_t) < r-\delta\}$$

and

$$\upsilon_\varepsilon = \inf\{t>0\colon \Theta^-_t \notin C_\varepsilon\cap\mathbb{S}^{d-1}\}.$$

Right-continuity of paths now ensures that the right-hand side above tends to zero as $\delta \to 0$.

On the other hand, from (15.5),

$$\int_{r-\delta\le|z|\le|y|} U_y^-(\mathrm{d}z) = \mathbb{P}_y(\tau^\oplus_{r-\delta}=\infty) = \mathbb{P}_{\frac{|y|}{(r-\delta)}1}(\tau^\oplus_1=\infty), \tag{15.16}$$

where we have used isotropy and scaling in the final equality. From (15.5) and (15.10), a rather tedious computation shows that

$$
\begin{aligned}
&\lim_{\eta\downarrow 1}\nu(\log\eta,\infty)\,\mathbb{P}_{\eta 1}(\tau^\oplus_1=\infty)\\
&= \lim_{\eta\downarrow 1}\frac{\alpha}{\Gamma(\alpha/2)\Gamma(1-\alpha/2)}\left(\int_{\log\eta}^\infty (1-\mathrm{e}^{-2v})^{-\frac{\alpha}{2}-1}\mathrm{e}^{-dv}\mathrm{d}v\right)\left(\int_0^{\eta^2-1}(u+1)^{-d/2}u^{\alpha/2-1}\mathrm{d}u\right)\\
&= \frac{1}{\Gamma(1+\alpha/2)\Gamma(1-\alpha/2)}.
\end{aligned}
$$

Hence

$$
\begin{aligned}
&\lim_{\delta\to 0}\sup_{|y|\in(r-\delta,r]} \frac{\int_{r-\delta\le|z|\le|y|} U_y^-(\mathrm{d}z)\mathbf{1}_{(z\notin C_{r,\delta,\varepsilon}(y))}}{\int_{r-\delta\le|z|\le|y|} U_y^-(\mathrm{d}z)}\\
&\le \lim_{\delta\to 0}\sup_{|y|\in(r-\delta,r]} \frac{\int_{r-\delta\le|z|\le|y|} U_y^-(\mathrm{d}z)\mathbf{1}_{(z\notin C_{r,\delta,\varepsilon}(y))}\frac{\nu(\log(|z|/(r-\delta)),\infty)}{\nu(\log(|z|/(r-\delta)),\infty)}}{\int_{r-\delta\le|z|\le|y|} U_y^-(\mathrm{d}z)}\\
&\le \lim_{\delta\to 0}\sup_{|y|\in(r-\delta,r]} \frac{\int_{r-\delta\le|z|\le|y|} U_y^-(\mathrm{d}z)\mathbf{1}_{(z\notin C_{r,\delta,\varepsilon}(y))}\nu(\log(|z|/(r-\delta)),\infty)}{\nu(\log(|y|/(r-\delta)),\infty)\mathbb{P}_y(\tau^{\oplus}_{r-\delta}=\infty)}\\
&\le \lim_{\delta\to 0}\sup_{1<\eta<1+\frac{\delta}{(r-\delta)}} \frac{\mathbb{P}_{r1}(\upsilon_\varepsilon<\sigma^-_{r-\delta})}{\nu(\log\eta,\infty)\mathbb{P}_{\eta 1}(\tau^{\oplus}_1=\infty)}\\
&=0,
\end{aligned}
$$

and thus plugging this back into (15.14) gives the result. □

With Lemma 15.3 in hand, noting in particular the representation (15.16), we can now return to (15.13) and note that, for each $\varepsilon>0$, we can choose δ sufficiently small such that (recall the notation in (15.12))

$$
\begin{aligned}
\Delta^\delta_r f(x) &= D(\varepsilon)\Delta^\delta_r 1(x)\\
&\quad+\frac{1}{\delta}C_{\alpha,d}|r^2-|x|^2|^{\alpha/2}\int_{|y|\in(r-\delta,r]}\mathrm{d}y\frac{|y-x|^{-d}}{|r^2-|y|^2|^{\alpha/2}}f(\arg(y))\mathbb{P}_y(\tau^{\oplus}_{r-\delta}=\infty),
\end{aligned}
$$

where, $|D(\varepsilon)|<\varepsilon$ and for $|x|>r$,

$$
\begin{aligned}
&\limsup_{\delta\to 0}|\Delta^\delta_r 1(x)|\\
&\le \limsup_{\delta\to 0}\left|\frac{1}{\delta}C_{\alpha,d}|r^2-|x|^2|^{\alpha/2}\int_{|y|\in(r-\delta,r]}\mathrm{d}y\frac{|y-x|^{-d}}{|r^2-|y|^2|^{\alpha/2}}\mathbb{P}_y(\tau^{\oplus}_{r-\delta}=\infty)\right|\\
&=\limsup_{\delta\to 0}\left|\frac{1}{\delta}\left(\mathbb{P}_x(\tau^{\oplus}_{r-\delta}=\infty)-\mathbb{P}_x(\tau^{\oplus}_r=\infty)\right)\right|\\
&=\frac{\Gamma(d/2)}{\Gamma((d-\alpha)/2)\Gamma(\alpha/2)}\left|\frac{\mathrm{d}}{\mathrm{d}v}\int_0^{(|x|^2/v^2)-1}(u+1)^{-d/2}u^{\alpha/2-1}\mathrm{d}u\right|_{v=r}\\
&=\frac{2\Gamma(d/2)}{\Gamma((d-\alpha)/2)\Gamma(\alpha/2)}\left(|x|^2-r^2\right)^{\alpha/2-1}r^{d-1-\alpha}|x|^{2-d},
\end{aligned}
$$

where in the third equality we have used Lemma 14.9.

We can now say that, *if the limit exists,*

$$
\begin{aligned}
&\lim_{\delta\to 0}\Delta^{\delta}_{r}f(x)\\
&= \lim_{\delta\to 0} C_{\alpha,d}|r^2-|x|^2|^{\alpha/2}\frac{1}{\delta}\int_{|y|\in(r-\delta,r]}\mathrm{d}y\frac{|y-x|^{-d}}{|r^2-|y|^2|^{\alpha/2}}f(\arg(y))\mathbb{P}_y(\tau^{\oplus}_{r-\delta}=\infty)\\
&= \lim_{\delta\to 0}\frac{C_{\alpha,d}|r^2-|x|^2|^{\alpha/2}}{}\delta\int_{r-\delta}^{r}\varrho^{d-1}\mathrm{d}\varrho\int_{\mathbb{S}^{d-1}(0,\varrho)}\sigma_{\varrho}(\mathrm{d}\theta)\frac{|\varrho\theta-x|^{-d}}{|r^2-\varrho^2|^{\alpha/2}}f(\theta)\mathbb{P}_{\varrho\theta}(\tau^{\oplus}_{r-\delta}=\infty)\\
&= \lim_{\delta\to 0}\frac{C_{\alpha,d}|r^2-|x|^2|^{\alpha/2}}{\delta}\int_{r-\delta}^{r}\varrho^{d-1}\mathrm{d}\varrho\frac{\mathbb{P}_{\varrho 1}(\tau^{\oplus}_{r-\delta}=\infty)}{|r^2-\varrho^2|^{\alpha/2}}\int_{\mathbb{S}^{d-1}(0,\varrho)}\sigma_{\varrho}(\mathrm{d}\theta)|\varrho\theta-x|^{-d}f(\theta),
\end{aligned}
\tag{15.17}
$$

where, in the second equality, we have switched from the d-dimensional Lebesgue measure to the generalised polar measure $\varrho^{d-1}\mathrm{d}\varrho\times\sigma_{\varrho}(\mathrm{d}\theta)$, so that $\varrho>0$ is the radial distance from the origin and $\sigma_{\varrho}(\mathrm{d}\theta)$ is the surface measure on $\mathbb{S}^{d-1}(0,\varrho)$, normalised to have unit mass. In the third equality, we have used isotropy to write $\mathbb{P}_{\varrho\theta}(\tau^{\oplus}_{r-\delta}=\infty)=\mathbb{P}_{\varrho 1}(\tau^{\oplus}_{r-\delta}=\infty)$ for $\theta\in\mathbb{S}^{d-1}$.

Noting the continuity of the integral $\int_{\mathbb{S}^{d-1}(0,\varrho)}\sigma_{\varrho}(\mathrm{d}\theta)|\varrho\theta-x|^{-d}f(\theta)$ in ϱ, the proof of Theorem 15.1 is complete as soon as we can evaluate the limit

$$
\lim_{\delta\to 0}\frac{1}{\delta}\int_{r-\delta}^{r}\varrho^{d-1}\mathrm{d}\varrho\frac{\mathbb{P}_{\varrho 1}(\tau^{\oplus}_{r-\delta}=\infty)}{|r^2-\varrho^2|^{-\alpha/2}}
\tag{15.18}
$$

in (15.17). To this end, we need a technical lemma.

Lemma 15.4 *Let* $D_{\alpha,d}=\Gamma(d/2)/\Gamma((d-\alpha)/2)\Gamma(\alpha/2)$. *Then*

$$
\lim_{\delta\to 0}\sup_{\varrho\in[r-\delta,r]}\left|(\varrho^2-(r-\delta)^2)^{-\alpha/2}r^{\alpha}\mathbb{P}_{\varrho 1}(\tau^{\oplus}_{r-\delta}=\infty)-\frac{2D_{\alpha,d}}{\alpha}\right|=0.
$$

Proof Appealing again to Lemma 14.9, we start by noting that

$$
\begin{aligned}
&\sup_{\varrho\in[r-\delta,r]}\left|D_{\alpha,d}\int_0^{\varrho^2/(r-\delta)^2-1}u^{\alpha/2-1}\mathrm{d}u-\mathbb{P}_{\varrho 1}(\tau^{\oplus}_{r-\delta}=\infty)\right|\\
&\qquad\le\sup_{\varrho\in[r-\delta,r]}D_{\alpha,d}\int_0^{\varrho^2/(r-\delta)^2-1}\left|(1+u)^{-d/2}-1\right|u^{\alpha/2-1}\mathrm{d}u\\
&\qquad\le\sup_{\varrho\in[r-\delta,r]}D_{\alpha,d}\int_0^{\varrho^2/(r-\delta)^2-1}\left|1-\frac{(r-\delta)^d}{\varrho^d}\right|u^{\alpha/2-1}\mathrm{d}u\\
&\qquad\le D_{\alpha,d}\left|1-\frac{(r-\delta)^d}{r^d}\right|\frac{2}{\alpha}\left(r^2-(r-\delta)^2\right)^{\alpha/2}(r-\delta)^{-\alpha},
\end{aligned}
\tag{15.19}
$$

which tends to zero as $\delta \to 0$. Furthermore,

$$\begin{aligned}
&\sup_{\varrho\in[r-\delta,r]}\left|D_{\alpha,d}\int_0^{\varrho^2/(r-\delta)^2-1}u^{\alpha/2-1}\mathrm{d}u-\frac{2D_{\alpha,d}}{\alpha}(\varrho^2-(r-\delta))^{\alpha/2}r^{-\alpha}\right|\\
&\quad=\sup_{\varrho\in[r-\delta,r]}\frac{2D_{\alpha,d}}{\alpha}(\varrho^2-(r-\delta)^2)^{\alpha/2}\left|(r-\delta)^{-\alpha}-r^{-\alpha}\right|\\
&\quad\le\frac{2D_{\alpha,d}}{\alpha}(r^2-(r-\delta))^{\alpha/2}\left|(r-\delta)^{-\alpha}-r^{-\alpha}\right|,
\end{aligned}\tag{15.20}$$

which also tends to zero as $\delta \to 0$. Summing (15.19) and (15.20) in the context of the triangle inequality and dividing by $r^{-\alpha}(r^2-(r-\delta))^{\alpha/2}$, we can also deduce that

$$\lim_{\delta\to 0}\sup_{\varrho\in[r-\delta,r]}\left|(\varrho^2-(r-\delta)^2)^{-\alpha/2}r^{\alpha}\mathbb{P}_{\varrho 1}(\tau^{\oplus}_{r-\delta}=\infty)-\frac{2D_{\alpha,d}}{\alpha}\right|=0,$$

and the lemma is proved. □

We are now ready to evaluate (15.18), and identify its limit, thereby completing the proof of Theorem 15.1. Appealing to Lemma 15.4, for all $\varepsilon > 0$, there exists a δ sufficiently small,

$$\begin{aligned}
&\left|\frac{1}{\delta}\int_{r-\delta}^{r}\mathrm{d}\varrho\frac{\mathbb{P}_{\varrho 1}(\tau^{\oplus}_{r-\delta}=\infty)}{(r^2-\varrho^2)^{\alpha/2}}-\frac{2D_{\alpha,d}r^{-\alpha}}{\alpha}\frac{1}{\delta}\int_{r-\delta}^{r}\frac{(\varrho^2-(r-\delta)^2)^{\alpha/2}}{(r^2-\varrho^2)^{\alpha/2}}\mathrm{d}\varrho\right|\\
&\qquad<\frac{\varepsilon}{\delta}\int_{r-\delta}^{r}\frac{(\varrho^2-(r-\delta)^2)^{\alpha/2}}{(r^2-\varrho^2)^{\alpha/2}}\mathrm{d}\varrho.
\end{aligned}\tag{15.21}$$

Next note that

$$\begin{aligned}
&\lim_{\delta\to 0}\frac{1}{\delta}\int_{r-\delta}^{r}\frac{(\varrho^2-(r-\delta)^2)^{\alpha/2}}{(r^2-\varrho^2)^{\alpha/2}}\mathrm{d}\varrho\\
&=\lim_{\delta\to 0}\frac{1}{\delta}\int_{r-\delta}^{r}\left[\frac{\varrho-(r-\delta)}{r-\varrho}\right]^{\alpha/2}\left[\frac{\varrho+(r-\delta)}{r+\varrho}\right]^{\alpha/2}\mathrm{d}\varrho\\
&=\lim_{\delta\to 0}\int_0^1\left[\frac{u}{1-u}\right]^{\alpha/2}\left[\frac{2r-2\delta+\delta u}{2r-\delta+\delta u}\right]^{\alpha/2}\mathrm{d}u\\
&=\int_0^1(1-u)^{-\alpha/2}u^{\alpha/2}\mathrm{d}u\\
&=\Gamma(1-\alpha/2)\Gamma(1+\alpha/2),
\end{aligned}\tag{15.22}$$

where we have used the substitution $\varrho = (r-\delta)+u\delta$ in the second equality and dominated convergence in the third.

Proof of Theorem 15.1 Putting the pieces together, we can take limits in (15.21) using (15.22) to deduce that

$$\lim_{\delta\to 0}\frac{1}{\delta}\int_{r-\delta}^{r} \mathrm{d}\varrho \frac{\mathbb{P}_{\varrho 1}(\tau^{\oplus}_{r-\delta}=\infty)}{(r^2-\varrho^2)^{\alpha/2}} = \frac{2}{\alpha}D_{\alpha,d}\Gamma(1-\alpha/2)\Gamma(1+\alpha/2)r^{-\alpha},$$

which, in turn, can be plugged into (15.17) and we find that

$$\begin{aligned}
&\lim_{\delta\to 0}\Delta^{\delta}_{r}f(x)\\
&=\frac{2}{\alpha}D_{\alpha,d}\Gamma(1-\alpha/2)\Gamma(1+\alpha/2)C_{\alpha,d}\\
&\qquad\times r^{d-\alpha-1}|r^2-|x|^2|^{\alpha/2}\int_{\mathbb{S}^{d-1}(0,r)}\sigma_r(\mathrm{d}\theta)|r\theta-x|^{-d}f(\theta)\\
&=\pi^{-d/2}\frac{\Gamma(d/2)^2}{\Gamma((d-\alpha)/2)\Gamma(\alpha/2)}r^{d-\alpha-1}|r^2-|x|^2|^{\alpha/2}\int_{\mathbb{S}^{d-1}(0,r)}\sigma_r(\mathrm{d}\theta)|r\theta-x|^{-d}f(\theta).
\end{aligned}$$

Now suppose that g is another bounded measurable function on $[0,\infty)$, then

$$\begin{aligned}
&\mathbb{E}_x[g(|X_{\mathsf{G}(\infty)}|)f(\arg(X_{\mathsf{G}(\infty)}))]\\
&=\pi^{-d/2}\frac{\Gamma(d/2)^2}{\Gamma((d-\alpha)/2)\Gamma(\alpha/2)}\int_0^{|x|}\int_{\mathbb{S}^{d-1}(0,r)} r^{d-1}\mathrm{d}r\,\sigma_r(\mathrm{d}\theta)\frac{|r^2-|x|^2|^{\alpha/2}}{r^{\alpha}|r\theta-x|^d}f(\theta)g(r)\\
&=\pi^{-d/2}\frac{\Gamma(d/2)^2}{\Gamma((d-\alpha)/2)\Gamma(\alpha/2)}\int_{|y|<|x|}\frac{||y|^2-|x|^2|^{\alpha/2}}{|y|^{\alpha}|y-x|^d}f(\arg(y))g(|y|),
\end{aligned}$$

which is equivalent to the statement of Theorem 15.1. □

Let us now return to the problem of first entrance into and first exit from a ball. From the proof of Theorem 14.7, we saw that these two problems are essentially the same thanks to the Riesz–Bogdan–Żak transform. In this section we aim to enrich one (and hence both) of these problems by looking at a more detailed distributional spatial analysis of how first passage occurs. Recall the first passage times

$$\tau^{\oplus}_a := \inf\{t>0\colon |X_t|<a\} \text{ and } \tau^{\ominus}_a := \inf\{t>0\colon |X_t|>a\}.$$

Theorem 15.5 (Triple law at first entrance/exit of a ball) *Fix $r>0$ and define, for $x,z,y,v\in\mathbb{R}^d\backslash\{0\}$,*

$$w_x(z,y,v) := \pi^{-3d/2}\frac{\Gamma((d+\alpha)/2)}{|\Gamma(-\alpha/2)|}\frac{\Gamma(d/2)^2}{\Gamma(\alpha/2)^2}\frac{||z|^2-|x|^2|^{\alpha/2}||y|^2-|z|^2|^{\alpha/2}}{|z|^{\alpha}|z-x|^d|z-y|^d|v-y|^{\alpha+d}}.$$

(i) Write

$$\mathsf{G}(\tau^{\oplus}_r-) = \inf\{s<\tau^{\oplus}_r\colon |X_s| = \inf_{u<\tau^{\oplus}}|X_u|\},$$

for the time of the point of closest reach from the origin immediately before first entry into $\mathbb{S}^{d-1}(0,r)$. *For* $|x| > |z| > r$, $|y| > |z|$ *and* $|v| < r$,

$$\mathbb{P}_x(X_{\mathtt{G}(\tau_r^\oplus-)} \in \mathrm{d}z,\ X_{\tau_r^\oplus-} \in \mathrm{d}y,\ X_{\tau_r^\oplus} \in \mathrm{d}v;\ \tau_r^\oplus < \infty) = w_x(z,y,v)\,\mathrm{d}z\,\mathrm{d}y\,\mathrm{d}v.$$

(ii) Define $\mathtt{H}(t) = \inf\{s \le 0\colon |X_s| = \sup_{u\le t}|X_u|\}$, $t \ge 0$, *and write*

$$\mathtt{H}(\tau_r^\ominus-) = \inf\{s < \tau_r^\ominus : |X_s| = \sup_{u<\tau_r^\ominus}|X_u|\}$$

for the point of furthest reach from the origin immediately before first exit from $\mathbb{S}^{d-1}(0,r)$. *For* $|x| < |z| < r$, $|y| < |z|$ *and* $|v| > r$,

$$\mathbb{P}_x(X_{\mathtt{H}(\tau_r^\ominus-)} \in \mathrm{d}z,\ X_{\tau_r^\ominus-} \in \mathrm{d}y,\ X_{\tau_r^\ominus} \in \mathrm{d}v) = w_x(z,y,v)\,\mathrm{d}z\,\mathrm{d}y\,\mathrm{d}v.$$

The reader will note that the dependency of the density w_x on r appears only in the restriction of the variable v. Indeed, it is easy to verify that the

$$w_{x/r}(z/r, y/r, v/r)\mathrm{d}(z/r)\mathrm{d}(y/r)\mathrm{d}(v/r) = w_x(z,y,v)\mathrm{d}z\mathrm{d}y\mathrm{d}v,$$

which agrees with the scaling property of the probabilities in the statement of Theorem 15.5.

Marginalising the first triple law in Theorem 15.1 to give the joint law of the pair $(X_{\mathtt{G}(\tau_r^\oplus-)}, X_{\tau_r^\oplus})$ or the pair $(X_{\tau_r^\oplus-}, X_{\tau_r^\oplus})$ is not necessarily straightforward (although arguably, could be done using some of the integral manipulations presented in the previous sections). Whist an analytical computation for the marginalisation should be possible, and likely complicated, we provide a probabilistic proof, appealing to fluctuation identities.

Corollary 15.6 (First entrance/exit and closest reach) *Fix* $r > 0$ *and define, for* $x, z, v \in \mathbb{R}^d\backslash\{0\}$,

$$w_x(z,\bullet,v) := \frac{\Gamma(d/2)^2}{\pi^d|\Gamma(-\alpha/2)|\Gamma(\alpha/2)}\frac{||z|^2-|x|^2|^{\alpha/2}}{||z|^2-|v|^2|^{\alpha/2}|z-v|^d|z-x|^d}.$$

(i) For $|x| > |z| > r$ $|v| < r$,

$$\mathbb{P}_x(X_{\mathtt{G}(\tau_r^\oplus-)} \in \mathrm{d}z,\ X_{\tau_r^\oplus} \in \mathrm{d}v;\ \tau_r^\oplus < \infty) = w_x(z,\bullet,v)\mathrm{d}z\,\mathrm{d}v.$$

(ii) For $|x| < |z| < r$ *and* $|v| > r$,

$$\mathbb{P}_x(X_{\mathtt{H}(\tau_r^\ominus-)} \in \mathrm{d}z,\ X_{\tau_r^\ominus} \in \mathrm{d}v) = w_x(z,\bullet,v)\,\mathrm{d}z\,\mathrm{d}v.$$

Corollary 15.7 (First entrance/exit and preceding position) *Fix* $r > 0$ *and define, for* $x, z, y, v \in \mathbb{R}^d\backslash\{0\}$,

$$w_x(\bullet,y,v) := \frac{\Gamma((d+\alpha)/2)\Gamma(d/2)}{\pi^d|\Gamma(-\alpha/2)|\Gamma(\alpha/2)^2}\left(\int_0^{\zeta_r(x,y)}(u+1)^{-d/2}u^{\alpha/2-1}\mathrm{d}u\right)\frac{|x-y|^{\alpha-d}}{|v-y|^{\alpha+d}}\,\mathrm{d}v\,\mathrm{d}y,$$

where

$$\zeta_r(x,y) := (|x|^2 - r^2)(|y|^2 - r^2)/r^2|x-y|^2.$$

(i) For $|x|, |y| > r$ $|v| < r$,

$$\mathbb{P}_x(X_{\tau_r^{\oplus}-} \in \mathrm{d}y,\ X_{\tau_r^{\oplus}} \in \mathrm{d}v;\ \tau_r^{\oplus} < \infty) = w_x(\bullet, y, v)\mathrm{d}y\,\mathrm{d}v.$$

(ii) For $|x|, |y| < r$ *and* $|v| > r$,

$$\mathbb{P}_x(X_{\tau_r^{\ominus}-} \in \mathrm{d}y,\ X_{\tau_r^{\ominus}} \in \mathrm{d}v) = w_x(\bullet, y, v)\,\mathrm{d}y\,\mathrm{d}v.$$

We need a number of preliminary results first before we can engage with the proof of the above theorem.

In a similar spirit to the role of the local time ℓ, we can construct the local time L for the process $\overline{\xi} - \xi$, where $\overline{\xi}$ is the running maximum of ξ. As such, (H^+, Θ^+) gives us the ascending ladder MAP, where $H^+ = \xi_{L^{-1}}$ and $\Theta^+ = \Theta_{L^{-1}}$. Note that there is no killing of the ascending ladder MAP as $\lim_{t\to\infty} \xi_t = \infty$.

We define

$$U_x^+(\mathrm{d}z) = \int_0^\infty \mathbf{P}_{\log|x|,\arg(x)}(\mathrm{e}^{H_t^+}\Theta_t^+ \in \mathrm{d}z)\mathrm{d}t, \qquad |z| \geq |x|. \tag{15.23}$$

Lemma 15.8 *For* $|x| > 0$,

$$U_x^+(\mathrm{d}z) = \pi^{-d/2}\frac{\Gamma(d/2)^2}{\Gamma((d-\alpha)/2)\Gamma(\alpha/2)}\frac{||z|^2 - |x|^2|^{\alpha/2}}{|z|^\alpha|x-z|^d}, \qquad |z| > |x|. \tag{15.24}$$

Proof The Riesz–Bogdan–Żak transform in Theorem 12.8 shows that, for $x \neq 0$, $(KX_{\eta(t)}, t \geq 0)$ under $\mathbb{P}_{Kx}$ is equal in law to $(X_t, t \geq 0)$ under $\mathbb{P}_x^\circ$, where

$$\left.\frac{\mathrm{d}\mathbb{P}_x^\circ}{\mathrm{d}\mathbb{P}_x}\right|_{\sigma(X_s, s\leq t)} = \frac{|X_t|^{\alpha-d}}{|x|^{\alpha-d}}, \qquad t \geq 0, \tag{15.25}$$

and $\eta(t) = \inf\{s > 0\colon \int_0^s |X_u|^{-2\alpha}\mathrm{d}u > t\}$. Moreover, in Theorem 12.5 we see that $\mathbb{P}_x^\circ$, $x \in \mathbb{R}^d\backslash\{0\}$ can be understood as the stable process conditioned to limit continuously to the origin. In Corollary 12.9 we also showed that the effect of the change of measure (15.25) on the underlying MAP (ξ, Θ), when issued from $(\log|x|, \arg(x))$, $x \in \mathbb{R}$, is that it is equal in law to $(-\xi, \Theta)$ issued from $(-\log|x|, \arg(x))$. The Riesz–Bogdan–Żak transform thus ensures that, for Borel $A \subseteq \{z \in \mathbb{R}^d\colon |z| < |x|\}$,

$$\frac{|z|^{\alpha-d}}{|x|^{\alpha-d}}\mathbf{P}_{\log|x|,\arg(x)}(\mathrm{e}^{H_t^+}\Theta_t^+ \in A) = \mathbf{P}_{-\log|x|,\arg(x)}(\mathrm{e}^{-H_t^-}\Theta_t^- \in KA, t < \ell_\infty),$$

where $KA = \{Kz\colon z \in A\}$. Hence, back in (15.23), we have, for $|x| > 0$,

$$U_x^+(\mathrm{d}z) = \pi^{-d/2}\frac{\Gamma(d/2)^2}{\Gamma((d-\alpha)/2)\Gamma(\alpha/2)}\frac{||z|^{-2}-|x|^{-2}|^{\alpha/2}}{|z|^{-\alpha}\left|(z/|z|^2)-(x/|x|^2)\right|^d}\frac{|x|^{\alpha-d}}{|z|^{\alpha-d}}\frac{\mathrm{d}z}{|z|^{2d}}$$

$$= \pi^{-d/2}\frac{\Gamma(d/2)^2}{\Gamma((d-\alpha)/2)\Gamma(\alpha/2)}\frac{||z|^2-|x|^2|^{\alpha/2}}{|z|^\alpha|x-z|^d}, \qquad |z| > |x|, \qquad (15.26)$$

where we have used the fact that

$$|Kx - Kz| = \frac{|x-z|}{|x||z|}, \qquad (15.27)$$

which is a special case of (14.5). □

One notices that the analytical structure of the potential measures $U_x^-(\mathrm{d}z)$ and $U_x^+(\mathrm{d}z)$ are identical albeit that the former is supported on $|z| < |x|$ and the latter on $|z| > |x|$.

Proposition 15.9 *For $\theta \in \mathbb{S}^{d-1}$, we have*

$$\mathbb{N}_\theta\left(\mathrm{e}^{\epsilon(\varsigma)}\Theta^\epsilon(\varsigma) \in \mathrm{d}y; \varsigma < \infty\right)$$
$$= \frac{\alpha\pi^{-d/2}}{2}\frac{\Gamma((d-\alpha)/2)}{\Gamma(1-\alpha/2)}|1-|y|^2|^{-\alpha/2}|\theta - y|^{-d}\mathrm{d}y, \qquad |y| \le 1.$$

Proof Take $|x| > r > r_0 > 0$ and suppose that $f\colon \mathbb{R}^d \mapsto [0,\infty)$ is continuous with support which is compactly embedded in the ball of radius r_0. We have, on the one hand, from Theorem 14.7, the identity

$$\mathbb{E}_x[f(X_{\tau_r^\oplus}); \tau_r^\oplus < \infty] = \pi^{-(d/2+1)}\Gamma(d/2)\sin\left(\frac{\pi\alpha}{2}\right)\int_{|y|<r}\frac{|r^2-|x|^2|^{\alpha/2}}{|r^2-|y|^2|^{\alpha/2}}|x-y|^{-d}f(y)\mathrm{d}y.$$

On the other hand, from (15.6), we also have

$$\mathbb{E}_x[f(X_{\tau_r^\oplus}); \tau_r^\oplus < \infty] = \int_{r<|z|<|x|} U_x^-(\mathrm{d}z)\int_{|y||z|<r}\mathbb{N}_{\arg(z)}(f(|z|e^{\epsilon(\varsigma)}\Theta^\epsilon(\varsigma)); \varsigma < \infty).$$

Note that, for each $z \in \mathbb{R}^d\backslash\{0\}$,

$$z \mapsto \mathbb{N}_{\arg(z)}(f(|z|e^{\epsilon(\varsigma)}\Theta^\epsilon(\varsigma)); \varsigma < \infty)$$

is bounded thanks to the fact that f is bounded and its support is compactly embedded in the ball or radius r_0. Indeed, there exists an $\varepsilon > 0$, which depends only on the support of f, such that $|z|e^{\epsilon(\varsigma)} < r_0 - \varepsilon$ if $f(|z|e^{\epsilon(\varsigma)}) > 0$. Hence, since $|z| > r$, we have $\epsilon(\varsigma) < \log((r_0-\varepsilon)/r)$ when $f(|z|e^{\epsilon(\varsigma)}) > 0$ and thus

$$\sup_{r<|z|<|x|}\left|\mathbb{N}_{\arg(z)}(f(|z|e^{\epsilon(\varsigma)}\Theta^\epsilon(\varsigma)); \varsigma < \infty)\right| \le \|f\|_\infty \nu(-\log((r_0-\varepsilon)/r), \infty) < \infty.$$

Moreover, since we can write

$$\mathbb{N}_{\arg(z)}(f(|z|e^{\epsilon(\varsigma)}\Theta^\epsilon(\varsigma)); \varsigma < \infty) = \mathbb{N}_1(f(|z|e^{\epsilon(\varsigma)}\Theta^\epsilon(\varsigma)\star\arg(z)); \varsigma < \infty), \quad (15.28)$$

where, for any $a \in \mathbb{S}^d$, the operation $\star a$ rotates the sphere so that the 'North Pole', $1 = (1, 0, \ldots, 0) \in \mathbb{S}^{d-1}$ moves to a, using a straightforward dominated convergence argument, we see that $\mathbb{N}_{\arg(z)}(f(|z|e^{\epsilon(\varsigma)}\Theta^\epsilon(\varsigma)); \varsigma < \infty)$ is continuous in z thanks to the continuity of f.

Appealing to Lemma 15.3, we thus have that

$$
\begin{aligned}
&\mathbb{N}_{\arg(x)}(f(|x|e^{\epsilon(\varsigma)}\Theta^\epsilon(\varsigma)); \varsigma < \infty)\\
&= \lim_{r \uparrow |x|} \frac{\int_{r<|z|<|x|} U_x^-(\mathrm{d}z) \int_{|y||z|<r} \mathbb{N}_{\arg(z)}(f(|z|e^{\epsilon(\varsigma)}\Theta^\epsilon(\varsigma)); \varsigma < \infty)}{\int_{r<|z|\le|x|} U_x^-(\mathrm{d}z)}\\
&= \lim_{r \uparrow |x|} \frac{\mathbb{E}_x[f(X_{\tau_r^\oplus}); \tau_r^\oplus < \infty]}{\mathbb{P}_x(\tau_r^\oplus = \infty)}.
\end{aligned}
$$

Substituting in the analytical form of the ratio on the right-hand side above using (15.28) and (14.32), we may continue with

$$
\begin{aligned}
&\mathbb{N}_{\arg(x)}(f(|x|e^{\epsilon(\varsigma)}\Theta^\epsilon(\varsigma)); \varsigma < \infty)\\
&= \lim_{r \uparrow |x|} \pi^{-d/2} \frac{\Gamma((d-\alpha)/2)}{\Gamma(1-\alpha/2)} \frac{(|x|^2 - r^2)^{\alpha/2} \int_{|y|<r} |r^2 - |y|^2|^{-\alpha/2} |x-y|^{-d} f(y)\mathrm{d}y}{\int_0^{(|x|^2-r^2)/r^2} (u+1)^{-d/2} u^{\alpha/2-1}\mathrm{d}u}\\
&= \pi^{-d/2} \frac{\Gamma((d-\alpha)/2)}{\Gamma(1-\alpha/2)} \int_{|y|<|x|} ||x|^2 - |y|^2|^{-\alpha/2} |x-y|^{-d} f(y)\mathrm{d}y\\
&\qquad\qquad \times \lim_{r \uparrow |x|} \frac{r^\alpha [(|x|^2 - r^2)/r^2]^{\alpha/2}}{\int_0^{(|x|^2-r^2)/r^2} (u+1)^{-d/2} u^{\alpha/2-1}\mathrm{d}u}\\
&= \frac{\alpha\pi^{-d/2}}{2} \frac{\Gamma((d-\alpha)/2)}{\Gamma(1-\alpha/2)} \int_{|y|<|x|} |x|^\alpha ||x|^2 - |y|^2|^{-\alpha/2} |x-y|^{-d} f(y)\mathrm{d}y, \qquad (15.29)
\end{aligned}
$$

where we have used that the support of f is compactly embedded in the ball of radius $|x|$ to justify the first term in the second equality. □

Proposition 15.10 *For $x \in \mathbb{R}^d \backslash \{0\}$, and continuous $g\colon \mathbb{R}^d \mapsto [0, \infty)$ whose support is compactly embedded in the exterior of the ball of radius $|x|$,*

$$
\mathbb{N}_{\arg(x)}\left(\int_0^\varsigma g(|x|e^{\epsilon(u)}\Theta^\epsilon(u))\mathrm{d}u\right) = 2^{-\alpha} \frac{\Gamma((d-\alpha)/2)^2}{\Gamma(d/2)^2} \int_{|x|<|z|} g(z) U_x^+(\mathrm{d}z).
$$

Proof Fix $0 < r < |x|$. Recall from the Lamperti–Kiu representation in Theorem 11.13 that $X_t = \exp\{\xi_{\varphi(t)}\}\Theta(\varphi(t))$, $t \ge 0$, where $\int_0^{\varphi(t)} \exp\{\alpha\xi_u\}\mathrm{d}u = t$. In particular, this implies that, if we write $s = \varphi(t)$, then

$$
e^{\alpha\xi_s}\mathrm{d}s = \mathrm{d}t, \qquad t > 0. \qquad (15.30)
$$

Splitting the occupation over individual excursions, we have with the help of (15.1) that

$$
\begin{aligned}
\mathbb{E}_x\left[\int_0^{\tau_r^\oplus} g(X_t)\mathrm{d}t\right]
&= \mathbb{E}_x\left[\int_0^\infty \mathbf{1}_{(\mathrm{e}^{\xi_s}>r)} g(\mathrm{e}^{\xi_s}\Theta_s)\mathrm{e}^{\alpha\xi_s}\mathrm{d}s\right] \\
&= \int_{r<|z|<|x|} U_x^-(\mathrm{d}z)\mathbb{N}_{\arg(z)}\left(\int_0^\varsigma g(|z|\mathrm{e}^{\epsilon(s)}\Theta^\epsilon(s))(|z|\mathrm{e}^{\epsilon(s)})^\alpha \mathrm{d}s\right). \qquad (15.31)
\end{aligned}
$$

Note that the left-hand side is necessarily finite as it can be upper bounded by $\mathbb{E}_x\left[\int_0^\infty g(X_t)\mathrm{d}t\right]$, which, considering Theorem 3.11, can easily be verified to be finite for the given assumptions on g.

Straightforward arguments, similar to those presented around (15.28), tell us that for continuous g with compact support that is compactly embeded in the exterior of the ball of radius $|x|$, we have that, for $r < |z| < |x|$,

$$
\mathbb{N}_{\arg(z)}\left(\int_0^\varsigma g(|z|\mathrm{e}^{\epsilon(s)}\Theta^\epsilon(s))\mathrm{e}^{\alpha\epsilon(s)}\mathrm{d}s\right) = \int_0^\infty \mathbb{N}_{\arg(z)}\left(g(|z|\mathrm{e}^{\epsilon(s)}\Theta^\epsilon(s))\mathrm{e}^{\alpha\epsilon(s)};\ s<\varsigma\right)\mathrm{d}s
$$

is a continuous function. Accordingly, we can again use Lemma 15.3 and Theorem 14.7 and write, for $x \in \mathbb{R}^d$,

$$
\begin{aligned}
&\mathbb{N}_{\arg(x)}\left(\int_0^\varsigma g(|x|\mathrm{e}^{\epsilon(s)}\Theta^\epsilon(s))(|x|\mathrm{e}^{\epsilon(s)})^\alpha\right) \\
&= \lim_{r\uparrow|x|} \frac{\int_{r<|z|<|x|} U_x^-(\mathrm{d}z)\mathbb{N}_{\arg(z)}\left(\int_0^\varsigma g(|z|\mathrm{e}^{\epsilon(s)}\Theta^\epsilon(s))(|z|\mathrm{e}^{\epsilon(s)})^\alpha\right)}{\int_{r<|z|\le|x|} U_x^-(\mathrm{d}z)} \\
&= \frac{\mathbb{E}_x\left[\int_0^{\tau_r^\oplus} g(X_s)\mathrm{d}s\right]}{\mathbb{P}_x(\tau_r^\oplus = \infty)} \\
&= 2^{-\alpha}\pi^{-d/2}\frac{\Gamma((d-\alpha)/2)}{\Gamma(\alpha/2)} \\
&\qquad \times \lim_{r\uparrow|x|} \frac{\int_{|x|<|z|} \mathrm{d}z\, \mathbf{1}_{(r<|z|)} g(z)|x-z|^{\alpha-d} \int_0^{\zeta_r(x,z)} (u+1)^{-d/2}u^{\alpha/2-1}\mathrm{d}u}{\int_0^{(|x|^2-r^2)/r^2} (u+1)^{-d/2}u^{\alpha/2-1}\mathrm{d}u} \\
&= 2^{-\alpha}\pi^{-d/2}\frac{\Gamma((d-\alpha)/2)}{\Gamma(\alpha/2)} \int_{|x|<|z|} g(z)|x-z|^{-d}(|z|^2-|x|^2)^{\alpha/2}\mathrm{d}z,
\end{aligned}
$$

where in the final equality we have used dominated convergence (in particular the assumption on the support of g). By inspection, we also note that the right-hand side above is equal to

$$2^{-\alpha}\frac{\Gamma((d-\alpha)/2)^2}{\Gamma(d/2)^2}\int_{|x|<|z|} g(z)|z|^\alpha U_x^+(\mathrm{d}z).$$

The proof is completed by replacing $g(z)$ by $g(z)|z|^{-\alpha}$. □

Proof of Theorem 15.5 (i) Appealing to the Lévy system compensation formula for the jumps of X, we have, on the one hand,

$$\begin{aligned}
&\mathbb{E}_x[f(X_{\mathtt{G}(\tau_r^\oplus-)})g(X_{\tau_r^\oplus-})h(X_{\tau_r^\oplus});\tau_r^\oplus<\infty]\\
&=\mathbb{E}_x\left[\sum_{t>0} f(X_{\mathtt{G}(t-)})g(X_{t-})h(X_t)\mathbf{1}_{(|X_{\mathtt{G}(t-)}|>r)}\mathbf{1}_{(|X_t|<r)}\right]\\
&=\mathbb{E}_x\left[\int_0^{\tau_r^\oplus} f(X_{\mathtt{G}(t)})g(X_t)k(X_t)\right],
\end{aligned}\qquad (15.32)$$

where continuous $\mathbb{R}^d$-valued functions f, g, h are such that the first two are compactly supported in $\{z\in\mathbb{R}^d\colon |z|>r\}$ and the third is compactly supported in the open ball of radius r and

$$k(y)=\int_{|y+w|<r}\Pi(\mathrm{d}w)h(y+w).$$

On the other hand, a calculation similar in spirit to (15.31), using (15.1), followed by an application of Proposition 15.10, tells us that

$$\begin{aligned}
&\mathbb{E}_x\left[\int_0^{\tau_r^\oplus} f(X_{\mathtt{G}(t)})g(X_t)k(X_t)\,\mathrm{d}t\right]\\
&=\int_{r<|z|<|x|} U_x^-(\mathrm{d}z)f(z)\mathbb{N}_{\arg(z)}\left(\int_0^\varsigma g(|z|e^{\epsilon(s)}\Theta^\epsilon(s))k(|z|e^{\epsilon(s)}\Theta^\epsilon(s))(|z|e^{\epsilon(s)})^\alpha\mathrm{d}s\right)\\
&=2^{-\alpha}\frac{\Gamma((d-\alpha)/2)^2}{\Gamma(d/2)^2}\int_{r<|z|<|x|} U_x^-(\mathrm{d}z)f(z)\int_{|z|<|y|} U_z^+(\mathrm{d}y)g(y)k(y)|y|^\alpha.
\end{aligned}$$

Putting the pieces together, we get

$$\begin{aligned}
&\mathbb{E}_x[f(X_{\mathtt{G}(\tau_r^\oplus-)})g(X_{\tau_r^\oplus-})h(X_{\tau_r^\oplus});\tau_r^\oplus<\infty]\\
&=2^{-\alpha}\frac{\Gamma((d-\alpha)/2)^2}{\Gamma(d/2)^2}\\
&\quad\times\int_{r<|z|<|x|}\int_{|z|<|y|}\int_{|w-y|<r} U_x^-(\mathrm{d}z)U_z^+(\mathrm{d}y)\Pi(\mathrm{d}w)f(z)g(y)|y|^\alpha h(y+w)\\
&=c_{\alpha,d}\int_{r<|z|<|x|}\int_{|z|<|y|}\int_{|v|<r}\frac{||z|^2-|x|^2|^{\alpha/2}||y|^2-|z|^2|^{\alpha/2}}{|z|^\alpha|z-x|^d|z-y|^d|v-y|^{\alpha+d}}f(z)g(y)h(v)\,\mathrm{d}y\,\mathrm{d}z\,\mathrm{d}v,
\end{aligned}$$

where

$$c_{\alpha,d} = \frac{\Gamma((d+\alpha)/2)}{|\Gamma(-\alpha/2)|}\frac{\Gamma(d/2)^2}{\pi^{3d/2}\Gamma(\alpha/2)^2}.$$

This is equivalent to the statement of part (i) of the theorem.

(ii) The second statement in the theorem is a straightforward application of the Riesz–Bogdan–Żak transformation, with computations in the style of those used to prove, for example, the second part of Theorem 14.10. For the sake of brevity, the proof is left as an exercise for the reader. □

Proof of Corollary 15.6 As in the proof of Theorem 15.5, we only prove (i). Part (ii) can be derived appealing to the Riesz–Bogdan–Żak transformation.

From (15.6) (albeit replacing the role of 1 by r), (15.11) and Proposition 15.9, more specifically (15.29), we have that for bounded measurable functions f, g on $\mathbb{R}^d$,

$$\begin{aligned}
&\mathbb{E}_x[g(X_{\mathtt{G}(\tau_r^\oplus-)})f(X_{\tau_r^\oplus}); \tau_r^\oplus < \infty] \\
&= \int_{r<|z|<|x|} U_x^-(\mathrm{d}z)\mathbb{N}_{\arg(z)}(f(|z|e^{\epsilon(\varsigma)}\Theta^\epsilon(\varsigma))\mathbf{1}_{(|z|e^{\epsilon(\varsigma)}<r)}; \varsigma < \infty)g(z) \\
&= \frac{\Gamma(d/2)^2\sin(\pi\alpha/2)}{\pi^d|\Gamma(-\alpha/2)|\Gamma(\alpha/2)}\int_{r<|z|<|x|}\int_{|v|<r}\frac{||z|^2-|x|^2|^{\alpha/2}}{||z|^2-|v|^2|^{\alpha/2}|z-v|^d|z-x|^d}f(v)g(z)\,\mathrm{d}z\,\mathrm{d}v.
\end{aligned}$$

The desired result now follows. □

Proof of Corollary 15.7 As with the previous proof, we only deal with (i). Setting $f \equiv 1$ in (15.32), we see with the help of Theorem 14.10 and the jump measure (3.23) that

$$\begin{aligned}
&\mathbb{E}_x[g(X_{\tau_r^\oplus-})h(X_{\tau_r^\oplus}); \tau_r^\oplus < \infty] \\
&= \mathbb{E}_x\left[\int_0^{\tau_r^\oplus} g(X_t)k(X_t)\,\mathrm{d}t\right] \\
&= \frac{2^\alpha\Gamma((d+\alpha)/2)}{\pi^{d/2}|\Gamma(-\alpha/2)|}\int_{|y|>r} g(y)\int_{|y+w|<r}\frac{1}{|w|^{\alpha+d}}\mathrm{d}w\,h(y+w)u_r(x,y)\mathrm{d}y \\
&= \frac{2^\alpha\Gamma((d+\alpha)/2)}{\pi^{d/2}|\Gamma(-\alpha/2)|}\int_{|y|>r}\int_{|v|<r} g(y)h(v)\frac{1}{|v-y|^{\alpha+d}}u_r(x,y)\,\mathrm{d}v\,\mathrm{d}y,
\end{aligned}$$

where we recall that

$$k(y) = \int_{|y+w|<r}\Pi(\mathrm{d}w)h(y+w)$$

and u_r is the potential density given in (14.44). The result now follows. □

15.3 Deep factorisation in d-dimensions

The manipulations we have made in the previous section, in particular in Proposition 15.10, are precisely what we need to demonstrate an analogue of the deep Wiener–Hopf factorisation. To this end, let us introduce some new notation. Define the resolvents

$$\mathbf{R}_z[f](\theta) = \mathbf{E}_{0,\theta}\left[\int_0^\infty e^{-z\xi_t} f(\Theta_t)\mathrm{d}t\right], \qquad \theta \in \mathbb{S}^{d-1}, \mathrm{Re}(z) \in (0, d-\alpha), \quad (15.33)$$

as well as

$$\hat{\rho}_z[f](\theta) = \mathbf{E}_{0,\theta}\left[\int_0^{\ell_\infty} e^{-zH_t^-} f(\Theta_t^-)\mathrm{d}t\right] = \int_{|y|<1} |y|^z f(\arg(y))U_\theta^-(\mathrm{d}y), \qquad (15.34)$$

for $\mathrm{Re}(z) > -(d-\alpha)$ and

$$\rho_z[f](\theta) = \mathbf{E}_{0,\theta}\left[\int_0^\infty e^{-zH_t^+} f(\Theta_t^+)\mathrm{d}t\right] = \int_{|y|>1} |y|^{-z} f(\arg(y))U_\theta^+(\mathrm{d}y), \qquad (15.35)$$

for $\mathrm{Re}(z) > 0$ and bounded measurable $f\colon \mathbb{S}^{d-1} \mapsto [0,\infty)$. Note, for the expression (15.35) the ladder MAP is not killed as $\lim_{t\to\infty}|X_t| = \infty$.

It is not immediately clear why the restrictions on $\mathrm{Re}(z)$ are thus imposed for each of $\mathbf{R}_z[f](\theta)$, $\hat{\rho}_z[f](\theta)$ and $\rho_z[f](\theta)$. Let us spend a little while justifying the need as such.

First, recall from Theorem 3.11, that the free potential measure of a stable process issued from $x \in \mathbb{R}^d$ has density given by

$$u(x,y) = \frac{\Gamma((d-\alpha)/2)}{2^\alpha \pi^{d/2}\Gamma(\alpha/2)}|y-x|^{\alpha-d}, \qquad y \in \mathbb{R}^d.$$

Accordingly, taking account of (15.30), it is straightforward to compute

$$\begin{aligned}\mathbf{R}_z[f](\theta) &= \mathbf{E}_{0,\theta}\left[\int_0^\infty e^{-(z+\alpha)\xi_s} f(\Theta_s)e^{\alpha\xi_s}\mathrm{d}s\right]\\ &= \mathbb{E}_\theta\left[\int_0^\infty |X_t|^{-(\alpha+z)} f(\arg(X_t))\mathrm{d}t\right]\\ &= \int_{\mathbb{R}^d} f(\arg(y))\frac{u(\theta,y)}{|y|^{\alpha+z}}\mathrm{d}y,\end{aligned}$$

where we have used the Lamperti–Kiu transform in the second equality. Converting the final integral to generalised polar coordinates, that is, $y = r\theta$, for $r > 0$ and $\theta \in \mathbb{S}^{d-1}$, we see that, for uniformly bounded f, the integral is well defined for large r as long as $\int^\infty r^{\alpha-d}r^{-(\alpha+z)}r^{d-1}\mathrm{d}r < \infty$, that is, $\mathrm{Re}(z) > 0$, and well defined for small r as long as $\int_{0+} r^{-(\alpha+z)}r^{d-1}\mathrm{d}r < \infty$, that is, $d-\alpha-\mathrm{Re}(z) > 0$.

Appealing to (15.26), we note that, for $\theta \in \mathbb{S}^{d-1}$ and $\mathrm{Re}(z) \geq 0$,

$$\boldsymbol{\rho}_z[f](\theta) = \pi^{-d/2}\frac{\Gamma(d/2)^2}{\Gamma((d-\alpha)/2)\Gamma(\alpha/2)}\int_{|y|>1} f(\arg(y))\frac{(|y|^2-1)^{\alpha/2}}{|y|^{\alpha+z}|y-\theta|^d}\mathrm{d}y.$$

Converting the above integral to generalised polar coordinates, it becomes apparent that the integral is well defined as soon as $\int^{\infty} r^{-z-1}\mathrm{d}r < \infty$, that is, $\mathrm{Re}(z) > 0$. Similarly appealing to Corollary 15.2, we also note that

$$\hat{\boldsymbol{\rho}}_z[f](\theta) = \pi^{-d/2}\frac{\Gamma(d/2)^2}{\Gamma((d-\alpha)/2)\Gamma(\alpha/2)}\int_{|y|<1} f(\arg(y))\frac{(|y|^2-1)^{\alpha/2}}{|y|^{\alpha-z}|y-\theta|^d}\mathrm{d}y.$$

Again converting to generalised polar coordinates, we see that the integral is well defined providing $\int_{0+} r^{d-\alpha+z-1}\mathrm{d}r < \infty$, that is, $\mathrm{Re}(z) > -(d-\alpha)$.

Theorem 15.11 (Deep factorisation of the d-dimensional isotropic stable process) *Suppose that $f\colon \mathbb{S}^{d-1} \mapsto [0,\infty)$ is bounded and measurable. Then*

$$\mathbf{R}_{c-\mathrm{i}\lambda}[f](\theta) = C_{\alpha,d}\,\boldsymbol{\rho}_{\mathrm{i}\lambda+d-\alpha-c}[\boldsymbol{\rho}_{c-\mathrm{i}\lambda}[f]](\theta), \qquad \theta \in \mathbb{S}^{d-1}, \lambda \in \mathbb{R}, c \in (0, d-\alpha), \tag{15.36}$$

where $C_{\alpha,d} = 2^{-\alpha}\Gamma((d-\alpha)/2)^2/\Gamma(d/2)^2$.

Proof From the second and third equalities of equation (15.31) (taking $r \to 0$) and Proposition 15.10, providing the left-hand side is finite, which it is for $c \in (0, d-\alpha)$ and $\lambda \in \mathbb{R}$, we have

$$\begin{aligned}\mathbf{R}_{c-\mathrm{i}\lambda}[f](\theta) &= \int_{|w|<1} U^-_\theta(\mathrm{d}w)\mathbb{N}_{\arg(w)}\left(\int_0^{\varsigma}(|w|\mathrm{e}^{\epsilon(s)})^{-c+\mathrm{i}\lambda}f(\Theta^\epsilon(s))\right)\\ &= 2^{-\alpha}\frac{\Gamma((d-\alpha)/2)^2}{\Gamma(d/2)^2}\int_{|w|<1}U^-_\theta(\mathrm{d}w)\int_{|w|<|y|}f(\arg(y))|y|^{-c+\mathrm{i}\lambda}U^+_w(\mathrm{d}y).\end{aligned} \tag{15.37}$$

Note that, by conditional stationary and independent increments, for any $w \in \mathbb{R}^d\backslash\{0\}$ and $c > 0$,

$$\begin{aligned}\int_{|w|<|y|}|y|^{-c+\mathrm{i}\lambda}f(\arg(y))U^+_w(\mathrm{d}y) &= \mathbf{E}_{\log|w|,\arg(w)}\left[\int_0^\infty \mathrm{e}^{-c+\mathrm{i}\lambda H^+_t}f(\Theta^+_t)\mathrm{d}t\right]\\ &= |w|^{-c+\mathrm{i}\lambda}\mathbf{E}_{0,\arg(w)}\left[\int_0^\infty \mathrm{e}^{-c+\mathrm{i}\lambda H^+_t}f(\Theta^+_t)\mathrm{d}t\right]\\ &= |w|^{-c+\mathrm{i}\lambda}\int_{1<|y|}|y|^{-c+\mathrm{i}\lambda}f(\arg(y))U^+_{\arg(w)}(\mathrm{d}y),\end{aligned} \tag{15.38}$$

where, by (15.26), converting to polar coordinates, the right-hand side is uniformly bounded by an integral of $O\left(\int^\infty r^{-c-1}\mathrm{d}r\right) < \infty$. Hence plugging (15.38) back in (15.37), we have

$$\mathbf{R}_{c-\mathrm{i}\lambda}[f](\theta) = 2^{-\alpha}\frac{\Gamma((d-\alpha)/2)^2}{\Gamma(d/2)^2}\hat{\rho}_{c-\mathrm{i}\lambda}[\rho_{c-\mathrm{i}\lambda}[f]](\theta), \qquad \lambda \in \mathbb{R}, c \in (0, d-\alpha).$$

Finally, we note from (15.26) that, making the change of variables $y = Kw$, so that $\arg(y) = \arg(w)$, $|y| = 1/|w|$ and $\mathrm{d}y = |w|^{-2d}\mathrm{d}w$ and, for $\theta \in \mathbb{S}^{d-1}$, $|\theta - Kw| = |\theta - w|/|w|$, we have

$$\begin{aligned}
\hat{\rho}_{c-\mathrm{i}\lambda}[f](\theta) &= \pi^{-d/2}\frac{\Gamma(d/2)^2}{\Gamma((d-\alpha)/2)\Gamma(\alpha/2)}\int_{|y|>1} |y|^{-c+\mathrm{i}\lambda} f(\arg(y))\frac{(|y|^2-1)^{\alpha/2}}{|y|^\alpha|\theta - y|^d}\mathrm{d}y \\
&= \pi^{-d/2}\frac{\Gamma(d/2)^2}{\Gamma((d-\alpha)/2)\Gamma(\alpha/2)}\int_{|w|<1} |w|^{c-\mathrm{i}\lambda+\alpha-d} f(\arg(w))\frac{(1-|w|^2)^{\alpha/2}}{|\theta-w|^d|w|^\alpha}\mathrm{d}w \\
&= \rho_{\mathrm{i}\lambda+d-\alpha-c}[f](\theta), \qquad \lambda \in \mathbb{R}, c \in (0, d-\alpha),
\end{aligned}$$

which is also uniformly bounded by an integral of $O\left(\int^\infty r^{c-(d-\alpha)-1}\mathrm{d}r\right) < \infty$.

□

The reader is reminded that Theorem 15.11 is stated for dimension $d \geq 2$ and for isotropic processes. However, if we are careful about assumptions, it is easily verified that the theorem also holds in dimension $d = 1$ and is consistent with (13.34). Recall that in dimension $d \geq 2$ we have $\lim_{t\to\infty}|X_t| = \infty$ almost surely. This is a crucial part of the proof of Theorem 15.11. Accordingly, when checking the proof in dimension $d = 1$, we should confine ourselves to the additional assumptions that the process is symmetric and $\alpha \in (0, 1)$.

The aforementioned case was discussed in Sections 13.2 and 13.5. In that case, Θ and Θ^+ live on the states $\{-1, 1\}$ and hence, in the definitions of the resolvents (15.33) and (15.34), we have $f\colon \{1, -1\} \mapsto [0, \infty)$. In effect we can thus reduce these resolvents to matrices. Accordingly we prefer to write (for all appropriate $z \in \mathbb{C}$), $\mathbf{R}_z(j,k) = \mathbf{R}_z[\mathbf{1}_k](j)$ and $\rho_z[\mathbf{1}_k](j) = \rho_z(j,k)$, for $j, k \in \{1, -1\}$, where $\mathbf{1}_k = \mathbf{1}_k(\cdot)$ is the indicator function of the state k. The factorisation (15.36) now reads (up to an unimportant multiplicative constant)

$$\begin{aligned}
\mathbf{R}_{c-\mathrm{i}\lambda}(j,k) &= \rho_{\mathrm{i}\lambda+d-\alpha-c}[\rho_{c-\mathrm{i}\lambda}[\mathbf{1}_k]](j) \\
&= \sum_{\ell=1,-1}\mathbf{E}_{0,j}\left[\int_0^\infty \mathrm{e}^{-(\mathrm{i}\lambda+d-\alpha-c)H_t^+}\mathbf{1}_\ell(\Theta_t^+)\mathrm{d}t\right]\rho_{c-\mathrm{i}\lambda}[\mathbf{1}_k](\ell) \\
&= \sum_{\ell=1,-1}\rho_{\mathrm{i}\lambda+d-\alpha-c}(j,\ell)\rho_{c-\mathrm{i}\lambda}(\ell,k), \qquad k,\ell \in \{1,-1\}.
\end{aligned} \tag{15.39}$$

In the same notation as Theorem 13.3, we have, for example, that

$$\begin{aligned}
\rho_{c-\mathrm{i}\lambda}(j,k) &= \mathbf{E}_{0,j}\left[\int_0^\infty \mathrm{e}^{-(c-\mathrm{i}\lambda)H_t^+}\mathbf{1}_k(\Theta_t^+)\mathrm{d}t\right] \\
&= \int_0^\infty \mathrm{e}^{-(c-\mathrm{i}\lambda)y}\boldsymbol{u}_{j,k}(y)\mathrm{d}y \\
&:= \boldsymbol{\kappa}_{i,j}^{-1}(c-\mathrm{i}\lambda),
\end{aligned}$$

where $(\boldsymbol{u}_{j,k}, j,k \in \{1,-1\})$ is the potential density of (H^+, Θ^+). We can also easily check that $\det(\Psi(c - \mathrm{i}\lambda)) \neq 0$ and hence $\Psi(c - \mathrm{i}\lambda)$ is invertible. By performing an analytical extension in the definition of $\boldsymbol{\Psi}$, we note that $\mathbf{R}_{c-\mathrm{i}\lambda}(j,k) = [\boldsymbol{\Psi}^{-1}(1-\alpha-c+\mathrm{i}\lambda)]_{j,k}$. As such, we see that (15.39) agrees with the matrix Wiener–Hopf factorisation in (13.34).

15.4 Radial reflection

In Section 13.1, we discussed the reflection of a stable process in its radial maximum for one-dimensional stable processes. Naturally, we can develop this notion to the setting of higher dimensions with reflection in the sphere of the radial maximum.

Define $M_t = \sup_{s\le t}|X_s|$, $t \ge 0$. In a similar spirit to the computations in Section 13.1, it is a straightforward exercise to show that $(X_t, X_t/M_t)$, $t \ge 0$ is a Markov process which lives on $\mathbb{R}^d \times \mathbb{B}(0,1)$, where $\mathbb{B}(0,1) = \{x \in \mathbb{R}^d : |x| \le 1\}$. Thanks to the transience of X, it is clear that $\lim_{t\to\infty} M_t = \infty$. As in the one-dimensional setting, thanks to repeated scaling of X by its radial maximum, we can expect that the $\lim_{t\to\infty} X_t/M_t$ exists as a weak limit.

Theorem 15.12 *For all bounded measurable $f\colon \mathbb{B}(0,1) \mapsto \mathbb{R}$ and $x \in \mathbb{R}\backslash\{0\}$,*

$$\begin{aligned}\lim_{t\to\infty} &\mathbb{E}_x[f(X_t/M_t)] \\ &= \pi^{-d/2}\frac{\Gamma((d+\alpha)/2)}{\Gamma(\alpha/2)}\int_{\mathbb{S}^{d-1}}\sigma_1(\mathrm{d}\phi)\int_{|w|<1} f(w)\frac{|1-|w|^2|^{\alpha/2}}{|\phi-w|^d}\mathrm{d}w, \qquad (15.40)\end{aligned}$$

where $\sigma_1(\mathrm{d}y)$ is the surface measure on $\mathbb{S}^{d-1}$, normalised to have unit mass.

Remark 15.13 Although we are dealing with the case $d \ge 2$, we can verify that the above limiting identity agrees with the stationary distribution for the radially reflected process in one-dimensional symmetric case, given in Theorem 13.2 when $\rho = 1/2$, if we set $d = 1$ and $\alpha \in (0,1)$ in (15.40). Indeed, note that $\mathbb{S}^{d-1} = \{-1, 1\}$ and hence $\sigma_1 = \delta_{-1} + \delta_1$, in which case,

$$\begin{aligned}&\int_{\mathbb{S}^{d-1}}\sigma_1(\mathrm{d}\phi)\int_{|w|<1} f(w)\frac{|1-|w|^2|^{\alpha/2}}{|\phi-w|^d}\mathrm{d}w \\ &= \int_0^1 f(u)(1-u^2)^{\alpha/2}(1-u)^{-1}\mathrm{d}u + \int_0^1 f(u)(1-u^2)^{\alpha/2}(1+u)^{-1}\mathrm{d}u \\ &= \int_0^1 f(u)\left((1+u)^{\alpha/2}(1-u)^{(\alpha/2)-1} + (1+u)^{(\alpha/2)-1}(1-u)^{\alpha/2}\right)\mathrm{d}u.\end{aligned}$$

Proof of Theorem 15.12 Recall from the description of the Riesz–Bogdan–Żak transform, in particular Corollary 12.9, that (ξ, Θ) under the change of measure in (15.25), or equivalently the MAP change of measure

$$\left.\frac{\mathrm{d}\mathbf{P}^\circ_{x,\theta}}{\mathrm{d}\mathbf{P}_{x,\theta}}\right|_{\sigma((\xi_s,\Theta_s),s\le t)} = \mathrm{e}^{(\alpha-d)(\xi_t - x)}, \qquad t \ge 0, \tag{15.41}$$

is equal in law to $(-\xi, \Theta)$. Recalling the excursion theory laid out in Section 15.1, we have for $q > 0$, $x \in \mathbb{R}^d\backslash\{0\}$ and bounded measurable g, whose support is compactly embedded in the ball of unit radius,

$$\begin{aligned}
&\mathbf{E}_{-\log|x|,\arg(x)}[g(\mathrm{e}^{-(\overline{\xi}_{\mathbf{e}_q}-\xi_{\mathbf{e}_q})}\Theta_{\mathbf{e}_q})]\\
&= \mathbf{E}_{\log|x|,\arg(x)}\left[\frac{\mathrm{e}^{(\alpha-d)\xi_{\mathbf{e}_q}}}{|x|^{\alpha-d}} g(\mathrm{e}^{-(\xi_{\mathbf{e}_q}-\underline{\xi}_{\mathbf{e}_q})}\Theta_{\mathbf{e}_q})\right]\\
&= |x|^{d-\alpha}\mathbf{E}_{\log|x|,\arg(x)}\left[\sum_{\mathrm{g}\in G}\mathbf{1}_{(\varsigma_{\mathrm{g}'}<\mathbf{e}_q^{\mathrm{g}'},\,\forall G i \mathrm{g}'<\mathrm{g})}\mathrm{e}^{(\alpha-d)\xi_{\mathrm{g}}}\mathrm{e}^{(\alpha-d)\epsilon_{\mathrm{g}}(\mathbf{e}_q^{\mathrm{g}})}\right.\\
&\qquad\qquad\left.\times\, g(\mathrm{e}^{-\epsilon_{\mathrm{g}}(\mathbf{e}_q^{\mathrm{g}})}\Theta^{\epsilon}_{\mathrm{g}}(\mathbf{e}_q^{\mathrm{g}}))\mathbf{1}_{(\mathbf{e}_q^{\mathrm{g}}<\varsigma_{\mathrm{g}})}\right]\\
&= |x|^{d-\alpha}\mathbf{E}_{\log|x|,\arg(x)}\left[\int_0^\infty \mathrm{e}^{-qt}\mathrm{e}^{(\alpha-d)\xi_t}\mathbb{N}_{\Theta_t}\left(\mathrm{e}^{(\alpha-d)\epsilon(\mathbf{e}_q)}g(\mathrm{e}^{-\epsilon(\mathbf{e}_q)}\Theta^{\epsilon}(\mathbf{e}_q));\mathbf{e}_q<\varsigma\right)\mathrm{d}\ell_t\right]\\
&= |x|^{d-\alpha}\mathbf{E}_{\log|x|,\arg(x)}\left[\int_0^\infty \mathrm{e}^{-q\ell_s^{-1}}\mathrm{e}^{-(\alpha-d)H_s^-}\right.\\
&\qquad\qquad\left.\times\,\mathbb{N}_{\Theta_s^-}\left(\mathrm{e}^{(\alpha-d)\epsilon(\mathbf{e}_q)}g(\mathrm{e}^{-\epsilon(\mathbf{e}_q)}\Theta^{\epsilon}(\mathbf{e}_q));\mathbf{e}_q<\varsigma\right)\mathrm{d}s\right],
\end{aligned}$$

where, for each $\mathrm{g} \in G$, $\mathbf{e}_q^{\mathrm{g}}$ are additional marks on the associated excursion which are independent and exponentially distributed with rate q. Hence, if we define

$$U_x^{(q),-}(\mathrm{d}y) = \int_0^\infty \mathrm{d}s\,\mathbf{E}_{\log|x|,\arg(x)}\left[\mathrm{e}^{-q\ell_t^{-1}};\,\mathrm{e}^{-H_s^-}\Theta_s^-\in\mathrm{d}y,\ s<\ell_\infty\right], \qquad |y|<|x|,$$

then

$$\begin{aligned}
&\mathbf{E}_{-\log|x|,\arg(x)}[g(\mathrm{e}^{-(\overline{\xi}_{\mathbf{e}_q}-\xi_{\mathbf{e}_q})}\Theta_{\mathbf{e}_q})]\\
&= \int_{(0,\infty)}\int_{|y|<|x|} qU_x^{(q),-}(\mathrm{d}y)\frac{|y|^{\alpha-d}}{|x|^{\alpha-d}}\mathbb{N}_{\arg(y)}\left(\int_0^\varsigma \mathrm{e}^{-qt}\mathrm{e}^{(\alpha-d)\epsilon(t)}g(\mathrm{e}^{-\epsilon(t)}\Theta(t))\mathrm{d}t\right).
\end{aligned} \tag{15.42}$$

Recall that there exists a local time L for the process $\overline{\xi}-\xi$, where $\overline{\xi}$ is the running maximum of ξ, which defines the ascending ladder MAP (H^+, Θ^+).

Here, $H^+ = \xi_{L^{-1}}$ and $\Theta^+ = \Theta_{L^{-1}}$. Moreover, L^{-1} is a subordinator (without reference to the accompanying modulation Θ^+) on account of the fact that it is the local time of a Lévy process. Suppose we denote its Laplace exponent by

$$\Lambda^+(q) := -\log \mathbf{E}_{0,\theta}\left[\exp\{-qL_1^{-1}\}\right], \qquad q \geq 0,$$

where $\theta \in \mathbb{S}^{d-1}$ is unimportant in the definition. Appealing again to the Riesz–Bogdan–Żak transform, we note that, for a bounded and measurable function h on $\mathbb{S}^{d-1}$, using obvious notation

$$\begin{aligned}
&\int_{|y|<|x|} \frac{|y|^{\alpha-d}}{|x|^{\alpha-d}} qU_x^{(q),-}(\mathrm{d}y)h(\arg(y)) \\
&\qquad = q\int_0^\infty \mathrm{d}s\, \mathbf{E}_{-\log|x|,\arg(x)}\left[\mathrm{e}^{-qL_s^{-1}} h(\Theta_s^+)\right] \\
&\qquad = \frac{q}{\Lambda^+(q)}\int_0^\infty \mathrm{d}s\, \Lambda^+(q)\mathrm{e}^{-\Lambda^+(q)s}\mathbf{E}^{(q)}_{-\log|x|,\arg(x)}\left[h(\Theta_s^+)\right] \\
&\qquad = \frac{q}{\Lambda^+(q)}\mathbf{E}^{(q)}_{-\log|x|,\arg(x)}\left[h\left(\Theta^+_{\mathbf{e}_{\Lambda^+(q)}}\right)\right],
\end{aligned} \tag{15.43}$$

where $\mathbf{P}^{(q)}_{-\log|x|,\arg(x)}$ appears as the result of a change of measure with martingale density $\exp\{-qL_s^{-1} + \Lambda^+(q)s\}$, $s \geq 0$ and $\mathbf{e}_{\Lambda^+(q)}$ is an independent exponential random variable with parameter $\Lambda^+(q)$.

Next, we want to take $q \downarrow 0$ in (15.42). To this end, we start by remarking that, as L is a local time for the Lévy process ξ, it is known from classical Wiener–Hopf factorisation theory that, up to a multiplicative constant, $c > 0$, which depends on the normalisation of the local times ℓ and L,

$$q = c\Lambda^+(q)\Lambda^-(q), \qquad q \geq 0,$$

where $\Lambda^-(q)$ is the Laplace exponent of the local time of ξ at its running infimum; see (2.33). On account of the fact that X is transient, that is, $\lim_{t\to\infty}|X_t| = \infty$, it follows that $\lim_{t\to\infty}\xi_t = \infty$ and hence we know that ℓ_∞ is exponentially distributed. Recall that we earlier normalised our choice of ℓ such that its rate, $\Lambda^-(0) = 1$; see the discussion above equation (15.4). This implies that $\lim_{q\downarrow 0} q/\Lambda^+(q) = c$.

Appealing to isotropy, the recurrence of $\{0\}\times\mathbb{S}^d$ for $(\overline{\xi}-\xi,\Theta)$ and weak convergence back in (15.43) as we take the limit with $q \downarrow 0$, to find that

$$\lim_{t\to\infty}\int_{|y|<|x|}\frac{|y|^{\alpha-d}}{|x|^{\alpha-d}}qU_x^{(q),-}(\mathrm{d}y)h(\arg(y)) = c\int_{\mathbb{S}^{d-1}}\sigma_1(\mathrm{d}\phi)h(\phi),$$

where we recall that $\sigma_1(\mathrm{d}\phi)$ is the surface measure on $\mathbb{S}^{d-1}$ normalised to have unit mass. Hence, back in (15.42), we have with the help of Proposition 15.10 and (15.26),

$$
\begin{aligned}
&\lim_{q\downarrow 0} \mathbf{E}_{-\log|x|,\arg(x)}[g(\mathrm{e}^{-(\bar{\xi}_{\mathbf{e}_q}-\xi_{\mathbf{e}_q})}\Theta_{\mathbf{e}_q})] \\
&= \int_{\mathbb{S}^{d-1}} \sigma_1(\mathrm{d}\phi)\mathbb{N}_\phi\left(\int_0^\varsigma \mathrm{e}^{(\alpha-d)\epsilon(t)} g(\mathrm{e}^{-\epsilon(t)}\Theta(t))\mathrm{d}t\right) \\
&= c\pi^{-d/2}2^{-\alpha}\frac{\Gamma((d-\alpha)/2)}{\Gamma(\alpha/2)}\int_{\mathbb{S}^{d-1}}\sigma_1(\mathrm{d}\phi)\int_{|z|>1} g(Kz)\frac{(|z|^2-1)^{\alpha/2}}{|z|^d|\phi-z|^d}\mathrm{d}z, \quad (15.44)
\end{aligned}
$$

where we recall that $Kz = z/|z|^2$.

Next, we follow the reasoning that led to the equality (13.32) for the stationary distribution of the radially reflected stable process in one dimension. Using the Lamperti–Kiu transform and (15.30), for bounded measurable f,

$$
f(X_t/M_t)\,\mathrm{d}t = f\left(\mathrm{e}^{-(\bar{\xi}_s-\xi_s)}\Theta_s\right)\mathrm{e}^{\alpha\xi_s}\mathrm{d}s,
$$

where $s = \varphi(t)$. Using the arguments alluded to above, that, we now have that, for $y \in \mathbb{R}^d\backslash\{0\}$,

$$
\lim_{t\to\infty}\mathbb{E}_y[f(X_t/M_t)] = \lim_{s\to\infty}\mathbf{E}_{\log|y|,\arg(y)}\left[f\left(\mathrm{e}^{-(\bar{\xi}_s-\xi_s)}\Theta_s\right)\mathrm{e}^{\alpha\xi_s}\right]. \quad (15.45)
$$

Putting this together with (15.44), for f and x as before, we conclude that,

$$
\begin{aligned}
&\lim_{t\to\infty}\mathbb{E}_{Kx}[f(X_t/M_t)] \\
&= \lim_{q\downarrow 0}\mathbf{E}_{-\log|x|,\arg(x)}[f(\mathrm{e}^{-(\bar{\xi}_{\mathbf{e}_q}-\xi_{\mathbf{e}_q})}\Theta_{\mathbf{e}_q})\mathrm{e}^{\alpha\xi_{\mathbf{e}_q}}] \\
&= c\pi^{-d/2}2^{-\alpha}\frac{\Gamma((d-\alpha)/2)}{\Gamma(\alpha/2)}\int_{\mathbb{S}^{d-1}}\sigma_1(\mathrm{d}\phi)\int_{|z|>1} f(Kz)\frac{|Kz|^\alpha(|z|^2-1)^{\alpha/2}}{|z|^d|\phi-z|^d}\mathrm{d}z \\
&= c\pi^{-d/2}2^{-\alpha}\frac{\Gamma((d-\alpha)/2)}{\Gamma(\alpha/2)}\int_{\mathbb{S}^{d-1}}\sigma_1(\mathrm{d}\phi)\int_{|w|<1} f(w)\frac{(1-|w|^2)^{\alpha/2}}{|\phi-w|^d}\mathrm{d}w, \quad (15.46)
\end{aligned}
$$

where we changed variables $w = Kz$, or equivalently $z = Kw$, and we used (15.27), that $|w| = 1/|z|$ and that $\mathrm{d}z = \mathrm{d}w/|w|^{2d}$.

In order to pin down the constant c, we need to ensure that, when $f \equiv 1$, the integral on the right-hand side of (15.46) is identically equal to 1. To do this, we recall the Poisson potential formula in (14.29), which tells us that

$$
(1-|w|^2)^{-1} = \int_{\mathbb{S}^{d-1}}|\phi-w|^{-d}\sigma_1(\mathrm{d}\phi), \qquad |w|<1.
$$

Writing $\sigma_r(\mathrm{d}\theta)$, $\theta \in \mathbb{S}^{d-1}(0,r)$ for the uniform surface measure on $\mathbb{S}^{d-1}(0,r)$ normalised to have total mass equal to one, it follows that

$$\int_{\mathbb{S}^{d-1}} \sigma_1(\mathrm{d}\phi) \int_{|w|<1} \frac{(1-|w|^2)^{\alpha/2}}{|\phi - w|^d} \mathrm{d}w$$
$$= \int_{|w|<1} (1-|w|^2)^{\frac{\alpha}{2}-1} \mathrm{d}w$$
$$= \frac{2\pi^{d/2}}{\Gamma(d/2)} \int_0^1 r^{d-1} \mathrm{d}r \int_{\mathbb{S}^{d-1}(0,r)} \sigma_r(\mathrm{d}\theta)(1-r^2)^{\frac{\alpha}{2}-1}$$
$$= \frac{\pi^{d/2}}{\Gamma(d/2)} \int_0^1 y^{\frac{d}{2}-1}(1-y)^{\frac{\alpha}{2}-1} \mathrm{d}y$$
$$= \pi^{d/2} \frac{\Gamma(\alpha/2)}{\Gamma((d+\alpha)/2)}.$$

This forces us to take

$$c = 2^\alpha \frac{\Gamma((d+\alpha)/2)}{\Gamma((d-\alpha)/2)},$$

and so we have

$$\lim_{t\to\infty} \mathbb{E}_{Kx}[f(X_t/M_t)]$$
$$= \pi^{-d/2} \frac{\Gamma((d+\alpha)/2)}{\Gamma(\alpha/2)} \int_{\mathbb{S}^{d-1}} \sigma_1(\mathrm{d}\phi) \int_{|w|<1} f(w) \frac{(1-|w|^2)^{\alpha/2}}{|\phi - w|^d} \mathrm{d}w, \qquad (15.47)$$

as required. □

We conclude this section with an observation regarding the analytical structure of the distributional limit in Theorem 15.12.

Corollary 15.14 *The limiting distribution in Theorem 15.12 is equal in law to the independent product of random variables* U × $\sqrt{\mathrm{B}}$, *where* U *is uniformly distributed on* $\mathbb{S}^{d-1}$ *and* B *is a Beta*($d/2, \alpha/2$) *distribution.*

Proof Indeed, suppose we take $f(w) = |w|^{2\gamma} g(\arg(w))$ for $\gamma > 0$, then we also see that

$$\lim_{t\to\infty} \mathbb{E}_x[f(X_t/M_t)]$$
$$= \frac{2\Gamma((d+\alpha)/2)}{\Gamma(d/2)\Gamma(\alpha/2)} \int_{\mathbb{S}^{d-1}} \sigma_1(\mathrm{d}\phi) \int_0^1 r^{2\gamma+d-1}(1-r^2)^{\alpha/2} \mathrm{d}r \int_{\mathbb{S}^{d-1}} \frac{g(\theta)}{|\phi - r\theta|^d} \sigma_1(\mathrm{d}\theta).$$

The Newton potential formula tells us that $(1-r^2)\int_{\mathbb{S}^{d-1}} |\phi - r\theta|^{-d} \sigma_1(\mathrm{d}\phi) = 1$, see Remark 14.5, and hence, after an application of Fubini's theorem for the two spherical integrals and change of variable,

$$\lim_{t\to\infty} \mathbb{E}_x[f(X_t/M_t)] = \frac{\Gamma((d+\alpha)/2)}{\Gamma(d/2)\Gamma(\alpha/2)} \int_0^1 u^{\gamma+\frac{d}{2}-1}(1-u)^{\frac{\alpha}{2}-1} \mathrm{d}u \times \int_{\mathbb{S}^{d-1}} g(\theta)\sigma_1(\mathrm{d}\theta),$$

verifying the claimed distributional decomposition. □

15.5 Comments

Although it is a special case of general excursion theory as prescribed by, for example, Maisonneuve [147], the radial excursion theory presented in Section 15.1 seems not to have appeared until very recently in the theory of stable processes. See Blumenthal [38] for further reading on the general theory of excursions from a set. Together with the Lamperti–Kiu transform and the limiting procedure in Lemma 15.3, this forms the key to the analysis in Sections 15.2, 15.3 and 15.4, all of which is based entirely on the very recent work in Kyprianou et al. [134]. As with the analogous result in one dimension, the justification of (15.45) comes from the use of Revuz measures; see the bottom of p. 240 in Walsh [212].

16

Windings and up-crossings of stable processes

For a planar isotropic stable process, written $X = (X_t, t \geq 0)$ as usual, we can take advantage of the fact that the plane is isomorphic to $\mathbb{C}$. As such, we may write the process X in polar coordinates, say

$$X_t = r_t \exp(\mathrm{i}\theta_t), \qquad t > 0. \tag{16.1}$$

Clearly, $r_t = |X_t|$ and the angular part, $\theta := (\theta_t, t \geq 0)$, is often referred to as its *winding number*. Roughly speaking θ tells us the net position of winding clockwise and unwinding anti-clockwise around the origin.

Although excluded from the current setting, if we momentarily assume that X is a planar Brownian motion issued away from the origin, a classical result concerning its winding number in (16.1), proved by Spitzer in 1958, states that

$$\frac{2}{\log t}\theta_t \underset{t\to\infty}{\overset{(w)}{\longrightarrow}} C, \tag{16.2}$$

where C is a standard Cauchy random variable. A classical time inversion equality for Brownian motion, that is, $(tX_{1/t}, t > 0)$ is equal in law to $(X_t, t > 0)$, means that (16.2) result also induces a similar winding result as $t \to 0$.

In this chapter, we will explore the natural development of Spitzer's classical result in the setting of isotropic planar stable processes, both for small and for large times. Our approach will make heavy use of the representation of planar stable processes as self-similar Markov processes and will lead us to results concerning the up-crossings of the origin of one-dimensional stable processes.

16.1 Polar decomposition of planar stable processes

As always, we will write $\mathbb{P}_x$, $x \in \mathbb{R}^2$ for the probabilities of X, reserving the special notation $\mathbb{P}$ as shorthand for $\mathbb{P}_0$. Recall from Section 11.5 that planar stable processes are also self-similar Markov processes. As such, they enjoy a representation as a space-time transformation of a Markov additive process given in Theorem 11.13. Taking account of the planar representation (16.1), it is natural that the generalised Lamperti representation may be written more conveniently in the form

$$X_t = \exp\left\{\xi_{\varphi(t)} + \mathrm{i}\chi_{\varphi(t)}\right\}, \qquad t \geq 0, \tag{16.3}$$

where

$$\varphi(t) = \inf\left\{s \geq 0\colon \int_0^s \mathrm{e}^{\alpha\xi_u}\mathrm{d}u > t\right\} = \int_0^t |X_s|^{-\alpha}\mathrm{d}s. \tag{16.4}$$

Here, $(\xi,\chi) = ((\xi_t,\chi_t)\colon t \geq 0)$ is such that (ξ,χ) is a strong Markov process with probabilities $(\mathbf{P}_{x,y},\ x,y \in \mathbb{R}^2)$ with the MAP property that, $(\xi_{t+s} - \xi_t, \chi_{t+s})$, $s \geq 0$, given $\sigma((\xi_u,\chi_u)\colon u \leq s)$, is equal in law to (ξ,χ) under $\mathbf{P}_{0,y}$ with $y = \chi_t$. We thus have the polar representation in terms of the underlying MAP

$$r_t = \mathrm{e}^{\xi_{\varphi(t)}} \text{ and } \theta_t = \chi_{\varphi(t)}, \qquad t \geq 0. \tag{16.5}$$

As our first result, we will show that we can discern more properties of the MAP (ξ,χ) other than those given in the standard definition above.

Lemma 16.1 *The pair (ξ,χ) is a two-dimensional Lévy process with no Gaussian component and Lévy measure given by*

$$\Pi_{\xi,\chi}(\mathrm{d}x,\mathrm{d}\vartheta) = \frac{\alpha 2^\alpha}{\pi}\frac{\Gamma(1+\alpha/2)}{\Gamma(1-\alpha/2)}\frac{\mathrm{e}^{2x}}{(\mathrm{e}^{2x}+1-2\mathrm{e}^x\cos(\vartheta))^{1+\alpha/2}}\,\mathrm{d}x\,\mathrm{d}\vartheta, \tag{16.6}$$

for $x \in \mathbb{R}, \vartheta \in (-\pi,\pi]$.

Proof Referring back to Theorem 12.3, in the notation used there, we have $\Theta_t = \exp\{\mathrm{i}\chi_t\}$. Hence, for all bounded and measurable f,

$$\mathbf{E}_{0,\theta}\left[\sum_{s>0} f(s,\Delta\xi_s,\Delta\chi_s)\right]$$
$$= 2c(\alpha)\mathbf{E}_{0,\theta}\left[\int_0^\infty \mathrm{d}s\int_0^\pi \mathrm{d}\vartheta\int_{\mathbb{R}}\mathrm{d}x\frac{\mathrm{e}^{2x}}{|\mathrm{e}^{2x}+1-2\mathrm{e}^x\cos\vartheta\,|^{(\alpha+2)/2}}f(s,x,\vartheta)\right],$$

where $\Delta\chi_t = \chi_t - \chi_{t-}$ and $\Delta\xi_t = \xi_t - \xi_{t-}$, $t \geq 0$. The jump measure $\Pi_{\xi,\chi}$ can be read directly out of the above equality. □

As already noted in Section 5.7 the isotropic property of X implies that $(|X_t|, t \geq 0)$ is a positive self-similar Markov process. In particular, when one considers ξ as a lone process, without information about χ, then it is a Lévy process. With an abuse of notation, we denote its probabilities by $\mathbf{P}_x$, $x \in \mathbb{R}$. The fact that $\lim_{t\to\infty}|X_t| = \infty$ (due to transience, see Section 3.6) implies that

$\lim_{t\to\infty} \xi_t = \infty$, almost surely. In Theorem 5.21, we derived the characteristic exponent of ξ. For $z \in \mathbb{R}$,

$$-\log \mathbf{E}_0[e^{iz\xi_1}] =: \Psi(z) = 2^\alpha \frac{\Gamma(\frac{1}{2}(-iz+\alpha))}{\Gamma(-\frac{1}{2}iz)} \frac{\Gamma(\frac{1}{2}(iz+2))}{\Gamma(\frac{1}{2}(iz+2-\alpha))}.$$

It is straightforward to see that this exponent can be analytically extended to the Laplace exponent $\psi(u) := -\Psi(-iu)$ for $-2 < u < \alpha$, which is convex, having roots at $u = 0$ and $u = \alpha - 2$ and exploding at $u = -2$ and $u = \alpha$.

The main objective in the remainder of this chapter is to analyse the winding number $(\theta_t, t \geq 0)$. From (16.5), we see that the winding number is a Lévy process subordinated by a time change that is correlated to the radial distance from the origin. Our approach is, therefore, to understand the individual behaviour of the Lévy process χ and the time change φ and then to consider their combined effect. An immediate observation we can make to this end is that the Lévy process χ is in the domain of attraction of a Brownian motion. This is a consequence of the fact that χ has jumps that are bounded in magnitude by π and hence is a Lévy process with finite second moments. Indeed, recalling Theorem 2.11, we have the following corollary.

Corollary 16.2 *As $r \to \infty$, $(r^{-1/2}\chi_{rt}, t \geq 0)$ converges weakly on the Skorokhod space to $(\sqrt{k(\alpha)}B_t, t \geq 0)$, where B is a standard Brownian motion and*

$$k(\alpha) = \frac{\alpha 2^\alpha}{\pi} \frac{\Gamma(1+\alpha/2)}{\Gamma(1-\alpha/2)} \int_0^\infty \int_{-\pi}^{\pi} \frac{r\theta^2}{(1+r^2-2r\cos\theta)^{1+\alpha/2}} \, r\,dr\,d\theta.$$

16.2 Windings at infinity

In this section, we shall consider a result for planar isotropic stable processes in the spirit of (16.2). Rather than providing a weak limit in time, we prove the slightly stronger result of weak functional convergence of the winding number. Most noticeable from the following result is that, unlike the Brownian case where winding scales with speed $\log t$, winding for stable processes occurs with the slightly slower scaling $\sqrt{\log t}$.

Theorem 16.3 (Planar stable windings at ∞) *Suppose that $(X_t, t \geq 0)$ is an isotropic planar α-stable process, with $\alpha \in (0, 2)$, that is issued from a point different from the origin. There exists a constant*

$$s(\alpha) = \frac{\alpha}{\pi} \int_0^\infty \int_{-\pi}^{\pi} \frac{r\theta^2}{(1+r^2-2r\cos\theta)^{1+\alpha/2}} \, r\,dr\,d\theta,$$

such that the process $(r^{-1/2}\theta_{\exp(rt)}, t \geq 0)$ *converges weakly on the Skorokhod space, as* $r \to \infty$, *to* $(\sqrt{s(\alpha)}B_t, t \geq 0)$, *where* $(B_t, t \geq 0)$ *is a standard one-dimensional Brownian motion issued from the origin.*

In particular, in contrast to Spitzer's winding result for Brownian motion, (16.2), *we have*

$$\frac{\theta_t}{\sqrt{s(\alpha)\log t}} \xrightarrow[t\to\infty]{(w)} \mathcal{N}(0,1),$$

where $\mathcal{N}(0, 1)$ *is a standard Normal random variable.*

Taking account of the fact that $\theta_t = \chi_{\varphi(t)}$, $t \geq 0$, Theorem 16.3 will follow from Corollary 16.2 if we can demonstrate suitable control over the growth of $\varphi(t)$, as $t \to \infty$. Indeed, the class of path functionals that are continuous with respect to the Skorokhod topology is closed under the operation of composition and hence it would suffice to show that

$$\frac{\varphi(\exp(t))}{t} \to 2^{-\alpha}\frac{\Gamma(1-\alpha/2)}{\Gamma(1+\alpha/2)} \tag{16.7}$$

in distribution, as $t \to \infty$, in order to prove Theorem 16.3. In fact, it turns out that something slightly stronger than (16.7) can be proved.

Proposition 16.4 *The limit* (16.7) *holds almost surely and hence Theorem 16.3 follows.*

Proof We start by defining

$$\tilde{X}_t = \exp(-t/\alpha)X_{\exp(t)}, \qquad t \geq 0,$$

and we claim that the process $(\tilde{X}_t, t \geq 0)$ is stationary and ergodic. This will allow us to invoke the Ergodic Theorem which, thanks to (16.4), will give us the almost sure limit

$$\begin{aligned}\lim_{t\to\infty}\frac{\varphi(\exp(t))}{t} &= \lim_{t\to\infty}\frac{\varphi(1)}{t} + \lim_{t\to\infty}\frac{1}{t}\int_1^{\exp t}|X_s|^{-\alpha}\mathrm{d}s\\ &= \lim_{t\to\infty}\frac{1}{t}\int_1^t|\tilde{X}_u|^{-\alpha}\mathrm{d}u\\ &= \mathbb{E}[|X_1|^{-\alpha}].\end{aligned} \tag{16.8}$$

The expectation above can be computed by recalling that X is equal in law to $(B_{2\gamma_t}, t \geq 0)$, where $(B_t, t \geq 0)$ is a standard 2-dimensional Brownian motion and $(\gamma_t, t \geq 0)$ is an independent stable subordinator with index $\alpha/2$. For the purpose of computing moments, we also note that $B_{2\gamma_1}$ is equal in distribution to $\sqrt{2\gamma_1}B_1$. We can thus simplify

$$\mathbb{E}[|X_1|^{-\alpha}] = 2^{-\alpha/2}\mathbb{E}[|B_1|^{-\alpha}]\mathbb{E}[\gamma_1^{-\alpha/2}]. \tag{16.9}$$

To compute the two expectations on the right-hand side, we can make clever use of the integral representation of the gamma function that leads to

$$q^{-c} = \frac{1}{\Gamma(c)} \int_0^\infty e^{-qt} t^{c-1} dt, \qquad q > 0, c \in (0, 1).$$

Indeed, we can now compute

$$\begin{aligned}
\mathbb{E}[\gamma_1^{-\alpha/2}] &= \frac{1}{\Gamma(\alpha/2)} \int_0^\infty \mathbb{E}[e^{-\gamma_1 t}] t^{-1+\alpha/2} dt \\
&= \frac{1}{\Gamma(\alpha/2)} \int_0^\infty e^{-t^{\alpha/2}} t^{-1+\alpha/2} dt \\
&= \frac{2}{\alpha\Gamma(\alpha/2)} \int_0^\infty e^{-s} ds \\
&= \frac{1}{\Gamma(1 + \alpha/2)}.
\end{aligned}$$

Similarly, we have

$$\begin{aligned}
\mathbb{E}[|B_1|^{-\alpha}] &= \frac{1}{\Gamma(\alpha/2)} \int_0^\infty \mathbb{E}[e^{-|B_1|^2 t}] t^{-1+\alpha/2} dt \\
&= \frac{1}{\Gamma(\alpha/2)} \int_0^\infty (1 + 2t)^{-1} t^{-1+\alpha/2} dt \\
&= 2^{-\alpha/2}\Gamma(1 - \alpha/2), \qquad (16.10)
\end{aligned}$$

where we have used the expression for the moment generating function of a chi-squared variable with index 2 in the second equality and the representation as an integral of the beta function from (A.18), in the Appendix, in the third equality. Combining these moments back in (16.9) gives us the required constant in the limit (16.7).

To complete the proof, we must thus prove that $\tilde{X}$ is stationary and ergodic. To this end, let us introduce $\tilde{\mathbb{P}}_x$ for the law of the process $\tilde{X}$ when X has law $\mathbb{P}_x$, $x \in \mathbb{R}^2$. Following the same connection for the latter probabilities, we write $\tilde{\mathbb{P}}$ in place of $\tilde{\mathbb{P}}_0$. Note, $\tilde{X}$ is Markovian on account of the fact that, for $s, t > 0$ and all bounded and measurable functions g on $\mathbb{R}^2$, the scaling property implies

$$\begin{aligned}
\tilde{\mathbb{E}}_x[g(\tilde{X}_{t+s}) | \sigma(\tilde{X}_u, u \le t)] &= \mathbb{E}_x[g(e^{-s/\alpha} e^{-t/\alpha} X_{e^t e^s}) | \sigma(X_u, u \le e^t)] \\
&= \mathbb{E}_{x' e^{-t/\alpha}}[g(e^{-s/\alpha} X_{e^s})]_{x' = X_{e^t}} \\
&= \tilde{\mathbb{E}}_{x''}[g(\tilde{X}_s)]_{x'' = \tilde{X}_t}. \qquad (16.11)
\end{aligned}$$

A similar calculation using scaling gives us that, for $x \in \mathbb{R}^2$ and $t \ge 0$,

$$\mathbb{E}_x[g(\tilde{X}_t)] = \mathbb{E}_{xe^{-t/\alpha}}[g(X_1)].$$

Letting t tend to infinity now shows that $\tilde{X}$ has a limiting stationary distribution corresponding to the law of $(X_1, \mathbb{P})$.

For ergodicity, we can think of the probability measure $\tilde{\mathbb{P}}$ as being supported on the Skorokhod space associated to $\mathbb{R}^2$ (see Section A.10 in the Appendix). We may now define the transformation $T_t[\omega(\cdot)] = \exp(-t/\alpha)\omega(e^t \cdot)$, for $\omega \in \mathcal{D}(\mathbb{R}^2)$, which is measure preserving thanks to the scaling property of $(X, \mathbb{P})$ and hence of $(X, \tilde{\mathbb{P}})$. For us to conclude that $(\tilde{X}_t, t \geq 0)$ is an ergodic sequence, from which we can conclude (16.8), it suffices to check the defining criterion of ergodicity. Specifically, we need to check that, for an event A that is invariant with respect to $(T_t, t \geq 0)$, $\tilde{\mathbb{P}}(A)$ is equal to 0 or 1.

To this end, note that, on the one hand, we have by the property of uniformly integrable martingales,

$$\mathbf{1}_A = \lim_{u\to\infty} \tilde{\mathbb{E}}[\mathbf{1}_A | \sigma(\tilde{X}_s, s \leq u)]. \tag{16.12}$$

On the other hand, by the Markov property,

$$\tilde{\mathbb{E}}[\mathbf{1}_A | \sigma(\tilde{X}_s, s \leq u)] = \tilde{\mathbb{E}}[\mathbf{1}_A | \sigma(\tilde{X}_s, s \leq u)] = g(\tilde{X}_u), \tag{16.13}$$

where $g(x) = \tilde{\mathbb{P}}_x(A)$, $x \in \mathbb{R}^2$. Now taking expectation again in (16.13) and taking limits as $u \to \infty$, recalling that the stationary distribution of $\tilde{X}$ is given by $(X_1, \mathbb{P})$, we observe that $g(x) = \mathbb{E}[g(X_1)]$ for all $x \in \mathbb{R}^2$, showing that g is a constant function. Now comparing the right-hand side of (16.13) with the indicator on the left-hand side of (16.12), it becomes clear that g is valued as either 0 or 1. □

16.3 Windings at the origin

In this section, we will develop an understanding of the windings of the stable process as it emerges from the origin. In order to do this, we will use a technique based on the relationship between $(X, \mathbb{P})$ and the singular law of X conditioned to limit continuously to the origin. In particular, we will show how they are related simply via time reversal.

Before going into technical details, the concept of winding at the origin of a stable process needs more attention. We need to be careful with the notation $\theta := (\theta_t, t \geq 0)$, which no longer makes sense when the process is issued from the origin as, by time t, the process has already undergone an infinite number of windings around the origin in both directions. Instead, we need to talk about angular displacements in relative, rather than absolute, terms. To this end, we shall henceforth work with $\theta_{[a,b]}$, $0 < a \leq b < \infty$, which is well defined as the winding of X over the time interval $[a, b]$. Of course, in the setting that X is issued from a point other than the origin, we can continue to write $\theta_t = \theta_{[0,t]}$. Our main result in this section, which complements Theorem 16.3, thus becomes the following.

Theorem 16.5 (Planar stable windings at 0) *Suppose that $(X_t, t \geq 0)$ is an isotropic planar α-stable process, with $\alpha \in (0,2)$, that is issued from the origin. The process $(r^{-1/2}\theta_{[\exp(-rt),1]}, t \geq 0)$ converges weakly on the Skorokhod space to $(\sqrt{s(\alpha)}B_t, t \geq 0)$ as $r \to \infty$, where $(B_t, t \geq 0)$ is a standard one-dimensional Brownian motion issued from the origin and $s(\alpha)$ is the same constant appearing in Theorem 16.3.*

In particular,

$$\frac{\theta_{[t,1]}}{\sqrt{s(\alpha)\log(1/t)}} \underset{t\to 0}{\overset{(w)}{\longrightarrow}} \mathcal{N}(0,1),$$

where $\mathcal{N}(0,1)$ is a standard Normal random variable.

Recall that $\mathbb{P}^\circ := (\mathbb{P}^\circ_x, x \neq 0)$ was constructed in d-dimensions via a limiting procedure in Section 12.2. More precisely, if we define $\mathcal{F}_t := \sigma(X_u\colon u \leq t)$, $t \geq 0$, then, for all $t \geq 0$, $A \in \mathcal{F}_t$ and $x \neq 0$,

$$\mathbb{P}^\circ_x(A, t < \zeta) := \lim_{\epsilon \downarrow 0} \mathbb{P}_x(A \cap \{t < \tau^{B_\epsilon}\} | \tau^{B_\epsilon} < \infty), \qquad |x| > 0, \tag{16.14}$$

where $\tau^{B_\epsilon} := \inf\{s > 0\colon |X_s| < \epsilon\}$ and ζ is the lifetime of the process under $\mathbb{P}^\circ$, which agrees with $\tau^{\{0\}} = \inf\{t > 0\colon X_t = 0\}$. Note, in the above expression for $\mathbb{P}^x$, $x \neq 0$, we have taken account of the fact that $\mathbb{P}_x(\tau^{\{0\}} = \infty) = 1$, for all $x \neq 0$. Moreover, for all $t \geq 0$, $x \neq 0$, we can equivalently express the conditioned process via the change of measure

$$\left.\frac{\mathrm{d}\mathbb{P}^\circ_x}{\mathrm{d}\mathbb{P}_x}\right|_{\mathcal{F}_t} = \frac{|X_t|^{\alpha-2}}{|x|^{\alpha-2}}\mathbf{1}_{t<\zeta}. \tag{16.15}$$

The change of measure (16.15) ensures that $(X, \mathbb{P}^\circ_x)$, $x \in \mathbb{R}^2\backslash\{0\}$ is again an isotropic self-similar Markov process and therefore has a decomposition in the spirit of (16.3). Indeed, let us write $X^\circ = (X^\circ_t, t \geq 0)$ to mean a canonical version of $(X, \mathbb{P}^\circ_x)$, $x \in \mathbb{R}^2\backslash\{0\}$. We shall write its polar decomposition as

$$X^\circ_t = \exp\{\xi^\circ_{\varphi^\circ(t)} + \mathrm{i}\chi^\circ_{\varphi^\circ(t)}\}, \qquad t \leq \zeta^\circ, \tag{16.16}$$

where

$$\varphi^\circ(t) = \inf\left\{s \geq 0\colon \int_0^s \mathrm{e}^{\alpha\xi^\circ_u}\mathrm{d}u > t\right\} = \int_0^t |X^\circ_s|^{-\alpha}\mathrm{d}s$$

and $\zeta^\circ = \inf\{s > 0\colon X^\circ_s = 0\}$ is the lifetime of the process. Once again, the process (ξ°, χ°) is a Markov additive process, where χ° is the underlying modulation to ξ°. Isotropy also ensures that the process $|X^\circ| := (|X^\circ_t|, t \geq 0)$ is again a positive self-similar Markov process and ξ°, when observed as a lone process, is a Lévy process. On account of the Doob h-transform of the law of X° with respect to X, one easily verifies that its characteristic exponent is an Esscher transform with respect to that of ξ. Moreover, the characteristic exponent, Ψ° of ξ° satisfies

$$\Psi^\circ(z) = \Psi(z - \mathrm{i}(\alpha-2)) = 2^\alpha \frac{\Gamma(\frac{1}{2}(-\mathrm{i}z+2))}{\Gamma(-\frac{1}{2}(\mathrm{i}z+\alpha-2))} \frac{\Gamma(\frac{1}{2}(\mathrm{i}z+\alpha))}{\Gamma(\frac{1}{2}(\mathrm{i}z))} = \Psi(-z), \quad (16.17)$$

for $z \in \mathbb{R}$. That is to say, ξ° is equal in law to $-\xi$. In fact, one can go a little further and use Corollary 12.9 to deduce that the pair (ξ°, χ°) is equal in law to the pair $(-\xi, \chi)$.

As alluded to above, our aim is to relate the process $(X, \mathbb{P}_0)$ to the process X° via time reversal.

Lemma 16.6 *For each $a > 0$, define $\mathtt{D}_a = \sup\{s \geq 0 \colon |X_s| \leq a\}$. Conditionally on the event $\{X_{\mathtt{D}_a-} = x\}$, where $|x| < a$, the process $(X_{(\mathtt{D}_a-t)-}, t \leq \mathtt{D}_a)$ under $\mathbb{P}_0$ is equal in law to X° issued from $X^\circ_0 = x$.*

Proof We will prove the reformulation above using the Hunt–Nagasawa duality theory, introduced in the Appendix. In the language of Nagasawa's Duality Theorem A.15, we will take the semigroup $(\hat{\mathcal{P}}_t[f], t \geq 0)$ to be that of the conditioned process X°. Accordingly, since $(\mathbb{P}^\circ_x, x \neq 0)$, can be written as a Doob h-transform with respect to $(\mathbb{P}_x, x \neq 0)$, the requirement **(B)** in Theorem A.15 is automatically satisfied by appealing, for example, to dominated convergence.

To deal with the condition **(A)** of Theorem A.15, let us introduce some notation. For $x, y \in \mathbb{R}^2$, we shall write $R(x, \mathrm{d}y)$ for the resolvent of X. Theorem 3.11 tells us that

$$R(x, \mathrm{d}y) = C(\alpha)|x-y|^{\alpha-2}\mathrm{d}y, \qquad x, y \in \mathbb{R}^2, \quad (16.18)$$

where $C(\alpha)$ is a constant depending on the index of stability α that is of no interest here. Taking account of the fact that X is issued from the origin, paraphrasing condition **(A)**, we need to check that, with

$$\varpi(\mathrm{d}x) := \int_{\mathbb{R}^2} \delta_0(\mathrm{d}a) R(a, \mathrm{d}x) = R(0, \mathrm{d}x) = |x|^{\alpha-2}\mathrm{d}x, \qquad x \in \mathbb{R}^2,$$

we have

$$p_t(x, \mathrm{d}y)\varpi(\mathrm{d}x) = p^\circ_t(y, \mathrm{d}x)\varpi(\mathrm{d}y), \qquad x, y \in \mathbb{R}^2\backslash\{0\}, t \geq 0. \quad (16.19)$$

Here, $p_t(x, \mathrm{d}y)$ and $p^\circ_t(y, \mathrm{d}x)$ represent the transition semigroups of X and X°, respectively.

We now see that (16.19) requires us to check that

$$p_t(x,\mathrm{d}y)|x|^{\alpha-2}\mathrm{d}x = \frac{|x|^{\alpha-2}}{|y|^{\alpha-2}}p_t(y,\mathrm{d}x)|y|^{\alpha-2}\mathrm{d}y, \qquad x,y \in \mathbb{R}^2\backslash\{0\}.$$

Hence, we require that $p_t(x,\mathrm{d}y)\mathrm{d}x = p_t(y,\mathrm{d}x)\mathrm{d}y$, $x,y \in \mathbb{R}^2\backslash\{0\}$. However, this is nothing more than the classical duality property for Lévy process semi-groups (and in particular for isotropic stable process semi-groups). □

The consequence of this last lemma is that we can study the windings of X backwards from last exit from the ball of radius a by considering instead the windings of X° as $t \uparrow \zeta^\circ$ from a randomised initial position, which we can control by conditioning on the distribution of the aforesaid last exit point. However, because of the nature of the scaling in the winding functional limit theorem and that only finite winding can occur over finite time horizons, knowledge of backward winding of X from $\mathtt{D}_a$ to the origin is sufficient to tell us about backward winding of X from time 1 to the origin. Indeed,

$$\theta_{[t,1]} = \theta_{[t,\mathtt{D}_a]} + \theta_{(\mathtt{D}_a,1]}\mathbf{1}_{(\mathtt{D}_a\le 1)} - \theta_{(1,\mathtt{D}_a]}\mathbf{1}_{(\mathtt{D}_a>1)},$$

and hence, when scaling by $r^{-1/2}$, the difference $|\theta_{[\mathrm{e}^{-rt},1]} - \theta_{[\mathrm{e}^{-rt},\mathtt{D}_a]}|$ becomes irrelevant.

Proof of Theorem 16.5 Suppose we write

$$\mathrm{Arg}(X_t^\circ) = \mathrm{e}^{\mathrm{i}\theta_t^\circ}, \qquad t \le \zeta^\circ.$$

From the representation (16.16), we have that, on $\{s < \zeta^\circ\}$,

$$\theta^\circ_{\zeta^\circ - s} = \chi^\circ_{\varphi^\circ(\zeta^\circ - s)}, \tag{16.20}$$

where

$$\int_0^{\varphi^\circ(\zeta^\circ - s)} \mathrm{e}^{\alpha\xi_u^\circ}\mathrm{d}u = \zeta^\circ - s.$$

Lemma 16.6 tells us that studying winding to the origin of $\theta_{[t,1]}$ as $t \downarrow 0$, is equivalent to studying the winding to the origin of $\theta^\circ_{\zeta^\circ - s}$ as $s \downarrow 0$.

For convenience, let us write $\hat{\varphi}^\circ(s) = \varphi^\circ(\zeta^\circ - s)$, providing $s \le \zeta^\circ$. Note in particular that $\hat{\varphi}^\circ(\zeta^\circ) = 0$ and that $\hat{\varphi}^\circ(0) = \infty$. We also have that

$$\int_{\hat{\varphi}^\circ(s)}^\infty \mathrm{e}^{\alpha\xi_u^\circ}\mathrm{d}u = \int_0^\infty \mathrm{e}^{\alpha\xi_u^\circ}\mathrm{d}u - \int_0^{\varphi^\circ(\zeta^\circ - s)} \mathrm{e}^{\alpha\xi_u^\circ}\mathrm{d}u = \zeta^\circ - (\zeta^\circ - s) = s.$$

Differentiating, we see that, on $\{s < \zeta^\circ\}$,

$$\frac{\mathrm{d}\hat{\varphi}^\circ(s)}{\mathrm{d}s} = -\mathrm{e}^{-\alpha\xi^\circ_{\hat{\varphi}^\circ(s)}} = -|X^\circ_{\zeta^\circ - s}|^{-\alpha}$$

and hence, after integrating, since $\hat{\varphi}^\circ(\zeta^\circ) = 0$, on $\{t < \zeta^\circ\}$,

$$\hat{\varphi}^\circ(t) = \hat{\varphi}^\circ(t) - \hat{\varphi}^\circ(\zeta^\circ) = \int_t^{\zeta^\circ} |X^\circ_{\zeta^\circ - s}|^{-\alpha} \mathrm{d}s.$$

Now define $\hat{X}^\circ_v = \mathrm{e}^{v/\alpha} X^\circ_{\zeta^\circ - \mathrm{e}^{-v}}$, $v \in \mathbb{R}$, so that, on $\{t < \zeta^\circ\}$,

$$\hat{\varphi}^\circ(t) = \int_{-\log \zeta^\circ}^{-\log t} |\hat{X}^\circ_v|^{-\alpha} \mathrm{d}v. \tag{16.21}$$

Next, we recall from Lemma 16.6 that $(X^\circ_{(\zeta^\circ - s)-}, s \le \zeta^\circ)$ under $\mathbb{P}^\circ_{\mu_a}$, where $\mu_a(\mathrm{d}x) = \mathbb{P}_0(X_{\mathtt{D}_a -} \in \mathrm{d}x)$, $|x| < a$, agrees with $(X_s, s < \mathtt{D}_a)$ under $\mathbb{P}_0$. It therefore follows that, under $\mathbb{P}^\circ_{\mu_a}$, $(\hat{X}^\circ_t, \mathrm{e}^{-t} < \zeta^\circ)$ is equal in law to $(\mathrm{e}^{t/\alpha} X_{\mathrm{e}^{-t}}, \mathrm{e}^{-t} < \mathtt{D}_a)$ under $\mathbb{P}_0$. Following the reasoning in the proof of Proposition 16.4, we note that $\hat{X}_t = \mathrm{e}^{t/\alpha} X_{\mathrm{e}^{-t}}$, $t \in \mathbb{R}$, is a stationary ergodic Markov process with distribution at each time equal to that of X_1 and, moreover, almost surely,

$$\lim_{r\to\infty} \frac{\hat{\varphi}^\circ(\exp(-r))}{r} = \mathbb{E}[|X_1|^{-\alpha}] = 2^{-\alpha} \frac{\Gamma(1-\alpha/2)}{\Gamma(1+\alpha/2)}. \tag{16.22}$$

We know from Theorem 16.3 that $(r^{-1/2}\chi_{\varphi(\exp(rt))}, t \ge 0)$ converges on the Skorokhod space to $(\sqrt{s(\alpha)}B_t, t \ge 0)$, as $r \to \infty$. Since by (16.20),

$$(r^{-1/2}\theta^\circ_{\zeta^\circ - \exp(-rt)}, t \ge 0) \overset{(\mathrm{d})}{=} (r^{-1/2}\chi^\circ_{\hat{\varphi}^\circ(\exp(-rt))}, t \ge 0),$$

by Corollary 12.9, $\chi^\circ \overset{(\mathrm{d})}{=} \chi$ and by (16.22), $\hat{\varphi}^\circ(\exp(-rt))$ has the same almost sure growth as $\varphi(\exp(rt))$, we conclude that $(r^{-1/2}\theta^\circ_{\zeta^\circ - \exp(-rt)}, t \ge 0)$ converges in the Skorokhod topology to $(\sqrt{s(\alpha)}B_t, t \ge 0)$, as $r \to \infty$. Recalling the remark earlier that winding to the origin of $\theta_{[t,1]}$, as $t \downarrow 0$, is equivalent to studying the winding to the origin of $\theta^\circ_{\zeta^\circ - s}$, as $s \downarrow 0$, the proof is complete. □

16.4 Upcrossings of one-dimensional stable processes

In this section, we turn our interest to the one-dimensional case. Hereafter, $X = (X_t, t \ge 0)$ will denote a one-dimensional stable process with stability index $\alpha \in (0,2)$ and probabilities $\mathbb{P}_x$, $x \in \mathbb{R}$. Note in particular that we do not insist that X is symmetric; however, we do insist that it undergoes jumps of both signs, that is, $\rho \in (0,1)$ for $\alpha \in (0,1)$ and $0 < \alpha\rho, \alpha\hat{\rho} < 1$ for $\alpha \in (1,2)$.

Recall that the long term behaviour of X can differ from its two-dimensional counter part depending on the value of α. When $\alpha \in (0,1)$, we know that $\lim_{t\to\infty} |X_t| = \infty$ and $\mathbb{P}_x(\tau^{\{0\}} = \infty) = 1$, $x \ne 0$, where $\tau^{\{0\}} = \inf\{s > 0 \colon X_s = 0\}$. When $\alpha = 1$, we have $\limsup_{t\to\infty} |X_t| = \infty$ and $\liminf_{t\to\infty} |X_t| = 0$ and

$\mathbb{P}_x(\tau^{\{0\}} = \infty) = 1$, $x \neq 0$. Finally, when $\alpha \in (1, 2)$, we have $\lim_{t\to\infty} |X_t| = \infty$ and $\mathbb{P}_x(\tau^{\{0\}} < \infty) = 1$, $x \neq 0$.

As we have assumed that X has both positive and negative jumps, it cannot creep upwards or downwards, and hence up-crossings into the positive half-line will always be by a jump. Accordingly, let $\mathtt{U}_{[a,b]}$, $0 < a \leq b < \infty$, be the number of up-crossings in the time interval $[a, b]$. That is to say,

$$\mathtt{U}_{[a,b]} = \sum_{a\leq s\leq b} \mathbf{1}_{(X_s>0,\, X_{s-}<0)}.$$

We write $\mathtt{U}_t = \mathtt{U}_{[0,t]}$, $t \geq 0$, providing that $X_0 \neq 0$.

We are interested in up-crossings both as time tends to zero and to infinity in the case $\alpha \in (0, 1]$ (in which regime the origin is polar). In the setting $\alpha \in (1, 2)$, we are, in principle, interested in up-crossings as time tends to zero and as time tends to the first hitting time of the origin. More precisely, we prove strong laws of large numbers for the up-crossing count.

Theorem 16.7 (Stable up-crossings) *Suppose that X is a one-dimensional stable process with two-sided jumps and with index $\alpha \in (0, 2)$.*

(i) If $\alpha \in (0, 1]$, then when X is issued from a point other than the origin,

$$\lim_{t\to\infty} \frac{\mathtt{U}_t}{\log t} = \frac{\sin(\pi\alpha\rho)\sin(\pi\alpha\hat{\rho})}{\alpha\pi\sin(\pi\alpha)}, \tag{16.23}$$

almost surely, with the understanding that the constant on the right-hand side above is equal to infinity when $\alpha = 1$.

(ii) If $\alpha \in (0, 1]$, then when X is issued from the origin,

$$\lim_{t\to 0} \frac{\mathtt{U}_{[t,1]}}{\log(1/t)} = \frac{\sin(\pi\alpha\rho)\sin(\pi\alpha\hat{\rho})}{\alpha\pi\sin(\pi\alpha)}, \tag{16.24}$$

almost surely, where, again, we understand the constant on the right-hand side above is equal to infinity when $\alpha = 1$.

(iii) If $\alpha \in (1, 2)$, then, when X is issued from a point other than the origin,

$$\lim_{t\to 0} \frac{\mathtt{U}_{\tau^{\{0\}}-t}}{\log(1/t)} = \frac{\sin(\pi\alpha\rho)\sin(\pi\alpha\hat{\rho})}{\alpha\pi|\sin(\pi\alpha)|},$$

almost surely, where $\tau^{\{0\}} = \inf\{t > 0 : X_t = 0\}$.

In the above theorem, when $\alpha \in (1, 2)$ and X is issued from the origin, the reader may expect to see a result for $\mathtt{U}_{[t,1]}$, as $t \to 0$. However, the question of counting up-crossings does not make sense any more. For this parameter regime, because X is issued from the origin, $\tau^{\{0\}} = 0$ almost surely. Moreover, over each time horizon $[0, \varepsilon)$, $\varepsilon > 0$, X enjoys a countable infinity of excursions from the origin; and within each excursion there are a countable infinity of up-crossings.

The key to the proof of Theorem 16.7 will again be the representation of the stable process as a self-similar Markov process. Recall that there exists a MAP, $(\xi, J) = ((\xi_t, J_t), t \geq 0)$, taking values in $\mathbb{R} \times \{-1, 1\}$, with matrix exponent, $\boldsymbol{\Psi}$, given by (12.1), such that, for $X_0 \neq 0$,

$$X_t = \exp(\xi_{\varphi(t)})J_{\varphi(t)}, \qquad t \leq \tau^{\{0\}}, \tag{16.25}$$

where

$$\varphi(t) = \inf\left\{s \geq 0 \colon \int_0^s \mathrm{e}^{\alpha\xi_u}\mathrm{d}u > t\right\} = \int_0^t |X_s|^{-\alpha}\mathrm{d}s. \tag{16.26}$$

When $\alpha \in [1, 2)$, $\tau^{\{0\}} = \infty$ almost surely so the decomposition holds for all times, otherwise, when $\alpha \in (0, 1)$, it only gives a pathwise decomposition up until first hitting of the origin.

Proof of Theorem 16.7 (i) Let $N := (N_t, t \geq 0)$ be the counting process of the number of jumps of the process J from -1 to 1 in the time interval $[0, t]$ when X is issued from a point other than the origin. That is to say,

$$N_t = \sum_{0 \leq s \leq t} \mathbf{1}_{(J_{s-}=-1,\, J_s=1)}, \qquad t \geq 0.$$

The processes N and U are related by the time change $\mathrm{U}_t = N_{\varphi(t)}$, $t \leq \tau^{\{0\}}$. In the case that $X_0 \neq 0$, we set $N_0 = 0$ and, for every $n \in \mathbb{N}$, we define

$$T_n = \inf\{t > 0 \colon N_t = n\}.$$

The random time between two consecutive up-crossings in the time-scale of the MAP is distributed as the sum of two independent exponential variables, the holding times of J between the transitions $1 \to -1$ and $-1 \to 1$, with respective rates $\boldsymbol{\Psi}(0)_{1,-1}$ and $\boldsymbol{\Psi}(0)_{-1,1}$. Standard Markov chain theory tells us that, for all $x \in \mathbb{R}$ and $i \in \{-1, 1\}$, we have $\mathbf{P}_{x,i}$-almost surely,

$$\begin{aligned}
\lim_{t\to\infty} \frac{N_t}{t} &= \lim_{n\to\infty} \frac{n}{T_n} \\
&= \frac{\boldsymbol{\Psi}(0)_{1,-1}\boldsymbol{\Psi}(0)_{-1,1}}{\boldsymbol{\Psi}(0)_{1,-1} + \boldsymbol{\Psi}(0)_{-1,1}} \\
&= \frac{\Gamma(\alpha)}{\Gamma(\alpha\hat{\rho})\Gamma(1 - \alpha\hat{\rho}) + \Gamma(\alpha\rho)\Gamma(1 - \alpha\rho)} \\
&= \frac{\Gamma(\alpha)\sin(\pi\alpha\rho)\sin(\pi\alpha\hat{\rho})}{\pi(\sin(\pi\alpha\rho) + \sin(\pi\alpha\hat{\rho}))}.
\end{aligned} \tag{16.27}$$

As in the two-dimensional setting, it suffices to prove that $\varphi(t)$ grows like $\log t$ in an almost sure sense. The method we use is similar to the analysis of the clock φ for planar stable processes in Theorem 16.3. In particular, it is straightforward to see that $\tilde{X}_t = \mathrm{e}^{-\alpha/t} X_{\mathrm{e}^t}$, $t \geq 0$, is a stationary ergodic Markov process with stationary distribution equal to that of X_1. Hence, from (16.26) we have

$$\begin{aligned}
\lim_{t\to\infty} \frac{\varphi(\exp(t))}{t} &= \lim_{t\to\infty} \frac{\varphi(\exp(t)) - \varphi(1)}{t} \\
&= \frac{1}{t}\int_1^{\exp(t)} |X_u|^{-\alpha} \mathrm{d}u \\
&= \frac{1}{t}\int_0^{t} |\tilde{X}_v|^{-\alpha} \mathrm{d}v \\
&= \mathbb{E}[|X_1|^{-\alpha}]. \qquad (16.28)
\end{aligned}$$

We can compute the expectation above explicitly when $\alpha \in (0, 1)$ by appealing to Theorem 1.13. In particular,

$$\mathbb{E}[|X_1|^{-\alpha}] = \frac{\sin(\pi\alpha\rho) + \sin(\pi\alpha\hat{\rho})}{\Gamma(1+\alpha)\sin(\pi\alpha)}.$$

Note, however, that this moment explodes when $\alpha = 1$. One may also verify directly from the density in Theorem 1.17 that, for the Cauchy process, $\mathbb{E}[|X_1|^{-1}] = \infty$.

The almost sure limit (16.23) when $\alpha \in (0, 1)$ now follows by combining the two strong laws of large numbers in (16.27) and (16.28). When $\alpha = 1$, we note that, for each $M > 0$, we have for all t sufficiently large that $\varphi(\exp(t))/t > M$. Using the monotonicity of the counting process N and the strong law of large numbers in (16.27), it now follows that

$$\liminf_{t\to\infty} \frac{N_{\exp(t)}}{t} \geq \liminf_{t\to\infty} M \frac{\mathrm{U}_{Mt}}{Mt} > M/2\pi.$$

Since M can be chosen arbitrarily large, the statement of the theorem also follows for Cauchy processes.

(ii) Now suppose that $X_0 = 0$ and we consider the up-crossings of X as $t \to 0$ for $\alpha \in (0, 1]$. Recall that the laws $\mathbb{P}^\circ_x$, $x \in \mathbb{R}\backslash\{0\}$, are determined by the Doob h-transform

$$\left.\frac{\mathrm{d}\mathbb{P}^\circ_x}{\mathrm{d}\mathbb{P}_x}\right|_{\mathcal{F}_t} = \frac{h(X_t)}{h(x)} \mathbf{1}_{(t<\zeta)}, \qquad (16.29)$$

where

$$h(x) = (\sin(\pi\alpha\hat{\rho})\mathbf{1}_{(x\geq 0)} + \sin(\pi\alpha\rho)\mathbf{1}_{(x<0)})\, |x|^{\alpha-1} \qquad (16.30)$$

and $\zeta = \inf\{t > 0 \colon X_t = 0\}$.

Appealing to the Hunt–Nagasawa method of duality, we can show that the analogue of Lemma 16.6 also holds here. Specifically, the time reversed process $(X_{(\tau^{\{0\}}-s)-}, s \leq \tau^{\{0\}})$ when issued from a randomised point with law $\mu^\circ_a(\mathrm{d}x) := \mathbb{P}^\circ_0(X^\circ_{\mathrm{D}^\circ_a-} \in \mathrm{d}x)$, where $\mathrm{D}^\circ_a = \sup\{t > 0 \colon |X^\circ_t| < a\}$ and $a > 0$, is equal in law to $(X^\circ_t, t < \mathrm{D}^\circ_a)$ under $\mathbb{P}^\circ_0$.

Indeed, the counterpart of (16.19) can be easily checked, recalling in particular that the resolvent $R(x, \mathrm{d}y)$, $x, y \in \mathbb{R}$, is known to satisfy $R(x, \mathrm{d}y) = h(y-x)\mathrm{d}y$, up to a multiplicative constant.

If we write X° as a canonical version of the real-valued self-similar Markov process $(X, \mathbb{P}^\circ)$, it is now the case that understanding $\mathrm{U}_{[t,1]}$, as $t \to 0$, is equivalent to understanding $U^\circ_{\zeta^\circ-s}$, as $s \to 0$, where U° is the number of up-crossings of X° and $\zeta^\circ = \inf\{s > 0 \colon X^\circ_s = 0\}$. Note that up-crossings of X corresponds to down-crossings of X°, however, every up-crossing is followed by a down-crossing and vice versa. At this point, we note that the MAP that underlies the process X° has the property that its modulating chain, say J°, has the same Q-matrix as J, albeit the roles of ρ and $\hat{\rho}$ are interchanged. We can see this by inspecting its matrix exponent which was given in (12.4). This has the effect that, if we write $(N^\circ_t, t \geq 0)$ for the process that gives the running count of the number of switches that J° makes from -1 to 1, then it also respects the same strong law of large numbers as (16.27). Note in particular that the right-hand side of (16.27) is invariant to exchanging the roles of ρ and $\hat{\rho}$.

The Lamperti–Kiu representation of X° tells us that, if $\varphi^\circ(t)$ is the time change associated with its underlying MAP, then

$$\varphi^\circ(t) = \int_0^t |X^\circ_s|^{-\alpha}\mathrm{d}s.$$

A computation similar to the one that leads to the equation (16.21) also tells us that $U^\circ_{\zeta^\circ-s} = N^\circ_{\gamma^\circ_s}$ and, for $s \leq \zeta^\circ$,

$$\gamma^\circ_t = \int_{-\log \zeta^\circ}^{-\log t} |\tilde{X}^\circ_v|^{-\alpha}\mathrm{d}v, \qquad t \leq \zeta^\circ,$$

where $\tilde{X}^\circ_v = \mathrm{e}^{v/\alpha} X^\circ_{\zeta^\circ-\mathrm{e}^{-v}}$, $v \in \mathbb{R}$. Continuing along the lines of the proof of (16.22), we find that, almost surely,

$$\lim_{s\to\infty} \frac{\gamma^\circ_{\mathrm{e}^{-s}}}{s} = \mathbb{E}[|X_1|^{-\alpha}].$$

Combining the strong law of large numbers for N° with the above almost sure limit, catering for the case $\alpha = 1$ in a similar way to the proof of part (i), we find that (16.24) holds.

(iii) Examining the proof of part (ii) above for the limit as $t \to 0$, one also essentially sees the proof for the up-crossings as $t \to \tau^{\{0\}}$ in the case $\alpha \in (1,2)$. Specifically, we note that $\mathrm{U}_{\tau^{\{0\}}-s} = N_{\gamma_s}$ where, for $s \le \tau^{\{0\}}$,

$$\gamma_t = \int_{-\log \tau^{\{0\}}}^{-\log t} |\tilde{X}_v|^{-\alpha} \mathrm{d}v, \qquad t \le \tau^{\{0\}},$$

where $\tilde{X}_v = \mathrm{e}^{v/\alpha} X_{\tau^{\{0\}}-\mathrm{e}^{-v}}$, $v \in \mathbb{R}$. Continuing again along the lines of the proof of (16.22), in particular noting the Hunt–Nagasawa duality of $(X_{\tau^{\{0\}}-s}, s \le \tau^{\{0\}})$ and X° (recall from Theorem 12.5 that X° now plays the role of the stable process conditioned to avoid the origin), we have the limit

$$\lim_{s\to\infty} \frac{\gamma_{\mathrm{e}^{-s}}}{s} = \mathbb{E}_0^\circ[|X_1|^{-\alpha}], \tag{16.31}$$

$\mathbb{P}_{\mu_a^\circ}$-almost surely, where $\mu_a^\circ(\mathrm{d}x) := \mathbb{P}_0^\circ(X^\circ_{\mathrm{D}_a^\circ -} \in \mathrm{d}x)$, providing that the expectation on the right-hand side makes sense.

For the expectation on the right-hand side of (16.31), we can appeal to Theorem 11.16, bearing in mind (13.8). In particular, we have

$$\begin{aligned}\mathbb{E}_0^\circ[f(X_1)] &= \Gamma(-\alpha)\frac{\sin(\pi\alpha\rho)}{\pi}\hat{\mathbf{E}}_{0,1}\left[f(I^{-1/\alpha})I^{-1}\right] \\ &\quad + \Gamma(-\alpha)\frac{\sin(\pi\alpha\hat{\rho})}{\pi}\hat{\mathbf{E}}_{0,-1}\left[f(-I^{-1/\alpha})I^{-1}\right],\end{aligned} \tag{16.32}$$

where $I = \int_0^\infty \exp\{\alpha\xi_u\}\mathrm{d}u$. For the special case that $f(x) = |x|^{-\alpha}$, it now follows rather easily from (16.32) that

$$\mathbb{E}_0^\circ[|X_1|^{-\alpha}] = \Gamma(-\alpha)\frac{\sin(\pi\alpha\rho) + \sin(\pi\alpha\hat{\rho})}{\pi}.$$

Combining with the strong law of large numbers in (16.27), which remains valid in the setting that $\alpha \in (1,2)$, the result now follows as, in the $\mathbb{P}_{\mu_a^\circ}$-almost sure sense,

$$\begin{aligned}\lim_{t\to\infty} \frac{N_{\varphi(\mathrm{e}^{-t})}}{t} &= \frac{\Gamma(\alpha)\sin(\pi\alpha\rho)\sin(\pi\alpha\hat{\rho})}{\pi(\sin(\pi\alpha\rho) + \sin(\pi\alpha\hat{\rho}))} \times \Gamma(-\alpha)\frac{\sin(\pi\alpha\rho) + \sin(\pi\alpha\hat{\rho})}{\pi} \\ &= \frac{\sin(\pi\alpha\rho)\sin(\pi\alpha\hat{\rho})}{-\alpha\pi\sin(\pi\alpha)}\end{aligned}$$

and hence, $\mathbb{P}_{\mu_a^\circ}$-almost surely,

$$\lim_{t\to 0} \frac{\mathrm{U}_{\tau^{\{0\}}-t}}{\log(1/t)} = \frac{\sin(\pi\alpha\rho)\sin(\pi\alpha\hat{\rho})}{\alpha\pi|\sin(\pi\alpha)|}, \tag{16.33}$$

where we have used the reflection formula for the gamma function.

Unlike before, we now have the problem that, because of the direction of time-reversal, we cannot use the same trick as in the remarks preceding the proof of Theorem 16.5 A way around this is to first show that μ_a° is absolutely continuous with respect to Lebesgue measure, with a strictly positive density. As we can vary the value of $a > 0$, this would give us (16.33), $\mathbb{P}_x$-almost surely, for almost every $x \in \mathbb{R}\backslash\{0\}$.

The missing Lebesgue-null set of starting points can be recovered by a simple trick. Suppose $x \neq 0$ is such a point. We can run the stable process until it first enters the interval $(-x/2, x/2)$, which it will do with probability 1. Noting that the first entrance into this interval is almost surely finite and the law of the first entry point is absolutely continuous with respect to Lebesgue measure with a strictly positive density (cf. Theorem 6.9), the Lebesgue a.e. behaviour in (16.33) now delivers the desired result.

We are thus left to prove that μ_a° is absolutely continuous with respect to Lebesgue measure with a strictly positive density. Invoking a simple scaling argument, similar to those that we have already seen, it suffices to show that μ_1° is absolutely continuous.

To this end, let us consider $b > 1$. Thanks to the identified duality, we have that, under $\mathbb{P}_{\mu_b^\circ}$, the random time

$$\sup\{t > 0\colon |X_{(\tau^{\{0\}}-t)-}| < 1\} = \inf\{t > 0\colon |X_t| < 1\} =: \tau^{(-1,1)}$$

is equal in law to $\mathtt{D}_1$ under $\mathbb{P}_0^\circ$ and hence the law of $X_{\tau^{(-1,1)}}$ is equal in law to μ_1°.

Note that $\mathbb{P}_0^\circ(\lim_{t\to\infty}|X_t^\circ| = \infty) = 1$. This follows on account of the fact that, if (ξ, J) is the MAP underlying X through the Lamperti–Kiu tranform, then $(-\xi, J)$ is the MAP underlying X°; see Corollary 12.9. As a consequence μ_b° converges weakly to the Dirac measure $\delta_{\pm\infty}$, where $\pm\infty := \{\infty\} \cup \{-\infty\}$ is seen as the one-point compactification of $\mathbb{R}$. Equivalently, for all bounded and measurable f on $(-1, 1)$,

$$\int_{(-1,1)} f(x)\mu_1^\circ(\mathrm{d}x) = \lim_{b\to\infty} \mathbb{E}_{\mu_b^\circ}[f(X_{\tau^{(-1,1)}})] = \lim_{|x|\to\infty} \mathbb{E}_x[f(X_{\tau^{(-1,1)}})], \tag{16.34}$$

where the first limit exists because the second one does. The limit on the right-hand side of (16.34) can be calculated thanks to Theorem 6.9. Indeed, by inspecting equation (6.5), or more conveniently (6.7), one may easily take the limit to see that

$$\lim_{|x|\to\infty} \mathbb{E}_x[f(X_{\tau^{(-1,1)}})] = \frac{2^{\alpha-1}\Gamma(2-\alpha)}{\Gamma(1-\alpha\hat{\rho})\Gamma(1-\alpha\rho)} \int_{-1}^{1} f(y)(1+y)^{-\alpha\rho}(1-y)^{-\alpha\hat{\rho}}\mathrm{d}y,$$

and hence, μ_1° is absolutely continuous with respect to Lebesgue measure with strictly positive density, as required. □

16.5 Comments

The idea to study windings for Brownian motion was introduced by Spitzer [199]. An extensive literature for winding in the diffusive setting follows; see for example [75, 149, 170, 191, 209, 216]. This chapter is almost entirely based on the two papers of Bertoin and Werner [25] and Kyprianou and Vakeroudis [136], who deal exclusively with the case of stable windings; see also Doney and Vakeroudis [64]. Examples of where windings appear can be found in the setting of planar polymers in Vakeroudis et al. [210] and neuroscience in Ditlevsen and Greenwood [58], to name but two.

Appendix

A.1 Useful results from complex analysis

Assume that the function $f(z)$ is analytic on a punctured disk $D = \{0 < |z - a| < \epsilon\}$ in $\mathbb{C}$. We define *the residue* of f at point a as a contour integral

$$\mathrm{Res}(f, a) := \frac{1}{2\pi i} \oint_{\mathsf{C}} f(z)\mathrm{d}z,$$

where C is a simple closed contour, oriented counter-clockwise and lying in D. If we write the Laurent expansion of f near point a in the form

$$f(z) = \sum_{n=-\infty}^{\infty} c_n (z - a)^n,$$

then $\mathrm{Res}(f, a) = c_{-1}$. If f has a simple pole at a, that is to, say $c_k = 0$ for $k \leq -2$, then one can check that

$$\mathrm{Res}(f, a) = \lim_{z \to a} (z - a) f(z). \tag{A.1}$$

The following proposition can be applied in a wide variety of examples.

Proposition A.1

(i) Assume that $f(z) = g(z)/h(z)$, where both functions g and h are analytic in the neighbourhood of a, and $h(a) = 0$, $h'(a) \neq 0$. Then

$$\mathrm{Res}(f, a) = \frac{g(a)}{h'(a)}. \tag{A.2}$$

(ii) Assume that $f = g(z)h(z)$, where g is analytic in the neighbourhood of a, and h has a simple pole at a. Then

$$\mathrm{Res}(f, a) = g(a) \times \mathrm{Res}(h, a). \tag{A.3}$$

Below we list a number of classical results that are used throughout this text. Before stating them, we introduce some notation. Let us define $\mathbb{C}^+ := \{z \in \mathbb{C} : \mathrm{Im}(z) \geq 0\}$ and $\mathbb{C}_0^+ := \{z \in \mathbb{C} : \mathrm{Im}(z) > 0\}$. We will write $\Lambda_{a,b} = \{z \in \mathbb{C} : a < \mathrm{Im}(z) < b\}$ for the open horizontal open strip with end points a and b, $\bar{\Lambda}_{a,b} = \{z \in \mathbb{C} : a \leq \mathrm{Im}(z) \leq b\}$, for the associated closed horizontal strip, and $\gamma_a := \{z \in \mathbb{C} : \mathrm{Im}(z) = a\}$, for the horizontal line at a. Schwartz's reflection principle implies the following result.

Proposition A.2 *Let G be a symmetric region, that is to say $z \in G$ if and only if $\bar{z} \in G$, and assume that $f : G \cap \mathbb{C}^+ \mapsto \mathbb{C}$ is a continuous function which is analytic on $G \cap \mathbb{C}_0^+$ and such that $f(z)$ is real for $z \in G \cap \mathbb{R}$. Then there exists an analytic function $g : G \mapsto \mathbb{C}$ such that $g(z) = f(z)$ for all $z \in G \cap \mathbb{C}^+$.*

Corollary A.3 *Assume that $a < b < c$ and $f(z)$ is a meromorphic function in $\Lambda_{a,c}$ which has no poles on γ_b and takes real values on this line. Then $f(z)$ can be analytically continued to a meromorphic function in $\Lambda_{a-h,a+h}$, where $h = \max(c - b, b - a)$.*

Theorem A.4 (Morera's Theorem) *Let D be a region in the complex plane $\mathbb{C}$. If $f : D \mapsto \mathbb{C}$ is a continuous function satisfying*

$$\oint_C f(z)\mathrm{d}z = 0,$$

for every closed contour C in D, then $f(z)$ is analytic in D.

Proposition A.5 (Cauchy's estimates) *Suppose that a function f is analytic in the neighbourhood of a disc of radius R centred at $z^* \in \mathbb{C}$. Suppose that*

$$M_R = \sup\{|f(z)| : |z - z^*| = R\},$$

which is necessarily finite, then the nth derivative of f satisfies

$$f^{(n)}(z^*) \leq \frac{n!M_R}{R^n}.$$

Theorem A.6 (Liouville's Theorem) *If f is a uniformly bounded entire function (i.e. analytic on $\mathbb{C}$), then it must be constant.*

A.2 Mellin and Laplace–Fourier inversion

Let $f : \mathbb{R} \mapsto \mathbb{R}$ be a measurable function. The Laplace–Fourier transform of f is given by

$$\mathcal{F}f(z) = \int_{\mathbb{R}} \mathrm{e}^{-zx} f(x)\mathrm{d}x, \tag{A.4}$$

for any $z \in \mathbb{C}$ such that the integral on the right-hand side exits. If $\int_{\mathbb{R}} |f(x)|\mathrm{d}x < \infty$, then there is no problem in dealing with the case that $z = -\mathrm{i}\theta$, $\theta \in \mathbb{R}$, in which case (A.4) corresponds to the Fourier transform of f. It is not uncommon that the support belongs to $[0, \infty)$, in which case, we can take $z = \lambda - \mathrm{i}\eta$, for $\lambda > 0$ and $\eta \in \mathbb{R}$ and (A.4) corresponds to the Laplace transform of f.

Also in the setting that $\mathrm{supp} f$ belongs to $[0, \infty)$, the Mellin transform of f is defined as

$$\mathcal{M}f(z) := \int_0^\infty x^{z-1} f(x)\mathrm{d}x, \tag{A.5}$$

whenever the integral on the right-hand side exists. Performing the substitution $x = \mathrm{e}^{-y}$ transforms the above integral into

$$\mathcal{M}f(z) = \int_{\mathbb{R}} f(\mathrm{e}^{-y})\,\mathrm{e}^{-yz}\mathrm{d}y, \tag{A.6}$$

and we see that the Mellin transform of f is the same as the Laplace–Fourier transform of $f(\mathrm{e}^{-y})$. In this sense, the Mellin transform inversion is very closely related to Laplace–Fourier inversion.

Assuming that the function $w \in \mathbb{R} \mapsto F(c + \mathrm{i}w)$ is integrable for some c in an open interval. In the case that $F = \mathcal{F}f$, the function f can be obtained via the inverse Laplace–Fourier transform

$$f(x) = \frac{1}{2\pi\mathrm{i}} \lim_{w\to\infty} \int_{c-\mathrm{i}w}^{c+\mathrm{i}w} \mathrm{e}^{zx} F(z)\mathrm{d}z.$$

If it is the case that $F = \mathcal{M}f$, then f can be obtained by the inverse Mellin transform

$$f(x) = \frac{1}{2\pi\mathrm{i}} \lim_{w\to\infty} \int_{c-\mathrm{i}w}^{c+\mathrm{i}w} x^{-z} F(z)\mathrm{d}z.$$

A.3 Gamma and beta functions

The gamma function is defined as

$$\Gamma(z) = \int_0^\infty t^{z-1}\mathrm{e}^{-t}\mathrm{d}t, \tag{A.7}$$

for $z \in \mathbb{C}$ such that $\mathrm{Re}(z) > 0$. Integration by parts means that this function satisfies

$$\Gamma(z+1) = z\Gamma(z), \tag{A.8}$$

and this identity can be used to provide an analytic continuation to the entire complex plane $\mathbb{C}$. Accordingly, $\Gamma(z)$ is a meromorphic function on $\mathbb{C}$, having simple poles at $-n$ for $n \in \mathbb{N} \cup \{0\}$.

Throughout this section, unless otherwise stated, all formulae will be valid for $z \in \mathbb{C}\backslash\{0, -1, -2, \cdots\}$.

There are many other representations of the gamma function which lead to numerous other identities other than the recursion formula (A.8). The first alternative representation we mention is that of the infinite product

$$\Gamma(z) = \frac{\mathrm{e}^{hz}}{z} \prod_{n=1}^{\infty} \left(1 + \frac{z}{n}\right)^{-1} \mathrm{e}^{z/n}, \tag{A.9}$$

where $h \approx 0.577216$ is the Euler–Mascheroni constant. This representation allows us to see that, as $z \to -n$, for $n = 0, 1, 2, \ldots$, we have

$$\Gamma(z) = \frac{(-1)^n}{z+n} + O(1), \tag{A.10}$$

which implies that

$$\mathrm{Res}(\Gamma, -n) = \frac{(-1)^n}{n!}. \tag{A.11}$$

The gamma function satisfies the reflection formula

$$\Gamma(1-z)\Gamma(z) = \frac{\pi}{\sin(\pi z)}, \tag{A.12}$$

as well as

$$\Gamma(-n+x) = (-1)^{n-1}\frac{\Gamma(-x)\Gamma(1+x)}{\Gamma(n+1-x)}, \qquad n \in \mathbb{N}. \tag{A.13}$$

In a similar spirit, we also have the duplication formula

$$\Gamma(z)\Gamma\left(z+\tfrac{1}{2}\right) = 2^{1-2z}\sqrt{\pi}\Gamma(2z). \tag{A.14}$$

For any $\epsilon \in (0, \pi)$, the gamma function satisfies the asymptotic relation

$$\log(\Gamma(z)) = \left(z - \tfrac{1}{2}\right)\log z - z + \frac{1}{2}\log(2\pi) + \frac{1}{12z} + O(z^{-3}), \tag{A.15}$$

as $|z| \to \infty$, uniformly in the sector $|\arg(z)| < \pi - \epsilon$. An important asymptotic relation can be derived from (A.15) is

$$|\Gamma(x+\mathrm{i}y)| = \sqrt{2\pi}\mathrm{e}^{-\frac{\pi}{2}|y|}|y|^{x-\frac{1}{2}}(1+o(1)), \qquad x \in \mathbb{R}, \tag{A.16}$$

as $y \to \infty$, uniformly in any finite interval $-\infty < a \le x \le b < \infty$. Another useful asymptotic result is

$$\frac{\Gamma(z+a)}{\Gamma(z)} = z^a(1+o(1)), \tag{A.17}$$

as $|z| \to +\infty$, uniformly in any sector $|\mathrm{Arg}(z)| < \pi - \epsilon$.

Either defined through gamma functions, or through one of two associated integrals, we have the beta function

$$B(x,y) := \frac{\Gamma(x)\Gamma(y)}{\Gamma(x+y)} = \int_0^1 s^{x-1}(1-s)^{y-1}\,\mathrm{d}s = \int_0^\infty \frac{t^{x-1}}{(1+t)^{x+y}}\,\mathrm{d}t, \tag{A.18}$$

for $\mathrm{Re}(x), \mathrm{Re}(y) > 0$. Another representation takes the form

$$B(x,y) = 2\int_0^{\pi/2} (\sin\theta)^{2x-1}(\cos\theta)^{2y-1}\mathrm{d}\theta, \qquad \mathrm{Re}(x), \mathrm{Re}(y) > 0. \tag{A.19}$$

A.4 Double gamma function

Much of the account of double gamma functions that we will give in this section originates from the work of Barnes [10]. The double gamma function was originally introduced as an infinite product in Weierstrass's form,

$$G(z;\tau) = \frac{z}{\tau}\mathrm{e}^{a\frac{z}{\tau}+b\frac{z^2}{2\tau}} \prod_{\substack{m,n\ge 0\\ m+n>0}} \left(1 + \frac{z}{m\tau+n}\right)\exp\left(-\frac{z}{m\tau+n} + \frac{z^2}{2(m\tau+n)^2}\right),$$

for $|\arg(\tau)| < \pi$ and $z \in \mathbb{C}$. Note that, by definition, $z \mapsto G(z;\tau)$ is an entire function and it has simple zeros on the lattice $m\tau + n$, $m \le 0$, $n \le 0$. Barnes also showed that $G(z;\tau)$ can also be expressed as a single infinite product:

$$G(z;\tau) = \frac{1}{\tau\Gamma(z)} e^{\frac{\tilde{a}z}{\tau} + \frac{\tilde{b}z^2}{2\tau^2}} \prod_{m\ge 0} \frac{\Gamma(m\tau)}{\Gamma(z+m\tau)} e^{z\psi(m\tau) + \frac{z^2}{2}\psi'(m\tau)}, \tag{A.20}$$

where $\psi(z) = \frac{\mathrm{d}}{\mathrm{d}z}\log(\Gamma(z))$ is the digamma function (see Gradshteyn and Ryzhik [82]). The constant $\tilde{a}$ and $\tilde{b}$ are related to a and b via

$$\tilde{a} = a - h\tau \qquad \text{and} \qquad \tilde{b} = b + \frac{\pi^2\tau^2}{6},$$

where $h = -\psi(1)$ is the Euler–Mascheroni constant. One of the most important properties of the double gamma function is that it is quasi-periodic with periods 1 and τ, namely

$$G(z+1;\tau) = \Gamma\left(\frac{z}{\tau}\right) G(z;\tau) \text{ and } G(z+\tau;\tau) = (2\pi)^{\frac{\tau-1}{2}} \tau^{-z+\frac{1}{2}} \Gamma(z) G(z;\tau), \tag{A.21}$$

provided that constants $\tilde{a}$ and $\tilde{b}$ are such that

$$\tilde{a} = \frac{\tau}{2}\log(2\pi\tau) + \frac{1}{2}\log(\tau) - \tau C(\tau),$$
$$\tilde{b} = -\tau\log(\tau) - \tau^2 D(\tau).$$

Here, $C(\tau)$ and $D(\tau)$ are certain transcendental functions of τ which can be computed as the following limits as $m \to \infty$:

$$C(\tau) = \sum_{k=1}^{m-1} \psi(k\tau) + \frac{1}{2}\psi(m\tau) - \frac{1}{\tau}\log\left(\frac{\Gamma(m\tau)}{\sqrt{2\pi}}\right) - \frac{\tau}{12}\psi'(m\tau) + \frac{\tau^3}{720}\psi^{(3)}(m\tau) + O(m^{-5}),$$

$$D(\tau) = \sum_{k=1}^{m-1} \psi'(k\tau) + \frac{1}{2}\psi'(m\tau) - \frac{1}{\tau}\psi(m\tau) - \frac{\tau}{12}\psi''(m\tau) + \frac{\tau^3}{720}\psi^{(4)}(m\tau) + O(m^{-6}).$$

It turns out that, with these choices of constants, we also have $G(1;\tau) = 1$. There exists a different and slightly simpler expression for these constants (see Lawrie and King [140]), however we decided to present the original Barnes formulas as they are more convenient for numerical calculations. It is also possible to give an integral representation for $\log(G(z;\tau))$, see Billingham and King [140], as well asymptotic expansions for this function (see Billingham and King [31]).

Next, we want to introduce two identities about the double gamma function that will are useful. To do so, we must first describe the q-Pochhammer symbol

$$(a:q)_n := \prod_{k=0}^{n-1}(1 - aq^k), \qquad \text{for} \quad n \in \mathbb{N},$$

and $(a:q)_0 = 1$. If $|q| < 1$, we define

$$(a:q)_\infty := \prod_{k\geq 0}(1 - aq^k).$$

The first identity we present is an analogue of the reflection formula for the gamma function. Specifically,

$$-2\pi i\tau G\left(\frac{1}{2} + z; \tau\right) G\left(\frac{1}{2} - z; -\tau\right) = \frac{\left(-e^{2\pi i z}; e^{2\pi i\tau}\right)_\infty}{\left(e^{2\pi i\tau}; e^{2\pi i\tau}\right)_\infty}, \tag{A.22}$$

for $\mathrm{Im}(\tau) > 0$. The second identity is the transformation formula

$$G(z;\tau) = (2\pi)^{\frac{z}{2}\left(1-\frac{1}{\tau}\right)} \tau^{\frac{z-z^2}{2\tau} + \frac{z}{2} - 1} G\left(\frac{z}{\tau}; \frac{1}{\tau}\right). \tag{A.23}$$

We complete this section by discussing the asymptotic behaviour of the double gamma function. Assume that $\tau > 0$. When $s \to \infty$ in the domain $|\arg(s)| < \pi - \epsilon < \pi$, we have

$$\log\left[\frac{G(a+s;\tau)}{G(s;\tau)}\right] = \frac{a}{\tau} s\log(s) - \frac{a}{\tau}(1 + \log(\tau))s + \frac{a}{2\tau}(a - 1 - \tau)\log(s) + O(1). \tag{A.24}$$

A.5 Double sine function

Let $\alpha > 0$ and define, for $z \in \mathbb{C}$,

$$S_2(z;\alpha) = (2\pi)^{(1+\alpha)/2-z} \frac{G(z;\alpha)}{G(1+\alpha-z;\alpha)}, \qquad z \in \mathbb{C}. \tag{A.25}$$

This is a meromorphic function, which has the Weierstrass product representation of the form

$$S_2(z;\alpha) = e^{A(\alpha)+B(\alpha)z+C(\alpha)z^2} \frac{z}{z-1-\alpha} \prod_{\substack{m,n\geq 0\\ m+n>0}} \frac{P(-z/(m\alpha+n))}{P(z/((m+1)\alpha+n+1))}, \tag{A.26}$$

where $P(z) := (1-z)\exp(z + z^2/2)$. When $\alpha = 1$, we obtain the simpler expression

$$S_2(z;1) = e^z \prod_{n=1}^{\infty}\left[\left(\frac{1-\frac{z}{n}}{1+\frac{z}{n}}\right)^n e^{2z}\right], \qquad z \in \mathbb{C}, \tag{A.27}$$

as well as

$$S_2(z;1) = (2\sin(\pi z))^z \exp\left((1/2\pi)\sum_{n=1}^{\infty}\frac{\sin(2\pi nz)}{n^2}\right), \qquad 0 < z < 1. \tag{A.28}$$

It is clear from (A.26) that $S_2(z;\alpha)$ is a meromorphic function of z, which has zeros at points $\{-m\alpha - n \;:\; m \geq 0, n \geq 0\}$ and poles at points $\{m\alpha + n \;:\; m \geq 1, n \geq 1\}$. All zeros and poles are simple if and only if α is irrational.

The double sine function satisfies the following identities

$$S_2(z+1;\alpha) = \frac{S_2(z;\alpha)}{2\sin(\pi z/\alpha)},$$
$$S_2(z+\alpha;\alpha) = \frac{S_2(z;\alpha)}{2\sin(\pi z)},$$
$$S_2(z;\alpha) = 1/S_2(1+\alpha-z;\alpha),$$
$$S_2(z;\alpha) = S_2(z/\alpha;1/\alpha).$$

We have special values

$$S_2(1;\alpha) = \sqrt{\alpha},$$
$$S_2(\alpha;\alpha) = 1/\sqrt{\alpha},$$
$$S_2(1/2;\alpha) = S_2(\alpha/2;\alpha) = \sqrt{2},$$
$$S_2((1+\alpha)/2;\alpha) = 1.$$

We also have the following asymptotic expression

$$\lim_{y\to\infty} |S_2(c+\mathrm{i}y;\alpha)| \exp(\pi|y|(2c-1-\alpha)/(2\alpha)) = 1, \tag{A.29}$$

which holds uniformly in c on any bounded interval of $\mathbb{R}$.

A.6 Hypergeometric functions

The class of hypergeometric functions are defined by the power series

$$_2F_1(a,b;c;z) = \sum_{k\geq 0} \frac{(a)_k(b)_k}{(c)_k}\frac{z^k}{k!}, \tag{A.30}$$

for $|z| \leq 1$, which converges if c is not negative when $|z| < 1$. If $|z| = 1$, we have convergence for $\mathrm{Re}(c-a-b) > 0$. Here, $(q)_n$ is the Pochhammer symbol satisfying

$$(q)_n = \begin{cases} 1 & \text{if } n = 0, \\ q(q+1)\cdots(q+n-1) & \text{if } n \in \mathbb{N}. \end{cases}$$

In the identities below, unless otherwise stated, we will assume that a, b, c, z respect the same constraints as given above.

We will make use of the integral representation

$$B(b,c-b)\ {}_2F_1(a,b,c;z) = \int_0^1 t^{b-1}(1-t)^{c-b-1}(1-zt)^{-a}\mathrm{d}t, \tag{A.31}$$

for $\mathrm{Re}(c) > \mathrm{Re}(b) > 0$ and $|z| < 1$ or $|z| = 1$, provided both the left- and right-hand sides converge.

The following combination of hypergeometric functions, leading to a simpler expression in terms of gamma functions, is one of an enormous catalogue of identities that can be proved. For $a, b > 0$,

$$_2F_1(a,a+b,a+1;-1)b + {}_2F_1(b,a+b,b+1;-1)a = \frac{\Gamma(a+1)\Gamma(b+1)}{\Gamma(a+b)}. \tag{A.32}$$

Another identity of use has the form

$$ {}_2F_1(c-a, c-b, c; z) = (1-z)^{a+b-c}\ {}_2F_1(a, b, c; z), \tag{A.33} $$

as well as

$$ {}_2F_1(a, b; b; x) = (1-z)^{-a}. \tag{A.34} $$

Finally, hypergeometric functions are related to incomplete beta integrals via the relation

$$ \int_0^x s^{a-1}(1-s)^{b-1}\mathrm{d}s = \frac{x^a}{a}\,{}_2F_1(a, 1-b; a+1; x), \tag{A.35} $$

where $x \in (0,1)$ and $\mathrm{Re}(a), \mathrm{Re}(b) > 0$.

A.7 Additive and subadditive functions

A real-valued function f is said to be additive if it satisfies the Cauchy functional equation

$$ f(x+y) = f(x) + f(y), \qquad \text{for} \quad x, y \in \mathbb{R}. \tag{A.36} $$

Obvious examples are of the form $f(x) = cx$, for $c \in \mathbb{R}$. Non-trivial examples of additive functions may exist but, subject to measurability constraints, additive functions are always of the form cx.

Theorem A.7 *If f is additive and measurable, then $f(x) = cx$, for some $c \in \mathbb{R}$.*

A function $g : (0, \infty) \to (0, \infty)$ is said to be multiplicative if

$$ g(x)g(y) = g(xy), \qquad \text{for} \quad x, y \geq 0. $$

If $f(x) = \log g(\mathrm{e}^x)$, then f is additive if and only if g is multiplicative, so the previous theorem translates to the following.

Theorem A.8 *If g is multiplicative and measurable, then $g(x) = x^c$, for $x > 0$, and some $c \in \mathbb{R}$.*

Now, let us consider a real-valued function f defined on a subset I and satisfying the condition

$$ f(x+y) \leq f(x) + f(y), \qquad \text{for} \quad x, y \in I \text{ and } x+y \in I. \tag{A.37} $$

Functions satisfying the above conditions are called subadditive.

It is clear that constants are subadditive functions and that the class of subadditive functions is closed under addition. Moreover, the supremum of subadditive functions is also subadditive. We also recall that the Cauchy functional equation (A.36) has non-measurable solutions in addition to the continuous solutions of the form cx. Since any real solution of (A.36) also satisfies (A.37), it is clear that there are non-measurable subadditive functions.

In what follows, all subadditive functions are assumed to be measurable.

Theorem A.9 *If f is subadditive and finite in (a, ∞) where $a > 0$, then f is bounded above in every interval $(2a + \delta, 2a + 1/\delta)$ and bounded below in $(a, a + 1/\delta)$ for $\delta > 0$.*

Said another way, the above theorem implies that a finite subadditive function is bounded in any finite closed interval, which is interior to its interval of definition. It may, however, become unbounded when t approaches either end point of its interval of definition (one of which is ∞). The following result describes the behaviour for large values.

Theorem A.10 *If f is subadditive and finite in (a, ∞), $a \geq 0$, then*

$$\lim_{t\to\infty} \frac{f(t)}{t} = \inf_{t>a} \frac{f(t)}{t} < \infty.$$

Proof Since f is finite in (a, ∞), we necessarily have

$$\beta = \inf_{t>a} \frac{f(t)}{t},$$

is either finite or $-\infty$. We only prove the result when β is finite, similar arguments allow us to deduce the case $\beta = -\infty$.

Let $x > a$ and $\epsilon > 0$, such that $f(x) < (\beta + \epsilon)x$ and take $t \in [(n+2)x, (n+3)x)$, then

$$\beta \leq \frac{f(t)}{t} \leq \frac{nx}{t}\frac{f(x)}{x} + \frac{f(t-nx)}{t} \leq \frac{nx}{t}(\beta + \epsilon) + \frac{f(t-nx)}{t}. \tag{A.38}$$

Since $t - nx \in [2x, 3x]$, we see that $f(t - nx)$ stays bounded according to Theorem A.9 and hence the right-hand side of (A.38) goes to $\beta + \epsilon$, as t increases. Since ϵ can be taken arbitrarily small, we deduce the result. □

We remark that the above result is true if we replace the assumption of measurability by boundedness in every compact subset of $(2a, \infty)$.

A.8 Random difference equations

Let us consider the following random equation

$$R \stackrel{d}{=} Q + VR, \tag{A.39}$$

where R is independent of the pair (Q, V). Observe that the latter identity implies

$$R \stackrel{d}{=} \sum_{k=1}^{\infty} Q_k \Pi_{k-1},$$

where $\Pi_n = \prod_{i=1}^{n} V_i$ and (Q_k, V_k), for $k \geq 1$, are independent with the same law as (Q, V).

It turns out that under suitable assumptions there is a unique random variable, R, satisfying (A.39) such that the behaviour of $\mathrm{P}(R > x)$ and $\mathrm{P}(R < -x)$, for large x, can be determined explicitly. In order to make this statement precise, let us assume that $V \geq 0$ and satisfies the following:

(i) the range of V is not arithmetic,
(ii) there exist $\kappa > 0$, such that $\mathrm{E}[V^\kappa] = 1$,
(iii) $\mathrm{E}[V^\kappa \ln^+(V)] < \infty$, where $\ln^+ x = 0 \vee \ln x$.

Theorem A.11 *Let Q and V be random variables on the same probability space. Suppose that*

$$\mathrm{E}[|Q|^{\kappa}] < \infty,$$

and V satisfies conditions (i), (ii) and (iii). Then there is a unique law R satisfying (A.39) *such that*

$$\mathrm{P}(R > x) \sim C_+ x^{-\kappa} \quad \text{and} \quad \mathrm{P}(R < -x) \sim C_- x^{-\kappa}, \quad \text{as} \quad x \to \infty,$$

where $m = \mathbb{E}[V^{\kappa} \ln(V)]$,

$$C_+ = \frac{\mathrm{E}[((Q + VR)^+)^{\kappa}] - \mathrm{E}[((VR)^+)^{\kappa}]}{\kappa m},$$

and

$$C_- = \frac{\mathrm{E}[((Q + VR)^-)^{\kappa}] - \mathrm{E}[((VR)^-)^{\kappa}]}{\kappa m}.$$

Here we use the standard notation $x^+ = 0 \wedge x$ and $x^- = 0 \vee x$.

A.9 A generalisation of the Borel–Cantelli Lemma

The Borel–Cantelli Lemma is a fundamental result in probability theory that characterises scenarios in which events (resp. independent events) fail (resp. succeed) to occur infinitely often almost surely. Here we present a more complex analogue of the aforementioned classical result.

Lemma A.12 *Let $\{E_k\}_{k\geq 1}$ be a sequence of events satisfying the condition that for $m \neq n$ there exists a constant $c > 0$ such that*

$$\mathrm{P}(E_m \cap E_n) \leq c\mathrm{P}(E_m)\mathrm{P}(E_n). \tag{A.40}$$

If

$$\sum_{k\geq 1} \mathrm{P}(E_k) = \infty,$$

then

$$\mathrm{P}\left(\limsup_{k\geq 1} E_k\right) \geq \frac{1}{c}.$$

Proof For each $n \geq 1$, we introduce the following random variable

$$X_n = \sum_{k=1}^{n} \mathbf{1}_{E_k},$$

and observe that

$$\mathrm{E}[X_n] = \sum_{k=1}^{n} \mathrm{P}(E_k) \quad \text{and} \quad \mathrm{E}[X_n^2] = \sum_{i,j=1}^{n} \mathrm{P}(E_i \cap E_j).$$

Next, for $\epsilon \in (0, 1)$, we introduce the set

$$F_n^{(\epsilon)} = \left\{X_k \geq \epsilon\mathrm{E}[X_k] \quad \text{for some } k \geq n\right\}.$$

From Cauchy–Schwarz's inequality, we deduce

$$\mathrm{P}(F_n^{(\epsilon)}) \geq \frac{\mathrm{E}\left[\mathbf{1}_{F_n^{(\epsilon)}} X_n\right]^2}{\mathrm{E}[X_n^2]} = \frac{\mathrm{E}\left[X_n\left(1 - \mathbf{1}_{(F_n^{(\epsilon)})^c}\right)\right]^2}{\mathrm{E}[X_n^2]}.$$

On the other hand, since $\mathrm{E}[\mathbf{1}_{((F_n^{(\epsilon)})^c)} X_n] \leq \epsilon \mathrm{E}[X_n]$, we get

$$\mathrm{P}(F_n^{(\epsilon)}) \geq (1-\epsilon)^2 \frac{\mathrm{E}\left[X_n\right]^2}{\mathrm{E}[X_n^2]}.$$

From our assumptions, the expectation $\mathrm{E}[X_n]$ goes to ∞, as n increases, which implies that

$$\mathrm{P}\left(\limsup_{k\geq 1} E_k\right) \geq \lim_{n\to\infty} \mathrm{P}(F_n^{(\epsilon)}) \geq (1-\epsilon)^2 \lim_{n\to\infty} \frac{\mathrm{E}\left[X_n\right]^2}{\mathrm{E}[X_n^2]}.$$

Since ϵ can be taken arbitrarily small and the sequence $\{E_k\}_{k\geq 1}$ satisfies (A.40), we deduce

$$\mathrm{P}\left(\limsup_{k\geq 1} E_k\right) \geq \lim_{n\to\infty} \frac{\mathrm{E}\left[X_n\right]^2}{\mathrm{E}[X_n^2]} = \lim_{n\to\infty} \frac{\sum_{k=1}^n \mathrm{P}(E_k)}{\sum_{i,j=1}^n \mathrm{P}(E_i \cap E_j)} \geq \frac{1}{c}.$$

This completes the proof. □

A.10 Skorokhod space

Throughout this section, we suppose that (E, d_E) is a locally compact Hausdorff space with a countable base. We need a canonical space on which we can build the laws of E-valued stochastic processes with path discontinuities. Our construction will allow for the possibility that E contains a distinguished state Δ, which serves as a cemetery (absorbing) state for killed processes.

Let $\mathcal{D}(E)$ be the space of (killed) trajectories consisting of mappings $\omega : [0,\infty) \to E$ which are right-continuous with left limits, having lifetime $\zeta(\omega) = \inf\{t > 0 : \omega(t) = \Delta\}$. Here continuity is understood with respect to the metric d_E.

We want to build a metric on $\mathcal{D}(E)$ from which we can define the Borel sigma-algebra on $\mathcal{D}(E)$, thus giving us a measurable space on which to assign probabilities. To this end, let us denote by Λ the class of strictly increasing continuous mappings of $[0,\infty)$ onto itself such that $\lambda \in \Lambda$ necessarily satisfies $\lambda(0) = 0$ and $\lim_{t\to\infty} \lambda(t) = \infty$. For $\lambda \in \Lambda$, we define

$$\|\lambda\|_{\mathrm{sk}} = \sup_{0\leq s\leq t<\infty} \left| \log\left(\frac{\lambda(t) - \lambda(s)}{t - s}\right)\right|.$$

We can now define the Skorokhod metric d_{Sk} such that, for $\omega, \omega' \in \mathcal{D}(E)$,

$$d_{\mathrm{Sk}}(\omega, \omega') = \sum_{k\geq 1} 2^{-k} \left(\inf_{\lambda\in\Lambda} \left\{ \|\lambda\|_{\mathrm{sk}} + \sup_{0\leq t\leq k} d_E\Big(\omega(t),\ \omega' \circ \lambda(t)\Big)\right\} \wedge 1\right).$$

The importance of the Skorokhod metric is that $(\mathcal{D}(E), d_{Sk})$ is a metric space that is complete and separable. Moreover, the metric d_{Sk} allows us to define a topology open sets on $\mathcal{D}(E)$, from which we understand the Borel sigma algebra on $\mathcal{D}(E)$ to mean the smallest sigma algebra containing all the d_{Sk}-open sets.

Another advantage of the Skorokhod metric is that it allows us to be clear about the notion of weak convergence for path-valued random variables. Formally speaking, we say that a sequence of probability measures $(\mu_n, n \geq 1)$ converges to μ with respect to the Skorokhod topology if, for all bounded and d_{Sk}-continuous $F\colon \mathcal{D}(E) \to [0, \infty)$,

$$\lim_{n\to\infty} \mu_n(F) = \mu(F).$$

Another way of saying this is that there is *weak convergence on the Skorokhod space.*

A.11 Feller processes

Suppose that $Z = (Z_t, t \geq 0)$ is an E-valued stochastic process on a filtered probability space with probabilities $\mathrm{P}_x(\cdot) = \mathrm{P}(\cdot\,|\,Z_0 = x)$ and filtration $\mathfrak{G} = (\mathfrak{G}_t, t \geq 0)$. For the purpose of this discussion, it will suffice that E is a locally compact Hausdorff space with a countable base, with metric written d_E. Note, as in Section A.10, our setting permits killed process, however, for convenience, we will always identify the cemetery state Δ as an additional point annexed to E (rather than included in E). The lifetime of Z is thus $\zeta = \inf\{t > 0 : Z_t = \Delta\}$.

Remark A.13 The filtration $\mathfrak{G}$ is often taken to be the *natural filtration*, that is, $\mathfrak{G}_t := \sigma(Z_s, s \leq t)$, $t \geq 0$. In some circumstances, it is preferable to work with the *natural enlargement* of a given filtration. To describe what this means, write $\mathrm{P}_\nu = \int_E \nu(\mathrm{d}a)\mathrm{P}_a$, for any probability measure ν on E. We say $\mathfrak{G}$ is naturally enlarged if, for each $t \geq 0$, $\mathfrak{G}_t$ is complete with respect to the null sets of $\mathrm{P}_\nu|_{\mathfrak{G}_t}$, for all possible ν, and there is right-continuity, in the sense that $\mathfrak{G}_t = \bigcap_{s>t} \mathfrak{G}_s$. Whereas we have introduced the notion of natural enlargement, it is commonplace in other literature to assume that the filtration $\mathfrak{G}$ satisfies *les conditions habituelles*, meaning that, for each $t \geq 0$, $\mathfrak{G}_t$ is complete with respect to all null sets of $\mathbb{P}_\nu$, for all possible ν, and right-continuous. This can create problems, for example, when looking at changes of measure. The reader is encouraged to read Warning 1.3.39. of Bichteler [29] for further investigation.

The process $Z = (Z_t, t \geq 0)$ possesses the Markov property if, for each bounded and measurable $f\colon E \to \mathbb{R}$, $x \in E$ and $s, t \geq 0$, on $\{t < \zeta\}$,

$$\mathrm{E}_x[f(Z_{t+s})\mathbf{1}_{(t+s<\zeta)}\,|\,\mathfrak{G}_t] = \mathcal{P}_s[f](Z_t),$$

where, for all $x \in E$ and $s \geq 0$, $\mathcal{P}_s[f](x) := \mathrm{E}_x[f(Z_s)\mathbf{1}_{(s<\zeta)}]$. We may think of $\mathcal{P} = (\mathcal{P}_t, t \geq 0)$ as a sequence of operators on the space of bounded and measurable functions on E. As such, $\mathcal{P}$ satisfies a *semigroup* property in the sense that, $\mathcal{P}_0[f](x) = f(x)$ and for $s, t \geq 0$, $\mathcal{P}_t[\mathcal{P}_s[f](\cdot)](x) = \mathcal{P}_{t+s}[f](x)$, $x \in \mathbb{R}$. The last equality is thanks to the Markov property.

We want to classify $\mathcal{P}$ according to how it behaves as an operator on the Banach space $C_0(E)$ of bounded measurable functions which decay to 0 as $d_E(x) \to \infty$, equipped with the supremum norm, $\|f\| = \sup_{x\in\mathbb{R}} |f(x)|$, for $f \in C_0(E)$.

Definition A.14 We say that $(\mathcal{P}_t, t \geq 0)$ is a *Feller semigroup* if it has the *Feller property*. That is, for all $f \in C_0(E)$,

(i) for all $t \geq 0$, $\mathcal{P}_t[f] \in C_0(E)$ and,
(ii) for all $x \in E$, $\lim_{t\downarrow 0} \mathcal{P}_t[f](x) = f(x)$.

(Note, it can be shown that the semigroup property together with (i) and (ii) imply the stronger continuity property that, for all $f \in C_0(E)$, $\lim_{t\to 0} \|\mathcal{P}_t[f] - f\| = 0$.)

An important consequence of the Feller property is that the Markov property also holds at certain types of random times. A non-negative random variable, say τ, is called a *stopping time* if

$$\{\tau \leq t\} \in \mathfrak{G}_t,$$

for all $t \geq 0$. It is possible that a stopping time may be infinite in value with positive probability. Associated to each stopping time τ is the sigma-algebra

$$\mathfrak{G}_\tau := \{A \in \mathfrak{G} : A \cap \{\tau \leq t\} \in \mathfrak{G}_t \text{ for all } t \geq 0\}.$$

(Note, it is a simple exercise to verify that $\mathfrak{G}_\tau$ is a sigma-algebra.) The process Z is said to satisfy the strong Markov property if, for each bounded, measurable f on E, $x \in E$, $s \geq 0$ and stopping time τ,

$$\mathrm{E}_\nu[f(Z_{\tau+s})\mathbf{1}_{(\tau+s<\zeta)}|\mathfrak{G}_\tau] = \mathcal{P}_s[f](Z_\tau) \text{ on } \{\tau < \zeta\}.$$

As alluded to above, Feller processes are also strong Markov processes. The Feller property offers more than the strong Markov property. Among other things, general theory allows us to conclude that every Feller process admits a modification with paths that are almost surely right-continuous with left limits. Moreover every such Feller process can additionally be shown to have quasi-left continuous paths. That is to say, paths which are left-continuous at increasing sequences of stopping times. A Feller process with all these properties is said to be a *regular Feller process*.

A.12 Hunt–Nagasawa duality

The duality discussed in Section 2.11 belongs to a much richer narrative for the setting of general Markov processes. We give some exposure to the general here, albeit being a thin slice of the much more expansive theory of the potential analytic/functional analytic view of Markov processes and their semigroups.

As defined in Section A.11 suppose, that $Z = (Z_t, t \geq 0)$ with probabilities P_x, $x \in E$, is a regular Feller process that lives on an open domain $E \subseteq \mathbb{R}^d$ with annexed cemetery state Δ and killing time $\zeta = \inf\{t > 0 : Z_t = \Delta\}$. Recall from the previous section that $\mathcal{P} := (\mathcal{P}_t, t \geq 0)$ is the associated semigroup and $\mathrm{P}_\nu = \int_{\mathbb{R}^d} \nu(\mathrm{d}a)\mathrm{P}_a$, for any probability measure ν on the state space of Z.

Suppose that $\mathfrak{G}$ is the σ-algebra generated by Z and write $\mathfrak{G}(\mathrm{P}_\nu)$ for its completion by the null sets of P_ν, for all possible probability distributions ν on E. Moreover, write $\overline{\mathfrak{G}} = \bigcap_\nu \mathfrak{G}(\mathrm{P}_\nu)$, where the intersection is taken over all probability measures on E. A finite random time $\mathbf{k}$ is called an L-time (generalised last exit time) if

(i) $\mathtt{k} \leq \zeta$ and $\mathtt{k}$ is measurable in $\overline{\mathfrak{G}}$,
(ii) $\{t + s < \mathtt{k}\} = \{t < \mathtt{k}\} \cap \{s < \vartheta_t \circ \mathtt{k}\}$, for all $t, s \geq 0$,

where ϑ_t, $t \geq 0$, is the usual Markov shift operator. The class of L-times also contains classical last exit times. For example, when B is an open set in $\mathbb{R}^d$, we say that

$$\ell^B := \sup\{t > 0 : Z_t \in B\},$$

where $\sup \emptyset := 0$, is the last passage time in B. The lifetime ζ is also an L-time.

Theorem 3.5 of Nagasawa [156], shows that, under suitable assumptions on the Markov process Z, the class of L-times are a family of 'good times' at which the pathwise time-reversal

$$\overleftarrow{Z}_t := Z_{(\mathtt{k}-t)-}, \qquad 0 \leq t \leq \mathtt{k},$$

is a Markov process. (Note that the left limit above is used to ensure that $\overleftarrow{Z}$ is a right-continuous process.) In order to state Nagasawa's aforementioned theorem, we need to state up front its two main conditions.

(A) For $a \in E$, the potential measure $U(a, \cdot)$ associated to $\mathcal{P}$, is defined by the relation

$$\int_E f(x)U(a, \mathrm{d}x) = \int_0^\infty \mathcal{P}_t[f](a)\mathrm{d}t = \mathbf{E}_a\left[\int_0^\infty f(X_t)\,\mathrm{d}t\right], \qquad \text{(A.41)}$$

for bounded and measurable f on E, is a σ-finite measure. For a σ-finite measure ν concentrated on E, if we put

$$\mu(A) = \int_E U(a, A)\,\nu(\mathrm{d}a), \qquad \text{for } A \in \mathcal{B}(E), \qquad \text{(A.42)}$$

assume that there exists a Markov transition semigroup, say $\hat{\mathcal{P}} := (\hat{\mathcal{P}}_t, t \geq 0)$ such that

$$\int_E g(x)\mathcal{P}_t[f](x)\mu(\mathrm{d}x) = \int_E f(x)\hat{\mathcal{P}}_t[g](x)\,\mu(\mathrm{d}x), \quad t \geq 0, \qquad \text{(A.43)}$$

for bounded, measurable test-functions f, g that are compactly supported in E.

In other words, (A.43) asks for the semigroup $\mathcal{P}$ to be *in weak duality to a semigroup* $\hat{\mathcal{P}}$ *with respect to the measure* μ taking the form (A.42).

(B) For any continuous test-function $f \in C_0(E)$, the space of continuous and compactly supported functions, and $a \in E$, assume that $\mathcal{P}_t[f](a)$ is right-continuous in t and assume that, for $q > 0$, $\hat{U}^{(q)}[f](\overleftarrow{Z}_t)$ is right-continuous in t, where, for bounded and measurable f on E,

$$\hat{U}^{(q)}[f](a) = \int_0^\infty \mathrm{e}^{-qt}\hat{\mathcal{P}}_t[f](a)\mathrm{d}t, \qquad a \in E$$

is the q-potential associated to $\hat{\mathcal{P}}$.

Theorem A.15 (Nagasawa's Duality Theorem) *Suppose that assumptions* **(A)** *and* **(B)** *hold. For the given starting probability distribution ν in* **(A)** *and any L-time,* $\mathtt{k}$, *the* *time-reversed process* $\overleftarrow{Z}$ *under* P_ν *is a time-homogeneous Markov process such that,* P_ν*-almost surely,*

$$\mathrm{P}_\nu(\overleftarrow{Z}_t \in A \mid \overleftarrow{Z}_r, 0 < r < s) = \mathrm{P}_\nu(\overleftarrow{Z}_t \in A \mid \overleftarrow{Z}_s) = \hat{p}(t-s, \overleftarrow{Z}_s, A), \tag{A.44}$$

for all $0 < s < t$ *and Borel* A *in* E*, where* $\hat{p}(u, x, A)$*,* $u \geq 0$*,* $x \in E$*, is the transition measure associated to the semigroup* $\hat{\mathcal{P}}$*.*

It is often the case that good probabilistic intuition serves as the best method of informing the paired choice of ν and $\overleftarrow{P}$ that fits (A.42) and (A.43). Nagasawa's Theorem A.15 builds on earlier work on Markov chains by Kolmogorov and, more formally, on the potential analytic framework of Hunt [91, 92]. A recent treatment can be found in the book of Chung and Walsh [50] as well as in material more immediately subsequent to Nagasawa's work, for example, [196, 212].

A.13 Poisson point processes

We give some brief notes on Poisson random measures and Poisson point processes here. Predominantly, this is to clarify two notational perspectives that are used in this text. We start with the formal definition of a Poisson random measure.

Definition A.16 (Poisson random measure) Let $(S, \mathcal{S}, \eta)$ be an arbitrary sigma-finite measure space and $(\Omega, \mathcal{F}, P)$ a probability space. Let $N : \Omega \times \mathcal{S} \to \{0, 1, 2, ...\} \cup \{\infty\}$ in such a way that the family $\{N(\cdot, A) : A \in \mathcal{S}\}$ are random variables defined on $(\Omega, \mathcal{F}, P)$. Henceforth, for convenience, we shall suppress the dependency of N on ω. Then N is called a *Poisson random measure* on $(S, \mathcal{S}, \eta)$ (or sometimes a Poisson random measure on S with intensity η) if

(i) for mutually disjoint $A_1, ..., A_n$ in $\mathcal{S}$, the variables $N(A_1), ..., N(A_n)$ are independent,
(ii) for each $A \in \mathcal{S}$, $N(A)$ is Poisson distributed with parameter $\eta(A)$ (here we allow $0 \leq \eta(A) \leq \infty$),
(iii) $N(\cdot)$ is a measure P-almost surely.

Definition A.17 (Poisson point process) Suppose now that $S = [0, \infty) \times \mathcal{U}$, $\mathcal{S} = \mathcal{B}[0, \infty) \times \mathcal{B}(\mathcal{U})$, the product of the natural Borel sigma-algebras on the product space, and $\eta(\mathrm{d}t, \mathrm{d}u) = \mathrm{d}t \times \mu(t, \mathrm{d}u)$ for some family of *intensity* kernels $(\mu(t, \cdot), t \geq 0)$ on the measurable space $(\mathcal{U}, \mathcal{B}(\mathcal{U}))$. The associated Poisson random measure, written $N(\mathrm{d}t, \mathrm{d}u)$, has support which is represented by a time-evolving system of points in $\mathcal{U}$. The system of points in $[0, \infty) \times \mathcal{U}$ is called a *Poisson point process.*

In the case that the intensity kernels $\mu(t, \cdot)$ are all the same and thus no longer time-dependent, say μ, the associated process is called a homogenous Poisson point process (although the word 'homogenous' is often omitted). The associated Poisson point process is sometimes written $((t, u_t), t \geq 0)$, where it is understood that we are only considering $(t, u_t) \in \operatorname{supp} N$. It is a consequence of the fact that the measure η is a product measure with one of them being diffuse, that is, Lebesgue measure, that there is at most one pair $(t, u_t) \in \operatorname{supp} N$ for each $t \geq 0$.

There are numerous examples of Poisson point processes that have been used throughout this text. Generally speaking, these occur in the context of the Lévy–Itô decomposition or through excursion theory. As such, there are a number of additive functionals that can be equivalently written

$$\int_{[0,t]}\int_{\mathcal{U}} f(s,u)N(\mathrm{d}s,\mathrm{d}u) = \sum_{s\le t} f(s,u_s), \tag{A.45}$$

for measurable $f\colon [0,\infty)\times\mathcal{U}\mapsto[0,\infty)$, where the sum on the right-hand side is understood to be over $(s,u_s)\in\operatorname{supp}N$ such that $s\le t$. Campbell's formula tells us that the expectation of (A.45) is equal to

$$E\left[\int_0^t\int_{\mathcal{U}} f(s,u)\mu(s,\mathrm{d}u)\mathrm{d}s\right].$$

The representation of the integral as a summation in (A.45) proves to be convenient when the integrand is additionally taken to be random and measurable in the appropriate sense.

Theorem A.18 *Suppose that $f\colon [0,\infty)\times\mathcal{U}\to[0,\infty)$ is a measurable function and that $(Z_s, s\ge 0)$ is a left-continuous stochastic process which is measurable with respect to the filtration $(\mathcal{H}_t, t\ge 0)$, where $\mathcal{H}_t=\sigma(N(A,B) : A\in\mathcal{B}[0,t]\times\mathcal{B}(\mathcal{U}))$. Then, for all $t\ge 0$,*

$$E_\theta\left[\sum_{s\le t} Z_s f(s,u_s)\right] = E_\theta\left[\int_0^t\int_{\mathcal{U}} Z_s f(s,u)\mu(s,\mathrm{d}u)\,\mathrm{d}s\right], \tag{A.46}$$

with the understanding that the right-hand side is infinite if and only if the left-hand side is.

The notion of a Markov additive Poisson point process is a generalisation of the setting described above. The basic idea is to replace the intensity measure μ on $\mathcal{U}$ by a family of intensity kernels μ_ϑ, for ϑ belonging to some domain E, and to randomise ϑ through time according to the evolution of a suitable Markov process on E.

To this end, suppose that $\vartheta := (\vartheta_t, t\ge 0)$ is an E-valued Feller process and that $\mu\colon E\times\mathcal{B}(\mathcal{U})\to[0,\infty]$ is a family of kernels. A denumerable sequence of points $((t_i,u_i), i\in\mathbb{N})$ in $[0,\infty)\times\mathcal{U}$ defines a counting measure

$$N(A,B) = \sum_{i\in\mathbb{N}} \mathbf{1}_{(t_i\in A, u_i\in B)}, \qquad A\in\mathcal{B}([0,\infty)), B\in\mathcal{B}(\mathcal{U}).$$

Note that the enumeration of $((t_i,u_i), i\in\mathbb{N})$ is not necessarily ordered according to the value of the elements t_i, which may not be possible, for example, if each finite time interval hosts an infinite number of points. When $\vartheta_0=\theta$, let us denote the joint law of the processes N and ϑ by P_θ.

Definition A.19 (Markov additive Poisson point process) The system (N,ϑ) is called a Markov additive Poisson point process if, for all $\theta\in E$ and $f\colon [0,\infty)\times\mathcal{U}\times E\to[0,\infty)$,

$$E_\theta\left[\exp\left(-\int_{[0,t]}\int_{\mathcal{U}} f(s,u,\vartheta_s)N(\mathrm{d}s,\mathrm{d}u)\right)\Big|(\vartheta_s, s\ge 0)\right]$$
$$= \exp\left(-\int_0^t\int_{\mathcal{U}}(1-\mathrm{e}^{-f(s,u,\vartheta_s)})\mu_{\vartheta_s}(\mathrm{d}u)\mathrm{d}s\right).$$

As in Theorem A.18, we can work with additive functionals with respect to N via (A.45). Thinking of the pair (t_i, u_i) as describing the time t_i at which a point u_i arrives in $\mathcal{U}$, we can associate the time evolving sigma algebra for both N and ϑ by $\mathcal{H}_t^\vartheta = \sigma((N(A,B) : A \in \mathcal{B}[0,t] \times \mathcal{B}(\mathcal{U})) \cup (\vartheta_s, s \le t))$ for $t \ge 0$.

Corollary A.20 *For $f\colon [0,\infty) \times \mathcal{U} \times E \to [0,\infty)$ and $(Z_s, s \ge 0)$ is a left-continuous stochastic process which is measurable with respect to the filtration $(\mathcal{H}_t^\vartheta, t \ge 0)$, we have*

$$E_\theta\left[\sum_{s\le t} Z_s f(s, u_s, \vartheta_s)\right] = E_\theta\left[\int_0^t \int_{\mathcal{U}} Z_s f(s, u, \vartheta_s)\mu_{\vartheta_s}(\mathrm{d}u)\,\mathrm{d}s\right], \qquad \text{(A.47)}$$

References

[1] L. Alili, L. Chaumont, Pi. Graczyk, and T. Żak. Inversion, duality and Doob h-transforms for self-similar Markov processes. *Electron. J. Probab.*, 22(20):18, 2017.

[2] L. Alili, W. Jedidi, and V. Rivero. On exponential functionals, harmonic potential measures and undershoots of subordinators. *ALEA Lat. Am. J. Probab. Math. Stat.*, 11(1):711–35, 2014.

[3] G. Alsmeyer. On the Markov renewal theorem. *Stochastic Process. Appl.*, 50(1):37–56, 1994.

[4] G. Alsmeyer. Quasistochastic matrices and Markov renewal theory. *J. Appl. Probab.*, 51A(Celebrating 50 Years of The Applied Probability Trust):359–76, 2014.

[5] P. Andrew. A proof from 'first principles' of Kesten's result for the probabilities with which a subordinator hits points. *Electron. Comm. Probab.*, 11:58–63 (electronic), 2006.

[6] D. Applebaum. *Lévy processes and stochastic calculus*, volume 116 of Cambridge Studies in Advanced Mathematics. Cambridge University Press, Cambridge, second edition, 2009.

[7] S. Asmussen. *Applied probability and queues*, volume 51 of Stochastic Modelling and Applied Probability. Springer-Verlag, New York, second edition, 2003.

[8] S. Asmussen and H. Albrecher. *Ruin probabilities*, volume 14 of Advanced Series on Statistical Science & Applied Probability. World Scientific, Hackensack, NJ, second edition, 2010.

[9] F. Avram, A. E. Kyprianou, and M. R. Pistorius. Exit problems for spectrally negative Lévy processes and applications to (Canadized) Russian options. *Ann. Appl. Probab.*, 14(1):215–38, 2004.

[10] E. W. Barnes. The genesis of the double gamma functions. *Proc. Lond. Math. Soc.*, 31:358–81, 1899.

[11] H. Bateman and A. Erdélyi. *Vysshie transtsendentnye funktsii. I: Gipergeometricheskaya funktsiya. Funktsii Lezhandra.* Izdat. 'Nauka', Moscow, unrevised edition, 1973. Translated from the English by N. Ja. Vilenkin.

[12] E. J. Baurdoux. Some excursion calculations for reflected Lévy processes. *ALEA Lat. Am. J. Probab. Math. Stat.*, 6:149–62, 2009.

[13] G. Baxter. An operator identity. *Pacific J. Math.*, 8:649–63, 1958.

[14] A. Behme. Exponential functionals of Lévy processes with jumps. *ALEA Lat. Am. J. Probab. Math. Stat.*, 12(1):375–97, 2015.

[15] A. Behme and A. Lindner. On exponential functionals of Lévy processes. *J. Theoret. Probab.*, 28(2):681–720, 2015.

[16] A. Behme, A. Lindner, and M. Maejima. On the range of exponential functionals of Lévy processes. In *Séminaire de Probabilités XLVIII*, volume 2168 of Lecture Notes in Mathematics, pages 267–303. Springer, Cham, 2016.

[17] V. Bernyk, R. C. Dalang, and G. Peskir. The law of the supremum of a stable Lévy process with no negative jumps. *Ann. Probab.*, 36(5):1777–89, 2008.

[18] J. Bertoin. *Lévy processes*, volume 121 of Cambridge Tracts in Mathematics. Cambridge University Press, Cambridge, 1996.

[19] J. Bertoin. On the first exit time of a completely asymmetric stable process from a finite interval. *Bull. London Math. Soc.*, 28(5):514–20, 1996.

[20] J. Bertoin. Regularity of the half-line for Lévy processes. *Bull. Sci. Math.*, 121(5):345–54, 1997.

[21] J. Bertoin. Subordinators: examples and applications. In *Lectures on probability theory and statistics (Saint-Flour, 1997)*, volume 1717 of Lecture Notes in Mathematics, pages 1–91. Springer, Berlin, 1999.

[22] J. Bertoin and M. E. Caballero. Entrance from 0+ for increasing semi-stable Markov processes. *Bernoulli*, 8(2):195–205, 2002.

[23] J. Bertoin and M. Savov. Some applications of duality for Lévy processes in a half-line. *Bull. Lond. Math. Soc.*, 43(1):97–110, 2011.

[24] J. Bertoin, K. van Harn, and F. W. Steutel. Renewal theory and level passage by subordinators. *Statist. Probab. Lett.*, 45(1):65–9, 1999.

[25] J. Bertoin and W. Werner. Stable windings. *Ann. Probab.*, 24(3):1269–79, 1996.

[26] J. Bertoin and M. Yor. On subordinators, self-similar Markov processes and some factorizations of the exponential variable. *Electron. Comm. Probab.*, 6:95–106 (electronic), 2001.

[27] J. Bertoin and M. Yor. The entrance laws of self-similar Markov processes and exponential functionals of Lévy processes. *Potential Anal.*, 17(4):389–400, 2002.

[28] J. Bertoin and M. Yor. Exponential functionals of Lévy processes. *Probab. Surv.*, 2:191–212, 2005.

[29] K. Bichteler. *Stochastic integration with jumps*, volume 89 of Encyclopedia of Mathematics and Its Applications. Cambridge University Press, Cambridge, 2002.

[30] J. D. Biggins. Uniform convergence of martingales in the branching random walk. *Ann. Probab.*, 20(1):137–51, 1992.

[31] J. Billingham and A. C. King. Uniform asymptotic expansions for the Barnes double gamma function. *Proc. Roy. Soc. London Ser. A*, 453(1964):1817–29, 1997.

[32] P. Billingsley. *Convergence of probability measures*. Wiley Series in Probability and Statistics. John Wiley & Sons, Inc., New York, second edition, 1999.

[33] I. Binder and M. Braverman. The rate of convergence of the Walk on Spheres algorithm. *Geom. Funct. Anal.*, 22(3):558–87, 2012.

[34] N. H. Bingham. Maxima of sums of random variables and suprema of stable processes. *Z. Wahrscheinlichkeitstheorie und Verw. Gebiete*, 26:273–96, 1973.

[35] N. H. Bingham. Fluctuation theory in continuous time. *Advances in Appl. Probab.*, 7(4):705–66, 1975.

[36] J. Bliedtner and W. Hansen. *Potential theory: An analytic and probabilistic approach to balayage*. Universitext. Springer-Verlag, Berlin, 1986.

[37] L. E. Blumenson. Classroom notes: a derivation of n-dimensional spherical coordinates. *Amer. Math. Monthly*, 67(1):63–6, 1960.

[38] R. M. Blumenthal. *Excursions of Markov processes*. Probability and its Applications. Birkhäuser Boston, Inc., Boston, MA, 1992.

[39] R. M. Blumenthal, R. K. Getoor, and D. B. Ray. On the distribution of first hits for the symmetric stable processes. *Trans. Amer. Math. Soc.*, 99:540–54, 1961.

[40] K. Bogdan and T. Żak. On Kelvin transformation. *J. Theoret. Probab.*, 19(1):89–120, 2006.

[41] A. A. Borovkov. *Stochastic processes in queueing theory*. Springer-Verlag, New York-Berlin, 1976. Translated from the Russian by Kenneth Wickwire, Applications of Mathematics, No. 4.

[42] J. Bretagnolle. Résultats de Kesten sur les processus à accroissements indépendants. In *Séminaire de Probabilités, V (Univ. Strasbourg, Année Universitaire 1969-1970)*, volume 191 of Lecture Notes in Mathematics, pages 21–36. Springer, Berlin, 1971.

[43] M. E. Caballero and L. Chaumont. Conditioned stable Lévy processes and the Lamperti representation. *J. Appl. Probab.*, 43(4):967–83, 2006.

[44] M. E. Caballero and L. Chaumont. Weak convergence of positive self-similar Markov processes and overshoots of Lévy processes. *Ann. Probab.*, 34(3):1012–34, 2006.

[45] M. E. Caballero, J. C. Pardo, and J. L. Pérez. Explicit identities for Lévy processes associated to symmetric stable processes. *Bernoulli*, 17(1):34–59, 2011.

[46] L. Chaumont, A. E. Kyprianou, J. C. Pardo, and V. Rivero. Fluctuation theory and exit systems for positive self-similar Markov processes. *Ann. Probab.*, 40(1):245–79, 2012.

[47] L. Chaumont, H. Panti, and V. Rivero. The Lamperti representation of real-valued self-similar Markov processes. *Bernoulli*, 19(5B):2494–523, 2013.

[48] L. Chaumont and J. C. Pardo. The lower envelope of positive self-similar Markov processes. *Electron. J. Probab.*, 11(49): 1321–41, 2006.

[49] K.-L. Chung and W. H. J. Fuchs. On the distribution of values of sums of random variables. *Mem. Amer. Math. Soc.*, (6):12, 1951.

[50] K.-L. Chung and J. B. Walsh. *Markov processes, Brownian motion, and time symmetry*, volume 249 of Grundlehren der Mathematischen Wissenschaften [Fundamental Principles of Mathematical Sciences]. Springer, New York, second edition, 2005.

[51] O. Chybiryakov. The Lamperti correspondence extended to Lévy processes and semi-stable Markov processes in locally compact groups. *Stochastic Process. Appl.*, 116(5):857–72, 2006.

[52] E. Çinlar. Markov additive processes. I, II. *Z. Wahrscheinlichkeitstheorie und Verw. Gebiete*, 24:85–93.

[53] E. Çinlar. Lévy systems of Markov additive processes. *Z. Wahrscheinlichkeitstheorie und Verw. Gebiete*, 31:175–85, 1974/75.

[54] E. Çinlar. Entrance-exit distributions for Markov additive processes. *Math. Programming Stud.*, (5):22–38, 1976.

[55] E. Çinlar. *Probability and stochastics*, volume 261 of Graduate Texts in Mathematics. Springer, New York, 2011.

[56] D. A. Darling. The maximum of sums of stable random variables. *Trans. Amer. Math. Soc.*, 83:164–9, 1956.

[57] S. Dereich, L. Döring, and A. E. Kyprianou. Real self-similar processes started from the origin. *Ann. Probab.*, 45(3):1952–2003, 2017.

[58] S. Ditlevsen and P. Greenwood. The Morris-Lecar neuron model embeds a leaky integrate-and-fire model. *J. Math. Biol.*, 67(2):239–59, 2013.

[59] R. A. Doney. On Wiener-Hopf factorisation and the distribution of extrema for certain stable processes. *Ann. Probab.*, 15(4):1352–62, 1987.

[60] R. A. Doney. *Fluctuation theory for Lévy processes*, volume 1897 of Lecture Notes in Mathematics. Springer, Berlin, 2007.

[61] R. A. Doney. A note on the supremum of a stable process. *Stochastics*, 80(2–3, IMS Lecture Notes – Monograph Series):151–5, 2008.

[62] R. A. Doney and A. E. Kyprianou. Overshoots and undershoots of Lévy processes. *Ann. Appl. Probab.*, 16(1):91–106, 2006.

[63] R. A. Doney and M. S. Savov. The asymptotic behavior of densities related to the supremum of a stable process. *Ann. Probab.*, 38(1):316–26, 2010.

[64] R. A. Doney and S. Vakeroudis. Windings of planar stable processes. In *Séminaire de Probabilités XLV*, volume 2078 of Lecture Notes in Mathematics, pages 277–300. Springer, Cham, 2013.

[65] T. Duquesne. Path decompositions for real Levy processes. *Ann. Inst. H. Poincaré Probab. Statist.*, 39(2):339–70, 2003.

[66] A. Dvoretzky and P. Erdös. Some problems on random walk in space. In *Proceedings of the Second Berkeley Symposium on Mathematical Statistics and Probability, 1950*, pages 353–67. University of California Press, Berkeley and Los Angeles, 1951.

[67] K. B. Erickson. The strong law of large numbers when the mean is undefined. *Trans. Amer. Math. Soc.*, 185:371–81, 1973.

[68] S. N. Ethier and T. G. Kurtz. *Markov processes: Characterization and convergence*. Wiley Series in Probability and Mathematical Statistics. John Wiley & Sons, New York, 1986.

[69] W. Feller. *An introduction to probability theory and its applications. Volume II.* John Wiley & Sons, New York-London-Sydney, second edition, 1971.

[70] R. Feng, A. Kuznetsov, and F. Yang. Exponential functionals of Lévy processes and variable annuity guaranteed benefits. *Stochastic Process. Appl.*, 129(2):604–25, 2019.

[71] P. J. Fitzsimmons. On the existence of recurrent extensions of self-similar Markov processes. *Electron. Comm. Probab.*, 11:230–41, 2006.

[72] S. Fourati. Inversion de l'espace et du temps des processus de Lévy stables. *Probab. Theory Relat. Fields*, 135(2):201–15, 2006.

[73] B. Fristedt. The behavior of increasing stable processes for both small and large times. *J. Math. Mech.*, 13:849–56, 1964.

[74] B. Fristedt. Sample functions of stochastic processes with stationary, independent increments. In *Advances in probability and related topics, Volume 3*, pages 241–396. Dekker, New York, 1974.

[75] J. Geiger and G. Kersting. Winding numbers for 2-dimensional, positive recurrent diffusions. *Potential Anal.*, 3(2):189–201, 1994.

[76] R. K. Getoor. Continuous additive functionals of a Markov process with applications to processes with independent increments. *J. Math. Anal. Appl.*, 13:132–53, 1966.

[77] I. I. Gikhman and A. V. Skorokhod. *The theory of stochastic processes. I.* Classics in Mathematics. Springer-Verlag, Berlin, 2004. Translated from the Russian by S. Kotz, Reprint of the 1974 edition.

[78] I. I. Gikhman and A. V. Skorokhod. *The theory of stochastic processes. II.* Classics in Mathematics. Springer-Verlag, Berlin, 2004. Translated from the Russian by S. Kotz, Reprint of the 1975 edition.

[79] I. I. Gikhman and A. V. Skorokhod. *The theory of stochastic processes. III.* Classics in Mathematics. Springer, Berlin, 2007. Translated from the Russian by Samuel Kotz, Reprint of the 1974 edition.

[80] B. V. Gnedenko and A. N. Kolmogorov. *Limit distributions for sums of independent random variables.* Translated from the Russian, annotated, and revised by K. L. Chung. With appendices by J. L. Doob and P. L. Hsu. Revised edition. Addison-Wesley, Reading, MA-London-Don Mills., Ont., 1968.

[81] P. Graczyk and T. Jakubowski. On Wiener-Hopf factors for stable processes. *Ann. Inst. Henri Poincaré Probab. Stat.*, 47(1):9–19, 2011.

[82] I. S. Gradshteyn and I. M. Ryzhik. *Table of integrals, series, and products.* Elsevier/Academic Press, Amsterdam, eighth edition, 2015. Translated from the Russian, Translation edited and with a preface by Daniel Zwillinger and Victor Moll, Revised from the seventh edition [MR2360010].

[83] P. Greenwood and J. Pitman. Fluctuation identities for Lévy processes and splitting at the maximum. *Adv. in Appl. Probab.*, 12(4):893–902, 1980.

[84] D. V. Gusak and V. S. Korolyuk. The combined distribution of a process with stationary increments and its maximum. *Teor. Verojatnost. i Primenen*, 14:421–30, 1969.

[85] D. Hackmann and A. Kuznetsov. A note on the series representation for the density of the supremum of a stable process. *Electron. Commun. Probab.*, 18(42): 5, 2013.

[86] P. Hall. A comedy of errors: the canonical form for a stable characteristic function. *Bull. London Math. Soc.*, 13(1):23–7, 1981.

[87] C. C. Heyde. On the maximum of sums of random variables and the supremum functional for stable processes. *J. Appl. Probab.*, 6:419–29, 1969.

[88] J. Horowitz. Semilinear Markov processes, subordinators and renewal theory. *Z. Wahrscheinlichkeitstheorie und Verw. Gebiete*, 24:167–93, 1972.

[89] E. L. Horton and A. E. Kyprianou. More on hypergeometric Lévy processes. *Adv. in Appl. Probab.*, 48(A):153–158, 2016.

[90] F. Hubalek and A. Kuznetsov. A convergent series representation for the density of the supremum of a stable process. *Electron. Commun. Probab.*, 16:84–95, 2011.

[91] G. A. Hunt. Markoff processes and potentials. I, II. *Illinois J. Math.*, 1:44–93, 316–69, 1957.

[92] G. A. Hunt. Markoff chains and Martin boundaries. *Illinois J. Math.*, 4:313–40, 1960.

[93] C-O. Hwang and M. Mascagni. Efficient modified 'walk on spheres' algorithm for the linearized Poisson–Bolzmann equation. *Appl. Phys. Lett.*, 78(6):787–9, 5 February 2001.

[94] C.-O. Hwang, M. Mascagni, and J. A. Given. A Feynman–Kac path-integral implementation for Poisson's equation using an h-conditioned Green's function. *Math. Comput. Simul.*, 62(3–6):347–55, 3 March 2003.

[95] J. Jacod and A. N. Shiryaev. *Limit Theorems for Stochastic Processes*, volume 288 of Grundlehren der Mathematischen Wissenschaften [Fundamental Principles of Mathematical Sciences]. Springer-Verlag, Berlin, second edition, 2003.

[96] A. Janicki and A. Weron. *Simulation and Chaotic Behavior of α-stable Stochastic Processes*, volume 178 of Monographs and Textbooks in Pure and Applied Mathematics. Marcel Dekker, New York, 1994.

[97] W. Jedidi and S. Vakeroudis. Windings of planar processes, exponential functionals and Asian options. *Adv. in Appl. Probab.*, 50(3):726–42, 2018.

[98] M. Kac. Some remarks on stable processes. *Publ. Inst. Statist. Univ. Paris*, 6:303–6, 1957.

[99] M. Kanda. Two theorems on capacity for Markov processes with stationary independent increments. *Z. Wahrscheinlichkeitstheorie und Verw. Gebiete*, 35(2):159–65, 1976.

[100] M. Kanda. Characterization of semipolar sets for processes with stationary independent increments. *Z. Wahrscheinlichkeitstheorie und Verw. Gebiete*, 42(2):141–54, 1978.

[101] M. Kanda. On the class of polar sets for a certain class of Lévy processes on the line. *J. Math. Soc. Japan*, 35(2):221–42, 1983.

[102] H. Kaspi. On the symmetric Wiener-Hopf factorization for Markov additive processes. *Z. Wahrsch. Verw. Gebiete*, 59(2):179–96, 1982.

[103] H. Kesten. *Hitting probabilities of single points for processes with stationary independent increments.* Memoirs of the American Mathematical Society, No. 93. American Mathematical Society, Providence, R.I., 1969.

[104] H. Kesten. Renewal theory for functionals of a Markov chain with general state space. *Ann. Probability*, 2:355–86, 1974.

[105] A. Khintchine. Sur la croissance locale des processus stochastiques homogènes à accroissements indépendants. *Bull. Acad. Sci. URSS. Sér. Math. [Izvestia Akad. Nauk SSSR]*, 1939:487–508, 1939.

[106] A. Ya. Khintchine. *Limit laws for sums of independent random variables.* ONTI, Moscow, Leningrad, 1938.

[107] A. Ya. Khintchine and P. Lévy. Sur les lois stables. *C. R. Acad. Sci. Paris Sér. A*, 202:374–6, 1936.

[108] I. Khoffman-Iënsen. Stable densities. *Teor. Veroyatnost. i Primenen.*, 38(2):470–6, 1993.

[109] D. Khoshnevisan, T. M. Lewis, and W. V. Li. On the future infima of some transient processes. *Probab. Theory Related Fields*, 99(3):337–60, 1994.

[110] J. F. C. Kingman. Recurrence properties of processes with stationary independent increments. *J. Austral. Math. Soc.*, 4:223–8, 1964.

[111] S. W. Kiu. Semistable Markov processes in $\mathbf{R}^n$. *Stochastic Process. Appl.*, 10(2):183–91, 1980.

[112] J. Klafter, S. C. Lim, and R. Metzler. *Fractional dynamics: recent advances*. World Scientific, Singapore 2011.

[113] R. Klages, G. Radons, and I. M. Sokolov. *Anomalous Transport: Foundations and Applications*. Wiley, Darmstadt 2008.

[114] J. Kuelbs. A representation theorem for symmetric stable processes and stable measures on *H*. *Z. Wahrscheinlichkeitstheorie und Verw. Gebiete*, 26:259–71, 1973.

[115] A. Kuznetsov. Wiener-Hopf factorization and distribution of extrema for a family of Lévy processes. *Ann. Appl. Probab.*, 20(5):1801–30, 2010.

[116] A. Kuznetsov. On extrema of stable processes. *Ann. Probab.*, 39(3):1027–60, 2011.

[117] A. Kuznetsov. On the density of the supremum of a stable process. *Stochastic Process. Appl.*, 123(3):986–1003, 2013.

[118] A. Kuznetsov, A. E. Kyprianou, and J. C. Pardo. Meromorphic Lévy processes and their fluctuation identities. *Ann. Appl. Probab.*, 22(3):1101–35, 2012.

[119] A. Kuznetsov, A. E. Kyprianou, J. C. Pardo, and K. van Schaik. A Wiener-Hopf Monte Carlo simulation technique for Lévy processes. *Ann. Appl. Probab.*, 21(6):2171–90, 2011.

[120] A. Kuznetsov, A. E. Kyprianou, J. C. Pardo, and A. R. Watson. The hitting time of zero for a stable process. *Electron. J. Probab.*, 19(30): 26, 2014.

[121] A. Kuznetsov and J. C. Pardo. Fluctuations of stable processes and exponential functionals of hypergeometric Lévy processes. *Acta Appl. Math.*, 123:113–39, 2013.

[122] A. E. Kyprianou. First passage of reflected strictly stable processes. *ALEA Lat. Am. J. Probab. Math. Stat.*, 2:119–23, 2006.

[123] A. E. Kyprianou. *Fluctuations of Lévy processes with applications*. Universitext. Springer, Heidelberg, second edition, 2014. Introductory lectures.

[124] A. E. Kyprianou. Deep factorisation of the stable process. *Electron. J. Probab.*, 21:Paper No. 23, 28, 2016.

[125] A. E. Kyprianou. Stable Lévy processes, self-similarity and the unit ball. *ALEA Lat. Am. J. Probab. Math. Stat.*, 15(1):617–90, 2018.

[126] A. E. Kyprianou, A. Osojnik, and T. Shardlow. Unbiased 'walk-on-spheres' Monte Carlo methods for the fractional Laplacian. *IMA J. Numer. Anal.*, 38(3):1550–78, 2018.

[127] A. E. Kyprianou, J. C. Pardo, and V. Rivero. Exact and asymptotic n-tuple laws at first and last passage. *Ann. Appl. Probab.*, 20(2):522–64, 2010.

[128] A. E. Kyprianou, J. C. Pardo, and A. R. Watson. The extended hypergeometric class of Lévy processes. *J. Appl. Probab.*, 51A(Celebrating 50 Years of The Applied Probability Trust):391–408, 2014.

[129] A. E. Kyprianou, J. C. Pardo, and A. R. Watson. Hitting distributions of α-stable processes via path censoring and self-similarity. *Ann. Probab.*, 42(1):398–430, 2014.

[130] A. E. Kyprianou, J. C. Pardo, and M. Vidmar. Double hypergeometric Lévy processes and self-similarity. *J. Appl. Probab.*, 58(1):254–73, 2021.

[131] A. E. Kyprianou and V. Rivero. Special, conjugate and complete scale functions for spectrally negative Lévy processes. *Electron. J. Probab.*, 13(57): 1672–701, 2008.

[132] A. E. Kyprianou, V. Rivero, and B. Şengül. Deep factorisation of the stable process II: Potentials and applications. *Ann. Inst. Henri Poincaré Probab. Stat.*, 54(1):343–62, 2018.

[133] A. E. Kyprianou, V. Rivero, B. Şengül, and T. Yang. Entrance laws at the origin of self-similar Markov processes in high dimensions. *Trans. Amer. Math. Soc.*, 373(9):6227–99, 2020.

[134] A. E. Kyprianou, V. Rivero, and W. Satitkanitkul. Deep factorisation of the stable process III: the view from radial excursion theory and the point of closest reach. *Potential Anal.*, 53(4):1347–75, 2020.

[135] A. E. Kyprianou, V. M. Rivero, and W. Satitkanitkul. Conditioned real self-similar Markov processes. *Stochastic Process. Appl.*, 129(3):954–77, 2019.

[136] A. E. Kyprianou and S. M. Vakeroudis. Stable windings at the origin. *Stochastic Process. Appl.*, 128(12):4309–25, 2018.

[137] A. E. Kyprianou and A. R. Watson. Potentials of stable processes. In *Séminaire de Probabilités XLVI*, volume 2123 of Lecture Notes in Mathematics, pages 333–43. Springer, Cham, 2014.

[138] S. P. Lalley. Conditional Markov renewal theory. I. Finite and denumerable state space. *Ann. Probab.*, 12(4):1113–48, 1984.

[139] J. Lamperti. Semi-stable Markov processes. I. *Z. Wahrscheinlichkeitstheorie und Verw. Gebiete*, 22:205–25, 1972.

[140] J. B. Lawrie and A. C. King. Exact solution to a class of functional difference equations with application to a moving contact line flow. *European J. Appl. Math.*, 5(2):141–57, 1994.

[141] J. Letemplier and T. Simon. Unimodality of hitting times for stable processes. In *Séminaire de Probabilités XLVI*, volume 2123 of Lecture Notes in Mathematics, pages 345–57. Springer, Cham, 2014.

[142] P. Lévy. Théorie des erreurs. La loi de Gaußet les lois exceptionnelles. *Bull. Soc. Math. France*, 52:49–85, 1924.

[143] P. Lévy. *Calcul de Probabilités.* Gauthier-Villars, Paris, 1925.

[144] P. Lévy. *Théorie de l'Addition des Variables Aléatoires.* Gauthier-Villars, Paris, 1937.

[145] Z. Li and W. Xu. Asymptotic results for exponential functionals of Lévy processes. *Stochastic Process. Appl.*, 128(1):108–31, 2018.

[146] W. Linde. *Probability Measures in Banach Spaces: Stable and Infinitely Divisible Distributions.* Wiley, New York, 1983.

[147] B. Maisonneuve. Exit systems. *Ann. Probability*, 3(3):399–411, 1975.

[148] K. Maulik and B. Zwart. Tail asymptotics for exponential functionals of Lévy processes. *Stochastic Process. Appl.*, 116(2):156–77, 2006.

[149] P. Messulam and M. Yor. On D. Williams' 'pinching method' and some applications. *J. London Math. Soc. (2)*, 26(2):348–64, 1982.

[150] J. L. Mijnheer. *Sample path properties of stable processes*. Mathematisch Centrum, Amsterdam, 1975. Doctoral dissertation, University of Leiden, Leiden, Mathematical Centre Tracts, No. 59.

[151] G. A. Mikhailov. Estimation of the difficulty of simulating the process of "random walk on spheres" for some types of regions. *USSR Computational Mathematics and Mathematical Physics*, 19(2):247–54, 1979.

[152] D. Monrad and M. L. Silverstein. Stable processes: sample function growth at a local minimum. *Z. Wahrsch. Verw. Gebiete*, 49(2):177–210, 1979.

[153] M. Motoo. Proof of the law of iterated logarithm through diffusion equation. *Ann. Inst. Stat. Math.*, 10:21–8, 1958.

[154] M. Motoo. Some evaluations for continuous Monte Carlo method by using Brownian hitting process. *Ann. Inst. Stat. Math.*, 11(1):49–54, February 1959.

[155] M. E. Muller. Some continuous Monte Carlo methods for the Dirichlet problem. *Ann. Math. Stat.*, 27(3):569–89, 1956.

[156] M. Nagasawa. Time reversions of Markov processes. *Nagoya Math. J.*, 24:177–204, 1964.

[157] S. Orey. Polar sets for processes with stationary independent increments. In *Markov Processes and Potential Theory (Proc. Sympos. Math. Res. Center, Madison, Wis., 1967)*, pages 117–26. Wiley, New York, 1967.

[158] S. Palau, J. C. Pardo, and C. Smadi. Asymptotic behaviour of exponential functionals of Lévy processes with applications to random processes in random environment. *ALEA Lat. Am. J. Probab. Math. Stat.*, 13(2):1235–58, 2016.

[159] J. C. Pardo. On the future infimum of positive self-similar Markov processes. *Stochastics*, 78(3):123–55, 2006.

[160] J. C. Pardo. The upper envelope of positive self-similar Markov processes. *J. Theoret. Probab.*, 22(2):514–42, 2009.

[161] J. C. Pardo, H. Panti, and V. Rivero. Recurrent extensions of real-valued self-similar markov processes. *Potential Analysis*, 53:899–920, 2020 2018. `arXiv:1808.00129 [math.PR]`.

[162] J. C. Pardo, P. Patie, and M. Savov. A Wiener-Hopf type factorization for the exponential functional of Lévy processes. *J. Lond. Math. Soc. (2)*, 86(3):930–56, 2012.

[163] J. C. Pardo, V. Rivero, and K. van Schaik. On the density of exponential functionals of Lévy processes. *Bernoulli*, 19(5A):1938–64, 2013.

[164] P. Patie. A few remarks on the supremum of stable processes. *Statist. Probab. Lett.*, 79(8):1125–8, 2009.

[165] P. Patie and M. Savov. Exponential functional of Lévy processes: generalized Weierstrass products and Wiener-Hopf factorization. *C. R. Math. Acad. Sci. Paris*, 351(9–10):393–6, 2013.

[166] P. Patie and M. Savov. Bernstein-gamma functions and exponential functionals of Lévy processes. *Electron. J. Probab.*, 23:Paper No. 75, 101, 2018.

[167] G. Peskir. The law of the hitting times to points by a stable Lévy process with no negative jumps. *Electron. Commun. Probab.*, 13:653–9, 2008.

[168] E. A. Pečerskiĭ and B. A. Rogozin. The combined distributions of the random variables connected with the fluctuations of a process with independent increments. *Teor. Verojatnost. i Primenen.*, 14:431–44, 1969.

[169] E. J. G. Pitman and J. Pitman. A direct approach to the stable distributions. *Adv. in Appl. Probab.*, 48(A):261–82, 2016.

[170] J. Pitman and M. Yor. Asymptotic laws of planar Brownian motion. *Ann. Probab.*, 14(3):733–79, 1986.

[171] S. C. Port. Hitting times and potentials for recurrent stable processes. *J. Analyse Math.*, 20:371–95, 1967.

[172] S. C. Port. The first hitting distribution of a sphere for symmetric stable processes. *Trans. Amer. Math. Soc.*, 135:115–25, 1969.

[173] S. C. Port and C. J. Stone. *Brownian motion and classical potential theory*. Academic Press [Harcourt Brace Jovanovich, Publishers], New York-London, 1978.

[174] N. U. Prabhu. *Queues and inventories. A study of their basic stochastic processes*. John Wiley & Sons, New York-London-Sydney, 1965.

[175] C. Profeta and T. Simon. On the harmonic measure of stable processes. In *Séminaire de Probabilités XLVIII*, volume 2168 of Lecture Notes in Mathematics, pages 325–345. Springer, Cham, 2016.

[176] P. E. Protter. *Stochastic integration and differential equations*, volume 21 of Stochastic Modelling and Applied Probability. Springer-Verlag, Berlin, second edition, 2005. Version 2.1, Corrected third printing.

[177] M. Riesz. Intégrales de Riemann-Liouville et potentiels. *Acta. Sci. Math. Szeged.*, 9:1–42, 1938.

[178] M. Riesz. Rectification au travail 'Intégrales de Riemann-Liouville et potentiels'. *Acta. Sci. Math. Szeged.*, 9:116–8, 1938.

[179] V. Rivero. A law of iterated logarithm for increasing self-similar Markov processes. *Stoch. Stoch. Rep.*, 75(6):443–72, 2003.

[180] V. Rivero. Recurrent extensions of self-similar Markov processes and Cramér's condition. *Bernoulli*, 11(3):471–509, 2005.

[181] V. Rivero. Recurrent extensions of self-similar Markov processes and Cramér's condition. II. *Bernoulli*, 13(4):1053–70, 2007.

[182] L. C. G. Rogers and D. Williams. *Diffusions, Markov processes, and martingales. Vol. 1*. Cambridge Mathematical Library. Cambridge University Press, Cambridge, 2000. Foundations, Reprint of the second (1994) edition.

[183] L. C. G. Rogers and D. Williams. *Diffusions, Markov processes, and martingales. Vol. 2*. Cambridge Mathematical Library. Cambridge University Press, Cambridge, 2000. Itô calculus, Reprint of the second (1994) edition.

[184] B. A. Rogozin. The local behavior of processes with independent increments. *Teor. Verojatnost. i Primenen.*, 13:507–12, 1968.

[185] B. A. Rogozin. Distribution of the position of absorption for stable and asymptotically stable random walks on an interval. *Teor. Verojatnost. i Primenen.*, 17:342–9, 1972.

[186] K. K. Sabelfeld. *Monte Carlo methods in boundary value problems*. Springer Series in Computational Physics. Springer, 1991.

[187] K. K. Sabelfeld and D. Talay. Integral formulation of the boundary value problems and the method of random walk on spheres. *Monte Carlo Methods Appl.*, 1(1):1–34, 1995.

[188] P. Salminen and L. Vostrikova. On moments of integral exponential functionals of additive processes. *Statist. Probab. Lett.*, 146:139–46, 2019.

[189] G. Samorodnitsky and M. S. Taqqu. *Stable Non-Gaussian Random Processes*. Chapman & Hall, New York, 1994.

[190] K-I Sato. *Lévy processes and infinitely divisible distributions*, volume 68 of Cambridge Studies in Advanced Mathematics. Cambridge University Press, Cambridge, 2013. Translated from the 1990 Japanese original, Revised edition of the 1999 English translation.

[191] Z. Shi. Windings of Brownian motion and random walks in the plane. *Ann. Probab.*, 26(1):112–31, 1998.

[192] M. F. Shlesinger, G. M. Zaslavsky, and U. Frisch, editors. Lévy Flights and Related Topics in Physics, volume 450 of Lecture Notes in Physics. Springer-Verlag, 1995.

[193] M. L. Silverstein. Classification of coharmonic and coinvariant functions for a Lévy process. *Ann. Probab.*, 8(3):539–75, 1980.

[194] A. V. Skorokhod. Limit theorems for stochastic processes. *Teor. Veroyatnost. i Primenen.*, 1:289–319, 1956.

[195] A. V. Skorokhod. Limit theorems for stochastic processes with independent increments. *Teor. Veroyatnost. i Primenen.*, 2:145–77, 1957.

[196] R. T. Smythe and J. B. Walsh. The existence of dual processes. *Invent. Math.*, 19:113–48, 1973.

[197] F. Spitzer. A combinatorial lemma and its application to probability theory. *Trans. Amer. Math. Soc.*, 82:323–39, 1956.

[198] F. Spitzer. The Wiener-Hopf equation whose kernel is a probability density. *Duke Math. J.*, 24:327–43, 1957.

[199] F. Spitzer. Some theorems concerning 2-dimensional Brownian motion. *Trans. Amer. Math. Soc.*, 87:187–97, 1958.

[200] F. Spitzer. The Wiener-Hopf equation whose kernel is a probability density. II. *Duke Math. J.*, 27:363–72, 1960.

[201] F. Spitzer. *Principles of random walk*. The University Series in Higher Mathematics. D. Van Nostrand, Princeton, N. J.-Toronto-London, 1964.

[202] E. S. Štatland. On local properties of processes with independent increments. *Teor. Verojatnost. i Primenen.*, 10:344–50, 1965.

[203] R. Stephenson. On the exponential functional of Markov additive processes, and applications to multi-type self-similar fragmentation processes and trees. *ALEA Lat. Am. J. Probab. Math. Stat.*, 15(2):1257–92, 2018.

[204] L. Takács. *Combinatorial methods in the theory of stochastic processes*. John Wiley & Sons, New York-London-Sydney, 1967.

[205] J. Takeuchi. A local asymptotic law for the transient stable process. *Proc. Japan Acad.*, 40:141–4, 1964.

[206] J. Takeuchi. On the sample paths of the symmetric stable processes in spaces. *J. Math. Soc. Japan*, 16:109–27, 1964.

[207] S. J. Taylor. Sample path properties of a transient stable process. *J. Math. Mech.*, 16:1229–46, 1967.

[208] V. V. Uchaikin and V. M. Zolotarev. *Chance and stability*. Modern Probability and Statistics. VSP, Utrecht, 1999. Stable distributions and their applications, With a foreword by V. Yu. Korolev and Zolotarev.

[209] S. Vakeroudis. On the windings of complex-valued Ornstein-Uhlenbeck processes driven by a Brownian motion and by a stable process. *Stochastics*, 87(5):766–93, 2015.

[210] S. Vakeroudis, M. Yor, and D. Holcman. The mean first rotation time of a planar polymer. *J. Stat. Phys.*, 143(6):1074–95, 2011.

[211] V. Vigon. *Simplifiez vos Lévy en titillant la factorisation de Wiener–Hopf.* PhD thesis, L'INSA de Rouen, 2002.

[212] J. B. Walsh. Markov processes and their functionals in duality. *Z. Wahrscheinlichkeitstheorie und Verw. Gebiete*, 24:229–46, 1972.

[213] T. Watanabe. Sample function behavior of increasing processes of class *L*. *Probab. Theory Related Fields*, 104(3):349–74, 1996.

[214] M. Winkel. Electronic foreign-exchange markets and passage events of independent subordinators. *J. Appl. Probab.*, 42(1):138–52, 2005.

[215] Y. Xiao. Asymptotic results for self-similar Markov processes. In *Asymptotic methods in probability and statistics (Ottawa, ON, 1997)*, pages 323–40. North-Holland, Amsterdam, 1998.

[216] M. Yor. Generalized meanders as limits of weighted Bessel processes, and an elementary proof of Spitzer's asymptotic result on Brownian windings. *Studia Sci. Math. Hungar.*, 33(1–3):339–43, 1997.

[217] V. M. Zolotarev. Analogue of the iterated logarithm law for semicontinuous stable processes. *Teor. Verojatnost. i Primenen.*, 9:512–3, 1964.

[218] V. M. Zolotarev. The first passage time of a level and the behavior at infinity for a class of processes with independent increments. *Teor. Verojatnost. i Primenen.*, 9:653–62, 1964.

[219] V. M. Zolotarev. The moment of first passage of a level and the behaviour at infinity of a class of processes with independent increments. *Teor. Verojatnost. i Primenen.*, 9:724–33, 1964.

[220] V. M. Zolotarev. One-dimensional stable distributions. *Amer. Math. Soc., Providence, RI.*, 1986.

Index

ascending ladder MAP, 294
  excursion measure, 294
  matrix Laplace exponent, 294

beta function, 431, 432
β-subordinator, 78
  Lamperti-stable, 81
Borel–Cantelli Lemma, 438
Brownian motion, 28
  hitting unit sphere, 363
  Spitzer's winding limt, 412
  subordination, 71
  walk-on-spheres algorithm, 381

Cauchy density, 16
Cauchy process, 63
  transition density, 63
Cauchy's estimates, 430
censored stable process, 136
  lower envelope, 277
  upper envelope, 278
characteristic exponent, 2
  Cauchy distribution, 3
  d-dimensional, 21
  isotropic stable, 24
  isotropic stable process, 71
  Lévy–Khintchine formula, 2
  normalised, 24
  normalised stable, 12
  stable distribution, 3
  stable process, 59
compensation formula, 444
creeping, 49

deep factorisation, 295, 330, 331, 403
  d-dimensional, 404
  in dimension $d \geq 0$, 403
  inverse ladder MAP exponent, 334
  inverse matrix factorisation, 332
  ladder MAP exponent, 348
  ladder MAP potential matrix, 335
Donsker-type convergence, 39
double gamma function, 96, 432
double sine function, 184, 434
drifting, 35
duality, 41
  duality lemma, 41
  Hunt–Nagasawa duality, 120, 442
  Markov additive process, 290
  q-resolvent, 42
  semigroup, 42

entrance law, 117
  censored stable process, 144
  duality, 119
  pssMp, 118
  radial part of isotropic stable process, 151
  recurrent extension, 122
  stable process conditioned to stay positive, 133
  time reversal, 120
Esscher transform
  Lévy process, 38
  Markov additive process, 291
excursion, 44
excursion theory
  excursion measure, 45, 177
  Markov additive process, 292
exit problem
  hitting points, 160
  interval, 154, 156, 159

exponential functional
  log moments, 253

Feller process, 441
  Lévy process, 34
  regular, 115, 121, 286, 300, 441
Feller semigroup, 441
filtration
  les conditions habituelles, 440
  MAP, 287, 300
  natural, 440
  natural enlargement, 122, 152, 176, 287, 300, 302, 308, 331, 440
  ssMp, 302
  stable, 308
  stable process, 122, 313, 331
first entrance from a ball
  triple law, 395
first entrance in a ball, 366
  triple law, 395
first entrance into a bounded interval, 162, 163, 318
first exit from a ball, 366
first exit from an interval, 154, 156, 159
first hitting of a sphere, 355
first hitting of a two-point set, 168
first passage problem, 50
  asymptotic overshoot subordinator, 52
  Lévy process, 50
  reflected stable process, 177
  stable process, 69
  stable subordinator, 68
  subordinator, 51
Fourier transform, 430

gamma function, 431
  duplication formula, 432
  infinite product representation, 431
  recursion formula, 431
  reflection formula, 432
generalised last exit time, 441
generalised polar coordinates, 301

hitting points, 42
  first hitting of a two-point set, 168
  in an interval for stable process, 160
  Mellin transform of hitting time for stable process, 174
  stable process, 65
hitting points in an interval, 160
Hunt–Nagasawa duality, 120, 419, 442
hypergeometric function, 435
  linear combination of, 435
  special formulae for, 436
hypergeometric Lévy process, 78, 84, 252
  admissible parameters, 84
  first passage problem, 89

infinitesimal generator, 35, 55
  Lévy process, 35

Jacobian
  sphere inversion, 362
  sphere inversion with reflection, 362

Lévy measure, 2
  isotropic stable, 25
  stable Cartesian coordinates, 25
  stable distribution, 5
  stable polar coordinates, 25
  stable process, 58
Lévy process, 27
  asymmetry, 31
  β-subordinator, 78
  Brownian motion, 28
  Cauchy process, 63
  characteristic exponent, 27, 54
  compound Poisson with drift, 29, 32
  creeping, 49
  definition, 27
  Donsker-type convergence, 39
  drifting, 35
  duality, 41, 55
  Esscher transform, 38
  excursion theory, 44
  exponential change of measure, 37
  exponential moments, 37
  Feller property, 34, 54
  first passage problem, 50
  hitting points, 42
  hypergeometric, 78, 84, 252
  in dimension $d \geq 2$, 54
  infinitesimal generator, 35, 55
  killing, 30
  Lévy–Itô decomposition, 29
  Lévy–Khintchine formula, 28
  Lamperti-stable, 87
  Markov property, 34, 35
  moments, 36, 55
  non-arithmetic, 228, 260
  oscillating, 35
  path variation, 31
  polarity of points, 55
  recurrence, 40, 55
  regularity, 44
  resolvent with killing at first passage, 52

semigroup, 34
spectrally one-sided, 33
strong Markov property, 34
subordinator, 33
transience, 40, 55
triple law, 50
Wiener–Hopf factorisation, 44, 47
Lévy–Khintchine formula, 2
in dimension $d \geq 2$, 54
isotropic stable, 24
regularisation function, 3
stable distribution, 3
Lamperti transform, 115
Lamperti–Kiu transform, 287
in dimension $d \geq 2$, 301
Lamperti-stable MAP, 307
dual process, 309
Lamperti-stable process, 87
admissible parameters, 88
subordinator, 81
Laplace exponent
spectrally negative stable process, 63
stable subordinator, 64
Laplace–Fourier transform, 430
last exit time, 442
generalised, 441
law of the iterated logarithm
radial maximum of a stable process, 221
spectrally negative stable process, 217
stable process conditioned to limit to 0 from above, 274
stable subordinator, 208
Liouville's theorem, 430
lower envelope
censored stable process, 277
pssMp, 235, 239
radial part of isotropic stable process, 280, 283
stable process conditioned to hit 0 continuously, 328
stable processes, 219
stable processes conditioned to limit to 0 from above, 273

MAP, *see* Markov additive process
Markov additive Poisson point process, 444
compensation formula, 445
Markov additive process, 287
ascending ladder MAP, 293
compensation formula, 311
deep factorisation, 295
dual ladder MAP, 337
dual to Lamperti-stable MAP, 309
Esscher transform, 291
excursion theory, 292
filtration, 287, 300
isotropic self-similar Markov process, 302
ladder MAP dual potential, 337
ladder MAP potential, 335, 387
Lamperti-stable MAP, 307
matrix exponent, 289, 291
matrix Laplace exponent, 294
modulator, 287
ordinator, 287
reflected, 292
strong law of large numbers, 292
time change, 291
Wiener–Hopf factorisation, 295, 403, 404
with discrete modulation, 287
Markov process
Hunt–Nagasawa duality, 442
Markov property
Lévy process, 34, 35
Mellin transform, 430
first hitting time of point for stable process, 174
Morera's Theorem, 430

Nagasawa duality, *see* Hunt–Nagasawa duality
natural filtration, 440
naturally enlarged filtration, 440
Newtonian Poisson formula, 363, 368

oscillating, 35

perpetuity, 437
tail distribution, 438
Pochhammer symbol, 435
point of closest reach, 166, 389
point of furthest reach, 166
Poisson point process, 443
Markov additive Poisson point process, 444
Poisson random measure, 443
positive self-similar Markov process, 115, 227, 260
censored stable process, 136
conservative, 117
entrance at 0, 117
envelopes, 227
last passage time, 228
non-conservative, 117
path functional distributions, 260
radial part of stable process, 144
radius of an isotropic self-similar Markov process, 303

recurrent extension, 121
stable process conditioned to limit to 0 from above, 133
positivity parameter, 61
pssMp, *see* positive self-similar Markov process
path decomposition, 227

radial excursion theory, 383
excursion measure, 385, 398
radial part of isotropic stable process
lower envelope, 280, 283
upper envelope, 281, 284
radial part of stable process, 144
radial reflection, 406
limiting distribution, 406
radially reflected stable process, 331
limiting distribution, 331
real self-similar Markov process, 287, 288
recurrence, 40, 55
criterion for Lévy processes, 40
stable process, 64
recurrent extension, 122
regular Feller process, 300
regularisation function, 3
regularity of half-line, 44
renewal measure, 50
residue, 429
resolvent, 41
adjusted resolvent, 74
entrance into a bounded interval, 165, 319
exit from a half-line, 159
exit from an interval, 157
first entry into ball, 370
first exit from ball, 370
first hitting of a sphere, 364
first hitting of point, 174
first passage of reflected process, 180
killed process, 42
Lévy process killed at first passage, 52
ladder height, 50
q-excessive, 43
q-resolvent, 41
stable process, 72
Riesz–Bogdan–Żak transform
in dimension $d \geq 2$, 324
one dimensional, 316
rssMp, *see* real self-similar Markov process

Schwartz's Reflection Principle, 430
self-similar Markov process, 287
censored stable process, 136
entrance at 0, 118, 303, 305
in dimension $d \geq 2$, 300
isotropic, 302
radial part of stable process, 144
stable process conditioned to avoid the origin, 314
stable process conditioned to hit 0 continuously, 314
stable process conditioned to limit to 0 from above, 133
stable process conditioned to stay positive, 128
stable process killed on entering $(-\infty, 0)$, 122
self-similarity, 60
stable process, 60
semigroup, 440
Feller property, 441
Lévy process, 34, 440
Skorokhod space, 118, 265, 304, 439
Skorokhod topology, 440
weak convergence, 440
spectrally negative stable process
law of the iterated logarithm, 217
sphere inversion, 351
with reflection, 354
ssMp, *see* self-similar Markov process
stable density, 14, 16
Cauchy, 16
power series, 16
stable distribution, 1
Cauchy, 3
Cauchy density, 16
definition, 1
density function, 14, 16
higher dimensions, 20
isotropic, 23
Lévy measure, 5
moment formula, 14
moments, 9, 14
normalised, 11
one dimension, 1
parameters, 12
positivity parameter, 11, 15
symmetric, 22
Zolotarev's duality, 15
stable process, 4, 58
adjusted resolvent, 74
admissible parameters, 61
Cauchy process, 63
censored, 136

characteristic exponent, 59, 71
conditioned to avoid the origin, 314
conditioned to hit 0 continuously, 314
conditioned to limit to 0 from above, 133
conditioned to stay positive, 128
deep factorisation, 404
drifting and oscillating, 64
excursion theory, 177
filtration, 122, 308, 313, 331
first entrance from a ball, 395
first entrance in a ball, 366, 395
first entrance into a bounded interval, 162, 163, 318
first exit from a ball, 366
first exit from an interval, 154, 156, 159
first hitting of a sphere, 355
first hitting of a two-point set, 168
first passage, 68
first passage problem, 69
hitting points, 65
hitting points in an interval, 160
in dimension $d \geq 2$, 70
isotropic, 70
killed on entering $(-\infty, 0)$, 122
Lévy measure, 58
moments, 64
normalised, 60
one dimension, 58
path variation, 62
point of closes and furthest reach, 166
polarity, 72
positivity parameter, 61
radial process, 144
radial reflection, 406
reflected, 176
regularity of half-line, 66
resolvent density, 72
Riesz–Bogdan–Żak transform, 316, 324
self-similarity, 60
spectrally negative, 63
subordinator, 64, 206
transience and recurrence, 64, 72
transition density, 60
up-crossings of origin, 422
Wiener–Hopf factorisation, 66
winding at infinity, 414
winding at origin, 418
stable process conditioned to hit 0 continuously
lower envelope, 328
upper envelope, 328
stable process conditioned to limit to 0 from above
law of the iterated logarithm, 274
lower envelope, 273
upper envelope, 274
stable process conditioned to stay positive
law of the iterated logarithm, 270
lower envelope, 266
upper envelope, 268, 269
stable subordinator, 206
law of the iterated logarithm, 208
stopping time, 441
strong Markov process, 441
subadditive functions, 436
submultiplicative function, 37
subordinator, 33
asymptotic first passage, 52
β-subordinator, 78
Lamperti-stable subordinator, 81
Laplace exponent, 33
stable subordinator, 64

transience, 40, 55
criterion for Lévy processes, 40
stable process, 64
triple law, 50

up-crossings of stable process, 422
upper envelope
censored stable process, 278
pssMp, 241, 248, 250
radial part of isotropic stable process, 281, 284
stable process, 214
stable process conditioned to hit 0 continuously, 328
stable process conditioned to limit to 0 from above, 274
stable process conditioned to stay positive, 268, 269

walk-on-spheres, 374
Brownian motion, 381
first exit problem, 375
first exit problem with resolvent, 379
Monte Carlo, 375
number of steps to completion, 375
stable process, 375
Wiener–Hopf factorisation, 47
censored stable, 143
deep factorisation, 295, 403
Doney classes, 196
Doney's factorisation, 196, 203

Doney–Kuznetsov factorisation, 184
Kuznetsov's factorisation, 183, 185
Lévy process, 47
radius of stable process, 150
space-time factorisation, 47, 48
spatial factorisation, 48
stable conditioned to limit to 0 from above, 135
stable conditioned to stay positive, 132
stable killed on exiting $(-\infty, 0)$, 125
stable process, 66
temporal factorisation, 47, 48
winding number, 412
winding of Brownian motion, 412
winding of stable process
at infinity, 414
at the origin, 418

Zolotarev's duality, 15